A Themed Issue in Honor of Prof. Dr. Vicente Rives

A Themed Issue in Honor of Prof. Dr. Vicente Rives

Editors

Miguel A. Vicente
Raquel Trujillano
Francisco Martín Labajos

Basel • Beijing • Wuhan • Barcelona • Belgrade • Novi Sad • Cluj • Manchester

Editors
Miguel A. Vicente
Universidad de Salamanca
Salamanca
Spain

Raquel Trujillano
Universidad de Salamanca
Salamanca
Spain

Francisco Martín Labajos
Universidad de Salamanca
Salamanca
Spain

Editorial Office
MDPI
St. Alban-Anlage 66
4052 Basel, Switzerland

This is a reprint of articles from the Special Issue published online in the open access journal *ChemEngineering* (ISSN 2305-7084) (available at: https://www.mdpi.com/journal/ChemEngineering/special_issues/honor_Vicente).

For citation purposes, cite each article independently as indicated on the article page online and as indicated below:

Lastname, A.A.; Lastname, B.B. Article Title. *Journal Name* **Year**, *Volume Number*, Page Range.

ISBN 978-3-0365-9576-4 (Hbk)
ISBN 978-3-0365-9577-1 (PDF)
doi.org/10.3390/books978-3-0365-9577-1

Cover image courtesy of Vicente Rives

© 2024 by the authors. Articles in this book are Open Access and distributed under the Creative Commons Attribution (CC BY) license. The book as a whole is distributed by MDPI under the terms and conditions of the Creative Commons Attribution-NonCommercial-NoDerivs (CC BY-NC-ND) license.

Contents

About the Editors . . . . **vii**

Miguel Angel Vicente, Raquel Trujillano and Francisco M. Labajos
A Themed Issue in Honor of Prof. Dr. Vicente Rives
Reprinted from: *ChemEngineering* **2023**, 7, 102, doi:10.3390/chemengineering7060102 . . . . **1**

Daiana A. Bravo Fuchineco, Angélica C. Heredia, Sandra M. Mendoza, Enrique Rodríguez-Castellón and Mónica E. Crivello
Esterification of Levulinic Acid to Methyl Levulinate over Zr-MOFs Catalysts
Reprinted from: *ChemEngineering* **2022**, 6, 26, doi:10.3390/chemengineering6020026 . . . . **4**

Fernando Cazaña, Zainab Afailal, Miguel González-Martín, José Luis Sánchez, Nieves Latorre, Eva Romeo, et al.
Hydrogen and CNT Production by Methane Cracking Using Ni–Cu and Co–Cu Catalysts Supported on Argan-Derived Carbon
Reprinted from: *ChemEngineering* **2022**, 6, 47, doi:10.3390/chemengineering6040047 . . . . **22**

Violeta Ureña-Torres, Gelines Moreno-Fernández, Juan Luis Gómez-Urbano, Miguel Granados-Moreno and Daniel Carriazo
Graphene-Wine Waste Derived Carbon Composites for Advanced Supercapacitors
Reprinted from: *ChemEngineering* **2022**, 6, 49, doi:10.3390/chemengineering6040049 . . . . **42**

Alexander Misol, Alejandro Jiménez and Francisco M. Labajos
Use of Ethylamine, Diethylamine and Triethylamine in the Synthesis of Zn,Al Layered Double Hydroxides
Reprinted from: *ChemEngineering* **2022**, 6, 53, doi:10.3390/chemengineering6040053 . . . . **53**

María Oset, Alejandro Moya, Guillermo Paulo-Redondo and Isaac Nebot-Díaz
Nanoparticle Black Ceramic Pigment Obtained by Hydrotalcite-like Compound Microwave Treatment
Reprinted from: *ChemEngineering* **2022**, 6, 54, doi:10.3390/chemengineering6040054 . . . . **72**

Omar José de Lima, Denis Talarico de Araújo, Liziane Marçal, Antonio Eduardo Miller Crotti, Guilherme Sippel Machado, Shirley Nakagaki, et al.
Photodegradation of Fipronil by Zn-AlPO$_4$ Materials Synthesized by Non-Hydrolytic Sol–Gel Method
Reprinted from: *ChemEngineering* **2022**, 6, 55, doi:10.3390/chemengineering6040055 . . . . **83**

Vladimir Sánchez, María Dolores González, Pilar Salagre and Yolanda Cesteros
Acid-Modified Clays for the Catalytic Obtention of 5-Hydroxymethylfurfural from Glucose
Reprinted from: *ChemEngineering* **2022**, 6, 57, doi:10.3390/chemengineering6040057 . . . . **105**

Jacinta García-Talegón, Adolfo Carlos Iñigo, Rosa Sepúlveda and Eduardo Azofra
Effect of Artificial Freeze/Thaw and Thermal Shock Ageing, Combined or Not with Salt Crystallisation on the Colour of Zamora Building Stones (Spain)
Reprinted from: *ChemEngineering* **2022**, 6, 61, doi:10.3390/chemengineering6040061 . . . . **118**

Alejandro Jiménez, Marta Valverde, Alexander Misol, Raquel Trujillano, Antonio Gil and Miguel Angel Vicente
Preparation of $Ca_2Al_{1-m}Fe_m(OH)_6Cl \cdot 2H_2O$-Doped Hydrocalumites and Application of Their Derived Mixed Oxides in the Photodegradation of Ibuprofen
Reprinted from: *ChemEngineering* **2022**, 6, 64, doi:10.3390/chemengineering6040064 . . . . **129**

Juan Pablo Zanin, German A. Gil, Mónica C. García and Ricardo Rojas
Drug-Containing Layered Double Hydroxide/Alginate Dispersions for Tissue Engineering
Reprinted from: *ChemEngineering* 2022, 6, 70, doi:10.3390/chemengineering6050070 147

David Caballero, Ruben Beltrán-Cobos, Fabiano Tavares, Manuel Cruz-Yusta, Luis Sánchez Granados, Mercedes Sánchez-Moreno and Ivana Pavlovic
The Inhibitive Effect of Sebacate-Modified LDH on Concrete Steel Reinforcement Corrosion
Reprinted from: *ChemEngineering* 2022, 6, 72, doi:10.3390/chemengineering6050072 162

Olga Yu. Golubeva, Yulia A. Alikina, Elena Yu. Brazovskaya and Nadezhda M. Vasilenko
Adsorption Properties and Hemolytic Activity of Porous Aluminosilicates in a Simulated Body Fluid
Reprinted from: *ChemEngineering* 2022, 6, 78, doi:10.3390/chemengineering6050078 175

Joel Silva, Cláudio Rocha, M. A. Soria and Luís M. Madeira
Catalytic Steam Reforming of Biomass-Derived Oxygenates for H_2 Production: A Review on Ni-Based Catalysts
Reprinted from: *ChemEngineering* 2022, 6, 39, doi:10.3390/chemengineering6030039 192

Didier Tichit and Mayra G. Álvarez
Layered Double Hydroxide/Nanocarbon Composites as Heterogeneous Catalysts: A Review
Reprinted from: *ChemEngineering* 2022, 6, 45, doi:10.3390/chemengineering6040045 243

David Suescum-Morales, José Ramón Jiménez and José María Fernández-Rodríguez
Review of the Application of Hydrotalcite as CO_2 Sinks for Climate Change Mitigation
Reprinted from: *ChemEngineering* 2022, 6, 50, doi:10.3390/chemengineering6040050 287

Raquel Trujillano
30 Years of Vicente Rives' Contribution to Hydrotalcites, Synthesis, Characterization, Applications, and Innovation
Reprinted from: *ChemEngineering* 2022, 6, 60, doi:10.3390/chemengineering6040060 305

E. I. García-López, L. Palmisano and G. Marcì
Overview on Photoreforming of Biomass Aqueous Solutions to Generate H_2 in the Presence of $g-C_3N_4$-Based Materials
Reprinted from: *ChemEngineering* 2023, 7, 11, doi:10.3390/chemengineering7010011 321

Is Fatimah, Ganjar Fadillah, Suresh Sagadevan, Won-Chun Oh and Keshav Lalit Ameta
Mesoporous Silica-Based Catalysts for Biodiesel Production: A Review
Reprinted from: *ChemEngineering* 2023, 7, 56, doi:10.3390/chemengineering7030056 338

About the Editors

Miguel A. Vicente

Miguel A. Vicente is Full Professor of Inorganic Chemistry at Faculty of Chemistry, Universidad de Salamanca, Spain. Professor Vicente earned his BSc and MSc in Chemistry at Universidad de Salamanca, and his PhD in Inorganic Chemistry at UNED University, Madrid (1994). He carried out postdoctoral research at the Unité de Catalyse et Chimie des Matériaux Divisés, Université Catholique de Louvain (Belgium) and at the Laboratoire de Réactivité de Surface, Université Pierre et Marie Curie (Paris, France). His research activity has been mainly focused on chemical modification of clays (acid activation, pillaring, functionalization) and on the applications of the obtained materials.

Raquel Trujillano

Raquel Trujillano is Full Professor of Inorganic Chemistry at Higher Polytechnical School, Universidad de Salamanca, Spain. Professor Trujillano earned her BSc, MSc and PhD (1997) in Pharmacy at Universidad de Salamanca. She carried out postdoctoral research at the Natural Resources and Agrobiology Institute, Spanish National Council of Scientific Research (CSIC) and at Laboratoire de Réactivité de Surface, Université Pierre et Marie Curie (Paris, France). Her research activity has been mainly focused on hydrotalcites and synthetic clays.

Francisco Martín Labajos

Francisco Martín Labajos is Full Professor of Inorganic Chemistry at the School of Industrial Engineering, Universidad de Salamanca, Spain. Professor Labajos earned his BSc, MSc and PhD (1993) in Chemistry at Universidad de Salamanca. He carried out postdoctoral research at Chemistry Department, Cambridge University (UK). His research activity has been mainly focused on hydrotalcites. His academic activity includes management positions, as Dean of the School of Industrial Engineering and Head of Extraordinary Chair "Iberdrola-USAL", funded by Iberdrola company.

Editorial

A Themed Issue in Honor of Prof. Dr. Vicente Rives

Miguel Angel Vicente *, Raquel Trujillano and Francisco M. Labajos

Recognized Research Group "QUESCAT", Departamento de Química Inorgánica, Universidad de Salamanca, E-37008 Salamanca, Spain; rakel@usal.es (R.T.); labajos@usal.es (F.M.L.)
* Correspondence: mavicente@usal.es

Professor Vicente Rives developed a very long and fruitful career as a teacher of Inorganic Chemistry and Materials Chemistry and has been a dedicated researcher in these and related fields. After obtaining his Ph.D. degree in Chemistry in 1978 at the University of Seville (Spain), Prof. Rives worked at the University of Salamanca for forty years. During this time, he worked on dozens of research projects; published more than 450 research papers; communicated to hundreds of scientific meetings; managed research as Editor of various Journals; received various research awards, etc. Most importantly, he created and headed an important group of researchers on Solid State Chemistry, Materials Chemistry and Heterogeneous Catalysis, establishing vital research collaborations with several groups from different countries.

Prof. Rives left the University of Salamanca in 2021 on his retirement. For this reason, as his disciples, colleagues and friends, we proposed to *ChemEngineering* the edition of a Special Issue devoted to him. The theme of this issue was opened to the research fields in which Prof. Rives worked along his career, namely, layered double hydroxides (LDHs), metal oxides, clay minerals, catalysis and photocatalysis, thermal analysis, and cultural heritage conservation, among others, mainly inviting (but not exclusively) to contribute researchers who had collaborated with Prof. Rives at any moment of his career.

This Special Issue has received eighteen contributions, six review papers [1–6], and twelve research papers [7–18]. Most of the papers originated from groups who have previously worked with Prof. Rives, but five of the contributions came from authors who had not previously worked with him [6,7,13,17,18]. The materials most studied by Prof. Rives throughout his career were LDHs. Therefore, it was not surprising that these materials were the basis of eight of the papers published in this Special Issue [2–4,10,11,15–17]. Other materials included Ni-based catalysts [1], g-C_3N_4 [5], mesoporous silica [6], MOFs [7], carbons [8,9], AlPO$_4$ [12], acid-modified clays [13], and porous aluminosilicates [17]. Finally, one of the papers was devoted to the study of cultural heritage conservation [14].

The review papers reported very interesting revisions on the use of different materials for certain applications. Thus, Soria et al. reported on the use of Ni-based catalysts on the steam reforming of oxygenated compounds derived from biomass for H_2 production [1]. Tichit and Alvarez reviewed the use of LDH–carbon nanocomposites as heterogeneous catalysts [2], and Fernández-Rodríguez et al. reviewed the use of LDHs in construction materials providing a sink for CO_2 [3]. Trujillano reviewed the very wide contribution of Prof. Rives to the study of LDHs, underlying his substantive results derived after more than 30 years of continuous research on these materials [4]. García-López et al. reviewed the use of g-C_3N_4-based materials to produce H_2 by the photoreforming of biomass [5], and Fatimah et al. reviewed [6] the production of biodiesel from catalysts based on mesoporous silica.

The research papers also reported on very interesting applications of the different materials studied. Rodríguez-Castellón et al. reported on the esterification of levulinic acid to methyl levulinate, catalyzed by Zr-MOFs [7]; Monzón et al. studied the production of H_2 and CNT from methane using bimetallic catalysts based on carbon [8], whilst Carriazo et al. used carbon composites for the preparation of advanced supercapacitors [9]. Labajos

Citation: Vicente, M.A.; Trujillano, R.; Labajos, F.M. A Themed Issue in Honor of Prof. Dr. Vicente Rives. *ChemEngineering* 2023, 7, 102. https://doi.org/10.3390/chemengineering7060102

Received: 17 October 2023
Accepted: 27 October 2023
Published: 31 October 2023

Copyright: © 2023 by the authors. Licensee MDPI, Basel, Switzerland. This article is an open access article distributed under the terms and conditions of the Creative Commons Attribution (CC BY) license (https://creativecommons.org/licenses/by/4.0/).

et al. explored new routes for the preparation of LDHs using various amines in the synthesis process [10], while Nebot-Díaz et al. used LDH as precursors of nanoparticle black pigments for the ceramic industry [11]. Ciuffi et al. used Zn-AlPO$_4$ as photocatalysts for the degradation of fipronil [12], and Cesteros et al. used acid-modified clays as catalysts for the preparation of 5-hydroxymethylfurfural from glucose [13]. Vicente et al. prepared Fe-doped hydrocalumites (a type of LDH) from aluminum slags and used them for the photodegradation of ibuprofen [15]. Rojas et al. prepared LDH–alginate composites as carriers of either ibuprofen or naproxen [16], while Pavlovic et al. proposed new corrosion inhibitors for reinforced concrete based on LDHs modified with sebacate anions [17]. In an area more related to biological applications, Golubeva et al. studied the adsorption and hemolytic behavior of porous aluminosilicates in a simulated body fluid [18]. On the other hand, related to cultural heritage conservation, García-Talegón et al. reported on the ageing of Spanish building stones under different physical agents, mainly studied by the evolution of their color [14].

Thus, this Special Issue contains a very interesting series of papers on the subjects out of which Prof. Rives developed his career, and will be of great interest for the researchers working in these fields.

Finally, we want to express our gratitude to *ChemEngineering* and MDPI for this Special Issue, and particularly to Ms. Camile Wang for her help and patience in all the processing steps required for this edition. We also thank the different researchers who have kindly reviewed these manuscripts.

Author Contributions: Conceptualized, written and reviewed by the three authors. All authors have read and agreed to the published version of the manuscript.

Funding: This research received no external funding.

Conflicts of Interest: The authors declare no conflict of interest.

References

1. Silva, J.; Rocha, C.; Soria, M.A.; Madeira, L.M. Catalytic Steam Reforming of Biomass-Derived Oxygenates for H$_2$ Production: A Review on Ni-Based Catalysts. *ChemEngineering* **2022**, *6*, 39. [CrossRef]
2. Tichit, D.; Álvarez, M.G. Layered Double Hydroxide/Nanocarbon Composites as Heterogeneous Catalysts: A Review. *ChemEngineering* **2022**, *6*, 45. [CrossRef]
3. Suescum-Morales, D.; Jiménez, J.R.; Fernández-Rodríguez, J.M. Review of the Application of Hydrotalcite as CO$_2$ Sinks for Climate Change Mitigation. *ChemEngineering* **2022**, *6*, 50. [CrossRef]
4. Trujillano, R. 30 Years of Vicente Rives' Contribution to Hydrotalcites, Synthesis, Characterization, Applications, and Innovation. *ChemEngineering* **2022**, *6*, 60. [CrossRef]
5. García-López, E.I.; Palmisano, L.; Marcì, G. Overview on Photoreforming of Biomass Aqueous Solutions to Generate H$_2$ in the Presence of g-C$_3$N$_4$-Based Materials. *ChemEngineering* **2023**, *7*, 11. [CrossRef]
6. Fatimah, I.; Fadillah, G.; Sagadevan, S.; Oh, W.-C.; Ameta, K.L. Mesoporous Silica-Based Catalysts for Biodiesel Production: A Review. *ChemEngineering* **2023**, *7*, 56. [CrossRef]
7. Bravo Fuchineco, D.A.; Heredia, A.C.; Mendoza, S.M.; Rodríguez-Castellón, E.; Crivello, M.E. Esterification of Levulinic Acid to Methyl Levulinate over Zr-MOFs Catalysts. *ChemEngineering* **2022**, *6*, 26. [CrossRef]
8. Cazaña, F.; Afailal, Z.; González-Martín, M.; Sánchez, J.L.; Latorre, N.; Romeo, E.; Arauzo, J.; Monzón, A. Hydrogen and CNT Production by Methane Cracking Using Ni–Cu and Co–Cu Catalysts Supported on Argan-Derived Carbon. *ChemEngineering* **2022**, *6*, 47. [CrossRef]
9. Ureña-Torres, V.; Moreno-Fernández, G.; Gómez-Urbano, J.L.; Granados-Moreno, M.; Carriazo, D. Graphene-Wine Waste Derived Carbon Composites for Advanced Supercapacitors. *ChemEngineering* **2022**, *6*, 49. [CrossRef]
10. Misol, A.; Jiménez, A.; Labajos, F.M. Use of Ethylamine, Diethylamine and Triethylamine in the Synthesis of Zn,Al Layered Double Hydroxides. *ChemEngineering* **2022**, *6*, 53. [CrossRef]
11. Oset, M.; Moya, A.; Paulo-Redondo, G.; Nebot-Díaz, I. Nanoparticle Black Ceramic Pigment Obtained by Hydrotalcite-like Compound Microwave Treatment. *ChemEngineering* **2022**, *6*, 54. [CrossRef]
12. de Lima, O.J.; de Araújo, D.T.; Marçal, L.; Crotti, A.E.M.; Machado, G.S.; Nakagaki, S.; de Faria, E.H.; Ciuffi, K.J. Photodegradation of Fipronil by Zn-AlPO$_4$ Materials Synthesized by Non-Hydrolytic Sol–Gel Method. *ChemEngineering* **2022**, *6*, 55. [CrossRef]
13. Sánchez, V.; González, M.D.; Salagre, P.; Cesteros, Y. Acid-Modified Clays for the Catalytic Obtention of 5-Hydroxymethylfurfural from Glucose. *ChemEngineering* **2022**, *6*, 57. [CrossRef]

14. García-Talegón, J.; Iñigo, A.C.; Sepúlveda, R.; Azofra, E. Effect of Artificial Freeze/Thaw and Thermal Shock Ageing, Combined or Not with Salt Crystallisation on the Colour of Zamora Building Stones (Spain). *ChemEngineering* **2022**, *6*, 61. [CrossRef]
15. Jiménez, A.; Valverde, M.; Misol, A.; Trujillano, R.; Gil, A.; Vicente, M.A. Preparation of $Ca_2Al_{1-m}Fe_m(OH)_6Cl \cdot 2H_2O$-Doped Hydrocalumites and Application of Their Derived Mixed Oxides in the Photodegradation of Ibuprofen. *ChemEngineering* **2022**, *6*, 64. [CrossRef]
16. Zanin, J.P.; Gil, G.A.; García, M.C.; Rojas, R. Drug-Containing Layered Double Hydroxide/Alginate Dispersions for Tissue Engineering. *ChemEngineering* **2022**, *6*, 70. [CrossRef]
17. Caballero, D.; Beltrán-Cobos, R.; Tavares, F.; Cruz-Yusta, M.; Granados, L.S.; Sánchez-Moreno, M.; Pavlovic, I. The Inhibitive Effect of Sebacate-Modified LDH on Concrete Steel Reinforcement Corrosion. *ChemEngineering* **2022**, *6*, 72. [CrossRef]
18. Golubeva, O.Y.; Alikina, Y.A.; Brazovskaya, E.Y.; Vasilenko, N.M. Adsorption Properties and Hemolytic Activity of Porous Aluminosilicates in a Simulated Body Fluid. *ChemEngineering* **2022**, *6*, 78. [CrossRef]

Disclaimer/Publisher's Note: The statements, opinions and data contained in all publications are solely those of the individual author(s) and contributor(s) and not of MDPI and/or the editor(s). MDPI and/or the editor(s) disclaim responsibility for any injury to people or property resulting from any ideas, methods, instructions or products referred to in the content.

Article

Esterification of Levulinic Acid to Methyl Levulinate over Zr-MOFs Catalysts

Daiana A. Bravo Fuchineco [1], Angélica C. Heredia [1], Sandra M. Mendoza [2], Enrique Rodríguez-Castellón [3,*] and Mónica E. Crivello [1]

1. Centro de Investigación y Tecnología Química (CITeQ), Consejo Nacional de Investigaciones Científicas y Técnicas (CONICET), Facultad Regional Córdoba, Universidad Tecnológica Nacional (UTN-FRC), Córdoba 5016, Santa Fe, Argentina; dbravo@frc.utn.edu.ar (D.A.B.F.); angelicacheredia@gmail.com (A.C.H.); mcrivello@frc.utn.edu.ar (M.E.C.)
2. Consejo Nacional de Investigaciones Científicas y Técnicas (CONICET), Facultad Regional Reconquista, Universidad Tecnológica Nacional (UTN-FRRQ), Reconquista 3560, Santa Fe, Argentina; smendoza@frrq.utn.edu.ar
3. Departamento de Química Inorgánica, Cristalografía y Mineralogía, Facultad de Ciencias, Universidad de Málaga, 29071 Málaga, Spain
* Correspondence: castellon@uma.es

Citation: Bravo Fuchineco, D.A.; Heredia, A.C.; Mendoza, S.M.; Rodríguez-Castellón, E.; Crivello, M.E. Esterification of Levulinic Acid to Methyl Levulinate over Zr-MOFs Catalysts. *ChemEngineering* 2022, 6, 26. https://doi.org/10.3390/chemengineering6020026

Academic Editors: Miguel A. Vicente, Raquel Trujillano and Francisco Martín Labajos

Received: 27 January 2022
Accepted: 23 March 2022
Published: 25 March 2022

Publisher's Note: MDPI stays neutral with regard to jurisdictional claims in published maps and institutional affiliations.

Copyright: © 2022 by the authors. Licensee MDPI, Basel, Switzerland. This article is an open access article distributed under the terms and conditions of the Creative Commons Attribution (CC BY) license (https://creativecommons.org/licenses/by/4.0/).

Abstract: At present, the trend towards partial replacement of petroleum-derived fuels by those from the revaluation of biomass has become of great importance. An effective strategy for processing complex biomass feedstocks involves prior conversion to simpler compounds (platform molecules) that are more easily transformed in subsequent reactions. This study analyzes the metal–organic frameworks (MOFs) that contain Zr metal clusters formed by ligands of terephthalic acid (UiO-66) and aminoterephthalic acid (UiO-66-NH$_2$), as active and stable catalysts for the esterification of levulinic acid with methanol. An alternative synthesis is presented by means of ultrasonic stirring at room temperature and 60 °C, in order to improve the structural properties of the catalysts. They were analyzed by X-ray diffraction, scanning electron microscopy, infrared spectroscopy, X-ray photoelectron spectroscopy, microwave plasma atomic emission spectroscopy, acidity measurement, and N$_2$ adsorption. The catalytic reaction was carried out in a batch system and under pressure in an autoclave. Its progress was followed by gas chromatography and mass spectrometry. Parameters such as temperature, catalyst mass, and molar ratio of reactants were optimized to improve the catalytic performance. The MOF that presented the highest activity and selectivity to the desired product was obtained by synthesis with ultrasound and 60 °C with aminoterephthalic acid. The methyl levulinate yield was 67.77% in batch at 5 h and 85.89% in an autoclave at 1 h. An analysis of the kinetic parameters of the reaction is presented. The spent material can be activated by ethanol washing allowing the catalytic activity to be maintained in the recycles.

Keywords: UiO-66; UiO-66-NH$_2$; levulinic acid; methyl levulinate; esterification kinetics; batch and pressure reactions

1. Introduction

In recent decades, due to the finite reserves and environmental pollution problems of fossil energy, governments have put forward the "energy strategy", stimulating the utilization of renewable energy and resources [1]. The development of new technologies for producing energy and chemicals from them has prompted biomass valorization to become an important area of research. Biomass provides an ideal alternative to fossil resources; indeed, biomass is the only sustainable source of organic compounds and has been proposed as the ideal equivalent to petroleum for the production of fuels and chemicals [2]. In general, biomass is defined as all organic matter, including crops, food, plants, and farming and forestry residues, which can be used as a source of energy [1,3]. Lignocellulose, the most abundant form of biomass, is composed of three primary components; cellulose (40–50 wt%), hemicellulose (25–30 wt%), and lignin (15–30 wt%), a phenolic polymer. Woody, herbaceous plants, and other crop residues such as sugarcane bagasse, and wheat straw, are suitable lignocellulosic feedstocks applied in the biorefinery process, to proffer valuable applications that can be derived from the petroleum-based source [2,4]. There are various available methods, through which biomass can be converted to different viable products that depend on the catalyst selected and the reaction conditions. For example, sorbitol is obtained by hydrogenation, fructose through isomerization, and 5-hydroxymethylfurfural via dehydration; this can be further transformed into levulinic acid (LA) and formic acid (FA) by hydrolysis [4,5].

Levulinic acid (LA), which is considered as one of the most promising platform molecules, has been hailed as the chemical bridge between biomass and petroleum [6,7], since it can be transformed into special chemicals such as fuels, solvents, monomers for polymers, plasticizers, surfactants, agrochemicals, and pharmaceuticals [8–10]. Within these-levulinate esters are unique potential value-added energy chemicals that have significantly attracted remarkable attention of global researchers. Possessing great commercial importance due to their applications as fuel-blending components, bio-lubricants, refining of mineral oils, chemicals synthesis/synthetic reagent, polymer precursor, foam comprising material, resin precursors, green solvents, plasticizer, food flavor agent, coating composition, pharmaceutical/cosmetics, degreasing surface agent/stain removal, and building blocks for polycarbonate and herbicides synthesis [11,12].

In particular, levulinate esters have specific properties which make them suitable additives for fuels; some of them are low toxicity, high lubricity, flash point stability, and moderate flow under low-temperature conditions. [13]. Levulinic esters can be prepared by acid-catalyzed alcoholysis and are usually synthesized using homogeneous acid catalysts such as H_2SO_4, HCl, and H_3PO_4 [9,14,15]; however, these liquid acids involved in homogeneous industrial catalytic processes generally suffer from high toxicity and corrosivity, high cost of regeneration or quenching as well as a large number of noxious byproducts and waste. A major challenge for the efficient preparation of levulinic esters is to develop robust, selective, inexpensive, and eco-friendly catalysts [6,12].

Acid solids are used as heterogeneous catalysts for the replacement of homogeneous catalysts, due to their low corrosivity, high selectivity, and easy separation from the reaction system. It is worth noting that MOFs, a new class of crystalline porous hybrid materials composed of metal ions/clusters and multi-topic organic linkers, have shown great potential as emerging porous materials due to their highly tunable textures, such as porosity, adjustable pore size, structural diversity, surface areas, and chemical functionalities [16–20].

MOF UiO-66 was synthesized for the first time in 2008 by scientists from the Universitetet i Oslo (UiO) using a solvothermal treatment at 120 °C in dimethylformamide as solvent [21,22]. The MOF UiO-66-NH_2 developed by Kandiah et al., 2010, is formed of metal clusters, made up of six zirconium atoms $Zr_6O_4(OH)_4$, joined together by μ_3-O and μ_3-OH groups, coming from the organic 2-aminoterephthalate ligand [23,24].

Given the Lewis acid character of UiO-66 type materials and their noticeable thermal, chemical, and mechanical stability, these materials could be good candidates for the esterification of levulinic acid using various alcohols [25–28]. In particular, it has been shown that the activity of UiO-66-NH_2 in the esterification of levulinic acid (LA) produces a reaction enhancing effect due to the close position of the NH_2 group to the metal center, allowing simultaneous activation of both LA such as alcohol. UiO-66-NH_2 behaves as a bifunctional acid–base catalyst [29,30].

The aim of this work is to evaluate the activity of organic–metal frameworks (MOFs) in fine chemistry reactions, such as the esterification of levulinic acid with methanol. For this purpose, the MOFs UiO-66 and UiO-66-NH_2 were synthesized by the solvothermal method, varying certain conditions such as their stirring method. These reactions were carried out in two systems, one batch and the other under pressure.

2. Materials and Methods

2.1. Chemicals and Reagents

Analytical grade reagents were used for the preparation of catalysts and catalytic reactions. $ZrCl_4$ ($\geq$98%, Merck, Darmstadt, Alemania), terephthalic acid (BDC, 98%, Aldrich, Saint Louis, MO, USA), aminoterephthalic acid (NH_2-BDC, 98%, Aldrich, Saint Louis, MO, USA), N,N-dimetilformamide (DMF $\geq$ 99.8%, Biopack, Buenos Aires, Argentina), Acetone (99.5% Sintorgan, Buenos Aires, Argentina), Methanol ($\geq$99.5%, Biopack, Buenos Aires, Argentina), and Levulinic Ácid (98%, Aldrich, Saint Louis, MO, USA) were employed as received.

2.2. Synthesis of MOF UiO-66 and UiO-66-NH_2

MOFs UiO-66 and UiO-66-NH_2 were obtained by solvothermal method [21,31,32]. Two different binding agents of terephthalic acid (BDC) or aminoterephthalic acid (NH_2-BDC) were used. The synthesis was carried out by magnetic and ultrasonic stirring, the latter at room temperature or 60 °C. The samples were named as M_X, M_X-U, or M_X-UT, where "X" can be "A" related to synthesis with BDC or "B" related to synthesis with NH_2-BDC. The letter U indicates the synthesis on ultrasound at room temperature. The letters U-T indicate the synthesis on ultrasound at 60 °C of temperature.

The synthesis consisted of dissolving 1.042 g in 50 mL g of $ZrCl_4$ in 50 mL of the solvent (DMF); the solution was magnetic or ultrasonic stirred for 5 min. In the case of ultrasound synthesis, the container is placed inside ultrasound equipment, which works at room temperature or at 60 °C. Then, 0.708 g of terephthalic acid or 0.788 g of aminoterephthalic acid was incorporated. The mixture was kept under stirring (magnetic or ultrasonic) for 30 min. The gel was transferred to a Teflon-lined stainless-steel autoclave and kept in an oven at 120 °C for 24 h. The material obtained was immersed in DMF and then washed with acetone with a lower boiling point and easily removed. The precipitate was separated by centrifugation. Finally, the solid was dried at 90 °C for 24 h, obtaining ~1.404 g and ~1.735 g of a white (A) or yellow (B) solid powder, respectively.

2.3. Characterization of MOF UiO-66 and UiO-66-NH_2

XRD powder patterns were collected on an X'pert diffractometer (PANanalytical, Netherlands) using monochromatized Cu Kα radiation (λ = 1.54 Å) at a scan speed of 0.25° min^{-1} in 2θ.

Infrared analyses were carried out on a spectrophotometer Smartomi-Transmission Nicolet iS10 Thermo Scientific (Thermo Scientific, Waltham, MA, USA) in a range of 4000–400 cm^{-1}.

The micrographs of the mixed oxides were obtained by SEM instrument model JSM-6380 LV (JEOL, Japan) equipped with a Supra 40 (Carl Zeiss, Oberkochen, Germany), the samples were metalized with chromium.

The specific surface area (SSA), analysis was carried out in an ASAP 2020 instrument (Micromeritics, Norcross, GA, USA) and was calculated by the Brunauer–Emmett–Teller (BET) method.

X-ray photoelectron spectra (XPS) were recorded with a PHI VersaProbe II Scanning XPS Microprobe (Physical Electronics, Chanhassen, MN, USA) with scanning monochromatic X-ray Al Kα radiation as the excitation source (100 µm area analyzed, 52.8 W, 15 kV, 1486.6 eV), and a charge neutralizer. The pressure in the analysis chamber was maintained below 2.0 $\times$ 10^{-6} Pa. High-resolution spectra were recorded at a given take-off angle of 45 by a multi-channel hemispherical electron analyzer operating in the constant pass energy mode at 29.35 eV. Spectra were charge referenced with the C 1s of adventitious carbon at 284.8 eV. The energy scale was calibrated using Cu $2p_{3/2}$, Ag $3d_{5/2}$, and Au $4f_{7/2}$ photoelectron lines at 932.7, 368.2, and 83.95 eV, respectively. The Multipack software version 9.6.0.15 was employed to analyze in detail the recorded spectra. The obtained spectra were fitted using Gaussian–Lorentzian curves to more accurately extract the binding energies of the different element core levels.

In order to carry out the elemental analysis, samples were acid-digested in a closed-vessel microwave oven (SCP Science, Baie-D'Urfe, QC, Canada). The chemical quantification was performed by microwave plasma atomic emission spectroscopy (MP-AES), using an Agilent 4200 MP AES instrument (Agilent, Santa Clara, CA, USA).

To study surface acidity, the adsorbed CO molecules were measured by FTIR analysis at two different temperatures (50 °C and 100 °C) in a Nicolet iS10 instrument. Initially, samples were treated under vacuum and then were put in contact with CO molecules during 2 h at room temperature.

2.4. Catalytic Esterification Reaction

The synthesized MOFs were used for the catalytic esterification of levulinic acid and methanol (Figure 1), with a 1:15 molar ratio and 0.05 g of the solid [25,33–36]. Two reaction systems were used. On the one hand, the reaction was carried out in a glass batch reactor with magnetic stirring at 65 °C, coupled to a reflux condenser. The total time was 5 h and 0.15 mL of samples were collected from the reaction medium at 0, 1, 2, 3, 4, and 5 h. The time t = 0 h was the time at which the temperature of the reaction medium reached 65 °C. Samples were collected using a microsyringe equipped with a filter (Whatman paper filter n° 5) to remove catalyst particles. On the other hand, a pressure system was used constituted of an autoclave reactor. The reactor was placed inside an oven at 130 °C with magnetic stirring at 400 rpm. It was previously purged with N_2 and then pressurized to 30 bar. The reaction products were analyzed by gas chromatography and mass spectrometry. It was used on an Agilent Technologies 7820A instrument equipped with a HP-20M column and a Perkin Elmer Clarus 560 instrument.

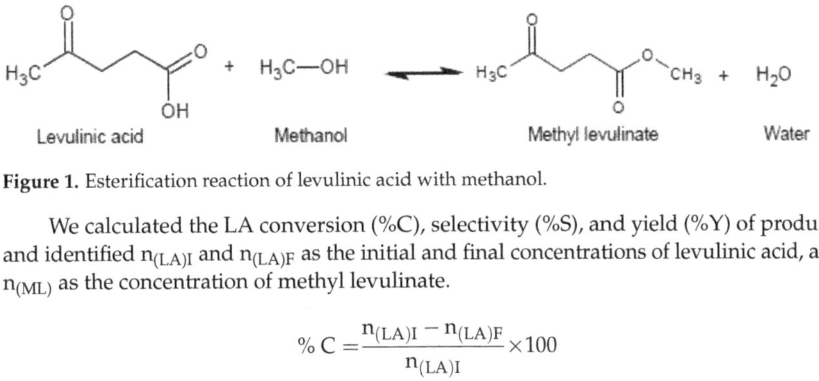

Figure 1. Esterification reaction of levulinic acid with methanol.

We calculated the LA conversion (%C), selectivity (%S), and yield (%Y) of products and identified $n_{(LA)I}$ and $n_{(LA)F}$ as the initial and final concentrations of levulinic acid, and $n_{(ML)}$ as the concentration of methyl levulinate.

$$\% C = \frac{n_{(LA)I} - n_{(LA)F}}{n_{(LA)I}} \times 100 \qquad (1)$$

$$\% S = \frac{n_{(ML)}}{n_{(LA)I} - n_{(LA)F}} \times 100 \qquad (2)$$

$$\% Y = \frac{C \times S}{100} \qquad (3)$$

3. Results

3.1. Physicochemical Characterization

3.1.1. X-ray Diffraction

Figure 2 shows the X-ray diffraction patterns of the MOFs. Two peaks located close to 2θ of 7.4, 8.5, and 25.6° are associated with the diffraction by (1 1 1), (2 0 0), and (600) planes characteristic of the MOF UiO-66-NH$_2$. [22,37–39]. The M_A-U and M_A-UT samples show an exchange in the intensity of the main peaks (111) and (002) and the appearance of a new peak at 2θ angle of ~17.6° related to a phase change, is the same as that reported in the literature [22]. This behavior could be promoted by temperature and ultrasound applied in the synthesis. On the other hand, the XRD patterns in series B show peaks of similar intensity with good crystalline order, regardless of the applied treatment [40].

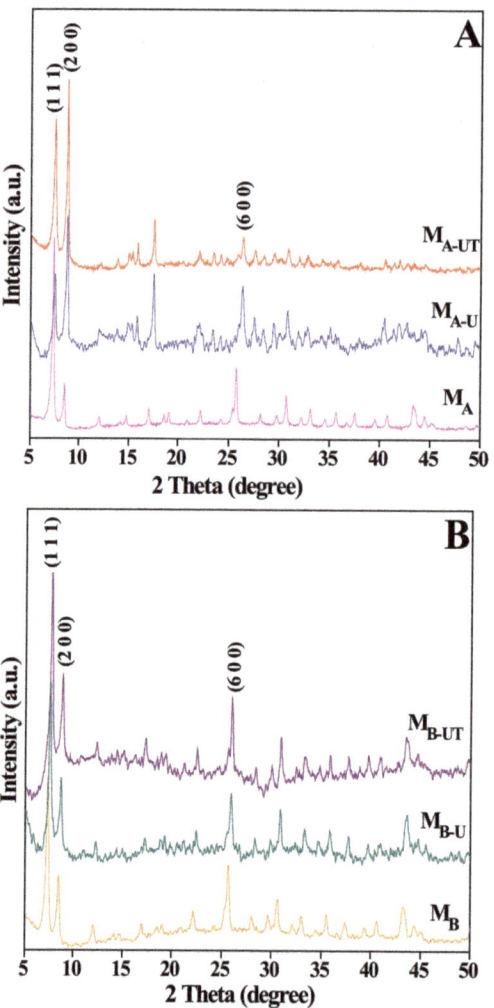

Figure 2. X-ray diffraction patterns. (**A**) UiO-66 magnetic stirring (M_A), UiO-66 ultrasound stirring (M_A-U), and ultrasound stirring at 60 °C (M_A-UT). (**B**) UiO-66 magnetic stirring (MB), UiO-66 ultrasound stirring (M_B-U), and ultrasound stirring at 60 °C (M_B-UT).

3.1.2. FTIR Spectroscopy

The chemical states and functional groups of the different samples were investigated by FT-IR (Figure 3). The peaks at 1586 and 1395 cm^{-1} are attributed to the asymmetric and symmetric stretching vibrations of the COO^- groups in the terephthalic acid, and the peak at 1506 cm^{-1} is the typical peaks of the C=C peak of the vibration of the aromatic ring. The two lowest frequencies at 488 cm^{-1} belong to the stretching vibrations of Zr–O in the Zr_6 cluster, and the band at 551 cm^{-1} is assigned to the asymmetric stretching vibration of Zr-(OC). For the case of materials with the BDC-NH_2 ligand, the signals at 3449 and 3348 cm^{-1} corresponded with the amine symmetric and asymmetric stretching bands, while the signals at 1258 and 1386 cm^{-1} were attributed to C-N binding absorption [41–43].

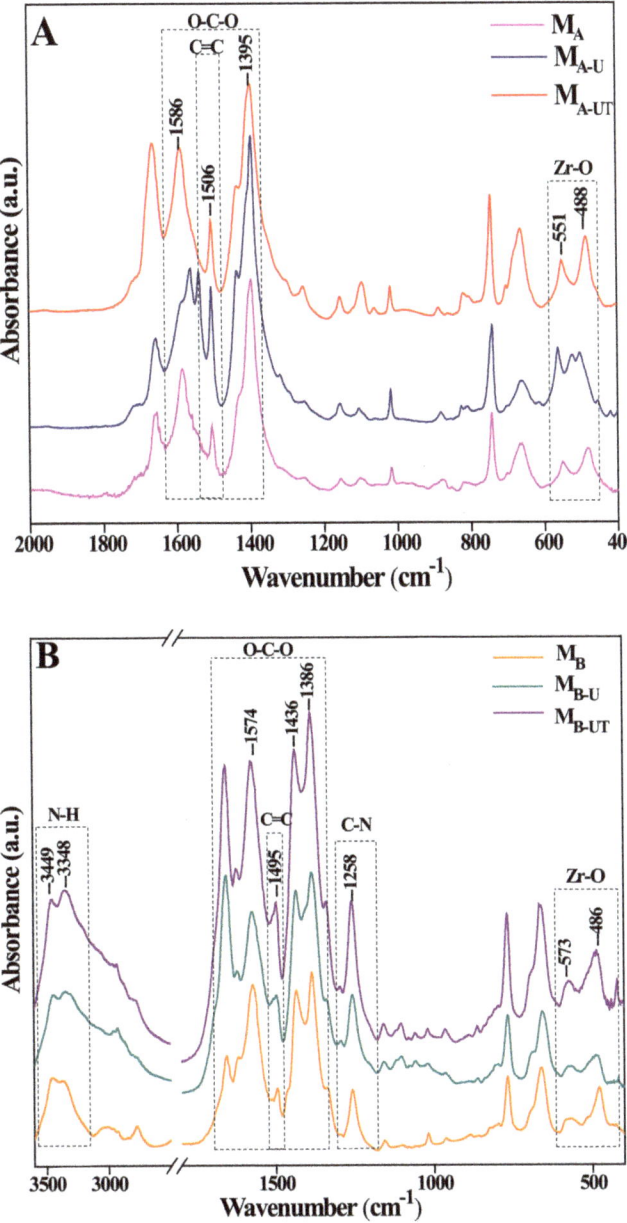

Figure 3. FTIR spectra. (**A**) UiO-66 magnetic stirring and ultrasound at room temperature and at 60 °C; (**B**) UiO-66-NH$_2$ magnetic stirring and ultrasound at room temperature and at 60 °C.

3.1.3. SEM-EDS Analysis

The morphology was characterized by a scanning electron microscope (SEM). Figure 4 shows the synthesized UiO-66 of regular octahedral structure, smooth surface, and good dispersity.

The UiO-66 samples (series A) exhibit a narrow crystal size distribution ranging between 285.7 nm and 488.5 nm, while the UiO-66-NH$_2$ samples (series B) exhibit smaller particles with a mean crystal size between 209.7 and 406.0 nm (Figure S1).

The Zr, C, O, and N elemental mapping images (Figure S2), show dispersion and homogeneous distribution of these elements on the catalyst surface [44,45].

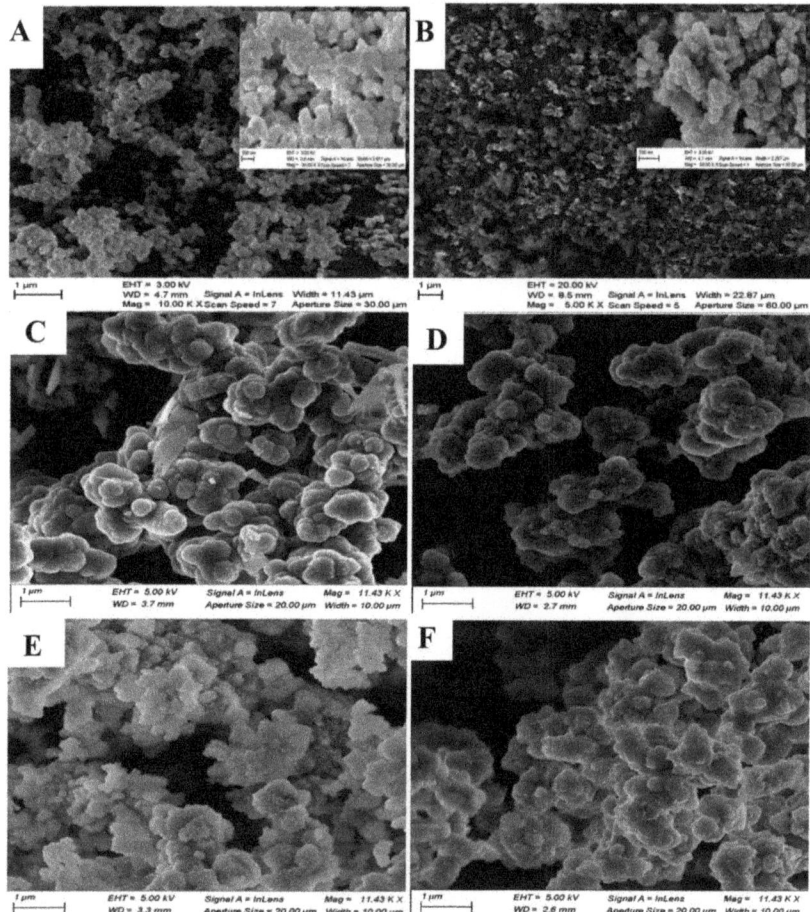

Figure 4. SEM image of (**A**) M_A, (**B**) M_B, (**C**) M_A-U, (**D**) M_B-U, (**E**) M_A-UT, and (**F**) M_B-UT.

3.1.4. Microwave Plasma Atomic Emission Spectrometry (MP-AES) and the Surface Area by BET

Through elemental analysis, the percentage of Zr present in the bulk of the catalyst was determined (Table 1). It is observed that the samples that had a temperature in the synthesis managed to anchor a greater amount of the Zr. Furthermore, comparing the binding agents, those with BDC have a higher % Zr than the others.

To investigate the textural characteristics, N_2 adsorption−desorption isotherms for UiO-66 and UiO-66-NH$_2$ were measured and these are shown in Figure 5. All materials exhibit type Ib isotherms, where the pore size distributions are in a range including micropores wider than ∼1 nm and narrow mesopores (of width < ∼2.5 nm) [46]. The N_2 uptake at the low-pressure range ($P/P_0 < 0.3$) suggests the existence of micropores, while the uptake at the middle-pressure range ($P/P_0 > 0.5$) with hysteresis between adsorption and desorption branches is characteristic of mesopores. For the sample M_A, which was synthesized without ultrasonic and temperature treatment, no hysteresis loop was observed on the isotherm, indicating a typical microporous structure. The rest of the samples show H1 type hysteresis loops [46], which at low and high relative pressure belong to disparate types of isotherms characteristic of solids with microporous windows and partial mesoporous cages [27,47]. Ultrasonic and temperature treatment applied, have decreased the pore size from 2.64 nm to 2.30 nm for series A and from 2.36 nm to 1.07 nm for series B. These comparative results suggest that the ultrasonic and temperature treatments have a significant influence on the pore structures, promoting the presence of a narrow mesopore and micropore distribution. Pore size and surface area in the series B samples have been largely reduced compared with that of series A. This might be because of the introduction of amino groups (-NH_2) in the BDC-NH_2 blocking the pores [48]. The surface area reported by the literature indicates higher values than reported in this work, along with lower pore volume values. Wang et al. [49], Su et al. [50], and Hu et al. [48], were reported a surface area of 1110 $m^2\ g^{-1}$, 967 $m^2\ g^{-1}$, and 1525 $m^2\ g^{-1}$, respectively, while the pore volume reported was 0.60 $cm^3\ g^{-1}$, 0.57 $cm^3\ g^{-1}$, and 0.66 $cm^3\ g^{-1}$ corresponding to micropore size distribution. The lower values of pore volume reported in the literature could be attributed to the different time and temperature autoclave steps, as well as the different ways and times stirring was applied in this work.

Table 1. Composition and surface analysis.

Catalyst	Zr (w %) Experimental	BET ($m^2\ g^{-1}$)	Pore Size (nm)	Pore Volume ($cm^3\ g^{-1}$)
M_A	28.4 ± 0.6	683	2.64	0.248
M_A-U	16.5 ± 0.3	640	2.46	0.163
M_A-UT	17.2 ± 0.3	658	2.30	0.153
M_B	20.5 ± 0.4	400	2.36	0.164
M_B-U	14.8 ± 0.3	300	1.21	0.099
M_B-UT	16.2 ± 0.3	312	1.07	0.095

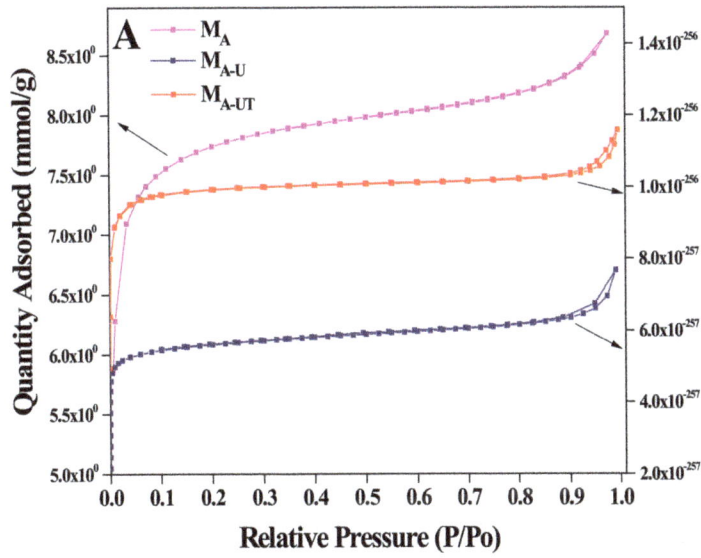

Figure 5. *Cont.*

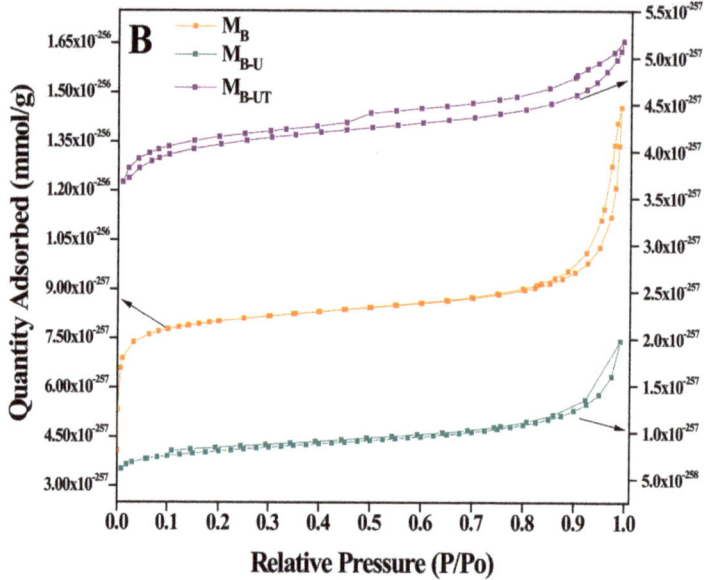

Figure 5. N_2 adsorption–desorption isotherms (**A**) UiO-66 magnetic stirring and ultrasound at room temperature and at 60 °C; (**B**) UiO-66-NH_2 magnetic stirring and ultrasound at room temperature and at 60 °C.

3.1.5. XPS-Spectroscopy

The surface composition of the studied catalysts was studied by XPS. Table 2 shows this composition in atomic concentration %. As expected, all samples exhibit a relatively high concentration of C, typical of MOFs materials. The % of C slightly decreases upon using ultrasound and after thermal treatment. Samples of series A show a small % of N due to the use of DMF in the synthesis, without being part of the structure. The % of Zr increases upon using ultrasound. In the case of the catalysts of series B, as expected, the % of N increases due to the presence of the amino group. This percentage increases when using ultrasound.

Table 2. Surface chemical composition (in atomic concentration %) of the studied catalysts determined by XPS.

Sample	C%	Zr%	O%	N%
M_A	62.59	5.59	31.51	0.31
M_A-U	57.94	7.02	34.72	0.33
M_A-UT	56.03	7.76	35.29	0.92
M_B	60.54	4.97	29.87	4.62
M_B-U	57.95	5.63	31.60	4.83
M_B-UT	59.52	5.42	29.96	5.09

The XPS spectra are shown in the Supplementary Material (Figure S3), and in Table 3, the binding energy values in eV for the C 1s, Zr $3d_{5/2}$, O 1s, and N 1s signals are included. The catalysts of series A show C 1s core-level spectra composed of two contributions at 284.8 and 288.8 eV. The former and more intense is assigned to –C-C- and –C=C- bonds of adventitious carbon and mainly to the aromatic ring, and the latter to the presence of carboxylic groups. The Zr $3d_{5/2}$ maxima appear at about 182.8 eV, typical of Zr(IV). The O 1s core-level spectra can be decomposed into two contributions at about 530.1 eV and

531.8 eV. The latter and more intense is assigned to carboxylic oxygen, and the former to lattice oxygen bonded to Zr. The observed higher relative intensity of the contribution at 531.8 eV for samples synthesized with ultrasound is concomitant with the increase in the %Zr (see Table 2). The catalysts of series B show N 1s binding energy values typical of amino groups. The C 1s core-level spectra of these catalysts with amino functional groups show a new contribution at 286.0 eV due to C-N bonds [51].

Table 3. Binding energy values (in eV) of different elements for the studied catalysts. Area percentages are indicated in parentheses.

Sample	C 1s	Zr $3d_{5/2}$	O 1s	N 1s
M_A	284.8 (86) 288.8 (14)	182.8	530.4 (14) 531.8 (86)	400.6
M_A-U	284.8 (82) 288.8 (18)	182.8	530.1 (18) 531.8 (82)	401.4
M_A-UT	284.8 (83) 288.8 (17)	182.7	530.2 (23) 531.8 (77)	400.7
M_B	284.8 (64) 286.1 (18) 288.7 (18)	182.9	530.3 (12) 532.0 (88)	399.8
M_B-U	284.8 (65) 286.1 (17) 288.8 (18)	182.9	530.5 (12) 531.9 (88)	399.5
M_B-UT	284.8 (65) 286.0 (17) 288.8 (18)	182.8	530.6 (17) 531.9 (83)	399.4

3.1.6. Acid Analysis by FTIR-CO

The acid sites were studied by FTIR CO adsorbed at different temperatures (50 and 100 °C). The Lewis and Brønsted acid sites were measured by the relative area percentage (Table 4). The sites measured at 50 °C are defined as weak, while those measured at 100 °C are defined as medium [52,53]. The graphs are attached in the Supplementary Material (Figure S4). Making an analysis of the acid sites present, it can be observed that when using ultrasound in the synthesis, an increase in the amount of Lewis acid sites is produced, which could be related to the better catalytic results obtained with said materials. As the acid–base duality is also needed, those with BDC-NH$_2$ are the ones that produce the best conversion [54].

Table 4. FTIR of CO absorbed.

Sample	% Area (50 °C) Weak Sites		% Area (100 °C) Medium Sites		% Area Total Sites		L/B
	Lewis	Brønsted	Lewis	Brønsted	Lewis	Brønsted	
M_A	45.87	54.13	53.40	46.60	49.64	50.36	0.99
M_A-U	57.38	42.62	52.68	47.32	55.03	44.97	1.22
M_A-UT	60.43	39.57	54.87	45.13	57.65	42.35	1.36
M_B	42.97	57.03	42.74	57.26	42.85	57.15	0.75
M_B-U	45.40	54.60	46.74	53.26	46.07	53.93	0.85
M_B-UT	48.25	51.75	45.75	54.25	47.00	53.00	0.89

3.2. Catalytic Activity: Esterification for Methyl Levulinate Production

Esterification products were identified by gas chromatography with FID and mass spectrometry. The main esterification product was methyl levulinate (ML) and the byproduct was β-lactone from the dehydration of levulinic acid [55–59]. A control experiment was carried out in the absence of catalyst, which gave a negligible methyl levulinate yield of only 1.56%, attributed to the auto-catalysis reaction due to the acidity of levulinic acid [60].

The M_B-UT presented the best yield to the ML (Table 5). These esterification reactions with the presence of the amino group, occur through a dual activation mechanism, related to an acid–base site, in which the Zr sites (acid site) interact with the adsorbed LA while the amino group (basic site), near the Zr, forms an adduct linked to hydrogen in the alcohol molecule, which increases the nucleophilic character of the O atom, favoring the first reaction step [30,33–35]. Due to this behavior, series A without the basic sites ($-NH_2$), showed a lower yield than series B.

Table 5. Catalytic activity. T = 65 °C, 0.05 g catalyst.

Catalyst	Conversion %	Selectivity ML %	Yield %
M_A	16.73	13.41	2.24
M_A-U	42.03	58.31	24.51
M_A-UT	53.85	78.02	42.02
M_B	21.08	73.10	15.41
M_B-U	58.48	84.88	49.64
M_B-UT	63.57	92.22	58.62

By XPS, M_B-U showed a higher Zr (5.63%) content indicating Lewis acid sites, while M_B-UT contains a higher content of N (5.09%) by $-NH_2$ groups, which would indicate more number of basic sites. Better catalytic results are shown in M_B-UT, probably due to the excess of Lewis acid sites as shown in the relation L/B in Table 4 and related with the Zr at the surface as shown in the XPS result. That would ensure that each basic site ($-NH_2$) is close to a Zr, generating a synergistic effect by acid–base duality in the catalyst.

To improve the catalytic performance, some reaction parameters were modified. Prior to it, the catalyst was washed with acetone using ultrasound, centrifuged, and dried at 90 °C, to take out the solvent present in the pores. This allowed the conversion and yield rate to be increased to 70.87% and 67.77%, respectively.

3.2.1. Effect of Reaction Parameters on the LA Esterification with Methanol Catalyzed by M_B-UT

In order to determine the optimal conditions in the catalytic activity of M_B-UT for esterification from LA to ML under reflux conditions, the roles of temperature and reaction time, catalyst loading, and molar ratio of LA to methanol were studied.

Effect of Catalyst Loading

The catalyst mass was varied based on 0.05 g. The catalytic reaction was tested with half, double, and triple of said mass (Table 6). As the catalyst mass increases, the conversion also increases, but this slight increase is not economically justified. Once all the active surface area of the catalyst is occupied by the reactants, the addition of a greater quantity of mass blocks part of its active sites. For this reason, a catalyst mass of 0.05 g was taken as the optimal value.

Table 6. Catalytic activity.

Catalyst Mass	Conversion %	Yield %	ΔY/ΔC
0.025 g	56.79	45.22	—
0.05 g	63.57	58.62	536
0.1 g	67.55	63.31	93.8
0.15 g	70.00	65.72	48.2

Effect of Temperature

The original reaction was performed at a temperature close to the boiling point of methanol, so that it does not evaporate while it occurs. Then, the reaction temperature was modified to 55 °C, 60 °C, and 70 °C (Table 7). Although the reaction at 70 °C has the highest conversion, the temperature increase of 5 °C was not justified to increase only 3% of conversion; therefore, the optimal working temperature is 65 °C, a temperature with which a good conversion to the desired product is ensured and also the evaporation of one of the reagents is avoided, which could change the molar ratio between them.

Table 7. Catalytic activity.

Temperature	Conversion %	Selectivity ML %	Yield %	ΔY/ΔT
55 °C	48.20	70.29	33.88	- - - -
60 °C	62.86	88.79	55.81	4.26
65 °C	70.87	89.99	63.77	1.59
70 °C	73.79	91.23	67.32	0.71

Effect of Molar Ratio

Regarding the molar ratio of the reagents, we initially worked with a 1:15 ratio of LA: methanol. Variations were performed at a more concentrated point (1:10) and a more dilute point (1:20) (Table 8). In this case, the optimal value corresponds to a ratio of 1:15.

Table 8. Catalytic activity.

Molar Ratio	Conversion %	Yield %	ΔY/Δmol [M]
1:10	59.97	54.19	- - -
1:15	70.87	63.77	1.92
1:20	67.28	64.47	0.14

3.2.2. Kinetic Model and Estimation of Kinetic Parameters

The esterification reaction of levulinic acid (LA) in the presence of methanol (M) to produce methyl levulinate (ML) and water (W) is given in the reaction below:

$$LA + M \leftrightarrow ML + W$$

The adjustment was made with the experimental data obtained in the batch reaction catalyzed by the M_{B-UT} material. To determine the reaction kinetics, an irreversible reaction and a pseudo-homogeneous model were considered [25,61–63]. Since the alcohol (methanol) is in excess, a zero-order reaction was assumed for it. According to this, the reaction rate is expressed in Equation (5), where [LA] and [M] are the concentrations of levulinic acid and methanol, respectively, $[LA_0]$ is the initial concentration of levulinic acid, k is the rate constant for the forward reactions.

$$-\frac{dC_A}{dt} = k[LA][M] \quad (4)$$

where [M] ≫ [A] so [M] = ctes

$$-\frac{dC_A}{dt} = k[LA] \rightarrow -kt = Ln\left[\frac{LA_0}{LA_0-x}\right] \rightarrow -kt = Ln\frac{[LA_0] \text{ (inicial)}}{[LA_t] \text{ (final)}} \quad (5)$$

$$Ln[LA] = -kt + Ln[LA_0] \rightarrow y = mx + b \text{ (equation of a line)} \quad (6)$$

Furthermore, with the variation of the reaction temperature and by using the Arrhenius Equation (7), it was possible to determine the activation energy and the heat of the reaction [61].

$$k = A\, e^{-\left(\frac{E_a}{R_g T}\right)} \quad (7)$$

$$Ln(k) = \frac{-E_a}{R_g} \cdot \frac{1}{T} + Ln(A) \rightarrow y = mx + b \text{ (equation of a line)} \quad (8)$$

where Ea, A, and Rg are the activation energy (J/mol), constant "pre-exponential or frequency factor" and ideal gas constant (8.3143 J/mol K), respectively. Both the frequency factor A and the activation energy Ea, are obtained by non-linear regression. The plot of ln k vs. 1/T is plotted by a straight line (Figure S6), the activation energy (Ea) is the slope of the line, and the pre-exponential factor (A) is the ordinate at the origin.

The experimental data were fitted, with pseudo-first-order kinetics, through Equation (6). With the adjustment, a constant kinetic value was obtained k = 3.57 × 10^{-3} min^{-1} with an R^2 of 0.98. As for the activation energy (Ea), adjusted with Equation (8), it yielded a value of 48.99 kJ/mol with an R^2 of 0.98. Additional data have been included as electronic Supplementary Material (Tables S1 and S2, Figures S5 and S6).

3.3. Esterification of Levulinic Acid to Methyl Levulinate in Pressure System, Study of the Stability

For the pressure system, the values obtained are shown in Table 9. It is observed that the conversion and yield results at 1 h of reaction were better for all materials, compared to the batch system at 5 h. The presence of pressure and high temperature in the autoclave system promotes the diffusion of the substrates through the pores of the material and their coordination with the active acid–basic sites, favoring the catalytic performance [41,64–67].

Table 9. Catalytic activity. T = 130 °C, P = 30 bars, 0.05 g catalyst.

Catalyst	Conversion %	Selectivity ML %	Yield %
M_A	50.18	58.63	29.42
M_A-U	50.08	67.64	33.87
M_A-UT	72.05	87.82	63.27
M_B	73.06	83.11	60.73
M_B-U	77.85	100	77.85
M_B-UT	85.89 (4) *	100	85.89
M_B-UT R1	70.18 (24) *	100	70.18
M_B-UT R2	50.46 (5) *	100	50.46
M_B-UT R3	40.20 (12) *	63.12	25.37
Cl_4Zr	98.24	100	98.24

* Zr content (ppm) in the supernatant, determined by MP-AES.

To study the stability of the catalyst, three recycles of the reaction were carried out with sample M_{B-UT} (M_B-UT R1, M_B-UT R2, M_B-UT R3); in each recycle, the catalyst was separated from the reaction mixture and dried at 90 °C. It was observed that the yield decreases while the selectivity is maintained in the first two recycles and falls in the third (Table 9).

By MP-AES, the concentration of Zr in the recycles supernatant was measured, finding values between 5 ppm and 24 ppm, that values represent ~(0.5–2.9)% Zr p/p leached from the solid in each recycle. The activity of the reaction in a homogeneous phase in the presence of Cl_4Zr was also measured, obtaining 98.24% yield to ML. This indicates that the presence of Zr leached in the supernatant did not generate a homogeneous phase reaction or increase the conversion in the recycles as shown in Table 9.

The decrease in catalytic activity can be mainly attributed to the partial blocking of Zr^{+4} sites by electrostatic attraction with ML molecules which cannot be removed by methanol in successive recycles. To support this explanation, after the third recycle catalyst, two reactions with ethanol (M_B-UT R4, M_B-UT R5) were carried out, obtaining an increase in the conversion (58.66–76.03%) with 100% selectivity (Figure 6).

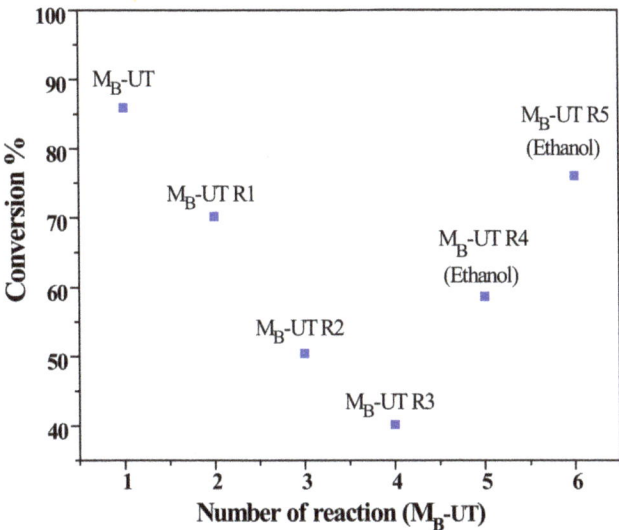

Figure 6. Catalitic activity of the M_B-UT reused with methanol (M_B-UT R1, M_B-UT R2, M_B-UT R3) and ethanol (M_B-UT R4, M_B-UT R5).

This behavior is attributed to the removal of methyl levulinate molecules on the surface catalyst due to the greater affinity and solubility of ethanol molecules, releasing the Zr^{+4} acid sites.

On the other hand, the M_B-UT R3 catalyst was washed with ethanol and ultrasonic stirring, then dried at 90 °C and analyzed by FTIR. Figure 7 shows FTIR spectra of M_B-UT, $M_{B\text{-}UT}$ used, M_B-UT R3, M_B-UT R3 supernatant, and M_B-UT R3 washed.

The characteristic peaks associated at COO–, C=C, Zr–O, Zr-(OC) groups were maintained in $M_{B\text{-}UT}$ used and M_B-UT R3 catalyst.

A peak at ~1716 cm^{-1} associated with ML in the supernatant M_B-UT R3, is correlated with the peak in the M_B-UT used and M_B-UT R3 indicating the ML adsorbed on the catalyst after use. A decrease in the intensity of this peak after M_B-UT R3 washed with ethanol indicates the removal of ML molecules coordinated with acid sites Zr^{+4} on the surface.

XRD patterns of the catalysts (M_B-UT, M_B-UT used, and M_B-UT R3) are shown in Figure S7. The intensity and the crystalline order of the MOFs phase were maintained after each recycle and mainly after the third one.

Finally, a decrease in catalytic activity with the successive reuse of the catalyst could be mainly related to the blockade of Zr^{+4} acid sites by ML and to a lesser extent by the leachate of this metal.

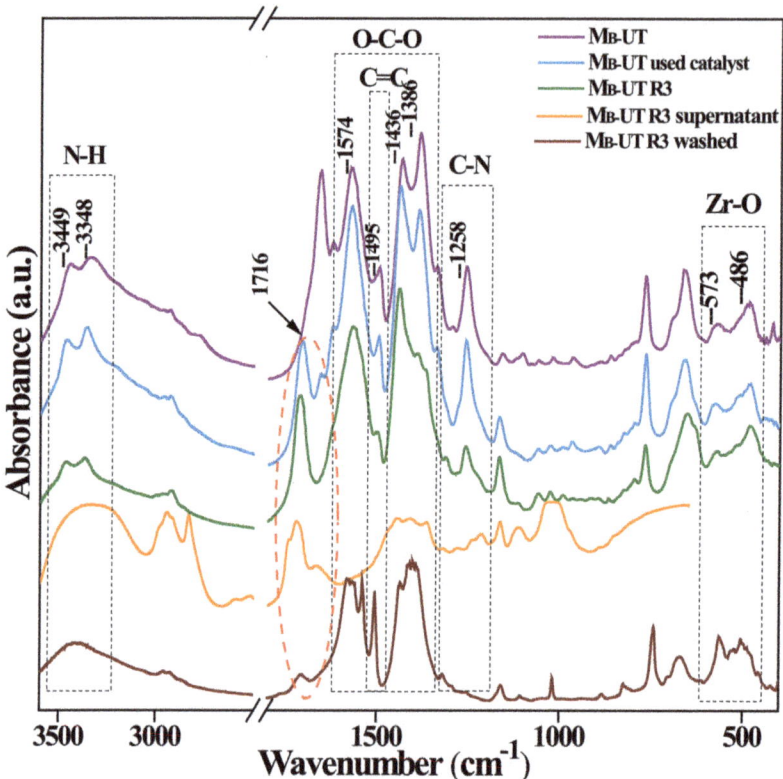

Figure 7. FTIR spectra for the fresh, used catalysts and supernatant reaction.

4. Conclusions

The synthesized materials presented outstanding crystallinity and porosity properties. The implementation of ultrasound and temperature in the synthesis promoted the formation of a micro/mesoporous structure, with crystal sizes between ~209.7 and ~488.5 nm and a homogeneous distribution of Zr, C, O, and N on the surface, which can be corroborated by BET and SEM analyses. Regarding the catalytic evaluation in a batch system, the materials of each series synthesized with ultrasound at 60 °C exhibited the highest conversions, where the M_B-UT sample is the one that presented the best activity, with a yield of 67.77% for the methyl levulinate. The optimal working conditions for this were 65 °C, 0.05 g of the catalyst and a mol ratio of reactants of 1:15 (LA:M). These results are associated with the greater dispersion of Lewis acid sites available to catalyze the reaction and also with the acid–base duality due to the -NH$_2$ group. This dual mechanism involves LA interacting with Zr (Lewis acid site) while the alcohol forms a hydrogen bond with the amino group (basic site). This simultaneous acid–base interaction favors the first reaction step.

A pseudo-homogeneous model was adjusted to first-order kinetics with a value of the reaction constant $k = 3.57 \times 10^{-3}$ min^{-1} and activation energy of 48.99 kJ/mol, with an R^2 of 0.98 for both parameters.

Improvements in the catalytic results were achieved, using more severe conditions, through the use of pressure reactors with greater selectivity to the desired product and reduction in the reaction time from 5 to 1 h. The deactivated catalyst observed in the successive recycles is mainly related to the blocking of the Zr^{+4} acid sites by electrostatic attraction of the ML molecules and, to a lesser extent, to the leaching of this metal. The spent material can be activated by ethanol washing allowing the catalytic activity to be maintained in the recycles.

Finally, it can be said that the synthesis of levulinate compounds from levulinic acid derived from biomass using heterogeneous catalysts such as UiO-66 and UiO-66-NH$_2$ is a highly viable and efficient alternative, which allows eco-compatible reactions to be carried out with the medium, with easy separation of the products of interest.

Supplementary Materials: The following supporting information can be downloaded at: https://www.mdpi.com/article/10.3390/chemengineering6020026/s1, Additional information about the XPS graphics, the particle size, the EDS images, the XRD pattern for the used catalyst, the CO-FTIR absorption spectra, the kinetics of the reaction, and the adjustments made have been included in the file supplied as Supplementary Materials.

Author Contributions: Conceptualization, A.C.H. and M.E.C.; methodology, D.A.B.F.; validation, A.C.H., M.E.C. and D.A.B.F.; formal analysis, E.R.-C., S.M.M., A.C.H., M.E.C. and D.A.B.F.; investigation, E.R.-C., S.M.M., A.C.H., M.E.C. and D.A.B.F.; writing—original draft preparation, D.A.B.F.; writing—review and editing, A.C.H., M.E.C. and E.R.-C.; supervision, M.E.C. and E.R.-C. All authors have read and agreed to the published version of the manuscript.

Funding: Projects PID-UTN-SCyT (MATCBCO 008094TC and MAIAICO 6570TC) Universidad Tecnológica Nacional. Projects RTI2018-099668-BC22 of Ministerio de Ciencia, Innovación y Universidades, and UMA18-FEDERJA- 126 and P20_00375 of Junta de Andalucía and FEDER funds.

Data Availability Statement: Not applicable.

Acknowledgments: Financial support from the Consejo Nacional de Investigaciones Científicas y Tecnológicas (CONICET) and Universidad Tecnológica Nacional–Facultad Regional Córdoba (UTN-FRC). E.R.C. thanks to project RTI2018-099668-BC22 of Ministerio de Ciencia, Innovación y Universidades, and projects UMA18-FEDERJA-126 and P20_00375 of Junta de Andalucía and FEDER funds.

Conflicts of Interest: The authors declare no conflict of interest.

References

1. Han, X.; Guo, Y.; Liu, X.; Xia, Q.; Wang, Y. Catalytic conversion of lignocellulosic biomass into hydrocarbons: A mini review. *Catal. Today* **2019**, *319*, 2–13. [CrossRef]
2. Climent, M.J.; Corma, A.; Iborra, S. Conversion of biomass platform molecules into fuel additives and liquid hydrocarbon fuels. *Green Chem.* **2014**, *16*, 516–547. [CrossRef]
3. Romanelli, G.P.; Ruiz, D.M.; Pasquale, G.A. *Química de la Biomasa y los Biocombustibles*; Editorial de la Universidad Nacional de La Plata (EDULP): Buenos Aires, Argentina, 2020. [CrossRef]
4. Adeleye, A.T.; Louis, H.; Akakuru, O.U.; Joseph, I.; Enudi, O.C.; Michael, D.P. A Review on the conversion of levulinic acid and its esters to various useful chemicals. *AIMS Energy* **2019**, *7*, 165–185. [CrossRef]
5. Corma Canos, A.; Iborra, S.; Velty, A. Chemical routes for the transformation of biomass into chemicals. *Chem. Rev.* **2007**, *107*, 2411–2502. [CrossRef]
6. Yu, Z.; Lu, X.; Xiong, J.; Ji, N. Transformation of levulinic acid to valeric biofuels: A review on heterogeneous bifunctional catalytic systems. *ChemSusChem* **2019**, *12*, 3915–3930. [CrossRef]
7. Serrano-Ruiz, J.C.; Pineda, A.; Balu, A.M.; Luque, R.; Campelo, J.M.; Romero, A.A.; Ramos-Fernández, J.M. Catalytic transformations of biomass-derived acids into advanced biofuels. *Catal. Today* **2012**, *195*, 162–168. [CrossRef]
8. Dutta, S.; Bhat, N.S. Recent advances in the value addition of biomass-derived levulinic acid: A review focusing on its chemical reactivity patterns. *ChemCatChem* **2021**, *13*, 3202–3222. [CrossRef]
9. Omoruyi, U.; Page, S.; Hallett, J.; Miller, P.W. Homogeneous catalyzed reactions of levulinic acid: To Γ-valerolactone and beyond. *ChemSusChem* **2016**, *9*, 2037–2047. [CrossRef]
10. Jeong, H.; Park, S.Y.; Ryu, G.H.; Choi, J.H.; Kim, J.H.; Choi, W.S.; Lee, S.M.; Choi, J.W.; Choi, I.G. Catalytic conversion of hemicellulosic sugars derived from biomass to levulinic acid. *Catal. Commun.* **2018**, *117*, 19–25. [CrossRef]
11. Badgujar, K.C.; Bhanage, B.M. Thermo-chemical energy assessment for production of energy-rich fuel additive compounds by using levulinic acid and immobilized lipase. *Fuel Process. Technol.* **2015**, *138*, 139–146. [CrossRef]
12. Bhat, N.S.; Mal, S.S.; Dutta, S. Recent advances in the preparation of levulinic esters from biomass-derived furanic and levulinic chemical platforms using heteropoly acid (HPA) catalysts. *Mol. Catal.* **2021**, *505*, 111484. [CrossRef]
13. Ogino, I.; Suzuki, Y.; Mukai, S.R. Esterification of levulinic acid with ethanol catalyzed by sulfonated carbon catalysts: Promotional effects of additional functional groups. *Catal. Today* **2018**, *314*, 62–69. [CrossRef]
14. Démolis, A.; Essayem, N.; Rataboul, F. Synthesis and applications of alkyl levulinates. *ACS Sustain. Chem. Eng.* **2014**, *2*, 1338–1352. [CrossRef]

15. Bart, H.J.; Reidetschläger, J.; Schatka, K.; Lehmann, A. Kinetics of esterification of succinic anhydride with methanol by homogeneous catalysis. *Int. J. Chem. Kinet.* **1994**, *26*, 1013–1021. [CrossRef]
16. Gong, W.; Liu, Y.; Li, H.; Cui, Y. Metal-organic frameworks as solid Brønsted acid catalysts for advanced organic transformations. *Coord. Chem. Rev.* **2020**, *420*, 213400. [CrossRef]
17. Qu, H.; Liu, B.; Gao, G.; Ma, Y.; Zhou, Y.; Zhou, H.; Li, L.; Li, Y.; Liu, S. Metal-organic framework containing BrØnsted acidity and Lewis acidity for efficient conversion glucose to levulinic acid. *Fuel Process. Technol.* **2019**, *193*, 1–6. [CrossRef]
18. Sun, Y.; Zhou, H.C. Recent progress in the synthesis of metal-organic frameworks. *Sci. Technol. Adv. Mater.* **2015**, *16*, 54202. [CrossRef]
19. Czaja, A.U.; Trukhan, N.; Müller, U. Industrial applications of metal–organic frameworks. *Chem. Soc. Rev.* **2009**, *38*, 1284–1293. [CrossRef]
20. Stock, N.; Biswas, S. Synthesis of metal-organic frameworks (MOFs): Routes to various MOF topologies, morphologies, and composites. *Chem. Rev.* **2012**, *112*, 933–969. [CrossRef]
21. Lillerud, K.P.; Cavka, J.H.; Lamberti, C.; Guillou, N.; Bordiga, S.; Jakobsen, S.; Olsbye, U. A new zirconium inorganic building brick forming metal organic frameworks with exceptional stability. *J. Am. Chem. Soc.* **2008**, *130*, 13850–13851. [CrossRef]
22. Rahmawati, I.D.; Ediati, R.; Prasetyoko, D. Synthesis of UiO-66 using solvothermal method at high temperature. *IPTEK J. Proc. Ser.* **2014**, *1*, 42–46. [CrossRef]
23. Abid, H.R.; Shang, J.; Ang, H.M.; Wang, S. Amino-functionalized Zr-MOF nanoparticles for adsorption of CO_2 and CH_4. *Int. J. Smart Nano Mater.* **2013**, *4*, 72–82. [CrossRef]
24. Bai, Y.; Dou, Y.; Xie, L.H.; Rutledge, W.; Li, J.R.; Zhou, H.C. Zr-based metal-organic frameworks: Design, synthesis, structure, and applications. *Chem. Soc. Rev.* **2016**, *45*, 2327–2367. [CrossRef] [PubMed]
25. Cirujano, F.G.; Corma, A.; Llabrés I Xamena, F.X. Zirconium-containing metal organic frameworks as solid acid catalysts for the esterification of free fatty acids: Synthesis of biodiesel and other compounds of interest. *Catal. Today* **2015**, *257*, 213–220. [CrossRef]
26. Wei, R.; Gaggioli, C.A.; Li, G.; Islamoglu, T.; Zhang, Z.; Yu, P.; Farha, O.K.; Cramer, C.J.; Gagliardi, L.; Yang, D.; et al. Tuning the properties of Zr_6O_8 nodes in the metal organic framework UiO-66 by selection of node-bound ligands and linkers. *Chem. Mater.* **2019**, *31*, 1655–1663. [CrossRef]
27. Li, H.; Chu, H.; Ma, X.; Wang, G.; Liu, F.; Guo, M.; Lu, W.; Zhou, S.; Yu, M. Efficient heterogeneous acid synthesis and stability enhancement of UiO-66 impregnated with ammonium sulfate for biodiesel production. *Chem. Eng. J.* **2021**, *408*, 127277. [CrossRef]
28. Herbst, A.; Janiak, C. MOF catalysts in biomass upgrading towards value-added fine chemicals. *CrystEngComm* **2017**, *19*, 4092–4117. [CrossRef]
29. Caratelli, C.; Hajek, J.; Cirujano, F.G.; Waroquier, M.; Llabrés i Xamena, F.X.; Van Speybroeck, V. Nature of active sites on UiO-66 and beneficial influence of water in the catalysis of Fischer esterification. *J. Catal.* **2017**, *352*, 401–414. [CrossRef]
30. Cirujano, F.G.; Corma, A.; Llabrés i Xamena, F.X. Conversion of levulinic acid into chemicals: Synthesis of biomass derived levulinate esters over Zr-containing MOFs. *Chem. Eng. Sci.* **2015**, *124*, 52–60. [CrossRef]
31. Garibay, S.J.; Cohen, S.M. Isoreticular synthesis and modification of frameworks with the UiO-66 topology. *Chem. Commun.* **2010**, *46*, 7700–7702. [CrossRef]
32. Lozano, L.A.; Iglesias, C.M.; Faroldi, B.M.C.; Ulla, M.A.; Zamaro, J.M. Efficient solvothermal synthesis of highly porous UiO-66 nanocrystals in dimethylformamide-free media. *J. Mater. Sci.* **2018**, *53*, 1862–1873. [CrossRef]
33. Ramli, N.A.S.; Zaharudin, N.H.; Amin, N.A.S. Esterification of renewable levulinic acid to levulinate esters using amberlyst-15 as a solid acid catalyst. *J. Teknol.* **2017**, *79*, 137–142. [CrossRef]
34. Liang, X.; Fu, Y.; Chang, J. Sustainable production of methyl levulinate from biomass in ionic liquid-methanol system with biomass-based catalyst. *Fuel* **2020**, *259*, 116246. [CrossRef]
35. Di, X.; Zhang, Y.; Fu, J.; Yu, Q.; Wang, Z.; Yuan, Z. Biocatalytic upgrading of levulinic acid to methyl levulinate in green solvents. *Process Biochem.* **2019**, *81*, 33–38. [CrossRef]
36. Chaffey, D.R.; Bere, T.; Davies, T.E.; Apperley, D.C.; Taylor, S.H.; Graham, A.E. Conversion of levulinic acid to levulinate ester biofuels by heterogeneous catalysts in the presence of acetals and ketals. *Appl. Catal. B Environ.* **2021**, *293*, 120219. [CrossRef]
37. Rubio-Martinez, M.; Batten, M.P.; Polyzos, A.; Carey, K.C.; Mardel, J.I.; Lim, K.S.; Hill, M.R. Versatile, high quality and scalable continuous flow production of metal-organic frameworks. *Sci. Rep.* **2014**, *4*, 5443. [CrossRef]
38. Lu, A.X.; McEntee, M.; Browe, M.A.; Hall, M.G.; Decoste, J.B.; Peterson, G.W. MOFabric: Electrospun nanofiber mats from PVDF/UiO-66-NH2 for chemical protection and decontamination. *ACS Appl. Mater. Interfaces* **2017**, *9*, 13632–13636. [CrossRef]
39. Kandiah, M.; Nilsen, M.H.; Usseglio, S.; Jakobsen, S.; Olsbye, U.; Tilset, M.; Larabi, C.; Quadrelli, E.A.; Bonino, F.; Lillerud, K.P. Synthesis and stability of tagged UiO-66 Zr-MOFs. *Chem. Mater.* **2010**, *22*, 6632–6640. [CrossRef]
40. Ragon, F.; Horcajada, P.; Chevreau, H.; Hwang, Y.K.; Lee, U.H.; Miller, S.R.; Devic, T.; Chang, J.S.; Serre, C. In situ energy-dispersive X-ray diffraction for the synthesis optimization and scale-up of the porous zirconium terephthalate UiO-66. *Inorg. Chem.* **2014**, *53*, 2491–2500. [CrossRef]
41. Arrozi, U.S.F.; Wijaya, H.W.; Patah, A.; Permana, Y. Efficient acetalization of benzaldehydes using UiO-66 and UiO-67: Substrates accessibility or Lewis acidity of zirconium. *Appl. Catal. A Gen.* **2015**, *506*, 77–84. [CrossRef]
42. Luu, C.L.; Van Nguyen, T.T.; Nguyen, T.; Hoang, T.C. Synthesis, characterization and adsorption ability of UiO-66-NH2. *Adv. Nat. Sci. Nanosci. Nanotechnol.* **2015**, *6*, 025004. [CrossRef]

43. Ding, Y.; Wei, F.; Dong, C.; Li, J.; Zhang, C.; Han, X. UiO-66 based electrochemical sensor for simultaneous detection of Cd(II) and Pb(II). *Inorg. Chem. Commun.* **2021**, *131*, 108785. [CrossRef]
44. Han, Y.; Liu, M.; Li, K.; Zuo, Y.; Wei, Y.; Xu, S.; Zhang, G.; Song, C.; Zhang, Z.; Guo, X. Facile synthesis of morphology and size-controlled zirconium metal-organic framework UiO-66: The role of hydrofluoric acid in crystallization. *CrystEngComm* **2015**, *17*, 6434–6440. [CrossRef]
45. Hou, J.; Luan, Y.; Tang, J.; Wensley, A.M.; Yang, M.; Lu, Y. Synthesis of UiO-66-NH2 derived heterogeneous copper (II) catalyst and study of its application in the selective aerobic oxidation of alcohols. *J. Mol. Catal. A Chem.* **2015**, *407*, 53–59. [CrossRef]
46. Thommes, M.; Kaneko, K.; Neimark, A.V.; Olivier, J.P.; Rodriguez-Reinoso, F.; Rouquerol, J.; Sing, K.S.W. Physisorption of gases, with special reference to the evaluation of surface area and pore size distribution (IUPAC Technical Report). *Pure Appl. Chem.* **2015**, *87*, 1051–1069. [CrossRef]
47. Ye, C.; Qi, Z.; Cai, D.; Qiu, T. Design and synthesis of ionic liquid supported hierarchically porous zr metal-organic framework as a novel Brønsted-Lewis acidic catalyst in biodiesel synthesis. *Ind. Eng. Chem. Res.* **2019**, *58*, 1123–1132. [CrossRef]
48. Hu, Z.; Peng, Y.; Kang, Z.; Qian, Y.; Zhao, D. A modulated hydrothermal (MHT) approach for the facile synthesis of UiO-66-type MOFs. *Inorg. Chem.* **2015**, *10*, 4862–4868. [CrossRef] [PubMed]
49. Wang, F.; Chen, Z.; Chen, H.; Goetjen, T.A.; Li, P.; Wang, F.X.; Alayoglu, S.; Ma, K.; Chen, Y.; Wang, T.; et al. Interplay of Lewis and brønsted acid sites in Zr-based metal-organic frameworks for efficient esterification of biomass-derived levulinic acid. *ACS Appl. Mater. Interfaces* **2019**, *11*, 32090–32096. [CrossRef] [PubMed]
50. Su, Y.; Zhang, Z.; Liu, H.; Wang, Y. $Cd_{0.2}Zn_{0.8}S$@UiO-66-NH2 nanocomposites as efficient and stable visible-light-driven photocatalyst for H2 evolution and CO_2 reduction. *Appl. Catal. B Environ.* **2017**, *200*, 448–457. [CrossRef]
51. Moulder, J.F.; Stickle, W.F.; Sobol, P.E.; Bomben, K.D. *Handbook of X-ray Photoelectron Spectroscopy*; Perkin-Elmer Corp., Eden Prairie: Chichester, UK, 1992; ISBN 9780470014226.
52. Wiersum, A.D.; Soubeyrand-Lenoir, E.; Yang, Q.; Moulin, B.; Guillerm, V.; Yahia, M.B.; Bourrelly, S.; Vimont, A.; Miller, S.; Vagner, C.; et al. An evaluation of UiO-66 for gas-based applications. *Chem.-Asian J.* **2011**, *6*, 3270–3280. [CrossRef]
53. Timofeeva, M.N.; Panchenko, V.N.; Jun, J.W.; Hasan, Z.; Matrosova, M.M.; Jhung, S.H. Effects of linker substitution on catalytic properties of porous zirconium terephthalate UiO-66 in acetalization of benzaldehyde with methanol. *Appl. Catal. A Gen.* **2014**, *471*, 91–97. [CrossRef]
54. Strauss, I.; Chakarova, K.; Mundstock, A.; Mihaylov, M.; Hadjiivanov, K.; Guschanski, N.; Caro, J. UiO-66 and UiO-66-NH2 based sensors: Dielectric and FTIR investigations on the effect of CO_2 adsorption. *Microporous Mesoporous Mater.* **2020**, *302*, 110227. [CrossRef]
55. Caretto, A.; Perosa, A. Upgrading of levulinic acid with dimethylcarbonate as solvent/reagent. *ACS Sustain. Chem. Eng.* **2013**, *1*, 989–994. [CrossRef]
56. Langlois, D.P.; Wolff, H. Pseudo esters of levulinic acid. *J. Am. Chem. Soc.* **1948**, *70*, 2624–2626. [CrossRef]
57. Lima, C.G.S.; Monteiro, J.L.; de Melo Lima, T.; Weber Paixão, M.; Corrêa, A.G. Angelica lactones: From biomass-derived platform chemicals to value-added products. *ChemSusChem* **2018**, *11*, 25–47. [CrossRef]
58. Al-Shaal, M.G.; Ciptonugroho, W.; Holzhäuser, F.J.; Mensah, J.B.; Hausoul, P.J.C.; Palkovits, R. Catalytic upgrading of α-angelica lactone to levulinic acid esters under mild conditions over heterogeneous catalysts. *Catal. Sci. Technol.* **2015**, *5*, 5168–5173. [CrossRef]
59. Ramli, N.A.S.; Sivasubramaniam, D.; Amin, N.A.S. Esterification of levulinic acid using ZrO_2-supported phosphotungstic acid catalyst for ethyl levulinate production. *Bioenergy Res.* **2017**, *10*, 1105–1116. [CrossRef]
60. Nandiwale, K.Y.; Sonar, S.K.; Niphadkar, P.S.; Joshi, P.N.; Deshpande, S.S.; Patil, V.S.; Bokade, V.V. Catalytic upgrading of renewable levulinic acid to ethyl levulinate biodiesel using dodecatungstophosphoric acid supported on desilicated H-ZSM-5 as catalyst. *Appl. Catal. A Gen.* **2013**, *460–461*, 90–98, ISBN 9120259026. [CrossRef]
61. Zubir, M.I.; Chin, S.Y. Kinetics of modified Zirconia-catalyzed heterogeneous esterification reaction for biodiesel production. *J. Appl. Sci.* **2010**, *10*, 2584–2589. [CrossRef]
62. Jrad, A.; Abu Tarboush, B.J.; Hmadeh, M.; Ahmad, M. Tuning acidity in zirconium-based metal organic frameworks catalysts for enhanced production of butyl butyrate. *Appl. Catal. A Gen.* **2019**, *570*, 31–41. [CrossRef]
63. Emel'yanenko, V.N.; Altuntepe, E.; Held, C.; Pimerzin, A.A.; Verevkin, S.P. Renewable platform chemicals: Thermochemical study of levulinic acid esters. *Thermochim. Acta* **2018**, *659*, 213–221. [CrossRef]
64. Feng, J.; Li, M.; Zhong, Y.; Xu, Y.; Meng, X.; Zhao, Z.; Feng, C. Hydrogenation of levulinic acid to γ-valerolactone over Pd@UiO-66-NH2 with high metal dispersion and excellent reusability. *Microporous Mesoporous Mater.* **2020**, *294*, 109858. [CrossRef]
65. Cao, W.; Lin, L.; Qi, H.; He, Q.; Wu, Z.; Wang, A.; Luo, W.; Zhang, T. In-situ synthesis of single-atom Ir by utilizing metal-organic frameworks: An acid-resistant catalyst for hydrogenation of levulinic acid to Γ-valerolactone. *J. Catal.* **2019**, *373*, 161–172. [CrossRef]
66. Sosa, L.F.; da Silva, V.T.; de Souza, P.M. Hydrogenation of levulinic acid to γ-valerolactone using carbon nanotubes supported nickel catalysts. *Catal. Today* **2020**, *381*, 86–95. [CrossRef]
67. Tulchinsky, M.L.; Briggs, J.R. One-pot synthesis of alkyl 4-alkoxypentanoates by esterification and reductive etherification of levulinic acid in alcoholic solutions. *ACS Sustain. Chem. Eng.* **2016**, *4*, 4089–4093. [CrossRef]

Article

Hydrogen and CNT Production by Methane Cracking Using Ni–Cu and Co–Cu Catalysts Supported on Argan-Derived Carbon

Fernando Cazaña [1], Zainab Afailal [2], Miguel González-Martín [1], José Luis Sánchez [2], Nieves Latorre [1], Eva Romeo [1], Jesús Arauzo [2] and Antonio Monzón [1,*]

[1] Departamento de Ingeniería Química y Tecnologías del Medio Ambiente, Instituto de Nanociencia y Materiales de Aragón, CSIC-Universidad de Zaragoza, 50018 Zaragoza, Spain; fcazana@unizar.es (F.C.); m.gonzalezmartin@posta.unizar.es (M.G.-M.); nlatorre@unizar.es (N.L.); evaromeo@unizar.es (E.R.)

[2] Departamento de Ingeniería Química y Tecnologías del Medio Ambiente, Instituto de Ingeniería de Aragón (I3A), Universidad de Zaragoza, 50018 Zaragoza, Spain; zainabafailal@unizar.es (Z.A.); jlsance@unizar.es (J.L.S.); jarauzo@unizar.es (J.A.)

* Correspondence: amonzon@unizar.es

Abstract: The 21st century arrived with global growth of energy demand caused by population and standard of living increases. In this context, a suitable alternative to produce CO_x-free H_2 is the catalytic decomposition of methane (CDM), which also allows for obtaining high-value-added carbonaceous nanomaterials (CNMs), such as carbon nanotubes (CNTs). This work presents the results obtained in the co-production of CO_x-free hydrogen and CNTs by CDM using Ni–Cu and Co–Cu catalysts supported on carbon derived from Argan (Argania spinosa) shell (ArDC). The results show that the operation at 900 °C and a feed-ratio CH_4:H_2 = 2 with the Ni–Cu/ArDC catalyst is the most active, producing 3.7 g_C/g_{metal} after 2 h of reaction (equivalent to average hydrogen productivity of 0.61 g H_2/g_{metal}·h). The lower productivity of the Co–Cu/ArDC catalyst (1.4 g_C/g_{metal}) could be caused by the higher proportion of small metallic NPs (<5 nm) that remain confined inside the micropores of the carbonaceous support, hindering the formation and growth of the CNTs. The TEM and Raman results indicate that the Co–Cu catalyst is able to selectively produce CNTs of high quality at temperatures below 850 °C, attaining the best results at 800 °C. The results obtained in this work also show the elevated potential of Argan residues, as a representative of other lignocellulosic raw materials, in the development of carbonaceous materials and nanomaterials of high added-value.

Keywords: CO_x-free hydrogen; CNTs; CDM; methane; argan-derived carbon; Ni–Cu; Co–Cu

1. Introduction

The potential applications of the carbonaceous nanomaterials (CNMs), for example, carbon nanotubes (CNTs), carbon nanofibers (CNFs) or graphene, due to their unique mechanical, electronic, chemical and physical properties, have motivated the enormous research effort carried out during the last few decades in almost all fields of nanoscience and nanotechnology [1–8].

Among the currently available technologies for the production of CNTs, the catalytic decomposition of light hydrocarbons in the gas (or vapor) phase stands as the most interesting method to satisfy the large potential demand, due to its easy scalability for production at large-scale and low-cost [9,10]. If the carbon source used is methane, the co-production of pure hydrogen is an additional and very relevant advantage due to the increasing necessity of the CO_x-free hydrogen in the actual energetic, environmental and political scenario [11–14]. Significant efforts were made to find the most suitable operating conditions and catalyst compositions to provide both, high carbon and hydrogen yields, and the necessary selectivity and quality for the desired carbon structure. However, the application of this process is still limited because of the rapid deactivation of the catalysts commonly employed [13,15,16].

Several metallic and also non-metallic (generally carbon materials) components of the catalysts were tested in the reaction of methane cracking. Thus, catalysts based on transition metals such as Ni, Fe, and Co are the most active, can operate at moderate temperature, and are also able to produce valuable CNMs as co-product [17–22]. The incorporation of a second metal as a catalytic promoter was reported to have positive effects on the activity and stability [23–26]. In previous studies, we demonstrated that an appropriate design of the catalyst composition and an adequate selection of the operating conditions (mainly temperature and CH_4/H_2 feed-ratio) allows not only the control of hydrogen yield in the catalytic decomposition of methane (CDM) reaction but also of the selectivity towards the desired carbonaceous nanomaterials [27]. Thus, reaction temperatures below 800–850 °C favor the formation of CNTs [27], while temperatures above 900–950 °C are more prone to produce graphene-related materials (GRMs) [24,28], depending on the catalyst composition.

Moreover, the increasing concern for the environment is leading to the study of new processes to obtain high added-value products and materials using renewable natural bioresources [11,29–31]. In this concern, one of the natural resources that can be used is the shells of argan (Argania Spinosa), a waste coming from the use of the kernels of their nuts for the production of argan oil, one of the most expensive vegetable oils in the world. The argan fruits are principally produced in Morocco and Tunisia, with an estimated production of 80,000 tons per year [32].

In this regard, the use of vine shoots wastes and cellulose as raw materials for the preparation of carbon-based supports of metallic catalysts is being investigated in our group [24,30,33,34]. These catalysts, prepared by controlled thermal decomposition of the impregnated raw materials (e.g., argan shells) with the metallic precursors exhibited high catalytic performance due to the good dispersion of metallic nanoparticles and the controlled textural properties of the support [35–37]. This technique easily allows converting hierarchical structures formed by a biological process into inorganic materials with high potential for different applications due to their porous texture developed during the preparation [38–42]. In addition, as the natural raw resource is previously impregnated with metallic catalytic precursors, the catalyst is obtained in one step [24]. The possibility of using different raw materials and metals makes this method a very useful and versatile tool to prepare mono or multimetallic supported catalysts with a wide range of compositions [24,35,37].

Thus, the ultimate goal is to demonstrate that it is possible to prepare proper catalytic supports using lignocellulosic residua ("Biomass Derived Carbons—BDC"), without very low commercial value, such as the "Argan Derived Carbon—ArDC" used here. The objective is to use these sustainable supports as an alternative to the traditional metallic oxides, such as silica or alumina, which have a large environmental and economic impact. In this context, we present here the results of the use of Ni–Cu and Co–Cu catalysts supported by Argan-Derived Carbon (ArDC) on the reaction of catalytic decomposition of methane to produce CO_x-free hydrogen and carbon nanomaterials.

The composition of the active phases (Ni–Cu and Co–Cu) was chosen based on previous works with the aim of obtaining different materials such as carbon nanotubes [27], or even graphene-related materials (GRMs) [24]. As a means to optimize the productivity and also the quality of the carbonaceous material formed, the effect of the main operating conditions during the reaction, i.e., feed composition and reaction temperature, was investigated.

2. Materials and Methods

2.1. Catalysts Preparation

Ni–Cu and Co–Cu/ArDC catalysts were prepared by thermal decomposition under reductive atmosphere of milled Argan shells, as described elsewhere [36]. First, the dried Argan shell was impregnated by incipient wetness with the appropriate amount of an aqueous solution containing the precursor salts ($Ni(NO_3)_2 \cdot 6H_2O$ provided by Alfa Aesar, $Co(NO_3)_2 \cdot 6H_2O$ and $Cu(NO_3)_2 \cdot 3H_2O$ provided by Sigma-Aldrich), to obtain a nominal weight composition of 5%Ni–1.35%Cu (atomic ratio Ni/Cu = 4) and 5%Co–1.35%Cu (atomic ratio Co/Cu = 4) with respect to the initial amount of Argan shell. After impregnation, the solid was dried at 100 °C overnight under 100 mL/min N_2 and thermally decomposed at 800 °C under reductive atmosphere (15% H_2/85% N_2) for 75 min with a heating rate of 50 °C/min. Finally, the catalyst was milled and sieved obtaining a particle size distribution ranging from 80 μm to 200 μm. These values are selected to avoid any internal diffusion limitations. The presence of the external limitations is also minimized by using high flow rates (700 mL/min, equivalent to 1680 mL/gcat.h).

2.2. Catalytic Decomposition of Methane

The catalytic decomposition of methane (CDM) was used to obtain carbon nanotubes (CNTs) at atmospheric pressure in a quartz thermobalance (CI Precision Ltd., Salisbury, UK, model MK2, https://www.ciprecision.com/ accessed on 20 February 2022) operated as a continuous fixed-bed differential reactor. In order to ensure the integrity of the equipment, and avoid any corrosive damage to the head of the balance, a continuous flow of inert gas was used, N_2 in our case. On the other part, the objective of introducing hydrogen into the feed mixture is to investigate its effect on the activity and stability of the catalysts during the reaction. In absence of H_2, the deactivation of the catalyst is very rapid.

Carbon mass evolution and temperature reaction were usually recorded during 120 min of reaction. The temperature range studied was 750–950 °C while the feed gas composition was varied from 0.5 to 3 of CH_4:H_2 ratio, using N_2 as balance gas (total flow rate 700 mL/min). In a characteristic experiment, 25 mg of the catalyst was placed into a copper mesh sample holder and then, the sample was heated at 10 °C/min under N_2 flow (700 mL/min) until reaching the reaction temperature. Once the desired reaction temperature was achieved, the reactive gas mixture CH_4/H_2/N_2 (700 mL/min) was fed into the reactor keeping constant the temperature during the reaction. Finally, the sample was cooled down to room temperature under N_2.

2.3. Catalysts and Carbonaceous Nanomaterials Characterization

The metal content of both catalysts was calculated by thermogravimetric analysis under oxidative atmosphere (air, 50 mL/min) in a Mettler Toledo TGA/SDTA 851 equipment. The desired amount of catalyst (~10 mg) was heated from 35 °C to 1000 °C with a heating rate of 10 °C/min. The final percentage of Ni–Cu and Co–Cu was determined considering the nominal metal quantity incorporated and the resulting mixture of metal oxides obtained upon combustion of the ArDC support during the TGA-air experiment.

Specific surface area and porosity values for both catalysts were obtained from N_2 adsorption–desorption isotherms measured at −196 °C using a TriStar 3000 instrument. Previous to the analysis, the catalysts were degassed at 200 °C for 8 h. BET specific surface area was calculated in the relative pressure range of P/P_0 = 0.01–0.10. Total pore volume was obtained at the maximum relative pressure reached by the adsorption branch (P/P_0 > 0.985), while the micropore volume was estimated by the t-plot method. Crystalline phase identification of the fresh catalysts was performed by X-ray diffraction (XRD) in a Rigaku D/Max 2500 apparatus from 5° to 90° 2θ degrees using Cu Kα radiation (λ = 1.5406 Å).

Morphological and structural information on the carbonaceous nanomaterials grown was obtained by electron microscopy. Transmission electron microscopy (TEM) images were acquired using an FEI Tecnai T-20 microscope operated at 200 kV. Scanning electron microscopy (SEM) and energy-dispersive X-ray spectroscopy (EDS) analysis were carried out in an FEI Inspect F50 microscope.

The carbonaceous structure of the Argan-derived carbon support and the quality of the CNTs obtained were characterized by Raman spectroscopy in a WiTec Alpha300 confocal Raman microscope using a 532 nm laser excitation beam. The intensity ratios I_D/I_G, I_{2D}/I_G

of the characteristic D (~1350 cm^{-1}), G (~1580 cm^{-1}), and 2D (~2690 cm^{-1}) bands from different sample spots (5 spectra) were used to evaluate the defects in the structure of the carbonaceous nanomaterials obtained.

3. Results

3.1. Fresh Catalyst Characterization

The Ni–Cu/ArDC and Co–Cu/ArDC catalysts were synthesized with nominal contents (wt%) of 5% Ni and 1.35% Cu and 5% Co and 1.35% Cu, respectively, with respect to the initial amount of milled Argan shell. After synthesis, the amounts of metal, calculated from the TGA-air data, were ca. Ni(16%)–Cu(4%) and Co(18%)–Cu(5%). This increment in the metal content is caused by the weight loss of carbonaceous support during the thermal decomposition of the natural source, a decisive stage to control the dispersion and the final content of the metal nanoparticles in the synthesized catalysts [24].

Information about the crystalline phase of the active metals before the reaction was addressed by the XRD technique in Figure 1. For the Ni–Cu/ArDC catalyst, it is observed that the peaks appearing at 44.3°, 51.6° and 76.0° do not correspond to the metallic Ni (45-1027 JCPDS) or Cu (04–0806 JCPDS) patterns. The shift observed in the 2θ value for these peaks can be associated with the formation of a Ni–Cu alloy [43]. However, the XRD pattern depicted in Figure 1 for Co–Cu/ArDC shows that both Co and Cu were present in metallic form (15–0806 JCPDS for Co^0 and 04–0806 JCPDS for Cu^0), not observing the formation of an alloy in this case. In addition, the Ni–Cu/ArDC and Co–Cu/ArDC patterns do not show any peak associated with the presence of oxidized species, confirming that the catalysts are completely in the reduced state.

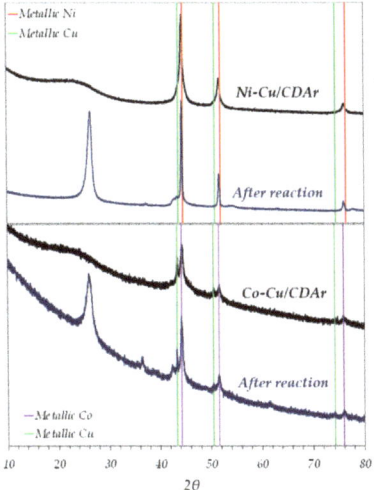

Figure 1. XRD patterns of fresh and used catalysts (800 °C; CH$_4$/H$_2$/N$_2$(%): 28.6/14.3/57.1).

The wide diffraction peak at about 2θ = 26°, attributed to the (002) plane of the hexagonal graphite structure (75–2078 JCPDS), confirms the amorphous nature of the carbonaceous ArDC support.

As it is shown in Figure 1, XRD results indicate that the metallic phases of both catalysts (Ni–Cu and Co–Cu) do not suffer relevant modifications during the reaction. The new peaks at 2θ = 26° correspond to the CNMs formed in each case, being more intense in the Ni–Cu sample, in agreement with the Raman and TEM results presented in the next paragraph.

Morphology of the bulk Ni–Cu/ArDC and Co–Cu/ArDC catalysts was addressed by electron microscopy images presented in Figure 2. The SEM images show the macrostructure and smoothness characteristic of this type of carbonized material for both catalysts prepared. In addition, TEM images obtained for the two fresh catalysts indicate that the metal nanoparticles were well distributed on the ArDC support owing to their characteristic biomorphic textural structure. TEM images were employed to estimate the particle size of the incorporated metals by measuring not less than 500 particles from different sample areas. The Ni–Cu/ArDC sample shows a trimodal metal particle size distribution, finding relatively small particles (~7 nm), intermediate particles (~17 nm) and large particles (~35 nm). Moreover, the Co–Cu/ArDC catalyst shows a bimodal metal particle size distribution with a large number of spherical nanoparticles about 5 nm and a low number of nanoparticles about 31 nm. In both cases, these metal size distributions are a consequence of the heating rate and high temperature used during the catalyst synthesis [36].

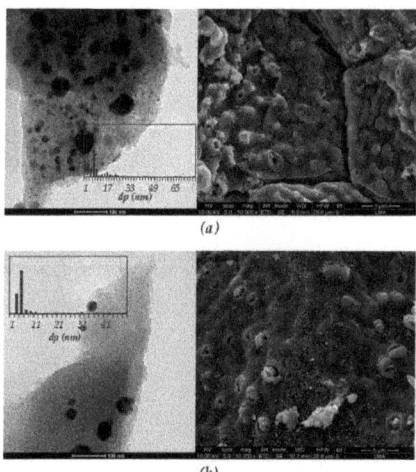

Figure 2. Electron microscopy study of the fresh catalysts: (**a**) Ni–Cu/ArDC (**b**) Co–Cu/ArDC.

Table 1 summarizes the textural properties of the fresh catalysts before reaction. In this Table 1, the values corresponding to the same type of metallic catalysts (Ni–Cu and Co–Cu), but supported on other carbon-based material, carbon derived from cellulose (CDC), and prepared using the same method are also included [24,27]. The main difference was found in the higher microporosity of the catalysts supported by the argan-derived carbon, especially in the case of the Ni–Cu catalyst. In this case, the Ni–Cu/ArDC has a 63% less pore volume but its microporosity is 2.5 times higher than the Co–Cu/ArDC one. Comparing the two catalysts supported by ArDC, both present similar textural properties corresponding to highly microporous materials, although the Ni-based sample has a slightly less developed porosity.

Table 1. Textural properties of the fresh catalysts.

Sample	BET Area (m^2/g)	Pore Vol. [1] (cm^3/g)	μpore Vol. [2] (cm^3/g)	μpore Vol. (%)
Ni–Cu/ArDC	404	0.168	0.138	82
Co–Cu/ArDC	433	0.182	0.164	90
Ni–Cu/CDC [24]	343	0.451	0.148	33
Co–Cu/CDC [27]	438	0.206	0.160	78

[1] Total pore volume at $P/P_0 > 0.989$. [2] Estimated with the t-plot method.

The carbonaceous structure of the fresh catalysts was also characterized by Raman spectroscopy shown in Figure 4. For both catalysts, the characteristic G band attributed to the in-plane vibration of the C sp^2-hybridized bonds was observed at ~1590 cm^{-1}, while the D band was associated with defects within the C sp^2 network of the carbonaceous support was detected at ~1350 cm^{-1} [44]. The wideness of these peaks (D and G bands) is a consequence of the different structural contributions of the several kinds of graphitic domains created during the thermal decomposition stage [36,45]. The wide and weak peak at about 2700 cm^{-1}, known as the overtone of the D band, indicates the amorphous nature of the catalyst support, which is in agreement with the XRD results shown in Figure 1.

3.2. Catalytic Decomposition of Methane

3.2.1. Influence of Reaction Temperature

The effect of the reaction temperature on the activity, productivity and morphology of the CNMs grown on both catalysts was evaluated in the interval from 750 °C to 950 °C, using a CH$_4$/H$_2$ ratio equal to 2 (28.6%CH$_4$/14.3%H$_2$/57.1%N$_2$). Figure 3 shows the evolution along time of the carbon concentration, m_C (g$_C$/g$_{metal}$), at different reaction temperatures for Ni–Cu/ArDC (Figure 3a) and Co–Cu/ArDC (Figure 3b). In addition, the summary of the results of activity and productivity after 2 h of reaction time for both catalysts is presented in Table 2.

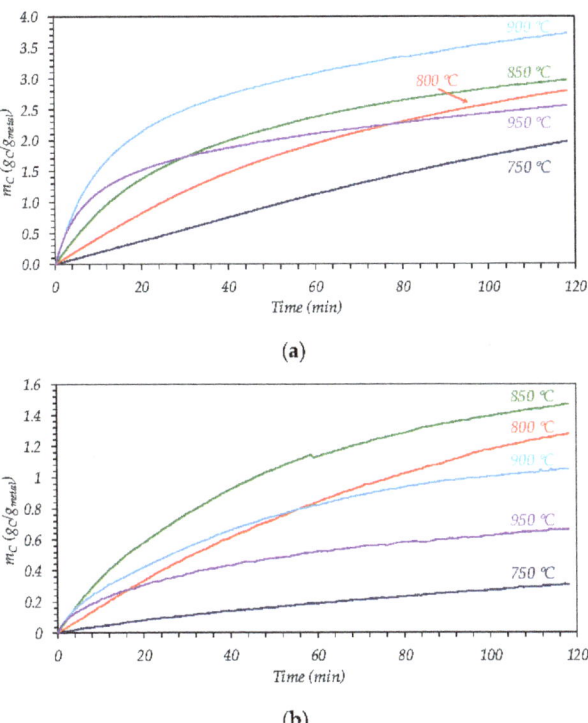

Figure 3. Cont.

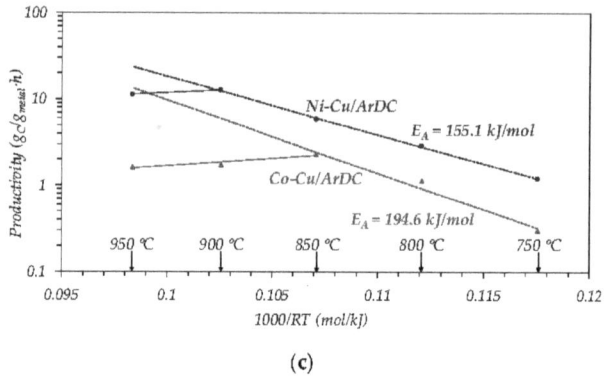

(c)

Figure 3. Influence of reaction temperature on evolution of carbon concentration, m_C: (**a**) Ni–Cu/ArDC; (**b**) Co–Cu/ArDC; (**c**) Final productivity at 2 h (Dashed line: Arrhenius plot).

Table 2. Influence of temperature on the initial reaction rate and carbon productivity *.

Temperature (°C)	Ni–Cu/ArDC			Co–Cu/ArDC		
	r_{C0} ($g_C/g_{metal}·min$)	Carbon Product. ($g_C/g_{metal}·h$) *	Carbon Product./r_{C0}	r_{C0} ($g_C/g_{metal}·min$)	Carbon Product. ($g_C/g_{metal}·h$) *	Carbon Product./r_{C0}
750	1.2	0.99	0.83	0.3	0.16	0.52
800	2.9	1.40	0.49	1.1	0.64	0.56
850	5.9	1.49	0.25	2.3	0.74	0.32
900	12.6	1.87	0.15	1.7	0.53	0.30
950	11.22	1.29	0.11	1.6	0.34	0.21

* Carbon productivity after 2 h of reaction.

For both samples, an increase in the carbon productivity (or space–time carbon yield, calculated as the carbon concentration, m_C, at a given time divided by that reaction time) is observed, and also in the initial carbon growth rate (r_{C0}, measured from the initial slope of the m_C vs. *time* curves in Figure 3) as the temperature passes from 750 °C to 900 °C, see Table 2. Thus, in the case of Ni-based catalyst, both the carbon productivity at 2 h and the initial reaction rate reached their maximum values, 1.87 and 12.60 $g_C/g_{metal}·h$, respectively, at 900 °C. At higher temperatures, the intense deactivation suffered for the catalyst after an initial short period of time of around 5 min, see Figure 3a, caused a continuous decrease in the reaction rate. Consequently, the final productivity of the catalyst is severely decreased in these conditions, attaining a value of 1.29 $g_C/g_{metal}·h$ at 950 °C.

In fact, the results shown in Figure 3 indicates that the reaction rate is continuously decreasing along the time on stream for all the conditions studied and also that the Co-based catalyst is less active than the Ni-based. Therefore, the decay of the catalyst activity occurs in all the cases and during all the time on stream. The value of carbon productivity attained at the end of one experiment of 2 h is the average of the decreasing reaction rates along the time, consequently, the carbon productivities are always lower than r_{C0}, see Table 2. The ratio between the carbon productivity and r_{C0} can be taken as an indicative index of the severity of the deactivation suffered in a given experiment.

As regards the effect of reaction temperature, considering that both phenomena, the reaction and the deactivation, are activated processes; an increase in the temperature will augment both rates, see Figure 3c. However, the lower value carbon productivity attained at high temperatures, e.g., 950 °C for the Ni–Cu catalyst, is a consequence of the prevalence in this case of the deactivation over the main reaction, due to the higher value of the activation energy of the deactivation process [27]. This fact implies that the observed activation energy becomes apparently negative (ca. −30 kJ/mol), as a consequence of the deactivation of the catalyst which indicates the change in the slope of the line drawn in the high-temperature zone.

This fact is confirmed for both catalysts by the continuous decrease in the temperature of the dimensionless ratio "Carbon productivity/r_{C0}" shown in Table 2. In addition, it is also interesting to note that for both catalysts, at the higher temperatures studied, the initial reaction rate, which corresponds to a fresh (not deactivated) catalyst situation, does not follow the expected trend of an activated phenomenon, and the value obtained at 950 °C for Ni–Cu, 11.22 $g_C/g_{metal} \cdot h$ is even lower than that obtained at 900 °C, see Table 2 and Figure 3c. This fact can be explained by a change in the catalyst selectivity towards the formation of a different type of carbon nanomaterial in these conditions, as can be seen in the characterization results section [27,30,46,47].

The effect of reaction temperature on the performance of Co–Cu/ArDC is qualitatively similar to the Ni–Cu catalyst, but the maximum value of carbon productivity (0.74 $g_C/g_{metal} \cdot h$) and r_{CO} (2.28 $g_C/g_{metal} \cdot h$)) appears at a lower temperature, 850 °C, and therefore the transition of the selectivity to the formation of different carbonaceous products is produced at lower temperatures. In this regard, the apparent activation energies of the initial reaction rate, r_{C0}, are 155 kJ/mol for Ni–Cu/ArDC and 194 kJ/mol for Co–Cu/ArDC, see Figure 3c. These results obtained considering only the values of r_{C0} in Table 2 below the maximum, confirm the better performance of the Ni–Cu/ArDC catalyst in this reaction. Furthermore, these values of the apparent activation energies are both lower than those obtained in this reaction with a Co–Cu/CDC catalyst in the same interval of temperatures, 221.7 kJ/mol.

The presence of Cu is always beneficial in increasing the activity and lifetime of the monometallic catalysts based on Ni or Co [25]. In comparison with the metals, Cu presents quite lower solubility of the carbon atoms, and also a lower methane dissociation rate (hydrogen atom abstraction from the adsorbed methane molecules). These facts reduce the formation of amorphous carbon deposits (i.e., deactivating carbon) explaining the increase in the net rate of reaction in presence of Cu.

On the other side, although the procedure of preparation of the catalyst is the same, the behavior of the different metals, Ni–Cu vs. Co–Cu, during the stage of thermal decomposition under a reductive atmosphere is clearly different. Thus, the results in Table 1 (Textural properties) and in Figures 1 and 2 reveal some important differences. During the thermal decomposition of the Argan shells impregnated with the metallic precursors, the process that really occurs is fast catalytic pyrolysis that decomposes both the metallic and the carbon precursors, producing an in situ pyrolysis–gasification of the evolving raw material (Argan). The different catalytic activity of Ni for the pyrolysis compared to Co, and also the ability of the Ni and Cu to form an alloy, explain the different structural properties of the carbon formed, and also the different nanoparticle size distributions presented in Figure 2. Consequently, the higher activity of Ni–Cu catalyst with respect to the Co–Cu sample can be due to the different metallic nanoparticle size distribution of these catalysts, shown on the inserts in Figure 2. Thus, the Co–Cu/ArDC sample, Figure 2b, contains a larger amount of small metal nanoparticles (<5 nm), while the quantity of the small particles is significantly lower for the Ni-based catalyst. The catalyst synthesis method used here determines the porous structure of the carbonaceous support developed and therefore the metallic particle size distribution [36,48]. If the microporosity of the support formed is very high, as in the case of the Co–Cu, a large fraction of the smaller metallic nanoparticles is confined to the internal structure of these micropores, remaining embedded inside the carbon matrix. In these conditions, the diffusion of the methane molecules to the exposed surface of the metallic nanoparticles is hindered, and consequently, the growth of the CNMs is slowed down [36]. Moreover, it was demonstrated that very small metallic NPs yield low growth rates and fast deactivation. On the contrary, if the NPs are very large, they have low activity due to the low exposed area, indicating the existence of an optimum size to maximize the catalyst activity [49]. The formation of an alloy in the case of the Ni–Cu catalyst extends the lifetime of the catalyst in this reaction, minimizing the formation of amorphous carbon deposits, which finally contribute to the decay of the catalyst activity [50–52]. The coupling of all these factors explains the higher activity of Ni-based catalysts in comparison to the Co catalysts.

In general, the amount of carbon formed should increase with the metal loading, but as obtained here, this result also depends on the nanoparticle size distribution, the potential formation of an alloy, and the textural properties of the support. Thus, the Co-based catalyst has a higher metal loading (18% of Co + 5% of Cu, wt%) than the Ni sample (16% of Ni + 5% of Cu, wt%) and yet it is much less productive. Therefore, the effect of metal loading on catalyst productivity is clearly modified by the particle size distribution and the textural properties of the support, which are controlled during the catalyst preparation stage. Furthermore, the higher solubility of carbon atoms on the Ni–Cu nanoparticles also explains the higher productivity of this catalyst, but the lower selectivity for the formation of pristine CNTs, as in the case of the Co-based catalyst.

In terms of hydrogen productivity at 2 h, the values attained are 0.61 g $H_2/g_{metal} \cdot h$ with Ni–Cu/ArDC at 900 °C; and 0.25 g $H_2/g_{metal} \cdot h$ at 850 °C for Co–Cu/ArDC. All these results demonstrate the great potential of the argan residue in the development of proper catalytic supports for hydrogen production by the decomposition of methane.

Figure 4a,b show the Raman spectra of the samples after reaction at increasing temperatures, from 750 °C to 950 °C, using a feed-ratio $CH_4/H_2 = 2$ (28.6 % CH_4, 14.3 % H_2, 57.1 % N_2). For both catalysts, it was observed the appearance of a maximum value of I_G/I_D ratio. In the case of Ni–Cu catalyst, this maximum is more intense and occurs at around 850 °C. The variation of this ratio in the case of the Co–Cu is less intense, and the maximum appears at around 900 °C.

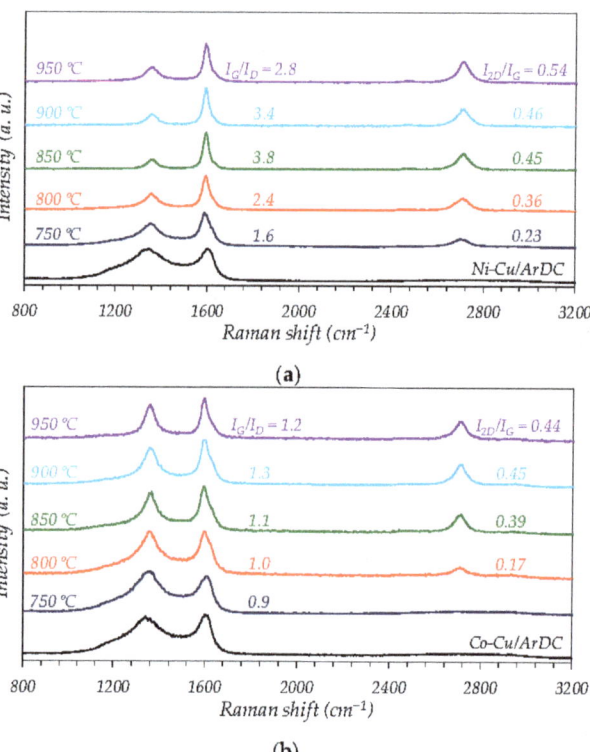

Figure 4. Cont.

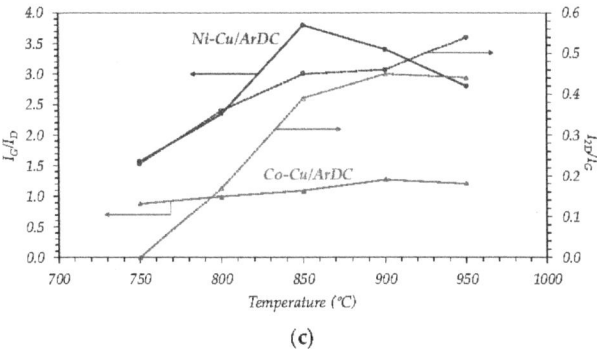

Figure 4. Influence of temperature on the Raman results of the used catalysts: (**a**) Ni–Cu/ArDC; (**b**) Co–Cu/ArDC, (**c**) I_G/I_D (solid symbols) and I_{2D}/I_G (open symbols) ratios.

These results are in agreement with the variation of the kinetic regime observed in Figure 3c. At low temperatures (till 850 °C), there is an increase in the intensity of the G band, related to the increase in the productivity, and a decrease in the D band, as a consequence of an increase in the size and crystallinity (i.e., fewer defects) of the graphitic domains of the obtained carbon material.

Above this point, the change of tendency in the evolution of the I_G/I_D ratio is due to a dramatic modification of the nature of carbon formed in agreement with the kinetic results, see Figure 3, and TEM observations, see Figures 5 and 6. On the other hand, as temperature augments, the separation between the D and G bands becomes clearer, and the intermediate shoulder between both bands (corresponding to the original carbon support) almost disappeared as a result of the increase in the surface coverage by the CNMs grown, which was risen with the reaction temperature (Figure 3).

As regards the evolution of the I_{2D}/I_G ratio, for the Ni–Cu/ArDC sample, the rise of this ratio suggests that the materials obtained at increasing temperatures are of nature more graphitic, with larger in-plane crystallite sizes (L_a) [24]. The values obtained, $I_{2D}/I_G < 0.50$, suggest that such graphitic nanostructures, which encapsulate the metallic NPs, have more than five stacked layers of graphene [53].

Figures 5 and 6 show the images of electron microscopy, SEM and TEM, of both catalysts after the reaction at the different temperatures studied. As was detected by the kinetic experiments, Figure 3, and the Raman results, Figure 4, these SEM and TEM images confirm the change in the structure and morphology observed as the temperature is increased. For the Ni–Cu/ArDC, the transition temperature is placed in the interval of 800–850 °C, and for the Co–Cu/ArDC, the change is between 800 and 850 °C. In the case of the Ni-based catalyst, Figure 5, until 800 °C the CNMs are mainly composed of CNTs and CNFs with a high proportion of defects as was also indicated by the Raman results. At temperatures above this point, the materials produced are of graphitic nature, that are encapsulating the metal nanoparticles, which are embedded inside these thick and short nanofibers. For the Co-based catalyst, Figure 6, at 750 °C the formation of pristine CNTs is still incipient, and these types of materials are formed, maintaining excellent quality, till 800 °C. From this temperature, the TEM and Raman results show a slight decrease in the quality, but the majority of products still are the CNTs. These results demonstrate that metallic catalysts supported on carbon derived from lignocellulosic residues, such as argan shells, are excellent candidates to produce CNTs of good quality, comparable to that obtained using commercial cellulose [27].

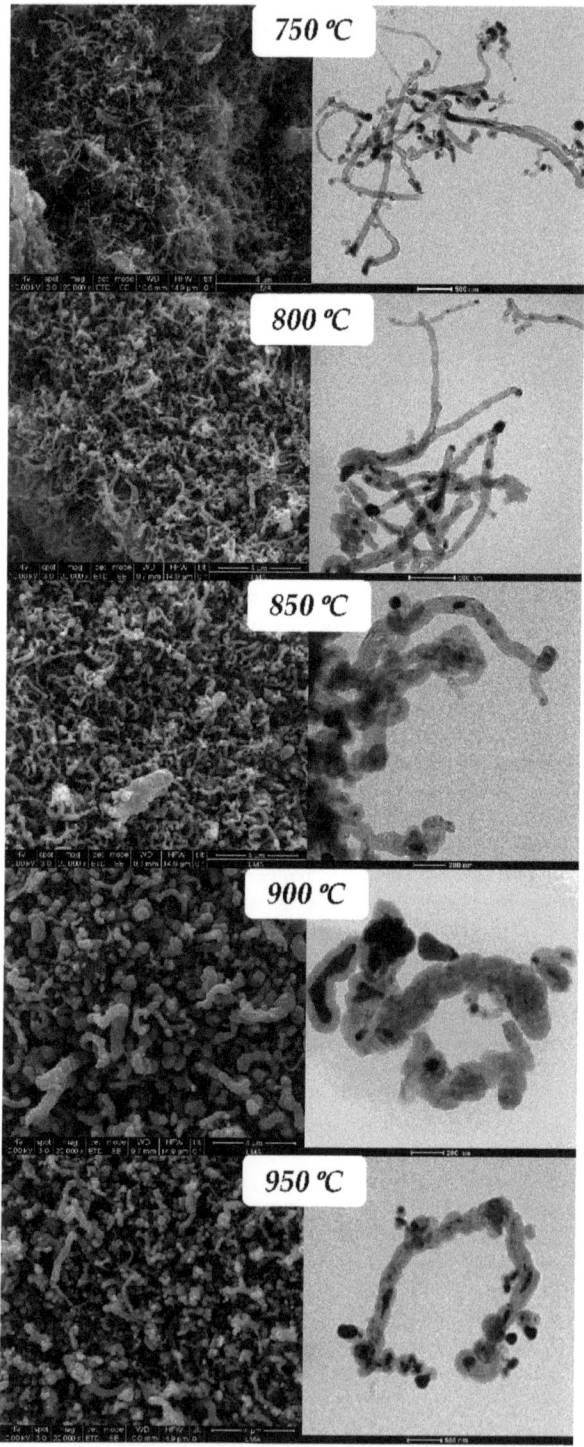

Figure 5. SEM and TEM images of the carbonaceous nanomaterials grown at different reaction temperatures for Ni–Cu/ArDC. Feed composition(%): $CH_4/H_2/N_2$:28.6/14.3/57.1.

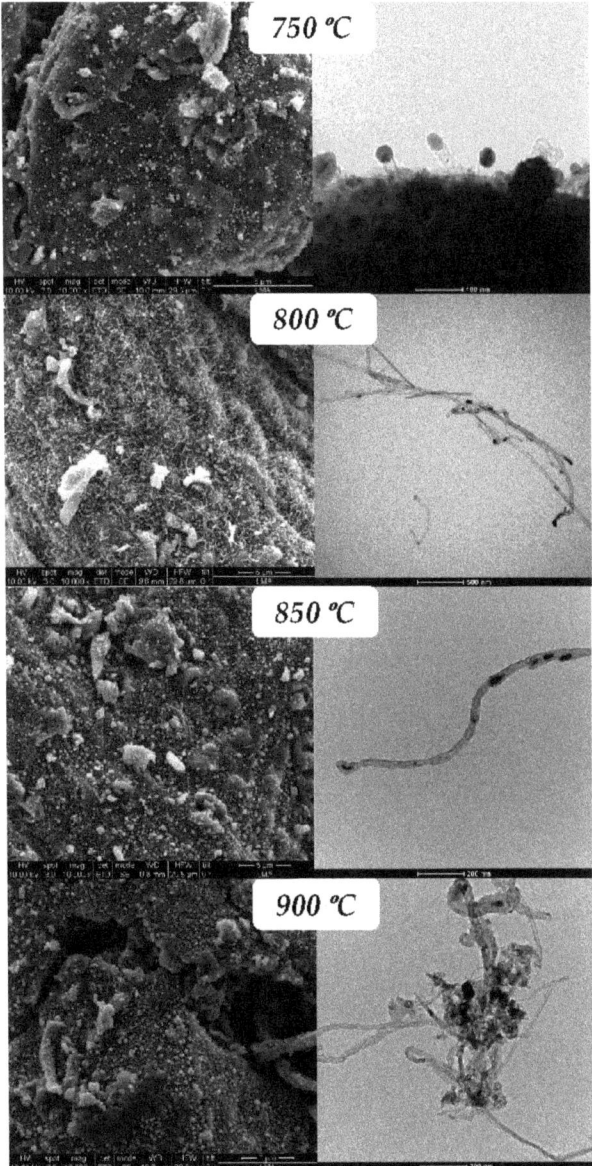

Figure 6. SEM and TEM images of the carbonaceous nanomaterials grown at different reaction temperatures for Co–Cu/ArDC. Feed composition (%):$CH_4/H_2/N_2$: 28.6/14.3/57.1.

3.2.2. Influence of Feed Composition

The effect of the feed gas composition on the carbon production and morphology of the grown carbonaceous materials was evaluated by varying the $CH_4:H_2$ ratio from 0.5 to 3 for both catalysts. All the experiments were carried out at 800 °C given that at this temperature the selectivity towards CNTs was the highest, as just described in the previous section. Figure 7 and Table 3 show that both the reaction rate and the carbon productivity after 2 h of reaction increase with the $CH_4:H_2$ ratio in the interval studied.

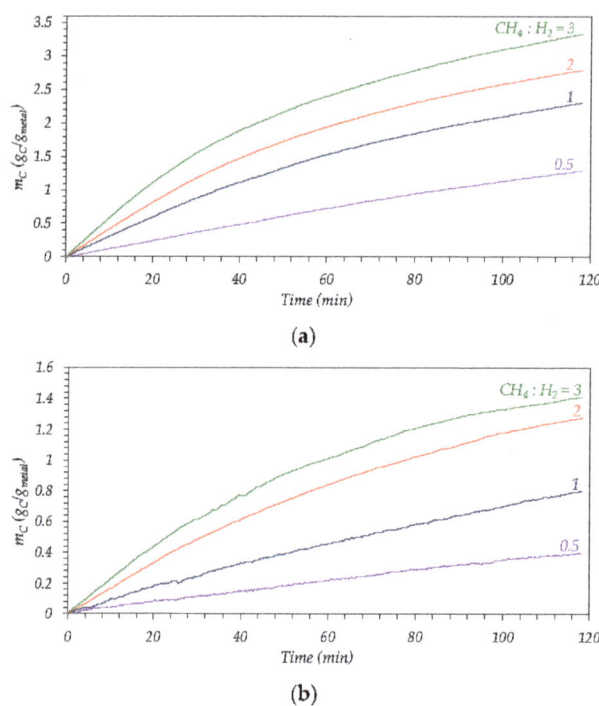

Figure 7. Influence of CH$_4$/H$_2$ feed-ratio on the evolution of carbon concentration along time: (a) Ni–Cu/ArDC; (b) Co–Cu/ArDC.

Table 3. Influence of CH$_4$/H$_2$ feed-ratio on the initial reaction rate and carbon productivity.

	Ni–Cu/ArDC			Co–Cu/ArDC		
CH$_4$:H$_2$	r_{C0} (g$_C$/g$_{metal}$·min)	Carbon Product. (g$_C$/g$_{metal}$·h) *	Carbon Product./r_{C0}	r_{C0} (g$_C$/g$_{metal}$·min)	Carbon Product. (g$_C$/g$_{metal}$·h) *	Carbon Product./r_{C0}
0.5	0.8	0.65	0.83	0.2	0.20	0.83
1	2.0	1.16	0.57	0.6	0.40	0.67
2	2.9	1.40	0.49	1.1	0.64	0.56
3	4.1	1.67	0.41	1.6	0.71	0.45

* Carbon productivity after 2 h of reaction.

As in the study of the effect of reaction temperature, the Ni-based catalyst is more active and productive that the Co catalyst at all the ratios studied. As was previously discussed, the higher proportion of small metallic NPs (<5 nm) in the Co–Cu/ArDC catalyst leaves this fraction of the metal buried in the micropores of the support, diminishing the activity of the catalyst. In addition, these small nanoparticles are less active to catalyze the formation of the CNTs. Therefore, these results give the clue for the optimization of the metallic particle size distribution of the catalyst, which is determined by the duration and the final temperature used during the thermal decomposition stage [36,48,54,55].

The maximum values of carbon productivity, 1.67 g$_C$/g$_{metal}$·h for Ni–Cu/ArDC and 0.71 g$_C$/g$_{metal}$·h for Co–Cu/ArDC, are obtained at the highest ratio CH$_4$:H$_2$, see Table 3. However, the quotient "carbon productivity/r_{C0}" continuously decreases as the CH$_4$:H$_2$ rises, indicating that the enhancement of both processes (carbon growth and catalyst deactivation) due to the methane content, is higher for the deactivation rate than for the carbon growth rate. That means that the apparent kinetic order with respect to methane is higher in the case of the deactivation rate.

These facts are explained considering that during the reaction of methane decomposition, the increase in methane amount in the feed enhances the carburization of the metallic nanoparticles, growing the number of carbon atoms dissolved in them and thus favoring the carbon precipitation at the metal–support interface [24,56,57]. At the same time, the presence of large amounts of carbon atoms on the exposed surface of the metallic nanoparticle, which is the situation corresponding to high methane concentrations, also favors the formation of encapsulating graphitic nanostructures that cause the catalyst deactivation. The balance between the rates of reaction and of the deactivation is quite similar for both catalysts, as is shown by the evolution of the ratio carbon productivity/r_{C0}, see Table 3, indicating the apparent orders of reaction for CH_4 and H_2 are similar for both catalysts.

As regards the results with the Co-based catalyst, the Raman spectra in Figure 8b are indicating the high selectivity of this catalyst for the formation of carbon nanotubes, and there are no significant differences among the spectra obtained at the different $CH_4:H_2$ ratios. However, for the case of $CH_4:H_2 = 0.5$, the Raman result is very similar to that obtained with the fresh catalyst, indicating the low coverage of the catalyst surface by the CNTs grown due to the low carbon productivity obtained under this reaction condition, see Figure 7 and Table 3.

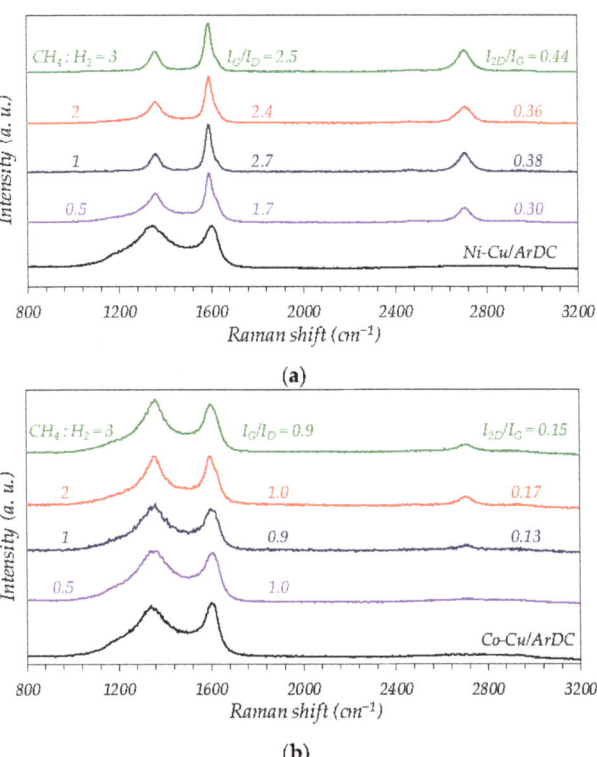

Figure 8. Raman spectra obtained for (**a**) Ni–Cu/ArDC and (**b**) Co–Cu/ArDC. Influence of feed composition ($CH_4:H_2$ ratio). Reaction temperature: 800 °C.

Figures 9 and 10 show the electron microscopy, SEM and TEM, results of the carbonaceous nanomaterials obtained at 800 °C using different $CH_4:H_2$ ratios for both catalysts. In the case of the Ni–Cu/ArDC, and in agreement with the kinetic results (see Figure 7a), the images in Figure 9 show that the increase in the methane concentration boosts the quantity, length and especially the thickness of the carbon materials grown, that in this case are mainly CNFs.

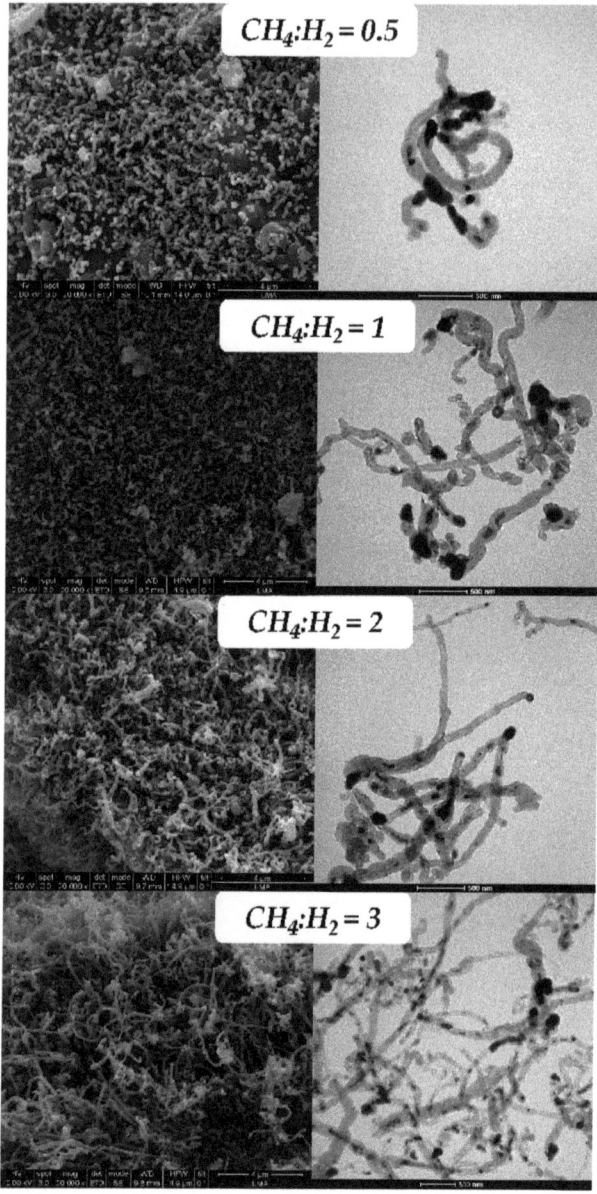

Figure 9. SEM and TEM images of the carbonaceous nanomaterials grown at different $CH_4:H_2$ ratios for Ni–Cu/ArDC. Reaction temperature: 800 °C.

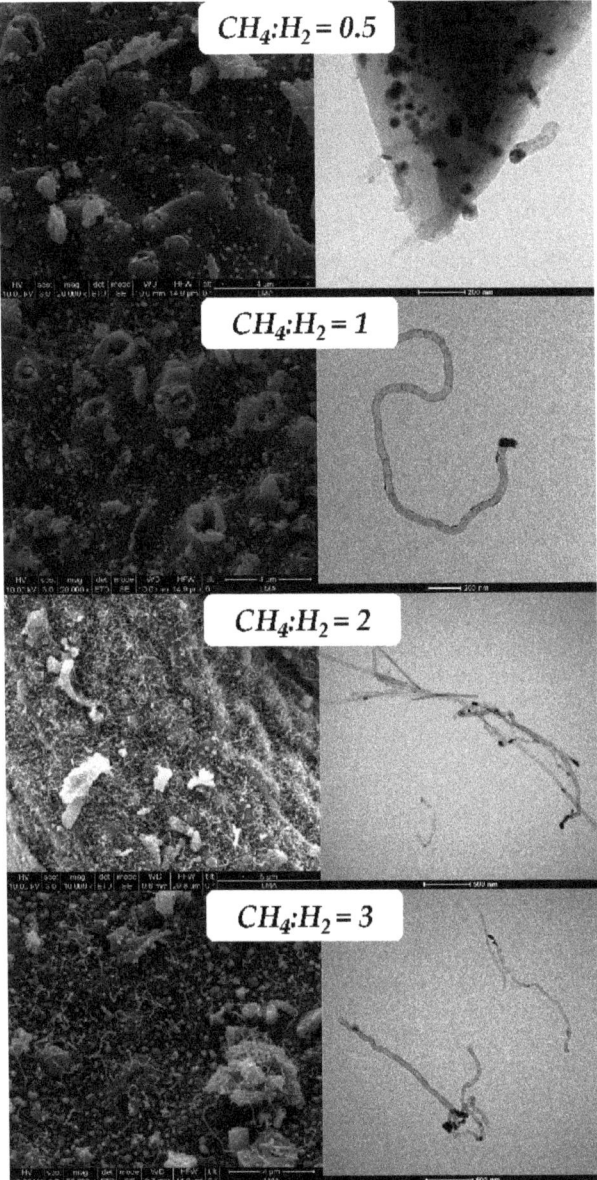

Figure 10. SEM and TEM images of the carbonaceous nanomaterials grown at different $CH_4:H_2$ ratios for Co–Cu/ArDC. Reaction temperature: 800 °C.

As was discussed previously, a large proportion of methane in the feed promotes the carburization step of the metallic NPs, growing the number of carbon atoms dissolved in them and thus favoring the carbon precipitation and growth. However, at the reaction temperature used (800 °C), the main part of carbon produced is in form of CNFs instead of CNTs, even at the low methane content experiment ($CH_4:H_2 = 0.5$).

In the case of Co-based catalyst, SEM and TEM results presented in Figure 10, are in total agreement with the Raman results, obtaining the selective formation of CNTs. Thus, at low methane contents ($CH_4:H_2 = 0.5$) the formation of CNTs is very incipient, see Figure 10, in accordance with the corresponding Raman spectra in Figure 8b. In these conditions of low partial pressure of methane, the low chemical potential for the carburization of the metallic NPs slows down the diffusion and precipitation of CNTs and therefore the amount and length of the CNTs formed are very low, being only detectable by TEM, see Figure 10.

4. Conclusions

In this work, Ni(16%)–Cu(4%) and Co(18%)–Cu(5%) catalysts supported on carbon derived from Argan shells (ArDC) were proved to be active and selective for the production of carbonaceous nanomaterials and hydrogen via catalytic decomposition of methane. The most active catalyst was the Ni-based, reaching maximum carbon productivity of 1.87 gC/gmetal·h at 900 °C, while the best productivity of the Co-based catalyst was 0.74 gC/gmetal·h at 850 °C, in both cases with a $CH_4:H_2$ =2. These results are comparable to those obtained with catalysts supported on commercial cellulose-derived carbon.

The higher activity of the Ni catalyst is explained in terms of the different particle size distribution and the average size of the metallic NPs. Thus the Co-based catalyst presents a large proportion of small NPs, which are less active due to the effect of the diffusional restrictions caused by the presence of these smaller NPs in the micropores of the support. On the other hand, the formation of an alloy in the case of the Ni–Cu catalyst extends the lifetime of the catalyst minimizing the formation of amorphous carbon deposits, which finally cause the decay of the catalyst activity.

The kinetic results indicate that both catalysts present a maximum in the CNMs productivity with the reaction temperature. This maximum is given at around 900 °C for the catalyst based on Ni, and at 850 °C for the catalysts based on Co. In both cases, this behavior can be attributed to the fact that an increment in the reaction temperature promotes a rapid diffusion of the carbon atoms through the metal nanoparticles. This fact increases the carbon precipitation rate at the metal catalyst support interface, favoring the formation of the carbonaceous nanomaterials. At temperatures above the maximum, the deactivation of the catalyst by the formation of encapsulating graphitic structures is boosted, decreasing the carbon productivity. On the other hand, the increment in the $CH_4:H_2$ ratio produces an increase in the CNMs productivity for the two catalysts. The increase in methane partial pressure enhances the carburization of the exposed surface of the metallic nanoparticles, increasing the number of carbon atoms dissolved in them and thus favoring the carbon precipitation at the metal–support interface.

Regarding the type of carbonaceous nanomaterials obtained in the function of the operational conditions, at temperatures below 800 °C, the Ni–Cu/ArDC catalyst is selective mainly for the production of CNFs. SEM, TEM and Raman results indicate that the highest graphitic character of these CNFs was obtained at 800 °C using a $CH_4:H_2$ = 3. In the case of the Co–Cu catalyst, operating at temperatures below 850 °C, the main solid product obtained was carbon nanotubes.

The amount of carbon formed should increase with the metal loading, but as obtained here, this result also depends on the nanoparticle size distribution, the potential formation of an alloy, and the textural properties of the support. Thus, the Co-based catalyst has a higher metal loading (18% of Co + 5% of Cu, wt%) than the Ni sample (16% of Ni + 5% of Cu, wt%) and yet it is less productive. Therefore, the effect of metal loading on catalyst productivity is clearly modified by these factors, which are controlled during the catalyst preparation stage. Furthermore, the higher solubility of carbon atoms on the Ni–Cu nanoparticles also explains the higher productivity of this catalyst, but the lower selectivity for the formation of pristine CNTs, as in the case of the Co-based catalyst.

In summary, although the Co-based catalyst is less productive than the Ni-based one, their high selectivity to the CNTs formation confirms the elevated potential of the Co-based catalyst supported on carbon derived from renewable lignocellulosic residues, such as argan shells, for the production of valuable CNMs by CCVD of methane. However, in terms of productivity of hydrogen and carbon, the Ni–Cu/ArDC is the more active and therefore the more interesting for the production of CO_x-free hydrogen.

Author Contributions: Conceptualization, J.L.S., N.L., E.R., J.A. and A.M.; methodology, F.C., Z.A., M.G.-M., J.L.S., N.L., E.R., J.A. and A.M.; validation, F.C., J.L.S., N.L., E.R., J.A. and A.M.; formal analysis, J.A. and A.M.; investigation, F.C., Z.A., M.G.-M., J.L.S., N.L., E.R., J.A. and A.M.; resources, N.L., E.R., J.A. and A.M.; data curation, F.C., N.L., E.R. and A.M.; writing—original draft preparation, F.C. and A.M.; writing—review and editing, N.L., E.R., J.A. and A.M.; visualization, F.C., E.R., J.A. and A.M.; supervision, J.L.S., N.L., E.R., J.A. and A.M.; project administration, J.A. and A.M.; funding acquisition, N.L., J.A. and A.M. All authors have read and agreed to the published version of the manuscript.

Funding: Grant PID2020-113809RB-C31 funded by MCIN/AEI/10.13039/501100011033; grant PLEC2021-008086 funded by MCIN/AEI/10.13039/501100011033 and by the European Union NextGenerationEU/PRTR; grant PRE2018-086557 funded by MCIN/AEI/10.13039/501100011033 and by ESF Investing in your future.

Data Availability Statement: Not applicable.

Acknowledgments: The authors would like to acknowledge the use of Servicio General de Apoyo a la Investigación-SAI and Laboratorio de Microscopías Avanzadas-LMA, Universidad de Zaragoza.

Conflicts of Interest: The authors declare no conflict of interest.

References

1. Dresselhaus, M.S.; Dresselhaus, G.; Hone, J. (Eds.) *Carbon Nanotubes: Synthesis, Structure, Properties and Applications*; Springer: Berlin/Heidelberg, Germany, 2001.
2. Serp, P.; Corrias, M.; Kalk, P. Carbon nanotubes and nanofibers in catalysis. *Appl. Catal. A* **2003**, *253*, 337–358. [CrossRef]
3. Terrones, M. Science and Technology of the Twenty-First Century: Synthesis, Properties, and Applications of Carbon Nanotubes. *Annu. Rev. Mater. Res.* **2003**, *33*, 419–501. [CrossRef]
4. Ding, S.; Xiang, Y.; Ni, Y.Q.; Kumar Thakur, V.; Wang, X.; Han, B.; Ou, J. In-situ synthesizing carbon nanotubes on cement to develop self-sensing cementitious composites for smart high-speed rail infrastructures. *Nano Today* **2022**, *43*, 101438. [CrossRef]
5. Zhou, X.; Wang, Y.; Gong, C.; Liu, B.; Wei, G. Production, structural design, functional control, and broad applications of carbon nanofiber-based nanomaterials: A comprehensive review. *Chem. Eng. J.* **2020**, *402*, 126189. [CrossRef]
6. Bai, Y.; Yue, H.; Wang, J.; Shen, B.; Sun, S.; Wang, S.; Wang, H.; Li, X.; Xu, Z.; Zhang, R.; et al. Super-durable ultralong carbon nanotubes. *Science* **2020**, *369*, 1104–1106. [CrossRef]
7. Hills, G.; Lau, C.; Wright, A.; Fuller, S.; Bishop, M.D.; Srimani, T.; Kanhaiya, P.; Ho, R.; Amer, A.; Stein, Y.; et al. Modern microprocessor built from complementary carbon nanotube transistors. *Nature* **2019**, *572*, 595–602. [CrossRef]
8. Kumar, S.; Saeed, G.; Zhu, L.; Nam Hui, K.; Hoon Kim, N.; Hee Lee, J. 0D to 3D carbon-based networks combined with pseudocapacitive electrode material for high energy density supercapacitor: A review. *Chem. Eng. J.* **2021**, *403*, 126352. [CrossRef]
9. Rao, R.; Pint, C.L.; Islam, A.E.; Weatherup, R.S.; Hofmann, S.; Hart, A.J. Carbon Nano-tubes and Related Nanomaterials: Critical Advances and Challenges for Synthesis toward Mainstream Commercial Applications. *ACS Nano* **2018**, *12*, 11756–11784. [CrossRef]
10. Pinilla, J.L.; de Llobet, S.; Moliner, R.; Suelves, I. Ni-Co bimetallic catalysts for the simultaneous production of carbon nanofibres and syngas through biogas decomposition. *Appl. Catal. B Environ.* **2017**, *200*, 255–264. [CrossRef]
11. Gómez-Pozuelo, G.; Pizarro, P.; Botas, J.A.; Serrano, D.P. Hydrogen production by catalytic methane decomposition over rice husk derived silica. *Fuel* **2021**, *306*, 121697. [CrossRef]
12. Fan, Z.; Weng, W.; Zhou, J.; Gu, D.; Xiao, W. Catalytic decomposition of methane to produce hydrogen: A review. *J. Energy Chem.* **2021**, *58*, 415–430. [CrossRef]
13. Dipu, A.L. Methane decomposition into CO_x-free hydrogen over a Ni-based catalyst: An overview. *Int. J. Energy Res.* **2021**, *45*, 9858–9877. [CrossRef]
14. Qian, J.X.; Chen, T.W.; Enakonda, L.R.; Liu, D.B.; Basset, J.M.; Zhou, L. Methane decomposition to pure hydrogen and carbon nano materials: State-of-the-art and future perspectives. *Int. J. Hydrogen Energy* **2020**, *45*, 15721–15743. [CrossRef]
15. Sanchez-Bastardo, N.; Schlogl, R.; Ruland, H. Methane Pyrolysis for CO_2-Free H_2 Production: A Green Process to Overcome Renewable Energies Unsteadiness. *Chem. Ing. Technik.* **2020**, *92*, 1596–1609. [CrossRef]
16. Jia, J.; Wang, Y.; Tanabe, E.; Shishido, T.; Takehira, K. Carbon fibers prepared by pyrolysis of methane over Ni/MCM-41 catalyst. *Micropor. Mesopor. Mat.* **2003**, *57*, 283–289. [CrossRef]
17. Jourdain, V.; Bichara, C. Current understanding of the growth of carbon nanotubes in catalytic chemical vapour deposition. *Carbon* **2013**, *58*, 2–39. [CrossRef]
18. Helveg, S.; López-Cartes, C.; Sehested, J.; Hansen, P.L.; Clausen, B.S.; Rostrup-Nielsen, J.R.; Abild-Pedersen, F.; Nørskov, J.K. Atomic-scale imaging of carbon nanofibre growth. *Nature* **2004**, *427*, 426–429. [CrossRef]
19. Pudukudy, M.; Yaakob, Z. Methane decomposition over Ni, Co and Fe based monometallic catalysts supported on sol gel derived SiO_2 microflakes. *Chem. Eng. J.* **2015**, *262*, 1009–1021. [CrossRef]

20. Alves Silva, J.; Oliveira Santos, J.B.; Torres, D.; Pinilla, J.L.; Suelves, I. Natural Fe-based catalysts for the production of hydrogen and carbon nanomaterials via methane decomposition. *Int. J. Hydrogen Energy* **2021**, *46*, 35137–35148. [CrossRef]
21. Zhou, L.; Reddy Enakonda, L.; Li, S.; Gary, D.; Del-Gallo, P.; Mennemann, C.; Basset, J.M. Iron ore catalysts for methane decomposition to make CO_x free hydrogen and carbon nano material. *J. Taiwan Inst. Chem. Eng.* **2018**, *87*, 54–63. [CrossRef]
22. Awadallah, A.E.; Aboul-Enein, A.A.; Kandil, U.F.; Reda Taha, M. Facile and large-scale synthesis of high quality few-layered graphene nano-platelets via methane decomposition over unsupported iron family catalysts. *Mater. Chem. Phys.* **2017**, *191*, 75–85. [CrossRef]
23. Latorre, N.; Cazaña, F.; Martínez-Hansen, V.; Royo, C.; Romeo, E.; Monzón, A. Ni-Co-Mg-Al catalysts for hydrogen and carbonaceous nanomaterials production by CCVD of methane. *Catal. Today* **2011**, *172*, 143–151. [CrossRef]
24. Cazaña, F.; Latorre, N.; Tarifa, P.; Labarta, J.; Romeo, E.; Monzón, A. Synthesis of graphenic nanomaterials by decomposition of methane on a Ni-Cu/biomorphic carbon catalyst. Kinetic and characterization results. *Catal. Today* **2018**, *299*, 67–79. [CrossRef]
25. Lin, X.; Zhu, H.; Huang, M.; Wan, C.; Li, D.; Jiang, L. Controlled preparation of Ni–Cu alloy catalyst via hydrotalcite-like precursor and its enhanced catalytic performance for methane decomposition. *Fuel Process. Technol.* **2022**, *233*, 107271. [CrossRef]
26. Tezel, E.; Eren Figen, H.; Baykara, S.Z. Hydrogen production by methane decomposition using bimetallic Ni–Fe catalysts. *Int. J. Hydrogen Energy* **2019**, *44*, 9930–9940. [CrossRef]
27. Henao, W.; Cazaña, F.; Tarifa, P.; Romeo, E.; Latorre, N.; Sebastian, V.; Delgado, J.J.; Monzón, A. Selective synthesis of carbon nanotubes by catalytic decomposition of methane using Co-Cu/cellulose derived carbon catalysts: A comprehensive kinetic study. *Chem. Eng. J.* **2021**, *404*, 126103. [CrossRef]
28. Cazaña, F.; Latorre, N.; Tarifa, P.; Royo, C.J.; Sebastián, V.; Romeo, E.; Centeno, M.A.; Monzón, A. Performance of AISI 316L-stainless steel foams towards the formation of graphene related nanomaterials by catalytic decomposition of methane at high temperature. *Catal. Today* **2022**, *383*, 236–246. [CrossRef]
29. Elmouwahidi, A.; Zapata-Benabithe, Z.; Carrasco-Marín, F.; Moreno-Castilla, C. Activated carbons from KOH-activation of argan (*Argania spinosa*) seed shells as supercapacitor electrodes. *Bioresour. Technol.* **2012**, *111*, 185–190. [CrossRef]
30. Azuara, M.; Latorre, N.; Villacampa, J.I.; Sebastián, V.; Cazaña, F.; Romeo, E.; Monzón, A. Use of Ni Catalysts Supported on Biomorphic Carbon Derived From Lignocellulosic Biomass Residues in the Decomposition of Methane. *Front. Energy Res.* **2019**, *7*, 34. [CrossRef]
31. Dahbi, M.; Kiso, M.; Kubota, K.; Horiba, T.; Chafik, T.; Hida, K.; Matsuyama, T.; Komaba, S. Synthesis of hard carbon from argan shells for Na-ion batteries. *J. Mater. Chem. A* **2017**, *5*, 9917. [CrossRef]
32. Harhar, H.; Gharby, S.; Ghanmi, M.; El Monfalouti, H.; Guillaume, D.; Charrouf, Z. Composition of the Essential Oil of Argania spinosa (*Sapotaceae*) Fruit Pulp. *Nat. Prod. Commun.* **2010**, *5*, 1934578X1000500626. [CrossRef]
33. Santos, J.L.; Mäki-Arvela, P.; Monzón, A.; Murzin, D.Y.; Centeno, M.A. Metal catalysts supported on biochars: Part I synthesis and characterization. *Appl. Catal B Environ.* **2020**, *268*, 118423. [CrossRef]
34. Santos, J.L.; Mäki-Arvela, P.; Wärnå, J.; Monzón, A.; Centeno, M.A.; Murzin, D.Y. Hydrodeoxygenation of vanillin over noble metal catalyst supported on biochars: Part II: Catalytic behaviour. *Appl. Catal B Environ.* **2020**, *268*, 118425. [CrossRef]
35. Cazaña, F.; Jimaré, M.T.; Romeo, E.; Sebastián, V.; Irusta, S.; Latorre, N.; Royo, C.; Monzón, A. Kinetics of liquid phase cyclohexene hydrogenation on Pd-Al/biomorphic carbon catalysts. *Catal. Today* **2015**, *249*, 127–136. [CrossRef]
36. Cazaña, F.; Galetti, A.; Meyer, C.; Sebastián, V.; Centeno, M.A.; Romeo, E.; Monzón, A. Synthesis of Pd-Al/biomorphic carbon catalysts using cellulose as carbon precursor. *Catal. Today* **2018**, *301*, 226–238. [CrossRef]
37. Tarifa, P.; Megías-Sayago, C.; Cazaña, F.; González-Martín, M.; Latorre, N.; Romeo, E.; Delgado, J.J.; Monzón, A. Highly Active Ce- and Mg-Promoted Ni Catalysts Supported on Cellulose-Derived Carbon for Low-Temperature CO_2 Methanation. *Energy Fuels* **2021**, *35*, 17212–17224. [CrossRef]
38. Mann, S. *Biomineralization: Principles and Concepts in Bioinorganic Materials Chemistry*; Oxford University Press: Oxford, UK, 2001.
39. Will, J.; Zollfrank, C.; Kaindl, A.; Sieber, H.; Grei, P. Biomorphic ceramics: Technologies based on nature. *Keram. Z* **2010**, *62*, 114.
40. Zuo, C.-Y.; Li, Q.S.; Peng, G.R.; Xing, G.Z. Manufacture of biomorphic Al_2O_3 ceramics using filter paper as template. *Prog. Nat. Sci. Mater. Int.* **2011**, *21*, 455–459. [CrossRef]
41. Luo, J.; Xu, H.; Liu, Y.; Chu, W.; Jiang, C.; Zhao, X. A facile approach for the preparation of biomorphic $CuO–ZrO_2$ catalyst for catalytic combustion of methane. *Appl. Catal. A Gen.* **2012**, *423*, 121–129. [CrossRef]
42. Chakraborty, R.; RoyChowdhury, D. Fish bone derived natural hydroxyapatite-supported copper acid catalyst: Taguchi optimization of semibatch oleic acid esterification. *Chem. Eng. J.* **2013**, *215*, 491–499. [CrossRef]
43. Wu, Q.; Duchstein, L.D.L.; Chiarello, G.L.; Christensen, J.M.; Damsgaard, C.D.; Elkjær, C.F.; Wagner, J.B.; Temel, B.; Grunwaldt, J.D.; Jensen, A.D. In Situ Observation of Cu–Ni Alloy Nanoparticle Formationby X-Ray Diffraction, X-Ray Absorption Spectroscopy, and Transmission Electron Microscopy: Influence of Cu/Ni Ratio. *Chem. Cat. Chem.* **2014**, *6*, 301–310.
44. Malard, L.M.; Pimenta, M.A.; Dresselhaus, G.; Dresselhaus, M.S. Raman spectroscopy in graphene. *Phys. Rep.* **2009**, *473*, 51–87. [CrossRef]
45. Rhim, Y.R.; Zhang, D.; Fairbrother, D.H.; Wepasnick, K.A.; Livi, K.J.; Bodnar, R.J.; Nagle, D.C. Changes in electrical and microstructural properties of microcrystalline cellulose as function of carbonization temperature. *Carbon* **2010**, *48*, 1012–1024. [CrossRef]

46. Latorre, N.; Cazaña, F.; Sebastián, V.; Royo, C.; Romeo, E.; Monzón, A. Effect of the Operating Conditions on the Growth of Carbonaceous Nanomaterials over Stainless Steel Foams. Kinetic and Characterization Studies. *Int. J. Chem. React. Eng.* **2017**, *15*, 20170121. [CrossRef]
47. Ahmad, M.; Silva, S.R.P. Low temperature growth of carbon nanotubes—A review. *Carbon* **2020**, *158*, 24–44. [CrossRef]
48. Galetti, A.; Barroso, M.N.; Monzón, A.; Abello, M.C. Synthesis of Nickel Nanoparticles Supported on Carbon Using a Filter Paper as Biomorphic Pattern for Application in Catalysis. *Mater. Res.* **2015**, *18*, 1278–1283. [CrossRef]
49. Chen, D.; Christensen, K.O.; Ochoa-Fernández, E.; Yu, Z.; Tøtdal, B.; Latorre, N.; Monzón, A.; Holmen, A. Synthesis of carbon nanofibers: Effects of Ni crystal size during methane decomposition. *J. Catal.* **2005**, *229*, 82–96. [CrossRef]
50. Zhu, J.; Jia, J.; Kwong, F.L.; Ng, D.H.L. Synthesis of bamboo-like carbon nanotubes on a cop-per foil by catalytic chemical vapor deposition from ethanol. *Carbon* **2012**, *50*, 2504–2512. [CrossRef]
51. Jia, Y.; Wu, P.Y.; Fang, F.; Zhou, S.S.; Peng, D.Y. Synthesis and characterization of un-branched and branched multi-walled carbon nanotubes using Cu as catalyst. *Solid State Sci.* **2013**, *18*, 71–77. [CrossRef]
52. Mohana Krishna, V.; Somanathan, T. Efficient strategy to Cu/Si catalyst into vertically aligned carbon nanotubes with Bamboo shape by CVD technique. *Bull. Mater. Sci.* **2016**, *39*, 1079–1084. [CrossRef]
53. Calizo, I.; Bejenari, I.; Rahman, M.; Liu, G.; Balandin, A.A. Ultraviolet Raman microscopy of single and multilayer graphene. *J. Appl. Phys.* **2009**, *106*, 043509. [CrossRef]
54. Santos, J.L.; Centeno, M.A.; Odriozola, J.A. Reductant atmospheres during slow pyrolysis of cellulose: First approach to obtaining efficient char-based catalysts in one pot. *J. Anal. Appl. Pyrolysis* **2020**, *148*, 104821. [CrossRef]
55. Ramirez, A.; Gevers, L.; Bavykina, A.; Ould-Chikh, S.; Gascon, J. Metal Organic Framework-Derived Iron Catalysts for the Direct Hydrogenation of CO_2 to Short Chain Olefins. *ACS Catal.* **2018**, *8*, 9174–9182. [CrossRef]
56. Monzón, A.; Lolli, G.; Cosma, S.; Mohamed, S.B.; Resasco, D.E. Kinetic Modeling of the SWNT Growth by CO Disproportionation on CoMo Catalysts. *J. Nanosci. Nanotechnol.* **2008**, *8*, 6141–6152. [CrossRef]
57. Latorre, N.; Romeo, E.; Cazaña, F.; Ubieto, T.; Royo, C.; Villacampa, J.I.; Monzón, A. Carbon Nanotube Growth by Catalytic Chemical Vapor Deposition: A Phenomenological Kinetic Model. *J. Phys. Chem. C* **2010**, *114*, 4773–4782. [CrossRef]

Article

Graphene-Wine Waste Derived Carbon Composites for Advanced Supercapacitors

Violeta Ureña-Torres [1], Gelines Moreno-Fernández [1,*], Juan Luis Gómez-Urbano [1,2], Miguel Granados-Moreno [1] and Daniel Carriazo [1,3,*]

1. Centre for Cooperative Research on Alternative Energies (CIC energiGUNE), Basque Research and Technology Alliance (BRTA), Alava Technology Park, Albert Einstein 48, 01510 Vitoria-Gasteiz, Spain; violetauat@gmail.com (V.U.-T.); jlgomezurbano@cicenergigune.com (J.L.G.-U.); magranados@cicenergigune.com (M.G.-M.)
2. Departamento de Química Orgánica e Inorgánica, Facultad de Ciencia y Tecnología, Universidad del País Vasco, UPV/EHU, 48080 Bilbao, Spain
3. IKERBASQUE, Basque Foundation for Science, 48013 Bilbao, Spain
* Correspondence: mamoreno@cicenergigune.com (G.M.-F.); dcarriazo@cicenergigune.com (D.C.); Tel.: +34-945-297-108 (G.M.-F. & D.C.)

Abstract: In this work, we investigate the potential of a novel carbon composite as an electrode for high-voltage electrochemical double-layer capacitors. The carbon composite was prepared following a sustainable synthetic approach that first involved the pyrolysis and then the activation of a precursor formed by winery wastes and graphene oxide. The composite prepared in this way shows a very high specific surface area (2467 $m^2 \cdot g^{-1}$) and an optimum pore size distribution for their use in supercapacitor electrodes. Graphene-biowaste-derived carbon composites are tested as active electrode materials in two different non-aqueous electrolytes, the ammonium salt-based conventional organic electrolyte and one imidazolium-based ionic liquid (1 M Et_4NBF_4/ACN and EMINTFSI). It was found that the presence of graphene oxide led to significant morphological and textural changes, which result in high-energy and power densities of ~27 $W \cdot h \cdot kg^{-1}$ at 13,026 $W \cdot kg^{-1}$. Moreover, the devices assembled retain above 70% of the initial capacitance after 6000 cycles in the case of the organic electrolyte.

Keywords: biowastes; EDLC; 2-D carbons; advanced electrolytes

Citation: Ureña-Torres, V.; Moreno-Fernández, G.; Gómez-Urbano, J.L.; Granados-Moreno, M.; Carriazo, D. Graphene-Wine Waste Derived Carbon Composites for Advanced Supercapacitors. *ChemEngineering* **2022**, *6*, 49. https://doi.org/10.3390/chemengineering6040049

Academic Editor: Miguel A. Vicente

Received: 26 May 2022
Accepted: 24 June 2022
Published: 29 June 2022

Publisher's Note: MDPI stays neutral with regard to jurisdictional claims in published maps and institutional affiliations.

Copyright: © 2022 by the authors. Licensee MDPI, Basel, Switzerland. This article is an open access article distributed under the terms and conditions of the Creative Commons Attribution (CC BY) license (https://creativecommons.org/licenses/by/4.0/).

1. Introduction

The rapid growth undergone by our society in the last decades has pushed us to look for new strategies in order to achieve sustainable development. This mainly involves the change of our current energy model based on the combustion of fossil fuels to a more eco-friendly one based on the use of renewable and greener sources, such as wind or solar energies. To attain this objective, it is necessary to develop new energy storage technologies that allow the use of these intermittent energies when they are not available [1]. Electrochemical capacitors, commonly known as supercapacitors, together with batteries, are the most widely used storage systems in portable devices, and their use is already beginning to be spread into larger applications such as electric vehicles or stationary energy storage stations [2]. Unlike batteries, where the charge is stored by faradaic processes, in supercapacitors, the charge is stored through an electrostatic process at the interface between the electrode and the electrolyte [3,4]. This mechanism does not involve ions insertion into the material's bulk; thus, its kinetics are not limited by diffusion processes, being able to provide higher power and a longer life-time than batteries [5]. These features make supercapacitors very attractive in a certain number of applications where reliability, wide operating temperature range, long-term stability or high power is required.

Electrical double-layer capacitors (EDLC) are typically formed by two similar electrodes made of highly porous carbon material impregnated in a liquid electrolyte and separated by a semipermeable membrane that allows the diffusion of ions but avoids the direct contact between electrodes preventing the short-circuit of the cell [6]. Taking into account that the energy stored in EDLC is proportional to the capacitance and to the square voltage (1/2 CV^2), most of the strategies are focused on the capacitance increasing through the optimization of the textural properties of active materials or to the operating voltage window extension through the use of highly electrochemically stable electrolytes [7].

In this regard, activated carbons (ACs) have been the preferred choice as electrodes in high-performing EDLCs due to their large specific surface area and low cost, as well as superior physical and chemical stability. Additionally, ACs can be readily produced from worldwide abundant and easily accessible wastes [8–11]. On the other hand, the incorporation of graphene has been found to be useful for increasing the electronic conductivity of the composites by tailoring their particle morphology and textural properties [12]. Nevertheless, most of the reported works rely on the use of aqueous-based electrolytes, which considerably limits the energy density of the EDLCs due to their narrower voltage window [13–15]. Thus, the evaluation of graphene-AC composites in non-aqueous electrolytes (organic, ionic liquids) is needed to subtract its maximal potential as electrodes for advanced EDLCs [16,17]. Within this context, ionic liquids have received great interest due to their low vapor pressure, wide temperature stability range and high electrochemical stability, which allow them to operate in a stable way in a wide voltage window, thus significantly increasing the energy density of the devices [18].

Herein we have investigated the potential that industrial winery biowastes combined with graphene have as carbonaceous precursor for their use as active electrode materials for high-voltage electrical double-layer capacitors. The results point out that the presence of graphene tunes the textural properties of the final carbon and leads to an improvement in the electrochemical performances of these materials measured in both ionic liquid and conventional organic electrolytes in terms of energy and power density. However, the higher reactivity of the graphene sheets can compromise the long-term stability of the systems, especially when tested in wide operating voltage windows.

2. Materials and Methods

2.1. Materials Synthesis

Wine wastes of the "Tempranillo" variety collected from a vineyard located in Navarra (Spain) were chosen as the carbon precursor, and graphene oxide (GO) from Graphenea (4 mg·mL^{-1}) was selected as the graphene source. First, wine wastes composed of skin, seeds, branches and leaves were dried at 80 °C for 4 consecutive days to subsequently crush them and convert them into waste powder. Then, 3 g of wine waste powder were mixed with 25 mL of graphene oxide and vigorously stirred for 3 h. Then, the suspension was freeze-dried for 4 days to produce the dry GO-wine waste precursor.

After that, the dried powder was pre-carbonized at 400 °C for 3 h using a heating ramp of 5 °C·min^{-1} under a dynamic Ar atmosphere in a tubular furnace. Finally, to obtain the activated carbon, the sample was ground together with KOH in a mortar using 1:4 mass ratio (C:KOH) and further heated at 800 °C for 1 h under Ar atmosphere using a heating ramp of 5 °C·min^{-1}. The resulting material was washed once with a diluted solution of HCl and several times with hot deionized water and further dried to obtain the final activated carbon powder. The reduced graphene oxide-wine waste-derived activated carbon sample is hereafter denoted as rGO-WW. For the sake of comparison, an activated carbon, denoted as WW, was also prepared following the same route but in the absence of graphene oxide.

2.2. Physicochemical Characterization

Morphological characterization was performed by scanning electron microscopy (SEM) in a Quanta200 FEI (3 kV, 30 kV) microscope. The composition of the samples was characterized by elemental analysis in a Flash 2000 Thermo Scientific (CHNS/O), where the samples were burned at 900 °C with an excess of oxygen. The textural properties of the composites were obtained by nitrogen adsorption/desorption isotherms that were registered at −196 °C in an ASAP 2460 from Micromeritics. The specific surface area values were calculated using the Brunauer–Emmet–Teller (BET) equation in the relative pressure range between 0.05 and 0.25, and the pore size distributions were calculated by the 2D-NLDFT model applied from the data of the adsorption branches using the SAIEUS software. Raman spectra were recorded with a Renishaw spectrometer (Nanonics multiview 2000). A laser beam with an excitation wavelength of 532 nm and 10 s exposition time was used for the spectrum acquisition.

2.3. Electrochemical Characterization

Carbon-based electrodes were processed by mixing 85% of active materials together with 10% of Super P C65 (Imerys Graphite & Carbon, Bironico, Switzerland) and 5% of polytetrafluoroethylene (PTFE) binder in the presence of ethanol. Then, the mixture was kneaded until plasticity and rolled to obtain a film with a thickness of ~100 μm. Circular-shaped electrodes of 11 mm diameter and 3.5 ± 1.5 mg were punched from the film and dried at 120 °C under vacuum overnight. The electrochemical performances of these electrodes were evaluated as symmetric EDCLs using a two-electrode Swagelok-type cell. The cells contained D-type glass fiber discs as a separator. The chosen electrolytes were 1 M tetraethylammonium tetrafluoroborate (Et_4NBF_4) in acetonitrile (ACN), and 1-ethyl-3-methylimidazolium bis(trifluoromethanesulfonyl) imide (EMINTFSI). Cyclic voltammetry (CV) and galvanostatic charge/discharge were carried out in a multichannel VMP3 generator from Biologic. The operating voltage window of each cell was selected from cyclic voltammetry measurements following the criteria reported elsewhere [19]. Applied current density (I_g) was calculated with respect to the total mass of active material (85%). Specific capacitance of the cells was obtained from the discharge branch of galvanostatic plots using Equation (1):

$$C_s = 2 \times \frac{I_g \times t_d}{\Delta V} \quad (1)$$

where t_d and ΔV are the discharge time and the operational voltage window, respectively, once the total resistance drop is subtracted. Gravimetric energy (E) and power densities (P) were calculated according to Equations (2) and (3):

$$E = \frac{1}{3.6}\left[\frac{1}{8}C_s \times \left(V_{max}^2 - V_{min}^2\right)\right] \quad (2)$$

$$P = \frac{E}{t_d} \quad (3)$$

where V_{max} and V_{min} are the maximum and the minimum of the cell potential, once the corresponding resistance drop is discarded.

3. Results and Discussion

As previously explained in the experimental section, an activated carbon composite comprising reduced graphene oxide/winery wastes (rGO-WW) has been synthesized as an electrode for supercapacitors. The winery wastes were firstly dispersed into a graphene oxide suspension, freeze-dried, pre-carbonized and then submitted to the activation process to increase their specific surface area and get the most appropriate balance between pore distribution, morphology and surface area [10]. On the other hand, for the sake of comparison, a pristine activated carbon derived from winery wastes (WW) was directly pre-carbonized and submitted to the activation process. SEM images registered for the materials are shown in Figure 1. It can be observed that the WW sample (Figure 1a,c) is formed by large polyhedral-shaped particles of 30–40 μm with cavities of two different sizes, 4–6 and 1.5 μm. On the other side, rGO-WW (Figure 1b,d) clearly shows the presence of graphene sheets homogeneously distributed within the carbon particles, and the particle size is clearly reduced to 7–21 µm. The morphological differences between the two samples are immediately recognizable at higher magnification images (Figure 1c,d).

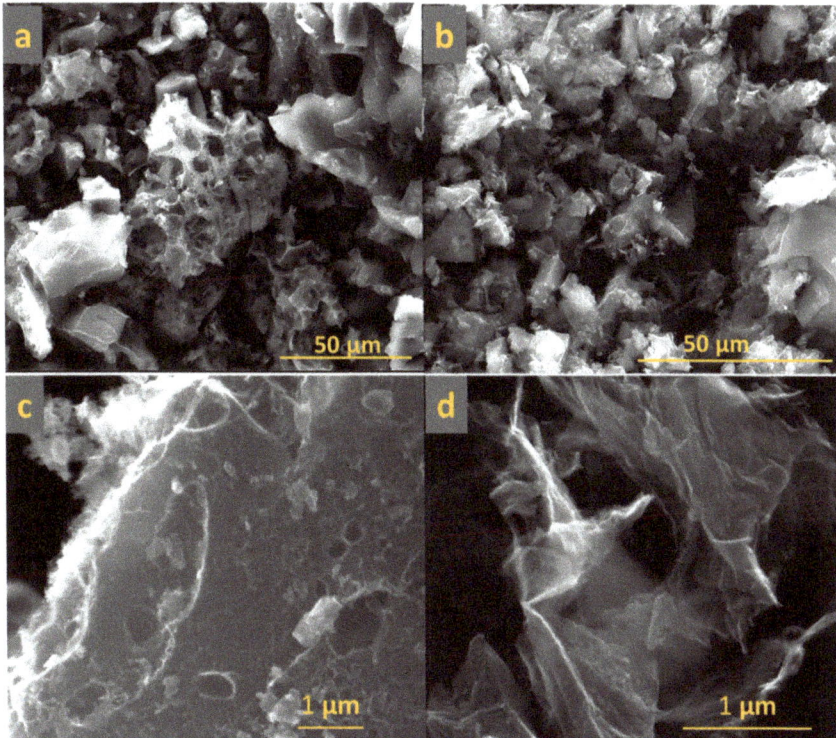

Figure 1. SEM images registered for WW (**a,c**) and rGO-WW (**b,d**) samples.

The chemical composition measured by the elemental analysis of the ACs prepared in this work is summarized in Table 1. As expected, the elemental analysis reveals that the main element in both materials is carbon (~90%). Other elements such as nitrogen, oxygen or hydrogen are present in much lower amounts. The presence of nitrogen could be, in this case, attributed to the alcoholic fermentation of grapes.

Table 1. Elemental composition of the activated carbons.

Material	% C	% O	% N	% H
WW	87.5	11.2	0.6	0.7
rGO-WW	91.2	8.3	0.2	0.2

Raman spectra were recorded to evaluate the structural differences between the two carbonaceous samples and to confirm the presence of rGO in the rGO-WW after the activation step. The Raman spectra presented in Figure 2a show the characteristics D and G bands in both WW and rGO-WW samples at ca. 1343 and ca. 1590 nm, respectively. The D and G bands are related to the presence of structural defects and graphitic domains. As expected, the addition of GO results in a smaller A_D/A_G ratio, pointing out a lower disorder degree. Additionally, the Raman spectrum of rGO-WW shows a band at ca. 2680 nm corresponding with the 2D band. This band is related to the 2D morphology of the rGO and confirms the proper conservation of the rGO sheets after the activation step.

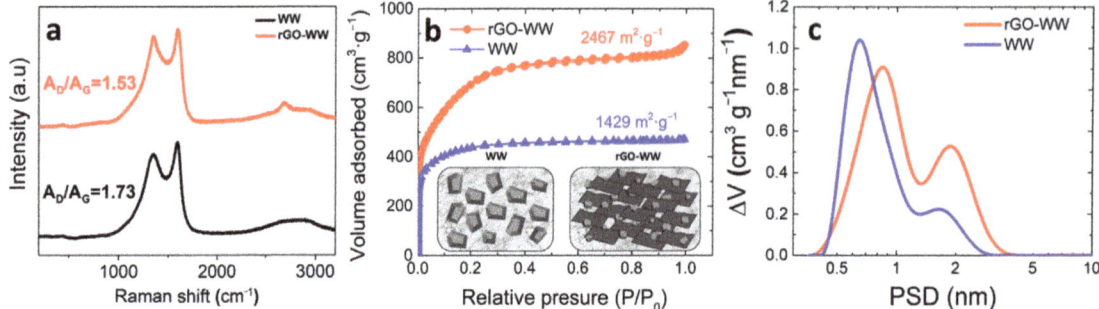

Figure 2. Raman spectra (**a**), nitrogen adsorption/desorption isotherms (**b**) and pore size distributions (**c**) measured for indicated samples.

Nitrogen adsorption/desorption isotherms were registered in order to assess the textural properties of the materials. Figure 2b,c include the isotherms registered for these samples and the pore size distributions associated with each of them. Both materials show similar adsorption profiles matching with type-I isotherms according to the IUPAC classification, which corresponds to porous samples with pores within the microporous range (<2 nm) [20]. The BET-specific surface area calculated for rGO-WW and WW gives values of 2467 and 1429 $m^2 \cdot g^{-1}$, respectively. A schematic representation of the particle shape and size of the samples has been included in Figure 2b. The 2D characteristic of the rGO sheets and the effect of the carbon particles acting as spacers are responsible for the specific surface area increment, also modifying the pore size of rGO-WW. Regarding the pore size distribution, the sample containing graphene oxide (rGO-WW) contains medium-size micropores centered at 0.9 nm and large-sized micropores of around 2 nm, while the sample without graphene oxide (WW) contains small-size pores of 0.7 nm and a few larger ones at ca. 1.6 nm.

According to these results, it seems that the presence of the graphene sheets within the rGO-WW sample helps to maximize the activation process giving rise to carbons with much larger specific surface areas and wider pores compared to the graphene-free sample (WW). It is worth remarking that the specific surface area achieved for the rGO-WW composite surpasses most of the carbons derived from winery wastes in the reported literature [13–15,21–25].

These homemade carbonaceous materials were evaluated as active materials in a two-electrode symmetric configuration using the conventional organic electrolyte 1 M Et_4NBF_4/ACN and the bare ionic liquid EMINTFSI. Figure 3 shows the CV curves registered for the supercapacitor cells at different sweep rates in the previously determined operating voltage window. From these CV curves, a much better electrochemical stability can be assessed for those systems measured in ionic liquid, which enables the expansion of the operating voltage window compared to those measured in the organic electrolyte. Moreover, the graphene-free sample (WW) seems to be more stable electrochemically than the composite (rGO-WW). Specifically, operating voltage windows of 2.9 and 3.1 V were determined for the WW sample (Figure 3a,c) and 2.7 and 2.9 V for the composite rGO-WW (Figure 3b,d) in the electrolytes 1 M Et_4NBF_4/ACN and EMINTFSI, respectively. Both materials show typical rectangular-shaped profiles, which are characteristic of capacitive-type charge storage mechanisms [26]. The graphene-containing sample rGO-WW outperforms the pristine WW whenever the electrolyte is tested in terms of specific capacitance and capacitance retention. This can be ascribed to the enhancement of the specific surface area and the pore widening upon graphene addition to the winery wastes, as previously explained.

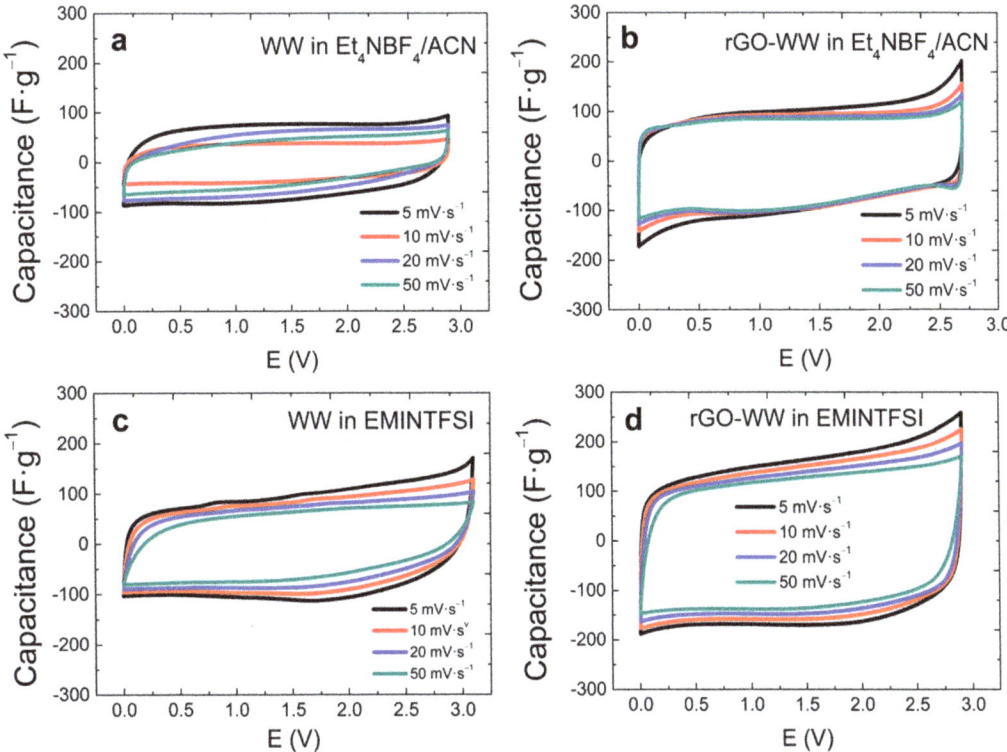

Figure 3. Cyclic voltammetry curves registered for supercapacitors assembled using WW in 1 M Et$_4$NBF$_4$ (**a**), and EMINTFSI (**c**); and corresponding measurements for rGO-WW (**b**,**d**).

Galvanostatic charge/discharge curves at different current densities registered for these samples at 0.5 and 5 A·g^{-1} are depicted in Figure 4. It can be observed that all samples show linear voltage increase/decrease during charge and discharge with the typical triangular shape of capacitive-type charge storage mechanisms. At the low current density of 0.5 A·g^{-1}, both ACs show very small ohmic drop (Figure 4a,c), which are progressively enlarged as the current rate is increased to 5 A·g^{-1} (Figure 4b,d). It is worth highlighting that this behavior is more pronounced for the EDLCs measured in the ionic liquid (EMINTFSI) due to its lower ionic conductivity and its higher viscosity compared to the organic electrolyte (Et$_4$NBF$_4$) (see Table 2). The sample without graphene oxide exhibits a larger ohmic drop compared to the graphene-containing one, which could be associated to its lower electronic conductivity.

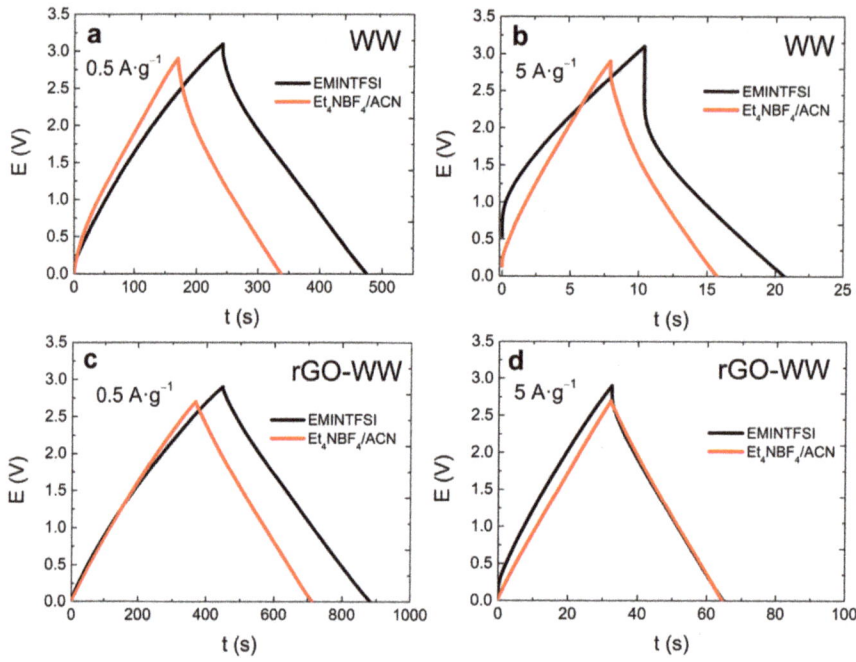

Figure 4. Galvanostatic charge-discharge curves registered in both electrolytes for WW sample at 0.5 (**a**) and 5 A·g^{-1} (**b**); and corresponding measurements for rGO-WW (**c**,**d**).

Table 2. Physical properties of the selected electrolytes.

Electrolyte	Viscosity (mPa·s)	Conductivity (mS·cm^{-1})	Cation Size (nm)	Anion Size (nm)	Ref.
1 M Et$_4$NBF$_4$/ACN	0.57	63	0.69	0.46	[27]
EMINTFSI	35.55	9	0.76	0.79	[28]

The evolution of the specific capacitance values at different current rates for the two electrolytes, calculated from the discharge branch of galvanostatic curves, are shown in Figure 5a. In good agreement with the CV results, the specific capacitance of the composite rGO-WW is larger than that of the pristine WW regardless of the electrolyte used. In fact, at low current densities, rGO-WW doubles the specific capacitance of WW either in the IL (160 F·g^{-1} vs. 79 F·g^{-2}) or in the organic electrolyte (130 F·g^{-1} vs. 65 F·g^{-1}). It is also worth highlighting that at low current densities, the specific capacitance measured for both samples is significantly larger in the ionic liquid compared to the organic electrolyte. This has been previously explained by the improved ion confinement in the pores of EMINTFSI due to the geometry of the ions [29]. In contrast, samples measured in the organic electrolyte exhibit better capacitance retention due to its lower viscosity and higher ionic conductivity (Table 2) (Figure 5b). Furthermore, at high current densities, the diffusion to the pores is not favored, and this takes special relevancy when the electrolytes are formed by large size ions, as in the case of ionic liquids (Table 2). Nevertheless, regardless of the electrolyte used, the graphene-containing sample exhibits much higher capacitance retention than the pristine (86% for rGO-WW vs. 72% for WW in 1 M Et$_4$NBF$_4$/ACN and 40% for rGO-WW vs. 11% for WW in EMINTFSI) (Figure 5b).

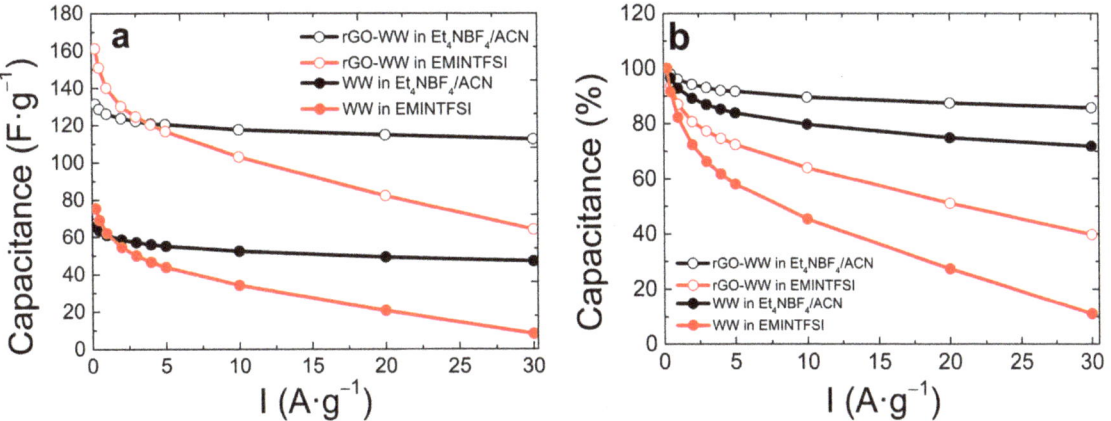

Figure 5. Capacitance evolution vs. current density (**a**) and capacitance retention (**b**) of noted samples.

The energy-to-power density plot for these home-designed symmetric cells is shown in Figure 6. Ragone plots confirm the larger energy density values and better power performance of graphene-wine waste (rGO-WW) electrode-based systems due to its larger specific capacitance and improved electronic conductivity. The systems measured using the ionic liquid EMINTFSI as electrolyte show maximized energy density values at low power densities due to the wider operating voltage window (46 W·h·kg^{-1} at 180 W·kg^{-1} for rGO-WW and 24 W·h·kg^{-1} at 191 W·kg^{-1} for WW) compared to the conventional organic electrolyte 1 M Et$_4$NBF$_4$/ACN (33.1 W·h·kg^{-1} at 160 W·kg^{-1} for rGO-WW and 19.1 W·h·kg^{-1} at 181 W·kg^{-1} for WW). However, EDLCs measured in the IL show a much more prominent decay at high power densities. In fact, 82% and 65% of the energy is still delivered for the supercapacitors rGO-WW, and WW assembled in the organic electrolyte at the maximum power rate, while only 38% and 7% are delivered for those using the pure ionic liquid.

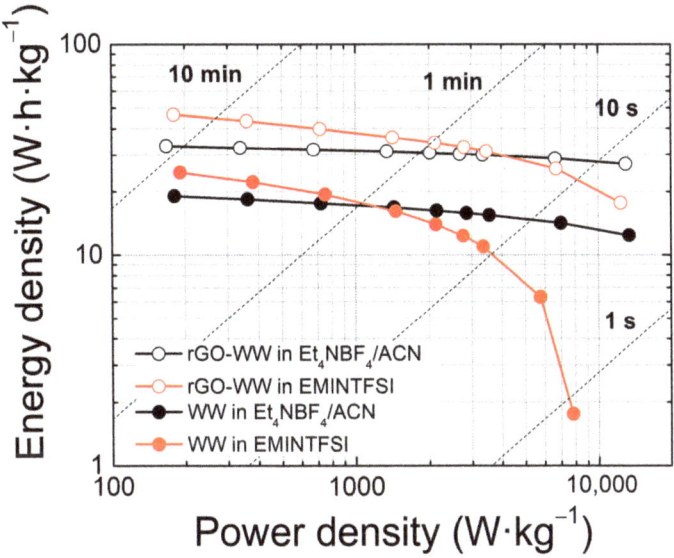

Figure 6. Ragone plot of indicated samples in both electrolytes.

These results are in line with most representative biowaste-derived carbons reported in the literature using organic and ionic liquid electrolytes. It is true that the specific capacitance of our reduced graphene oxide/winery waste-derived carbon surpasses most of these materials, and the energy delivered is similar or even lower [10,30–33]. This could be related to the higher reactivity of the graphene sheets, especially in the ionic liquid, which could represent a handicap narrowing the operational voltage as well as compromising the cycling stability. Taking this into account and the exceptional electrochemical performance of both AC in the organic electrolyte, the stability of these EDLC systems was evaluated by monitoring the specific capacitance value from galvanostatic charge/discharge measurements at 10 $A·g^{-1}$ at regular intervals (Figure 7).

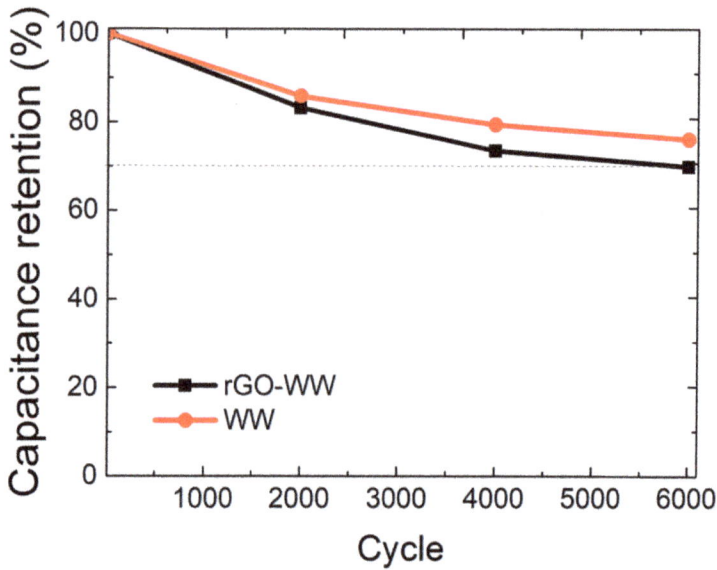

Figure 7. Capacitance evolution along cycles of indicated samples measured at 10 $A·g^{-1}$ in the organic electrolyte.

The cyclability measurement shows that WW still retains 78% of the initial capacitance after 6000 cycles compared to the 70% of capacitance retention measured for the rGO-WW. Once again, this can be ascribed to the higher reactivity of samples containing graphene oxide that may arise as a consequence of their larger concentration of oxygen groups of the graphene oxide that react with the electrolyte and promote its degradation.

4. Conclusions

Winery wastes were demonstrated as promising raw material for its revalorization as a porous carbon. The optimization of different parameters, such as particle size, electronic conductivity and specific surface areas plays a significant role and have a huge impact on the final electrochemical performance of these materials as electrodes for symmetric supercapacitors.

The results show that the reduced graphene oxide notably improved the electrochemical properties of these devices. The most favorable results in terms of energy, power density and stability are obtained for supercapacitors operating with an organic electrolyte (~27 $W·h·kg^{-1}$ at 13,026 $W·kg^{-1}$). The choice of ionic liquid as the electrolyte can also provide some important advantages under some specific operating conditions, specifically at low power densities.

In summary, the reutilization of agricultural waste as a precursor of active material for electrodes in double-layer supercapacitors is a sustainable, innovative solution. Furthermore, the presence of reduced graphene oxide improved the energy and power density, but due to their higher reactivity, the stability can be compromised in the long term. These supercapacitors still retain more than 70% of their initial capacitance after more than 6000 charge-discharge cycles.

Author Contributions: Conceptualization, D.C.; Funding acquisition, D.C.; Investigation, V.U.-T., G.M.-F., J.L.G.-U. and M.G.-M.; Methodology, V.U.-T., G.M.-F., J.L.G.-U. and M.G.-M.; Supervision, G.M.-F. and D.C.; Validation, J.L.G.-U. and M.G.-M.; Writing—original draft, G.M.-F.; Writing—review and editing, V.U.-T., J.L.G.-U., M.G.-M. and D.C. All authors have read and agreed to the published version of the manuscript.

Funding: This research was funded by the European Union through the project Graphene Flagship, Core 3, Grant number 881603; by the Spanish ministry of Science and Innovation through the project MICINN/FEDER (RTI2018-096199-B-I00); and by the Spanish Ministry of Education, Science and Universities through the FPU grant (16/03498).

Institutional Review Board Statement: Not applicable.

Data Availability Statement: Data are available upon request to the authors.

Acknowledgments: The authors thank the European Union (Graphene Flagship, Core 3, Grant number 881603) and the Spanish Ministry of Science and Innovation (MICINN/FEDER) (RTI2018-096199-B-I00) for the financial support of this work. J.L.G.-U. is very thankful to the Spanish Ministry of Education, Science and Universities (MICINN) for the FPU grant (16/03498). We also want to acknowledge the company GRAPHENEA for supplying the graphene oxide used in this work.

Conflicts of Interest: The authors declare no conflict of interest. The funders had no role in the design of the study; in the collection, analyses, or interpretation of data; in the writing of the manuscript, or in the decision to publish the results.

References

1. Koohi-Fayegh, S.; Rosen, M.A. A Review of Energy Storage Types, Applications and Recent Developments. *J. Energy Storage* **2020**, *27*, 101047. [CrossRef]
2. Hannan, M.A.; Hoque, M.M.; Mohamed, A.; Ayob, A. Review of Energy Storage Systems for Electric Vehicle Applications: Issues and Challenges. *Renew. Sustain. Energy Rev.* **2017**, *69*, 771–789. [CrossRef]
3. Li, M.; Lu, J.; Chen, Z.; Amine, K. 30 Years of Lithium-Ion Batteries. *Adv. Mater.* **2018**, *30*, 1800561. [CrossRef] [PubMed]
4. Poonam; Sharma, K.; Arora, A.; Tripathi, S.K. Review of Supercapacitors: Materials and Devices. *J. Energy Storage* **2019**, *21*, 801–825. [CrossRef]
5. Simon, P.; Gogotsi, Y.; Dunn, B. Where Do Batteries End and Supercapacitors Begin? *Science* **2014**, *343*, 1210–1211. [CrossRef]
6. Zhao, J.; Burke, A.F. Review on Supercapacitors: Technologies and Performance Evaluation. *J. Energy Chem.* **2021**, *59*, 276–291. [CrossRef]
7. González, A.; Goikolea, E.; Barrena, J.A.; Mysyk, R. Review on Supercapacitors: Technologies and Materials. *Renew. Sustain. Energy Rev.* **2016**, *58*, 1189–1206. [CrossRef]
8. Díez, N.; Botas, C.; Mysyk, R.; Goikolea, E.; Rojo, T.; Carriazo, D. Highly Packed Graphene–CNT Films as Electrodes for Aqueous Supercapacitors with High Volumetric Performance. *J. Mater. Chem. A* **2018**, *6*, 3667–3673. [CrossRef]
9. Moreno-Fernández, G.; Gómez-Urbano, J.L.; Enterría, M.; Rojo, T.; Carriazo, D. Flat-Shaped Carbon–Graphene Microcomposites as Electrodes for High Energy Supercapacitors. *J. Mater. Chem. A* **2019**, *7*, 14646–14655. [CrossRef]
10. Gómez-Urbano, J.L.; Moreno-Fernández, G.; Granados-Moreno, M.; Rojo, T.; Carriazo, D. Nanostructured Carbon Composites from Cigarette Filter Wastes and Graphene Oxide Suitable as Electrodes for 3.4 V Supercapacitors. *Batter. Supercaps* **2021**, *4*, 1749–1756. [CrossRef]
11. Gómez-Urbano, J.L.; Moreno-Fernández, G.; Arnaiz, M.; Ajuria, J.; Rojo, T.; Carriazo, D. Graphene-Coffee Waste Derived Carbon Composites as Electrodes for Optimized Lithium Ion Capacitors. *Carbon* **2020**, *162*, 273–282. [CrossRef]
12. Cheng, F.; Yang, X.; Zhang, S.; Lu, W. Boosting the Supercapacitor Performances of Activated Carbon with Carbon Nanomaterials. *J. Power Sources* **2020**, *450*, 227678. [CrossRef]
13. Guardia, L.; Suárez, L.; Querejeta, N.; Pevida, C.; Centeno, T.A. Winery Wastes as Precursors of Sustainable Porous Carbons for Environmental Applications. *J. Clean. Prod.* **2018**, *193*, 614–624. [CrossRef]
14. Suárez, L.; Centeno, T.A. Unravelling the Volumetric Performance of Activated Carbons from Biomass Wastes in Supercapacitors. *J. Power Sources* **2020**, *448*, 227413. [CrossRef]
15. Jiménez-Cordero, D.; Heras, F.; Gilarranz, M.A.; Raymundo-Piñero, E. Grape Seed Carbons for Studying the Influence of Texture on Supercapacitor Behaviour in Aqueous Electrolytes. *Carbon* **2014**, *71*, 127–138. [CrossRef]
16. Díez, N.; Mysyk, R.; Zhang, W.; Goikolea, E.; Carriazo, D. One-Pot Synthesis of Highly Activated Carbons from Melamine and Terephthalaldehyde as Electrodes for High Energy Aqueous Supercapacitors. *J. Mater. Chem. A* **2017**, *5*, 14619–14629. [CrossRef]

17. Moreno-Fernández, G.; Boulanger, N.; Nordenström, A.; Iakunkov, A.; Talyzin, A.; Carriazo, D.; Mysyk, R. Ball-Milling-Enhanced Capacitive Charge Storage of Activated Graphene in Aqueous, Organic and Ionic Liquid Electrolytes. *Electrochim. Acta* **2021**, *370*, 137738. [CrossRef]
18. Brandt, A.; Pohlmann, S.; Varzi, A.; Balducci, A.; Passerini, S. Ionic Liquids in Supercapacitors. *MRS Bull.* **2013**, *38*, 554–559. [CrossRef]
19. Bahdanchyk, M.; Hashempour, M.; Vicenzo, A. Evaluation of the Operating Potential Window of Electrochemical Capacitors. *Electrochim. Acta* **2020**, *332*, 135503. [CrossRef]
20. Thommes, M.; Kaneko, K.; Neimark, A.V.; Olivier, J.P.; Rodriguez-Reinoso, F.; Rouquerol, J.; Sing, K.S.W. Physisorption of Gases, with Special Reference to the Evaluation of Surface Area and Pore Size Distribution (IUPAC Technical Report). *Pure Appl. Chem.* **2015**, *87*, 1051–1069. [CrossRef]
21. Guardia, L.; Suárez, L.; Querejeta, N.; Vretenár, V.; Kotrusz, P.; Skákalová, V.; Centeno, T.A. Biomass Waste-Carbon/Reduced Graphene Oxide Composite Electrodes for Enhanced Supercapacitors. *Electrochim. Acta* **2019**, *298*, 910–917. [CrossRef]
22. Zuo, X.; Chang, K.; Zhao, J.; Xie, Z.; Tang, H.; Li, B.; Chang, Z. Bubble-Template-Assisted Synthesis of Hollow Fullerene-like MoS$_2$ Nanocages as a Lithium Ion Battery Anode Material. *J. Mater. Chem. A* **2016**, *4*, 51–58. [CrossRef]
23. Hoffmann, V.; Jung, D.; Zimmermann, J.; Rodriguez Correa, C.; Elleuch, A.; Halouani, K.; Kruse, A. Conductive Carbon Materials from the Hydrothermal Carbonization of Vineyard Residues for the Application in Electrochemical Double-Layer Capacitors (EDLCs) and Direct Carbon Fuel Cells (DCFCs). *Materials* **2019**, *12*, 1703. [CrossRef] [PubMed]
24. Ramanathan, S.; Moorthy, S.; Ramasundaram, S.; Rajan, H.K.; Vishwanath, S.; Selvinsimpson, S.; Duraraj, A.; Kim, B.; Vasanthkumar, S. Grape Seed Extract Assisted Synthesis of Dual-Functional Anatase TiO$_2$ Decorated Reduced Graphene Oxide Composite for Supercapacitor Electrode Material and Visible Light Photocatalytic Degradation of Bromophenol Blue Dye. *ACS Omega* **2021**, *6*, 14734–14747. [CrossRef] [PubMed]
25. Zhang, J.; Chen, H.; Bai, J.; Xu, M.; Luo, C.; Yang, L.; Bai, L.; Wei, D.; Wang, W.; Yang, H. N-Doped Hierarchically Porous Carbon Derived from Grape Marcs for High-Performance Supercapacitors. *J. Alloys Compd.* **2021**, *854*, 157207. [CrossRef]
26. Babu, B.; Simon, P.; Balducci, A. Fast Charging Materials for High Power Applications. *Adv. Energy Mater.* **2020**, *10*, 2001128. [CrossRef]
27. Pohlmann, S.; Kühnel, R.-S.; Centeno, T.A.; Balducci, A. The Influence of Anion-Cation Combinations on the Physicochemical Properties of Advanced Electrolytes for Supercapacitors and the Capacitance of Activated Carbons. *ChemElectroChem* **2014**, *1*, 1301–1311. [CrossRef]
28. Singh, R.; Rajput, N.N.; He, X.; Monk, J.; Hung, F.R. Molecular Dynamics Simulations of the Ionic Liquid [EMIM+][TFMSI−] Confined inside Rutile (110) Slit Nanopores. *Phys. Chem. Chem. Phys.* **2013**, *15*, 16090. [CrossRef]
29. Merlet, C.; Péan, C.; Rotenberg, B.; Madden, P.A.; Daffos, B.; Taberna, P.-L.; Simon, P.; Salanne, M. Highly Confined Ions Store Charge More Efficiently in Supercapacitors. *Nat. Commun.* **2013**, *4*, 2701. [CrossRef]
30. Momodu, D.; Sylla, N.F.; Mutuma, B.; Bello, A.; Masikhwa, T.; Lindberg, S.; Matic, A.; Manyala, N. Stable Ionic-Liquid-Based Symmetric Supercapacitors from Capsicum Seed-Porous Carbons. *J. Electroanal. Chem.* **2019**, *838*, 119–128. [CrossRef]
31. Cui, Y.; Wang, H.; Mao, N.; Yu, W.; Shi, J.; Huang, M.; Liu, W.; Chen, S.; Wang, X. Tuning the Morphology and Structure of Nanocarbons with Activating Agents for Ultrafast Ionic Liquid-Based Supercapacitors. *J. Power Sources* **2017**, *361*, 182–194. [CrossRef]
32. Tian, W.; Gao, Q.; Tan, Y.; Yang, K.; Zhu, L.; Yang, C.; Zhang, H. Bio-Inspired Beehive-like Hierarchical Nanoporous Carbon Derived from Bamboo-Based Industrial by-Product as a High Performance Supercapacitor Electrode Material. *J. Mater. Chem. A* **2015**, *3*, 5656–5664. [CrossRef]
33. Wang, H.; Li, Z.; Tak, J.K.; Holt, C.M.B.; Tan, X.; Xu, Z.; Amirkhiz, B.S.; Harfield, D.; Anyia, A.; Stephenson, T.; et al. Supercapacitors Based on Carbons with Tuned Porosity Derived from Paper Pulp Mill Sludge Biowaste. *Carbon* **2013**, *57*, 317–328. [CrossRef]

Article

Use of Ethylamine, Diethylamine and Triethylamine in the Synthesis of Zn,Al Layered Double Hydroxides

Alexander Misol, Alejandro Jiménez and Francisco M. Labajos *

GIR-QUESCAT, Departamento de Química Inorgánica, Facultad de Ciencias Químicas, Universidad de Salamanca, 37008 Salamanca, Spain; alex_aspa6@usal.es (A.M.); alejm@usal.es (A.J.)
* Correspondence: labajos@usal.es

Abstract: Amines with two carbon atoms in the organic chain [ethylamine (EA), diethylamine (DEA), triethylamine (TEA)] have been used as precipitant agents to obtain a hydrotalcite-like compound with Zn (II) and Al (III) as layered cations and with nitrate anions in the interlayered region to balance the charge. This Layered Double Hydroxide was prepared following the coprecipitation method, and the effect on the crystal and particle sizes was studied. Also, the effect of submitting the obtained solids to hydrothermal post-synthesis treatment by conventional heating and microwave assisted heating were studied. The obtained solids were exhaustively characterized using several instrumental techniques, such as X-ray diffraction, Thermal Analysis (DTA and TG), Chemical Analysis, Infrared Spectroscopy (FT-IR), determination of Particle Size Distribution and BET-Surface area. Well crystallized solids were obtained showing two possible LDH phases, depending on the orientation of the interlayer anion with respect to the brucite-like layers. The results indicated that there is a certain influence of the amine, when used as a precipitating agent, and as a consequence of the degree of substitution, on the crystallinity and particle size of the final solid obtained. The LDHs obtained using TEA exhibited higher crystallinity, which was improved after a long hydrothermal treatment by conventional heating. Regarding the shape of the particles, the formation of aggregates in the former solid was detected, which could be easily disintegrated using ultrasound treatments, producing solid powder with high crystallinity and small particle size, with homogeneous size distribution.

Keywords: LDHs; hydrotalcite; amines; coprecipitation method; hydrothermal treatment; crystallinity

Citation: Misol, A.; Jiménez, A.; Labajos, F.M. Use of Ethylamine, Diethylamine and Triethylamine in the Synthesis of Zn,Al Layered Double Hydroxides. *ChemEngineering* 2022, 6, 53. https://doi.org/10.3390/chemengineering6040053

Academic Editor: Dmitry Yu. Murzin

Received: 3 June 2022
Accepted: 4 July 2022
Published: 6 July 2022

Publisher's Note: MDPI stays neutral with regard to jurisdictional claims in published maps and institutional affiliations.

Copyright: © 2022 by the authors. Licensee MDPI, Basel, Switzerland. This article is an open access article distributed under the terms and conditions of the Creative Commons Attribution (CC BY) license (https://creativecommons.org/licenses/by/4.0/).

1. Introduction

The term Layered Double Hydroxides (LDHs) is used to name synthetic or natural hydroxides with a layered structure and with at least two types of metal cations in the main layers, which are positively charged, and contain anionic species in the interlayer space. This large family of compounds is also called anionic clays, in comparison with cationic clays, which, in their interlayer region, contain cations to balance the negative charge of the layers [1]. They are also known as hydrotalcite-like compounds, because hydrotalcite is the most abundant mineral with this layered structure in Nature. These materials are not as abundant in Nature as the analogous cationic clays, but they are very easy to synthesize with a tuned composition and are generally not very expensive.

The layered structure of the LDH consists of brucite-like layers with divalent cations occupying octahedral spaces formed by OH^- ions $[M(OH)_6]$, in which an isomorphic, and partial, substitution of divalent cations by trivalent cations has taken place, leaving the layers positively charged. The electroneutrality of the compound is achieved by the incorporation of anions in the interlayer space [2]. The chemical composition of LDHs is described by the chemical formula: $\left[M^{II}_{1-x}M^{III}_x(OH)_2\right]^{x+}\left[A^{m-}\right]_{\frac{x}{m}} \cdot nH_2O$, where M^{II} and M^{III} are the divalent and trivalent metal cations, respectively, and A^- is the interlayer anion, with x defined as the $M^{III}/\left(M^{II}+M^{III}\right)$ ratio [1,3]. Both organic and inorganic anions can be incorporated into the interlayer space, in a wide range of sizes and charges [1,4]. In addition, in the interlayer space there are randomly arranged water molecules, and this interlayer region has quasi-liquid behavior. The wide range of anions and divalent and trivalent cations that can be used to prepare LDH, provide them with a diversity of compositions.

In recent decades, LDHs have established themselves as promising materials, due to their properties and applications in a number of fields, such as water remediation [5–7], catalysis [8–10], drug delivery [11,12], electroactivity [13], biomedicine [14,15], and others. Their applications as anion exchanging solids depend on the interlayer anion and its affinity for the LDH layers. Thus, Miyata et al. [16] have reported an order of anionic selectivity for MgAl hydrotalcite-like materials, which could be applicable to other combinations of elements [17]. According to this, for applications involving anion exchanging, LDHs containing nitrate have a higher ion exchange facility than those containing carbonate in the interlayer space.

As above mentioned, the synthesis of LDHs in the laboratory is relatively simple and cheap, and they could be synthesized by many different methods. The method selected has an impact on the properties of the final solid and, therefore, on its subsequent application. Among the different methods described, the most widely used is the coprecipitation method, due to its great ease of use and reproducibility [3]. This method is based on precipitation by the slow dropwise addition of a solution containing the mixture of M^{II} and M^{III} salts solution in a fixed ratio and the anion to an alkaline solution, working at constant pH. The addition of a second alkaline solution allows the pH of the precipitation medium to be maintained during the precipitation of the cations [3]. Many parameters are relevant to monitoring the process, such as the type and concentration of cations and anions, the precipitation medium, the pH and the temperature [18,19]. The optimal pH depends on the nature of the cations to be incorporated into the structure; thus, Kloprogge et al. [20] reported that the best crystallinity for samples of Zn/Al LDH were exhibit in the pH range 11–12. Moreover, when the coprecipitation method is carried out, the incorporation of the carbonate anion into the interlayer space is very difficult to avoid, due to the fact that it is the anion with the highest affinity for the LDH layers. However, its incorporation can be prevented by using decarbonated water and bubbling N_2.

LDH synthetized by the coprecipitation method generally exhibits low crystallinity with a high degree of aggregation and a wide particle size distribution. The most common way to obtain more uniform particle properties, with an improved crystallinity is by an aging process. For this reason, the coprecipitation method is generally followed by a long aging period, from 10 to 80 h, and often longer [19]. Also, post-synthesis treatment heating at moderate temperatures is used; for instance, by gentle-to-gentle reflux or by hydrothermal treatment. The aging presumably occurs through the Ostwald ripening process, in which larger and more perfect crystallites grow at the expense of smaller particles in solution by dissolution/precipitation processes [21]. Therefore, microwave-assisted hydrothermal treatment, or hydrothermal treatment by conventional heating, have been widely used with the aim of improving structural and textural properties [22–26].

In our previous work [27], Zn/Al-LDH in molar ratio 2:1 was synthesized using amines with one carbon atom in the organic chain [methylamine (MMA), dimethylamine (DMA) and trimethylamine (TMA)], such as precipitant agents. Furthermore, the evolution on the properties caused by hydrothermal treatment of samples synthetized with amines using the following two ways of heating was studied: conventional heating and microwave assisted heating. Highly crystalline LDH with nitrate anion in the interlayer was obtained using DMA as the precipitant agent and, after submitting the solid to conventional hydrothermal treatment. In the present work, we reported the use of amines with two carbon atoms in the organic chain [ethylamine (EA), diethylamine (DEA), triethylamine (TEA)] as precipitant agents in the synthesis of Zn,Al-LDH in molar ratio 2:1. Furthermore, the solids were treated hydrothermally by conventional heating and microwave-assisted heating to improve the crystallinity and properties of the solids. The effect of the treatment methods was analyzed for each amine-assisted synthesis condition.

2. Materials and Methods

2.1. Materials

$Zn(NO_3)_2 \cdot 6H_2O$ (98–102%), $Al(NO_3)_3 \cdot 9H_2O$ (98–102%) and NaOH (98%) were purchased from Panreac and used as received. An aqueous solution of ethylamine (70% in H_2O) was purchased from Alfa Aesar (Ward Hill, MA, USA). Diethylamine (99.5%) was purchased from Panreac (Barcelona, Spain). Triethylamine ($\geq$99%) was purchased from Sigma Aldrich (Burlington, MA, USA).

2.2. Synthesis

The coprecipitation method was used to prepare the desired solids [3]. In order to avoid the intercalation of carbonate anions, decarbonated water solutions and nitrogen atmosphere were used during the synthesis. In order to prepare the solution of the metal cations, 0.3 L in 2.5 M concentration of their nitrate salts in a M^{II}/M^{III} molar ratio 2:1 was prepared. For the precipitation medium, an aqueous solution of 4.5 M concentration of the desired amine was prepared. The metal cation solution was added dropwise to the amine solution, which was maintained under vigorous magnetic stirring. The pH of the precipitation media was kept at a preselected pH value of 10 by adding the required amount of a 2 M NaOH solution using a 240 CRISON pH-burette. After complete addition, the aqueous suspension of the precipitated solid was stirred for 1 h at room temperature and, then, the slurry obtained was subjected to different ageing treatments: (i) a portion of the sample without any hydrothermal treatment was kept as a reference; (ii) a portion was subjected to a hydrothermal treatment by microwave-assisted heating (MW) for 60 or 300 min at 90 °C; (iii) a portion was subjected to a hydrothermal treatment by conventional heating (HT) for 1 or 7 days at 90 °C using a home-made stainless-steel bomb lined with Teflon. Hydrothermal treatment by microwave-assisted heating was carried out in a MILESTONE ETHOS PLUS microwave oven, where the aqueous suspension was placed in Teflon digestion vessels, sealed and mounted on a turntable inside the oven. The programmed temperature was controlled by a thermocouple immersed in a reference vessel and software provided by the manufacturer. The different portions of the solid suspensions were separated and washed by centrifugation with distilled water until reaching a pH close to 7, in order to eliminate the equilibrium ions of the starting salts. Finally, the solids were dried at 40 °C in an oven under air atmosphere.

The samples were labelled according to the preparation procedure: reaction medium, hydrothermal treatment method and time period to which they had been subjected. So, a sample labeled as ZA2XYt, ZA2 represents the cations (Zn–Al, molar ratio 2/1); X represents the precipitation media (depending on the amine used EA, DEA or TEA); Y represents the aging treatment (STH for the reference sample, HT or MW); and t the treatment duration (in minutes for MW, in days for HT).

2.3. Characterization

A Yobin Ivon Ultima II apparatus at NUCLEUS (University of Salamanca, Salamanca, Spain) was used for elemental chemical analysis of Zn and Al by ICP-OES.

A Siemens D-5000 instrument was used to record Powder X-ray diffraction (PXRD) patterns using Cu-Kα radiation (λ = 1.54050 Å) with a scanning rate of 2°/min from 5° to 70° (2θ). The Scherrer equation was used to calculate the crystallite sizes from the FWHM (Full Width at Half Maximum) of the diffraction maximum (00l). The Warren correction for instrumental line broadening was taken into account, but the possible contribution of disorder effects and/or lattice strains to the peak broadening was ignored.

A Perkin-Elmer Spectrum One instrument was used to record the FT-IR spectra by transmission with a nominal resolution of 2 cm^{-1} from 4000 cm^{-1} to 450 cm^{-1}, using KBr pressed pellets.

SDT Q600 equipment from TA Instruments was used to carried out the thermogravimetric (TG) and differential thermal analyses (DTA). The thermal analyses were carried out by heating from room temperature to 900 °C at a rate of 10 °C/min under continuous oxygen (L'Air Liquide, 99.995%) flow (50 mL/min).

A Micromeritics Gemini VII 2390t apparatus was used to record the nitrogen (L'Air Liquide, 99.999%) adsorption–desorption isotherms at -196 °C, and to calculate the specific surface area and porosity data. The apparatus was calibrated with He (L'Air Liquide, 99.999%). Before measurements, the samples were pretreated at 110 °C for 2 h under a stream of N_2 in a Micromeritics FlowPrep 060 Sample Degass System.

A Diffraction Mastersizer 2000 equipment from Malvern Instruments was used to determine the particle size distribution (PSD) by Laser Diffraction. Using the dispersion unit Hydro 2000 from Malvern Instruments, the solid was dispersed in water at 25 °C (approx. 0.05 vol.%), and, after measuring the PSD for the dispersed samples, ultrasounds were applied in situ to disaggregate the particles.

3. Results and Discussion

3.1. Element Chemical Analysis

Table 1 gives the Zn/Al molar ratio values and chemical formulae of the samples synthesized in the presence of the different amines used as precipitant agents. The chemical formulae were deduced from the results of elemental chemical analysis and thermogravimetric analysis (see below).

Table 1. Element chemical analysis results and the chemical formulae of each sample.

Sample	Al [a]	Zn [a]	Zn/Al [b]	x [c]	Formulae
ZA2EASTH	8.32	40.27	2.00	0.33	$[Zn_{0.67}Al_{0.33}(OH)_2](NO_3)_{0.33} \cdot 0.44\ H_2O$
ZA2EAMW60	7.98	38.63	2.00	0.33	$[Zn_{0.67}Al_{0.33}(OH)_2](NO_3)_{0.33} \cdot 0.48\ H_2O$
ZA2EAMW300	7.96	38.30	1.99	0.33	$[Zn_{0.67}Al_{0.33}(OH)_2](NO_3)_{0.33} \cdot 0.50\ H_2O$
ZA2EAHT1	9.84	46.51	1.95	0.34	LDH * + ZnO + Al_2O_3
ZA2EAHT7	11.05	52.01	1.94	0.34	LDH * + ZnO + Al_2O_3
ZA2DEASTH	7.51	38.44	2.11	0.32	$[Zn_{0.68}Al_{0.32}(OH)_2](NO_3)_{0.32} \cdot 0.57\ H_2O$
ZA2DEAMW60	7.67	38.87	2.09	0.32	$[Zn_{0.68}Al_{0.32}(OH)_2](NO_3)_{0.32} \cdot 0.55\ H_2O$
ZA2DEAMW300	7.64	38.31	2.07	0.33	$[Zn_{0.67}Al_{0.33}(OH)_2](NO_3)_{0.33} \cdot 0.55\ H_2O$
ZA2DEAHT1	7.73	38.47	2.05	0.33	$[Zn_{0.67}Al_{0.33}(OH)_2](NO_3)_{0.33} \cdot 0.51\ H_2O$
ZA2DEAHT7	8.37	41.88	2.07	0.33	$[Zn_{0.67}Al_{0.33}(OH)_2](NO_3)_{0.33} \cdot 0.42\ H_2O$
ZA2TEASTH	7.64	37.56	2.03	0.33	$[Zn_{0.67}Al_{0.33}(OH)_2](NO_3)_{0.33} \cdot 0.55\ H_2O$
ZA2TEAMW60	7.62	37.62	2.04	0.33	$[Zn_{0.67}Al_{0.33}(OH)_2](NO_3)_{0.33} \cdot 0.51\ H_2O$
ZA2TEAMW300	7.60	37.39	2.03	0.33	$[Zn_{0.67}Al_{0.33}(OH)_2](NO_3)_{0.33} \cdot 0.52\ H_2O$
ZA2TEAHT1	7.67	37.82	2.04	0.33	$[Zn_{0.67}Al_{0.33}(OH)_2](NO_3)_{0.33} \cdot 0.50\ H_2O$
ZA2TEAHT7	8.13	41.16	2.09	0.32	$[Zn_{0.68}Al_{0.32}(OH)_2](NO_3)_{0.32} \cdot 0.43\ H_2O$

[a] Mass percentage. [b] Molar ratio. [c] Al/(Al + Zn) molar ratio. * It was not possible to determine the chemical formula of the LDH phase.

In all cases, the Zn/Al molar ratio approached the value of 2, suggesting a complete precipitation of the existing cations in the synthesis medium. In some cases, a small deviation could be observed, but never more than 5%.

For the determination of the chemical formula of each of the samples, the amount of nitrate anion was calculated from the Al/(Al + Zn) molar ratio, assuming that it was the only interlayered anion, as observed by FT-IR spectroscopy, neutralizing the positive charge excess of the layers.

Unlike samples synthesized without amines or in the presence of methylamine, for dimethylamine or trimethylamine samples synthesized in the presence of EA, DEA or TEA have water content per chemical formula which is generally lower, as reported in our previous work [27] The more regular stacking of the octahedral layers, as observed by PXRD, leads to a decrease in the number of water molecules per unit formula, and this fact is more evident in the samples having HT hydrothermal treatment.

On the other hand, samples ZA2EAHT1 and ZA2EAHT7 were mostly or completely formed by ZnO, as deduced from the PXRD analysis. However, after chemical analysis, the presence of Al in the solid sample could be determined, indicating the formation of an amorphous phase containing aluminum [28], not observed by X-ray diffraction. Therefore, as there was a mixture of phases in the solid sample, it was not possible to determine the amount of Zn and Al forming the LDH structure (if its collapse was not complete); thus, making it difficult to determine its chemical formula.

3.2. Powder X-ray Diffraction (PXRD)

The samples synthesized using EA, DEA and TEA as modifiers of the precipitation medium were also obtained in the form of microcrystalline powder. Figure 1 shows the PXRD diagrams of these samples without hydrothermal treatment. The positions and relative intensities of the recorded diffraction peaks revealed a layered structure of the solids, characteristic of an ordered $3R_1$ polytype of solids with the LDH structure (JCPDS: 22-0700) [29–31]. In all cases, the most intense diffraction peak, attributed to the diffraction plane (003) of the crystal structure, was recorded at a position 10.0° (2θ), with a spacing of 8.93 Å. This spacing is in agreement with the values reported by Miyata et al. [16] for LDH with nitrate as the interlayer anion arranged in a perpendicular orientation to the brucite-like layers and with a M^{2+}/M^{3+} molar ratio close to 2. Confirming the layered structure, diffraction peaks corresponding to crystallographic planes (006) and (009) were recorded at values close to 19.9° (2θ) and 30.0° (2θ), respectively, and with spacings of 4.46 Å and 2.98 Å. Reflections corresponding to diffraction planes (110) and (113) were recorded at 60.3° (2θ) and 61.3° (2θ), with spacings of 1.53 Å and 1.51 Å, respectively.

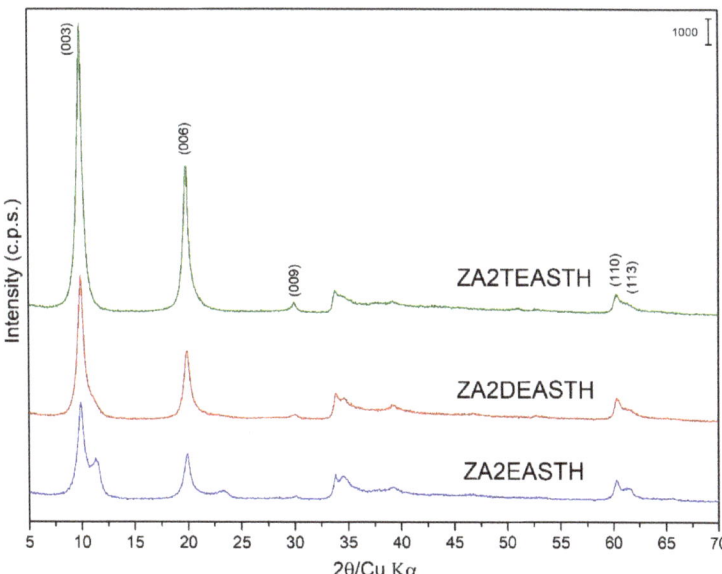

Figure 1. Powder X-ray diffraction diagrams of the samples prepared in the presence of EA, DEA and TEA with no hydrothermal treatment.

As can be seen in Figure 1, the solids prepared using DEA or TEA as precipitant agents showed a single crystallographic phase. However, in the case of the sample prepared in the presence of EA, diffraction peaks corresponding to a second LDH phase could be distinguished in the PXRD diagram. Thus, it is possible to distinguish the most characteristic peak of this secondary phase, which was recorded close to 11.33° (2θ) and ascribed to the diffraction peak (003), with a spacing of 7.81 Å. In this case, the diffraction angle for this diffraction peak was slightly higher than that found for the sample synthesized in the absence of amines [27], where, unlike the shoulder observed for the sample prepared in the absence of amines, a well-defined maximum could be clearly distinguished. In this second phase with a spacing of 7.81 Å for the peak, due to the (003) planes, the interlaminar nitrate anions were arranged with their molecular plane parallel to the plane of the brucite-like layers [32–34].

Observing the width and profile of the peaks ascribed to the diffraction plane (003) close to 10° (2θ), the crystallinity of these samples decreased in the order: ZA2TEASTH > ZA2DEASTH > ZA2EASTH. Again, it could be observed how the use of amines (and the nature of these) in the precipitation medium modifies the crystallinity of the solids, to the point of two phases coexisting, with nitrate anion in the interlayer space depending on the amine used.

The samples synthesized in the presence of EA, DEA and TEA were also subjected to hydrothermal treatments, using two heating routes: microwave-assisted heating (MW) and conventional oven heating (HT). Figure 2 includes the PXRD plots of these samples with different MW hydrothermal treatment periods. Observing the profile of the diffraction peaks and their relative intensities as the treatment time increased, it can be seen how, unlike the samples prepared in our previous work [27], the application of MW hydrothermal treatment increased the crystallinity of the solids as the MW treatment time increased. Comparing the width of the diffraction peaks (003) of the samples subjected to a longer treatment time, that is, the samples with 300 min of treatment, the following decreasing order of crystallinity could be established as a function of the amine used: ZA2TEAMW300 > ZA2DEAMW300 > ZA2EAMW300. It is noteworthy that, as the MW hydrothermal treatment time was prolonged on the samples synthesized with EA, the relative intensity of the diffraction peak (003) of the phase with the nitrate anion parallel to the brucite-like layers did not increase as the treatment time increased, going from being a well-defined peak in sample ZA2EASTH to being a shoulder of the peak (003) in sample ZA2EAMW300.

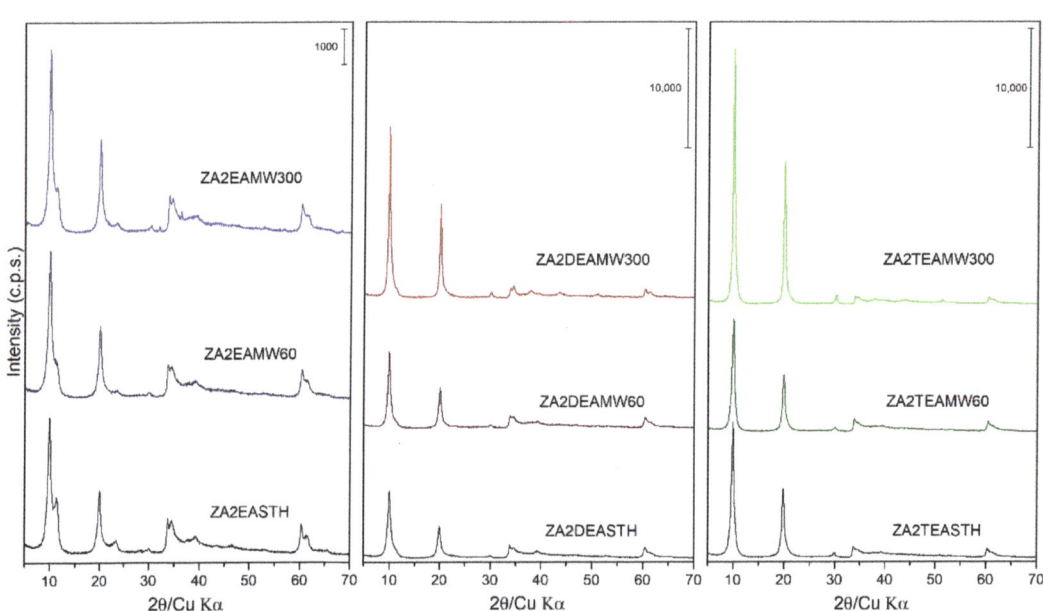

Figure 2. PXRD of the samples prepared in the presence of EA, DEA and TEA with 60 and 300 min of MW treatment.

The application of a conventional hydrothermal treatment (HT) resulted in a greater increase in the crystallinity of the solids synthesized in the presence of DEA and TEA.

However, the same effect is not observed for the samples synthesized in the presence of EA, where the application of HT treatment resulted in the collapse of the layered structure, as can be observed in Figure 3. Thus, the PXRD recorded for samples synthetized using EA showed the formation of a zinc oxide (ZnO) like zincite phase (JCPDS: 00-036-1451 [29]). This ZnO phase is identified mainly by the reflections recorded in the range of 28°–38° (2θ). While with 1 day of HT treatment the collapse of the LDH phase was not complete, as can be seen from its corresponding PXRD diagram, after 7 days of treatment the collapse was complete. So, the main diffraction peaks of sample ZA2EAHT7 corresponded to the ZnO phase and, with very little intensity, the diffraction peaks at 11.6° and 23.4° (2θ) corresponded to the crystallographic planes (003) and (006), respectively, of the LDH phase with the interlayer nitrate arranged parallel to the plane of the brucite-like layers.

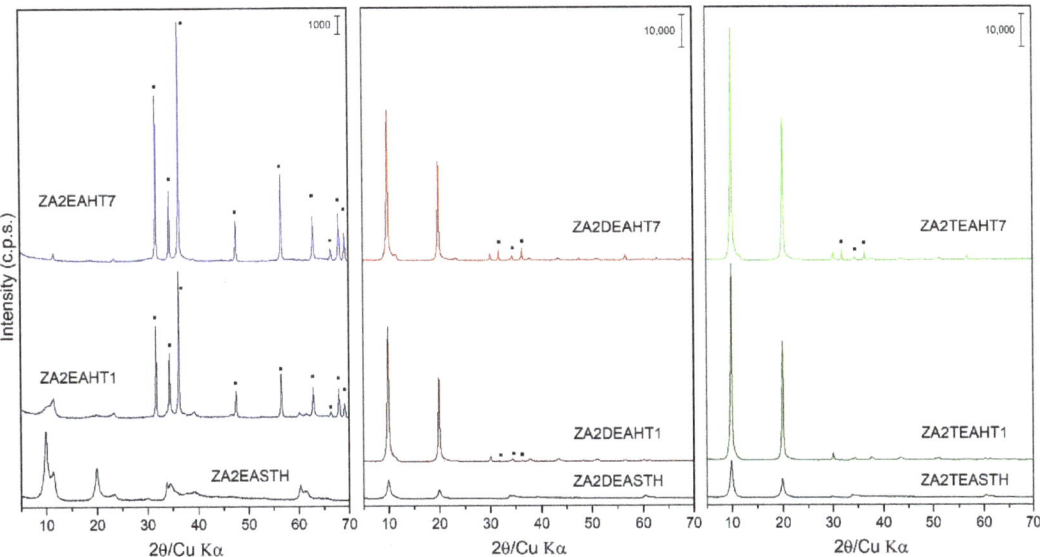

Figure 3. PXRD of the samples prepared in the presence of EA, DEA and TEA with 1 and 7 days of HT treatment. (■) ZnO phase.

In the case of the samples synthetized using DEA or TEA, segregation of a small amount of the zincite phase (ZnO) could also be observed as the HT treatment time increased. Thus, from the profile of the diffraction peaks (003) and their relative intensities for these two series of samples, it can be observed how with 1 day of treatment, solids of high crystallinity were obtained without segregation of the ZnO phase when TEA was used as the precipitant agent, and with a very small amount of this phase when DEA was used. Therefore, the segregation of ZnO decreased as the degree of substitution of the amino group in the compound used as precipitating agent increased. Comparing the width of the diffraction peaks (003) of the samples subjected to a longer treatment time (samples with 7 days of treatment) the following decreasing order of crystallinity can be established as a function of the amine used: ZA2TEAHT7 > ZA2DEAHT7. Table 2 shows the amount of zincite phase present in the sample calculated from the calibration line with the ratio of the areas of the characteristic peaks of the zincite phase and the LDH phase [35]. Thus, it is shown how the amount of zincite phase increased as the HT treatment time increased, finding an amount of approximately 2% of ZnO in the sample ZA2DEAHT7. By extrapolation of the calibration line to samples synthesized using EA, for the ZA2EAHT1 sample, 67% was ZnO and for longer treatments it was close to 100% (Table 2).

Table 2. ZnO content in the Zn and Al samples prepared in the presence of EA, DEA and TEA with HT treatment.

Sample	(101) Peak Area (ZnO) [a]	(003) Peak Area (LDH) [a]	Area Ratio (101)/(003)	ZnO Content [b]
ZA2EAHT1	1807.0	1424.0	1.268961	67
ZA2EAHT7	2776.0	-	-	≈100
ZA2DEAHT1	122.1	18,339.8	0.006659	0.32
ZA2DEAHT7	790.7	19,232.1	0.041114	2.14
ZA2TEAHT1	-	-	-	-
ZA2TEAHT7	502.6	26,627.0	0.018876	0.97

[a] a.u. [b] Mass percentage.

The lattice parameters of the prepared solids were calculated from the positions of the diffraction peaks due to the (003) and (110) planes [30]; being, $c = 3 \cdot d(003) \approx 26.6–26.8$ Å, and $a = 2 \cdot d(110) \approx 3.069–3.077$ Å. In addition, from the value of the Full Width at Half Maximum (FWHM) of reflection 003, the crystallite size (D) in the c direction was calculated using the Scherrer equation, $D = k\lambda / \beta \cos\theta$ [36,37], where k is a constant, taken in this case as 0.9; λ is the wavelength of the radiation used; β the FWHM and θ the diffraction angle; correction due to instrumental broadening was not applied. The values calculated for the samples obtained are included in Table 3, together with the calculated values for the number of stacked layers.

Table 3. Lattice parameters c and a, average crystal size D and number of stacked layers for the samples obtained.

Sample	c (Å)	a (Å)	D (Å)	Number of Stacked Layers
ZA2EASTH	26.80	3.0697	111	12
ZA2EAMW60	26.80	3.0697	118	13
ZA2EAMW300	26.80	3.0720	135	15
ZA2EAHT1	-	-	-	-
ZA2EAHT7	-	-	-	-
ZA2DEASTH	26.67	3.0673	126	14
ZA2DEAMW60	26.80	3.0697	148	17
ZA2DEAMW300	26.80	3.0720	210	23
ZA2DEAHT1	26.67	3.0743	284	32
ZA2DEAHT7	26.80	3.0766	334	37
ZA2TEASTH	26.80	3.0697	150	17
ZA2TEAMW60	26.80	3.0697	142	16
ZA2TEAMW300	26.80	3.0697	218	24
ZA2TEAHT1	26.80	3.0766	319	36
ZA2TEAHT7	26.67	3.0743	322	36

The similarity of the lattice parameter a value, with differences of less than 1%, are coherent with the homogeneity of metals composition in the samples.

The crystallite size (D) values highlight how the conventional hydrothermal treatment led to a greater increase in the crystallinity of the solids. When the MW treatment was applied, it was observed that, as the treatment time increased the crystal size increased, resulting in the obtaining of, in both the DEA and TEA series, crystal sizes around 200 Å. However, when HT treatment was applied, it could be observed that the samples synthesized in the presence of TEA with only one day of treatment reached a crystal size close to 320 Å, which was practically maintained, even if the HT treatment time increased. In the case of samples prepared in the presence of DEA, with one day of HT treatment a crystal size higher than 280 Å was obtained; reaching a crystal size of 334 Å when the treatment was extended to 7 days.

3.3. FT-IR Spectroscopy

FT-IR spectra of the synthesized solids are plotted in Figure 4. The broad band at 3460 cm^{-1} is ascribed to the stretching vibration modes of the hydroxyl groups and the water molecules of the interlayer space. At lower wavenumbers it was possible to observe a bands at 615 and 556 cm^{-1} caused by M-OH vibration modes. The vibration ascribed to the bending mode of the interlayer water molecules was recorded at 1624 cm^{-1} [18,38,39]. The bands at 2396 and 2428 cm^{-1} could be attributed to atmospheric CO_2 weakly bonded on the LDH surface.

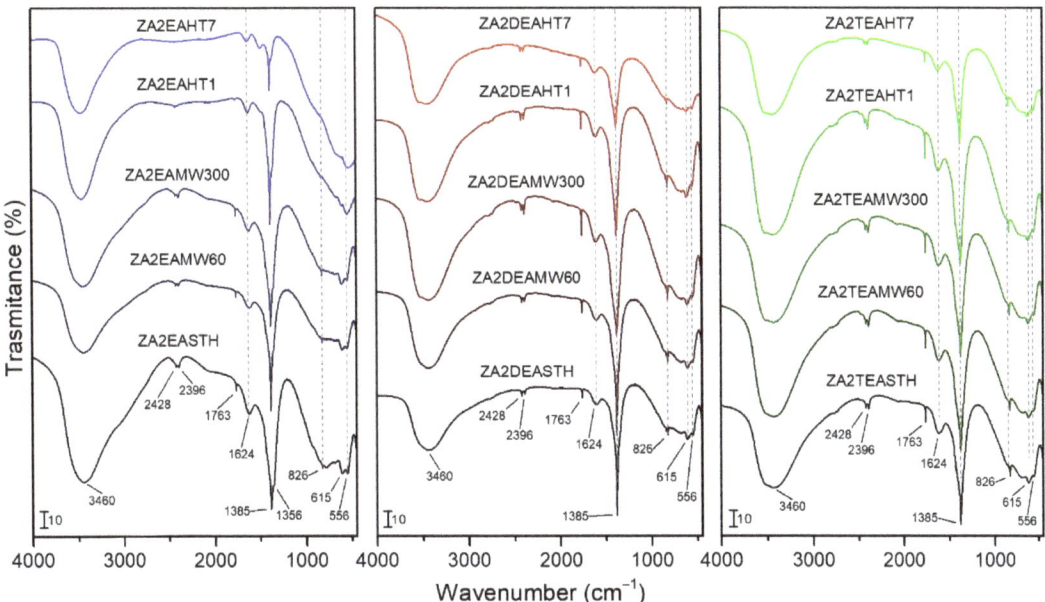

Figure 4. FT-IR spectra of the samples prepared in the presence of EA, DEA and TEA with no hydrothermal treatment and with MW and HT treatments.

In all cases, the characteristic bands of the vibrational modes of the nitrate anion molecules were recorded, confirming its presence as an interlayer anion. Thus, bands at 1385 cm^{-1} and 826 cm^{-1} can be observed, assigned to the ν3(E′) and ν2 (A2″) vibrational modes, respectively, of NO_3^- with a D_{3h} symmetry [19,38]. At 1763 cm^{-1} a narrow band can be observed, corresponding to the combination of the vibrational modes ν1(A1′) at 1068 cm^{-1} and ν4(E′) at 692 cm^{-1} of nitrate, the latter not clearly observed in the infrared spectra. For sample ZA2EASTH, a shoulder can be seen at 1356 cm^{-1} (Figure 4), which could be due to nitrate anions in parallel orientation in the second LDH phase observed by PXRD. As observed by PXRD for the samples synthesized in the presence of EA after MW hydrothermal treatment, a crystallinity increase of the phase with the anions in perpendicular orientation took place in detriment of the phase with the anions in parallel orientation to the layers. This effect was reflected in the FT-IR spectra, where the shoulder at 1356 cm^{-1} became less evident (Figure 4). On the other hand, when DEA or TEA were used as precipitaton agents, such a shoulder was also not observed at 1356 cm^{-1}.

When the sample synthesized in the presence of EA was subjected to a HT hydrothermal treatment process, as observed in the PXRD studies, the structure collapsed segregating the zinc oxide in its zincite phase, being almost complete with long treatment times; this caused the FT-IR spectra of these samples to change slightly. In Figure 4 the bands attributed to the vibrational modes of the hydroxyl groups and water molecules (band positions already mentioned above) can be identified. In addition, the vibration band at

1385 cm^{-1}, attributed to the presence of the nitrate anion of the LDH phase recorded with low intensity in the PXRD diagrams of the sample, can be clearly observed. As a result of the formation of ZnO in the HT-treated samples, a band at the limit of the spectrum, around 470 cm^{-1}, attributed to the stretching vibrational mode of ZnO, was observed in the FT-IR spectra [40].

In the spectra of the samples synthesized in the presence of DEA and TEA, no different bands were observed for the hydrothermally treated samples, both MW and HT, with respect to that of the samples that received no treatment. It should be noted that the higher crystallinity of the samples resulted in a better ordering of both the interlaminar anions and the layered structure, giving rise to a regularity that was reflected in a subtle increase in intensity and narrowing of the vibration bands.

Neither in the FT-IR spectra recorded for the samples without hydrothermal treatment, nor for the samples subjected to MW or HT hydrothermal treatments, were bands corresponding to the vibrational modes of the amines used during the synthesis observed.

3.4. Thermal Analysis

The thermal analysis of the samples prepared using EA, DEA or TEA as precipitan agents were carried out to determine their stability and evolution to mixed oxides. During the thermal analysis, mass spectrometry (MS) of the gases and vapors formed during the process (EGA, evolved gas analysis) was carried out. To identify the masses of the generated species, a complete mass spectrum was initially recorded and, in a second step, the MS analysis was performed by fixing these masses and following the change in their intensities throughout the analysis. The reference MS of the expected evolved gases was also taken into account. The signals monitorized corresponded to H_2O (m/z = 18), N_2 and CO (m/z = 28), NO (m/z = 30), N_2O (m/z = 44), NO_2 (m/z = 45), EA (fragment at m/z = 30, 44 and 45), DEA (fragments at m/z = 58 and 73) and TEA (fragments at m/z = 58 and 101).

TG curves of the samples without hydrothermal treatment (STH) are included in Figure 5, together with the tracked masses of the gases generated during the process. In all curves, the typical decomposition stages of LDH compounds can be identified. Three decomposition stages could be identified in all TG curves. First, below 180 °C, water removal was observed, as shown by the MS peak at m/z = 18. The second stage of decomposition was observed up to around 300 °C, and corresponded to the release of water from the condensation of hydroxyl groups of the brucite-like layers. Finally, a steady mass loss was observed between 300 and 700 °C, which corresponded to the removal of interlayer nitrate species. The MS signals recorded in this temperature range corresponded to formation of species such as NO, NO_2, and N_2O, from nitrate decomposition. In all the curves, the process of elimination of the hydroxyl groups practically overlapped with the last stage of decomposition/removal of the interlaminar anion. However, there was better differentiation of the first stage from the second decomposition stage as the degree of substitution of the amino group in the compound used as precipitating agent increased. Thus, for the ZA2TEASTH sample a small plateau could be observed around 200 °C.

From the tracking of the m/z signals of each amine, the absence of amine residues in the final solids could be concluded, as could also be observed by FT-IR spectroscopy. Only in the case of the sample prepared in the presence of EA, could the mass at 45 m/z be attributed to that amine. However, some of the masses associated with EA overlap with the signals of the decomposition products of the nitrate anion (N_2O, NO and NO_2). The absence of EA was confirmed because the mass tracking curves had the same profile as those of NOx gas formation in the solids synthesized in the presence of DEA and TEA.

For the hydrothermally treated samples, both MW and HT, had similar TG curves recorded, in which the aforementioned plateau was more evident as a result of the increase in the crystallinity of the samples.

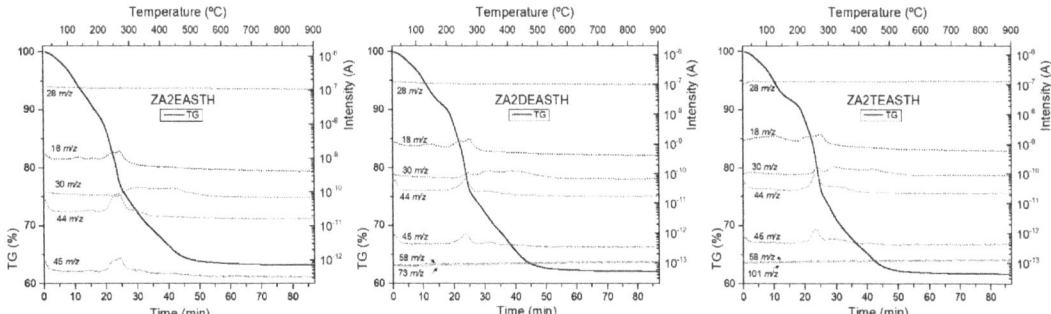

Figure 5. TG curves, in O_2 atmosphere, of the samples prepared in the presence of EA, DEA and TEA with no hydrothermal treatment and tracking of the characteristic m/z signals.

On the other hand, it can also be observed that, as the crystalline regularity of the solids increases, the total mass loss decreased (Table 4), due to the lower amount of water retained in the solids. Table 4 shows that as the hydrothermal treatment time increased, the total mass loss was lower, being in all cases between 30 and 40%. Only in the case of the samples synthesized in the presence of EA and with HT hydrothermal treatment was a total mass loss less than 30% observed. This was due to the formation of ZnO in these samples, where after 7 days of treatment practically the whole structure had collapsed and the mass loss observed in the TG curve was due to the small amount of the LDH phase, the retained water and the decomposition of the nitrate anion.

Table 4. Total weight loss and H_2O molecules per chemical formula calculated for each sample.

Sample	Weight Loss (%)	H_2O Molecules Per Chemical Formula (n)
ZA2EASTH	36.6	0.44
ZA2EAMW60	36.7	0.48
ZA2EAMW300	38.0	0.50
ZA2EAHT1	22.6	-
ZA2EAHT7	14.6	-
ZA2DEASTH	38.1	0.57
ZA2DEAMW60	37.8	0.55
ZA2DEAMW300	38.2	0.55
ZA2DEAHT1	37.3	0.51
ZA2DEAHT7	31.0	0.42
ZA2TEASTH	38.7	0.55
ZA2TEAMW60	38.3	0.51
ZA2TEAMW300	38.5	0.52
ZA2TEAHT1	37.9	0.50
ZA2TEAHT7	32.8	0.43

The DTA curves of the samples both without hydrothermal treatment and with hydrothermal treatments, MW or HT, are included in Figure 6. In all cases, endothermic minima associated with the different decomposition processes of the samples can be observed.

The minimum recorded at 120–130 °C in the DTA curves was associated with the process of release of the water retained in the interlayer space. However, in the case of the samples synthesized in the presence of EA, another minimum could be observed at 176 °C, which could correspond to the elimination of water molecules retained more strongly in the structure of the layered solid. On the other hand, at temperatures above 200 °C a minimum associated with the process of elimination of hydroxyl groups in the form of water vapor and the decomposition of the nitrate anion was found. In many cases this minimum presented a shoulder at lower temperature, or was even dissociated into two clearly distinguishable minima. This dissociation and, therefore, differentiation in the decomposition processes, became more evident as the crystalline regularity in the solids increased after the application of a hydrothermal treatment. On the other hand, it is worth mentioning that in all cases the main minimum was found at higher temperatures than in the case of the solids synthesized in our previous work using methylamine, dimethylamine or trimethylamine as precipitant agents, and even in the case of the solids synthesized in the absence of amines [27]. Thus, for the sample synthesized in the presence of DEA, this minimum was found at 260 °C, while for the sample synthesized in the presence of TEA, the minimum shifted to 272 °C.

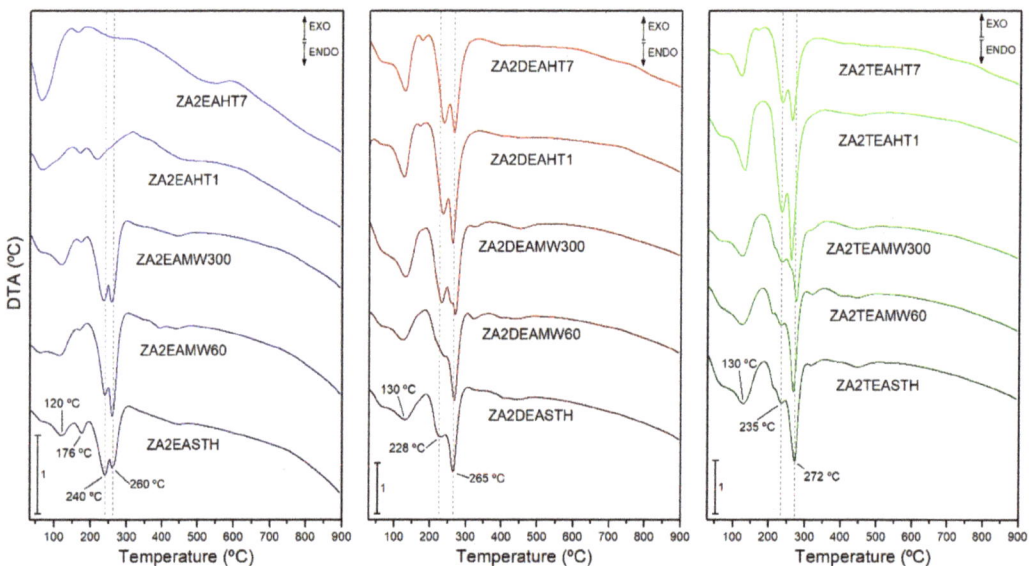

Figure 6. DTA curves, in O_2 atmosphere, of the samples prepared in the presence of EA, DEA and TEA with no hydrothermal treatment and with MW and HT treatments.

3.5. Specific Surface Area and Porosity

The textural properties of the synthesized solids were studied from the N_2 adsorption–desorption isotherms at −196 °C. Table 5 includes the values of the specific surface areas calculated by the BET (S_{BET}) method [41,42], the pore volume (V_{pore}) and the average pore diameter calculated by the BJH method [42,43] for the synthesized samples. For the samples prepared in the presence of DEA and TEA, both without hydrothermal treatment and with MW treatment, the adsorption capacity was below the confidence limit of the equipment used. Only for the samples with HT hydrothermal treatment (and sample ZA2DEAMW300) did the adsorption measurements present confidence values. This behavior was not observed for samples prepared in the presence of EA, although the samples presented low S_{BET} values, close to the detection limit. Figures 7 and 8 include the corresponding adsorption–desorption isotherms, which correspond to type II according to the IUPAC classification [44,45], corresponding to adsorption on non-porous or mesoporous adsorbents, where adsorption can occur without monolayer-multilayer restrictions. Moreover, it can be observed how all of them presented a hysteresis cycle, corresponding to the H3 type according to the IUPAC classification [42], indicating that adsorption took place in slit-shaped pores formed by layer-like particles.

Table 5. BET specific surface area, pore volume and average pore diameter of the samples prepared.

Sample	S_{BET} (m²/g)	V_{pore} (mm³/g)	BJH Desorption Average Pore Diameter (nm)
ZA2EASTH	6.1	14.9	9.1
ZA2EAMW60	3.4	9.8	9.6
ZA2EAMW300	4.5	11.2	9.6
ZA2EAHT1	74.3	73.7	4.6
ZA2EAHT7	118.5	83.2	3.0
ZA2DEASTH	-	-	-
ZA2DEAMW60	-	-	-
ZA2DEAMW300	6.4	16.0	6.9
ZA2DEAHT1	15.3	32.2	7.6
ZA2DEAHT7	28.2	42.2	5.3
ZA2TEASTH	-	-	-
ZA2TEAMW60	-	-	-
ZA2TEAMW300	-	-	-
ZA2TEAHT1	9.0	23.3	9.9
ZA2TEAHT7	21.0	32.4	5.7

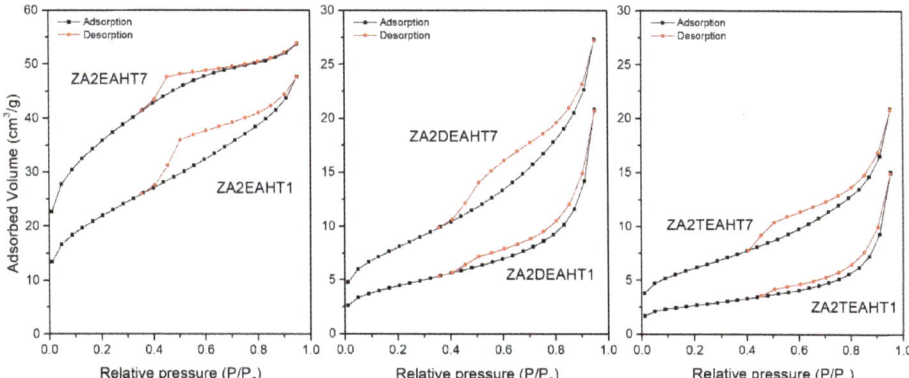

Figure 7. Nitrogen adsorption-desorption isotherms for HT treated samples prepared in the presence of EA, DEA and TEA.

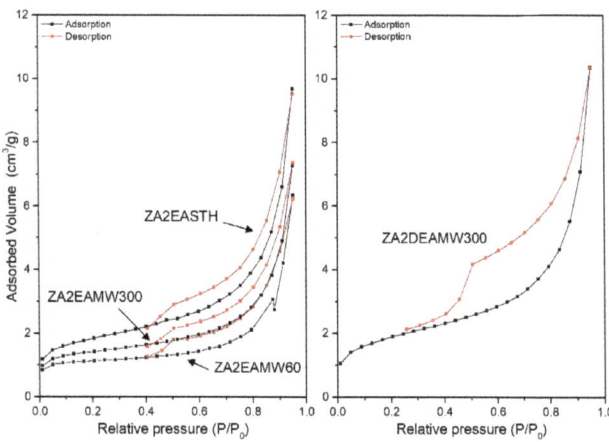

Figure 8. Nitrogen adsorption-desorption isotherms for MW treated samples prepared in the presence of EA and DEA.

In all cases S_{BET} values were higher than those obtained for the samples obtained with the amines used in our previous work and, also, the values were higher than those found for the samples synthesized in the absence of amines [27]. On the contrary, smaller pore diameter sizes were recorded.

In view of the results included in Table 5, the segregation of ZnO in the samples synthesized in the presence of EA with HT treatment led to a substantial increase in the specific surface area. Solids with higher porosity were obtained, where the pore volume increased, with smaller pore diameter sizes with respect to the sample without hydrothermal treatment.

In the case of the solids synthesized in the presence of DEA and TEA, after the application of a HT hydrothermal treatment, according to the PXRD results, an increase in the crystallinity took place, as a consequence of better ordering of the brucite-like layers. Together with the increase in the crystallinity of the solids, an increase in the S_{BET} could be observed and adsorption–desorption curves could be recorded in both cases. Moreover, the prolongation of the HT treatment resulted in higher crystallinity linked to an increase of the S_{BET} value, where, in both series of samples, the S_{BET} value for sample with 7 days of treatment was twice that for the sample with one day of treatment. However, it is important to remember the presence of approximately 2% of ZnO in sample ZA2DEAHT7, which could justify the difference in specific surface area with respect to that of sample ZA2TEAHT7.

Figure 8 shows the adsorption–desorption curves for the samples without hydrotermal treatment and with MW treatment. While the sample without hydrothermal treatment had an S_{BET} value close to 6 m^2/g, for samples synthetized using EA as precipitant agent, when a MW hydrothermal treatment was applied the surface area decreased. In the three curves a similar behavior against the desorption process can be observed, with slightly larger pore sizes when MW treatment was applied. On the other hand, when DEA or TEA were used as precipitant agents, only the adsorption–desorption curve for sample ZA2DEAMW300 was recorded. The S_{BET} value for this sample was 6.4 m^2/g, higher than that found for the analogous sample synthesized in the presence of EA.

3.6. Particle Size Distribution

Figure 9 shows the particle size distribution curves of the samples synthesized in the presence of EA, DEA and TEA without hydrothermal treatment and after the application of the longest periods of both hydrothermal treatments: MW and HT. For each of the samples two distribution curves are represented: (i) the distribution curve of the sample in aqueous suspension, black curve, and (ii) the distribution curve of the sample in aqueous suspension subjected to sonication treatment for 15 min, directly in the particle size analyzer, red curve. Sonication treatment is often used to disaggregate primary particles, changing the distribution curves.

The samples without hydrothermal treatment presented a monomodal size distribution centered between 300 and 400 μm. Although, as the degree of amino group substitution increased, a small shoulder could be observed at lower size values, approximately at 20 μm, as observed in Figure 9a–c. The application of ultrasound for 15 min did not have a great impact on the distribution curves; only in the case of sample ZA2TEASTH was the aforementioned shoulder accentuated at 20 μm. This fact indicates that samples prepared in the presence of EA and DEA give rise to robust particles that are more difficult to disaggregate than particles obtained using TEA as the precipitant agent.

The particle size distribution curves for the samples subjected to MW hydrothermal treatment for 300 min are included in Figure 9d–f. While the sample synthesized in the presence of EA presented the same profile as its version without hydrothermal treatment, broader distribution curves were observed for samples ZA2DEAMW300 and ZA2TEAMW300. After the application of ultrasound, practically no changes in the distribution were observed for sample ZA2EAMW300, indicating the difficulty in desaggregating its particles, showing the same robustness as the STH sample. In the case of samples prepared in the presence of DEA and TEA, after the application of ultrasound, disaggregation to smaller particle size took place, resulting in bimodal distribution curves with maxima at 40 µm. Moreover, after the application of ultrasound, it can be observed how these distributions became wider.

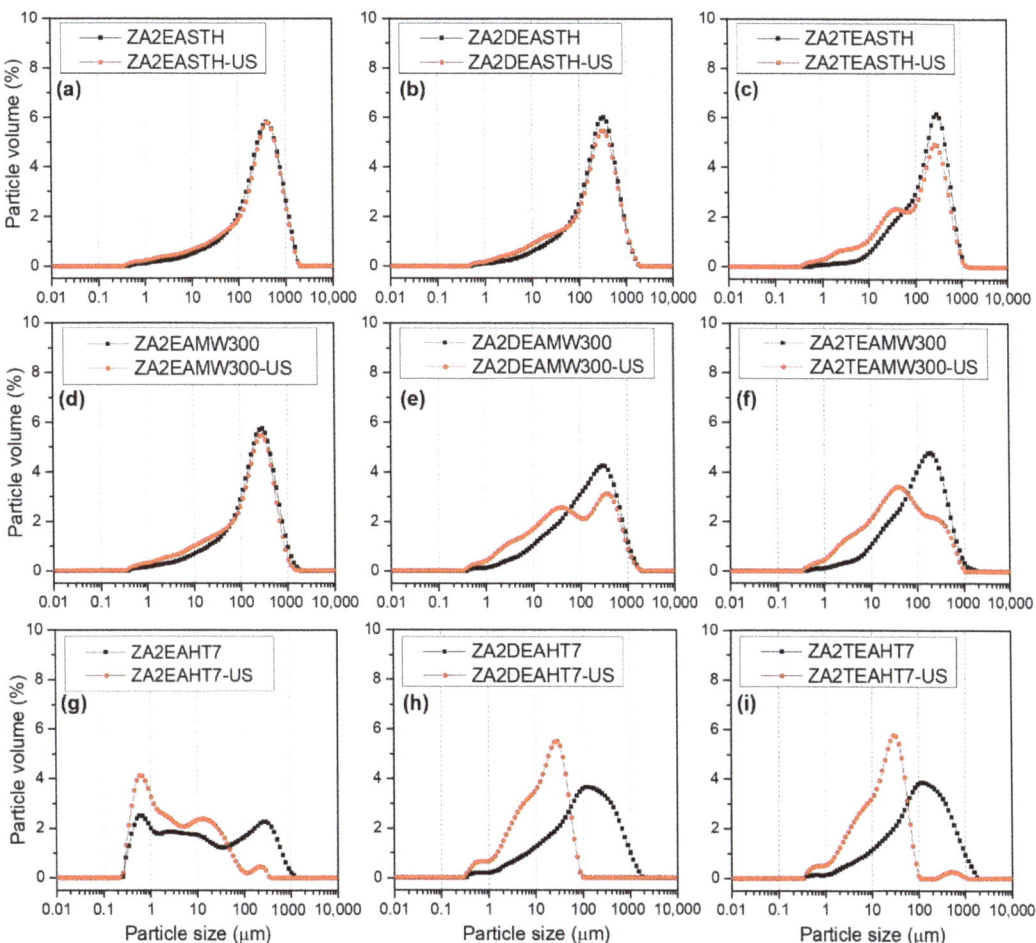

Figure 9. Particle size distribution before (black) and after (red) sonication in water suspension of synthetized samples (**a**) ZA2EASTH, (**b**) ZA2DEASTH, (**c**) ZA2TEASTH, (**d**) ZA2EAMW300, (**e**) ZA2DEAMW300, (**f**) ZA2TEAMW300, (**g**) ZA2EAHT7, (**h**) ZA2DEAHT7, (**i**) ZA2TEAHT7.

The distribution curves for the samples subjected to HT hydrothermal treatment for 7 days are included in Figure 9g–i, where it is observed that the samples synthesized in the presence of DEA and TEA presented the same behavior as the samples with MW hydrotermal treatment over 300 min, obtaining monomodal particle size distributions that covered a wide range of sizes. However, when an ultrasound treatment was applied, the particles disaggregated, obtaining monomodal distributions at smaller particle sizes with a maxima between 20 and 30 µm, which presented, in both cases, a shoulder at even smaller particle sizes, around 5 µm. Furthermore, in both samples it was observed how the size distributions became narrower. For example, the particle size distribution of sample ZA2DEAHT7 decreased from a maximum at 130 µm to a maximum at 30 µm, with a shoulder at 6 µm. In contrast, the curve of sample ZA2EAHT7 presented slightly different behavior, where a multimodal distribution was obtained over a wide range of particle sizes, with maxima at 0.6, 2.7, 10 and 300 µm. After the application of ultrasound, the disintegration of the larger particles occurred, also resulting in a multimodal distribution with maxima at 0.6, 2.5, 12 and 225 µm, with a higher percentage of smaller particles. The different behavior of sample ZA2EAHT7 was due to fact that in this sample the hydrotalcite-type structure had completely collapsed and the only phase detected by PXRD was ZnO (Figure 3).

Table 6 includes the volume-weighted mean particle diameter, D[4,3], values for each of the samples, both before and after ultrasound treatment [46,47]. In general, the lowest particle size values were obtained for the samples synthesized in the presence of TEA, obtaining values even lower than those of the samples obtained without amines in the reaction medium (reported in our previous work [27]). For all series of samples, similar D[4,3] values were found, regardless of the MW treatment time, whereas, when HT hydrothermal treatment was applied, as the treatment time increased, the D[4,3] value slightly decreased. The smallest particle size, 114 µm, was found for sample ZA2EAHT7, where, as previously commented, the structure had collapsed to zinc oxide.

Table 6. Volume-weighted mean particle diameter, D[4,3], values of prepared samples.

Sample	D[4,3] Before Sonication	D[4,3] After Sonication
ZA2EASTH	428	399
ZA2EAMW60	280	258
ZA2EAMW300	298	248
ZA2EAHT1	327	230
ZA2EAHT7	114	16
ZA2DEASTH	347	324
ZA2DEAMW60	268	263
ZA2DEAMW300	275	210
ZA2DEAHT1	272	94
ZA2DEAHT7	232	22
ZA2TEASTH	274	212
ZA2TEAMW60	221	219
ZA2TEAMW300	201	112
ZA2TEAHT1	253	79
ZA2TEAHT7	231	36

The application of a sonication treatment had a great impact on the D[4,3] value of the samples subjected to HT hydrothermal treatment, as can be deduced from the data collected in Table 6. Thus, the lowest value was obtained for sample ZA2DEAHT7, with a mean particle diameter of 36 µm.

It can be concluded that the samples subjected to MW treatment were formed by tightly bound particle agglomerates, deduced by the low disintegration after the sonication treatment. Only for samples with long periods of MW treatment was a slight disintegration of the agglomerates observed, obtaining the lowest D[4,3] value for the sample ZA2TEAMW300, with 112 µm.

4. Conclusions

The effect on the properties of Zn-Al-NO$_3$ LDHs, prepared using amines with two carbon atoms in the organic chain, together with the application of hydrothermal treatments using different energy sources, conventional or microwave heating, was studied. Well crystallized compounds were obtained, the crystallinity of which improved a lot after prolonged conventional hydrothermal treatment when diethylamine or triethylamine were used as precipitant agents. When ethylamine was used in the synthesis media, two LDH phases were obtained, in which the nitrate anion had two different orientations in the interlayer space. Furthermore, the LDH structure of solids prepared using ethylamine collapsed easily with a conventional hydrothermal treatment to form zinc oxide. When the samples were subjected to conventional hydrothermal treatment, the formation of ZnO was also observed when DEA or TEA were used, although to a smaller amount. So, the ZnO content is lower as the degree of substitution of the amino group of the compound used as precipitating agent increases. The results also showed the formation of aggregates which could be disaggregated by sonication. So, it was possible to obtain solids with high crystal sizes and low particle size distributions when a conventional treatment was used.

Author Contributions: Conceptualization, F.M.L.; methodology, A.M., A.J. and F.M.L.; investigation, A.M.; resources, F.M.L.; writing—original draft preparation, A.M.; writing—review and editing, A.M. and F.M.L.; visualization, A.M. and A.J.; supervision, F.M.L.; funding acquisition, F.M.L. All authors have read and agreed to the published version of the manuscript.

Funding: Financial support from Universidad de Salamanca (Plan I-B3) and Proyect RTC-2014-1908-3, Ministerio de Economía y Competitividad.

Institutional Review Board Statement: Not applicable.

Informed Consent Statement: Not applicable.

Data Availability Statement: Not applicable.

Acknowledgments: A. Misol thanks Junta de Castilla y León and ERDF for a predoctoral contract. A. Jiménez thanks Universidad de Salamanca and Banco Santander for a predoctoral contract. The authors want to express the great pleasure and privilege in collaborating with this paper for this special issue of the ChemEngineering in honor of our mentor and friend Vicente Rives, Professor of Inorganic Chemistry at University of Salamanca.

Conflicts of Interest: The authors declare no conflict of interest.

References

1. Cavani, F.; Trifirò, F.; Vaccari, A. Hydrotalcite-type anionic clays: Preparation, properties and applications. *Catal. Today* **1991**, *11*, 173–301. [CrossRef]
2. Rives, V.; Labajos, F.M.; Herrero, M. Layered Double Hydroxides as Nanofillers of Composites and Nanocomposite Materials Based on Polyethylene. In *Polyethylene-Based Blends, Composites and Nanocomposites*; Visakh, P.M., Morlanes, M.J.M., Eds.; Wiley: Beverly, MA, USA, 2015; pp. 163–200.
3. Rives, V. *Layered Double Hydroxides: Present and Future*; NOVA Science Publishers, Inc.: New York, NY, USA, 2001; ISBN 978-1-61209-289-8.
4. Wang, Q.; O'hare, D. Recent advances in the synthesis and application of layered double hydroxide (LDH) nanosheets. *Chem. Rev.* **2012**, *112*, 4124–4155. [CrossRef] [PubMed]
5. Alonso-de-Linaje, V.; Mangayayam, M.C.; Tobler, D.J.; Dietmann, K.M.; Espinosa, R.; Rives, V.; Dalby, K.N. Sorption of chlorinated hydrocarbons from synthetic and natural groundwater by organo-hydrotalcites: Towards their applications as remediation nanoparticles. *Chemosphere* **2019**, *236*, 124369. [CrossRef]
6. Dietmann, K.M.; Linke, T.; Trujillano, R.; Rives, V. Effect of chain length and functional group of organic anions on the retention ability of mgal-layered double hydroxides for chlorinated organic solvents. *ChemEngineering* **2019**, *3*, 89. [CrossRef]
7. Alonso-de-Linaje, V.; Mangayayam, M.C.; Tobler, D.J.; Rives, V.; Espinosa, R.; Dalby, K.N. Enhanced sorption of perfluorooctane sulfonate and perfluorooctanoate by hydrotalcites. *Environ. Technol. Innov.* **2021**, *21*, 101231. [CrossRef]
8. Trujillano, R.; Nájera, C.; Rives, V. Activity in the Photodegradation of 4-Nitrophenol of a Zn,Al Hydrotalcite-Like Solid and the Derived Alumina-Supported ZnO. *Catalysts* **2020**, *10*, 702. [CrossRef]
9. Karásková, K.; Pacultová, K.; Jirátová, K.; Fridrichová, D.; Koštejn, M.; Obalová, L. K-Modified Co–Mn–Al Mixed Oxide—Effect of Calcination Temperature on N_2O Conversion in the Presence of H_2O and NOx. *Catalysts* **2020**, *10*, 1134. [CrossRef]
10. Li, P.; Yu, F.; Altaf, N.; Zhu, M.; Li, J.; Dai, B.; Wang, Q. Two-Dimensional Layered Double Hydroxides for Reactions of Methanation and Methane Reforming in C1 Chemistry. *Materials* **2018**, *11*, 221. [CrossRef]
11. Rives, V.; Del Arco, M.; Martín, C. Layered double hydroxides as drug carriers and for controlled release of non-steroidal antiinflammatory drugs (NSAIDs): A review. *J. Control. Release* **2013**, *169*, 28–39. [CrossRef]
12. Choi, G.; Choy, J. Recent progress in layered double hydroxides as a cancer theranostic nanoplatform. *WIREs Nanomed. Nanobiotechnol.* **2021**, *13*, 1–19. [CrossRef]
13. Patel, R.; Park, J.T.; Patel, M.; Dash, J.K.; Gowd, E.B.; Karpoormath, R.; Mishra, A.; Kwak, J.; Kim, J.H. Transition-metal-based layered double hydroxides tailored for energy conversion and storage. *J. Mater. Chem. A* **2018**, *6*, 12–29. [CrossRef]
14. Saifullah, B.; Hussein, M.Z. Inorganic nanolayers: Structure, preparation, and biomedical applications. *Int. J. Nanomed.* **2015**, *10*, 5609–5633. [CrossRef]

15. Mishra, G.; Dash, B.; Pandey, S. Layered double hydroxides: A brief review from fundamentals to application as evolving biomaterials. *Appl. Clay Sci.* **2018**, *153*, 172–186. [CrossRef]
16. Miyata, S. Anion-Exchange Properties of Hydrotalcite-Like Compounds. *Clays Clay Miner.* **1983**, *31*, 305–311. [CrossRef]
17. Inayat, A.; Klumpp, M.; Schwieger, W. The urea method for the direct synthesis of ZnAl layered double hydroxides with nitrate as the interlayer anion. *Appl. Clay Sci.* **2011**, *51*, 452–459. [CrossRef]
18. Abderrazek, K.; Frini Srasra, N.; Srasra, E. Synthesis and Characterization of [Zn-Al] Layered Double Hydroxides: Effect of the Operating Parameters. *J. Chin. Chem. Soc.* **2017**, *64*, 346–353. [CrossRef]
19. Bukhtiyarova, M.V. A review on effect of synthesis conditions on the formation of layered double hydroxides. *J. Solid State Chem.* **2019**, *269*, 494–506. [CrossRef]
20. Kloprogge, J.T.; Hickey, L.; Frost, R.L. The effects of synthesis pH and hydrothermal treatment on the formation of zinc aluminum hydrotalcites. *J. Solid State Chem.* **2004**, *177*, 4047–4057. [CrossRef]
21. Galvão, T.L.P.; Neves, C.S.; Caetano, A.P.F.; Maia, F.; Mata, D.; Malheiro, E.; Ferreira, M.J.; Bastos, A.C.; Salak, A.N.; Gomes, J.R.B.; et al. Control of crystallite and particle size in the synthesis of layered double hydroxides: Macromolecular insights and a complementary modeling tool. *J. Colloid Interface Sci.* **2016**, *468*, 86–94. [CrossRef]
22. Benito, P.; Herrero, M.; Barriga, C.; Labajos, F.M.; Rives, V. Microwave-assisted homogeneous precipitation of hydrotalcites by urea hydrolysis. *Inorg. Chem.* **2008**, *47*, 5453–5463. [CrossRef]
23. He, J.; Wei, M.; Li, B.; Kang, Y.; Evans, D.G.; Duan, X. Preparation of Layered Double Hydroxides. In *Layered Double Hydroxides. Structure and Bonding*; Duan, X., Evans, D.G., Eds.; Springer: Berlin/Heidelberg, Germany, 2006; Volume 119, pp. 89–119.
24. Labajos, F.M.; Rives, V.; Ulibarri, M.A. Effect of hydrothermal and thermal treatments on the physicochemical properties of Mg-Al hydrotalcite-like materials. *J. Mater. Sci.* **1992**, *27*, 1546–1552. [CrossRef]
25. Ezeh, C.I.; Tomatis, M.; Yang, X.; He, J.; Sun, C. Ultrasonic and hydrothermal mediated synthesis routes for functionalized Mg-Al LDH: Comparison study on surface morphology, basic site strength, cyclic sorption efficiency and effectiveness. *Ultrason. Sonochem.* **2018**, *40*, 341–352. [CrossRef] [PubMed]
26. Zadaviciute, S.; Baltakys, K.; Bankauskaite, A. The effect of microwave and hydrothermal treatments on the properties of hydrotalcite. *J. Therm. Anal. Calorim.* **2017**, *127*, 189–196. [CrossRef]
27. Misol, A.; Labajos, F.M.; Morato, A.; Rives, V. Synthesis of Zn,Al layered double hydroxides in the presence of amines. *Appl. Clay Sci.* **2020**, *189*, 105539. [CrossRef]
28. Kooli, F.; Depège, C.; Ennaqadi, A.; De Roy, A.; Besse, J.P. Rehydration of Zn-Al layered double hydroxides. *Clays Clay Miner.* **1997**, *45*, 92–98. [CrossRef]
29. De La Rosa-Guzmán, M.Á.; Guzmán-Vargas, A.; Cayetano-Castro, N.; Del Río, J.M.; Corea, M.; Martínez-Ortiz, M.D.J. Thermal stability evaluation of polystyrene-Mg/zn/Al LDH nanocomposites. *Nanomaterials* **2019**, *9*, 1528. [CrossRef]
30. Drits, V.A.; Bookin, A.S. Crystal Structure and X-Ray Identification of Layered Double Hydroxides. In *Layered Double Hydroxides: Present and Future*; Rives, V., Ed.; NOVA Science Publishers, Inc.: New York, NY, USA, 2001; pp. 41–100.
31. Thomas, G.S.; Radha, A.V.; Kamath, P.V.; Kannan, S. Thermally induced polytype transformations among the Layered Double Hydrodides (LDHs) of Mg Zn with Al. *J. Phys. Chem. B* **2006**, *110*, 12365–12371. [CrossRef]
32. Marappa, S.; Radha, S.; Kamath, P.V. Nitrate-Intercalated Layered Double Hydroxides–Structure Model, Order, and Disorder. *Eur. J. Inorg. Chem.* **2013**, *2013*, 2122–2128. [CrossRef]
33. Karthikeyan, J.; Fjellvåg, H.; Bundli, S.; Sjåstad, A.O. Efficient Exfoliation of Layered Double Hydroxides; Effect of Cationic Ratio, Hydration State, Anions and Their Orientations. *Materials* **2021**, *14*, 346. [CrossRef]
34. Wang, S.L.; Wang, P.C. In situ XRD and ATR-FTIR study on the molecular orientation of interlayer nitrate in Mg/Al-layered double hydroxides in water. *Coll. Surf. A Physicochem. Eng. Asp.* **2007**, *292*, 131–138. [CrossRef]
35. Misol, A.; Jiménez, A.; Morato, A.; Labajos, F.M.; Rives, V. Quantification by Powder X-ray Diffraction of Metal Oxides Segregation During Formation of Layered Double Hydroxides. *Eur. J. Eng. Technol. Res.* **2020**, *5*, 1243–1248. [CrossRef]
36. Brown, J.G. *X-rays and Their Applications*; Plenum/Ros.; Plenum Publishing Corporation: New York, NY, USA, 1966; ISBN 0-306-20021-X.
37. Jenkins, R.; de Vries, J.L. *Worked Examples in X-ray Analysis*, 2nd ed.; Springer: New York, NY, USA, 1970; ISBN 978-1-4899-2649-4.
38. Nakamoto, K. *Infrared and Raman Spectra of Inorganic and Coordination Compounds. Part A: Theory and Applications in Inorganic Chemistry*, 6th ed.; John Wiley and Sons Inc.: Hoboken, NJ, USA, 2009; ISBN 9780471743392.
39. Zhang, Y.; Wang, L.; Zou, L.; Xue, D. Crystallization behaviors of hexagonal nanoplatelet MgAlCO$_3$ layered double hydroxide. *J. Cryst. Growth* **2010**, *312*, 3367–3372. [CrossRef]
40. Feng, W.; Chen, J.; Hou, C.-Y. Growth and characterization of ZnO needles. *Appl. Nanosci.* **2014**, *4*, 15–18. [CrossRef]
41. Brunauer, S.; Emmett, P.H.; Teller, E. Adsorption of Gases in Multimolecular Layers. *J. Am. Chem. Soc.* **1938**, *60*, 309–319. [CrossRef]
42. Lowell, S.; Shields, J.E.; Thomas, M.A.; Thommes, M. *Characterization of Porous Solids and Powders: Surface Area, Pore Size and Density*; Springer: Berlin/Heidelberg, Germany, 2010.
43. Barrett, E.P.; Joyner, L.G.; Halenda, P.P. The Determination of Pore Volume and Area Distributions in Porous Substances. I. Computations from Nitrogen Isotherms. *J. Am. Chem. Soc.* **1951**, *73*, 373–380. [CrossRef]
44. Thommes, M.; Kaneko, K.; Neimark, A.V.; Olivier, J.P.; Rodriguez-Reinoso, F.; Rouquerol, J.; Sing, K.S.W. Physisorption of gases, with special reference to the evaluation of surface area and pore size distribution (IUPAC Technical Report). *Pure Appl. Chem.* **2015**, *87*, 1051–1069. [CrossRef]

45. Brunauer, S.; Deming, L.S.; Deming, W.E.; Teller, E. On a Theory of the van der Waals Adsorption of Gases. *J. Am. Chem. Soc.* **1940**, *62*, 1723–1732. [CrossRef]
46. *A Basic Guide to Particle Characterization*; Malvern Instruments Limited: Malvern, UK, 2015; pp. 1–24. Available online: https://www.cif.iastate.edu/sites/default/files/uploads/Other_Inst/Particle%20Size/Particle%20Characterization%20Guide.pdf (accessed on 30 June 2022).
47. A Guidebook to Particle Size Analysis. *Horiba Sci.* **2019**, 1–32. Available online: https://www.horiba.com/aut/scientific/products/particle-characterization/particle-size-essentials-guidebook/ (accessed on 30 June 2022).

Article

Nanoparticle Black Ceramic Pigment Obtained by Hydrotalcite-like Compound Microwave Treatment

María Oset, Alejandro Moya, Guillermo Paulo-Redondo and Isaac Nebot-Díaz *

Escola Superior de Ceràmica de L'Alcora, ESCAL-ISEACV, 12110 L'Alcora, Castellón, Spain; maria.oset@escal.es (M.O.); alejandro.moya@escal.es (A.M.); guillermo.paulo@escal.es (G.P.-R.)
* Correspondence: isaac.nebot@escal.es; Tel.: +34-964-399-450

Abstract: Development of ceramic pigments with controlled particle sizes below 1 μm is essential for the preparation of ceramic inks used in inkjet digital decoration that is currently being applied in the ceramics sector. A black ceramic pigment based on NiCoCrFe composition has been prepared using thermal decomposition of hydrotalcite-like compounds. The stoichiometry ratio between different cations was studied to obtain the blackest pigment, giving $Ni_{0.5}Co_{0.5}CrFeO_4$ the better cation ratio, also the thermal treatment, comparing traditional firing in an electric furnace with microwave treatment. Samples have been characterized by X-ray diffraction, Scanning Electron Microscopy and Lab colour measurement. Microwave treatment showed the best way to obtain a pigment with spinel-type structure and a homogeneous size distribution near to 150 nm, with a high intensity and colorimetric data, reducing drastically the temperature and energy consumption to obtain a black ceramic pigment suitable to be utilized in digital ceramic inks.

Keywords: hydrotalcite; ceramic pigment; hydrotalcite-like compounds; ceramic ink

1. Introduction

A ceramic pigment is a crystalline structure that gives ceramic the chromatic properties that it does not possess [1]. In the case of black ceramic pigments, the main crystalline structure used is spinel, due to its stability and inalterability when it is introduced into the composition of ceramic glazes [2]. The traditional way of preparing these pigments is the traditional route or route in the solid state, where the corresponding proportions of metal oxides and mineralizing agents are mixed, and subsequently subjected to high calcination temperatures so that the corresponding reaction occurs [3].

The use of alternative routes for the preparation of ceramic pigments has been studied in recent decades to improve reactivity, reduce the high temperatures and times used in calcination, and eliminate the use of mineralizing agents, which are harmful to the environment [4–9].

Thermal decomposition of hydrotalcite-type compounds can produce spinel [10]. An adequate formulation of these compounds allows obtaining suitable products such as ceramic pigments, with a spinel structure, lowering the temperature at which they are obtained, as well as eliminating the use of any mineralizing agent [11].

The use of new technologies for digital decoration in ceramics through inkjet printing requires the use of inks where the solid component (pigments or glazes) has a size of less than 1 micron [12]. By means of the traditional preparation of the pigments, the size obtained after the calcination process is very high (>20 microns), for which a very energetic and expensive grinding stage is necessary. Use of different soft-chemistry synthesis routes has been used to develop particle size and to avoid or minimize the grinding stage [13–15], also the use of thermal decomposition of hydrotalcite-type compounds reduces this size in the final product, thus minimizing the energy consumption of the grinding process because of the adequate particle size obtained [11].

Finally, the use of microwave technology in the preparation of inorganic solids has had a great boom in recent years due to the short time it takes to reach the cooking temperature, which favours the low aggregation of the material and the obtaining of very small particle sizes with the appropriate crystalline conditions [16–19].

These advantages can be used together with decomposition of hydrotalcite-like compounds to obtain well-formed and small-sized pigments suitable to be use into ceramic inks for digital ink decoration.

2. Materials and Methods

$Co_{1-x}Ni_xFe_{2-y}Cr_yO_4$, with x and y values of 0.5 and 1, respectively, were prepared using traditional solid-state reaction and thermal decomposition of Hydrotalcite-like compound (HTLC) precursor.

For solid-state reaction, different metal oxides (Co_3O_4, NiO, Fe_2O_3 and Cr_2O_3) were used in industrial grade (minimum 98.5% purity), provided by Al-Farben company (Torrecid Group). Also, H_3BO_3 (Al-Farben 99.3%) in 5% w/w was used as flux agent

The HTLC precursors were prepared according to Kanezaki [20] procedure using 0.28 M solutions for divalent metallic cations and 0.07 M solutions for trivalent metallic cations. Solutions of NaOH 1 M and Na_2CO_3 0.25 M as buffer agent were also used. A variation using 0.5 M of all solutions was also studied to test the concentration effect. The solutions were prepared using $CoCl_2·6H_2O$, $NiCl_2·6H_2O$, $CrCl_3·6H_2O$ and $FeCl_3$ (Merck, 99%) and $Na_2CO_3·10H_2O$ (Merck, 99.5%).

Firing treatments were performed using a traditional kiln with 500 °C, 700 °C and 1000 °C with 10 °C/min heating rate and 30 min or 2 h at maximum temperature. Also, a domestic microwave oven (800 W) was used to heat an Al_2O_3 capsule painted inside with SiC (Figure 1). Into this capsule, different samples were located and microwaves with 800 W power were applied for times of 5, 15 and 30 min, reaching a maximum temperature of 1038 °C. Temperature reached was determined using process temperature control rings (PTCR) type ETH (temperature range 850 °C–1100 °C) from FERRO.

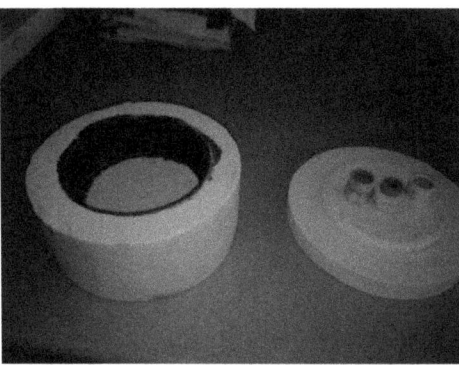

Figure 1. Capsule for microwave kiln painted with SiC.

Analysis and characterization of the samples were performed using XRD for crystalline development (BRUKER AXS, EndeavorD4, determination angles 5–70° 2θ with acquisition time 2 s. 0.05° 2θ), TG-DTA for thermal analysis (BÄHR STA503 with Pt crucible and heating rate 5 °C/min), SEM for morphological and size distribution (JEOL7001, FED warm cathode, 15 kV and carbon sputtering) and spectrophotometer for determination of colorimetric L*a*b* coordinates (KONICA MINOLTA, CM-3600A, SpectraMagicNX d65 illumination and 2° observer).

Colour tests were prepared with an industrial transparent glossy enamel (10% ZnO in composition) with an addition in 1% w/w of synthesized ceramic pigment and fired in an industrial cycle of 1100 °C with 5 min at maximum temperature and 50 min of total firing time (cold to cold).

Table 1 summarizes the references and synthesis conditions of different samples prepared.

Table 1. References of prepared samples.

Reference	Synthesis Method	Firing	Temperature (°C)	Remaining Time at Max. T (min)
C-crude	Solid state	No firing	-	-
C-M1	Solid state	Normal kiln	500	30
C-M2	Solid state	Normal kiln	700	30
C-M3	Solid state	Normal kiln	1000	30
C-M4	Solid state	Normal kiln	500	120
C-M5	Solid state	Normal kiln	700	120
C-M6	Solid state	Normal kiln	1000	120
C-MW1	Solid state	Microwave	<850	5
C-MW2	Solid state	Microwave	<850	15
C-MW3	Solid state	Microwave	1038	30
H-crude	HTLC	No firing		
H-M1	HTLC	Normal kiln	300	30
H-M2	HTLC	Normal kiln	400	30
H-M3	HTLC	Normal kiln	500	30
H-M4	HTLC	Normal kiln	700	30
H-M5	HTLC	Normal kiln	1000	30
H-M6	HTLC	Normal kiln	500	120
H-M7	HTLC	Normal kiln	700	120
H-M8	HTLC	Normal kiln	1000	120
H-MW1	HTLC	Microwave	<850	5
H-MW2	HTLC	Microwave	<850	15
H-MW3	HTLC	Microwave	1038	30

3. Results and Discussion

3.1. HTLC Preparation

Figure 2 shows DTA/TG of the HTLC prepared (H-crudo). Two endothermal peaks were observed at 168 °C and 327 °C with a mass loss associated (27.7%) corresponding with H_2O hydration and carbonate loss, respectively. In fact, 12.3% corresponds with water loss and 15.4% with carbonate loss [11].

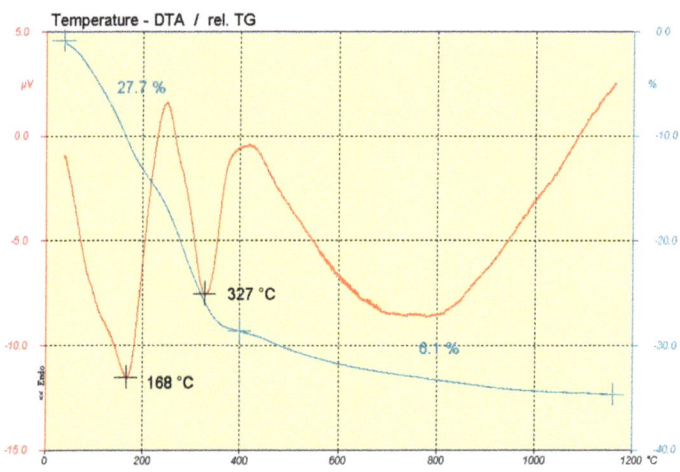

Figure 2. DTA (red line) and TG (blue line) of HTLC sample.

Figure 3 shows XRD analysis of HTLC sample (H-crude). It shows peaks corresponding with hydrotalcite, with a very low definition and crystallinity, because stoichiometry ratio used in sample preparation was M(II)/M(III) 1:2, far from the 4:1 ration in theoretical hydrotalcite, and there was no ageing time after precipitation.

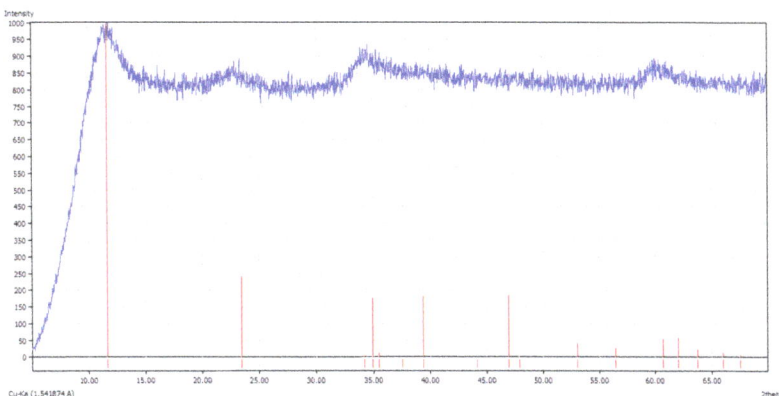

Figure 3. XRD of HTLC sample prepared (red lines correspond to hydrotalcite XRD pattern JCDP 010890460).

The poor crystallization and lower temperature of decarbonation respect to theoretical hydrotalcite can be explained because hydrotalcite can incorporate a maximum quantity of trivalent cation. $M^{II}(OH)_6$ octahedra are electrically neutral, but $M^{III}(OH)_6$ octahedra are positively charged, and the presence of this type of octahedra sharing edges is unstable; therefore, part of trivalent cation cannot be incorporated into the hydrotalcite structure, giving this structural deformation [10].

3.2. Thermal Treatment

In conventional solid state, at least 1000 °C and 120 min firing conditions are necessary to obtain a spinel phase (sample C-M6). The other firing conditions were not adequate to develop this crystalline phase. When microwave firing was used, 30 min were necessary to achieve a temperature of 1038 °C. When shorter calcination times were used, the spinel phase did not develop, since the temperatures reached were in all cases below 850 °C.

Figure 4 shows XRD analysis of sample C-M6 and C-MW3. The results were practically identical in both cases, theoretically obtaining the crystalline phase of spinel $Co_{0.5}Ni_{0.5}FeCrO_4$. In fact, the crystalline phase identified in DRX pattern corresponds to spinel $CoCr_2O_4$ (JCDPS 0801668) and $NiFe_2O_4$ (JCDPS 0742081). Up to our knowledge, the reference pattern for the spinel theoretically prepared ($Co_{0.5}Ni_{0.5}FeCrO_4$) does not exist, but it was clear that the solid prepared had the spinel structure and coincided with the patterns of the extreme solids of the corresponding series.

SEM micrographies (Figure 5) of these samples show well-formed aggregates where individual size was under 1 micron.

When calcining HTLCs, spinel phase formation occurred at much lower temperatures. Figure 6 shows XRD of H-M3 and H-MW1. The development of crystallization was very similar in both cases; therefore, the temperature reached with the microwave was possibly close to 500 °C. Several phases have been identified that could correspond to these peaks, namely NiO, FeO, CoO and $NiCr_2O_4$. Obviously, the peaks did not correspond to these phases, but were close to them, which gives an idea of possible intermediates made up of the different elements of the prepared composition.

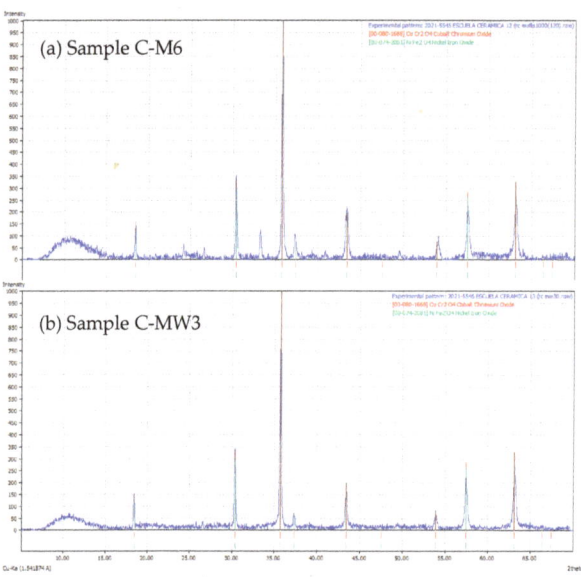

Figure 4. XRD of samples C-M6 (**a**) and C-MW3 (**b**) (red lines correspond to $CoCr_2O_4$ XRD pattern (JCDPS 0801668) and light blue ones to $NiFe_2O_4$ (JCDPS 0742081)).

Figure 5. SEM micrograph of samples C-M6 (**a**) and C-MW3 (**b**).

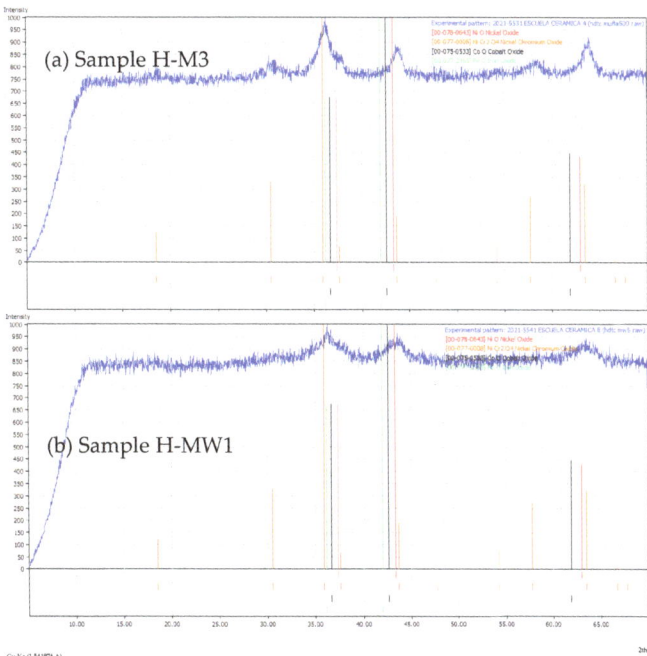

Figure 6. XRD of samples H-M3 (**a**) and H-MW1 (**b**) (red lines correspond to NiO XRD pattern (JCDPS 0780643), orange to $NiCr_2O_4$ (JCDPS 0770008), dark blue to CoO (JCDPS 0750533) and light green to FeO (JCDPS 0772355)).

Particle sizes were very low (under 100 nm) and there was no aggregation of particles, as shown in Figure 7 for sample H-MW1.

Figure 7. SEM micrograph of sample H-MW1.

When temperatures higher than 500 °C were used, the crystalline phase of spinel $Co_{0.5}Ni_{0.5}FeCrO_4$ can be observed with good crystallinity.

Figure 8 shows the XRD patterns of HM-4, HM-5, and H-M8 samples, while Figure 9 shows the XRD patterns of H-MW2 and H-MW3, respectively. As occurred in the solid-state samples calcined at 1000 °C, the target spinel was developed, being identified by the similarity with $CoCr_2O_4$ and $NiFe_2O_4$ crystalline phases, as explained above. In these cases, this phase was obtained at lower temperatures and with shorter retention times than those needed in the solid-state reaction, and with a very similar crystalline development in both cases, traditional firing and microwave treatment. Higher temperature led to better crystallization, but in all cases showing spinel formation.

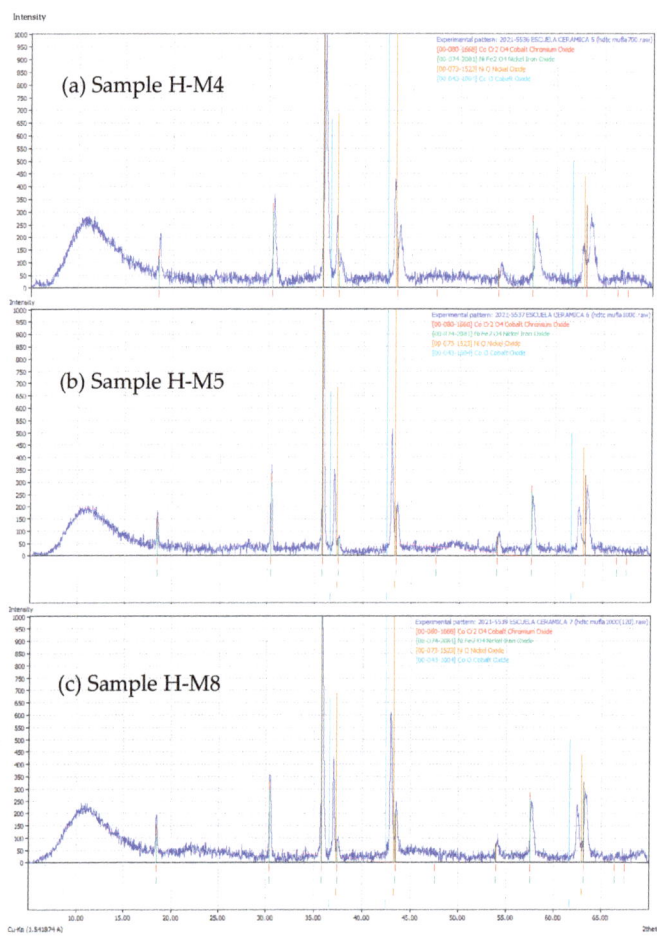

Figure 8. XRD patterns of samples H-M4 (**a**), H-M5 (**b**) and H-M8 (**c**) (red lines correspond to $CoCr_2O_4$ XRD pattern (JCDPS 0801668), light green to $NiFe_2O_4$ (JCDPS 0742081), and orange to NiO (JCDPS 0731523)).

In addition to the spinel phase, the mixed oxide of nickel and cobalt also appeared. This may be since these metals have the narrowest working pH ranges in which the mixed hydroxide is formed (7 to 10). Thus, it is possible that during the synthesis of hydrotalcite, the pH of the solution was below 7, giving rise to the dissolution of these metals. Subsequently, when the solution is basified again, a fractional precipitation may take place, giving rise to a mixed cobalt-nickel hydroxide. Finally, during the calcination, the spinel was obtained separately together with this mixed oxide. In all cases, it can be observed that the particle size obtained was close to 100 nm, forming aggregates in normal kiln, although with low retention time, as shown in Figure 10.

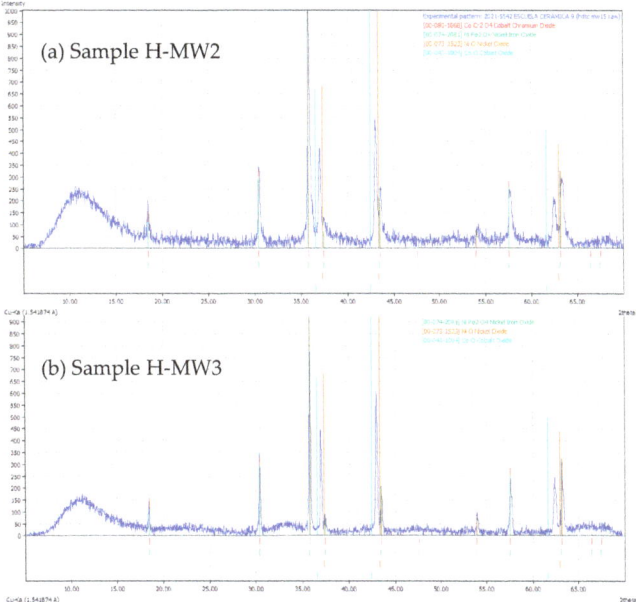

Figure 9. XRD patterns of samples H-MW2 (**a**) and H-MW3 (**b**) (red lines correspond to $CoCr_2O_4$ XRD pattern (JCDPS 0801668), light green to $NiFe_2O_4$ (JCDPS 0742081), and orange to NiO (JCDPS 0731523)).

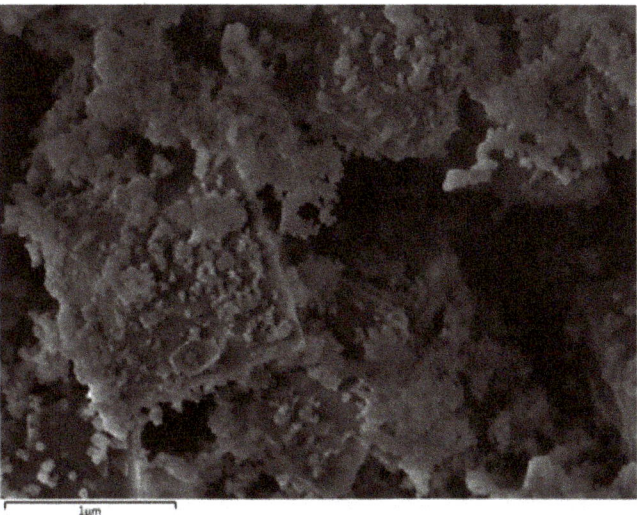

Figure 10. SEM micrograph of sample H-M5.

In case of using microwave kiln, there were no aggregates with a very well size distribution thanks to fast heating rates and minimum retention time during firing process, Figure 11.

Figure 11. SEM micrograph of sample H-MW2.

Thanks to this small particle size and the non-agglomeration of these particles, a ceramic ink could be prepared for inkjet application. Preliminary test has been developed, showing the availability to obtain ceramic inks, without clogging nozzles problems.

3.3. Colour Measurement

All the prepared pigments have been applied in ceramic glazes in 1% w/w, using a transparent glaze in an industrial firing cycle of porous single firing tiles "monoporosa" pieces (1100 °C with a total cycle duration of 45 min).

The pure black colour is the one that corresponds to chromatic coordinates L = 0, a = 0 and b = 0. Of course, these values should not be reached, so the closest values will be the ones that will provide the best colour development.

The results have been compared with a commercial black pigment (Table 2), obtaining in most cases in which HTLC synthesis has been used, a higher colour yield than the commercial one. When using "colour yield" term, it is meant that the colour is closer to pure black colour (Lower values for L coordinate and a, b values closest to 0).

Table 2. L*a*b* colorimetric coordinates of different pigments studied.

Reference	L	a	b
C-crude	44.47	−2.72	7.6
C-M1	42.85	−1.88	6.74
C-M2	38.99	−0.84	4.91
C-M3	36.36	0.14	3.01
C-M4	42.78	−1.97	6.51
C-M5	41.02	−1.79	5.44
C-M6	37.19	1.13	3.70
C-MW1	43.74	−2.47	6.98
C-MW2	36.72	1.85	3.03
C-MW3	36.74	1.65	1.71
H-crude	36.62	1.12	1.47
H-M1	33.37	0.93	0.76
H-M2	35.75	0.55	−0.18
H-M3	35.36	0.54	−0.16
H-M4	32.29	0.93	0.72
H-M5	31.30	1.38	−0.03
H-M6	36.65	0.76	1.12
H-M7	35.17	0.98	0.94
H-M8	29.40	1.15	0.04
H-MW1	39.09	1.04	0.61
H-MW2	32.12	1.12	0.32
H-MW3	32.25	1.00	−2.32
Commercial	36.04	1.36	−1.01

All the pigments prepared from the HTLC synthesis gave a very acceptable black colour performance; even the uncalcined sample had a very acceptable colour performance. In all cases where spinel has not formed after heat treatment, the developed black colour was good. This may be because the precursor obtained finished developing the spinel phase during the firing of the glazed ceramic piece, which allowed this synthesis to be used at low temperatures to obtain nanoparticulate pigments.

4. Conclusions

The main conclusions that can be drawn in this work are the following:

- Pigments made by non-conventional routes, namely using hydrotalcite as a precursor, have better/higher colour performance than pigments made by ceramic route, when similar firing conditions are compared. In all cases, using hydrotalcite as a precursor (except crude hydrotalcite), colour performance is better than when commercial black pigment is used.
- The use of microwave treatment reduces the final size of the primary particles due to high energy and low time firing process. When hydrotalcite precursor is used, it is enough applying 800 W for 15 min to obtain the spinel phase with the best black colour performance.
- For samples obtained by the non-conventional synthesis route, the development of the spinel phase begins at 700 °C, while in the conventional route it is necessary to reach 1000 °C to be able to observe the development of the spinel, showing the higher reactivity of hydrotalcite precursor in front of the traditional synthesis route.
- The developed method allows obtaining nanoparticulate ceramic pigments with a spinel structure under very favourable synthesis conditions for their industrial development.

Author Contributions: Investigation, M.O. and A.M.; Project administration, I.N.-D.; Supervision, G.P.-R. and I.N.-D.; Writing—review & editing, I.N.-D. All authors have read and agreed to the published version of the manuscript.

Funding: This research received no external funding.

Institutional Review Board Statement: Not applicable.

Informed Consent Statement: Not applicable.

Data Availability Statement: Not applicable.

Acknowledgments: Isaac Nebot-Díaz would like to thank Vicente Rives for all his teaching in the field of hydrotalcites. Without his help, the research line for the development of ceramic pigments from the thermal decomposition of hydrotalcite-type compounds would never have come true.

Conflicts of Interest: The authors declare no conflict of interest.

References

1. Nebot-Díaz, I. Estudio y Caracterización de Compuestos Tipo Espinela $M^{II}Al_2O_4$ Mediante Rutas de Síntesis no Convencionales. Ph.D. Thesis, University Jaume I, Castellón, Spain, 2001.
2. Monrós, G.; Badenes, J.A.; García, A.; Tena, M.A. El Color en los Materiales Cerámicos. In *El Color de la Cerámica*; Publicacions de la Universitat: Valencia, Spain, 2003; pp. 82–89.
3. Hohemberger, J.M.; Todorova, I.; Marchal, M. Pigmentos cerámicos. In *Esmaltes y Pigmentos Cerámicos*; Escribano, P., Carda, J.B., Cordoncillo, E., Eds.; Faenza Editrice Ibérica: Castellón, Spain, 2001; pp. 189–242.
4. Monrós, G.; Tena, M.A.; Escribano, P.; Cantavella, V.; Carda, J.B. Classical ceramic colours through coloidal and from alkoxides gels. *J. Sol.-Gel. Sci. Tech.* **1994**, *2*, 377–380. [CrossRef]
5. Chavarriaga, E.A.; Jaramillo, L.J.; Restrepo, O.J. Ceramic pigments with spinel structure obtained by low temperature methods. In *Characterization of Minerals, Metals and Materials*; Wiley (John Wiley & Sons, Inc: Hoboken, NJ, USA, 2012; pp. 155–162.
6. Ma, P.; Geng, Q.; Gao, X.; Yang, S.; Liu, G. $CuCr_2O_4$ spinel ceramic pigments synthesized by sol-gel self-combustion method for solar absorber coatings. *J. Mater. Eng. Perf.* **2016**, *25*, 2814–2823. [CrossRef]
7. Betancour-Granados, N.; Restrepo-Baena, O.J. Flame spray pyrolysis of ceramic nanopigments $CoCr_2O_4$: The effect of key variables. *J. Eur. Cer. Soc.* **2017**, *37*, 5051–5056. [CrossRef]

8. El Jabbar, Y.; Lakhlifi, H.; El Ouatib, R.; Er-Rakho, L.; Guillemet-Fritsch, S.; Durand, B. Preparation and characterization of green nano-sized cramic pigmens with the spinel structure AB_2O_4 (A=Co, Ni and B=Cr, Al). *Solid State Commu.* **2021**, *334*, 114394. [CrossRef]
9. Paborji, F.; Afarini, M.S.; Arabi, A.M.; Ghahari, M. Solution combustion synthesis of $FeCr_2O_4$ powders for pigment applications: Effect of fuel type. *Int. J. Appl. Cer. Tech.* 2022; *in press*. [CrossRef]
10. Nebot-Díaz, I.; Rives, V.; Rocha, J.; Carda, J.B. Thermal decomposition study of hydrotalcite-like compounds. *Bol. Soc. Esp. Cer. Vidr.* **2002**, *41*, 411–414. [CrossRef]
11. Rives, V.; Pérez-Bernal, M.E.; Ruano-Casero, R.J.; Nebot-Díaz, I. Development of a black pigment form non stoichiometric hydrotalcites. *J. Eur. Cer. Soc.* **2012**, *32*, 975–987. [CrossRef]
12. Nebot-Díaz, I.; Dal Corso, P.L. *Digital Ceramic Decoration, an Introduction*; ATC: Castellón, Spain, 2017.
13. Li, X.; Wang, Q.K.; Wang, C.; Zhang, W.J.; Yang, Y.L.; Liu, K.; Wankg, Y.Q.; Chang, Q.B. Ultrafine $Z-ZrSiO_4$ pigment prepared by a bottom-up approach: Particle size evolution and chromatic properties. *Adv. Powder Technol.* **2021**, *32*, 3934–3942. [CrossRef]
14. Molinari, C.; Conte, S.; Zanelli, C.; Ardite, M.; Cruciani, G.; Dondi, M. Ceramic pigments and dyes beyond the inkjet revolution: From technological requirements to constraints in colorant design. *Ceram. Int.* **2020**, *46*, 21839–21872. [CrossRef]
15. Tang, Q.; Zhu, H.X.; Chen, C.; Wang, Y.X.; Zhu, Z.G.; Wu, J.Q.; Shis, W.H. Preparation and characterization of nanoscale cobalt blue pigment for ceramic inkjet printing by sol-gel self-propagating combustion. *Mater. Res. Ibero-Am. J. Mater.* **2017**, *20*, 1340–1344. [CrossRef]
16. Obata, S.; Kato, M.; Yokohama, H.; Iwata, Y.; Kikumoto, M.; Sakurada, O. Synthesis of nano $CoAl_2O_4$ pigment for ink-jet printing to decorate porcelain. *J. Cer. Soc. Jpn.* **2011**, *119*, 208–213. [CrossRef]
17. Veronesi, P.; Leonelli, C.; Bondioli, F. Energy efficiency in the microwave-assisted solid-state synthesis of cobalt-aluminate pigment. *Technologies* **2017**, *5*, 42. [CrossRef]
18. Trujillano, R.; Nieto, D.; Rives, V. Microwave-assisted synthesis of Ni, Zn layered double hydroxysalts. *Microporous Mesoporous Mater.* **2017**, *253*, 129–136. [CrossRef]
19. Trujillano, R.; González-García, I.; Morato, A.; Rives, V. Controlling the synthesis conditions for tuning the properties of hydrotalcite like materials at the nano scale. *Chemengineering* **2018**, *2*, 31. [CrossRef]
20. Kanezaki, E. Thermal behaviour of the hydrotalcite-like layered structure of Mg and Al layered double hydroxides with interlayer carbonate by means of in situ powder HTXRD and DTA/TG. *Solid State Ion.* **1998**, *196*, 279–284. [CrossRef]

Article

Photodegradation of Fipronil by Zn-AlPO$_4$ Materials Synthesized by Non-Hydrolytic Sol–Gel Method

Omar José de Lima [1], Denis Talarico de Araújo [1], Liziane Marçal [1], Antonio Eduardo Miller Crotti [2], Guilherme Sippel Machado [3], Shirley Nakagaki [3], Emerson Henrique de Faria [1,*] and Katia Jorge Ciuffi [1,*]

[1] Grupo Sol-Gel, Universidade de Franca-UNIFRAN, Av. Dr. Armando Salles Oliveira, n° 201, Pq. Universitário, Franca 14404-600, SP, Brazil; omarjlima@yahoo.com.br (O.J.d.L.); denistalarico@gmail.com (D.T.d.A.); liziane.silva@unifran.edu.br (L.M.)
[2] Faculdade de Filosofia Ciências e Letras–USP, Av. Bandeirantes, 3900, Bairro Monte Alegre-Ribeirão Preto 14040-901, SP, Brazil; millercrotti@ffclrp.usp.br
[3] Departamento de Química, Universidade Federal do Paraná-UFPR, Rua XV de Novembro, 1299, Curitiba 80060-000, PR, Brazil; guimachado@ufpr.br (G.S.M.); shirleyn@ufpr.br (S.N.)
* Correspondence: emerson.faria@unifran.edu.br (E.H.d.F.); katia.ciuffi@unifran.edu.br (K.J.C.)

Abstract: In recent decades, the increasing use of pesticides to improve food productivity has led to the release of effluents that contaminate the environment. To prepare a material that may help to treat effluents generated during agricultural practice, we used a new method based on the non-hydrolytic sol-gel route to obtain zinc photocatalysts in aluminophosphate matrixes. IR spectroscopy, X-ray diffraction, thermal analysis, differential scanning electron microscopy, energy dispersion spectroscopy, and specific surface area and pore volume determined from the nitrogen adsorbed were used to characterize materials treated at different temperatures. X-ray analysis showed how heat-treatment affected the structure of the material: Zn-AlPO$_4$ in the trigonal and orthorhombic phase was obtained at 750 and 1000 °C, respectively. These phases directly influenced the ability of the material to generate OH radicals. The capacity of the materials to treat effluents was tested in the photodegradation of the pesticide Fipronil. The photocatalytic reactions were monitored by ultraviolet-visible spectroscopy and gas chromatography-mass spectrometry analyses. Zn-AlPO$_4$ treated at 750 °C showed better photodegradation results–it removed 80% of the pesticide in 2 h when higher mass (150 mg) was tested. Long-time treatment of the effluent with Zn-AlPO$_4$ treated at 750 °C completely photodegraded Fipronil. GC-MS analysis confirmed the photodegradation profile, and only traces of Fipronil were observed after photocatalytic reaction for 120 min in the presence of Zn-AlPO$_4$ treated at 750 °C under UV radiation.

Keywords: non hydrolytic sol–gel; advanced oxidation process; heterogeneous photocatalysis; photochemistry

Citation: de Lima, O.J.; de Araújo, D.T.; Marçal, L.; Crotti, A.E.M.; Machado, G.S.; Nakagaki, S.; de Faria, E.H.; Ciuffi, K.J. Photodegradation of Fipronil by Zn-AlPO$_4$ Materials Synthesized by Non-Hydrolytic Sol-Gel Method. ChemEngineering 2022, 6, 55. https://doi.org/10.3390/chemengineering6040055

Academic Editors: Ilenia Rossetti, Miguel A. Vicente, Raquel Trujillano and Francisco Martín Labajos

Received: 15 June 2022
Accepted: 11 July 2022
Published: 13 July 2022

Publisher's Note: MDPI stays neutral with regard to jurisdictional claims in published maps and institutional affiliations.

Copyright: © 2022 by the authors. Licensee MDPI, Basel, Switzerland. This article is an open access article distributed under the terms and conditions of the Creative Commons Attribution (CC BY) license (https://creativecommons.org/licenses/by/4.0/).

1. Introduction

The intensive use of agricultural land worldwide, along with the large-scale development of the agrochemical industry, has dramatically increased the variety and quantity of agrochemicals in both continental and marine waters. Contaminated water poses risks to human and animal health, and the removal of these contaminants has been a challenge for researchers in recent years [1].

Technologies known as advanced oxidation processes (AOPs) have emerged as strategies to remediate waters contaminated with agrochemicals [2,3]. These technologies allow pesticides to be removed up to mineralization, and include methods such as O_3, O_3/UV-B, H_2O_2/UV-C, Fenton, photo-Fenton, TiO_2-ZnO/UV-A, and TiO_2-ZnO/UV-B [4–6]. These methods share a common factor: all of them generate hydroxyl radicals (OH•).

Zinc oxide (ZnO) is a semiconducting species with a band gap similar to the band gap of TiO$_2$, but it offers a major advantage: ZnO absorbs more than one fraction of the UV spectrum, and has a threshold of 425 nm.

It is importante remark also that the UV emission corresponds to the near band edge (NBE) emission assigned to to the radiative annihilation of excitons and the visible emission is commonly referred to as a deep-level or trap-state emission [7–9]. The relative strength of NBE to deep level defect emissions exhibits a dramatic threshold dependence on surface roughness, and this point involves directly the method employed to preare the semicoductor. Surface optical emission efficiency increases over the roughness decreases to unit cell dimensions, highlighting the coupled role of surface morphology and near-surface defects for high efficiency ZnO emitters [7–9].

In addition, compared to TiO2, materials containing the Zn(II) ion constitute more effective photocatalysts to remediate water, because they generate H_2O_2 more efficiently [7–9], afford higher mineralization rates, contain a larger number of active sites for photocatalysis, and exhibit greater surface reactivity.

Incorporation of photoactive elements into molecular sieves provides materials with highly dispersed active species and enhanced photocatalytic activity [10,11]. In general, the high dispersion of photoactive sites in the network of the molecular sieves, along with the effective separation between electrons and holes, improves the photocatalytic activity of molecular sieves containing metallic transition elements. Photoactive elements (e.g., Zr and Ti) incorporated into the zeolite MCM-41 network cleave H_2O at least 80 times more effectively than ZrO_2 [12–14]. Hence, isomorphous substitution of Zn(II) ions into aluminophosphates seems to be a good strategy to make this semiconductor more efficient, and to avoid the corrosion that free Zn(II) species commonly experience under electrochemical conditions, acidic media, and UV light [15].

Fipronil, with the minimum formula $C_{12}H_4Cl_2F_6N_4OS$ (Figure 1a, is an insecticide of the phenylpyrazole series used to control pests of corn, cotton, and rice in several parts of the world [16]. It is effective against insects with resistance to pyrethroids, cyclodiene, organophosphorus, and carbamate insecticides [17]. Its mechanism of action involves the selective blocking of the passage of chloride ions through the γ-aminobutyric acid (GABA)-regulated chloride channel, which disrupts the activity of the central nervous system and, at sufficient doses, kills insects with a favorable safety factor between insects and mammals [18]. In aquatic environments, fipronil has been found at levels between 0.5 and 9 µg L^{-1} in surface waters and downstream of treated rice-cultivation fields, and up to 12.6 µg L^{-1} in residential areas [16]. In the Brazilian southeast region, concentrations between 6 and 465 µg L^{-1} were found [16].

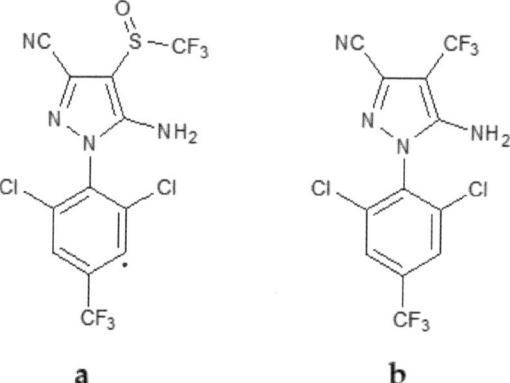

Figure 1. Chemical structure of dipronil (**a**) and its desulfurated derivative desthiofipronil (**b**).

In this context, the sol–gel process, which involves the hydrolysis and polycondensation of molecular precursors and formation of oxo bridges, is an important methodology to design and synthesize materials that are potentially applicable in catalysis and photocatalysis [19–21]. The sol–gel chemistry has the following advantages: it provides materials with (i) a high degree of purity; (ii) strictly controlled composition, structure, and homogeneity at the molecular level; and (iii) regulated texture. In recent years, authors have suggested many modifications to the sol–gel routes. In the modified methodology, the oxo bonds originate from other oxygen donors other than water—for example, alkoxides, ethers, or alcohols—giving rise to the term "non-hydrolytic sol-gel" route coined by Acosta et al. in 1994 [22].

This work aimed to synthesize new materials to treat sewage and wastewater from agrochemical activities. More specifically, it describes the synthesis of zinc aluminophosphate matrices via the non-hydrolytic sol–gel process, and their application in the photocatalytic degradation of fipronil—an insecticide that is widely used worldwide. Techniques such as X-ray powder diffraction, infrared spectroscopy, thermal analysis, and scanning electron microscopy aided the characterization of the new materials, calcined at different temperatures.

2. Materials and Methods

2.1. Reagents

All of the reagents were analytical grade, and were purchased from Aldrich, Sigma, or Merck, and treated when necessary. Fipronil was acquired from Sigma-Aldrich (St. Louis, MO, USA).

2.2. Preparation of Zn-AlPO$_4$ Matrices via the Non-Hydrolytic Sol–Gel Route (NHG)

The material was synthesized by modifying the methods described by Acosta et al. [22], Bourget et al. [23], and de Lima et al. [19]. In a two-necked round-bottomed flask, aluminum chloride (9.4×10^{-2} mol), isopropyl ether (iPr$_2$O, 2.8 mol), and zinc(II) chloride (2.7×10^{-2} mol) were mixed and refluxed at 110 °C, for 4 h, under an argon 5.0 quality ($\geq$99.999%) atmosphere. The condenser was adapted to a thermostatic bath and kept at -8 °C. After the gel merged, H$_3$PO$_4$ (1.5×10^{-1} mol) was added to the reaction system. After reflux, the system was cooled to room temperature and aged overnight in the mother liquor. The solvent was eliminated in a rotary evaporator, and the solid material was washed with different solvents, in the following order: dichloromethane, acetonitrile, and methanol. This procedure generated a non-hydrolytic gel (NHG) during the process, namely, an aluminum halide condensed with isopropyl ether upon cleavage of the O-R bond, which later produced an alkyl halide [22,24]. In the presence of aluminum alkoxides or chloroalkoxides, aluminum chloride exhibits stretching of structures of the µ-Cl and µ-OR type, which elicits two types of reaction: (a) nucleophilic attack of chlorine and electrophilic attack of aluminum, and (b) formation of Al-O-Al, Al-O-PO$_4$, and -P-O-Zn-OH-Al-O bonds due to nucleophilic attack of chlorine and electrophilic attack of carbon [25,26].

Figure 2 depicts a schematic representation of the prepared material, designated Zn-AlPO$_4$. Contact of the solid with the mother liquor is an important step during aging of the gel. Indeed, a previous work demonstrated [19] that longer aging periods give better yields, and promote anchoring of a larger amount of metallic ions onto the matrix. Up to this step, the network is flexible; additional condensation reactions and new crosslinkages occur during aging in the mother liquor and solvent removal.

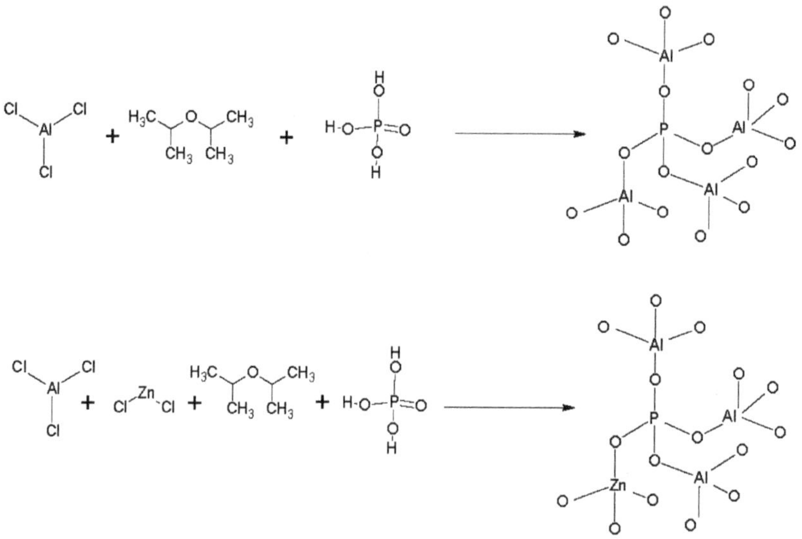

Figure 2. Schematic representation of the synthesis of aluminum phosphate (Zn-AlPO$_4$) via the non-hydrolytic sol–gel process.

Parts of the solid material were thermally treated at 260, 400, 750, or 1000 °C, which afforded the samples named Zn-AlPO$_4$-260, Zn-AlPO$_4$-400, Zn-AlPO$_4$-750, and Zn-AlPO$_4$-1000, respectively. As a reaction control (blank), heat-treated materials were prepared under the same conditions, but in the absence of ZnCl$_2$, which yielded the control samples AlPO$_4$-260, AlPO$_4$-400, AlPO$_4$-750, and AlPO$_4$-1000, respectively.

2.3. Characterizations

Thermogravimetric and differential thermal data (TG/DTA) were acquired using a thermal analyzer from TA Instruments (SDT Q600). Simultaneous DTA–TGA was carried out between ~25 and 1000 °C, at a heating rate of 20 °C min^{-1} and air flow of 100 mL min^{-1}.

The X-ray diffractograms were registered on a Rigaku MiniFlex II DESKTOP X-ray diffractometer, at room temperature; CuKα radiation (λ = 1.54 Å) was used. A 0.04° s^{-1} path was employed for 2θ values ranging from 17 to 80°.

Infrared spectroscopy was carried out on a Fourier-transform infrared spectrometer (Fourier ABB, Bomem, model ME 100), using KBr pellets prepared by mixing 2% of the sample in weight with KBr. The spectra were recorded after 20 scans, from 4000 to 400 cm^{-1}.

Specific surface areas were determined according to the BET method [27], by analyzing the nitrogen adsorption isotherms. A Micrometrics ASAP 2020 physical adsorption analyzer was employed.

Transmission electron microscopy (TEM) was performed using a JEOL JEM CX 100 II microscope.

Scanning electron microscopy (SEM) and energy-dispersive X-ray experiments were conducted using a Philips model XL30 scanning electron microscope, equipped with an EDAX energy-dispersive X-ray detector.

The electronic spectra were recorded on a Hewlett-Packard 8453 diode-array spectrophotometer coupled with an HP KAYAK-XA microcomputer; the program provided by the manufacturer was employed.

2.4. Photocatalytic Reactions

The photocatalytic reactions involving fipronil as a substrate were conducted in a sealed chamber containing a 95-watt lamp emitting short-wavelength UV-C radiation with a peak at 253.7 nm and germicidal action. This chamber was designed on the basis of other models described in the literature (Figure 3) [28,29], with a total volume of 2000 cm^3. A device that allowed us to accomplish simple and cost-effective photodegradation reactions was utilized. The lamp was surrounded by a quartz tube, which enabled us to use it inside the solution containing the substance to be degraded. The chamber also relied on a system mounted to disperse the solution and the photocatalyst, on a temperature sensor, and on an electrode coupled with a potentiometer (Hanna Instruments HI 9321) placed outside the chamber to measure pH (Orion® semi-micro pH 911600). An aqueous solution of fipronil with concentrations ranging from 2830 to 3370 µg L^{-1} was employed. The influence of exposure time and the mass of the photocatalysts was studied herein, with 0.100 or 0.153 g of photocatalyst in 1000 cm^3 of fipronil solution at two different concentrations—2830 or 3370 µg L^{-1} (quantified by calibration curves from UV–Vis spectroscopy)—and exposed to artificial ultraviolet radiation under constant stirring and placed in a thermostatic bath maintained at 25 ± 1 °C.

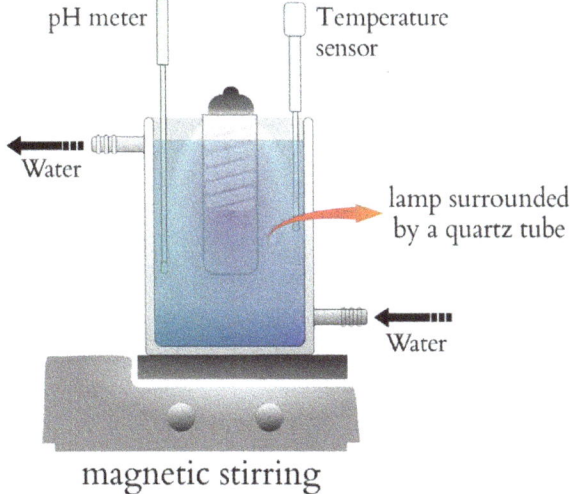

Figure 3. Photochemical reactor bench illustration.

Measurement of light intensity: The amount of light emission under artificial light and sunlight was quantified by a INSTRUTEMP model LM801 luximeter (Instrutemp, São Paulo, Brazil) with a 0 to 500,000 lux range, and UV light was quantified by an Instrutherm model MRU-201 UV light meter (Instrutemp, São Paulo, Brazil), from 290 nm to 390 nm, and with a quantification range from 4000 to 19,990 µW/cm^2 (distance varying from 5–38 cm).

Extraction protocol for GC–MS analysis: Photodegradation products were extracted using a typical extraction method. Briefly, 150 mL of the supernatant was placed into conical flasks and extracted with an organic solvent (150 mL of ethyl acetate). The samples were filtered into bottles and allowed to settle for 24 h. The filtrates from each photocatalytic reaction were then concentrated in vacuo at 40 °C using a rotary evaporator, and proceeded to analysis under GC–MS.

The photocatalyzed reactions were monitored by UV–Vis spectroscopy and gas chromatography–mass spectrometry (GC–MS) analyses. For the UV–Vis analysis, a 2000 µL aliquot was removed from the reaction medium and centrifuged at 3500 rpm for 15 min, followed by analysis on a Hewlett-Packard 8453 diode-array UV–Vis spectrometer, without previous treatment. For the GC–MS analysis, the reaction medium was extracted with ethyl acetate and analyzed using a Shimadzu.

A GCMS 2010 Plus (Shimadzu Corporation, Kyoto, Japan) system equipped with an AOC-20i autosampler was utilized; the column consisted of an Rtx-5MS (Restek Co., Bellefonte, PA, USA) fused silica capillary column (30 m length × 0.25 mm i.d. × 0.25 µm film thickness). The electron ionization mode was used at 70 eV. Helium (99.999%), with positive ionization, was employed as the carrier gas at a constant flow of 1.0 mL/min. The injection volume was 0.1 µL (split ratio of 1:10). The injector and the ion-source temperatures were set at 240 and 280 °C, respectively. The column temperature was programmed to rise from 80 to 240 °C at 3 °C/min, and was then held at 240 °C for 5 min. Mass spectra were taken with a scan interval of 0.5 s, in the mass range from 40 to 600 Da. The products from the photodegradation of fipronil were identified in the single-ion mode (SIM) chromatogram on the basis of retention times (R_t); the MS data were compared with the literature results [17,30].

3. Results

Figure 4 illustrates the thermogravimetric (TG/DTG and DTA) curves recorded for aluminophosphate samples synthesized in the presence of Zn(II) ions.

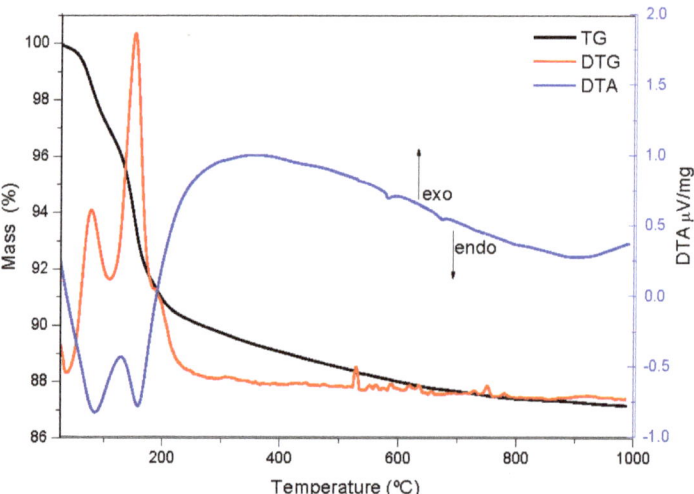

Figure 4. TG and DTA obtained for Zn-AlPO$_4$ under an oxidizing atmosphere at a heating rate of 20 °C min^{-1}.

Treatment of the samples up to 260 °C removed the liquid retained in the material after the synthesis. However, the AlPO$_4$ sample underwent successive mass losses from the onset of heating at 25 °C until 200 °C, assigned to volatile solvents employed on washing steps and adsorbed water that remained in the product. A similar phenomenon has already been verified for alumina matrices prepared via the same route, and it was attributed to the contact of the material with the room atmosphere after the synthetic procedure [19,31].

The mass losses taking place from 200 to 400 °C (around 12% wt) referred to loss of residual groups of alkyl halides from the precursors employed during the synthesis. In the DTA curve, endothermic peaks arose at 523, 589, 671, and 870 °C. These peaks were related to small mass losses, and indicated that the sample underwent successive structural arrangements and rearrangements under rising temperatures.

Regarding the Zn-AlPO$_4$ samples, they lost weakly bound or weakly adsorbed water between 63 and 200 °C (8.0% wt). As explained above for AlPO$_4$, these water molecules were adsorbed onto the material after the synthesis. Further mass loss took place between 200 and 600 °C, due to pyrolysis and oxidation of residual alkyl halide groups [32]. The DTA curve also displayed two endothermic peaks, at 578 and 670 °C—a region where mass loss did not occur; in fact, this region indicated a phase transition in the network.

Identification of the phases in Zn-AlPO$_4$ (Figure 5) relied on comparison of the peak positions and diffraction planes, as well as their intensities, with the corresponding files of the National Bureau Standards (NBSCAA-00-010-0423) for pure crystalline compounds.

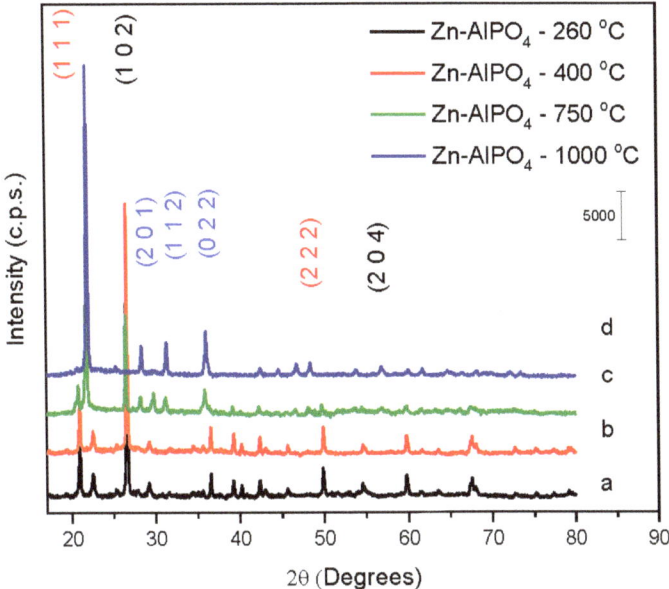

Figure 5. Powder X-ray diffractograms of the Zn-AlPO$_4$ samples heat-treated at (**a**) 260, (**b**) 400, (**c**) 750, and (**d**) 1000 °C.

The diffractograms of the Zn-AlPO$_4$ samples heat-treated at 260 and 400 °C demonstrated that the berlinite phase—a hexagonal system—predominated in these products. The materials heat-treated at 750 and 1000 °C exhibited different diffraction planes, suggesting that they had distinct configurations from those of the samples treated at lower temperatures. The sample heated at 750 °C had a trigonal phase, ascribed to AlZnP$_2$O$_8$ (ACHSE7-00-052-1506); the material treated at 1000 °C presented an orthorhombic phase, assigned to AlPO$_4$. This might have resulted from ZnO migration to regions outside the aluminophosphate network, corroborating formation of the AlPO$_4$ phase.

The hexagonal structure of the berlinite type prevailed in AlPO$_4$ (2θ = 26.43° and 54.34°), with diffraction planes typical of this phase: (hkl = 1,0,2) and (2,0,4) (NBSCAA-00-010-0423). The AlPO$_4$ berlinite structure presents the same atomic structure as quartz. It is obtained by alternately substituting Si for Al and P [33]. The presence of diffraction peaks due to the planes (1,0,0), (1,1,0), and (2,0,0) at larger angles also indicated that a hexagonal phase of the wurtzite type existed in the sample, suggesting that it was a polycrystalline material.

The AlPO$_4$ samples heat-treated at 260, 400, and 750 °C displayed the same diffraction planes as the material that did not undergo heat treatment; therefore, the berlinite phase prevailed in these cases as well. The AlPO$_4$ sample heat-treated at 1000 °C also contained berlinite, but structural defects gave rise to another non-identified phase.

The IR spectra (Figure 6) enabled us to compare the AlPO$_4$ and Zn-AlPO$_4$ samples heat-treated at different temperatures. The spectra of the AlPO$_4$ samples contained the typical vibrations of the berlinite structure [34–36], the bands associated with vibrations of the berlinite unit at 1220, 1096, 504, 468, and 418 cm^{-1}, and the bands related to the vibrations of the Al pseudo-network at 700, 690, 626, and 580 cm^{-1}. A band also appeared at 735 cm^{-1}, which corresponded to defects of the Al subnetwork in berlinite [35].

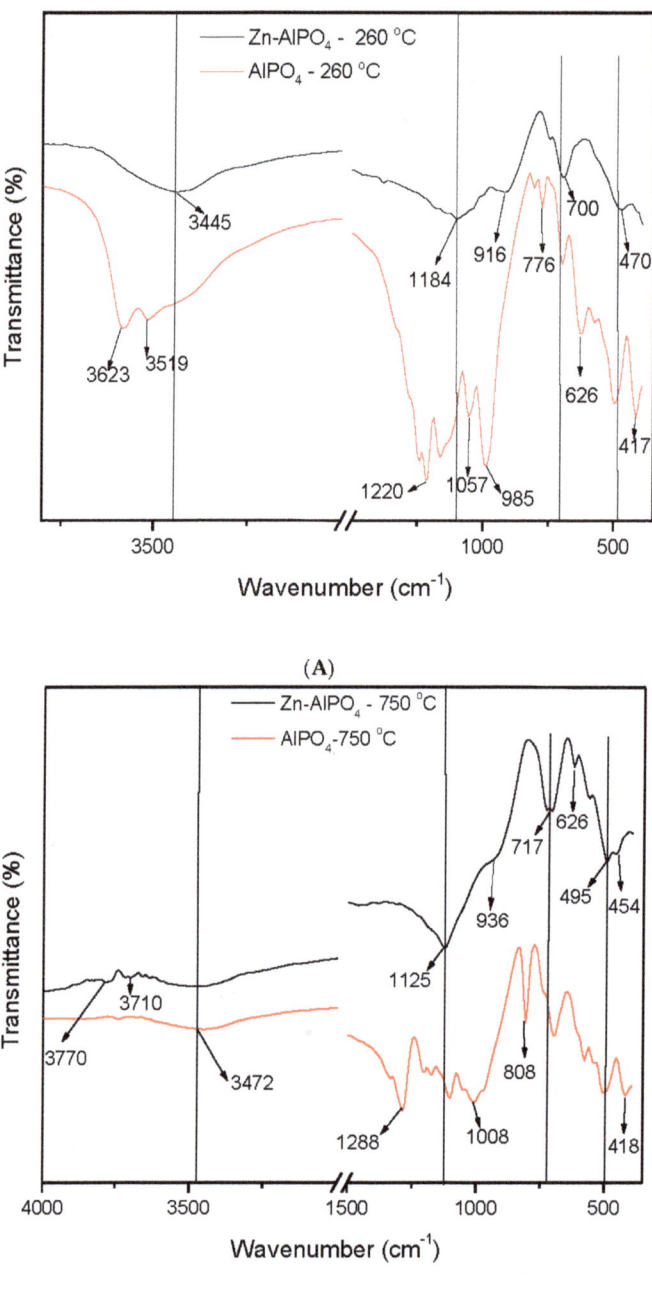

Figure 6. IR spectra of the Zn-AlPO$_4$ and AlPO$_4$ samples treated at (**A**) 260 °C and (**B**) 750 °C.

The Zn-AlPO$_4$ samples also displayed the bands detected for AlPO$_4$, although some of them shifted to lower wavenumbers as a result of the presence of Zn(II) in the berlinite network. Two additional vibrations appeared at 1009 and 936 cm^{-1}, as a result of Zn-OP vibrations [37]. ZnO IR vibrations usually occur below 600 cm^{-1}; in fact, a broad band in the

400–550 cm^{-1} region has been reported, constituting further evidence of Zn(II)'s incorporation into the network of berlinite [37]. The altered vibrations between 450 and 1290 cm^{-1}, along with an asymmetric band centered at 1125 cm^{-1}, clearly attested to the isomorphous substitution of Al(III) for Zn(II); the latter band intensified as the heat treatment temperature increased. According to the X-ray diffractometry results, this may have stemmed from formation of AlZnP$_2$O$_8$. Recent studies have described that both the activity and selectivity within a series of zeolites and molecular sieves—such as aluminophosphates—depend on total acidity. Apart from hydroxyl groups, other types of acid sites (for example, Lewis centers) contribute to the acidity of molecular sieves [26].

All of the Zn-AlPO$_4$ and AlPO$_4$ samples presented a broad band in the region of 3650–2900 cm^{-1} (centered at 3500 cm^{-1}), irrespective of the heat treatment temperature. Various authors have attributed this band to hydroxyl groups on the surface of free phosphorus; indeed, a hydrogen bond of the hydroxyl group on the surface of the compound disturbs phosphorus.

Hydroxyl vibrations appeared between 3000 and 4000 cm^{-1}. These bands allow for the detection of different types of OH groups, such as Al-OH, P-OH, and Zn-OH, which result from crystal defects and bridge hydroxyls, and confer Bronsted acidity to the material [32]. They indicate the type of metal incorporation into MeALPO and MeSAPO; that is, whether the metal isomorphically substitutes Al, P, or Si, or if only ion exchange occurs. Isomorphous substitution generates new OH groups that bind the metal to another tetrahedral ion in the structure of the zeolite or the molecular sieve. In contrast, ion exchange does not alter the typical structural vibrations. Hence, isomorphous substitution did take place in the synthesized materials.

Irrespective of the presence of Zn(II) in the matrix and the heat treatment temperature, all of the samples displayed bands between 3445 and 3500 cm^{-1}, characteristic of OH groups. These groups could correspond to water physisorbed onto the KBr pellet, or even onto the sample exposed to the environment after heat treatment. Indeed, thermogravimetry detected the presence of water due to handling of the material, as discussed earlier in this paper. The bands at 3583, 3623, 3710, and 3748 cm^{-1} corresponded to terminal -Al-OH and -P-OH groups [38]. The Zn-AlPO$_4$ samples may also present interactions of the -P-OH + Zn(OH)$_2$ type [38,39].

Table 1 lists data on the textural properties of the samples thermally treated at different temperatures, as determined by N$_2$ adsorption/desorption isotherms [27].

Table 1. Textural properties of the samples heat-treated at different temperatures.

Sample	Specific Surface Area (/m^{2}g^{-1})	Pore Volume (/cm^3 g^{-1})	Pore Size (/nm)
Zn-AlPO$_4$-260 °C	-	-	-
Zn-AlPO$_4$-400 °C	-	-	-
Zn-AlPO$_4$-750 °C	0.1	9.05 × 10^{-4}	44
Zn-AlPO$_4$-1000 °C	0.2	2.01 × 10^{-3}	49
AlPO$_4$-1000 °C	0.3	7.32 × 10^{-3}	12

* the traces (-) represent values of specific surface area lower than 1 m^{2}g^{-1}.

The Zn-AlPO$_4$ samples treated at 750 and 1000 °C had larger pore sizes. Increased heat treatment temperature rearranged the constituents in the matrix network, and subsequently eliminated organic residues, producing a more ordered crystalline structure. It is noteworthy that all of the AlPO$_4$ and Zn-AlPO$_4$ samples presented a lower surface area than those reported in the literature. This happened because the syntheses occurred in the absence of surfactants or any other molecular guides, to afford extremely dense materials.

The micrographs depicted in Figure 7A,B demonstrate that the samples were arranged as plaques, which generated a dense material and made it difficult for the electron beams to cross the structure. This reduced the TEM resolution, as observed for all of the AlPO$_4$ and Zn-AlPO$_4$ samples.

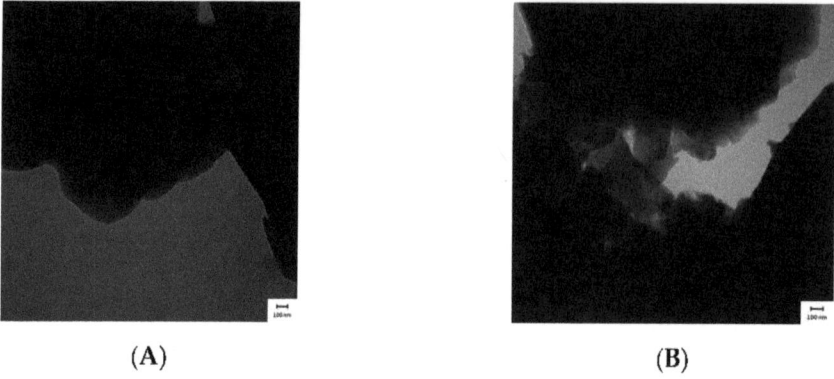

Figure 7. Transmission electron micrographs of (**A**) Zn-AlPO$_4$ and (**B**) AlPO$_4$ heat-treated at 400 °C.

AlPO$_4$ treated at 400 °C contained cavities, whereas Zn-AlPO$_4$ (Figure 7A,B), with high electron density, did not. In Zn-AlPO$_4$, Zn(II) may have been homogeneously dispersed in the aluminophosphate matrix, which justified the absence of cavities in the sample, and indicated that Zn(II) isomorphically substituted the Al(III) ions.

Zn-AlPO$_4$ thermally treated at 1000 °C (Figure 8B) bore agglomerates on the surface, which evidenced that the metal migrated to the surface at higher temperatures. As discussed earlier in the X-ray results confirmed the presence of ZnO and AlPO$_4$, which corroborated this migration.

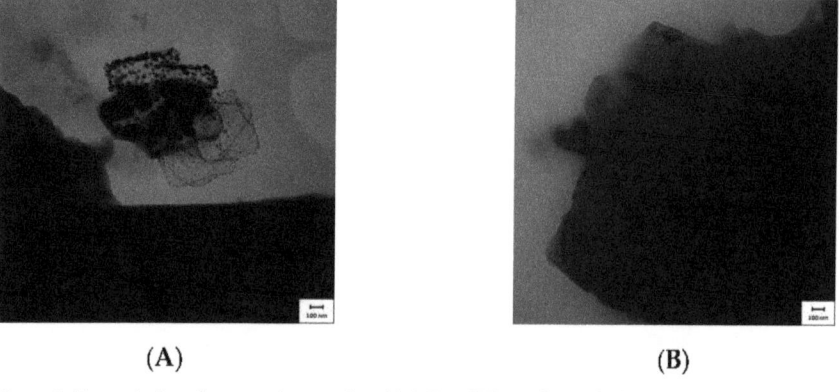

Figure 8. Transmission electron micrographs of (**A**) Zn-AlPO$_4$ and (**B**) AlPO$_4$ heat-treated at 1000 °C.

The SEM images depicted in Figures 9 and 10 show that the Zn(II) ions elicited changes in the features of the aluminophosphate matrix. Rising heat treatment temperature accentuated the differences between Zn-AlPO$_4$ and AlPO$_4$. Zn-AlPO$_4$ thermally treated at 400 °C had a smoother surface than AlPO$_4$; the latter material presented a more disordered structure, and granules of varied sizes.

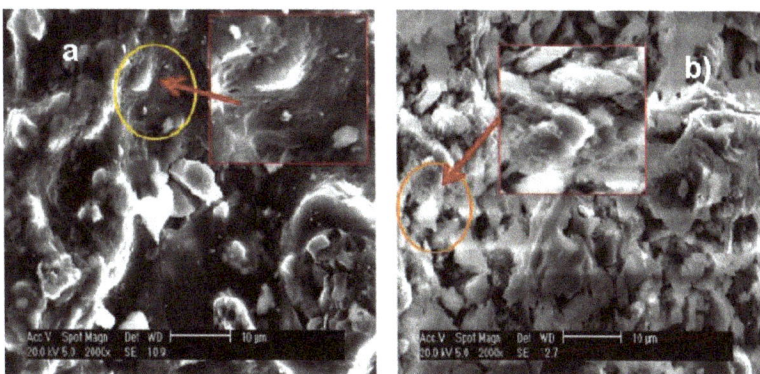

Figure 9. Scanning electron microscope images of (**a**) Zn-AlPO$_4$ and (**b**) AlPO$_4$ heat-treated at 400 °C.

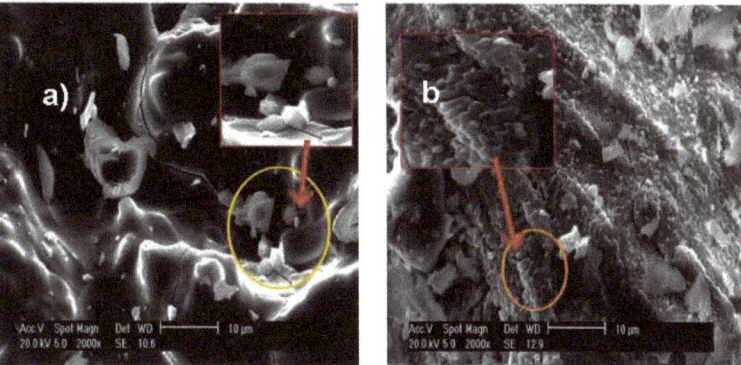

Figure 10. Scanning electron microscopies of (**a**) Zn-AlPO$_4$ and (**b**) AlPO$_4$ heat-treated at 1000 °C.

The differences between Zn-AlPO$_4$ and AlPO$_4$ became more pronounced upon heat treatment of the materials. The surface of the Zn-AlPO$_4$ sample contained some holes that resembled volcanoes, which may have originated during rising temperatures and the subsequent elimination of volatile compounds. Cracks also existed on the surface of Zn-AlPO$_4$. As for AlPO$_4$, the sample became whiter upon the removal of impurities. In the signaled area, and juxtaposed hexagonal plaques arose, which confirmed the X-ray and TEM results.

As for the Zn-AlPO$_4$ and AlPO$_4$ samples thermally treated at 1000 °C, Zn-AlPO$_4$ displayed a rough surface because the structure collapsed upon increasing heat treatment temperature. The AlPO$_4$ sample exhibited larger agglomerates and, according to the TEM results, this material consisted of juxtaposed plaques with extremities of lower thickness.

The surface was irregular. Synthetization seemed to culminate in low viscosity. The AlPO$_4$ surface presented larger agglomerates, and the particles resembled loose scales. These scales or plaques may also have originated from the collapse of the structures upon rising heat treatment temperature. The larger block was rough, and contained cracks along the surface.

To conduct a more detailed study on the composition of the samples analyzed by SEM, EDX analysis was carried out. Table 2 summarizes the percentage mass values of aluminum, phosphorus, zinc, oxygen, and carbon in the samples.

Table 2. Approximate percentages of the constituting elements in the Zn-AlPO$_4$ and AlPO$_4$ samples.

Sample	Zn (%)	Al (%)	P (%)	O (%)	C (%)
Zn-AlPO$_4$–260 °C	37	18	21	23.690	0.279
Zn-AlPO$_4$–400 °C	35	18	22	24.949	0.288
Zn-AlPO$_4$–750 °C	30	21	22	26.776	0.199
Zn-AlPO$_4$–1000 °C	28	23	23	25.823	0.466
AlPO$_4$–260 °C	-	20	33	46.321	0.847
AlPO$_4$–400 °C	-	19	34.918	44.550	1.461
AlPO$_4$–750 °C	-	21	35.355	42.798	1.138
AlPO$_4$–1000 °C	-	23	30.916	43.472	2.137

The amount of Zn(II) in the material decreased with the heat treatment temperature, because ZnO agglomerates arose, as verified by TEM. This provided further evidence that the ZnO agglomerates migrated to the surface of the polymeric matrix. Therefore, the samples heat-treated at higher temperatures should more efficient as photodegradation catalysts—the presence of these ZnO agglomerates on the surface of the matrix or around it should improve the quantum efficiency, consequently increasing the photogeneration of radicals that oxidize the target pesticide. As for AlPO$_4$, the ratios between the matrix constituents remained the same even after treatment at high temperatures. The exception was the sample treated at 1000 °C, which had better crystalline organization.

3.1. Photodegradation Studies

The photodegradation reactions lasted between 60 and 140 min, as monitored by UV spectroscopy (Figures 11 and 12). The typical bands of fipronil at 196, 219, and 284 nm disappeared as a function of the time that the sample was exposed to UV light; the temperature and pH values were kept constant. The influence of temperature and pH on photodegradation was also investigated. Finally, the optimized conditions were combined, and GC–MS was used to study how AlPO$_4$ and Zn-AlPO$_4$ affected the photodegradation of fipronil.

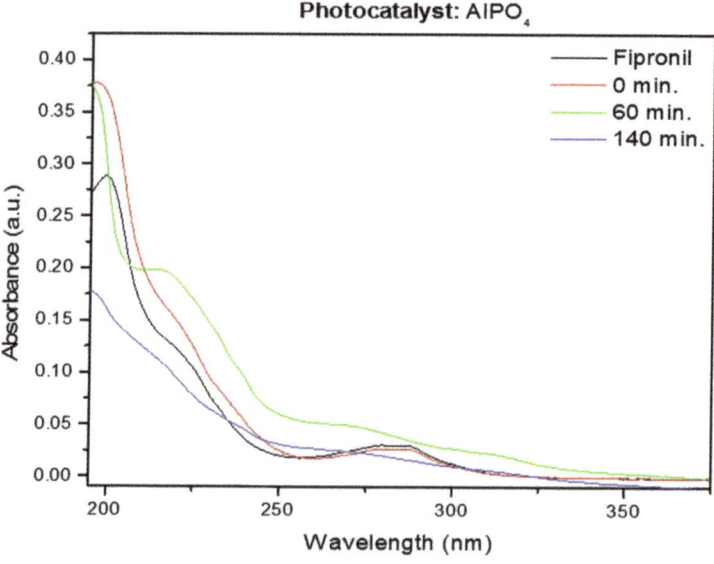

Figure 11. Fipronil (Ci = 2950 µg L^{-1}) absorption spectra in the presence of 153 mg of AlPO$_4$, under mechanical stirring and UV radiation (control reaction).

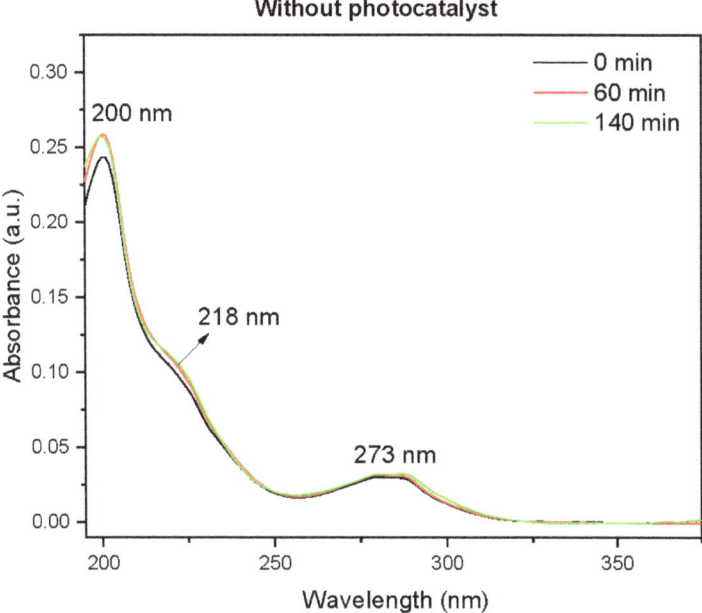

Figure 12. Absorption spectra of fipronil (Ci = 3370 µg L^{-1}) in the absence of Zn-AlPO$_4$, under mechanical stirring, in the absence of UV radiation.

According to previous studies, exposure to light prompts conversion of fipronil into its desulfurated derivative desthiofipronil, which has also been reported to act as an insecticide [16,40]. However, desthiofipronil is stable on the surface of leaves, which could culminate in persistence and non-target toxicity in the environment.

Analyses performed without a photocatalyst and in the presence of 153 mg of AlPO$_4$ (Figure 12)—the designated blank (control) reaction under irradiation—revealed that AlPO$_4$ samples without the semiconductor (Zn(II) species) photodegraded fipronil. The UV/AlPO$_4$ system initially formed a novel species, as observed through the new UV bands that emerged at 218 and 273 nm. After 140 min, these species had disappeared, as confirmed by the decreased UV absorption and complete change in the UV spectral profile. Reactive groups (F, N, and Cl) in the structure of fipronil probably favored the formation of reactive sites on the surface of the AlPO$_4$ matrix, which generated oxidizing radicals even in the absence of the photoconductor. Nevertheless, considering the same period of exposure to radiation, Zn-AlPO$_4$ materials photodegraded fipronil more efficiently.

As noted by Ngim et al. [30], photodegradation of fipronil under mechanical stirring and UV light radiation, in the absence of the catalyst, produced intermediate species. However, these intermediates seemed to regenerate the reactants throughout the experiment. These results suggested that photoinduced reactions did not occur simply due to UV light irradiation on the insecticide. Indeed, previous literature data demonstrate that total mineralization of certain pollutants exposed to radiation would require days or even months to occur [41]. The detailed discussion on the formation of intermediate species is presented in the section on CG–MS.

Experiments conducted in the absence and in the presence of UV radiation helped to evaluate how UV light affected the photodegradation of fipronil. In a typical adsorption experiment accomplished in a dark chamber and under mechanical stirring, the concentration of the pesticide did not change over time, suggesting that both the photocatalyst and light radiation were necessary to promote the photodegradation of fipronil via hydroxyl radicals [42].

The experiment accomplished in the presence of Zn-AlPO$_4$ but in the absence of UV light did not provide the typical degradation profile of the intermediate species, indicating that Zn-AlPO$_4$ only degraded the pesticide in the presence of UV artificial light source (253.7 nm). Fipronil clearly did not adsorb at the same intensity as it did throughout the exposure, because its concentration did not vary over the duration of the experiment. Indeed, BET textural analyses showed that the synthesized Zn-AlPO$_4$ material hardly adsorbed substances, which may explain why the pesticide did not adsorb onto the matrix containing the photosensitizer.

Because not even temporary modification of the chemical structure of fipronil took place during the experiment conducted in the absence of the photocatalyst, efficient decomposition of fipronil must have depended on the presence of hydroxyl groups generated on the surface of Zn-AlPO$_4$ under UV light, to yield the intermediate reactive species. These intermediate species allowed for complete mineralization of the pesticide.

3.1.1. Influence of the Zn-AlPO$_4$ Mass on the Photodegradation of Fipronil

Table 3 shows that photodegradation occurred as a function of the mass of Zn-AlPO$_4$ employed in the photodegradation experiment. The same conditions of photodegradation and an exposure time of 140 min did not degrade fipronil in the presence of 103 mg of the photocatalyst, regardless of the Zn-AlPO$_4$ heat treatment temperature. Nevertheless, the concentration of the pesticide decreased, confirming that photodegradation of fipronil occurred in the presence of catalysts treated at 400, 750, and 1000 °C.

Table 3. Evaluation of the photodegradation of fipronil as a function of the mass and calcination temperature of Zn-AlPO$_4$.

Sample	Zn-AlPO$_4$-260		Zn-AlPO$_4$-400		Zn-AlPO$_4$-750		Zn-AlPO$_4$-1000	
Mass of photocatalyst (mg)	103	152	103	152	103	152	103	152
Degradation of Fipronil (%)	-	37	-	17	17	80	70	10
Presence of byproduct (at the end of the reaction)	Yes	Yes	Yes	Yes	No	No	No	No

The catalyst calcined at 750 °C produced the best results, as shown in Figure 13. The sample heated at 750 °C had a trigonal phase, assigned to AlZnP$_2$O$_8$. Compared to the hexagonal phase, the trigonal phase probably contributed to more efficient generation of hydroxyl radicals due to the presence of larger amounts of ZnO in the AlZnP$_2$O$_8$ phase at the surface of the unit cell. The temperature at which Zn-AlPO$_4$ was treated also influenced the morphology of the material, as noted from TEM analysis. Indeed, heat treatment was directly related to the formation of particles with features that favored the generation of the electron–gap pair on the surface of the catalyst [43].

Among other factors, the efficiency of Zn-AlPO$_4$-750 may have been related to the treatment (higher temperatures) to which it was submitted after the synthesis. This treatment eliminated residual groups (e.g., chloride ions) that could interfere in the adsorption of the compound from which the reactive radicals originated [44]. The presence of such interfering agents would culminate in an effect known as scavenging—sweeping of the radicals from the surface of the photocatalyst, to favor recombination of the electron–gap pair, with subsequent release of energy as heat. This would harm AOPs, not to mention that these groups could affect adsorption of the substance that generates the radicals on the semiconductor matrix. The experiment depicted revealed that a smaller mass of catalyst (103 mg) and longer exposure time (140 min) decomposed almost 100% of the pesticide. A higher mass of catalyst and longer reaction time (Figures 13 and 14) indicated that Zn-AlPO$_4$ treated at 260 °C degraded the pesticide (Figure not shown).

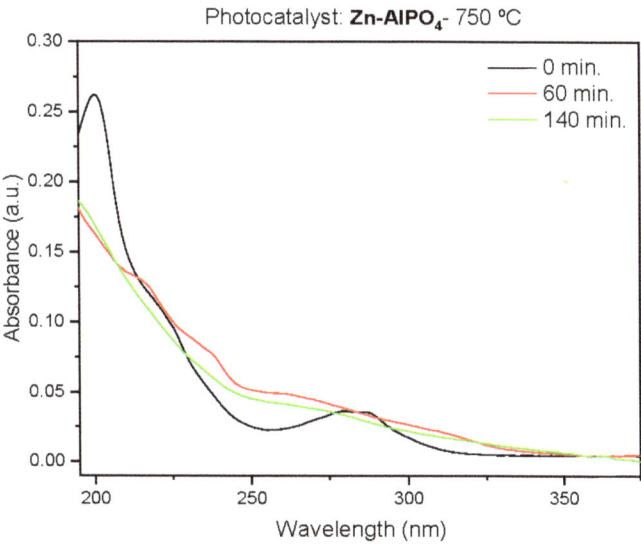

Figure 13. Absorption spectra of fipronil (Ci = 2880 µg L^{-1}) in the presence of Zn-AlPO$_4$ (103 mg) heat-treated at 750 °C, under magnetic stirring and UV radiation.

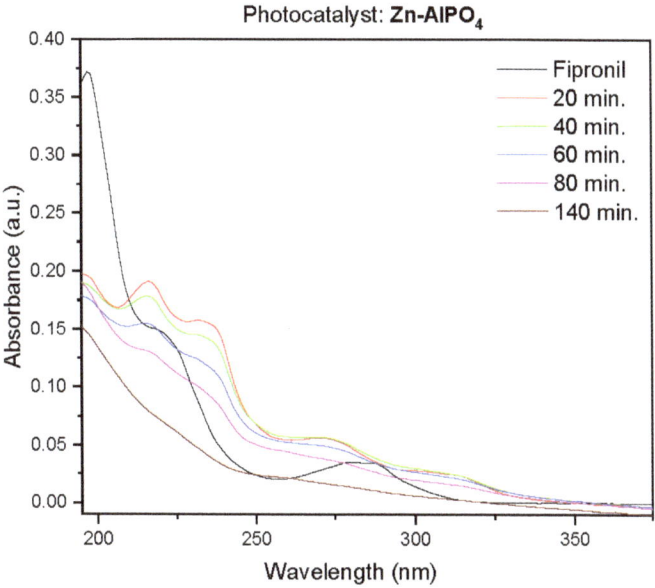

Figure 14. Absorption spectra of fipronil (Ci = 2920 µg L^{-1}) in the presence of Zn-AlPO$_4$ (103 mg) heat-treated at 1000 °C, under mechanical stirring and UV radiation.

A calcination temperature of 750 °C afforded the best photodegradation catalyst. All of the Zn-AlPO$_4$ materials prepared herein performed well during the degradation of fipronil. Degradation increased very quickly within a short time, especially for Zn-AlPO$_4$-750 and Zn-AlPO$_4$-1000. Eighty-percent photodegradation was achieved after only 2 h. Longer reaction times favored complete degradation. All of the calcined composites behaved similarly. These results resembled the results reported by Ngim et al. [30] and Bóbe et al. [18]. It is important to consider not only the photocatalytic nature of the composites, but also the amount of ZnO in the material when comparing the data obtained herein with the literature results. In some cases, our materials provided even better results than those reported by other authors. The photodegradation of fipronil as a function of time is shown in Table 4.

Table 4. Photodegradation of fipronil as a function of time in the presence of 153 mg of Zn-AlPO$_4$.

Time (min)	Zn-AlPO$_4$-260		Zn-AlPO$_4$-400		Zn-AlPO$_4$-750		Zn-AlPO$_4$-1000	
	(%)	Byproducts	(%)	Byproducts	(%)	Byproducts	(%)	Byproducts
0	-	Yes	0	Yes	0	Yes	0	Yes
15	-	Yes	2	Yes	5	Yes	-	Yes
30	-	Yes	3	Yes	12	Yes	-	Yes
60	-	Yes	5	Yes	15	Yes	-	Yes
140	37	Yes	17	Yes	80	No	10	No

Concerning Zn-AlPO$_4$ heat-treated at 1000 °C, a smaller mass of catalyst afforded better results. A series of studies have indicated that the photocatalytic degradation rate initially rises with larger photocatalyst mass, but later decreases significantly due to light scattering. A higher amount of the solid catalyst also favors agglomeration (particle–particle interaction); hence, lower surface area is available for light absorption, diminishing the rate of photocatalytic degradation. Although the number of active sites in the solution might increase with larger mass of the photocatalyst, the system reaches a point where excess particles no longer allow light to penetrate the catalyst. Therefore, there is an optimal mass that affords the highest efficiency possible. Further increase in the amount of the photocatalyst results in non-uniform distribution of light intensity, reducing the reaction yields despite the larger catalyst mass [45].

The results showed that the best photocatalytic conditions were obtained using the higher mass of photocatalyst heat-treated at 750 °C; this enhanced photocatalytic activity was assigned to trigonal-phase AlZnP$_2$O$_8$, producing 80% fipronil removal without byproducts; the material heat-treated at 1000 °C presented very low photocatalytic activity under the same conditions—only 10% photodegradation and the presence of byproducts were observed by UV–Vis spectroscopy; this difference could be assigned to the orthorhombic berlinite phase that presented lower photocatalytic activity; the sintering of pores and migration of Zn phases from matrix could also explain this difference.

3.1.2. Investigation into the Photodegradation of Fipronil Catalyzed by Zn-AlPO$_4$

GC–MS helped to investigate the potential of Zn-AlPO$_4$ to catalyze the photodegradation of fipronil. To this end, an aqueous solution of fipronil (1 µg/mL) was prepared. Then, the solution was placed in a UV chamber in the absence of Zn-AlPO$_4$ for 2 h, and in the presence of Zn-AlPO$_4$ for 120 min, and photodegradation was measured as shown in Figure 15.

Figure 15A shows the chromatogram of fipronil (**1**, R$_t$ = 11.9 min). Peaks with retention times at 11.7 (*m/z* 420), 13.1 (*m/z* 352), and 13.2 min (*m/z* 452) corresponded to fipronil–sulfide (**3**), 4-thiol–fipronil (**4**), and fipronil–sulfone (**5**), respectively (Figure 16). Compounds **3** and **4** corresponded to products from the photodegradation of fipronil [32,33,46]. Compound **5** referred to an intermediate that emerged during conversion of **1** to desulfinyl–fipronil (**2**) [18]. Table 5 summarizes the MS data of compounds **1** to **6**.

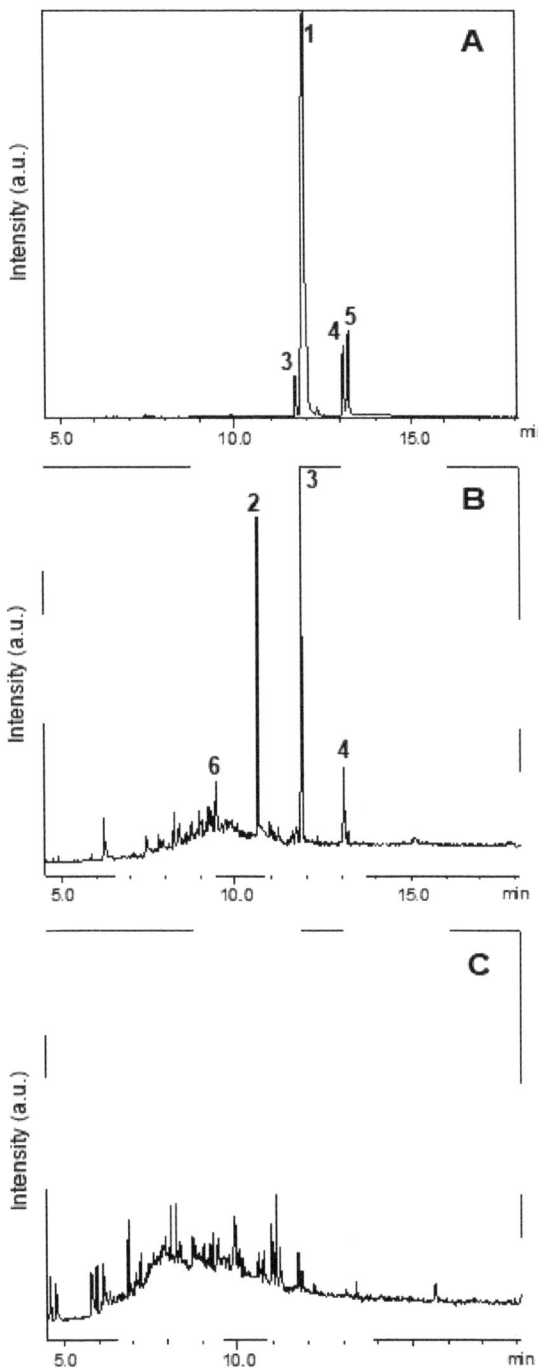

Figure 15. GC–MS chromatogram of (**A**) fipronil; (**B**) aqueous solution of fipronil after 120 min under exposure to UV light in the sealed chamber, in the absence of a catalyst; and (**C**) aqueous solution of fipronil after 8 h under exposure to UV light in the sealed chamber, in the presence of Zn-AlPO$_4$ (**C**). Fipronil (**1**), desthioderivative (**2**), fipronil–sulfide (**3**), 4-thiol–fipronil (**4**), fipronil–sulfone (**5**), and 2-[2-(2,6-dichloro-4-(trifluoromethyl) phenyl) diazenyl] acetonitrile (**6**).

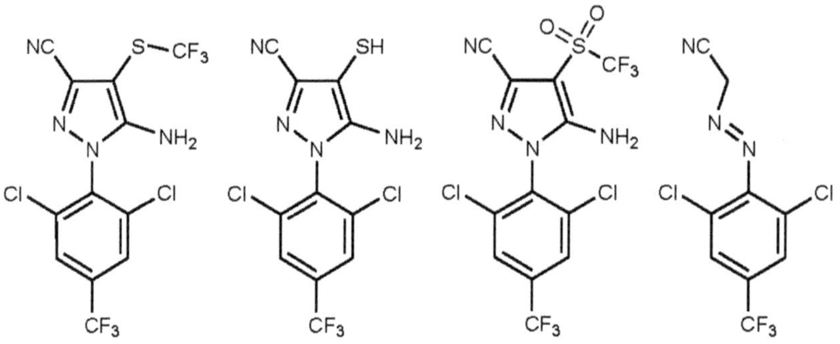

Figure 16. Chemical structures of the products from the photodegradation of fipronil identified in this study. Fipronil–sulfide (**3**), 4-thiol–fipronil (**4**), fipronil–sulfone (**5**), and 2-[2-(2,6-dichloro-4-(trifluoromethyl) phenyl) diazenyl] acetonitrile (**6**).

Table 5. GC retention times (Rt) and mass spectral data (EI-MS) for fipronil (1) and the degradation products 2–6.

Compound	R_t (min)	$M^{\bullet+}$	MS, m/z [a] (Relative Intensity, %)
1	11.9 ± 0.2	436	367 [$M^{\bullet+} - \bullet CF_3$] (100); 255 [$C_6H_2(Cl)_2(CF_3)(NCNH_2)$]$^+$ (17); 213 [$M^{\bullet+} - \bullet C_3N_2(CN)(F_3CSO)NH_2$] (25); 69 [$CF_3^+$] (11) Other ions: 353 (13), 351 (18), 257 (12), 215 (28), 77 (25)
2	10.6 ± 0.2	388	369 [$M^{\bullet+} - \bullet F$] (15); 333 [$M^{\bullet+} - \bullet CF_3$] (97); 278 [$C_6H_2(Cl)_2(CF_3)(C_3H_2N_2)$]$^+$ (11); 213 [$M^{\bullet+} - \bullet C_3N_2(CN)(F_3C)NH_2$] (38); 143 [$C_6H_2(CF_3)$]$^+$ (14); 69 [CF_3^+] (13) Other ions: 390 (50), 369 (15), 335 (34), 281 (28), 215 (12), 179 (13), 120 (15), 77 (18)
3	11.7 ± 0.2	420	351 [$M^{\bullet+} - \bullet CF_3$] (100); 255 [$C_6H_2(Cl)_2(CF_3)(NCNH_2)$]$^+$ (42); 213 [$M^{\bullet+} - \bullet C_3N_2(CN)(F_3CS)NH_2$] (18); 143 [$C_6H_2(CF_3)$]$^+$ (15); 69 [CF_3^+] (21). Other ions: 290 (27), 179 (11), 77 (18)
4	13.1 ± 0.2	352	283 [$M^{\bullet+} - \bullet CF_3$] (100); 255 [$C_6H_2(Cl)_2(CF_3)(NCNH_2)$]$^+$ (33); 213 [$M^{\bullet+} - \bullet C_3N_2(CN)(SH)NH_2$] (14); 143 [$C_6H_2(CF_3)$]$^+$ (14); 69 [CF_3^+] (30). Other ions: 113 (12), 112 (43), 99 (24), 83 (39)
5	13.2 ± 0.2	452	383 [$M^{\bullet+} - \bullet CF_3$] (100); 255 [$C_6H_2(Cl)_2(CF_3)(NCNH_2)$]$^+$ (52); 213 [$M^{\bullet+} - \bullet C_3N_2(CN)(F_3CSO_2)NH_2$] (44); 143 [$C_6H_2(CF_3)$]$^+$ (13); 69 [CF_3^+] (11). Other ions: 385 (72), 257 (35), 241 (30), 215 (28), 179 (11), 178 (18), 77 (28)
6	9.4 ± 0.2	281	255 [$M^{\bullet+} - \bullet CN$] (30); 254 [$M^{\bullet+} - HCN$] (43); 227 [$M^{\bullet+} - \bullet CN - N_2$] (100); 213 [$M^{\bullet+} - \bullet C_2H_2N_3$] (29); 143 [$C_6H_2(CF_3)$]$^+$ (28); 69 [CF_3^+] (59). Other ions: 215 (16), 230 (69), 203 (32), 201 (51), 157 (34), 88 (40), 61 (71)

[a] Ions with relative intensity lower than 10% are not shown.

Comparison of the GC–MS chromatogram of fipronil with the GC–MS chromatogram obtained from exposure of the aqueous solution of fipronil to UV radiation in the sealed chamber revealed that fipronil generated mainly fipronil–sulfide (**3**, R_t = 11.7 min) and desulfinyl–fipronil (**2**, R_t = 10.6 min) after 2 h. Compound **6** (R_t = 9.4 min)—a product from the degradation of the pyrazole ring, and a known product from the photodegradation of fipronil—also emerged. Conversion of fipronil to its desthioderivative **2** has been exhaustively investigated in the literature, mainly due to the increased stability of this product as compared to fipronil [46]. In this study, only traces of fipronil arose after 140 min of exposure of the aqueous solution of fipronil to UV radiation in the sealed chamber without Zn-AlPO$_4$.

The photodegradation of fipronil by Zn-AlPO$_4$ at 750 °C is confirmed in Figure 15C, which shows only lower intensities of the characteristic peaks of fipronil and its byproducts; the peak intensity is quantitatively proportional to the pollutant concentration.

Analysis of the GC–MS chromatogram obtained for the aqueous solution of fipronil after 140 min under exposure to UV light, in the presence of the photocatalyst Zn-AlPO$_4$, revealed that fipronil and its transformation products **2–6** underwent photodegradation. More detailed analysis of the MS data of each peak in this chromatogram evidenced total conversion of fipronil and its most commonly reported transformation products into a complex mixture of other photoproducts. Although it was not possible to identify these transformation products on the basis of GC–MS data alone, these results strongly suggested mineralization of fipronil and its transformation products. In summary, the results obtained in the photodegradation experiments reported herein suggest that the compound Zn-AlPO$_4$ is a promising photocatalyst for the degradation of fipronil.

4. Conclusions

The non-hydrolytic sol–gel route afforded new materials consisting of aluminophosphate and containing zinc in their structure (Zn-AlPO$_4$). Zinc isomorphically substituted Al and P in the AlPO$_4$ network, as attested by infrared spectroscopy. Thermal analyses indicated and X-ray diffraction later corroborated that Zn-AlPO$_4$ underwent various morphological transitions during thermal treatment, paving the way for future investigations into each of the generated phases. The berlinite phase predominated in all of the prepared catalysts.

X-ray powder diffraction showed the influence of heat treatment on structural changes of Zn-AlPO$_4$; trigonal and orthorhombic phases were obtained at 750 and 1000 °C, respectively; these phases directly affect the capability of OH radical generation. The capacity of the materials to produce effluent was tested in the photocatalyzed degradation of the pesticide fipronil. Heat-treated Zn-AlPO$_4$ at 750 °C showed better photodegradation results by removing 80% of the pesticide in 2 h, using the higher mass tested (150 mg). GC–MS confirmed the photodegradation profile, confirming that only traces of fipronil were observed after 140 min of photocatalytic reaction using a photocatalyst treated at 750 °C. On the other hand, the material treated at 1000 °C showed the presence of byproducts (quantified by GC–MS), in agreement with the XRD results.

The use of the non-hydrolytic sol–gel process enabled the construction of catalysts with structures that allowed the degradation of fipronil without the production of byproducts; we could highlight that this process provided high reproducibility and simplicity compared to different synthesis routes, without the use of complexes infrastructures resulting in very pure phases after heating treatments.

Zn-AlPO$_4$ heat-treated at 1000 °C showed the presence of agglomerates on the surface, evidencing that the Zn migrated to the surface at higher temperatures (reducing the catalytic activity) due to the increase in cluster size, along with the sintering of pores. As discussed earlier, X-ray results confirmed the presence of ZnO and AlPO$_4$, corroborating this migration.

The as-prepared Zn-AlPO$_4$ materials are potentially applicable in the photocatalytic degradation of the pesticide fipronil; they also have potential application in the degradation of other organic pollutants, other classes of pesticides, and dyes used in the textile industry. The non-hydrolytic sol–gel route is faster and simpler than other hydrothermal methods used to prepare aluminophosphates. In the future, this method could aid in the incorporation of other metals into these matrices, along with the preparation of photocatalysts with unique features.

Author Contributions: Conceptualization, O.J.d.L., L.M., G.S.M. and D.T.d.A.; methodology, O.J.d.L., L.M. and K.J.C.; validation, O.J.d.L., L.M., G.S.M., S.N. and E.H.d.F.; formal analysis, E.H.d.F., L.M., K.J.C. and S.N.; investigation, O.J.d.L., D.T.d.A., E.H.d.F. and L.M.; resources, K.J.C., S.N., A.E.M.C. and E.H.d.F.; data curation, E.H.d.F., K.J.C., L.M., A.E.M.C. and S.N.; writing—original draft preparation, O.J.d.L., L.M., D.T.d.A. and E.H.d.F.; writing—review and editing, L.M., E.H.d.F., S.N., A.E.M.C. and K.J.C.; supervision, E.H.d.F., K.J.C. and L.M.; project administration, K.J.C. and E.H.d.F.; funding acquisition, E.H.d.F., K.J.C., S.N. and A.E.M.C. All authors have read and agreed to the published version of the manuscript.

Funding: This group acknowledges the support from the research funding agencies the Fundação de Amparo à Pesquisa do Estado de São Paulo, FAPESP (2013/19523-3, 2017/15482-1 and 2020/06712-6), the Coordenação de Aperfeiçoamento de Pessoal de Nível Superior (CAPES)—finance code 001—and the Conselho Nacional de Desenvolvimento Científico e Tecnológico, CNPq (311767/2015-0, 303135/2018-2, 310151/2021-0, 305180/2019-3 and 405217/2018-8). The equipment of the Brazilian group was financed by FAPESP (1998/11022-3, 2005/00720-7, 2011/03335-8, 2012/11673-3 and 2016/01501-1).

Data Availability Statement: Not applicable.

Acknowledgments: This article is dedicated to Vicente Rives upon his retirement.

Conflicts of Interest: The authors declare no conflict of interest. The authors declare that they have no known competing financial interest or personal relationships that could have appeared to influence the work reported in this paper.

References

1. Kumar, A.; Shalini; Sharma, G.; Naushad, M.; Kumar, A.; Kalia, S.; Guo, C.; Mola, G.T. Facile hetero-assembly of superparamagnetic Fe3O4/BiVO4 stacked on biochar for solar photo-degradation of methyl paraben and pesticide removal from soil. *J. Photochem. Photobiol. A Chem.* **2017**, *337*, 118–131. [CrossRef]
2. Dewil, R.; Mantzavinos, D.; Poulios, I.; Rodrigo, M.A. New perspectives for Advanced Oxidation Processes. *J. Environ. Manag.* **2017**, *195*, 93–99. [CrossRef] [PubMed]
3. Oller, I.; Malato, S.; Sánchez Pérez, J.A. Combination of Advanced Oxidation Processes and biological treatments for wastewater decontamination—A review. *Sci. Total Environ.* **2011**, *409*, 4141–4166. [CrossRef] [PubMed]
4. Ribeiro, A.R.; Nunes, O.C.; Pereira, M.F.; Silva, A.M. An overview on the advanced oxidation processes applied for the treatment of water pollutants defined in the recently launched Directive 2013/39/EU. *Environ. Int.* **2015**, *75*, 33–51. [CrossRef]
5. Miklos, D.B.; Remy, C.; Jekel, M.; Linden, K.G.; Drewes, J.E.; Hübner, U. Evaluation of advanced oxidation processes for water and wastewater treatment—A critical review. *Water Res.* **2018**, *139*, 118–131. [CrossRef]
6. Sanches, S.; Crespo, M.T.B.; Pereira, V.J. Drinking water treatment of priority pesticides using low pressure UV photolysis and advanced oxidation processes. *Water Res.* **2010**, *44*, 1809–1818. [CrossRef]
7. Shi, X.; Liu, J.-B.; Hosseini, M.; Shemshadi, R.; Razavi, R.; Parsaee, Z. Ultrasound-aasisted photodegradation of Alprazolam in aqueous media using a novel high performance nanocomosite hybridation g-C3N4/MWCNT/ZnO. *Catal. Today* **2019**, *335*, 582–590. [CrossRef]
8. Ahmadi, M.; Samarbaf, S.; Golshan, M.; Jorfi, S.; Ramavandi, B. Data on photo-catalytic degradation of 4- chlorophenol from aqueous solution using UV/ZnO/persulfate. *Data Brief* **2018**, *20*, 582–586. [CrossRef]
9. Yang, Q.; Ma, Y.; Chen, F.; Yao, F.; Sun, J.; Wang, S.; Yi, K.; Hou, L.; Li, X.; Wang, D. Recent advances in photo-activated sulfate radical-advanced oxidation process (SR-AOP) for refractory organic pollutants removal in water. *Chem. Eng. J.* **2019**, *378*, 122149. [CrossRef]
10. Davari, N.; Farhadian, M.; Nazar, A.R.S.; Homayoonfal, M. Degradation of diphenhydramine by the photocatalysts of ZnO/Fe2O3 and TiO2/Fe2O3 based on clinoptilolite: Structural and operational comparison. *J. Environ. Chem. Eng.* **2017**, *5*, 5707–5720. [CrossRef]
11. Shinde, P.; Bhosale, C.; Rajpure, K. Zinc oxide mediated heterogeneous photocatalytic degradation of organic species under solar radiation. *J. Photochem. Photobiol. B: Biol.* **2011**, *104*, 425–433. [CrossRef] [PubMed]
12. Chien, Y.-C.; Wang, H.P.; Liu, S.-H.; Hsiung, T.; Tai, H.-S.; Peng, C.-Y. Photocatalytic decomposition of CCl4 on Zr-MCM-41. *J. Hazard. Mater.* **2008**, *151*, 461–464. [CrossRef] [PubMed]
13. Liu, S. Photocatalytic generation of hydrogen on Zr-MCM-41. *Int. J. Hydrogen Energy* **2002**, *27*, 859–862. [CrossRef]
14. Jafarzadeh, A.; Sohrabnezhad, S.; Zanjanchi, M.A.; Arvand, M. Synthesis and characterization of thiol-functionalized MCM-41 nanofibers and its application as photocatalyst. *Microporous Mesoporous Mater.* **2016**, *236*, 109–119. [CrossRef]

15. Han, J.; Qiu, W.; Gao, W. Potential dissolution and photo-dissolution of ZnO thin films. *J. Hazard. Mater.* **2010**, *178*, 115–122. [CrossRef]
16. Hidaka, H.; Tsukamoto, T.; Mitsutsuka, Y.; Takamura, T.; Serpone, N. Photochemical and Ga_2O_3-photoassisted decomposition of the insecticide Fipronil in aqueous media upon UVC radiation. *New J. Chem.* **2014**, *38*, 3939–3952. [CrossRef]
17. Goff, A.D.; Saranjampour, P.; Ryan, L.M.; Hladik, M.; Covi, J.A.; Armbrust, K.L.; Brander, S.M. The effects of fipronil and the photodegradation product fipronil desulfinyl on growth and gene expression in juvenile blue crabs, Callinectes sapidus, at different salinities. *Aquat. Toxicol.* **2017**, *186*, 96–104. [CrossRef]
18. Bobé, A.; Meallier, P.; Cooper, J.-F.; Coste, C.M. Kinetics and Mechanisms of Abiotic Degradation of Fipronil (Hydrolysis and Photolysis). *J. Agric. Food Chem.* **1998**, *46*, 2834–2839. [CrossRef]
19. De Lima, O.J.; de Aguirre, D.P.; de Oliveira, D.C.; da Silva, M.A.; Mello, C.; Leite, C.A.P.; Sacco, H.C.; Ciuffi, K.J. Porphyrins entrapped in an alumina matrix. *J. Mater. Chem.* **2001**, *11*, 2476–2481. [CrossRef]
20. Mutin, P.H.; Vioux, A. Nonhydrolytic Processing of Oxide-Based Materials: Simple Routes to Control Homogeneity, Morphology, and Nanostructure. *Chem. Mater.* **2009**, *21*, 582–596. [CrossRef]
21. Jusoh, N.; Jalil, A.A.; Triwahyono, S.; Setiabudi, H.D.; Sapawe, N.; Satar, M.; Karim, A.; Kamarudin, N.; Jusoh, R.; Jaafar, N.F.; et al. Sequential desilication–isomorphous substitution route to prepare mesostructured silica nanoparticles loaded with ZnO and their photocatalytic activity. *Appl. Catal. A Gen.* **2013**, *468*, 276–287. [CrossRef]
22. Acosta, S.; Corriu, R.; Leclercq, D.; Lefèvre, P.; Mutin, P.; Vioux, A. Preparation of alumina gels by a non-hydrolytic sol-gel processing method. *J. Non-Crystalline Solids* **1994**, *170*, 234–242. [CrossRef]
23. Bourget, L.; Corriu, R.; Leclercq, D.; Mutin, P.; Vioux, A. Non-hydrolytic sol–gel routes to silica. *J. Non-Crystalline Solids* **1998**, *242*, 81–91. [CrossRef]
24. Pal, N.; Bhaumik, A. Mesoporous materials: Versatile supports in heterogeneous catalysis for liquid phase catalytic transformations. *RSC Adv.* **2015**, *5*, 24363–24391. [CrossRef]
25. Yang, X.; Ma, H.; Xu, Z.; Xu, Y.; Tian, Z.; Lin, L. Hydroisomerization of n-dodecane over Pt/MeAPO-11 (Me=Mg, Mn, Co or Zn) catalysts. *Catal. Commun.* **2007**, *8*, 1232–1238. [CrossRef]
26. Zhou, L.; Lu, T.; Xu, J.; Chen, M.; Zhang, C.; Chen, C.; Yang, X.; Xu, J. Synthesis of hierarchical MeAPO-5 molecular sieves—Catalysts for the oxidation of hydrocarbons with efficient mass transport. *Microporous Mesoporous Mater.* **2012**, *161*, 76–83. [CrossRef]
27. Brunauer, S.; Deming, L.S.; Deming, W.E.; Teller, E. On a Theory of the van der Waals Adsorption of Gases. *J. Am. Chem. Soc.* **1940**, *62*, 1723–1732. [CrossRef]
28. Campos, M.L.A.M.; Mello, L.C.; Zanette, D.R.; de Souza Sierra, M.M.; Bendo, A. Construçao e Otimizaçao De Um Reator De Baixo Custo Para a Fotodegradaçao Da Materia Orgânica Em águas Naturais E Sua Aplicaçao No Estudo Da Especiaçao Do Cobre Por Voltametria. *Quim. Nova* **2001**, *24*, 12–16. [CrossRef]
29. Tiburtius, E.R.L.; Peralta-Zamora, P.; Emmel, A.; Leal, E.S. Degradação de BTXs via processos oxidativos avançados. *Quim. Nova* **2005**, *28*, 61–64. [CrossRef]
30. Ngim, K.K.; Mabury, S.A.; Crosby, D.G. Elucidation of Fipronil Photodegradation Pathways. *J. Agric. Food Chem.* **2000**, *48*, 4661–4665. [CrossRef]
31. Saltarelli, M.; de Faria, E.H.; Ciuffi, K.J.; Nassar, E.J.; Trujillano, R.; Rives, V.; Vicente, M.A. Aminoiron(III)–porphyrin–alumina catalyst obtained by non-hydrolytic sol-gel process for heterogeneous oxidation of hydrocarbons. *Mol. Catal.* **2018**, *462*, 114–125. [CrossRef]
32. Zhu, J.; Yang, J.; Bian, Z.-F.; Ren, J.; Liu, Y.-M.; Cao, Y.; Li, H.-X.; He, H.-Y.; Fan, K.-N. Nanocrystalline anatase TiO_2 photocatalysts prepared via a facile low temperature nonhydrolytic sol–gel reaction of $TiCl_4$ and benzyl alcohol. *Appl. Catal. B Environ.* **2007**, *76*, 82–91. [CrossRef]
33. Corà, F.; Alfredsson, M.; Barker, C.; Bell, R.G.; Foster, M.D.; Saadoune, I.; Simperler, A.; Catlow, C.A. Modeling the framework stability and catalytic activity of pure and transition metal-doped zeotypes. *J. Solid State Chem.* **2003**, *176*, 496–529. [CrossRef]
34. Pînzaru, S.C.; Onac, B.P. Raman study of natural berlinite from a geological phosphate deposit. *Vib. Spectrosc.* **2009**, *49*, 97–100. [CrossRef]
35. Rokita, M.; Handke, M.; Mozgawa, W. The $AlPO_4$ polymorphs structure in the light of Raman and IR spectroscopy studies. *J. Mol. Struct.* **2000**, *555*, 351–356. [CrossRef]
36. Rokita, M.; Handke, M.; Mozgawa, W. Spectroscopic studies of polymorphs of $AlPO_4$ and SiO_2. *J. Mol. Struct.* **1998**, *450*, 213–217. [CrossRef]
37. Pawlig, O.; Trettin, R. In-Situ DRIFT Spectroscopic Investigation on the Chemical Evolution of Zinc Phosphate Acid–Base Cement. *Chem. Mater.* **2000**, *12*, 1279–1287. [CrossRef]
38. Lourenço, J.; Ribeiro, M.; Borges, C.; Rocha, J.; Onida, B.; Garrone, E.; Gabelica, Z. Synthesis and characterization of new CoAPSO-40 and ZnAPSO-40 molecular sieves. Influence of the composition on the thermal and hydrothermal stability of $AlPO_4$-40-based materials. *Microporous Mesoporous Mater.* **2000**, *38*, 267–278. [CrossRef]
39. Finger, G.; Kornatowski, J.; Lutz, W.; Heidemann, D.; Schultze, D. Burning out of di-n-propylamine template from MeAPO-31 materials studied by thermal analysis. *Thermochim. Acta* **2004**, *409*, 49–54. [CrossRef]

40. Raveton, M.; Aajoud, A.; Willison, J.C.; Aouadi, H.; Tissut, M.; Ravanel, P. Phototransformation of the Insecticide Fipronil: Identification of Novel Photoproducts and Evidence for an Alternative Pathway of Photodegradation. *Environ. Sci. Technol.* **2006**, *40*, 4151–4157. [CrossRef]
41. Zhao, H.; Hiragushi, K.; Mizota, Y. 27Al and 29Si MAS–NMR studies of structural changes in hybrid aluminosilicate gels. *J. Eur. Ceram. Soc.* **2002**, *22*, 1483–1491. [CrossRef]
42. Fenoll, J.; Hellín, P.; Flores, P.; Garrido, I.; Navarro, S. Fipronil decomposition in aqueous semiconductor suspensions using UV light and solar energy. *J. Taiwan Inst. Chem. Eng.* **2014**, *45*, 981–988. [CrossRef]
43. Ahmed, S.; Rasul, M.; Brown, R.; Hashib, M. Influence of parameters on the heterogeneous photocatalytic degradation of pesticides and phenolic contaminants in wastewater: A short review. *J. Environ. Manag.* **2011**, *92*, 311–330. [CrossRef] [PubMed]
44. Gaya, U.I.; Abdullah, A.H.; Hussein, M.Z.; Zainal, Z. Photocatalytic removal of 2,4,6-trichlorophenol from water exploiting commercial ZnO powder. *Desalination* **2010**, *263*, 176–182. [CrossRef]
45. Choi, J.-H.; Kim, Y.-H. Reduction of 2,4,6-trichlorophenol with zero-valent zinc and catalyzed zinc. *J. Hazard. Mater.* **2009**, *166*, 984–991. [CrossRef]
46. Ngim, K.K.; Crosby, D.G. Abiotic processes influencing fipronil and desthiofipronil dissipation in California, USA, rice fields. *Environ. Toxicol. Chem.* **2001**, *20*, 972–977. [CrossRef]

Article

Acid-Modified Clays for the Catalytic Obtention of 5-Hydroxymethylfurfural from Glucose

Vladimir Sánchez, María Dolores González, Pilar Salagre and Yolanda Cesteros *

Departament de Química Física i Inorgànica, Universitat Rovira i Virgili, C/Marcel·lí Domingo 1, 43007 Tarragona, Spain; vladimir.sanchez@urv.cat (V.S.); mdolores.gonzalez@urv.cat (M.D.G.); pilar.salagre@urv.cat (P.S.)
* Correspondence: yolanda.cesteros@urv.cat; Tel.: +34-977-559571

Abstract: 5-hydroxymethylfurfural (5-HMF) is an important platform molecule for the synthesis of high-added value products. Several synthesized clay materials, such as mesoporous hectorite and fluorohectorite, in addition to commercial montmorillonite K-10, have been acid modified by different methodologies to be applied as catalysts for the obtention of 5-HMF from glucose. The effects of the Brønsted and/or Lewis acidity, the reaction temperature and time, and the catalyst/glucose ratio on the conversion but especially on the selectivity to 5-HMF have been studied. By comparing the synthesized clays, the best selectivity to 5-HMF (36%) was obtained at 140 °C for 4 h with H-fluorohectorite because of the presence of strong Brønsted acid sites, although its conversion was the lowest (33%) due to its low amounts of Lewis acid sites. Different strategies, such as physical mixtures of montmorillonite K10, which contains high amounts of Lewis acid centers, with Amberlyst-15, which has high amounts of Brønsted acid sites, or the incorporation of rhenium compounds, were carried out. The best selectivity to 5-HMF (62%) was achieved with a mixture of 44 wt % Amberlyst-15 and 56 wt % of montmorillonite K10 for a 56% of conversion at 140 °C for 4 h. This proportion optimized the amount of Brønsted and Lewis acid sites in the catalyst under these reaction conditions.

Keywords: 5-hydroxymethylfurfural; clays; glucose; acidity; Brønsted; Lewis

Citation: Sánchez, V.; González, M.D.; Salagre, P.; Cesteros, Y. Acid-Modified Clays for the Catalytic Obtention of 5-Hydroxymethylfurfural from Glucose. *ChemEngineering* **2022**, *6*, 57. https://doi.org/10.3390/chemengineering6040057

Academic Editors: Miguel A. Vicente, Raquel Trujillano and Francisco Martín Labajos

Received: 18 June 2022
Accepted: 18 July 2022
Published: 26 July 2022

Publisher's Note: MDPI stays neutral with regard to jurisdictional claims in published maps and institutional affiliations.

Copyright: © 2022 by the authors. Licensee MDPI, Basel, Switzerland. This article is an open access article distributed under the terms and conditions of the Creative Commons Attribution (CC BY) license (https://creativecommons.org/licenses/by/4.0/).

1. Introduction

The development of technologies to produce energy and chemicals from renewable resources, as an alternative to petroleum-derived products, has prompted biomass valorization [1–3]. Lignocellulose, the most abundant renewable biomass, is considered the main raw material in a biorefinery concept. Lignocellulose consists of three types of polymers: cellulose (40–50%), hemicellulose (25–35%) and lignin (15–20%). The conversion of biomass by fractionation into functionalized platform molecules, which includes sugars (glucose, xylose), polyols, furans (furfural, 5-hydroxymethylfurfural) and acids (levulinic) allows the production of a wide range of biofuels and chemicals [1].

Among the different platform molecules that can be obtained from biomass, 5-hydroxymethylfurfural (5-HMF) is one of the most versatile due to the presence in the same molecule of an aromatic aldehyde group, an alcohol and a furan ring that can be converted to high-added value products, such as bioplastics, biofuels or other chemicals [4]. 5-HMF is usually synthesized from glucose through two consecutive acid catalyzed reactions: isomerization of glucose to fructose catalyzed by Lewis acid sites and dehydration of fructose to 5-HMF catalyzed by Brønsted acid sites (Scheme 1) [5,6].

For the first isomerization step, which is the most difficult due to the stability of the six-carbon glucose ring, enzymatic, basic and Lewis acid catalysts have been tested [7]. Regarding basic catalysts, the use of as-synthesized, calcined and rehydrated Mg/Al-hydrotalcites led to high conversion values of glucose (73%) [8]. However, these catalytic systems required low glucose concentration and a long time of reaction. With respect to Lewis acid catalysts, $CrCl_2$ in a liquid ionic solvent resulted in 68% of yield at 100 °C after 3 h, while other Cu(II) or Fe(II) salts presented worse results [9].

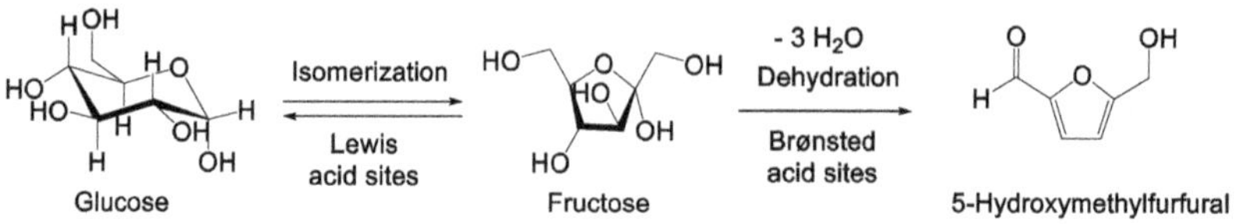

Scheme 1. Mechanism for the catalytic obtention of 5-HMF from glucose.

Heterogeneous catalysts have been also applied as an alternative to homogeneous catalysts due to their reusability. Sn-Beta zeolite presented interesting results related to the presence of SnO_2 as extra-framework species [10]. Davis et al. found that Sn-beta could efficiently catalyze the isomerization of glucose to fructose and then convert fast the generated fructose to 5-HMF with HCl at low pH [10,11]. The authors concluded that Sn^{4+} ions were active as Lewis acid centers for the glucose isomerization to fructose, while the partially hydrolyzed Sn–OH groups acted as Brønsted acid sites for the dehydration of fructose. A total amount of 5% Sn/SAPO-34 zeolite prepared by impregnation showed 98.5% of glucose conversion with a 64.4% of 5-HMF yield at 150 °C after 1.5 h of reaction [12]. $Nb_2O_5 \cdot nH_2O$ catalysts led to 49% yield of 5-HMF at 160 °C after 110 min of reaction time when H_3PO_4 was added to the catalytic system because of the presence of Brønsted acid sites [13]. In this way, the 76.3% yield of 5-HMF obtained by using $Ag_3PW_{12}O_{40}$ at 130 °C after 1.5 h of reaction was attributed to the synergistic effect between the Lewis and Brønsted acid sites present in the catalyst [14,15]. A carbon-based solid catalyst prepared from crystalline cellulose by carbonization and later sulfonation has been tested, achieving good catalytic results (73% of glucose conversion and 65% of selectivity to 5-HMF) [16]. Moreover, one sulfonated carbon obtained from an active carbon resulted in total glucose conversion and 93% of selectivity to 5-HMF at 160 °C after 3 h of reaction [17].

More recently, Pd nanoparticles supported in a highly acidic ZrO_2 exhibited 55% of glucose conversion with 74% of selectivity to 5-HMF after 3 h of reaction at 160 °C. The high catalytic activity was related to the balance between Brønsted and Lewis acid sites, in combination with the intrinsic activity of Pd species [18]. X. Li et al. concluded that the increase in Lewis acid sites promoted glucose dehydration, while Brønsted acid sites had a detrimental effect on glucose isomerization, achieving selectivity to 5-HMF of 71% for 92.6% of conversion using silica–alumina composite catalysts at optimized conditions [19]. However, a recent review on the topic remarked that more studies are needed to analyze and understand the role of Brønsted/Lewis acid sites present in the catalytic systems on the reaction [20].

Cationic clay minerals are layered microporous materials containing negatively charged aluminosilicate sheets, which further contain cations in their interlayer space to balance the charge and water molecules. Hectorites are cationic clays of the smectite group, with formula $M^{n+}_{x/n}[(Mg_{6-x}Li_x)Si_8O_{20}(OH)_4]$. When M^{n+} is replaced by H^+, the resulting clay has Brønsted acidity. The acidity of hectorites can be improved by delamination, which increases the accessibility to the acid sites [21], or by the substitution of the –OH groups by –F, leading to fluorohectorite [22]. There are few references about the use of clays for the obtention of 5-HMF from glucose. Sn-montmorillonite and Nb-montmorillonite led to high conversion, high yield and moderate-high selectivity values to 5-HMF (54.4% and 71.2%, respectively) [23,24]. The catalytically active sites in the Nb–OH groups favored glucose isomerization and subsequent dehydration with a 70.52% yield of 5-HMF at 170 °C for 3 h. Attapulgite modified by phosphoric acid showed 56.6% of 5-HMF yield in a 2-butanol-water biphasic system at 170 °C after 3 h of reaction. The results were again attributed to the proper amount of Brønsted and Lewis acid sites [25].

The aim of this work was to study the catalytic activity of several synthesized clay materials, such as mesoporous hectorite and fluorohectorite, in addition to commercial montmorillonite K-10, which were acid modified by different methodologies, including the preparation of physical mixtures with commercial sulfonic macroporous resin Amberlyst-15, or the incorporation of rhenium cations, to be applied as catalysts for the catalytic obtention of 5-hydroxymethylfurfural from glucose. The effect of the BrØnsted and/or Lewis acidity, the reaction temperature and time, and the ratio catalyst/glucose on the conversion and selectivity results were studied.

2. Experimental Section

2.1. Materials

MgO (97%), SiO_2 (99%), trimethyldodecylammonium chloride (98%) NH_4NO_3 (98%) and LiOH (98%) were supplied by Sigma Aldrich, LiF (98%) by Acros Oganics, $MgCl_2$ (99%), MgF_2 (99%) and NH_4ReO_4 (99%) by Alfa Aesar and 1-butanol (99.5%) by Scharlau.

Commercial montmorillonite K10 (Si/Al = 2.7) and Amberlyst-15 were supplied by Sigma-Aldrich (St. Louis, MO, USA).

2.2. Preparation of Catalysts

Na^+ mesoporous hectorite was synthesized as delaminated hectorite (Na-DH) using trimethyldodecylammonium chloride as template, following a method previously developed by our research group [21,26].

Fluorohectorite (Li-FH), with formula $Li_{0.7}[(Mg_{5.3}Li_{0.7})Si_8O_{20}F_4]$, was synthesized by mixing sintered SiO_2, sintered MgO, commercial LiF and commercial MgF_2 in a 8:4:2:2 molar ratio, respectively [22]. The mixture was homogenized following a sequence of different methods: grounding, suspending in acetone under an ultrasounds bath (Selecta) for 20 min and finally evaporating the solvent and drying the solid in an oven at 80 °C overnight. The resulting solid was then heated in a conventional muffle furnace at 800 °C for 3 h. The product was purified by selective sedimentation in a centrifuge (Hettich Zentrifugen Rotofix 32 A). The solid was suspended in about 50 mL of water and stirred overnight at room temperature. Then, it was centrifuged at 600 rpm for 6 min, and the supernatant was centrifuged at 4000 rpm for 30 min. The settled solid was dried and stored at room temperature.

Commercial montmorillonite K10 (MK10) was also tested as a catalyst for comparison.

H^+ delaminated hectorite (H-DH) and H^+ Fluorohectorite (H-FH) were prepared by the ionic exchange of previously synthesized Na-DH and Li-FH, respectively, with 1 M NH_4NO_3 aqueous solution by refluxing for 1 h. Then, the suspension was centrifuged, and the resultant solid was calcined at 540 °C for 3 h and stored at room temperature.

Several physical mixtures of montmorillonite K10 (MK10) and Amberlyst-15 (A) were prepared with different A-MK10 wt ratios: 0.4 (28 wt % A: 72 wt % MK10), 0.8 (44 wt % A: 56 wt % MK10) and 1 (50 wt % A: 50 wt % MK10).

MK10 was impregnated with an appropriate aqueous-ethanol solution of NH_4ReO_4 (with the minimum amount of water) to obtain 5 wt % of rhenium in the final sample. After rota evaporation of the solvents, the solid was calcined at 300 °C for 3 h (Re_2O_7/H-DH, Re_2O_7/MK10). The temperature of calcination of NH_4ReO_4 was chosen to prevent sublimation of NH_4ReO_4 at temperatures above 300 °C [27]. Other groups of samples were then prepared by mixing Amberlyst-15 with Re_2O_7/MK10 at the same A-Re_2O_7/MK10 ratios as those of A-MK10 mixtures. Finally, Re_2O_7, obtained by calcining NH_4ReO_4 at 300 °C for 3 h, was also tested alone for comparison.

Table 1 summarizes the different methodologies used for the acid modification of the clays.

Table 1. Methodologies for the acid modification of the clays.

Starting Clay	Acid Modification	Sample
Na-DH	NH_4^+ cation exchange + calcination	H-DH
Li-FH	NH_4^+ cation exchange + calcination	H-FH
MK10	Physical mixtures with Amberlyst-15 (A)	28 wt % A: 72 wt % MK10 44 wt % A: 56 wt % MK10 50 wt % A: 50 wt % MK10
MK10	Impregnation with NH_4ReO_4 to obtain 5 wt % of Re + calcination	Re_2O_7/MK10
H-DH	Impregnation with NH_4ReO_4 to obtain 5 wt % of Re + calcination	Re_2O_7/H-DH
Re_2O_7/MK10	Physical mixtures with Amberlyst-15 (A)	28 wt % A: 72 wt % Re_2O_7/MK10 44 wt % A: 56 wt % Re_2O_7/MK10 50 wt % A: 50 wt % Re_2O_7/MK10

2.3. Characterization of Catalysts

X-ray diffraction was used to identify and quantify the crystalline phases present in the catalytic precursors and catalysts. The experiments were carried out with a Siemens D5000 diffractometer (Bragg–Brentano parafocusing geometry and vertical θ–θ goniometer) fitted with a curved graphite diffracted-beam monochromator and diffracted-beam Soller slits, a 0.06° receiving slit and scintillation counter as a detector. The angular 2θ diffraction range was between 5 and 70°. The sample was dusted onto a low background Si (510) sample holder. The data were collected with an angular step of 0.05° at 3 s per step and sample rotation. CuKα radiation was obtained from a copper X-ray tube operated at 40 kV and 30 mA. The JCPDS files used for the identification of the crystalline phases were 00-003-0168, 01-075-0909 and 01-073-5680, for hectorite, anthophyllite and fluorohectorite, respectively.

BET surface areas were calculated from the nitrogen adsorption isotherms at −196 °C using a Quantachrome Quadrasorb SI surface analyzer and a value of 0.164 nm^2 for the cross-section of the nitrogen molecule.

The Brønsted acid capacity of the physical mixtures and their starting compounds was measured through the determination of cation exchange capacities using aqueous sodium chloride (2 M) solutions as a cationic exchange agent. The released protons were then potentiometrically titrated [28].

2.4. Catalytic Activity

Catalytic isomerization of commercial glucose (Sigma-Aldrich) was tested in a 100 mL stainless-steel autoclave equipped with an electronic temperature controller and a mechanical stirrer. Several reaction parameters were studied: temperature and time of reaction and catalyst/glucose ratio. The reaction conditions tested are summarized in Table 2.

Table 2. Reaction conditions tested for the obtention of 5-HMF from glucose.

Reaction Parameter	Reaction Conditions
Temperature	100, 120, 140, 160 and 180 °C
Time	1, 4 and 24 h
Catalyst wt /glucose wt (ratio)	0.6 g/2.4 g (0.25) 0.3 g/2.4 g (0.125) 0.15 g/1.2 g (0.125) 0.3 g/0.6 g (0.5) 0.15 g/0.6 g (0.25) 0.075 g/0.6 g (0.125)

In a typical experiment, glucose was dissolved in 50 mL of a mixture of THF:H_2O (35:15), and a fresh catalyst was transferred to the reactor. The temperature of the reaction was set with stirring of 600 rpm. At the end of the reaction, the reaction mixture was cooled, and the reaction products were separated from the catalyst by microfiltration.

Kit fructose/glucose assay, provided by Cygic Biocon SL, was used for determining the conversion of glucose and the selectivity to fructose.

Conversion of glucose (%) = (Number of moles of converted glucose)/(Number of moles of starting glucose).

Selectivity to fructose (%) = (Number of moles of glucose converted to fructose)/(Number of moles of converted glucose).

GC measurements were performed on Shimadzu GC-2010A series equipped with AOC-20i Series autoinjector and FID. The column was a Suprawax-280 (60 m × 0.25 mm × 0.50 µm). 1-Butanol was the internal standard. The quantification of products was determined based on GC data using the internal standard method in order to determine the selectivity to the desired product calculated as indicated below.

Selectivity to 5-HMF (%) = (Number of moles of glucose converted to 5-HMF)/(Number of moles of converted glucose).

3. Results and Discussion

3.1. Characterization of Catalysts

Table 3 summarizes the main characterization results obtained for the synthesized clays and for commercial montmorillonite K10.

Table 3. Characterization of the clays.

Catalyst	Crystalline Phases (XRD)	BET Area [a] (m^2/g)	Acidity [b] (meq/g)
MK10	Montmorillonite	233	—
Na-DH	Hectorite	327	0.90
H-DH	Hectorite	334	0.67
Li-FH	Fluorohectorite, anthophyllite	17	0.23
H-FH	Fluorohectorite, anthophyllite	21	0.53

[a] Calculated from N_2 physisorption results. [b] Obtained by NH_3-TPD. Data from [22].

The XRD patterns of the mesoporous hectorites (Na-DH and H-DH) showed the presence of the peaks corresponding to the crystalline hectorite with the exception of the (0 0 1) reflection, related to layer stacking, as expected, due to delamination (e.g., Figure 1a). The high BET surface areas were similar to those obtained for delaminated hectorites previously prepared in our research group [21]. The mesopores correspond to the interparticle space of lamellar-shaped crystallites, aggregated by edge-to-face bonding.

Li-FH and its protonated form (H-FH) presented the main crystalline phase, identified as fluorohectorite, with a well-defined (001) reflection at $2\Theta = 7°$ indicating a high order of the lamellar material in the stacking direction. In addition, anthophyllite and quartz phases were detected in lower amounts. After purification, quartz was eliminated, and anthophyllite was reduced, obtaining practically pure fluorohectorite (e.g., Figure 1b). These samples had much lower surface areas (17–21 m^2/g) than delaminated hectorites due to their microporosity, high crystallinity and the sinterization suffered during calcination at high temperature. Additionally, the residual anthophyllite phase could partially block the pores of fluorohectorite samples, contributing to lowering their surface area.

For the samples prepared with rhenium, it was not possible to detect the presence of rhenium compounds by XRD, probably due to the low amount loaded (5 wt %). Considering the calcination temperature used, the presence of Re_2O_7 should be expected, although according to other studies found in the literature, there could also be a mixture of Re^{6+}/Re^{7+} oxides [29].

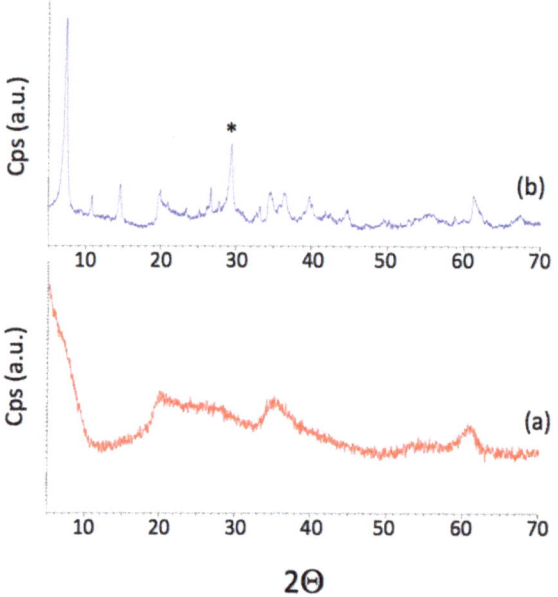

Figure 1. XRD patterns of several synthesized clays (**a**) delaminated hectorite (Na-DH) and (**b**) Li-fluorohectorite (Li-FH). * anthophyllite.

In a previous work, we evaluated the acidity (amount and strength) of this type of clay materials by NH_3-TPD [22]. The amount of acid sites is shown in Table 3. Na-DH had a higher amount of acid sites than H-DH because of the presence of higher amounts of Lewis acid sites (Na^+) but lower acidity strength, since the NH_3 desorption temperature was lower for Na-DH (190 °C) than for H-DH (224 °C) [22]. This confirms the higher strength of the Brønsted acid sites of the H-DH sample. The amount of acid sites of Li-FH was the lowest and can be mainly attributed to the Lewis acid sites (Li^+). H-FH showed a higher number of acid sites than Li-FH (Table 3), and, interestingly, very strong acidity, since the NH_3 desorption temperature for its main desorption peak was 721° C [22]. The effect of the partial substitution of –OH groups by F in the hectorite structure favored an increase in the acidity of the Brønsted acid sites (H^+) of the H-FH through an inductive effect of fluorine.

Commercial montmorillonite K10 is a montmorillonite, which was submitted to an acid treatment. This involves: (a) some dealumination of the structure, and therefore, the presence of Lewis acid sites due to extra framework cations; (b) higher surface area than montmorillonite because of the generation of some mesoporosity; and (c) the presence of some amount of Brønsted acid sites due to the acidic medium used during the treatment. For the Re-containing samples, the rhenium cations act as Lewis acid centers.

The Brønsted acidity of the A-MK10 and A-Re_2O_7/MK10 physical mixtures was evaluated by titration and compared to their starting compounds alone. The results are shown in Table 4.

Amberlyst-15 showed the highest amount of Brønsted acid sites, as expected, due to the sulfonic groups present in this macroporous resin, while Re_2O_7 and MK10 had the lowest, since they mainly had Lewis acidity, as commented above. The physical mixtures increased the values of Brønsted acid sites with respect to MK10 or Re_2O_7/MK10 due to the presence of Amberlyst-15. These results will be later correlated with the catalytic results.

Table 4. Characterization of the Brønsted acidity of the physical mixtures and starting compounds.

Catalyst	Brønsted Acidity (meq H^+/100g)
Re_2O_7	5.0
MK10	15.4
Re_2O_7/ MK10	27.2
A	478.8
28 wt % A: 72 wt % MK10	113.6
44 wt % A: 56 wt % MK10	190.4
50 wt % A: 50 wt % MK10	251.6
28 wt % A: 72 wt % Re2O7/ MK10	137.7
44 wt % A: 56 wt % Re2O7/ MK10	229.2
50 wt % A: 50 wt % Re2O7/ MK10	232.8

3.2. Catalytic Activity

First, the effect of the reaction temperature and the catalyst/glucose weight ratio were studied with commercial montmorillonite K-10 (MK10) in order to find the optimized conditions to check the rest of the catalysts. All studies were conducted with a solvent mixture of THF:H_2O in a volume ratio of 35:15. From several preliminary experiments with different solvents and mixtures of solvents, this was found to be the best solvent to favor the obtention of 5-HMF. Figure 2 shows the effect of the reaction temperature using 2.4 g of glucose and 0.3 g of MK10 catalyst for 1 h of reaction.

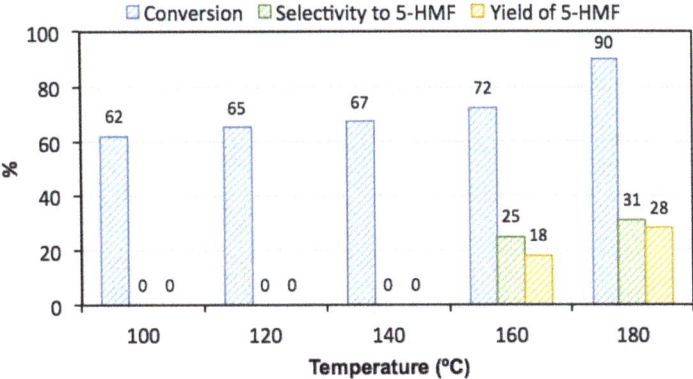

Figure 2. Effect of the reaction temperature on the catalytic activity of the catalyst MK10. Reaction conditions: 2.4 g of glucose, 0.3 g of catalyst, catalyst/glucose wt ratio = 0.125, solvent volume ratio (THF:H_2O) = 35:15 and reaction time = 1 h.

The results showed an increase in the conversion values with the temperature, as expected, and the formation of the desired product, 5-hydroxymethylfurfural (5-HMF), from 160 °C. No other reaction products, such as levulinic acid or formic acid, were detected by gas chromatography. Therefore, the other reaction products should be related to the polymerization of glucose, fructose and 5-HMF that result in the formation of condensation products. These undesired products easily formed from 100 °C, while isomerization reaction was slower and needed higher temperatures or, as concluded below, longer times.

The effects of using different catalyst/glucose weight ratios on the conversion and selectivity to 5-HMF values were then studied with the MK10 catalyst at 180 °C for 1 h (Figure 3).

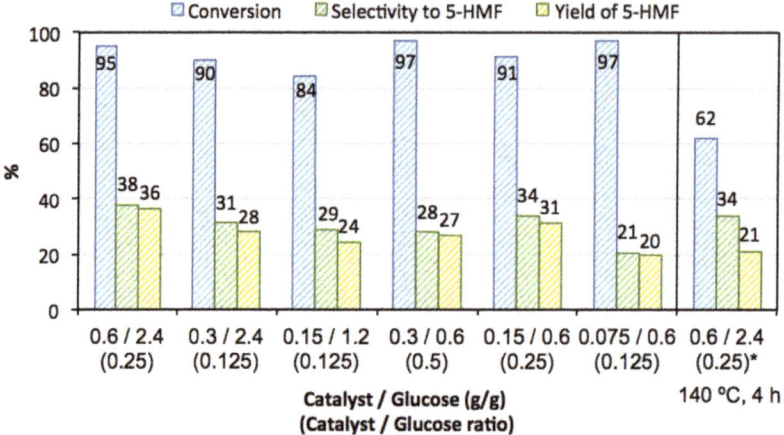

Figure 3. Effect of the different catalyst/glucose wt ratios on the catalytic activity of the catalyst MK10. Reaction conditions: solvent volume ratio (THF:H$_2$O) = 35:15, temperature = 180 °C and reaction time = 1 h. * Reaction temperature = 140 °C and Reaction time = 4 h.

On the whole, there are not significant differences between the catalytic tests. The highest selectivity to 5-HMF (38%) and the highest yield of 5-HMF (36%) were observed when using 0.6 g of catalyst and 2.4 g of glucose with a catalyst/glucose ratio of 0.25. This was the optimized ratio, and the second-best result was achieved with the same ratio using lower amounts of catalyst and glucose. One more catalytic test with catalyst MK10 was performed at the optimized catalyst/glucose ratio, decreasing the reaction temperature (140 °C) and increasing the reaction time (4 h) (Figure 3). The selectivity to 5-HMF was maintained (34%) at moderate conversion (62%).

We selected these reaction conditions—0.6 g of catalyst, 2.4 g of glucose, 140 °C and 4 h—to test catalysts with different total acidity and different presence of Brønsted and/or Lewis acid sites. The corresponding results are shown in Figure 4.

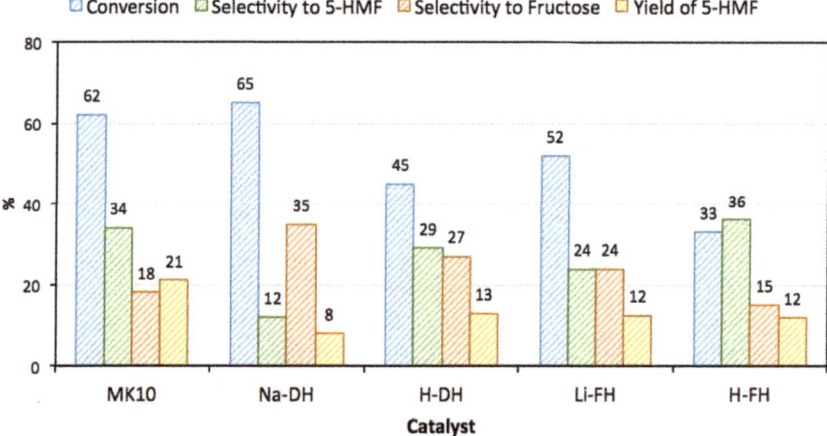

Figure 4. Catalytic activity of the commercial montmorillonite K10 and the synthesized and modified cationic clays. Reaction conditions: 2.4 g of glucose, 0.6 g of catalyst, catalyst/glucose wt ratio = 0.25, solvent volume ratio (THF:H$_2$O) = 35:5, temperature = 140 °C and reaction time = 4 h.

The catalysts with higher amounts of acid centers, especially Lewis acid centers, and higher surface areas (MK10 and Na-DH) (Table 3) showed higher conversion values, as expected, due to the importance of Lewis centers for the first isomerization step from glucose to fructose (Scheme 1). By comparing the selectivity to 5-HMF and to fructose obtained for these two catalysts, the lower amount of Brønsted acid sites of Na-DH justifies its lower selectivity to 5-HMF but higher selectivity to fructose, while the Brønsted acid sites remaining in the K10 montmorillonite, after the acid treatment to which it was submitted to obtain it, favored the transformation of fructose to 5-HMF, the second step of the reaction (Scheme 1).

H-DH had lower amounts of Lewis acid sites and higher amounts of Brønsted acid sites than Na-DH, and consequently, lower conversion but higher selectivity to 5-HMF was obtained when compared to Na-DH. For the fluorohectorite (Li-DH), the conversion was lower, and the selectivity to 5-HMF was higher than for the Na-DH catalyst. The lower amount of acid centers of Li-DH (Table 3) justifies its lower conversion, although it was higher than for H-DH, probably because of the higher amounts of Lewis acid centers due to cations Li^+. The higher selectivity to 5-HMF compared to Na-DH should be related to the presence of some amounts of Brønsted acid sites due to the hydrolysis of the Li^+ cations. Interestingly, the highest selectivity to 5-HMF for this group of catalysts was achieved with the protonated fluorohectorite (H-FH) due to its stronger Brønsted acid sites, as commented above, although its conversion was the lowest due to the low amounts of Lewis acid sites of this catalyst.

Taking into account that the montmorillonite K10 showed the best yield of 5-HMF and one of the best selectivity values to 5-HMF, in order to improve these results, we planned to modify its acidity by preparing several physical mixtures of MK10 with sulfonic macroporous resin Amberlyst-15 (A), which had higher amounts and stronger Brønsted acidity, in order to favor the second step of the reaction, the formation of 5-HMF from fructose (Scheme 1). Figure 5 shows the catalytic activity results obtained for these A-MK10 mixtures with different wt % ratios compared to the catalytic results of A and MK10 catalysts. The amount of Brønsted acid sites (meq H^+/100 g), determined by titration, is also indicated in the figure.

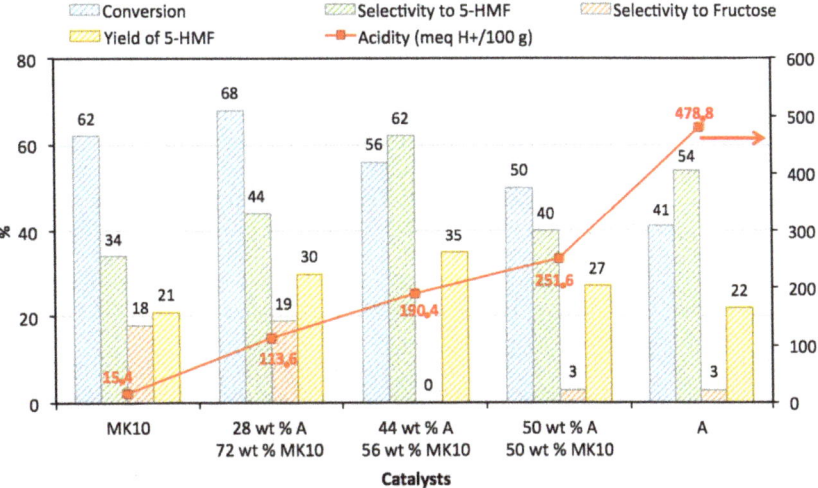

Figure 5. Catalytic activity of physical mixtures A-MK10 in different proportions compared to A and MK10 catalysts. Reaction conditions: 2.4 g of glucose, 0.6 g of catalyst mixture, catalyst/glucose wt ratio = 0.25, solvent volume ratio (THF:H_2O) = 35:15, temperature = 140 °C and reaction time = 4 h.

The catalytic activity of Amberlyst-15 (A) showed lower conversion but higher selectivity to 5-HMF and lower selectivity to fructose than that of catalyst MK10. This confirms again that Lewis acid sites are more active for the overall transformation of glucose, and

Brønsted acid sites are responsible for the selective conversion of fructose to 5-HMF. By mixing physically both catalysts in different wt % proportions, a synergetic effect was clearly observed. Thus, the presence of 28 wt % of Amberlyst-15 favored both an increase in conversion and selectivity to 5-HMF due to the stronger Brønsted acid sites provided by Amberlyst-15 (Table 4) but with a still significant contribution of the Lewis acid centers of MK10. On the other hand, the mixed catalyst with 50 wt % of A led to lower conversion than those of MK10 and the mixed catalyst with 28 wt % of A, and a slight increase in selectivity to 5-HMF with respect to MK10 but lower than for the mixed catalyst with 28 wt % of A (Figure 5). These results can be explained because of the covering of part of the Lewis acid sites of MK10 by the Brønsted acid sites of A. Then, an optimized A-MK10 ratio was searched, and it was found at the proportion 44 wt % of A and 56 wt % of MK10, obtaining a moderate glucose conversion of 56% and the highest selectivity to 5-HMF of 62%.

Another way to modify the acidity of the clay MK10 was to load a 5 wt % of rhenium to increase Lewis acidity and then to prepare mixtures with Amberlyst-15, with high amounts of strong Brønsted acidity. The catalytic results are shown in Figure 6.

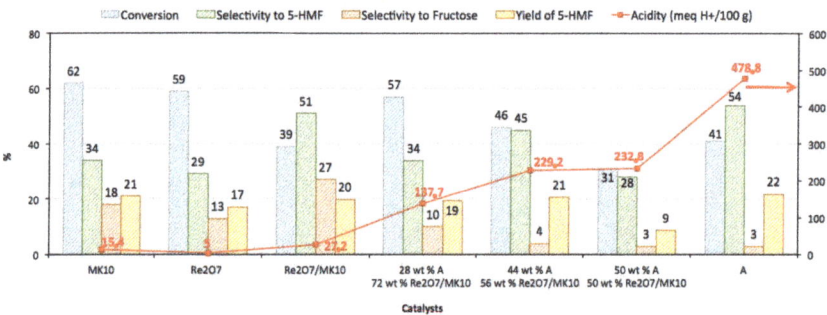

Figure 6. Catalytic activity of physical mixtures A-Re$_2$O$_7$/MK10 in different wt % proportions compared to A, MK10, Re$_2$O$_7$ and Re$_2$O$_7$/MK10 catalysts. Reaction conditions: 2.4 g of glucose, 0.6 g of catalyst mixture, catalyst/glucose wt ratio = 0.25, solvent volume ratio (THF:H$_2$O) = 35:15, temperature = 140 °C and reaction time = 4 h.

The incorporation of rhenium in MK10 increased the selectivity to fructose, with respect to the MK10 or Re$_2$O$_7$ alone, as expected due to its higher amount of Lewis acid sites, and it led to higher selectivity to 5-HMF (51%) than MK10 (34%) and Re$_2$O$_7$ (29%). This can be explained by the higher amount of Brønsted acidity of catalyst Re$_2$O$_7$/MK10, as determined by titration (Table 4, Figure 6), which could be related to some hydrolysis of the rhenium cations during the reaction. Finally, the decrease in conversion can be attributed to a decrease in the surface area of the catalyst because of the impregnation-calcination procedure employed.

From these results, the following attempt to improve the selectivity to 5-HMF was to mix this Re$_2$O$_7$/MK10 catalyst in different proportions with Amberlyst-15. Independently of the proportions used, the selectivity to 5-HMF did not improve the value obtained for catalyst Re$_2$O$_7$/MK10 (Figure 6). The mixture with the highest selectivity to 5-HMF (45%) was 44 wt % of Re$_2$O$_7$/MK10 and 56 wt % of Amberlyst-15. A possible explanation for this catalytic behavior is that the acid properties of Re$_2$O$_7$ were modified when mixed with Amberlyst-15. It is well known that rhenium cations can be protonated. This should decrease the amount and strength of Lewis acid sites of the mixtures.

Finally, the effect of the reaction time was studied for the best catalyst (44 wt % of A and 56 wt % of MK10) by comparing its catalytic activity at 4 and 24 h of reaction with those of its precursors, A and MK10, and also with catalyst Re$_2$O$_7$/MK10 (Figure 7).

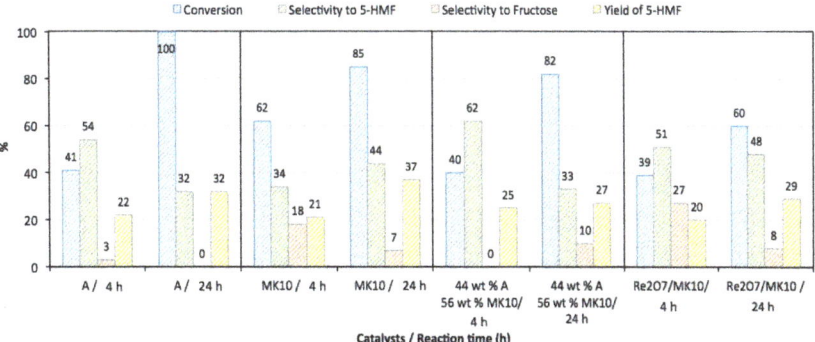

Figure 7. Effect of the reaction time, 4 h and 24 h, for several catalysts. Reaction conditions: 2.4 g of glucose, 0.6 g of catalyst mixture, catalyst/glucose wt ratio = 0.25, solvent volume ratio (THF:H_2O) = 35:15, temperature = 140 °C.

All catalysts showed an increase in conversion with the reaction time but, with the exception of MK10, this increase in conversion did not involve an increase in the selectivity to 5-HMF, and in general, a decrease in fructose was also observed. This means that the increase in conversion was at the expense of the formation of other reaction products, probably polymerization products, especially more favored for the catalysts that had higher amounts of stronger Brønsted acid sites due to Amberlyst-15. Interestingly, catalyst MK10 showed an increase of 10% in the selectivity to 5-HMF with time, and for Re_2O_7/MK10, only a slight decrease was observed (Figure 7). This behavior could be explained by the low amounts of acid centers present in these catalysts, which do not favor polymerization as much, and in the case of MK10 favor the slow transformation to 5-HMF with time.

Figure 8 shows the reproducibility of the catalytic results of the best catalysts, which were checked thrice. Higher standard deviation was observed when the physical mixture of Amberlyst-15 and montmorillonite K10 was tested. This can be related to its preparation procedure.

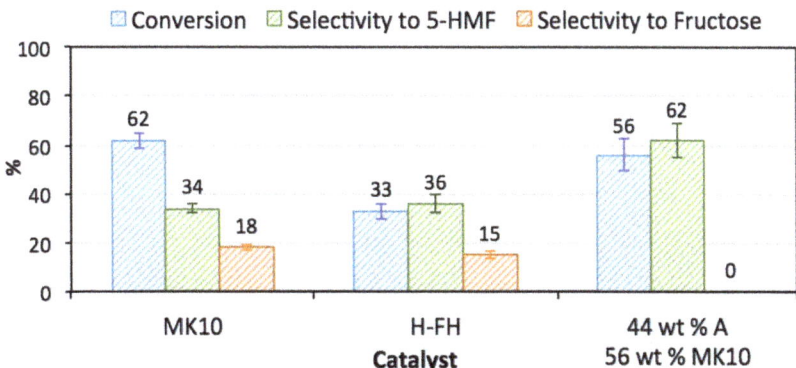

Figure 8. Reproducibility of the catalytic results of the best catalysts. Reaction conditions: 2.4 g of glucose, 0.6 g of catalyst mixture, catalyst/glucose wt ratio = 0.25, solvent volume ratio (THF:H_2O) = 35:15, temperature = 140 °C and reaction time = 4 h.

4. Conclusions

Several synthesized clays, such as delaminated Na-hectorite and Li-fluorohectorite, in addition to commercial montmorillonite MK-10, were acid modified by different methodologies: the incorporation of protons (H-hectorite and H-fluorohectorite), physical mixtures of montmorillonite MK-10 with Amberlyst-15 to combine Lewis and Brønsted acidity and the addition of rhenium compounds with Lewis acidity. The catalytic results confirmed the importance of combining Lewis and Brønsted acid centers in the appropriate amounts to favor the transformation of glucose to fructose by the Lewis acid sites and the subsequent conversion of fructose to 5-HMF by the Brønsted acid sites. The optimized reaction conditions were established in a reaction temperature of 140 °C, reaction time of 4 h, using 0.6 g of catalyst and 2.4 g of glucose, with a catalyst/glucose ratio of 0.25.

Regarding the synthesized clays, H-fluorohectorite showed the highest selectivity value to 5-HMF (36%) at 140 °C for 4 h that can be attributed to the strong Brønsted acid sites of this catalyst due to the incorporation of fluorine in the hectorite structure during fluorohectorite preparation. For the physical mixtures between Amberlyst-15, with high amounts of strong Brønsted acid sites, and montmorillonite K10, with Lewis acid centers, the best selectivity to 5-HMF (62%) was achieved with a mixture of 44 wt % Amberlyst-15 and 56 wt % of montmorillonite K10 for a 56% conversion. The incorporation of rhenium in MK10 increased the selectivity to fructose, with respect to the MK10 or Re_2O_7 alone, due to the higher amount of Lewis acid sites and led to higher selectivity to 5-HMF (51%) than MK10 (34%) and Re_2O_7 (29%). This has been related to the generation of Brønsted acidity in catalyst Re_2O_7/MK10 due to some hydrolysis of the rhenium cations during reaction. However, the physical mixtures of Re_2O_7/MK10 with Amberlyst-15 did not improve the selectivity to 5-HMF. Catalyst MK10 showed an increase of 10% in the selectivity toward 5-HMF when the reaction time was increased to 24 h. This was related to low amounts of acid centers in this catalyst, which, with time, do not favor polymerization as much and continue forming 5-HMF.

Author Contributions: Conceptualization, Y.C., P.S. and M.D.G.; validation, Y.C. and M.D.G.; investigation, V.S.; resources, Y.C. and M.D.G.; writing—original draft preparation, V.S., M.D.G. and Y.C.; writing—review and editing, Y.C.; supervision, Y.C., P.S. and M.D.G.; funding acquisition, Y.C. All authors have read and agreed to the published version of the manuscript.

Funding: This research was funded by the project PID2019-110735RB-C22 funded by MCIN/AEI/10.13039/501100011033.

Institutional Review Board Statement: Not applicable.

Informed Consent Statement: Not applicable.

Data Availability Statement: Not applicable.

Conflicts of Interest: The authors declare no conflict of interest.

References

1. Alonso, D.M.; Bond, J.Q.; Dumesic, J.A. Catalytic Conversion of Biomass to Biofuels. *Green Chem.* **2010**, *12*, 1493–1513.
2. Climent, M.J.; Corma, A.; Iborra, S. Conversion of Biomass Platform Molecules into Fuel Additives and Liquid Hydrocarbon Fuels. *Green Chem.* **2014**, *16*, 516–547.
3. Sheldon, R.A. Green and Sustainable Manufacture of Chemicals from Biomass: State of the Art. *Green Chem.* **2014**, *16*, 950–963.
4. Van Putten, R.J.; Van Der Waal, J.C.; De Jong, E.; Rasrendra, C.B.; Heeres, H.J.; De Vries, J.G. Hydroxymethylfurfural, a Versatile Platform Chemical Made from Renewable Resources. *Chem. Rev.* **2013**, *113*, 1499–1597.
5. Mukherjee, A.; Dumont, M.J.; Raghavan, V. Review: Sustainable Production of Hydroxymethylfurfural and Levulinic Acid: Challenges and Opportunities. *Biomass Bioenergy* **2015**, *72*, 143–183.
6. Rosatella, A.A.; Simeonov, S.P.; Frade, R.F.M.; Afonso, C.A.M. 5-Hydroxymethylfurfural (HMF) as a Building Block Platform: Biological Properties, Synthesis and Synthetic Applications. *Green Chem.* **2011**, *13*, 754–793.
7. Assary, R.S.; Curtiss, L.A. Comparison of Sugar Molecule Decomposition through Glucose and Fructose: A High-Level Quantum Chemical Study. *Energy Fuels* **2012**, *26*, 1344–1352.
8. Yu, S.; Kim, E.; Park, S.; Song, I.K.; Jung, J.C. Isomerization of Glucose into Fructose over Mg-Al Hydrotalcite Catalysts. *Catal. Commun.* **2012**, *29*, 63–67.
9. Pidko, E.A.; Degirmenci, V.; Hensen, E.J.M. On the Mechanism of Lewis Acid Catalyzed Glucose Transformations in Ionic Liquids. *ChemCatChem* **2012**, *4*, 1263–1271.
10. Nikolla, E.; Román-Leshkov, Y.; Moliner, M.; Davis, M.E. "One-Pot" Synthesis of 5-(Hydroxymethyl)Furfural from Carbohydrates Using Tin-Beta Zeolite. *ACS Catal.* **2011**, *1*, 408–410.

11. Moliner, M.; Roman-Leshkov, Y.; Davis, M.E. Tin-containing zeolites are highly active catalysts for the isomerization of glucose in water. *Proc. Natl. Acad. Sci. USA* **2010**, *107*, 6164–6168.
12. Song, X.; Yue, J.; Zhu, Y.; Wen, C.; Chen, L.; Liu, Q.; Ma, L.; Wang, C. Efficient Conversion of Glucose to 5-Hydroxymethylfurfural over a Sn-Modified SAPO-34 Zeolite Catalyst. *Ind. Eng. Chem. Res.* **2021**, *60*, 5838–5851.
13. Nakajima, K.; Baba, Y.; Noma, R.; Kitano, M.; Kondo, J.N.; Hayashi, S.; Hara, M. $Nb_2O_5 \cdot nH_2O$ as a Heterogeneous Catalyst with Water-Tolerant Lewis Acid Sites. *J. Am. Chem. Soc.* **2011**, *133*, 4224–4227.
14. Fan, C.; Guan, H.; Zhang, H.; Wang, J.; Wang, S.; Wang, X. Conversion of Fructose and Glucose into 5-Hydroxymethylfurfural Catalyzed by a Solid Heteropolyacid Salt. *Biomass Bioenergy* **2011**, *35*, 2659–2665.
15. Wang, J.; Xi, J.; Xia, Q.; Liu, X.; Wang, Y. Recent Advances in Heterogeneous Catalytic Conversion of Glucose to 5-Hydroxymethylfurfural via Green Routes. *Sci. China Chem.* **2017**, *60*, 870–886.
16. Zou, B.; Chen, X.; Zhou, C.; Yu, X.; Ma, H.; Zhao, J.; Bao, X. Highly-efficient and Low-cost Synthesis of 5-hydroxymethylfurfural from Monosaccharides. *Can. J. Chem. Eng.* **2018**, *96*, 1337.
17. Nahavandi, M.; Kasanneni, T.; Yuan, Z.S.; Xu, C.C.; Rohani, S. Efficient Conversion of Glucose into 5-Hydroxymethylfurfural Using a Sulfonated Carbon-Based Solid Acid Catalyst: An Experimental and Numerical Study. *ACS Sustain. Chem. Eng.* **2019**, *7*, 11970–11984.
18. Goyal, R.; Abraham, B.M.; Singh, O.; Sameer, S.; Bal, R.; Mondal, P. One-pot transformation of glucose into hydroxymethyl furfural in water over Pd decorated acidic ZrO_2. *Renew. Energy* **2022**, *183*, 791–801.
19. Li, X.; Peng, K.; Liu, X.; Xia, Q.; Wang, Y. Comprehensive Understanding of the Role of Brønsted and Lewis Acid Sites in Glucose Conversion into 5-Hydromethylfurfural. *ChemCatChem* **2017**, *9*, 2739–2746.
20. Megías-Sayago, C.; Navarro-Jaén, S.; Drault, F.; Ivanova, S. Recent Advances in the Brønsted/Lewis Acid Catalyzed Conversion of Glucose to HMF and Lactic Acid: Pathways toward Bio-Based Plastics. *Catalysts* **2021**, *11*, 1395.
21. Sánchez, T.; Salagre, P.; Cesteros, Y.; Bueno-López, A. Use of delaminated hectorites as supports of copper catalysts for the hydrogenolysis of glycerol to 1,2-propanediol. *Chem. Eng. J.* **2012**, *179*, 302–311.
22. Sánchez, V.; Dafinov, A.; Salagre, P.; Llorca, J.; Cesteros, Y. Microwave-Assisted Furfural Production Using Hectorites and Fluorohectorites as Catalysts. *Catalysts* **2019**, *9*, 706.
23. Wang, J.; Ren, J.; Liu, X.; Xi, J.; Xia, Q.; Zu, Y.; Lu, G.; Wang, Y. Direct Conversion of Carbohydrates to 5-Hydroxymethylfurfural Using Sn-Mont Catalyst. *Green Chem.* **2012**, *14*, 2506–2512.
24. Qiu, G.; Huang, C.; Sun, X.; Chen, B. Highly active niobium-loaded montmorillonite catalysts for the production of 5-hydroxymethylfurfural from glucose. *Green Chem.* **2019**, *21*, 3930–3939.
25. Yang, F.L.; Weng, J.S.; Ding, J.J.; Zhao, Z.Y.; Qin, L.Z.; Xia, F.F. Effective conversion of saccharides into hydroxymethylfurfural catalyzed by a natural clay, attapulgite. *Renew. Energy* **2020**, *151*, 829–836.
26. Sánchez, T.; Salagre, P.; Cesteros, Y. Ultrasounds and microwave-assisted synthesis of mesoporous hectorites. *Microporous Mesoporous Mater.* **2013**, *171*, 24–34.
27. Plaza de los Reyes, J.; Cid, R.; Pecchi, G.; Reyes, P. Platinum-Rhenium Catalysts. I. Thermal Study of Precursor Salts. *J. Chem. Res.* **1983**, *12*, 318.
28. Melero, J.A.; Stucky, G.D.; van Grieken, R.; Morales, G. Direct Syntheses of Ordered SBA-15 Mesoporous Materials Containing Arenesulfonic Acid Groups. *J. Mater. Chem.* **2002**, *12*, 1664–1670.
29. Chuseang, J.; Nakwachara, R.; Kalong, M.; Ratchahat, S.; Koo-Amornpattana, W.; Klysubun, W.; Khemthong, P.; Faungnawakij, K.; Assabumrungrat, S.; Itthibenchapong, V.; et al. Selective Hydrogenolysis of Furfural into Fuel-Additive 2-Methylfuran over a Rhenium-Promoted Copper Catalyst. *Sustain. Energy Fuels* **2021**, *5*, 1379–1393.

Article

Effect of Artificial Freeze/Thaw and Thermal Shock Ageing, Combined or Not with Salt Crystallisation on the Colour of Zamora Building Stones (Spain)

Jacinta García-Talegón [1,*], Adolfo Carlos Iñigo [2], Rosa Sepúlveda [3] and Eduardo Azofra [4]

1. Departamento de Geología, Universidad de Salamanca, 37008 Salamanca, Spain
2. Instituto de Recursos Naturales y Agrobiología de Salamanca (IRNASA), CSIC, 37008 Salamanca, Spain
3. Departamento de Estadística, Universidad de Salamanca, 37007 Salamanca, Spain
4. Departamento Historia del Arte, Universidad de Salamanca, 37001 Salamanca, Spain
* Correspondence: talegon@usal.es; Tel.: +34-666-58-56-59

Citation: García-Talegón, J.; Iñigo, A.C.; Sepúlveda, R.; Azofra, E. Effect of Artificial Freeze/Thaw and Thermal Shock Ageing, Combined or Not with Salt Crystallisation on the Colour of Zamora Building Stones (Spain). *ChemEngineering* **2022**, *6*, 61. https://doi.org/10.3390/chemengineering6040061

Academic Editors: Akira Otsuki, Alírio E. Rodrigues, Miguel A. Vicente, Raquel Trujillano and Francisco Martín Labajos

Received: 5 June 2022
Accepted: 28 July 2022
Published: 4 August 2022

Publisher's Note: MDPI stays neutral with regard to jurisdictional claims in published maps and institutional affiliations.

Copyright: © 2022 by the authors. Licensee MDPI, Basel, Switzerland. This article is an open access article distributed under the terms and conditions of the Creative Commons Attribution (CC BY) license (https://creativecommons.org/licenses/by/4.0/).

Abstract: After subjecting Zamora building stones to accelerated ageing tests, colour changes were studied, namely: (a) freezing/thawing and thermal shock (gelifraction and thermoclasty), and (b) combination of freezing/thawing plus thermal shock and salt crystallisation (sulphates or phosphates) (gelifraction, thermoclasty and haloclasty). Zamora building stones are silicified conglomerates (silcretes) from the Cretaceous that show marked colour changes due to the remobilisation of iron oxyhydroxides. In this work, four varieties were: white stone; ochreous stone; white and red stone; and purple stone Their micromorphological characterization (skeleton, weathering plasma and porosity/cutan) is formed of grains and fragments of quartz and quartzite as well as by accesory minerals muscovite and feldspar (more or less altered), and some opaque. Quartz, feldspar and illite/mica were part of the skeleton; kaolinite, iron oxyhydroxides, and CT opal were part of the weathering plasma or cutans; their porosity were 11.7–8.7%. Their chromatic data have been statistically analysed (MANOVA-Biplot). They showed higher variations in ΔE^*, ΔL^*, Δa^* and Δb^* on combined freezing/thawing plus thermal shock and sulphates crystallisation leading to rapid alteration of the building stones. Chromatic differences between the other two artificial ageing tests were less evident and were not detected in all samples. The global effect of ageing on the Zamora building stones darkened them and reduced their yellowing. The ochreous stone suffered the least variation and the purple stone the most. This study of the colour by statistical analyse may be of interest for the evaluation and monitoring of stone decay, which is an inexpensive, simple, easy and non-destructive technique.

Keywords: colour; building stones; silicified conglomerates; artificial ageing test; freezing/thawing; thermoclasty; salt crystallisation; MANOVA-Biplot

1. Introduction

Zamora is a "Romanesque city" with 24 Romanesque churches, the Castle, the Cathedral and City Walls from the 12th and 13th centuries built in silicified sandstones and conglomerates called Zamora building stones. These stones have been used in monuments of Zamora, in the masonry wall, and in the lower parts of religious and civil monuments of the cities of Salamanca and Avila, which were designated World Heritage Cities by UNESCO in 1988 and 1985 respectively [1]. Building stone monuments represent a significant part of the World Heritage, enclosing a historical, aesthetic and constructive material value. The architectural heritage of a city is strongly influenced by its geological context and by the place where the building stones were quarried in the surroundings of the urban centre. Thus, cities are often characterised by distinctive chromatic features that reflect those of the locally available building stones, as is the case with Villamayor sandstone in the city of Salamanca, whose natural alteration in monuments is accompanied by the development of a golden-reddish patina. Hence, the term 'golden sandstone' is one of the signatures of monuments and buildings constructed with this building stone [2,3].

The durability of building stone is its behaviour against decay processes that depend on the mineralogical composition, texture and structure of the stone (intrinsic factors), its location in the monument and environmental conditions (extrinsic factors). Its prediction is complex, given the variety of factors involved, and the great forms of decay that can occur [4]. Foreseeing the expected decay processes of stone materials is of paramount importance to anticipate problems arising from the accelerated ageing test; and currently climatic chambers under controlled conditions are used for this. In this work, different accelerated ageing tests simulating the conditions of the city of Zamora with a Mediterranean climate with a continental tendency and low polluted, with a mean annual temperature of 11–12 °C were carried out. The range of daily temperature fluctuations may be as large as 30 °C. The absolute maximum temperature measured based on a historical data series (15 years) is 37 °C, while the absolute minimum is -19 °C. Average annual rainfall is about 395 mm, mainly in spring and autumn, with July and August typically characterised by severe droughts [5–8].

On the exposed exteriors of monuments, thermoclasty and gelifraction are responsible for the decay of Zamora building stones (silicified conglomerates) leading to microfissures, plates, flakes, and surface arenisation. Additionally, outwards areas of the lower parts of the monuments, affected by capillary damping of polluted water, are subject to a synergistic effect between these factors and salt crystallization (haloclasty/gelifraction/thermoclasty). Inwards of monuments, the most extensive and commonly decay of silicified conglomerates is due to the presence of salts in microenvironments, which crystallise inside the pore network (sub-efflorescences) or outside it (efflorescences) [5–8].

The main water-soluble salts studied in the efflorescences and/or subefflorescences of historical stones of the buildings are composed of the following anions Cl^-, NO_3^-, SO_4^{2-}, CO_3^{2-}, PO_4^{3-} etc. The source of the salts may be multiple, from the mortar of the wall and from the outside e.g., bird droppings which, being very soluble, migrate easily to the surface. In the high parts of the building it is due to infiltration by rain off, which dissolves the soluble species in mortars, while in the lower parts the source of salts come from capillary damping (polluted urban groundwater) [9].

One of the aesthetic parameters of a building stone is its colour, which contributes greatly to its ornamental value. As with other properties, colour monitoring is of great importance to evaluate treatment effectiveness when the stone is subjected to conservation or restoration treatment [10–16], to assess changes by accelerated ageing tests [16–22] and to in situ deterioration in the building [3,23–29]. In most of the works cited, stone building materials, after the application of conservation treatments (waterproofing, consolidation, desalination, etc.), accelerated artificial ageing, or deterioration in situ in monuments, produce a greater darkness and more reddish and/or yellowish tones, depending on the case.

The purpose of this study was to quantify the colour changes in building stones that were submitted to three different accelerated ageing treatments (freezing/thawing and two combined freezing/thawing and salt crystallisation), to allow a rationale use of these building stones and to anticipate their behaviour in advance, in order to succeed in restoration. In the combined tests with salt crystallisation, sulphates were selected on the one hand, because they are soluble in water, which increases more in volume when crystallising, and their deterioration is greater through the so-called wedge effect (physical); and on the other hand, phosphates, because they give a very alcaline medium to the dissolution (pH > 11), and although the decay process is much slower, it would be by chemical dissolution, and the wedge effect (physical) would be lesser.

2. Materials and Methods

2.1. Zamora Building Stone

The Zamora building stones (silicified conglomerates and sandstones) correspond to the upper part of Salamanca Sandstone Formation (SSF). Remnants of silicifications appear in the SW border of the Tertiary Duero basin, on both the Iberian Hercynian basement deeply weathered and/or a sedimentary cover of siderolithic nature. Silicification represents an alteration of the parent minerals which gave rise to a relative concentration of silica as CT opal, a partial release of aluminum, the almost complete leaching of the other components and local concentrations of minerals of the alunite group This silicification process produces a silcrete, that is a variety of highly indurated duricrust formed as a result of the near-surface accumulation of silica within a soil, sediment, rock or weathered material [30–32]. The strength of the sedimentary cover increases towards the E and NE, thus separating vertically from the mantle of weathering and silicification process. According to their lithological characteristics, they are arranged in three lithostratigraphic units, the differentiating elements being the presence of iron and silica, colour and sequential organization. The upper section (Zamora building stones), presents a clearly prograding stratigraphic grain and strato-growth architecture generated by the activity of alluvial fan systems. Sedimentation took place within wide, shallow and highly mobile channels through which streams flowed with very high bottom loads and fine suspended material during a Mesozoic period [30–32].

This stone shows marked changes in colour due to the remobilization of iron oxyhydroxides under ancient hydromorphic and acid conditions, with local precipitation of sulphates of the allunite-jarosite group [30,31].

The Zamora building stone shows a chemical composition of the major elements with high contents in SiO_2, together with the presence of Al_2O_3 and H_2O.

In this work, four natural silicified conglomerates (Zamora building stones) from Zamora quarries (Figure 1) have been tested: Z1, white stone, microconglomerate; Z2, ochreous stone conglomerate; Z3B, white and red stone conglomerate and Z3R, purple stone conglomerate.

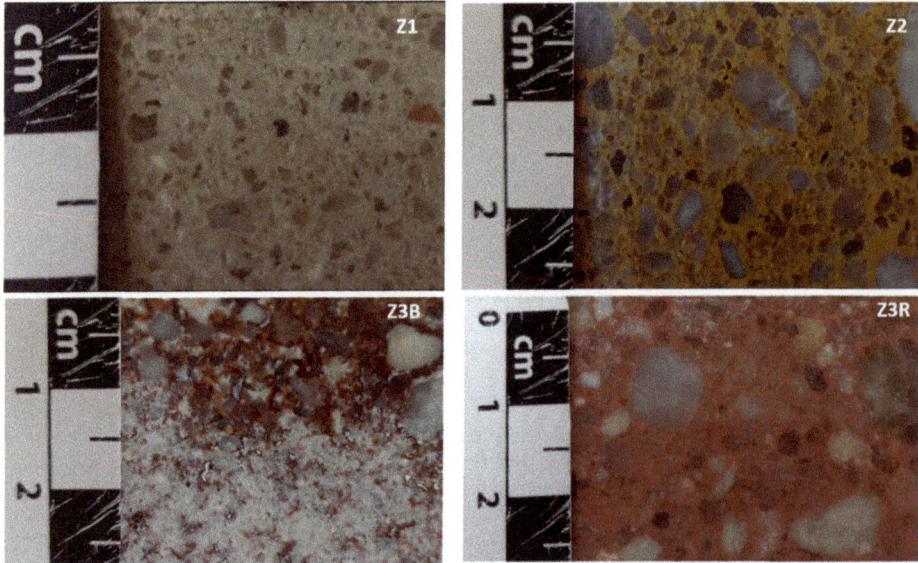

Figure 1. Macroscopic view of the Zamora Building Stones. (**Z1**) white stone, silicified microconglomerate. (**Z2**) ochreous stone, silicified conglomerate. (**Z3B**) white and red stone, silicified conglomerate. (**Z3R**) purple stone, silicified conglomerate.

Figure 2 shows the most remarkable micromorphology results of the four selected varieties, following the methodology published by the authors [32,33]. In all the samples studied, quartz is the dominant mineral in the skeleton, presenting many of its grains, corrosion gulfs on their edges (Figure 2a). Likewise, more or less altered potassium feldspars, muscovites flakes with loss of interference colours (Figure 2c,d), opaque (Fe and Ti oxyhydroxides, Figure 2e), tourmalines, epidote group minerals, and few zircons may appear. The appearance of an isotropic mass quite called the weathering plasma (Figure 2a). Iron oxyhydroxides can appear attached to the 1:1 phyllosilicate (Figure 2c,d), called weathering plasma, giving ocher and/or reddish tones and it is common to find minerals such as small accordion-shaped kaolinites (1:1 phyllosilicates) that are considered neoformation (Figure 2b) By visual samples, the concentration of these iron oxyhydroxydes gives zones with shades from intense red to purple.

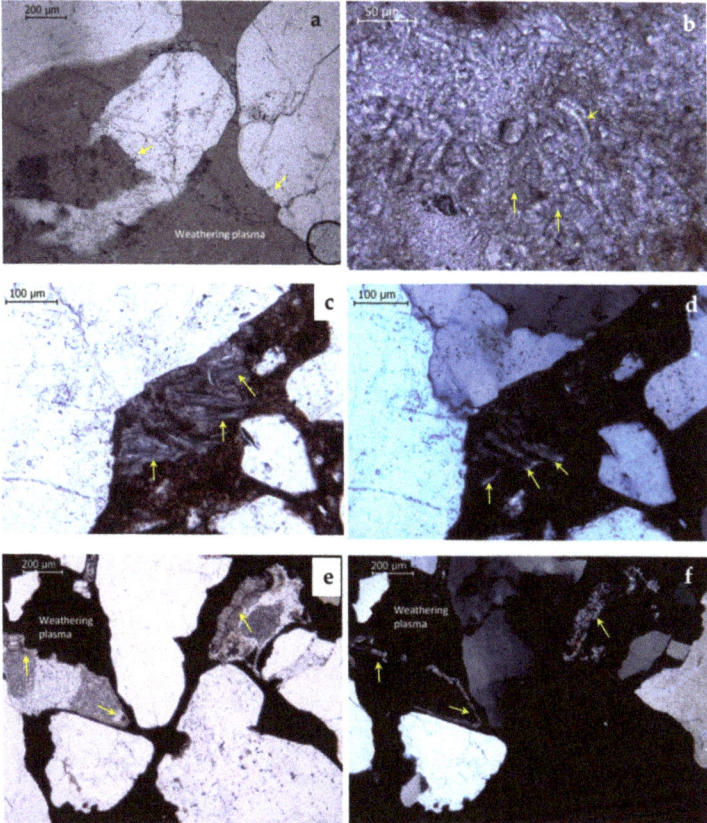

Figure 2. (a) Natural Light (NL) microphotograph. Skeleton of polycrystalline quartz grains (yellow arrows) with corrosion gulfs set in a weathering plasma of CT opal and 1:1 phyllosilicate (birefringent, silcretes); (b) Microphotograph (NL). Small accordion-shaped kaolinites (1:1 phyllosilicates of neoformation(yellow arrows); (c) Microphotograph (NL) and (d) microphotograph (crossed nicols, CN) of muscovites flakes with loss of interference colour (yellow arrows); (e) Microphotograph (NL) and (f) microphotograph (CN) of isotropic mass called weathering plasma formed by CT opal and iron oxyhydroxides (which gives a purple hue to the Zamora building stone), in addition, the partial filling of the porous system (cutan, to designate a modification of the texture, structure or fabric at natual surfaces in soil materials, due to concentration of particular soil constituents or in situ modification of the plasma) by fibrous silica or CT opal (yellow arrows), which have originated by traces of roots (bioturbation) in the tropical palaeosoil (pedogenic silcretes).

Study of thin samples of silicified conglomerates has been observed that: (i) the walls of some cavities appear covered by fibrous silica with their optic axes in different positions in relation to the elongation of the fibers called cutan (Figure 2e,f), and, (ii) most of the CT opal are located lining the walls of cavities (Figure 2e,f), these being interpreted as the traces of roots (bioturbation) of old palaeosols (pedogenic silcretes). In strongly silicified samples, quartz grains encompassed within a brown and more or less anisotropic mass compose of clay, iron oxyhidroxides and CT opal [30–34].

Table 1 shows the average results obtained from the physical properties of the four varieties of Zamora building stones, following the methodology published by the authors [35]. It is observed that the real density varies between 2.56 and 2.60 g/cm^3 and the apparent density between 2.28 and 2.33 g/cm^3. On the other hand, the order of the values obtained for free and total porosity, imbibition coefficient and capillary absorption and permeability in the four varieties studied is as follows: Z1 > Z2 ≈ Z3B ≈ Z3R. However, the order of values in the absorption coefficient is Z1 < Z2 ≈ Z3B ≈ Z3R.

Table 1. Physical characterisation of Zamora building stones by water.

Samples	FP (%)	TP (%)	AC (%)	RD (g/cm^3)	AD (g/cm^3)	IC (%)	CAC (g/cm^{2}S$^{\frac{1}{2}}$)	P (Kg/m^{2}s)
Z1	11.7	14.3	82	2.57	2.28	5.4	0.001075	0.000221
Z2	9.6	10.1	95	2.60	2.33	3.8	0.000844	0.000195
Z3B	8.7	9.2	95	2.57	2.33	4.2	0.000927	0.000179
Z3R	9.1	9.2	99	2.56	2.33	3.7	0.000866	0.000164

FP = Free porosity, TP = Total porosity, AC = Absorption coefficient, RD = Reel density, AD = Apparent density, IC = Imbibition coefficient, CAC = Capillary adsorption coefficient, P = Permeability.

2.2. Experimental and Statistical Methods

The durability of four Zamora building stones (ZBS) was determined by accelerated artificial ageing treatments. Two cubic samples (6 × 6 × 6 × 6 cm) were selected for each of the ZBS varieties, the chromatic coordinates of five of the faces were measured, and they were subjected to the following accelerated ageing treatments in a simulation chamber under controlled conditions. Five cycles were performed:

T1: Freez/thaw and cool/heat cycles (−20 to 110 °C) according to the following procedure: After a drying period at 60 °C to reach a constant weight, the blocks were immersed in distilled water for 16 h (the rocks were saturated), after which they were cooled to −20 °C and kept at that temperature for 3 h. The temperature was then raised to 110 °C (rate = 2 °C/min), and the blocks were kept at that temperature for 3 h. Finally, the blocks were left for 2 h at room temperature and the process was restarted [16].

T2: A combined freez/thaw treatment with sulphates crystallisation, following the method in T1 [16], but using a 14% (weight) solution of $Na_2SO_4 \times 10\ H_2O$ instead of distilled water.

T3: Same as T1 but using a 1% (weight) solution of $Na_3PO_4 \times 10\ H_2O$ instead of the distilled water [16].

Colour was measured with a MINOLTA MODEL CR-310 colourimeter for solids using the (L*,a*,b*) colour coordinates. Total colour difference (ΔE^*_{ab}) defined by the equation [36,37]: $\Delta E_{ab}^* = [(\Delta L^*)^2 + (\Delta a^*)^2 + (\Delta b^*)^2]^{1/2}$.

To study changes in the colour, increments ΔE^*, ΔL^*, Δa^*, and Δb^* have been defined as the difference of the parameter values between that for the untreated sample and those of the sample after each ageing treatment.

The statistical analysis was performed using the MANOVA-Biplot technique (Multivariate Analysis of Variance). The MANOVA-Biplot is a multivariate analysis method, complementary to MANOVA which permits to obtain simultaneous plots of the different groups to be compared, and the different variables being analysed, by also involving the specific Biplot characteristics [38,39]. The results are summarised on several factorial planes, where the variables are represented as vectors that start out from a hypothetical origin and the means of the different groups as star markers in the same reference system. If two variables are represented with a very small angle, then the variables are highly correlated, and are inversely correlated if they are opposite. Additionally, if the angle is close to perpendicular, their correlation is minimal. When projecting all the star markers perpendicularly onto the directions of any of the variables, the order of the projections in the direction of those variables is equivalent to the value that the population means take on for that variable. These interpretations are subject to a series of measurements of the quality of representation for the different planes (inertia absorption of the planes, the goodness of the projections of the measurements on the variables for the dimensions selected, etc.). Depending on these qualities of representation, it is or is not possible to interpret the position of the variables (or the means of the different groups) in the corresponding factorial plane.

The MANOVA-Biplot analysis was applied to a matrix consisting of 16 variables (4 for each chromatic parameter, i.e., ΔE^*, ΔL^*, Δa^*, and Δb^*) and 120 rows in 12 groupings accounting for the different combinations of silicified conglomerates with the three ageing treatments applied.

The different groupings of samples were named as follows:
* Type of sample: Four types of samples tested (Z1, Z2, Z3B and Z3R).
* Artificial ageing, corresponding to aged (T1, T2 and T3).

The 12 groupings have been labelled by combining the four types of stones and the different ageing treatments: ZWTY, where W = 1, 2, 3B and 3R (type of sample), and Y = 1, 2, and 3 (type of artificial ageing treatment). The average values for ΔE^*, ΔL^*, Δa^*, and Δb^* for each aging cycle are included in Tables 2–5.

Table 2. Average values for ΔE^* for every ageing treatment cycles.

Sample and Treatment	$\Delta E1^*$		$\Delta E2^*$		$\Delta E3^*$		$\Delta E4^*$		$\Delta E5^*$	
	Mean	(S.E.)	Mean	(S.E.)	Mean	(S.E.)	Mean	(S.E.)	Mean	(S.E.)
Z1, T1	0.505	0.133	0.657	0.175	0.918	0.172	1.199	0.194	1.011	0.198
Z1, T2	1.032	0.197	1.054	0.193	1.836	0.248	1.475	0.272	1.316	0.185
Z1, T3	0.643	0.236	0.835	0.296	0.962	0.329	0.721	0.349	0.914	0.208
Z2, T1	0.410	0.065	0.595	0.101	0.628	0.107	0.743	0.148	0.691	0.123
Z2, T2	0.539	0.061	0.623	0.113	0.948	0.102	1.533	0.155	2.079	0.607
Z2, T3	0.549	0.228	0.621	0.179	0.831	0.267	0.794	0.299	0.818	0.266
Z3B, T1	0.653	0.065	0.694	0.072	0.835	0.090	0.953	0.086	0.821	0.083
Z3B, T2	1.189	0.128	1.181	0.167	1.228	0.242	1.317	0.178	2.423	0.279
Z3B, T3	0.695	0.169	0.548	0.166	0.891	0.278	0.640	0.230	0.760	0.135
Z3R, T1	0.666	0.061	0.577	0.065	0.854	0.149	0.621	0.096	0.695	0.077
Z3R, T2	1.377	0.130	0.621	0.081	1.251	0.149	1.092	0.101	1.687	0.066
Z3R, T3	0.735	0.154	0.645	0.193	0.618	0.177	0.655	0.152	0.725	0.132

S.E. = Standard Error.

Table 3. Average values for ΔL^* for every ageing treatment cycles.

Sample and Treatment	$\Delta L1^*$		$\Delta L2^*$		$\Delta L3^*$		$\Delta L4^*$		$\Delta L5^*$	
	Mean	(S.E.)	Mean	(S.E.)	Mean	(S.E.)	Mean	(S.E.)	Mean	(S.E.)
Z1, T1	−0.295	0.098	−0.409	0.118	−0.703	0.124	−0.892	0.123	−0.703	0.094
Z1, T2	−0.587	0.224	0.375	0.319	0.914	0.660	0.376	0.444	−0.005	0.211
Z1, T3	−0.346	0.219	−0.394	0.303	−0.588	0.332	−0.445	0.309	−0.428	0.261
Z2, T1	−0.324	0.061	−0.397	0.102	−0.528	0.101	−0.506	0.124	−0.482	0.108
Z2, T2	−0.329	0.093	−0.006	0.026	−0.549	0.114	−0.569	0.097	−1.017	0.570
Z2, T3	−0.082	0.253	−0.020	0.263	−0.265	0.358	−0.101	0.355	−0.154	0.318
Z3B, T1	−0.534	0.090	−0.614	0.092	−0.673	0.125	−0.828	0.088	−0.698	0.104
Z3B, T2	−1.028	0.181	−0.932	0.183	−0.751	0.310	−0.972	0.170	−2.177	0.280
Z3B, T3	−0.610	0.195	−0.449	0.132	−0.763	0.260	−0.502	0.168	−0.676	0.123
Z3R, T1	−0.580	0.080	−0.516	0.082	−0.616	0.116	−0.534	0.085	−0.573	0.083
Z3R, T2	−1.310	0.128	−0.332	0.166	−0.983	0.286	−0.573	0.220	−1.347	0.191
Z3R, T3	−0.420	0.236	−0.177	0.294	−0.189	0.288	−0.151	0.279	−0.161	0.263

S.E. = Standard Error.

Table 4. Average values for Δa* for every ageing treatment cycles.

Sample and Treatment	Δa1*		Δa2*		Δa3*		Δa4*		Δa5*	
	Mean	(S.E.)	Mean	(S.E.)	Mean	(S.E.)	Mean	(S.E.)	Mean	(S.E.)
Z1, T1	0.000	0.019	0.096	0.024	0.251	0.027	0.326	0.030	0.166	0.026
Z1, T2	−0.080	0.032	−0.013	0.049	0.067	0.068	0.055	0.047	−0.008	0.037
Z1, T3	−0.190	0.050	−0.067	0.061	0.056	0.038	−0.064	0.031	−0.124	0.030
Z2, T1	0.017	0.025	0.079	0.029	0.080	0.029	0.095	0.032	0.033	0.039
Z2, T2	0.009	0.025	0.040	0.049	0.135	0.030	0.291	0.039	0.451	0.079
Z2, T3	0.099	0.043	0.147	0.031	0.208	0.042	0.170	0.052	0.108	0.068
Z3B, T1	0.176	0.028	0.121	0.028	0.182	0.034	0.140	0.030	0.077	0.035
Z3B, T2	0.350	0.110	0.473	0.108	0.499	0.135	0.515	0.119	0.660	0.097
Z3B, T3	−0.107	0.042	0.009	0.036	0.164	0.048	0.044	0.035	0.070	0.050
Z3R, T1	0.093	0.034	0.040	0.025	0.262	0.030	0.039	0.030	0.000	0.040
Z3R, T2	−0.240	0.106	−0.320	0.075	−0.305	0.154	−0.232	0.125	−0.108	0.165
Z3R, T3	−0.305	0.068	−0.214	0.083	−0.135	0.041	−0.287	0.050	−0.322	0.075

S.E. = Standard Error.

Table 5. Average values for Δb* for every ageing treatment cycles.

Sample and Treatment	Δb1*		Δb2*		Δb3*		Δb4*		Δb5*	
	Mean	(S.E.)	Mean	(S.E.)	Mean	(S.E.)	Mean	(S.E.)	Mean	(S.E.)
Z1, T1	0.143	0.135	0.260	0.172	0.345	0.159	0.619	0.180	0.504	0.218
Z1, T2	0.759	0.148	0.684	0.243	0.607	0.402	1.092	0.290	1.222	0.218
Z1, T3	0.000	0.264	−0.108	0.351	−0.233	0.357	0.098	0.320	−0.303	0.332
Z2, T1	0.114	0.059	0.251	0.089	0.146	0.082	0.362	0.129	0.146	0.133
Z2, T2	0.367	0.081	0.613	0.111	0.733	0.083	1.385	0.139	1.535	0.469
Z2, T3	−0.142	0.229	−0.058	0.229	−0.177	0.278	−0.131	0.323	−0.186	0.339
Z3B, T1	−0.170	0.034	0.105	0.040	0.234	0.051	0.397	0.045	0.204	0.069
Z3B, T2	−0.041	0.173	0.218	0.216	0.142	0.336	0.541	0.216	0.653	0.242
Z3B, T3	−0.014	0.116	0.107	0.175	0.073	0.228	0.366	0.171	−0.007	0.172
Z3R, T1	−0.106	0.050	0.015	0.038	0.288	0.149	0.205	0.071	0.006	0.092
Z3R, T2	−0.251	0.068	0.094	0.121	0.319	0.132	0.725	0.132	0.797	0.193
Z3R, T3	−0.301	0.092	−0.219	0.129	−0.292	0.086	−0.133	0.138	−0.299	0.138

S.E. = Standard Error.

3. Results and Discussion

The global analysis of the MANOVA-Biplot provides a lambda (Wilks) value of 5.6205, with $p < 0.01$, indicating that significant differences among the different groupings exist. The inertia absorption of the first five factorial axes is 87.47%. The effect of artificial ageing on the global analysis of colour changes is important because great angles between the variables are observed, Figures 3 and 4 (ΔE^*_1–ΔE^*_5, ΔL^*_1–ΔL^*_5, Δa^*_1–Δa^*_5 and Δb^*_1–Δb^*_5).

Figure 3 shows that in Z1 silicified conglomerate, T2 ageing gives rise to higher values in ΔE^* and ΔL^*, bringing about to the samples greater darkness. With T3 ageing, darkening also occurs, but of lesser magnitude. Furthermore, in Z2 silicified conglomerate, T2 ageing gives rise to higher values ΔE^*, compared to T1 and T3, with no differences in ΔL^*. In the Z3B silicified conglomerate, T3 ageing has no variations in ΔL^* or in ΔE^* and in T1 and T2 ageing the effect on ΔL^* is similar, these aging produce higher clarity. In the Z3R silicified conglomerate, T2 ageing produces higher values in ΔE^*, with T1 and T3 exhibiting similar outcomes, but in the opposite direction. At T1 and T2 there is no variation in ΔL^*, but at T3 ΔL^* is higher, resulting in a darkening. In general terms, for ΔL^* the order among all the ageing treatments is T2 > T3 > T1. Moreover, it is observed that the ageing test that produces the greatest variations in the different silicified conglomerates studied is T2. The Zamora building stone that suffers less variation is the Z2 silicified conglomerate and the one that suffers more variations is the Z3R silicified conglomerate.

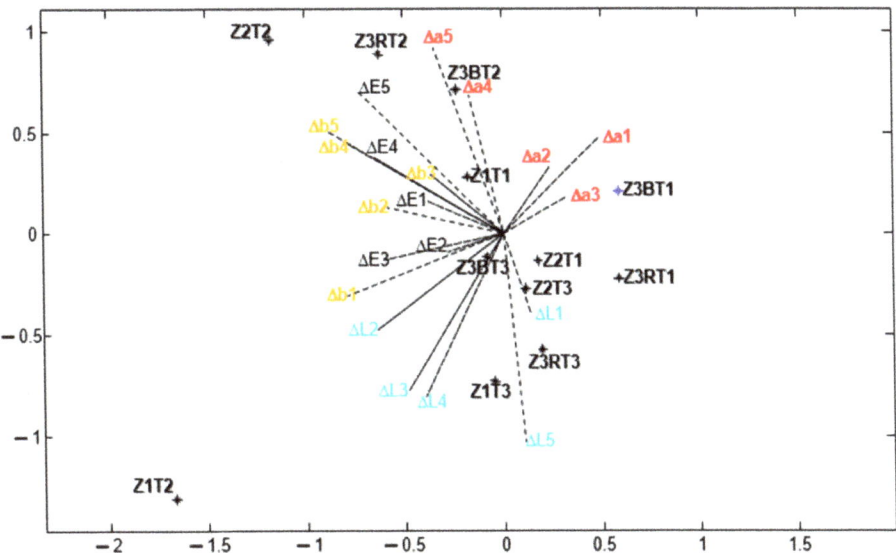

Figure 3. MANOVA-Biplot representation of the different Zamora building stones studied on the plane 1–2.

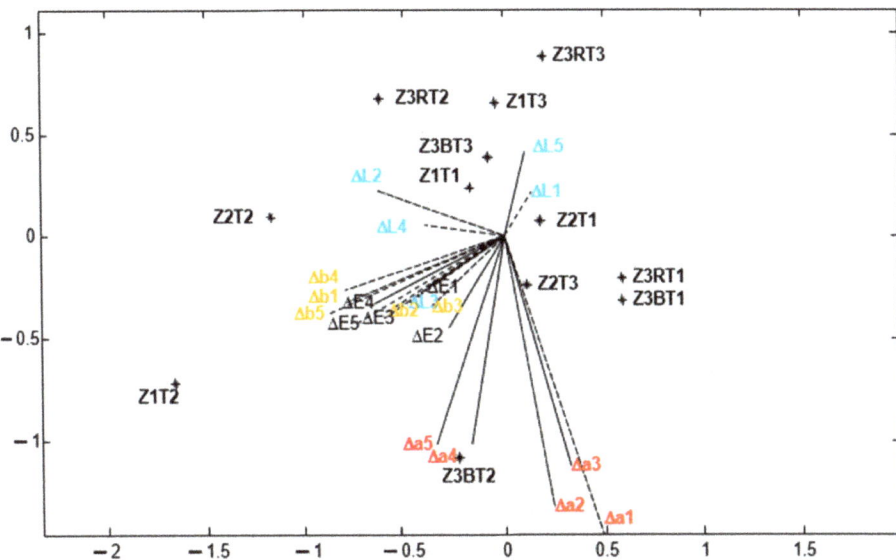

Figure 4. MANOVA-Biplot representation of the different Zamora building stones studied on the plane 1–4.

In the plane 1–2 (Figure 3) the variables Δa* and Δb* have low qualities of representation, they can be interpreted in the plane 1–4 (Figure 4).

In the plane 1–4 (Figure 4), in all silicified conglomerates studied, T2 ageing gives rise to higher values in Δb*, than T1 and T3, leading to a decrease of the b* parameter. Between T1 and T3 ageing there are no differences in Δb* values, but the samples aged with the T3 become more reddish than those treated with T1. On the other hand, in Z1 silicified conglomerate, T2 ageing gives rise to higher Δb* and Δa* values, than T1 and T3, thus giving rise to samples with less reddeningness and yellowing. The order among all ageing treatments would be T2 > T3 ≈ T1. In the Z2 silicified conglomerate, T2 ageing gives rise to higher Δb^* values compared to T1 and T3, giving rise to less yellowish samples, with no great differences in Δa^* between the three ageing treatments. In the Z3B silicified conglomerate, the T2 ageing produces higher in Δa^* and Δb^* values compared to T1 and T3. The order of the Δa^* values, among all ageing treatments would be T2 > T1 > T3, while for Δb^* the order would be T2 > T3 ≈ T1. In the Z3R silicified conglomerate, the T2 ageing produces higher values in Δb^* than T1 and T3, giving rise to less yellow samples, while the variable Δa^* has higher values at T3 and T2, but with a more reddish colour. This same figure also shows that the ageing that produces the greatest variations in the different silicified conglomerates studied is T2. The Zamora building stone that suffers the least variations is Z2 and the one that suffers the most variations is Z3R.

All these data indicate that the samples treated by a combined freezing/thawing and cooling/heating treatments with sulphates crystallisation (T2 ageing) undergo larger colour changes than with the other artificial ageings T1 (freezing/thawing and cooling/heating) and T3 (freezing/thawing treatment with phosphates crystallisation), due to the T2 ageing is the most aggressive of the three artificial ageings tested [6]. Furthermore, the increase in volume of sulphates (haloclasty process) when crystallising from solutions that contain them is greater than in the case of water (gelifraction process) and phosphates (haloclasty process), respectively.

Sulphate crystallisation produces the complete arenisation of almost all the cubic samples in the fifth cycle, but this does not happen for the other ageing treatments. The T2 treatment accelerates the changes in colour on the surface of Z1, Z3B and Z3R stones, because it develops a larger diameter size porosity which can be partly filled up by sulphates. All this due to successive crystallisation/dissolution processes of these salts in each of the cycles performed.

The effect of the T1 artificial ageing cycles (freezing/thawing and cooling/heating) on the chromatic parameters is very weak, as already reported by [2,20].

4. Conclusions

Changes in colour can determine the trends of the different groupings according to the positions from the variables in each artificial ageing cycle.

A significant interaction has been found between some of the groupings, due to T2 artificial ageing (a combined freezing/thawing and thermoclasty treatment with sulphate crystallisation), although some sort of interaction has also been observed in T1 and T3 artificial ageing.

The overall effect of T1, T2 and T3 ageings on the four Zamora building stones turned them darker and with less yellowing for most of the samples studied. The changes produced in ΔE^*, ΔL^*, Δa^* and Δb^* depend on the type of building stone variety, the ochreous stone suffered the least variation and the purple stone the most. There are following variations according to accelerated ageing tests:

(a) Z1: ΔE^* (T2 > T1 ≈ T3), ΔL^* (T2 > T3 > T1), Δa^* and Δb^* (T2 > T1 ≈ T3)
(b) Z2: ΔE^* (T2 > T1 ≈ T3) and Δb^* (T2 > T1 ≈ T3)
(c) Z3B: ΔL^* (T2 ≈ T1 > T3), Δa^* (T2 > T1 > T3) and Δb^* (T2 > T1 ≈ T3)
(d) Z3R: ΔE^* (T2 > T1 ≈ T3), ΔL^* (there are only differences in T3), Δa^* (T2 ≈ T3 > T1) and Δb^* (T2 > T1 ≈ T3).

This study of the colour by statistical analyse may be of interest for the evaluation and monitoring of stone decay, which is an inexpensive, simple, easy and non-destructive technique.

Author Contributions: Conceptualization, J.G.-T. and A.C.I.; methodology, J.G.-T. and A.C.I.; software, J.G.-T. and A.C.I.; validation, J.G.-T. and A.C.I.; formal analysis, J.G.-T. and A.C.I.; investigation, J.G.-T. and A.C.I.; resources, J.G.-T. and A.C.I.; data curation, J.G.-T., A.C.I. and R.S.; writing—original draft preparation, J.G.-T. and A.C.I.; writing—review and editing, J.G.-T. and A.C.I.; visualization, J.G.-T. and A.C.I.; supervision, J.G.-T. and A.C.I.; project administration, E.A., J.G.-T. and A.C.I.; funding acquisition, E.A. and J.G.-T. All authors have read and agreed to the published version of the manuscript.

Funding: This research was funded by the Ministry of Science, Innovation and Universities (PGC2018-098151-B-I00).

Data Availability Statement: Not Applicable.

Acknowledgments: Authors are grateful for financial support for this work from the Ministry of Science, Innovation and Universities (PGC2018-098151-B-I00) and Project "CLU-2019-05—IRNASA/CSIC Unit of Excellence". The authors want to express their gratitude in collaborating with this paper, in the Special Issue of the ChemEngineering in honor of Vicente Rives.

Conflicts of Interest: The authors declare no conflict of Interest.

References

1. García-Talegón, J.; Iñigo, A.C.; Alonso-Gavilán, G.; Vicente-Tavera, S. Villamayor Stone (Golden Stone) as a Global Heritage Stone Resource from Salamanca (NW of Spain). *Geol. Soc. Lond.* **2015**, *407*, 109–120. [CrossRef]
2. García-Talegon, J.; Vicente, M.A.; Vicente-Tavera, S.; Molina-Ballesteros, E. Assessment of chromatic changes due to artificial ageing and/or conservation treatments of sandstones. *Color Res. Appl.* **1998**, *23*, 46–51. [CrossRef]
3. Occhipinti, R.; Stroscio, A.; Belfiore, C.M.; Barone, G.; Mazzoleni, P. Chemical and colorimetric analysis for the characterization of degradation forms and surface colour modification of building stone materials. *Constr. Build. Mater.* **2021**, *302*, 124356. [CrossRef]
4. Rives, V.; Talegon, J.G. Decay and Conservation of Building Stones on Cultural Heritage Monuments. *Mater. Sci. Forum* **2006**, *514–516*, 1689–1694. [CrossRef]
5. Iñigo, A.C.; Rives, V.; Vicente, M.A. Reproducción en cámara climática de las formas de alteración más frecuentes detectadas en materiales graníticos, en clima de tendencia continental. *Mater. Construcc.* **2000**, *50*, 57–60. [CrossRef]
6. Iñigo, A.C.; Vicente-Tavera, S. Different degrees of stone decay on the inner and outer walls of a Cloister. *Build Environ.* **2001**, *36*, 911–917. [CrossRef]
7. Iñigo, A.C.; Vicente-Tavera, S. Surface-inside (10 cm) thermal gradients in granitic rocks: Effect of environmental conditions. *Build. Environ.* **2002**, *37*, 101–108. [CrossRef]
8. Iñigo, A.; García-Talegón, J.; Vicente-Palacios, V.; Vicente-Tavera, S. Canonical Biplot as a tool to detect microclimates in the inner and outer parts of El Salvador Church in Seville, Spain. *Measurement* **2019**, *136*, 745–760. [CrossRef]
9. García-Talegón, J.; Vicente, M.A.; Molina, E. Decay of granite monuments due to salt crystallization in a non-polluted urban environment. *Mater. Construcc.* **1999**, *49*, 17–27. [CrossRef]
10. Fort, R.; López-de Azcona, M.C.; Mingarro, F. Assessment of protective treatments based on their chromatic evolution: Limestone and granite in the Royal Palace of Madrid, Spain. In *5th International Symposium on the Conservation of Monuments in the Mediterranean Basin*; Galán, E., Zezza, F., Eds.; Swets & Zeitlinger B.V.: Lisse, The Netherlands, 2002; pp. 437–441.
11. Carmona-Quiroga, P.; Martinez-Ramirez, S.; de Rojas, M.S.; Blanco-Varela, M.T. Surface water repellent-mediated change in lime mortar colour and gloss. *Constr. Build. Mater.* **2010**, *24*, 2188–2193. [CrossRef]
12. García, O.; Rz-Maribona, I.; Gardei, A.; Riedl, M.; Vanhellemont, Y.; Santarelli, M.L.; Strupi-Suput, J. Comparative study of the variation of the hydric properties and aspect of natural stone and brick after the application of 4 types of anti-graffiti. *Mater. Construcc.* **2010**, *60*, 68–82.
13. Rivas, T.; Iglesias, J.; Taboada, J.; Vilán, J.A. Sulphide oxidation in ornamental slates: Protective treatment with siloxanes. *Mater. Construcc.* **2011**, *61*, 115–130. [CrossRef]
14. La Russa, M.F.; Barone, G.; Belfiore, C.M.; Mazzoleni, P.; Pezzino, A. Application of protective products to "Noto" calcarenite (south-eastern Sicily): A case study for the conservation of stone materials. *Environ. Earth Sci.* **2011**, *62*, 1263–1272. [CrossRef]
15. Pelin, V.; Sandu, I.; Gurlui, S.; Brânzilă, M.; Vasilache, V.; Borş, E.; Sandu, I.G. Preliminary investigation of various old geomaterials treated with hydrophobic pellicle. *Color Res. Appl.* **2016**, *41*, 317–320. [CrossRef]
16. Iñigo, A.C.; García-Talegón, J.; Vicente-Palacios, V.; Vicente-Tavera, S. Measuring the Effectiveness and Durability of Silicified Sandstones and Conglomerates from Zamora, Spain Subject to Silico-organic Treatments and/or Freezing/Thawing Processes. *Rock Mech. Rock Eng.* **2021**, *54*, 2697–2705. [CrossRef]
17. Grossi, C.M.; Brimblecombe, P.; Esbert, R.M.; Alonso, F.J. Color changes in architectural limestones from pollution and cleaning. *Color Res. Appl.* **2007**, *32*, 320–331. [CrossRef]
18. Alonso, F.J.; Vázquez, P.; Esbert, R.M.; Ordaz, J. Ornamental granite durability: Evaluation of damage caused by salt crystallization test. *Mat. Construcc.* **2008**, *58*, 191–201.
19. Rivas, T.; Prieto, B.; Silva, B. Artificial weathering of granite. *Mater. Construcc.* **2008**, *58*, 179–189.

20. Vazquez, P.; Luque, A.; Alonso, F.J.; Grossi, C.M. Surface changes on crystalline stones due to salt crystallisation. *Environ. Earth Sci.* **2013**, *69*, 1237–1248. [CrossRef]
21. Iñigo, A.C.; García-Talegón, J.; Vicente-Tavera, S. Canonical biplot statistical analysis to detect the magnitude of the effects of phosphates crystallization aging on the color in siliceous conglomerates. *Color Res. Appl.* **2014**, *39*, 82–87. [CrossRef]
22. Aly, N.; Gomez-Heras, M.; Hamed, A.; De Buergo, M.; Soliman, F. The influence of temperature in a capillary imbibition salt weathering simulation test on Mokattam limestone. *Mater. Construcc.* **2015**, *65*, e044. [CrossRef]
23. Navarro, R.; Catarino, L.; Pereira, D.; de Sá Campos Gil, F.P. Effect of UV radiation on chromatic parameters in serpentinites used as dimension stones. *Bull. Eng. Geol. Environ.* **2019**, *78*, 5345–5355. [CrossRef]
24. Martin, J.; Feliu, M.J.; Edreira, M.C.; Villena, A.; Calleja, S.; Pérez, F.; Barros, J.R.; Ortega, P. The original colour of the building facades from "El Pópulo" an old quarter of Cádiz. In *5th International Symposium on the Conservation of Monuments in the Mediterranean Basin*; Galan, E., Zezza, F., Eds.; Swets & Zeitlinger B.V.: Lisse, The Netherlands, 2002; pp. 649–653.
25. Zezza, F. Non-destructive technique for the assessment of the deterioration processes of prehistoric rock art in karstic caves: The paleolithic paintings of Altamira (Spain). In *5th International Symposium on the Conservation of Monuments in the Mediterranean Basin*; Galan, E., Zezza, F., Eds.; Swets & Zeitlinger B.V.: Lisse, The Netherlands, 2002; pp. 377–388.
26. Zezza, F. Inland dispersion of marine spray and its effects on monument stone. In *5th International Symposium on the Conservation of Monuments in the Mediterranean Basin*; Galán, E., Zezza, F., Eds.; Swets & Zeitlinger B.V.: Lisse, The Netherlands, 2002; pp. 23–39.
27. Grossi, C.M.; Brimblecombe, P. Past and future colouring patterns of historic stone buildings. *Mater. Construcc.* **2008**, *58*, 143–160.
28. Aparecida-del Lama, E.; Kazumi-Dehira, L.; Grossi, D.; Kuzmickas, L. The colour of the granite that built the city of São Paulo, Brazil. *Color Res. Appl.* **2015**, *41*, 241–245. [CrossRef]
29. Pelin, V.; Rusu, O.; Sandu, I.; Vasilache, V.; Gurlui, S.; Sandu, A.V.; Cazacu, M.M.; Sandu, I.G. Approaching on Colorimetric Change of Porous Calcareous Rocks Exposed in Urban Environmental Conditions from Iasi–Romania. *IOP Conf. Ser. Mater. Sci. Eng.* **2017**, *209*, 012080. [CrossRef]
30. García-Talegón, J.; Iñigo, A.C.; Vicente-Tavera, S.; Molina-Ballesteros, E. Silicified Granites (Bleeding Stone and Ochre Granite) as Global Heritage Stones Resources from Avila (Central of Spain). *Geosci. Can.* **2016**, *43*, 53–62. [CrossRef]
31. García Talegón, J.; Molina, E.; Vicente, M.A. Nature and characteristics of 1:1 phyllosilicates from weathered granite. Central Spain. *Clay Miner.* **1994**, *29*, 727–734.
32. Thiry, M.; Milnes, A.R.; Rayot, V.; Simon-Coinçon, R. Interpretation of palaeoweathering features and successive silicifications in the Tertiary regolith of inland Australia. *J. Geol. Soc. Lond.* **2006**, *163*, 723–736. [CrossRef]
33. Delvigne, J.E. Atlas of micromorphology of mineral alteration and weathering. *Mineral. Mag.* **1998**, *64*, 369–370.
34. Bauluz, B.; Mayayo, M.J.; Yuste, A.; López, J.M.G. Genesis of kaolinite from Albian sedimentary deposits of the Iberian Range (NE Spain): Analysis by XRD, SEM and TEM. *Clay Miner.* **2008**, *43*, 459–475. [CrossRef]
35. Iñigo, A.C.; Supit, J.F.; Prieto, O.; Rives, V. Change in Microporosity of Granitic Building Stones upon Consolidation Treatments. *J. Mater. Civ. Eng.* **2007**, *19*, 437–440. [CrossRef]
36. Robertson, A.R. The CIE 1976 Color-Difference Formulae. *Color Res. Appl.* **1977**, *2*, 7–11. [CrossRef]
37. Sève, R. New formula for the computation of CIE 1976 Hue difference. *Color Res. Appl.* **1991**, *16*, 217–218. [CrossRef]
38. Amaro, I.R.; Vicente-Villardón, J.L.; Galindo-Villardón, M.P. MANOVA Biplot para arreglos de tratamientos con dos factores basado en modelos lineales generales multivariantes. *Interciencia* **2004**, *29*, 26–32.
39. Vicente-Villardón, J.L. MULTBIPLOT: Multivariate Analysis Using Biplots. 2016. Available online: http://biplot.usal.es (accessed on 31 March 2016).

Article

Preparation of $Ca_2Al_{1-m}Fe_m(OH)_6Cl \cdot 2H_2O$-Doped Hydrocalumites and Application of Their Derived Mixed Oxides in the Photodegradation of Ibuprofen

Alejandro Jiménez [1], Marta Valverde [1], Alexander Misol [1], Raquel Trujillano [1], Antonio Gil [2] and Miguel Angel Vicente [1,*]

[1] Departamento de Química Inorgánica, GIR—QUESCAT, Universidad de Salamanca, E-37008 Salamanca, Spain
[2] INAMAT[2], Departamento de Ciencias, Edificio de los Acebos, Campus de Arrosadía, Universidad Pública de Navarra, E-31006 Pamplona, Spain
* Correspondence: mavicente@usal.es

Abstract: Aluminum from saline slags generated during the recycling of this metal, extracted under reflux conditions with aqueous NaOH, was used in the synthesis of hydrocalumite-type solids with the formula $Ca_2Al_{1-m}Fe_m(OH)_6Cl \cdot 2H_2O$. The characterization of the obtained solids was carried out by powder X-ray diffraction, infrared spectroscopy, thermal analysis, element chemical analysis, N_2 adsorption-desorption at $-196\,°C$ and electron microscopy. The results showed the formation of Layered Double Hydroxide-type compounds whose characteristics varied as the amount of incorporated Fe^{3+} increased. These solids were calcined at 400 °C and evaluated for the catalytic photodegradation of ibuprofen, showing promising results in the elimination of this drug by advanced oxidation processes. The CaAl photocatalyst (without Fe) showed the best performance under UV light for the photodegradation of ibuprofen.

Keywords: hydrocalumite; aluminum saline slag recovery; photocatalysis; ibuprofen degradation

1. Introduction

Technological progress in recent years requires the use of increasingly sophisticated materials for specific applications, and these materials must also be eco-friendly, avoiding as much as possible the generation of polluting wastes. Aluminum is the second most widely used metal in the world, after iron [1,2]. Its excellent properties make it an ideal material for various sectors such as military, machinery, aerospace, building or food [3–6]. The combination of the Bayer and Hall-Héroult processes allows for obtaining aluminum from natural bauxite, with the disadvantage of high electricity consumption and generation of several wastes, including *red mud*. This is known as *Primary Aluminum Production* (PAP) [2,4,6–10]. Another way to obtain it is *Secondary Aluminum Production* which is based on the recycling of the metal. In this case, the process requires less electricity consumption and the addition of salts (mainly NaCl and KCl) for melting the aluminum, generating an important residue, known as *Salt Cake* or *Saline Slags* [2,4,6–10]. According to the European Normative [11], salt cake is considered hazardous waste. Several valorization procedures were proposed, among which the following stand out: (a) its direct use as an adsorbent [12–14] or (b) extraction of aluminum [4,5,9] and its subsequent use in the preparation of Al^{3+}-based materials such as alumina [15], zeolites [7,16], or layered double hydroxides (LDHs) [8,17–20], among others.

LDHs are a family of compounds whose structure is derived from brucite ($Mg(OH)_2$) and whose general formula is $[M(II)_{1-x}M(III)_x(OH)_2]^{x+}[A_{x/n}]^{n-} \cdot mH_2O$ (M(II) and M(III) are divalent and trivalent cations and A is the interlayer anion) [21]. A large number of divalent cations (Ni, Co, Cu, Mg, Ca or Zn) form LDH with several trivalent cations

(Al, Fe, Cr, Y or Ga) [21], and the Li–Al LDH was also described [22,23]. Hydrocalumite, $Ca_2Al(OH)_6Cl \cdot 2H_2O$, belongs to the LDH family [21]. Among its most important applications are: adsorbent [24], antacid [25], ion-exchanger [26], and basic heterogeneous catalyst [27–31]. We recently reported the preparation of hydrocalumite from saline slags [8].

Ibuprofen (IBU) is a non-steroidal drug widely used as an anti-inflammatory, antipyretic, and analgesic [32]. Its chemical structure consists of an alkylbenzene ring with a carboxylic acid functional group, whose pK_a is 4.8 [33]. The presence of IBU and other pharmaceuticals and personal care products (PPCP) in surface waters was reported by several authors, their elimination being not possible by classical biological treatments and including these drugs among the emerging pollutants [34–44].

Advanced oxidation processes (AOPs), especially heterogeneous photocatalysis, are presented as a promising technology for wastewater purification and treatment [36]. This methodology achieves the degradation and/or mineralization of pollutants. Photocatalysis, ozonation or the Fenton process are AOP [41]. Heterogeneous photocatalysis is a versatile and environmentally friendly technique based on the use of light and a photocatalyst. Light is used to activate the catalyst, which is a semiconductor that generates highly reactive radicals that degrade the pollutant [41]. The par-excellence photocatalyst is titanium dioxide (TiO_2), while zinc oxide (ZnO), iron (III) oxide (Fe_2O_3) or vanadium (V) oxide (V_2O_5) are also widely used [45–47]. The performance of the process can be improved by dispersing the photocatalyst on a support such as mixed metal oxides (MMO) [48,49]. In this sense, LDH can be excellent precursors of MMO by calcination through topological transformation [21,47,50]. Di et al. have recently reported that bifunctional ZnFe-MMOs prepared from LDH show high performance in IBU degradation under simulated solar irradiation [51].

Although the Fe–Al substitution in hydrocalumites should be expected, to the best of our knowledge, only a few works have explored the study of this type of LDH compound. Thus, Phillips and Vandeperre have investigated the adsorption of nitrate, chloride and carbonate by LDH-CaAlFe-type solids [52], while Lu et al. have applied these solids as heterogeneous catalysts in the production of biodiesel from soybean oil by transesterification reaction [53], and Szabados et al. have studied the preparation of LDH-CaAlFe by a combination of dry-milling and ultrasonic irradiation in aqueous solution [54]. Although some work was carried out using other LDH solids, mainly hydrotalcite-type (an LDH family similar to hydrocalumite), hydrocalumite solids are very scarcely used for investigating the degradation of emerging pollutants. Sánchez-Cantú et al. have used hydrocalumite-type compounds as catalyst precursors in the photodegradation of 2,4-dichlorophenoxyacetid acid [55]. Thus, in this work, hydrocalumites with the theoretical formula $Ca_2Al_{1-m}Fe_m(OH)_6Cl \cdot 2H_2O$, incorporating variable amounts of Fe^{3+}, were prepared following the methodology recently reported by Jiménez et al. for the preparation of hydrocalumite from aluminum saline slags [8]. These CaAlFe hydrocalumites were calcined at 400 °C and the resulting solids were used in the removal of ibuprofen, evaluating their adsorption and catalytic photodegradation capacities. Therefore, this work can be framed within the circular economy, as waste from aluminum recycling was used to synthesize LDH-CaAlFe which, after being calcined at 400 °C, showed high efficiency in the removal of ibuprofen by means of an AOP.

2. Materials and Methods
2.1. Materials

Aluminum saline slag was kindly supplied by IDALSA (Ibérica de Aleaciones Ligeras S.L., Pradilla de Ebro, Zaragoza, Spain). NaOH (pharma grade), HCl (pharma grade, 37%) $CaCl_2$ (anhydrous, 95%), $FeCl_3 \cdot 6H_2O$ (97–102%) were from Panreac, while $CaCl_2 \cdot 2H_2O$ (ACS 99–105%) was supplied by Alfa Aesar and ibuprofen sodium salt (98%) was supplied by Sigma Aldrich, St. Louis, MO, USA. All were used as received, without any treatment.

2.2. Preparation of CaAlFe Mixed Metal Oxides

The methodology recently reported by some of us in [8] to prepare hydrocalumite was followed. Saline slags were ground and washed until obtaining a chloride-free solid, the fraction smaller than 0.4 mm was extracted with NaOH, as reported elsewhere [9], and the solution was treated with HCl to precipitate silicon-containing species as SiO_2; addition of $CaCl_2 \cdot 2H_2O$ and $FeCl_3 \cdot 6H_2O$ at pH 11.5 (fixed using NaOH) led to the formation of a precipitate, that was submitted to treatment under microwave (MW) radiation in a Milestone Ethos Plus Microwave oven for 2 h at 125 °C, leading to the formation of hydrocalumite. These as-prepared samples were named $CaAl_{1-m}Fe_m$, where m represents the amount of Fe^{3+} incorporated into the hydrocalumite-type solid, as fraction of the trivalent positions. In order to evaluate the photocatalytic activity, samples CaAl, $CaAl_{0.90}Fe_{0.10}$, $CaAl_{0.80}Fe_{0.20}$ and CaFe were calcined at 400 °C, this temperature was selected taking into account the results shown by the thermal analysis of the solids and those reported in previous studies for hydrocalumites and hydrotalcites, as at this calcination temperature this sort of solids exhibit the largest specific surface area values, while higher temperatures lead to crystallization of the individual and mixed oxides [27,56–58]. For the calcined samples, "−400" was added to the name of the samples, denoting the calcination temperature, in Celsius. Considering hydrocalumite as the parent solid, it was doped by substituting Al^{3+} with Fe^{3+}, in order to investigate the extent of the isomorphous substitution, and how the increase in Fe content influenced the catalytic performance. Thus, samples with m = 0, 0.1, 0.2, and 1 were selected in order to evaluate their photocatalytic activity in the degradation of ibuprofen.

2.3. Characterization Techniques

A Siemens D-5000 equipment was used to record the powder X-ray diffraction (PXRD) patterns of the samples (λ = 0.154 nm Cu–Kα radiation, fixed divergence, 5°–70° (2θ), scanning rate 2° (2θ)/min, steps of 0.05°, 1.5 s/step). The crystalline phases formed were identified by comparison with the JCPDS-International Centre for Diffraction Data (ICDD®) database [59].

The thermogravimetric (TG) curves were recorded in an SDT Q600 apparatus (TA Instruments, New Castle, DE, USA) at a heating rate of 10 °C/min up to 900 °C and under oxygen (Air Liquide, Madrid, Spain, 99.999%) flow (50 mL/min).

The infrared spectra, FT–IR, were recorded in a Perkin-Elmer Spectrum Two instrument with a nominal resolution of 4 cm^{-1} from 4000 to 400 cm^{-1}, using KBr (Merck, grade IR spectroscopy, Kenilworth, NJ, USA) pressed pellets and averaging 12 scans to improve the signal-to-noise ratio.

Element chemical analyses for several elements were carried out by Inductively Coupled Plasma Optical Emission Spectrometry (ICP-OES) in a Yobin Ivon Ultima II apparatus (Nucleus Research Platform, University of Salamanca, Salamanca, Spain).

N_2 adsorption–desorption isotherms were recorded at −196 °C using a Micromeritics Gemini VII 2390T. Prior to analysis, N_2 flowed through the sample (ca 0.1 g) at 110 °C for 2 h to remove weakly adsorbed species. Specific surface areas were calculated by the Brunauer–Emmet–Teller (BET) method and the average pore diameter by the Barrett–Joyner–Halenda (BJH) method [60].

Scanning electron microscopy (SEM) images were obtained using a JEOL IT500 Scanning Electron Microscope at the Nucleus Research Platform (University of Salamanca, Spain).

2.4. Photodegradation Studies

The study of the catalytic performance of the solids was carried out on an MPDS-Basic system from Peschl Ultraviolet, with a PhotoLAB Batch-L reactor and a TQ150-Z0 lamp (power 150 W), integrated into a photonCABINET. Its spectrum is continuous, with the main peaks at 366 nm (radiation flux, Φ 6.4 W) and 313 nm (4.3 W). For this purpose, 750 mL of a solution of ibuprofen sodium salt in distilled water of concentration 50 ppm was introduced. An amount of 0.75 g of photocatalyst was added and magnetically stirred in the dark for

35 min to ensure the adsorption–desorption equilibrium [51], then the UV lamp was turned on. For taking samples for analyses, the illumination was cut off and the suspension was allowed to decant in order to reduce the losses of catalyst as much as possible. The samples taken were filtered with a filter Macherey-Nagel CHROMAFIL Xtra PA-20/25 of 0.22 µm. The solutions were analyzed by an ultraviolet-visible spectrophotometer coupled to a computer with UV WINLAB 2.85 software, following the evolution of the absorption band of ibuprofen at 222 nm. The reproducibility of the experiments was tested by duplicating some experiments, the difference in the values obtained always being lower than 1%.

In order to determine the by-products generated during UV degradation, selected solutions were analyzed after several reaction times by mass spectrometry (MS). The equipment used was an Agilent 1100 HPLC mass spectrometer coupled to an ultraviolet detector and an Agilent Trap XCT mass spectrometer. These analyses were performed at the Servicio Central de Análisis Elemental, Cromatografía y Masas (Universidad de Salamanca, Salamanca, Spain).

3. Results

3.1. Extraction of Aluminum

The aluminum content in the final extraction solution was 13,937 mg/L, while other elements were not found. The extraction and synthesis conditions described in our previous work [8,9] allowed us to obtain a pure aluminum solution, which was successfully used for the synthesis of the doped hydrocalumite solids.

3.2. Characterization of the Solids

3.2.1. Hydrocalumite Type Solids

Figure 1A shows the diffractograms of the synthesized solids. All the diffractograms showed the characteristic peaks of the hydrocalumite-type layered structure (ICDD card 01-072-4773). In an octahedral environment, the Fe^{3+} cation has an ionic radius of 55 pm while Al^{3+} has a radius of 54 pm [61], the similarity in size between the two cations allowed their isomorphic substitution in a similar system without important structural changes, although as the amount of Fe increased, the crystallinity of the final solid decreased [18], in spite of the fact that all the samples were submitted to a microwave aging process at 125 °C for 2 h. One of the peculiarities of hydrocalumite is its high degree of crystallinity compared to other LDHs; this is due to the fact that the Ca and Al octahedra are not randomly distributed in the sheets but are perfectly ordered due to the larger radius of Ca^{2+} compared to Mg^{2+} [8,30,62–65]. On the other hand, the similarity of radii between the divalent cation and the trivalent cation produces an increase in the disorder of the octahedra in the sheets [62,65]. In hydrocalumite, the large size difference between Ca^{2+} (100 pm) and Al^{3+} (54 pm) [8,61] contributes to an increase in the degree of ordering of the octahedra in the sheets [8,62,65]. As mentioned above, the radius of Fe^{3+} is slightly larger than the radius of Al^{3+}. When Al^{3+} is isomorphically substituted by Fe^{3+}, the Ca^{2+}/M^{3+} radius ratio decreases, which implies an increase in the degree of disorder of the octahedra in the sheets with respect to hydrocalumite. Other phases were not detected by PXRD.

As described above, LDHs consist of $[M(II)_{1-x}M(III)_x(OH)_2]^{x+}$ octahedral sheets. In the case of hydrocalumite, M(II) is Ca^{2+} and M(III) is Al^{3+}, the main difference between hydrocalumite with other LDHs is that the Ca and Al octahedra are not randomly arranged but are ordered [30,63]. This higher order implies more intense and narrower diffraction peaks than in other LDH, and peaks due to diffraction by planes (003), (006), (110) and (009) are present in the diffractograms of both hydrocalumite and other LDHs [64,65]. As shown in Figure 1A, the samples showed a clear preferred orientation, which coincided with the direction of sheet stacking (crystallographic direction c), in agreement with the high intensity of the (003) and (006) peaks. The main difference between the CaAl and CaFe samples was observed in the (110) plane peak, this diffraction peak shifted to lower diffraction angles when all Al^{3+} were replaced by Fe^{3+} (m = 1) (Figure 1B,C). The interlayer spacing from the (110) diffraction peak allowed us to calculate the value of the lattice parameter a $\left(a = 2d_{(110)}\right)$ and from the (003) and (006) spacing values the parameter c was calculated $\left(c = 3/2[d_{(003)} + 2d_{(006)}]\right)$ [62]. These cell parameters were calculated for all the synthesized solids (Table 1); the literature reported that the a and c parameters for hydrocalumite are 0.575 nm and 2.349 nm, respectively (ICDD 01–072–4773), while for $Ca_2Fe(OH)_6Cl \cdot 2H_2O$ these parameters are $a = 0.587$ pm and $c = 2.336$ pm [65]. Cell parameters of a varied between 0.574 and 0.586 nm in the prepared solids, all values of a were found to be within the range of theoretical values reported in the literature. Although the size difference between Fe^{3+} and Al^{3+} is very small, ionic radii 55 pm and 54 pm, respectively, it affected the a parameter, which was larger when the trivalent cation was entirely Fe^{3+} than when the trivalent cation was Al^{3+}. Figure 2 shows the variation of the a parameter with composition, as the amount of Fe^{3+} in the structure increased, the parameter a also increased, according to Vegard's Law. Thus, the amount of incorporated Fe^{3+} could be estimated from the lattice parameter a determined by PXRD by the expression $y = 0.0012x + 5.742$, where y represents the lattice parameter a and x represents the percentage of Fe^{3+} as a trivalent cation. On the other hand, lattice parameter c varied between 2.356 and 2.331 nm, very close to the expected value. However, a small decrease was observed as the Fe^{3+} content increased. The lattice parameters a and c showed an opposite trend. The parameter c depends on the charge, size and orientation of the interlayer anions [65,66]; in this case, the interlayer anion was chloride; therefore, these small variations found could be attributed to differences in the degree of hydration as shown in Table 1. Moreover, due to the basic character of these solids, it was possible that they fixed atmospheric CO_2, incorporating traces of carbonate anion in the interlayer space, influencing on the value of this parameter. The values of parameter c were similar to those reported in the literature for $Ca_2Al(OH)_6Cl \cdot 2H_2O$ ($c_{theorical} = 2.349$ pm and $c_{experimental} = 2.356$ pm) and $Ca_2Fe(OH)_6Cl \cdot 2H_2O$ ($c_{theorical} = 2.336$ pm and $c_{experimental} = 2.331$ pm).

Table 1. Parameters determined for the synthesized solids.

Sample	a (nm) *	c (nm) **	Fe/M^{3+} (%) ***	Ca^{2+}/M^{3+} Molar Ratio	$D_{(003)}$ (nm)	$D_{(001)}$ (nm)	S_{BET} (m^2/g)	Average Pore Diameter (nm)	Hydration Water (wt. %) ****
CaAl	0.574	2.356	0.00	2.05	37	82	12	8.7	12.3
CaAl$_{0.95}$Fe$_{0.05}$	0.575	2.354	5.91	1.96	33	66	15	9.2	12.9
CaAl$_{0.90}$Fe$_{0.10}$	0.576	2.354	8.87	1.83	38	45	13	6.2	12.1
CaAl$_{0.80}$Fe$_{0.20}$	0.576	2.355	19.77	1.83	32	60	20	8.2	15.7
CaAl$_{0.60}$Fe$_{0.40}$	0.577	2.350	38.71	1.90	35	21	22	8.9	11.9
CaAl$_{0.40}$Fe$_{0.60}$	0.579	2.346	60.68	1.54	34	21	28	9.0	11.7
CaAl$_{0.20}$Fe$_{0.80}$	0.582	2.341	82.65	1.26	33	26	32	8.2	11.5
CaFe	0.586	2.331	100	1.68	37	75	27	9.3	11.1

* a theorical hydrocalumite = 0.575 pm and a_{CaFe} theoretical = 0.587 pm [65]. ** c theorical hydrocalumite = 2.349 pm and c_{CaFe} = 2.336 pm [65]. *** Percentage of Fe^{3+} occupying trivalent positions. **** End of the first mass loss (~200 °C).

The crystallite sizes of the synthesized solids along the packing directions (003) ($d_{(003)}$) and (110) ($d_{(110)}$) are also shown in Table 1. The crystal size (D) was calculated from Scherrer's equation (D = $\frac{k\lambda}{\beta \cos \theta}$; k = 0.94; λ = 0.154 nm; θ = Bragg diffraction angle; $\beta = \sqrt{B^2 - b^2}$; B = FWHM$_{(hkl)}$ (rad); b = instrument width (rad)) [67]. The crystal size along the sheet stacking direction ($d_{(003)}$) remained practically constant, showing values between 32 and 38 nm. However, in the (110) direction there was a decrease in the crystal size as the amount of Fe^{3+} in the structure increased, except for the CaFe sample, whose value approached that of the CaAl sample again.

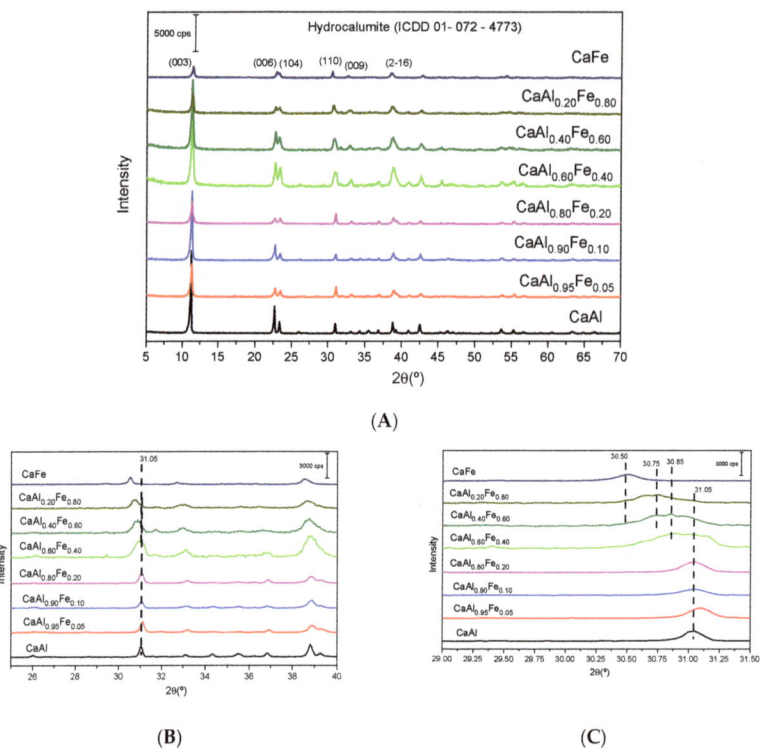

Figure 1. X-ray patterns of samples synthesized: (**A**) shows X-ray patterns of all samples, while (**B**,**C**) show in detail the variation in the position of the peak due to the (110) plane diffraction with the increasing of the Fe^{3+} content in the solids.

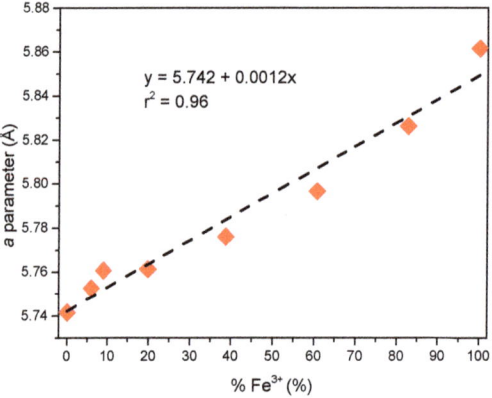

Figure 2. Variation of parameter *a* with the amount of Fe^{3+} incorporated. Red points = experimental values; Dashed line: straight adjustment.

Table 1 also shows the amount of Fe^{3+} present in the solids. All amounts agreed with the theoretical amount of Fe^{3+} predicted from the composition of the original solutions. Regarding the Ca^{2+}/M^{3+} molar ratio, it moved away from the theoretical value of 2 for Ca/Al hydrocalumite as the Fe^{3+} content increased.

Figure 3 shows the FT–IR spectra of the samples. There were few differences between all the solids. All of them showed bands between 3600 and 3400 cm^{-1} that corresponded to the stretching vibration of the O–H bonds. The band at 3632 cm^{-1} was assigned to the stretching vibration of the AlO–H bonds, while the band due to the stretching vibration of the CaO–H bonds was located at 3471 cm^{-1}. On the other hand, as the Fe^{3+} content increased in the samples, a band appeared at 3591 cm^{-1}, which could be assigned to the stretching vibration of the FeO–H bonds [54,57]. In this zone, the bands corresponding to the hydroxyl groups of the water molecules located in the interlayer space were also found [54,68,69]. The band at 1615 cm^{-1} was due to the bending vibration of water, which confirmed its presence in the interlayer space. The presence of CO$_3$$^{2-}$ ion was confirmed by the band at 1414 cm^{-1}. During the synthesis of the solids, special care was taken to work in an inert atmosphere and decarbonated water was used to avoid the formation of calcite, however, it was possible that the samples fixed atmospheric CO$_2$ during their handling, producing the incorporation of traces of carbonate. The band located at 1414 cm^{-1} increased its intensity as the Fe^{3+} content increased, in such a way that for the CaAl sample this band was practically not observed. The presence of CO$_3$$^{2-}$ in the interlayer space was confirmed by the existence of bands at 1506 cm^{-1} and 870 cm^{-1} in the samples with higher Fe^{3+} content (CaFe, CaAl$_{0.20}$Fe$_{0.80}$ and CaAl$_{0.40}$Fe$_{0.60}$) [31,70]. This seemed to contradict the PRXD results, since the larger the size of the interlayer anion, the higher the value of the reticular parameter c should be. The diameter of CO$_3$$^{2-}$ is 378 pm [61], while that of Cl$^-$ is 336 pm [61]; according to these values, the samples that present a band at 1414 cm^{-1} should present a higher value of the reticular parameter c. However, according to the values in Table 1, no increase in parameter c was observed with the presence of carbonate in the samples, so the differences in parameter c observed can be attributed to small variations in the degree of hydration of the anions located in the interlayer space. The carbonate may then be adsorbed on the surface or even in the form of very small particles not detectable by powder X-ray diffraction. The bands at 797 cm^{-1}, 739 cm^{-1}, 575 cm^{-1} and 421 cm^{-1} were assigned to M–O bonds, where M is Ca^{2+}, Al^{3+} or Fe^{3+} [54,68,69,71].

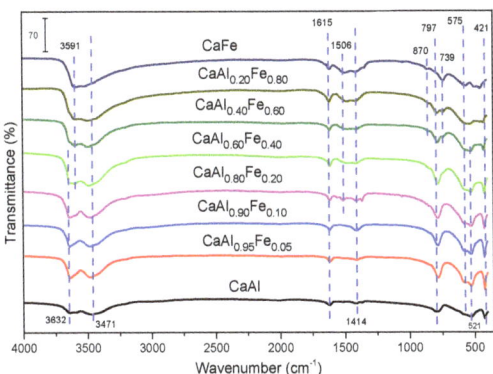

Figure 3. FT—IR spectra of the samples.

The TG and DTG curves of the samples are shown in Figure 4, all of them were very similar to each other, showing thermal decomposition processes characteristic of LDH [26,66,72], as in all cases the phases identified by PXRD corresponded to LDH and other phases were not detected. In all cases, three main mass loss steps were observed, with a total mass loss between 28 and 36% (theoretical total mass loss for complete dehydration/dehydroxylation of Ca$_2$Al(OH)$_6$Cl·2H$_2$O would be 32.1% and for Ca$_2$Fe(OH)$_6$Cl·2H$_2$O it would be 29.1%). The first one was located at a temperature between 140–150 °C, corresponding to the loss of hydration water located in the interlayer space or chemically bound to LDH [26,66,72]. The second mass loss appeared close to 300 °C and corresponded to the dehydroxylation of the layers. As the Fe^{3+} content increased, the temperature at which the process took place decreased, from 328 °C for the CaAl sample to 270 °C for the CaFe sample, which indicated that the Fe–OH bonds were weaker than the Al–OH bonds, which was in agreement with the FT–IR results. The third and last mass loss was located close to 700 °C and could be associated with the decarbonization of the samples. However, in our previous work, this last stage was shown to be compatible with the elimination of a water molecule, thus completing the dehydroxylation process to form the corresponding oxychloride [57]. In addition, in some samples ($CaAl_{0.80}Fe_{0.20}$ or CaFe) an additional mass loss was observed at a temperature below 100 °C, due to the elimination of physisorbed water [63,73]. Considering this thermal behavior, 400 °C was considered the ideal temperature for the calcination of the solids.

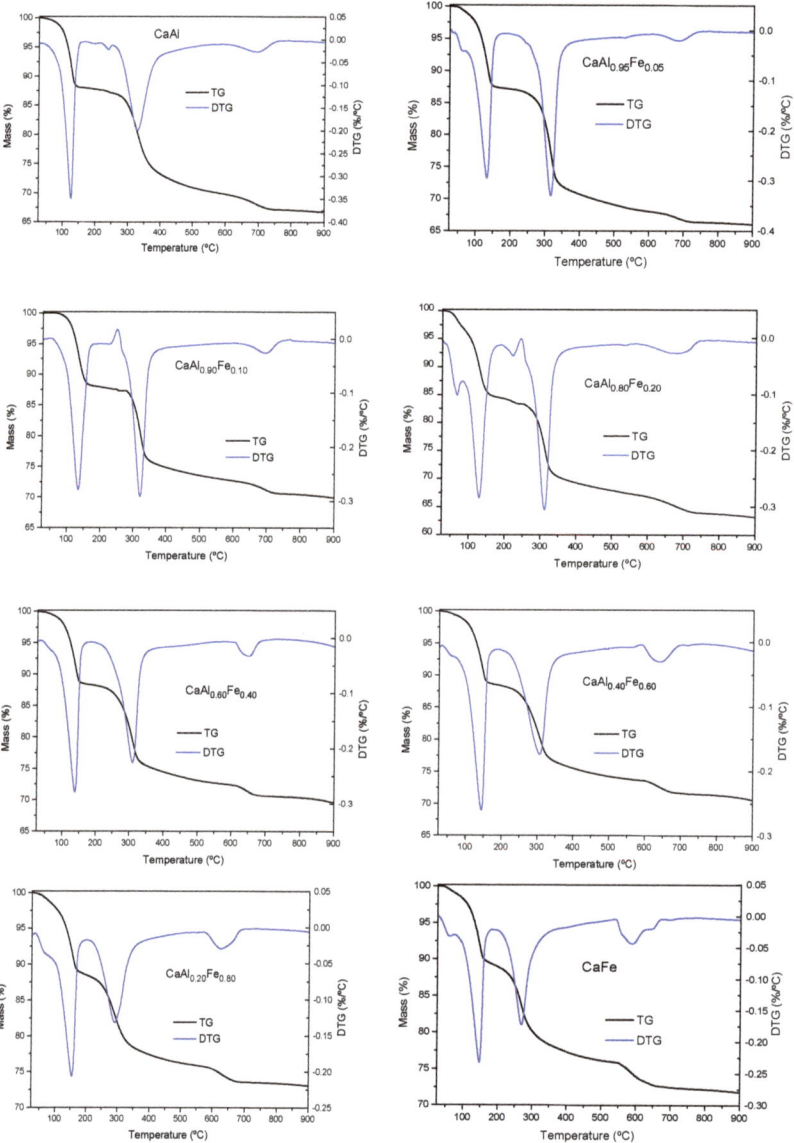

Figure 4. TG and DTG curves of the solids synthesized.

According to the IUPAC isotherm classification criteria [60], all synthesized solids showed type II N_2 adsorption isotherms (Figure 5). In addition, some of them showed a small hysteresis loop of type H3 after recording the desorption branch, which corresponded to the presence of plate-like particle aggregates leading to slit-like pores [60]. All samples presented low values of BET-specific surface area (Table 1), with a maximum of 32 m^2/g for sample CaAl$_{0.20}$Fe$_{0.80}$ and with an average pore width in the mesopore range. In general, as the amount of Fe^{3+} in the solid increased, there was an increase in S$_{BET}$ values, although, for values $m \geq 0.60$, the increase in Fe^{3+} content did not cause an increase in S$_{BET}$, with CaAl$_{0.40}$Fe$_{0.60}$, CaAl$_{0.20}$Fe$_{0.80}$ and CaFe showing similar S$_{BET}$ values (all of them higher than the those of the samples with lower Fe^{3+} content). This was in agreement with PXRD data because as the Fe^{3+} content increased, the crystallinity of the samples decreased.

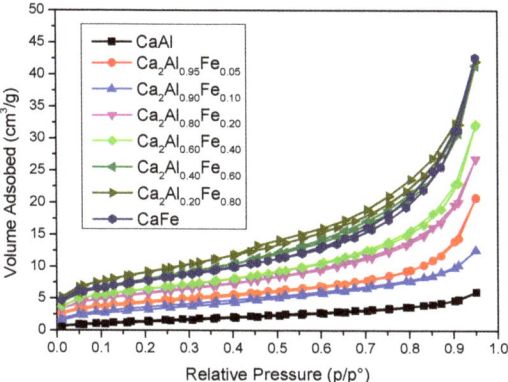

Figure 5. Nitrogen adsorption–desorption isotherms of the samples.

3.2.2. Solids Calcined at 400 °C

Figure 6 shows the diffractograms of the samples calcined at 400 °C. With respect to the LDH-type solids before calcination, the layered structure completely disappeared, giving rise to the formation of amorphous mixed oxides. In the CaFe-400 sample, the appearance of diffraction peaks corresponding to magnetite (Fe$_3$O$_4$, ICDD card 01–072–6170) and hematite (Fe$_2$O$_3$, ICDD card 01–084–0311) phases was observed. In the samples with lower Fe^{3+} or no Fe^{3+} content at all, no crystalline phases were observed.

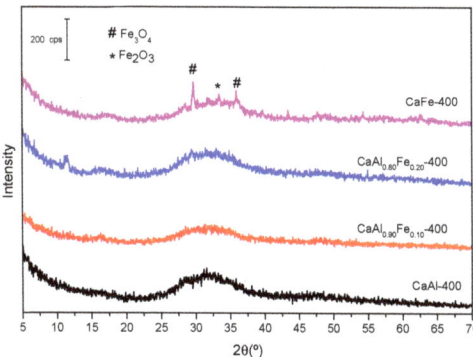

Figure 6. X-ray patterns of samples calcined at 400 °C.

The FT–IR spectra of the samples calcined at 400 °C are shown in Figure 7. In the samples with low Fe^{3+} content (CaAl-400, CaAl$_{0.90}$Fe$_{0.10}$ and CaAl$_{0.80}$Fe$_{0.20}$) a broad band centered at 3565 cm^{-1} was observed, due to the stretching vibrational mode of the hydroxyl groups. This band was broad and presented a shoulder centered at 3428 cm^{-1}, being composed of the superposition of hydroxyl group bands coming from several environments [58]. The presence of water was confirmed by its bending band at 1624 cm^{-1}. On the other hand, the characteristic bands of carbonate appeared again at 1419 cm^{-1}, 1496 cm^{-1} and 879 cm^{-1} [31,58,70]. This carbonate species may be formed during the precipitation of the starting solid or adsorbed due to the basic character of the hydrocalumites, in both cases by fixation of atmospheric CO_2. Finally, the band located at 596 cm^{-1} as well as others located in the region between 800 cm^{-1} and 400 cm^{-1} (not labeled in Figure 7) can be attributed to M–OH bonds, where M can be Ca^{2+}, Al^{3+} or Fe^{3+} [58]. The band at 814 cm^{-1} showed a shoulder on its left-hand side which could be attributed to the presence of carbonate in the samples, and it can overlap with the band at 870 cm^{-1} with characteristics of the carbonate ion. However, this band was not present in the spectrum of the CaFe-400 sample, suggesting that it can be attributed to the Al–OH bond, but band assignment in this region is complicated due to band overlapping and broadness of the bands.

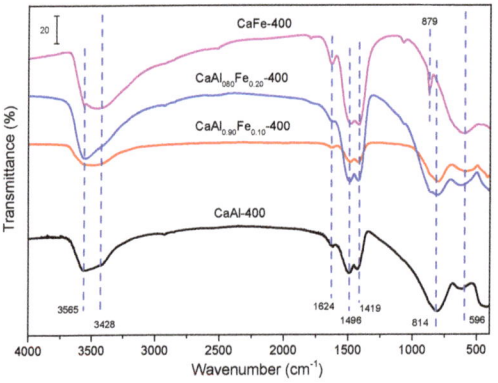

Figure 7. FT–IR spectra of the samples calcined at 400 °C.

Figure 8 shows the nitrogen adsorption–desorption isotherms of the solids calcined at 400 °C. All isotherms are type II, according to the IUPAC classification criteria [60]. Only the CaFe-400 sample showed a type H3 hysteresis loop [60]. Although the calcination should remove part of the interlayer water, the calcination process did not produce an increase in S_{BET} in the samples (Table 2), but a remarkable decrease, mainly associated with the loss of the layered structure (vide infra). Thus, for example, for the café sample, the S_{BET} decreased from 27 m^2/g to 16 m^2/g under calcination, probably by the loss of the layered structure and the formation under calcination of crystallites that may block access to the porosity. On the other hand, the average pore width remained in the mesopore range.

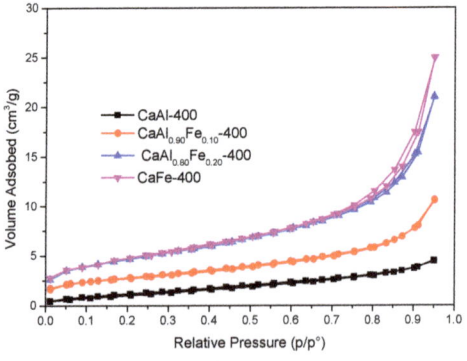

Figure 8. Nitrogen adsorption–desorption isotherms of samples calcined at 400 °C.

Table 2. S_{BET} and average pore diameter of samples calcined at 400 °C.

Sample	S_{BET} (m²/g)	Average Pore Diameter (nm)
CaAl-400	5	5.5
CaAl$_{0.90}$Fe$_{0.10}$-400	10	7.3
CaAl$_{0.80}$Fe$_{0.20}$-400	16	9.9
CaFe-400	16	10.1

Figure 9A,B show micrographs of the CaAl-400 sample. Aggregates of particles of size 70–80 µm are observed, consisting of hexagonal plate-shaped particles characteristic of LDH-type compounds [8]. However, unlike the uncalcined LDHs, the plate-shaped particles were aggregated into larger particles, showing that the layered structure was maintained at this calcination temperature [18,32,74]. However, sintering of the hexagonal plate-shaped particles with each other was observed. On the other hand, Figure 9C shows an SEM micrograph of the CaAl$_{0.80}$Fe$_{0.20}$-400 sample. Aggregates of the hexagonal plate-shaped particles were observed, although of considerably smaller size (5–10 µm) than in the case of the CaAl-400 sample and with an appearance of greater sponginess. Finally, Figure 9D,E show two SEM micrographs of the CaFe-400 sample. In this case, aggregates of particles with a spongy appearance were observed, totally different from those observed in the CaAl sample. This spongy aspect could justify the higher S_{BET} value for this solid. In the SEM micrograph of higher magnification (Figure 9E), particles in the form of a hexagonal plate remained, however, a higher degree of sintering for the CaAl-400 sample was observed.

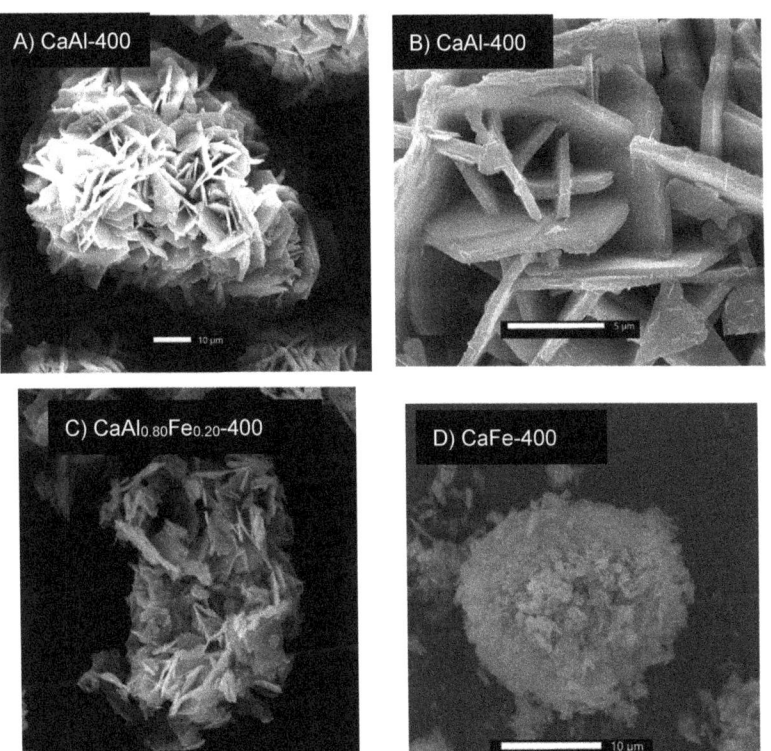

Figure 9. *Cont.*

Figure 9. SEM micrographs of samples calcined at 400 °C. (**A**,**B**) show micrographs of CaAl-400 sample. (**C**) belongs to CaAl$_{0.80}$Fe$_{0.20}$-400 sample and (**D**,**E**) show SEM micrographs of CaFe-400 sample.

3.3. Photocatalytic Application

Figure 10 shows the photodegradation of ibuprofen by the action of ultraviolet light in the absence and presence of LDHs calcined at 400 °C. Initially, the photodegradation of ibuprofen was evaluated in the absence of catalyst (photolysis) observing that 78% of the initial ibuprofen still remained after 152 min of irradiation with ultraviolet light in the photoreactor, i.e., in the presence of ultraviolet light but without photocatalyst, 22% of the initial ibuprofen was degraded, higher than the value reported in the literature [38]. When calcined LDH was added, the removal of the contaminant by adsorption was evaluated. For this purpose, the aqueous solution containing 50 ppm ibuprofen and 0.75 g of the photocatalyst was kept under stirring in the dark, and samples were taken at 0, 5, 10, 15, 20 and 35 min, finding that there was no elimination of ibuprofen by adsorption during this time. Subsequently, the ultraviolet light lamp was turned on and new aliquots were taken at various times. The concentration of the contaminant rapidly dropped to ca. 24% of the initial concentration after 42 min switching on the lamp (T$_{77}$, total time 77 min, 42 from the beginning of photocatalysis) with all the photocatalysts used. Up to this point, no major differences were observed in the percentage of ibuprofen degraded depending on the photocatalyst used; however, after this time of contact, the photocatalyst did not contain Fe^{3+} (CaAl-400) showed higher degradation percentages. Thus, 152 min after the lamp was switched on, the percentage of the initial ibuprofen remaining in the solution was only 4.7% for the CaAl sample and between 13% and 17% for the rest of the samples, those containing Fe^{3+}. Thus, for long times CaAl-400 photocatalyst showed the best performance and the presence of Fe^{3+} caused a low negative effect, this may be due to the formation of oxides or mixed oxides of Ca^{2+}, Fe^{3+} and Al^{3+} which did not possess photocatalytic properties, even though some of the properties of the photocatalysts were improved, such as S$_{BET}$ which increased with increasing the Fe^{3+} content. This behavior is similar to that reported on LDH-ZnFe systems, also calcined at 400 °C [51], in which ibuprofen degradation improved as the Zn/Fe ratio increased.

As indicated above, the photocatalyst obtained by calcination of LDHs at 400 °C that showed the best performance was CaAl-400; Figure 11A shows the ultraviolet-visible spectra of the aliquots taken at different times when this catalyst was used. During the "adsorption experiment", before the lamp was switched on (35 min) there was no variation in the characteristic band of ibuprofen at 222 nm [38,75], while after switching on the UV lamp the intensity of this band decreased, denoting photodegradation. Moreover, already for times T$_{42}$ and T$_{48}$ (7 min and 13 min after turning on the UV light, respectively), a new band appeared at 259 nm. This band may be due to the formation of a by-product of the degradation of ibuprofen; this band disappeared at longer times, indicating that this temporary intermediate was also photodegraded. For the rest of the photocatalysts, and also in the photolysis experiment, the evolution of the spectra was similar to that shown for the CaAl-400 sample.

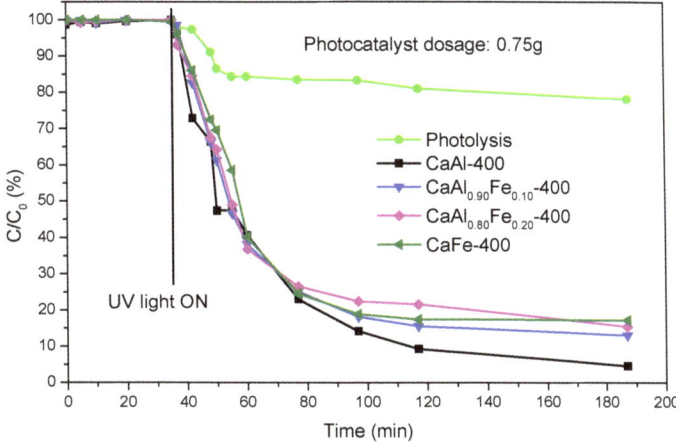

Figure 10. Ibuprofen degradation using samples calcined at 400 °C (photolysis is included for comparison).

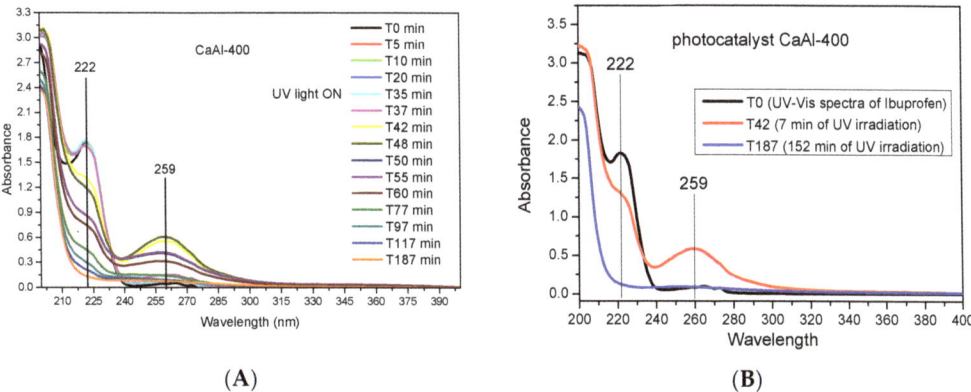

(A)   (B)

Figure 11. Ultraviolet-visible spectra of samples taken at different times when the CaAl-400 photocatalyst was used. (A) UV–Vis spectra of all samples taken at different times; (B) UV–Vis spectra of ibuprofen (T_0), aliquot at T_{42} and aliquot at T_{187}.

According to the most recent literature [40,44], the main photodegradation metabolites generated in the removal of ibuprofen by chemical methods (AOPs) were: 4-isobutylphenol ($C_{10}H_{14}O$), hydratropic acid ($C_9H_{10}O_2$), 4-(1-carboxyethyl)benzoic acid ($C_{10}H_{10}O_4$), 4-ethylbenzaldehyde ($C_9H_{10}O$), 2-[4-[4-(1-hydroxy-2-methylpropyl)phenyl] propanoic acid ($C_{13}H_{18}O_3$), 1-(4-isobutylphenyl)–1–ethanol ($C_{12}H_{18}O$), 4-acetylbenzoic acid ($C_9H_8O_3$), 1-isobutyl-4-vinylbenzene ($C_{12}H_{16}$) and 4-isobutylacetophenon ($C_{12}H_{16}O$). However, none of these metabolites were identified in the present work. As indicated, the spectrum of ibuprofen only showed the absorption maximum located at 222 nm (Figure 11B), but already for the aliquot taken 7 min after starting irradiation (T_{42}) an absorbance maximum at 259 nm was observed, and at the same time the HPLC–MS analysis showed the presence of two compounds: $C_{13}H_{18}O_2$ (ibuprofen) and a compound with formula $C_{13}H_{18}O_4$. This compound may correspond to a hydroxylated derivative of ibuprofen formed by a mechanism involving •OH radicals. In fact, the literature has listed various hydroxylated by-products in the photocatalytic degradation of ibuprofen [38,40,44,76,77], but the formation of one or another by-product is not clear and depends on the photocatalyst used. The band at 259 nm decreased as the irradiation time with UV light increased (Figure 11), so that for an irradiation time of 152 min this band practically did not exist and neither did the ibuprofen band, suggesting the disappearance of the aromatic ring. The aliquot of sample T_{187} was also analyzed by HPLC–MS, again finding a compound of formula $C_{13}H_{18}O_4$ and, although its concentration could not be quantified, both UV–Vis and HPLC–MS indicated a clear decrease in its presence from T_{42} to T_{187} aliquot. However, this did not mean that ibuprofen was completely and safely mineralized, as non-aromatic by-products may still remain in the solution. In this regard, further studies are required in order to better understand the performance of hydrocalumite-type solids and the reaction mechanism, also analyzing the mineralization of the pollutant by analysis of the total organic carbon.

4. Conclusions

Hydrocalumite LDH of formula $Ca_2Al_{1-m}Fe_m(OH)_6Cl \cdot 2H_2O$ were synthesized by the coprecipitation method, using as a source of aluminum one of the most hazardous wastes generated during aluminum recycling. The slags were treated in a basic medium under reflux conditions and the extraction solution was used as an aluminum source. The solids synthesized by the coprecipitation method were submitted to a microwave treatment at 125 °C for 2 h. In all cases, the corresponding LDH was obtained, the crystallinity decreasing as the Fe^{3+} content in the final solid increased, and the 110 diffraction peak shifted towards the lower two values. The lattice parameter a, determined from the (110) diffraction peak spacing, correlated with the amount of Fe^{3+} incorporated into the LDH, obeying Vegard's law. In spite of working in an inert atmosphere and using decarbonated water, the FT–IR results showed the presence of carbonate in the solids, due to the fixation of atmospheric CO_2 by the samples, due to their high basic character. As the amount of Fe^{3+} in the solids increased, the S_{BET} increased and the degree of hydration of the LDH decreased.

When LDH was calcined at 400 °C, a mixture of non-crystalline oxides was obtained, without crystalline phases, except in the CaFe-400 sample, where small diffraction peaks corresponding to hematite and magnetite were observed. The FT–IR results showed the presence of carbonate ions, possibly coming from the fixation of atmospheric CO_2, as well as the permanence of physisorbed water. The S_{BET} of the solids calcined at 400 °C was hardly modified for the samples with null or low Fe^{3+} content. However, for the CaFe-400 sample, a decrease in S_{BET} was observed.

The photocatalytic performance of the calcined solids was evaluated in the removal of ibuprofen. All the catalysts employed facilitated the degradation of this pollutant. However, as the Fe^{3+} content in the solid increased, the catalytic performance decreased. The catalyst that showed the best results was CaAl-400. In the same way, the HPLC–Mass and UV–Vis analyses of the aliquots taken at different times showed only the presence of a di-hydroxylated derivative, besides ibuprofen. This by-product reached its maximum formation after 7 min of irradiation, subsequently, it also degraded, although was still present in the solution after 152 min of irradiation with UV light. No other characteristic by-products of ibuprofen degradation were identified. More studies are required in order to investigate the effect of other parameters such as catalyst dosage, pH, etc., and mainly to elucidate the performance of hydrocalumite-type solids and the photodegradation mechanism.

Author Contributions: A.J.: Data curation, Formal analysis, Investigation, Methodology, Validation, Writing—original draft, Writing—review. M.V.: Data curation, Formal analysis, Investigation, Methodology, Validation. A.M.: Data curation, Formal analysis, Investigation, Methodology, Validation, Writing—original draft. R.T.: Conceptualization, Data curation, Formal analysis, Methodology, Supervision, Validation, Writing—review and editing. A.G.: Conceptualization, Data curation, Formal analysis, Methodology, Validation, Writing—review and editing. M.A.V.: Conceptualization, Data curation, Formal analysis, Methodology, Project administration, Resources, Supervision, Validation, Writing—review and editing. All authors have read and agreed to the published version of the manuscript.

Funding: This research was funded by Universidad de Salamanca (Plan I-B2).

Acknowledgments: This work belongs to the End-of-Degree work of Marta Valverde, supervised by Miguel Angel Vicente and Alejandro Jiménez, and is a section of the Ph. D. Thesis of Alejandro Jiménez, supervised by Miguel Angel Vicente and Vicente Rives. This article is included in the ChemEngineering Special Issue dedicated to Vicente Rives on his retirement (A Themed Issue in Honor of Vicente Rives, https://www.mdpi.com/journal/ChemEngineering/special_issues/honor_Vicente). With our participation in this Special Issue we want to express our immense gratitude to Vicente Rives for his help—as a teacher, as a researcher and as a friend—during all these years. A.J. thanks the Universidad de Salamanca and Banco Santander for a predoctoral contract.

Conflicts of Interest: The authors declare no conflict of interest.

References

1. World Bureau of Metal Statistics. Available online: http://www.world-bureau.com/ (accessed on 1 June 2022).
2. Gil, A. Management of the Salt Cake from Secondary Aluminum Fusion Processes. *Ind. Eng. Chem. Res.* **2005**, *44*, 8852–8857. [CrossRef]
3. He, L.; Shi, L.; Huang, Q.; Hayat, W.; Shang, Z.; Ma, T.; Wang, M.; Yao, W.; Huang, H.; Chen, R. Extraction of Alumina from Aluminum Dross by a Non-Hazardous Alkaline Sintering Process: Dissolution Kinetics of Alumina and Silica from Calcined Materials. *Sci. Total Environ.* **2021**, *777*, 146123. [CrossRef] [PubMed]
4. Tsakiridis, P.E. Aluminium Salt Slag Characterization and Utilization—A Review. *J. Hazard. Mater.* **2012**, *217–218*, 1–10. [CrossRef]
5. Tsakiridis, P.E.; Oustadakis, P.; Moustakas, K.; Agatzini, S.L. Cyclones and Fabric Filters Dusts from Secondary Aluminium Flue Gases: A Characterization and Leaching Study. *Int. J. Environ. Sci. Technol.* **2016**, *13*, 1793–1802. [CrossRef]
6. Mahinroosta, M.; Allahverdi, A. Hazardous Aluminum Dross Characterization and Recycling Strategies: A Critical Review. *J. Environ. Manag.* **2018**, *223*, 452–468. [CrossRef] [PubMed]
7. Jiménez, A.; Misol, A.; Morato, Á.; Rives, V.; Vicente, M.A.; Gil, A. Synthesis of Pollucite and Analcime Zeolites by Recovering Aluminum from a Saline Slag. *J. Clean. Prod.* **2021**, *297*, 126667. [CrossRef]
8. Jiménez, A.; Misol, A.; Morato, Á.; Rives, V.; Vicente, M.A.; Gil, A. Optimization of Hydrocalumite Preparation under Microwave Irradiation for Recovering Aluminium from a Saline Slag. *Appl. Clay Sci.* **2021**, *212*, 10217. [CrossRef]
9. Jiménez, A.; Rives, V.; Vicente, M.A.; Gil, A. A Comparative Study of Acid and Alkaline Aluminum Extraction Valorization Procedure for Aluminum Saline Slags. *J. Environ. Chem. Eng.* **2022**, *10*, 107546. [CrossRef]
10. Meshram, A.; Singh, K.K. Recovery of Valuable Products from Hazardous Aluminum Dross: A Review. *Resour. Conserv. Recycl.* **2018**, *130*, 95–108. [CrossRef]
11. EU Parliament. Directive 2010/75/EU of the European Parliament and of the Council of 24 November 2010 on Industrial Emissions (Integrated Pollution Prevention and Control). *Off. J. Eur. Union* **2010**, *334*, 17.
12. Gil, A.; Albeniz, S.; Korili, S.A. Valorization of the Saline Slags Generated during Secondary Aluminium Melting Processes as Adsorbents for the Removal of Heavy Metal Ions from Aqueous Solutions. *Chem. Eng. J.* **2014**, *251*, 43–50. [CrossRef]
13. Gil, A.; Korili, S.A. Management and Valorization of Aluminum Saline Slags: Current Status and Future Trends. *Chem. Eng. J.* **2016**, *289*, 74–84. [CrossRef]
14. Gil, A.; Arrieta, E.; Vicente, M.Á.; Korili, S.A. Application of Industrial Wastes from Chemically Treated Aluminum Saline Slags as Adsorbents. *ACS Omega* **2018**, *3*, 18275–18284. [CrossRef] [PubMed]
15. Das, B.R.; Dash, B.; Tripathy, B.C.; Bhattacharya, I.N.; Das, S.C. Production of η-Alumina from Waste Aluminium Dross. *Miner. Eng.* **2007**, *20*, 252–258. [CrossRef]
16. Yoldi, M.; Fuentes-Ordoñez, E.G.; Korili, S.A.; Gil, A. Zeolite Synthesis from Industrial Wastes. *Microporous Mesoporous Mater.* **2019**, *287*, 183–191. [CrossRef]
17. Gil, A.; Arrieta, E.; Vicente, M.A.; Korili, S.A. Synthesis and CO_2 Adsorption Properties of Hydrotalcite-like Compounds Prepared from Aluminum Saline Slag Wastes. *Chem. Eng. J.* **2018**, *334*, 1341–1350. [CrossRef]
18. Santamaría, L.; Vicente, M.A.; Korili, S.A.; Gil, A. Saline Slag Waste as an Aluminum Source for the Synthesis of Zn–Al–Fe–Ti Layered Double-Hydroxides as Catalysts for the Photodegradation of Emerging Contaminants. *J. Alloys Compd.* **2020**, *843*, 156017. [CrossRef]
19. Santamaría, L.; López-Aizpún, M.; García-Padial, M.; Vicente, M.A.; Korili, S.A.; Gil, A. Zn-Ti-Al Layered Double Hydroxides Synthesized from Aluminum Saline Slag Wastes as Efficient Drug Adsorbents. *Appl. Clay Sci.* **2020**, *187*, 105486. [CrossRef]
20. Santamaría, L.; García, L.O.; De Faria, E.H.; Ciuffi, K.J.; Vicente, M.A.; Korili, S.A.; Gil, A. M(II)-Al-Fe Layered Double Hydroxides Synthesized from Aluminum Saline Slag Wastes and Catalytic Performance on Cyclooctene Oxidation. *Miner. Eng.* **2022**, *180*, 107516. [CrossRef]
21. Rives, V. (Ed.) *Layered Double Hydroxides*; Nova Science Publishers, Inc.: New York, NY, USA, 2001.
22. Zhitova, E.S.; Pekov, I.V.; Chaikovskiy, I.I.; Chirkova, E.P.; Yapaskurt, V.O.; Bychkova, Y.V.; Belakovskiy, D.I.; Chukanov, N.V.; Zubkova, N.V.; Krivovichev, S.V.; et al. Dritsite, $Li_2Al_4(OH)_{12}Cl_2·3H_2O$, a New Gibbsite-Based Hydrotalcite Supergroup Mineral. *Minerals* **2019**, *9*, 492. [CrossRef]
23. Thiel, J.P.; Chiang, C.K.; Poeppelmeier, K.R. Structure of lithium aluminum hydroxide dihydrate ($LiAl_2(OH)_7·2H_2O$). *Chem. Mater.* **1993**, *2*, 297–304. [CrossRef]

24. Takaki, Y.; Qiu, X.; Hirajima, T.; Sasaki, K. Removal Mechanism of Arsenate by Bimetallic and Trimetallic Hydrocalumites Depending on Arsenate Concentration. *Appl. Clay Sci.* **2016**, *134*, 26–33. [CrossRef]
25. Linares, C.F.; Moscosso, J.; Alzurutt, V.; Ocanto, F.; Bretto, P.; González, G. Carbonated Hydrocalumite Synthesized by the Microwave Method as a Possible Antacid. *Mater. Sci. Eng. C* **2016**, *61*, 875–878. [CrossRef] [PubMed]
26. Murayama, N.; Maekawa, I.; Ushiro, H.; Miyoshi, T.; Shibata, J.; Valix, M. Synthesis of Various Layered Double Hydroxides Using Aluminum Dross Generated in Aluminum Recycling Process. *Int. J. Miner. Process.* **2012**, *110–111*, 46–52. [CrossRef]
27. Granados-Reyes, J.; Salagre, P.; Cesteros, Y. Effect of the Preparation Conditions on the Catalytic Activity of Calcined Ca/Al-Layered Double Hydroxides for the Synthesis of Glycerol Carbonate. *Appl. Catal. A Gen.* **2017**, *536*, 9–17. [CrossRef]
28. Granados-Reyes, J.; Salagre, P.; Cesteros, Y.; Busca, G.; Finocchio, E. Assessment through FT-IR of Surface Acidity and Basicity of Hydrocalumites by Nitrile Adsorption. *Appl. Clay Sci.* **2019**, *180*, 105180. [CrossRef]
29. Rosset, M.; Perez-Lopez, O.W. Cu–Ca–Al Catalysts Derived from Hydrocalumite and Their Application to Ethanol Dehydrogenation. *React. Kinet. Mech. Catal.* **2019**, *126*, 497–511. [CrossRef]
30. Souza Júnior, R.L.; Rossi, T.M.; Detoni, C.; Souza, M.M.V.M. Glycerol Carbonate Production from Transesterification of Glycerol with Diethyl Carbonate Catalyzed by Ca/Al-Mixed Oxides Derived from Hydrocalumite. *Biomass Convers. Biorefinery* **2020**. [CrossRef]
31. Gevers, B.R.; Labuschagné, F.J.W.J. Green Synthesis of Hydrocalumite (CaAl-OH-LDH) from $Ca(OH)_2$ and $Al(OH)_3$ and the Parameters That Influence Its Formation and Speciation. *Crystals* **2020**, *10*, 627. [CrossRef]
32. Fang, L.; Li, W.; Chen, H.; Xiao, F.; Huang, L.; Holm, P.E.; Hansen, H.C.B.; Wang, D. Synergistic Effect of Humic and Fulvic Acids on Ni Removal by the Calcined Mg/Al Layered Double Hydroxide. *RSC Adv.* **2015**, *5*, 18866–18874. [CrossRef]
33. Li, F.; Kong, Q.; Chen, P.; Chen, M.; Liu, G.; Lv, W.; Yao, K. Effect of Halide Ions on the Photodegradation of Ibuprofen in Aqueous Environments. *Chemosphere* **2017**, *166*, 412–417. [CrossRef] [PubMed]
34. Matamoros, V.; Duhec, A.; Albaigés, J.; Bayona, J.M. Photodegradation of Carbamazepine, Ibuprofen, Ketoprofen and 17α-Ethinylestradiol in Fresh and Seawater. *Water. Air. Soil Pollut.* **2009**, *196*, 161–168. [CrossRef]
35. Peuravuori, J.; Pihlaja, K. Phototransformations of Selected Pharmaceuticals under Low-Energy UVA-Vis and Powerful UVB-UVA Irradiations in Aqueous Solutions-the Role of Natural Dissolved Organic Chromophoric Material. *Anal. Bioanal. Chem.* **2009**, *394*, 1621–1636. [CrossRef] [PubMed]
36. Sá, A.S.; Feitosa, R.P.; Honório, L.; Peña-Garcia, R.; Almeida, L.C.; Dias, J.S.; Brazuna, L.P.; Tabuti, T.G.; Triboni, E.R.; Osajima, J.A.; et al. A Brief Photocatalytic Study of ZnO Containing Cerium towards Ibuprofen Degradation. *Materials* **2021**, *14*, 5891. [CrossRef] [PubMed]
37. Da Silva, J.C.C.; Teodoro, J.A.R.; Afonso, R.J.D.C.F.; Aquino, S.F.; Augusti, R. Photolysis and Photocatalysis of Ibuprofen in Aqueous Medium: Characterization of by-Products via Liquid Chromatography Coupled to High-Resolution Mass Spectrometry and Assessment of Their Toxicities against Artemia Salina. *J. Mass Spectrom.* **2014**, *49*, 145–153. [CrossRef]
38. Tian, H.; Fan, Y.; Zhao, Y.; Liu, L. Elimination of Ibuprofen and Its Relative Photo-Induced Toxicity by Mesoporous BiOBr under Simulated Solar Light Irradiation. *RSC Adv.* **2014**, *4*, 13061–13070. [CrossRef]
39. Li, F.H.; Yao, K.; Lv, W.Y.; Liu, G.G.; Chen, P.; Huang, H.P.; Kang, Y.P. Photodegradation of Ibuprofen under UV-VIS Irradiation: Mechanism and Toxicity of Photolysis Products. *Bull. Environ. Contam. Toxicol.* **2015**, *94*, 479–483. [CrossRef]
40. Arthur, R.B.; Bonin, J.L.; Ardill, L.P.; Rourk, E.J.; Patterson, H.H.; Stemmler, E.A. Photocatalytic Degradation of Ibuprofen over BiOCl Nanosheets with Identification of Intermediates. *J. Hazard. Mater.* **2018**, *358*, 1–9. [CrossRef]
41. Akkari, M.; Aranda, P.; Belver, C.; Bedia, J.; Ben Haj Amara, A.; Ruiz-Hitzky, E. ZnO/Sepiolite Heterostructured Materials for Solar Photocatalytic Degradation of Pharmaceuticals in Wastewater. *Appl. Clay Sci.* **2018**, *156*, 104–109. [CrossRef]
42. Gu, Y.; Yperman, J.; Carleer, R.; D'Haen, J.; Maggen, J.; Vanderheyden, S.; Vanreppelen, K.; Garcia, R.M. Adsorption and Photocatalytic Removal of Ibuprofen by Activated Carbon Impregnated with TiO_2 by UV–Vis Monitoring. *Chemosphere* **2019**, *217*, 724–731. [CrossRef]
43. Patterson, K.; Howlett, K.; Patterson, K.; Wang, B.; Jiang, L. Photodegradation of Ibuprofen and Four Other Pharmaceutical Pollutants on Natural Pigments Sensitized TiO_2 Nanoparticles. *Water Environ. Res.* **2020**, *92*, 1152–1161. [CrossRef] [PubMed]
44. Chopra, S.; Kumar, D. Ibuprofen as an Emerging Organic Contaminant in Environment, Distribution and Remediation. *Heliyon* **2020**, *6*, e04087. [CrossRef] [PubMed]
45. Bojer, C.; Schöbel, J.; Martin, T.; Ertl, M.; Schmalz, H.; Breu, J. Clinical Wastewater Treatment: Photochemical Removal of an Anionic Antibiotic (Ciprofloxacin) by Mesostructured High Aspect Ratio ZnO Nanotubes. *Appl. Catal. B Environ.* **2017**, *204*, 561–565. [CrossRef]
46. Kudo, A.; Miseki, Y. Heterogeneous Photocatalyst Materials for Water Splitting. *Chem. Soc. Rev.* **2009**, *38*, 253–278. [CrossRef] [PubMed]
47. Trujillano, R.; Nájera, C.; Rives, V. Activity in the Photodegradation of 4-Nitrophenol of a Zn,Al Hydrotalcite-Like Solid and the Derived Alumina-Supported ZnO. *Catalysts* **2020**, *10*, 702. [CrossRef]
48. Prince, J.; Tzompantzi, F.; Mendoza-Damián, G.; Hernández-Beltrán, F.; Valente, J.S. Photocatalytic Degradation of Phenol by Semiconducting Mixed Oxides Derived from Zn(Ga)Al Layered Double Hydroxides. *Appl. Catal. B Environ.* **2015**, *163*, 352–360. [CrossRef]
49. He, S.; Zhang, S.; Lu, J.; Zhao, Y.; Ma, J.; Wei, M.; Evans, D.G.; Duan, X. Enhancement of Visible Light Photocatalysis by Grafting ZnO Nanoplatelets with Exposed (0001) Facets onto a Hierarchical Substrate. *Chem. Commun.* **2011**, *47*, 10797–10799. [CrossRef]

50. Fan, G.; Li, F.; Evans, D.G.; Duan, X. Catalytic Applications of Layered Double Hydroxides: Recent Advances and Perspectives. *Chem. Soc. Rev.* **2014**, *43*, 7040–7066. [CrossRef]
51. Di, G.; Zhu, Z.; Zhang, H.; Zhu, J.; Lu, H.; Zhang, W.; Qiu, Y.; Zhu, L.; Küppers, S. Simultaneous Removal of Several Pharmaceuticals and Arsenic on Zn-Fe Mixed Metal Oxides: Combination of Photocatalysis and Adsorption. *Chem. Eng. J.* **2017**, *328*, 141–151. [CrossRef]
52. Phillips, J.D.; Vandeperre, L.J. Anion Capture with Calcium, Aluminium and Iron Containing Layered Double Hydroxides. *J. Nucl. Mater.* **2011**, *416*, 225–229. [CrossRef]
53. Lu, Y.; Zhang, Z.; Xu, Y.; Liu, Q.; Qian, G. CaFeAl Mixed Oxide Derived Heterogeneous Catalysts for Transesterification of Soybean Oil to Biodiesel. *Bioresour. Technol.* **2015**, *190*, 438–441. [CrossRef] [PubMed]
54. Szabados, M.; Pásztor, K.; Csendes, Z.; Muráth, S.; Kónya, Z.; Kukovecz, Á.; Carlson, S.; Sipos, P.; Pálinkó, I. Synthesis of High-Quality, Well-Characterized CaAlFe-Layered Triple Hydroxide with the Combination of Dry-Milling and Ultrasonic Irradiation in Aqueous Solution at Elevated Temperature. *Ultrason. Sonochem.* **2016**, *32*, 173–180. [CrossRef] [PubMed]
55. Sánchez-Cantú, M.; Barcelos-Santiago, C.; Gomez, C.M.; Ramos-Ramírez, E.; Ruiz Peralta, M.D.L.; Tepale, N.; González-Coronel, V.J.; Mantilla, A.; Tzompantzi, F. Evaluation of Hydrocalumite-Like Compounds as Catalyst Precursors in the Photodegradation of 2,4-Dichlorophenoxyacetic Acid. *Int. J. Photoenergy* **2016**, *2016*, 5256941. [CrossRef]
56. Gao, Y.; Zhang, Z.; Wu, J.; Yi, X.; Zheng, A.; Umar, A.; O'Hare, D.; Wang, Q. Comprehensive Investigation of CO_2 Adsorption on Mg-Al-CO_3 LDH-Derived Mixed Metal Oxides. *J. Mater. Chem. A* **2013**, *1*, 12782–12790. [CrossRef]
57. Jiménez, A.; Vicente, M.A.; Rives, V. Thermal Study of the Hydrocalumite—Katoite—Calcite System. *Thermochim. Acta* **2022**, *713*, 179242. [CrossRef]
58. Silva, J.M.; Trujillano, R.; Rives, V.; Soria, M.A.; Madeira, L.M. High Temperature CO_2 Sorption over Modified Hydrotalcites. *Chem. Eng. J.* **2017**, *325*, 25–34. [CrossRef]
59. *ICDD Database, JCPDS*; International Centre for Diffraction Data (ICDD®): Newtown Square, PA, USA, 2020.
60. Thommes, M.; Kaneko, K.; Neimark, A.V.; Olivier, J.P.; Rodriguez-Reinoso, F.; Rouquerol, J.; Sing, K.S.W. Physisorption of Gases, with Special Reference to the Evaluation of Surface Area and Pore Size Distribution (IUPAC Technical Report). *Pure Appl. Chem.* **2015**, *87*, 1051–1069. [CrossRef]
61. Lide, D.R. *CRC Handbook of Chemistry and Physics*, 76th ed.; CRC Press: Boca Raton, FL, USA, 1995.
62. Cavani, F.; Trifirò, F.; Vaccari, A. Hydrotalcite-Type Anionic Clays: Preparation, Properties and Applications. *Catal. Today* **1991**, *11*, 173–301. [CrossRef]
63. López-Salinas, E.; Serrano, M.E.L.; Jácome, M.A.C.; Secora, I.S. Characterization of Synthetic Hydrocalumite-Type [$Ca_2Al(OH)_6$]$NO_3 \cdot mH_2O$: Effect of the Calcination Temperature. *J. Porous Mater.* **1996**, *2*, 291–297. [CrossRef]
64. Radha, A.V.; Kamath, P.V.; Shivakumara, C. Mechanism of the Anion Exchange Reactions of the Layered Double Hydroxides (LDHs) of Ca and Mg with Al. *Solid State Sci.* **2005**, *7*, 1180–1187. [CrossRef]
65. Rousselot, I.; Taviot-Guého, C.; Leroux, F.; Léone, P.; Palvadeau, P.; Besse, J.P. Insights on the Structural Chemistry of Hydrocalumite and Hydrotalcite-like Materials: Investigation of the Series $Ca_2M^{3+}(OH)_6Cl \cdot 2H_2O$ (M^{3+}: Al^{3+}, Ga^{3+}, Fe^{3+}, and Sc^{3+}) by X-ray Powder Diffraction. *J. Solid State Chem.* **2002**, *167*, 137–144. [CrossRef]
66. Pérez-Barrado, E.; Pujol, M.C.; Aguiló, M.; Cesteros, Y.; Díaz, F.; Pallarès, J.; Marsal, L.F.; Salagre, P. Fast Aging Treatment for the Synthesis of Hydrocalumites Using Microwaves. *Appl. Clay Sci.* **2013**, *80–81*, 313–319. [CrossRef]
67. Jenkins, R.; de Vries, J.L. Worked Examples in X-ray Analysis. In *Part of the Philips Technical Library Book Series*; Springer: Berlin, Germany, 1978.
68. Nyquist, R.A.; Kagel, R.O. *Infrared Spectra of Inorganic Compounds*; Academic Press: New York, NY, USA, 2001.
69. Bastida, J.; Bolós, C.; Pardo, P.; Serrano, F.J. Análisis Microestructural Por DRX de CaO Obtenido a Partir de Carbonato Cálcico Molido (CCM). *Bol. Soc. Esp. Cerám. Vidr.* **2004**, *43*, 80–83. [CrossRef]
70. Pan, X.; Liu, J.; Wu, S.; Yu, H. Formation Behavior of Tricalcium Aluminate Hexahydrate in Synthetic Sodium Aluminate Solution with High Alkali Concentration and Caustic Ratio. *Hydrometallurgy* **2020**, *195*, 105373. [CrossRef]
71. Nakamoto, K. *Infrared and Raman Spectra of Inorganic and Coordination Compounds: Part A: Theory and Applications in Inorganic Chemistry*; Wiley: Hoboken, NJ, USA, 2008.
72. Granados-Reyes, J.; Salagre, P.; Cesteros, Y. Effect of Microwaves, Ultrasounds and Interlayer Anion on the Hydrocalumites Synthesis. *Micropor. Mesopor. Mater.* **2014**, *199*, 117–124. [CrossRef]
73. Domínguez, M.; Pérez-Bernal, M.E.; Ruano-Casero, R.J.; Barriga, C.; Rives, V.; Ferreira, R.A.S.; Carlos, L.D.; Rocha, J. Multiwavelength Luminescence in Lanthanide-Doped Hydrocalumite and Mayenite. *Chem. Mater.* **2011**, *23*, 1993–2004. [CrossRef]
74. Chen, G.; Qian, S.; Tu, X.; Wei, X.; Zou, J.; Leng, L.; Luo, S. Enhancement Photocatalytic Degradation of Rhodamine B on NanoPt Intercalated Zn-Ti Layered Double Hydroxides. *Appl. Surf. Sci.* **2014**, *293*, 345–351. [CrossRef]
75. Padilla Villavicencio, M.; Escobedo Morales, A.; de Ruiz Peralta, M.L.; Sánchez-Cantú, M.; Rojas Blanco, L.; Chigo Anota, E.; Camacho García, J.H.; Tzompantzi, F. Ibuprofen Photodegradation by Ag_2O and Ag/Ag_2O Composites Under Simulated Visible Light Irradiation. *Catal. Lett.* **2020**, *150*, 2385–2399. [CrossRef]

76. Liu, S.H.; Tang, W.T.; Chou, P.H. Microwave-Assisted Synthesis of Triple 2D g-C_3N_4/Bi_2WO_6/RGO Composites for Ibuprofen Photodegradation: Kinetics, Mechanism and Toxicity Evaluation of Degradation Products. *Chem. Eng. J.* **2020**, *387*, 124098. [CrossRef]
77. Miranda, M.O.; Cabral Cavalcanti, W.E.; Barbosa, F.F.; Antonio De Sousa, J.; Ivan Da Silva, F.; Pergher, S.B.C.; Braga, T.P. Photocatalytic Degradation of Ibuprofen Using Titanium Oxide: Insights into the Mechanism and Preferential Attack of Radicals. *RSC Adv.* **2021**, *11*, 27720–27733. [CrossRef]

Article

Drug-Containing Layered Double Hydroxide/Alginate Dispersions for Tissue Engineering

Juan Pablo Zanin [1,2,3,4], German A. Gil [3,4], Mónica C. García [5,6] and Ricardo Rojas [1,2,*]

1. Departamento de Fisicoquímica, Facultad de Ciencias Químicas, Universidad Nacional de Córdoba, Córdoba 5000, Argentina
2. Consejo Nacional de Investigaciones Científicas y Técnicas, CONICET, Instituto de Investigaciones en Fisicoquímica de Córdoba, INFIQC, Córdoba 5000, Argentina
3. Departamento de Química Biológica, Facultad de Ciencias Químicas, Universidad Nacional de Córdoba, Córdoba 5000, Argentina
4. Consejo Nacional de Investigaciones Científicas y Técnicas, CONICET, Centro de Investigaciones en Química Biológica de Córdoba, CIQUIBIC, Córdoba 5000, Argentina
5. Departamento de Ciencias Farmacéuticas, Facultad de Ciencias Químicas, Universidad Nacional de Córdoba, Córdoba 5000, Argentina
6. Consejo Nacional de Investigaciones Científicas y Técnicas, CONICET, Unidad de Investigación y Desarrollo en Tecnología Farmacéutica, UNITEFA, Córdoba 5000, Argentina
* Correspondence: ricardo.rojas@unc.edu.ar

Abstract: Alginate (Alg) is increasingly studied as a constitutive material of scaffolds for tissue engineering because of its easy gelation and biocompatibility, and the incorporation of drugs into its formulation allows for its functionality to be extended. However, Alg presents a low cell adhesion and proliferation capacity, and the incorporation of drugs may further reduce its biocompatibility. Layered double hydroxides (LDH) are promising fillers for Alg-based biomaterials, as they increase cell adhesion and interaction and provide drug storage and controlled release. In this work, LDH containing ibuprofen or naproxen were synthesized by coprecipitation at a constant pH and their properties upon their incorporation in Alg dispersions (LDH-Drug/Alg) were explored. Drug release profiles in simulated body fluid and the proliferation of pre-osteoblastic MC3T3-E1 cells by LDH-Drug/Alg dispersions were then evaluated, leading to results that confirm their potential as biomaterials for tissue engineering. They showed a controlled release with diffusive control, modulated by the in-situ formation of an Alg hydrogel in the presence of Ca^{2+} ions. Additionally, LDH-Drug/Alg dispersions mitigated the cytotoxic effects of the pure drugs, especially in the case of markedly cytotoxic drugs such as naproxen.

Keywords: scaffolding biomaterials; drug delivery systems; hydrogel; cytotoxicity

1. Introduction

Alginate (Alg) is a natural polymer that is extensively used in drug delivery and scaffolding applications. It is composed by monomers of α-L-guluronate and β-D-mannuronate in a ratio that varies depending on the type of seaweed from which it is extracted. Alg presents a high concentration of carboxylate groups that are charged at pH values above 3–4, being highly soluble under neutral and alkaline conditions. Alg is cross-linked by Ca^{2+} ions (Ba^{2+} and Zn^{2+} ions are also used to a lesser extent) to form hydrogels [1]. Alg is commonly used as excipient in the pharmaceutical industry, but it has also been used as a drug carrier, providing increased drug solubility, pH-triggered and/or controlled release rate [2]. More recently, Alg has been proposed for tissue engineering applications due to its similarity to other extracellular matrix components, as well as its biocompatibility, gel-forming ability, and water retention capacity, thus providing a suitable environment for the regeneration of tissues and organs, such as skeletal bone, skin, nerve, liver, and pancreas [3]. Particularly, Alg is extensively used to fabricate scaffolds for bone tissue engineering, either alone or in combination with other polymers, either hydrophilic (i.e., chitosan) or hydrophobic (i.e., polylactic acid and polycaprolactone), as well as (nano)particles [4–6]. Strategies such as cell seeding, and drug delivery have been applied to such scaffolds to expand and to optimize their bio-functionality.

Nevertheless, the applications of Alg in tissue engineering are hindered by its poor mechanical properties and low capacity for cell adhesion and proliferation. Furthermore, the leakage of drug loading during cross-linking of the Alg matrix and fast drug release under biological conditions decrease the effectiveness of approaches involving the incorporation of drugs and biomolecules [7,8]. Particles, particularly those of an inorganic nature, such as hydroxyapatite, graphene, clay and layered double hydroxides (LDH) have been used to reinforce the mechanical properties of Alg, increase its bifunctionality and control the release of the drugs included its formulations [9].

LDH are bidimensional solids with brucite ($Mg(OH)_2$)-like layers that present the isomorphic substitution of divalent by trivalent ions, which leads to anion intercalation and exchange capacity. LDH present a great versatility and customization capacity due to the different metal ions (either divalent or trivalent) that can be included in their layers, as well as their anion exchange and surface adsorption properties [10]. LDH, due to their lamellar structure and anion exchange capacity, as well as their biocompatibility and ability to interact with cells, are also extensively used for tissue engineering applications [11,12]. They also present the high drug loading capacity of acidic drugs between their layers, and have been proposed as nanocarriers for cell internalization and the release of drugs and genes. In this sense, several types of drugs have been intercalated into LDH interlayers, among which nonsteroidal anti-inflammatory drugs (NSAID) have shown a controlled release rate and improved solubility in acid media [13–17].

Due to the carboxylate anions in its structure, Alg presents electrostatic interactions with positive LDH surface charges. As a result, LDH particles attach to Alg chains [18,19], being retained upon the formation of Alg-Ca hydrogels [20]. LDHs intercalated with NSAID such as ibuprofen (Ibu), diclofenac and naproxen (Nap) have also been incorporated into Alg composites, and beads [7,21,22]. Silver sulfadiazine-loaded LDH have also been incorporated into Alg films [23] for wound-dressing applications. The inclusion of LDH particles produced controlled drug release and sustained antimicrobial activity while maintaining a low cytotoxicity in cells. Nevertheless, the drug release under body conditions, especially for non-gelled LDH/Alg dispersions, has scarcely been explored, and LDH's effect on the cytotoxicity of both Alg and the incorporated drugs is also poorly explored. These aspects are essential for their performance as biomaterials for tissue engineering, with non-gelled dispersions being especially suitable, as they allow for the easy incorporation of cells, allowing for a posterior in-situ gelation.

In this work, we prepared drug-loaded LDH (LDH-Drug) particlesin Alg dispersions (LDH-Drug/Alg) and explored their release performances and biocompatibility as scaffolding biomaterials for tissue engineering. With this aim, Mg-Al LDH intercalated with Ibu and Nap was synthesized by coprecipitation at a constant pH, and its properties upon incorporation in Alg dispersion were explored. The release profiles of intercalated drugs toward bio-relevant fluids and the proliferation capacity of these biomaterials were evaluated against pre-osteoblastic MC3T3-E1 cell cultures.

2. Materials and Methods

Ibu and Nap anhydrous acids ($\geq$98% purity, Parapharm®), $MgCl_2 \cdot 6H_2O$ (Baker®), $AlCl_3 \cdot 6H2O$ (Anedra®), NaOH (Baker®), NaOH granules (PA grade, Cicarelli®), 37% w w^{-1} HCl solution (PA grade, Cicarelli®), KH_2PO_4 and K_2HPO_4 (PA grade, Anedra®), NaCl (PA grade, Cicarelli®), $NaHCO_3$ (PA grade, Cicarelli®), KCl (PA grade, Anedra®), (PA grade, Cicarelli®), tris (hydroxymethyl)aminomethane (Tris buffer, PA grade, Biopack®), and Na_2SO_4 (PA grade, Baker, polyacrylate (PA, 40% w/w solution, molecular weight, MW = 8 kDa, Sigma Aldrich). Minimum Essential Medium Eagle without ascorbic acid (MEM, Thermo Fisher, Waltham, MA, USA), fetal bovine serum (FBS, Gibco/Thermo Fisher), Penicillin-

Streptomycin-Neomycin (PSN) Antibiotic Mixture (Thermo Fisher) and alamarBlue (Merck). Phosphate buffer solution (PBS) at pH 7.4 was prepared according to USP specifications [24]. Simulated body fluid (SBF) was prepared according to standardized parameters [25].

Deionized water (18 MΩ MilliQ, Millipore® System) was used in all experiments, which were conducted at room temperature (25 °C) unless otherwise stated.

2.1. Synthesis and Structural Characterization of LDH-Drug

Mg-Al LDH loaded with NSAID (LDH-Drug, either LDH-Ibu or LDH-Nap) were synthesized by the coprecipitation method at constant pH. A solution of the metal ions (0.4 mol L^{-1} $AlCl_3$ and 1.2 mol L^{-1} $MgCl_2$, 0.1 L) was added to a 0.1 L solution containing 0.06 mol of the corresponding drug, previously dissolved by addition of NaOH. The addition was performed dropwise, under constant stirring, in nitrogen atmosphere and at pH = 9, controlled by addition of a 2.0 mol L^{-1} NaOH solution. The addition of this solution was controlled by a Titrando 905 automatic titrator (Metrohm) coupled to a Metrohm 9.0262.100 combined pH electrode. The obtained slurries were centrifuged, washed, and finally dried at 50 °C until constant weight. For cell proliferation studies, a LDH intercalated with chloride (LDH-Cl) was synthesized with the same procedure, but replacing the drug with NaCl.

The powder X-ray diffraction (PXRD) patterns were recorded with a Phillips X'pert Pro instrument equipped with a Pixcell 1D detector and a CuKα lamp (λ = 1.5408 Å) at 40 kV and 40 mA. The scans were performed in continuous mode (10° min^{-1}) between 5 and 70°. Low angle measurements (2–10°) were performed in step mode (0.05°, 1.2 s) with a Xe detector coupled to a graphite monochromator. Fourier-transform infrared (FTIR) spectra were measured with a FTIR Bruker IFS28 instrument using KBr pellets (1:200 sample:KBr ratio) at a 4 cm^{-1} resolution and accumulating 32 scans. Scanning electron microscopy (SEM) images were obtained in a FE-SEM Σigma instrument on samples covered with a Cr layer. The Mg/Al ratio was determined by energy dispersive X-ray spectroscopy (EDS) in the same instrument. The drug content of the samples was determined in dispersions of samples (1 g L^{-1}) prepared in PBS, which were equilibrated until a constant drug concentration was reached. The drug concentration in the supernatants was determined by UV-Vis spectrophotometry (Agilent Technologies® Cary 60) at λ = 222 nm (Ibu) and 272 nm (Nap).

2.2. Dispersion of LDH–Drug Particles

The dispersion of the dried LDH–drug particles was essayed in different media and under different conditions to optimize their interaction with Alg and determine the effect on the particle distribution. Both LDH-Ibu and LDH-Nap (5 g L^{-1}), alone or together with Alg powder (2 % w w^{-1}), were dispersed in water and equilibrated for 24 h. These dispersions were then sonicated for 2 h and sterilized in an autoclave at 121 °C. Aliquots of the dispersions were taken after the sonication and the autoclave step and diluted 1:10 in water. The hydrodynamic apparent diameter (d) and zeta potential (ζ) of LDH–drug particles were determined by dynamic light scattering (DLS) and electrophoretic light scattering (ELS) measurements, respectively, using a Delsa Nano C instrument (Beckman Coulter). Drug released from LDH in Alg dispersions was determined after sonication and thermal treatment (TT). The dispersions were centrifuged and filtered and the free drug concentration in the supernatants was determined as previously described. A dispersion of the LDH–drug particles in a 2% polyacrylate (MW = 8 kDa) equilibrated for 24 h without further treatment was used for comparative purposes.

2.3. In Vitro Drug Release Studies

Release studies of both Ibu and Nap from 2% Alg dispersions of LDH–drug particles at two different concentrations (0.5 and 5 g L^{-1}, prepared as described in Section 2.3) were performed in bicompartmental diffusion devices (Franz cells). A semisynthetic acetate cellulose membrane (molecular cut-off 12 kDa, Sigma-Aldrich®) was mounted between the donor and the receptor compartments. The release profiles from 2% Alg dispersions with equivalent concentrations of the pure drugs were also analyzed. A total of 1 mL of each sample was carefully placed in the donor compartment and kept in contact with 16.5 mL of receptor medium (PBS and SBF) at 37.0 ± 0.5 °C. 1 mL aliquots of receptor medium were withdrawn at predetermined time intervals (5; 15; and 30 min; 1; 1,5; 2; 3; 4; 5; 6; 7; and 8 h) and replaced with equivalent volumes of preheated fresh medium. The concentration of each drug was determined by UV-Vis spectrophotometry as previously described, using calibration curves constructed for each drug at each receptor medium. All experiments were conducted in triplicate and the sink conditions were maintained.

The drug release profiles were statistically compared using the difference factor (f_1) and the similarity factor (f_2) (Equations (1) and (2), respectively). According to this methodology, an f_1 value above 15 and an f_2 value in the 0–49 range implies a difference between the release profiles [26].

$$f_1 = \frac{\sum_{t=1}^{n}|R_t - T_t|}{\sum_{t=1}^{n} Rt} \times 100 \tag{1}$$

$$f_2 = 50 \, log\left\{\left[1 + \frac{1}{n}\sum_{t=1}^{n}(R_t - T_t)^2\right]^{-0.5}\right\} \times 100 \tag{2}$$

where R_t and T_t are the cumulative percentages of drug released at each of the n time points of the reference and test sample, respectively.

The release profiles were fitted with common mathematical models [26]:

Zero order:

$$\% \, Drug = \% \, Drug_0 + k_Z t \tag{3}$$

Higuchi:

$$\% \, Drug = \% \, Drug_0 + k_H t^{0.5} \tag{4}$$

Korsmeyer–Peppas (K-P):

$$\% \, Drug = \% \, Drug_0 + k_P t^n \tag{5}$$

where % $Drug_0$ is the intercept, often referring to the initial amount of drug in the receptor or fast processes that are produced at the beginning of the release experiment (*burst* effect). k_0, k_H and k_P are the kinetic constants corresponding to zero-order, Higuchi, and K-P kinetic models, respectively. Finally, n parameter in the K-P model (Equation (5)) describes the release mechanism: when n = 0.5, the fraction of drug released is proportional to the square root of time (Higuchi model, Equation (4)) and the drug release is purely diffusion controlled; while, when n = 1, the equation is identical to that of zero-order (case-II transport, Equation (3)). Values of n between 0.5 and 1 indicate an anomalous process with contributions from different phenomena, such as ionic exchange and the relaxation of polymer chains, among others.

2.4. Cell Proliferation Assay

MC3T3-E1 cells were maintained in a complete medium (CM) prepared with MEM, 10% FBS, and 1 × PSN antibiotics at 37 °C in 5% CO_2 atmosphere. An MC3T3-E1 cell line was obtained from ATCC (Manassas, VA, USA) in 2020, cultured at 37 °C in 5% CO_2 atmosphere until passage 2–4, and then cultured for 3–4 additional passages if necessary. Cells were authenticated based on morphology and growth curve analysis. Mycoplasma detection was performed every 2 months by PCR and Hoechst 33,258 staining [27].

Cell proliferation was assessed using alamarBlue, a resazurin-based solutions (fluorometric/colorimetric) growth indicator that changes from an oxidized (nonfluorescent, blue) form to a reduced (fluorescent, red) form when reduced by mitochondrial respiration. Cells were seeded in 96-well plates (3000 cells/well) and maintained for 1 day until complete cell attachment. The growth medium was then discarded and replaced with medium containing the different samples, and 10% alamarBlue was added. Treated cells were incubated for 24 and 48 h and the fluorescence intensity was measured on a PlateReader (Biotek®, excitation/emission = 535 nm/590 nm). The performed tests were control (pure CM); Alg; LDH; LDH-Cl/Alg; pure drug (anionic); Drug/Alg; 5 g L^{-1} LDH-Drug; and 5 g L^{-1} LDH-Drug/Alg. The concentration of LDH-Cl and drug used for each control were equivalent to the amount contained in the 5 g L^{-1} LDH-Drug/Alg dispersions, while Alg concentration was 2% in all cases. The results were expressed as a percentage of the intensity recorded for the corresponding control experiment.

Data are expressed as mean ± standard deviation (SD) (n = 4). Statistical significance of comparisons of mean values was assessed by two-way ANOVA test, followed by Tukey's multiple comparison. p values above 0.05 were considered statically significant.

3. Results and Discussion

3.1. Structural Characterization of the LDH–Drug Samples

The proposed synthesis was effective in obtaining LDH phases that were completely intercalated with the corresponding drug. The PXRD patterns of the samples (Figure 1a) showed narrow and intense peaks corresponding to the (0 0 l) reflections of LDH structure at 2θ values lower than 30°. Wide and asymmetric peaks corresponding to (0 1 l) reflections and a peak due to a (1 1 0) plane were obtained at larger 2θ values. This last peak allowed for the a parameter of the rhombohedral cell of LDH structure (Table 1) to be obtained, which was similar to that commonly obtained for LDH phases in both cases [28] and to that obtained for the reference sample LDH-Cl. The (0 0 l) peaks were recorded at low 2θ, corresponding to the LDH intercalated with large anions, and led to large values of the c cell parameter, while LDH-Cl showed (0 0 l) peaks at larger 2θ values. The interlayer distances (c/3 = 22.2 and 22.4 Å for LDH-Ibu and LDH-Nap, respectively) gave a hint of the drug disposition between the layers. Thus, the Ibu anion presents a length, at its larger axis, of around 9.7 Å [29], which led to a theoretical interlayer distance (accounting on a 4.8 Å for the layer height) of 14.5 Å for a perpendicular, monolayer arrangement and 24.3 Å for a bilayer one. However, the most common values are around 22.5 Å, which was assigned to a tilted bilayer disposition that maximizes the interaction between carboxylate anions and the hydroxyl anions of the layers [29]. Although Nap is a larger anion (its length is around 12 Å), the interlayer spacings obtained for LDH-Nap are quite close to that of LDH-Ibu [30,31]. These lower spacings indicate that a higher tilting from a perpendicular arrangement to the LDH layers is produced for this anion.

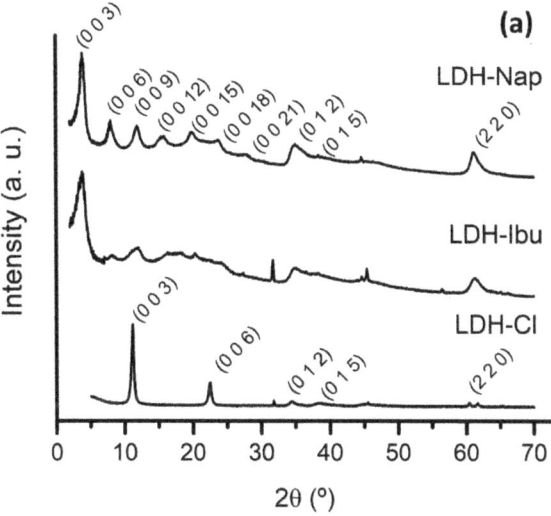

Figure 1. Cont.

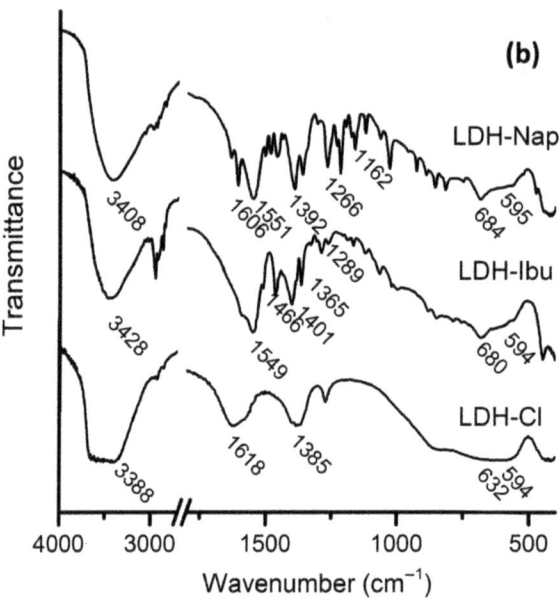

Figure 1. FT-IR spectra (**a**) and PXRD patterns (**b**) of LDH-Cl, LDH-Ibu and LDH-Nap samples.

Table 1. Chemical analysis and proposed formula for the synthesized LDH–drug samples.

Sample	% Mg	% Al	% D	% H$_2$O *	Chemical Formula	a(Å)	c (Å)
LDH-Ibu	11.3	6.5	47.9	9.5	Mg$_{0.66}$Al$_{0.34}$(OH)$_2$Ibu$_{0.34}$· 0.75 H$_2$O	3.02	66.5
LDH-Nap	12.2	6.2	46.9	9.3	Mg$_{0.69}$Al$_{0.31}$(OH)$_2$Nap$_{0.31}$· 0.72 H$_2$O	3.02	67.3

* Obtained from the weight loss at 200 °C.

The FT-IR spectra showed characteristic peaks in both the LDH structure and the intercalated drugs (Figure 1b). The band centered at 3428 and 3408 cm^{-1} for LDH-Ibu and LDH-Nap, respectively, was assigned to the O-H stretching mode (ν_{OH}) of OH$^-$ anions and H$_2$O water molecules. The bands recorded at wavenumbers below 1000 cm^{-1}, especially those at 594–595 cm^{-1}, were assigned to the lattice vibrations of the LDH layers [32,33]. The band corresponding to the bending vibration of structural water molecules between the layers registered at 1618 cm^{-1} for LDH-Cl was missing for LDH–drug samples due to the strong drug bands in the 1550–1650 cm^{-1} region. These bands were assigned to the ν_{asym} (at 1549 and 1551 cm^{-1} for LDH-Ibu and LDH-Nap, respectively) and ν_{sym} (1401 and 1392 cm^{-1}) of the carboxylate anions of the intercalated drugs. The difference between the maxima of these bands ($\Delta\nu = \nu_{asym} - \nu_{sym}$) is considered to be indicative of the interactions established between the carboxylate group and the hydroxylated layers [34]. The obtained values (148 and 159 cm^{-1} for LDH-Ibu and LDH-Nap, respectively) corresponded to weak interactions produced by electrostatic forces between the negatively charged carboxylate groups and the positively charged layers, as well as the hydrogen bonding between these groups and the hydroxyl anions of the layers. Other characteristic bands were registered at 1466 and 1365 cm^{-1} (δ (CH$_2$)) and at 1289 cm^{-1} (γ (OH)) for LDH-Ibu [35,36] and at 1266 (ν (C-O)), 1162 (ν (C-O-C)) and 1606 (aromatic ring) for LDH-Nap, among others [37,38]. Neither of the samples presented C = O stretching vibrations characteristic of the carboxylic group, which indicated that the acidic form of the drug was negligible in both cases.

Then, a single LDH phase, fully intercalated with the anionic form of the respective drugs, was obtained in both cases, as confirmed by the chemical analysis of the samples

(Table 1). The determined drug content (% D) was consistent with a 100% occupation of the exchange sites of LDH structure. Then, negligible anion excess was detected by the chemical analysis of the samples. Nevertheless, the presence of a slight anion excess in the surface of LDH-Ibu was suggested by the negative ζ of these particles (Figure 2). These values indicated that these anions were attached to LDH layers by specific interactions besides the electrostatic ones. In the case of anions attached exclusively by electrostatic interactions, the ζ values are positive, reversed only at high pH values [39,40]. In previous works, we even found a slight excess of Ibu$^-$ anions, caused by the additional stabilization assigned to hydrophobic interactions between the nonpolar sections of adjacent Ibu$^-$ anions. This excess was not produced for Nap$^-$ anions due to the presence of polar groups in its hydrophobic tail, which weakens the hydrophobic interactions [14,38].

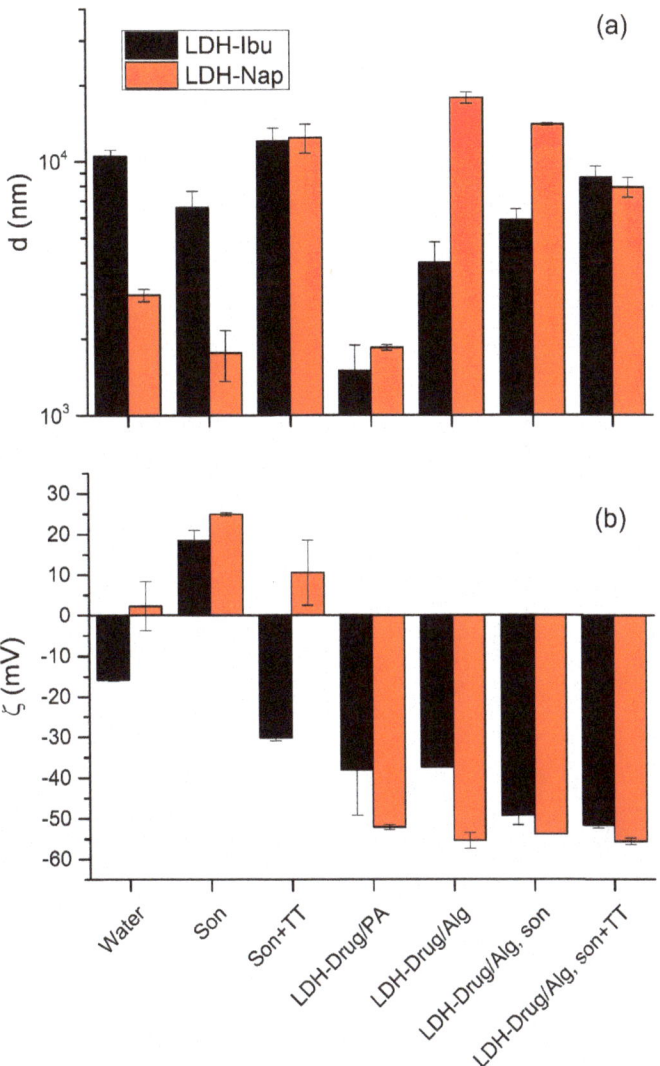

Figure 2. Hydrodynamic apparent diameter, d (**a**) and zeta potential, ζ (**b**) values obtained for LDH-Ibu and LDH-Nap dispersions (1 g L^{-1}) in different media and preparation conditions (son, sonicated; TT, thermal treatment; PA, 2% polyacrylate dispersion).

The morphology of LDH particles and agglomerates was studied using SEM images (Figure 3) and the d values registered by DLS (Figure 2a). The particles of both LDH-Ibu and LDH-Nap were heavily agglomerated, with the size of the agglomerates being larger for the former. Thus, LDH-Ibu particles in the SEM images completely lost the typical lamellar morphology of LDH particles, being merged in agglomerates with rounded edges and smooth surfaces. The size of LDH-Ibu agglomerates was variable, although most samples were above 10 μm, which was in accordance with the d values obtained by DLS measurements. The large aggregation of LDH-Ibu particles was assigned to the hydrophobic surface of the LDH-Ibu particles, as the anion drug is exposes its nonpolar tail to the aqueous side of the interface. Therefore, interactions between hydrophobic LDH particles was favored, while the surface tension with water is increased. The SEM images of LDH-Nap showed more defined particles and less aggregation, in good accordance with the d values obtained by DLS. The lower aggregation of LDH-Nap was assigned to the presence of a polar ether (C-O-C) group at the end of the hydrophobic tail. As a result, the interaction between the LDH-Nap particles is weaker and the surface tension with water is lower, allowing for a better dispersion of their particles and a lower agglomeration.

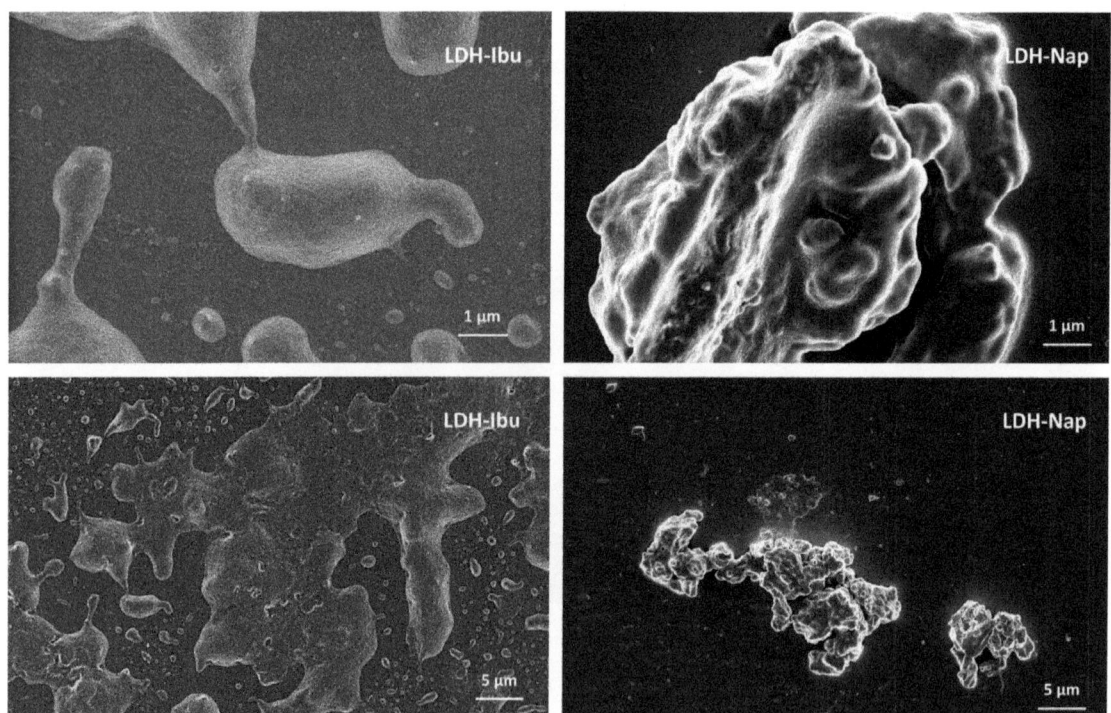

Figure 3. SEM images of LDH–drug samples shown at different magnifications.

3.2. Dispersion of LDH–Drug Particles

Both d and ζ values of the LDH particles were affected by dispersion conditions and the presence of Alg (Figure 2). Thus, the sonication of water dispersions produced a reversal of the ζ values of LDH-Ibu and an increase in the already positive ζ of LDH-Nap, which indicated that drug anions were detached from the particle surface. As a result, the surface hydrophilicity increased, which diminished aggregation, leading to smaller d values. In contrast, the TT sterilization of dispersions produced the opposite effect: the ζ of LDH-Ibu particles was reversed, that of LDH-Nap decreased and the d values of both LDH–drug dispersions increased. Then, the drug adsorption at the surface of LDH formed an equilibrium that was quite sensitive to the medium conditions, as reflected by the weak interaction of these anions with the particle surface, especially for those incorporated by specific interactions other than the electrostatic.

LDH–drug particles in Alg dispersions (LDH-Drug/Alg) impacted the properties on both d and ζ values. Nevertheless, this impact was greater in the latter, corresponding to polyelectrolytes with a relatively low affinity for LDH surface [19], which was decreased by the affinity and hydrophobic nature of the intercalated drugs. Thus, LDH-Drug/Alg and LDH-Drug/polyacrylate presented large and negative ζ values with relatively minor differences between both electrolytes, which indicated that LDH interacted with both polyelectrolytes. Nevertheless, the Alg interaction with LDH led to large d values in comparison with polyacrylate, which present a larger concentration of carboxylate anions and, consequently, stronger interactions [19,41,42]. Alg, on the other hand, presents a lower density of carboxylate anions, leading to a poor disaggregation in the case of LDH-Ibu/Alg, and even an increased particle size in the case of LDH-Nap/Alg. In the first case, the size diminution was related to the displacement of Ibu$^-$ anions of the LDH structure, diminishing the hydrophobicity of the surface, while the increased aggregation in the case of LDH-Nap particles was assigned to the bridging of LDH-Nap particles by Alg chains. In any case, d values for both LDH-Drug/Alg dispersions converged upon sonication and TT and the LDH-Drug/Alg dispersions in these conditions presented only a slightly lower aggregation than dispersions in pure water at the same conditions. Nevertheless, it should be considered that the results present a significant uncertainty, as the equipment used for the measurements presents significant random errors in this size range. Then, although LDH presents an interaction with Alg, this was not strong enough to produce a fine dispersion of the particles, similarly to that previously obtained for LDH-Cl nanoparticles [19]. On the other hand, interaction with a polyelectrolyte such as Alg produces a significant detachment of the drug from the particle surface [42]. Thus, 33% and 35% was released from the LDH in LDH-Ibu/Alg and LDH-Nap/Alg dispersions, respectively (values after sonication and TT).

3.3. In Vitro Drug Release Results

The release behavior of both drugs, Nap and Ibu, from LDH-Drug/Alg dispersions was studied to evaluate their performance as carrier systems (Figure 4). Two different LDH–drug particle concentrations (5 and 0.5 g L^{-1} LDH-Drug/Alg) were analyzed. The release profiles of Alg dispersions containing the pure drug in its anionic form (Drug/Alg) were obtained at a drug concentration equivalent to that of 0.5 g L^{-1} LDH-Drug/Alg. The release media were SBF, used to study the apatite-forming ability of implant materials [43], and PBS, a typical release medium to study drug release from delivery systems in simulated plasma conditions.

Controlled release towards both receptor media was achieved, but remarkable differences in the release profiles were observed depending on the release media. Thus, negligible *burst* effects (2.4 and 5.7% of the drug released at t = 15 min for LDH-Ibu/Alg and LDH-Nap/Alg, respectively) were observed toward PBS, lower than that of Drug/Alg (7.5% in both cases). This *burst* effect minimization was even more significant in SBF, in which the percentages of drug released at t = 15 min from 5 g L^{-1} LDH-Ibu/Alg and LDH-Nap/Alg were 1.5 and 3.2%, while those from Ibu/Alg and Nap/Alg were 13.9 and 12%, respectively.

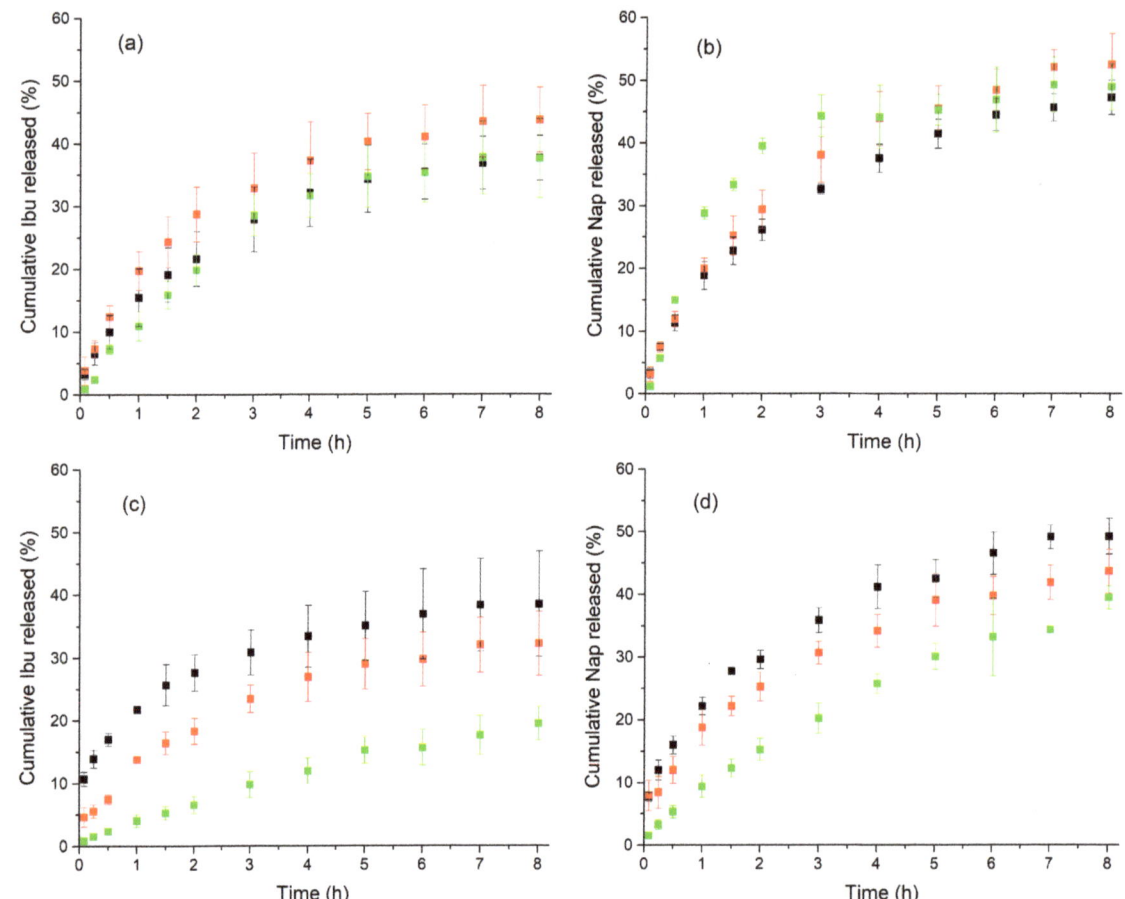

Figure 4. In vitro release profiles), using bicompartmental cells filled with PBS (**a**,**b**) and SBF (**c**,**d**), of Ibu (**a**,**c**) and Nap (**b**,**d**) from LDH-Drug/Alg dispersions at two different concentrations: 5 g L^{-1} (green), and 0.5 g L^{-1} (red), and Alg dispersions with the pure drug at a concentration equivalent to that of 0.5 g L^{-1} LDH-Drug/Alg (black).

For PBS, all release profiles, even those of pure drugs, were similar ($f_2 > 50$); however, the analysis of fitting curves and resulting kinetic parameters (Table 2) established slight but relevant differences. The best fittings were obtained when the K-P model was applied, with n coefficients near 0.5 for the profiles of Drug/Alg and 0. 5 g L^{-1} LDH-Drug/Alg samples and, accordingly, good fittings were also obtained with the Higuchi model. This indicated that the release rate was determined by diffusion processes. Higher n values (0.84 and 0.74 for Ibu and Nap, respectively) and poorer Higuchi fittings were obtained for LDH-Drug/Alg samples at 5 g L^{-1}, which indicate an anomalous behavior, assigned to the combined control of two main mechanisms, drug migration through the Alg dispersion and ion exchange from LDH particles. The main difference was a slightly slower release ($f_1 = 15$) obtained for Ibu (cumulative drug release at the end of the experiment between 38 and 43% in all cases) compared to Nap-containing samples at 0.5 g L^{-1} LDH-Drug/Alg (between 47 and 52%). This difference was related to the higher chemical affinity for the LDH interlayer of Ibu$^-$ anions compared to Nap$^-$ ones [14].

Table 2. Release kinetic data obtained from drug release studies using empirical equations: zero-order, Higuchi and K-P models, fittings with the best R^2 are marked with asterisks.

Sample	Receptor Media	Zero-Order			Higuchi			Korsmeyer-Peppas		
		% Drug$_0$	k_0	R^2	% Drug$_0$	k_H	R^2	k_P	n	R^2
LDH-Ibu/Alg, 5 g L^{-1}	PBS	6.6	4.8	0.88	−3.7	16.1	0.97 *	9.2	0.84	0.97 *
LDH-Ibu/Alg, 0.5 g L^{-1}		12.6	4.8	0.85	2.0	16.3	0.96	16.9	0.54	0.98 *
Ibu/Alg		9.9	4.2	0.88	0.8	14.3	0.98	14.1	0.54	0.99
LDH-Ibu/Alg, 5 g L^{-1}	SBF	1.5	2.4	0.98	−3.0	7.7	0.98	4.2	0.73	0.99 *
LDH-Ibu/Alg, 0.5 g L^{-1}		8.6	3.6	0.90	1.0	11.9	0.98 *	12.8	0.48	0.98 *
Ibu/Alg		17.0	3.3	0.86	9.8	11.2	0.97	21.8	0.30	0.99 *
LDH-Nap/Alg, 5 g L^{-1}	PBS	17.0	5.2	0.69	4.1	18.5	0.86 *	16.1	0.74	0.87 *
LDH-Nap/Alg, 0.5 g L^{-1}		12.0	6.1	0.89	−1.1	20.5	0.98 *	17.4	0.62	0.98 *
Nap/Alg		11.1	5.3	0.90	−0.23	18.0	0.99 *	16.3	0.58	0.99 *
LDH-Nap/Alg, 5 g L^{-1}	SBF	4.1	4.7	0.97	−5.0	15.2	0.99 *	9.0	0.73	0.99 *
LDH-Nap/Alg, 0.5 g L^{-1}		12.5	4.5	0.91	2.9	15.2	0.99 *	18.5	0.42	0.97
Nap/Alg		15.5	5.0	0.90	4.8	17.0	0.99 *	22.0	0.42	0.99 *

k_0, k_H and k_P expressed as % h^{-1}, % h$^{-0.5}$ and % h^{-n}, respectively. The concentration of pure drug experiments was calculated to match that of the 0.5 g L^{-1} LDH-Drug/Alg dispersions.

Release profiles in the SBF of both drugs were more influenced by their inclusion in LDH layers, with the release from both 0.5 and 5 g L^{-1} LDH-Drug/Alg being increasingly slower than that of Drug/Alg. Significant differences were observed in release profiles ($f_1 > 27$ when comparing Drug/Alg and 0.5 g L^{-1} LDH-Drug/Alg; $f_1 = 54$ and $f_2 = 48$ when comparing 0.5 and 5 g L^{-1} LDH-Drug/Alg). Nevertheless, and similarly to that observed in the PBS medium, the best fits were obtained with the K-P model (Table 2), with n values near 0.5 for all Drug/Alg and 0.5 g L^{-1} LDH-Drug/Alg and, in agreement, good fittings were also obtained with the Higuchi model. These results indicated that a diffusion-controlled mechanism was again controlling the drug release. In the case of 5 g L^{-1} LDH-Drug/Alg, the n values of K-P model increased (0.73 and 0.84 for Ibu and Nap, respectively), which indicated an anomalous behavior, like that produced towards PBS. Then, the mechanism of drug release was not so different in both media, and the main difference was the release rate, which decreased with increasing LDH–drug concentration. The cumulative drug release at the end of the assay reached 39, 32 and 19% for Ibu/Alg, 0.5 and 5 g L^{-1} LDH-Ibu/Alg, respectively, while 49, 44 and 39% were achieved for Nap-containing samples. This effect was more pronounced in the release from LDH-Ibu/Alg than for LDH-Nap/Alg, as corresponds to the above-mentioned affinity of Ibu$^-$ anions for the LDH interlayers.

The differences in the release profiles toward SBF compared to PBS were assigned to Alg gelation when exposed to the divalent calcium ions present in the SBF. This Ca-Alg gel included and fixed the LDH–drug particles, which crosslinked the Alg chains and performed as barrier for drug diffusion, decreasing the overall release rate. Effectively, a consistent gel, which was easily separated from the acetate cellulose membrane, was formed in the donor compartment of the Franz cells (not shown). This in-situ gel formation presents promising applications in tissue engineering [43]. This gel formation will also ease the fixation of the biomaterial to the insertion place and locate the action of loaded drug, which would be released in a controlled manner.

3.4. Cell Proliferation Assay

Biomaterials for bone regeneration must provide support for cell adhesion and proliferation, which can be compromised by cytotoxic effects caused by the inclusion of drugs such as Ibu and Nap. To explore the capacity of LDH to diminish these cytotoxic effects, the proliferation of pre-osteoblastic MC3T3-E1 cells in the presence of LDH-Drug/Alg dispersions was determined (Figure 5). Experiments with LDH-Cl, Alg and their mixture (LDH-Cl/Alg), as well as in the presence of the pure drugs, the LDH–drug particles, and Drug/Alg dispersions were also performed to differentiate between factors of their overall behavior.

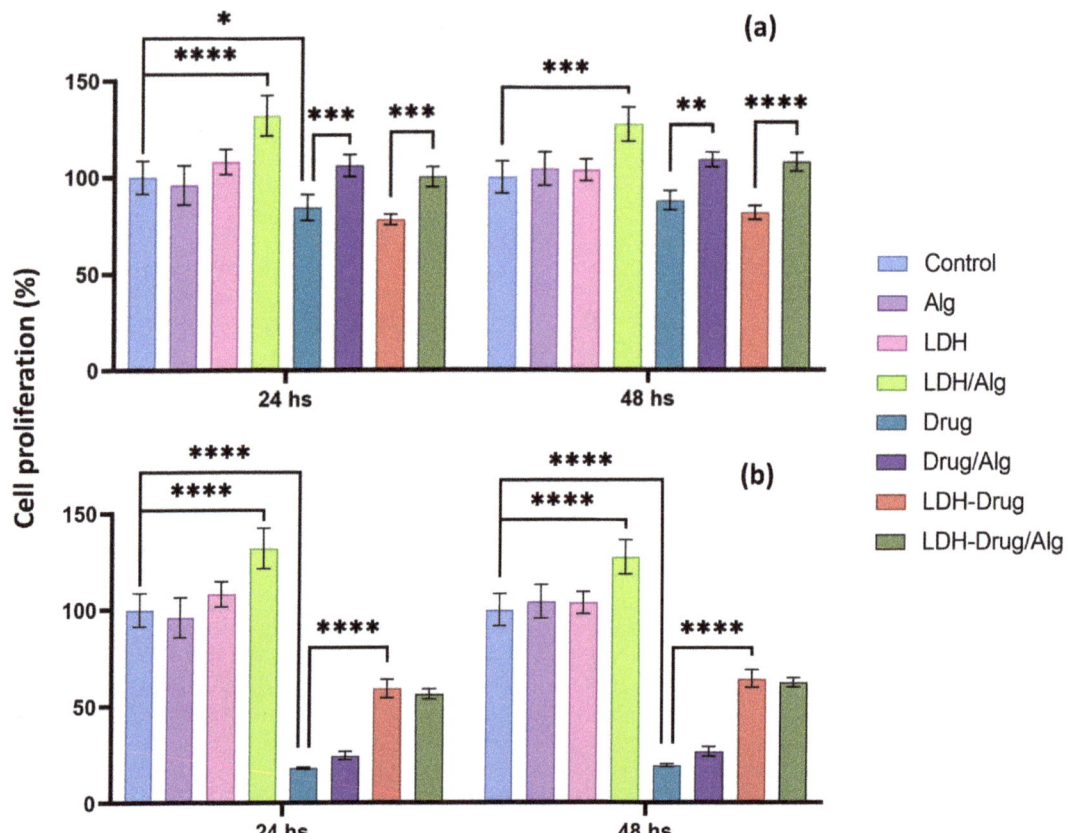

Figure 5. Anti-cytotoxic effect of 5 g L^{-1} LDH-Drug/Alg in MC3T3-E1 cell cultures. Proliferation of MC3T3-E1 cells cultured in CM (control), Alg, LDH, LDH/Alg, Drug (Ibu (**a**) and Nap (**b**)), Drug/Alg, LDH-Drug, LDH-Drug/Alg. Fluorescence of alamarBlue was measured at 24 and 48 h of and expressed as a percentage of control proliferation at that given time. Asterisks indicate significant differences between the indicated groups (****, $p < 0.0001$, ***, $p < 0.001$, **, $p < 0.01$ and *, $p < 0.05$).

In the absence of the drug, the separate presence of Alg or LDH-Cl did not produce significant differences in the proliferation of MC3T3-E1 cells compared to the control, which was in line with the low cytotoxicity of both materials [12,44]. Although LDH layers did not interfere with cell adhesion and proliferation, they have been proposed to have a positive effect on biomaterials such as polycaprolactone [45]. Comparable results were obtained for LDH-Cl/Alg, which produced a significant increase in cell activity ($p < 0.001$), which was related to the hydrophilicity and positive charge of the LDH surface, providing attachment sites due to electrostatic interactions with negatively charged cell membranes [12]. The presence of pure drugs decreased the percentage of cell proliferation compared to the control, with a reduction of approximately 81% and 13% for Nap ($p < 0.0001$) and Ibu, respectively, after 48 h, although the latter decrease was not considered significant according to ANOVA analysis. The cytotoxic effect of Ibu was significantly reduced when cultivated in the Alg matrix, showing an increase in proliferation of 24% compared to the pure drug ($p < 0.01$), while, in the case of Nap, the polymeric matrix did not significantly enhance the cell proliferation. On the other hand, the inclusion of Ibu in LDH did not show significant differences in cell proliferation compared to the pure drug, which was assigned to the low cytotoxicity of the drug, as well as the negative charge and hydrophobicity of the LDH-Ibu surface, while LDH-Nap reduced in 55% the cytotoxic effect of the pure drug ($p < 0.0001$). The results for the LDH-Drug/Alg dispersions were consistent with the previously exposed trends, as they showed a cellular protection that was dependent on the intercalated drug. In the case of Ibu, a significantly increased proliferation was obtained compared to pure Ibu and LDH-Ibu due to the protective action of Alg ($p < 0.001$). Instead, the LDH-Nap/Alg dispersion showed a highly significant anticytotoxic effect, like that of the LDH-Nap dispersion, which indicated that its inclusion in the LDH structure is mainly responsible for the protective effect ($p < 0.0001$). Therefore, LDH-Drug/Alg dispersions are a promising and versatile biomaterial for cell culturing that, depending on the physicochemical properties and composition of the included LDH, can provide cell adhesion centers and/or cytoprotective effects. This strategy can be extended to drugs that provide antimicrobial and anticancer properties of cell differentiation to biomaterials for tissue engineering applications.

4. Conclusions

In this work, the physicochemical properties of LDH/Alg dispersions were explored, aiming at their application in tissue engineering. LDH intercalated with Ibu and Nap were obtained and incorporated in Alg dispersions, with significant effects in their surface charge but minor effects on their aggregation. The sterilization of dispersions by TT did not produce significant changes in the aggregate size of LDH–drug particles. The drug release behavior was highly dependent on the release media, and particularly to the presence of Ca^{2+} ions present in the SBF. The formation of a gel layer upon exposure to SBF led to a release rate diminution with increasing LDH concentration. The main release mechanism was drug diffusion through the LDH/Alg dispersions, although anion exchange also influenced the overall rate at high LDH concentrations. Finally, LDH/Alg dispersions had a protective effect against the cytotoxic effects of drugs in cell culture. This protective effect was provided by Alg or LDH depending on the intercalated drug. LDH/Alg dispersions are then promising biomaterials for tissue engineering due to their capacity to provide drug release control and cell proliferation enhancement and/or protection. These dispersions can be included in scaffold formulations on their own, due to their in-situ gelling capacity, and can also be used, for example, as precursors of biomaterials gelled by ionic crosslinking with Ca^{2+} ions or lyophilized for subsequent rehydration with solutions containing stem cells or bone marrow aspirates.

Author Contributions: Conceptualization, R.R.; methodology, R.R., M.C.G., G.A.G.; validation, M.C.G., G.A.G.; formal analysis, R.R., M.C.G., G.A.G.; data curation, J.P.Z.; investigation, R.R., M.C.G., J.P.Z.; writing—original draft preparation, R.R., M.C.G., G.A.G., J.P.Z.; writing—review and editing, R.R., M.C.G.; visualization, R.R., M.C.G., J.P.Z.; supervision, R.R., M.C.G., G.A.G.; project administration, R.R.; funding acquisition, R.R., M.C.G., G.A.G. All authors have read and agreed to the published version of the manuscript.

Funding: This research was funded by the Agencia Nacional de Promoción Científica y Tecnológica—Fondo Nacional para la Investigación Científica y Tecnológica (Grant number FonCyT- PICT 2016-0986, PICT 2019-00048 and PICT-2020-SERIEA-01781), the Secretaría de Ciencia y Tecnología, Universidad Nacional de Córdoba (SeCyT-Formar Grant number 33820190100091CB SeCyT-ITT2019 and PRIMAR number 32520170100384CB) and the Ministerio de Ciencia y Tecnología de la Provincia de

Córdoba (MinCyT, Cordoba Innova Program). The SEM images were obtained at the Laboratorio de Microscopía Electrónica y Análisis por Rayos X (LAMARX).

Data Availability Statement: Not applicable.

Acknowledgments: J.P.Z. thanks CONICET for his scholarship grant. G.G., M.C.G. and R.R. are members of Argentinean National Council for Scientific and Technical Research (CONICET)'s scientific career. The authors wish to acknowledge the assistance of the CONICET and the National University of Cordoba (UNC, Argentina), both of which provided facilities for this work.

Conflicts of Interest: The authors declare no conflict of interest.

References

1. Hariyadi, D.M.; Islam, N. Current status of alginate in drug delivery. *Adv. Pharmacol. Pharm. Sci.* **2020**, *2020*, 8886095. [CrossRef] [PubMed]
2. Ray, P.; Maity, M.; Barik, H.; Sahoo, G.S.; Hasnain, M.S.; Hoda, M.N.; Nayak, A.K. Alginate-based hydrogels for drug delivery applications. In *Alginates in Drug Delivery*; Academic Press: Cambridge, MA, USA, 2020. [CrossRef]
3. Venkatesan, J.; Bhatnagar, I.; Manivasagan, P.; Kang, K.H.; Kim, S.K. Alginate composites for bone tissue engineering: A review. *Int. J. Biol. Macromol.* **2015**, *72*, 269–281. [CrossRef] [PubMed]
4. Sotome, S.; Uemura, T.; Kikuchi, M.; Chen, J.; Itoh, S.; Tanaka, J.; Tateishi, T.; Shinomiya, K. Synthesis and in vivo evaluation of a novel hydroxyapatite/collagen- alginate as a bone filler and a drug delivery carrier of bone morphogenetic protein. *Mater. Sci. Eng. C* **2004**, *24*, 341–347. [CrossRef]
5. Frankenberg, E. Controlled nucleation of hydroxyapatite on alginate scaffolds for stem cell-based bone tissue engineering. *Bone* **2012**, *23*, 1–7. [CrossRef]
6. Axpe, E.; Oyen, M.L. Applications of alginate-based bioinks in 3D bioprinting. *Int. J. Mol. Sci.* **2016**, *17*, 1976. [CrossRef]
7. Alcantara, A.C.S.; Aranda, P.; Darder, M.; Ruiz-Hitzky, E. Bionanocomposites based on alginate-zein/layered double hydroxide materials as drug delivery systems. *J. Mater. Chem.* **2010**, *20*, 9495–9504. [CrossRef]
8. Hasnain, M.S.; Ahmed, S.A.; Behera, A.; Alkahtani, S.; Nayak, A.K. Inorganic materials–alginate composites in drug delivery. In *Alginates in Drug Delivery*; Academic Press: Cambridge, MA, USA, 2020. [CrossRef]
9. Song, F.; Li, X.; Wang, Q.; Liao, L.; Zhang, C. Nanocomposite hydrogels and their applications in drug delivery and tissue engineering. *J. Biomed. Nanotechnol.* **2015**, *11*, 40–52. [CrossRef]
10. Rives, V. *Layered Double Hydroxides: Present and Future*; Nova Publishers: Hauppauge, NY, USA, 2001.
11. Izbudak, B.; Cecen, B.; Anaya, I.; Miri, A.K.; Bal-Ozturk, A.; Karaoz, E. Layered double hydroxide-based nanocomposite scaffolds in tissue engineering applications. *RSC Adv.* **2021**, *11*, 30237–30252. [CrossRef]
12. Rojas, R.; Mosconi, G.; Pablo, J.; Gil, G.A. Layered double hydroxide applications in biomedical implants. *Appl. Clay Sci.* **2022**, *224*, 106514. [CrossRef]
13. Rives, V.; del Arco, M.; Martín, C. Layered double hydroxides as drug carriers and for controlled release of non-steroidal antiinflammatory drugs (NSAIDs): A review. *J. Control. Release* **2013**, *169*, 28–39. [CrossRef]
14. Rojas, R.; Jimenez-Kairuz, A.F.; Manzo, R.H.; Giacomelli, C.E. Release kinetics from LDH-drug hybrids: Effect of layers stacking and drug solubility and polarity. *Colloids Surf. A Physicochem. Eng. Asp.* **2014**, *463*, 37–43. [CrossRef]
15. Szabados, M.; Gácsi, A.; Gulyás, Y.; Kónya, Z.; Kukovecz, Á.; Csányi, E.; Pálinkó, I.; Sipos, P. Conventional or mechanochemically-aided intercalation of diclofenac and naproxen anions into the interlamellar space of CaFe-layered double hydroxides and their application as dermal drug delivery systems. *Appl. Clay Sci.* **2021**, *212*, 106233. [CrossRef]
16. Bernardo, M.P.; Rodrigues, B.C.S.; de Olivera, T.D.; Guedes, A.P.M.; Batista, A.A.; Mattoso, L.H.C. Naproxen/layered double hydroxide composites for tissue-engineering applications: Physicochemical characterization and biological evaluation. *Clays Clay Miner.* **2020**, *68*, 623–631. Available online: https://link.springer.com/article/10.1007/s42860-020-00101-w (accessed on 22 February 2022). [CrossRef]
17. Yousefi, Y.; Tarhriz, V.; Eyvazi, S.; Dilmaghani, A. Synthesis and application of magnetic@layered double hydroxide as an anti-inflammatory drugs nanocarrier. *J. Nanobiotechnol.* **2020**, *18*, 155. [CrossRef]
18. Kang, H.; Shu, Y.; Li, Z.; Guan, B.; Peng, S.; Huang, Y.; Liu, R. An effect of alginate on the stability of LDH nanosheets in aqueous solution and preparation of alginate/LDH nanocomposites. *Carbohydr. Polym.* **2014**, *100*, 158–165. [CrossRef]
19. Vasti, C.; Borgiallo, A.; Giacomelli, C.E.; Rojas, R. Layered double hydroxide nanoparticles customization by polyelectrolyte adsorption: Mechanism and effect on particle aggregation. *Colloids Surf. A Physicochem. Eng. Asp.* **2017**, *533*, 316–322. [CrossRef]
20. Borgiallo, A.; Rojas, R. Reactivity and Heavy Metal Removal Capacity of Calcium Alginate Beads Loaded with Ca–Al Layered Double Hydroxides. *ChemEngineering* **2019**, *3*, 22. [CrossRef]
21. Zhang, J.P.; Wang, Q.; Xie, X.L.; Li, X.; Wang, A.Q. Preparation and swelling properties of pH-sensitive sodium alginate/layered double hydroxides hybrid beads for controlled release of diclofenac sodium. *J. Biomed. Mater. Res. Part B Appl. Biomater.* **2010**, *92*, 205–214. [CrossRef]
22. Viscusi, G.; Gorrasi, G. Facile preparation of layered double hydroxide (LDH)-alginate beads as sustainable system for the triggered release of diclofenac: Effect of pH and temperature on release rate. *Int. J. Biol. Macromol.* **2021**, *184*, 271–281. [CrossRef]

23. Munhoz, D.R.; Bernardo, M.P.; Malafatti, J.O.D.; Moreira, F.K.V.; Mattoso, L.H.C. Alginate films functionalized with silver sulfadiazine-loaded [Mg-Al] layered double hydroxide as antimicrobial wound dressing. *Int. J. Biol. Macromol.* **2019**, *141*, 504–510. [CrossRef]
24. U.S. Pharmacopoeial Convention. *United States Pharmacopoeia*; U.S. Pharmacopoeial Convention: Rockville, MD, USA, 2015.
25. Marques, M.R.C.; Loebenberg, R.; Almukainzi, M. Simulated biological fluids with possible application in dissolution testing. *Dissolut. Technol.* **2011**, *18*, 15–28. [CrossRef]
26. Costa, P.; Sousa Lobo, J.M. Modeling and comparison of dissolution profiles. *Eur. J. Pharm. Sci.* **2001**, *13*, 123–133. [CrossRef]
27. Castellaro, A.M.; Rodriguez-Baili, M.C.; Di Tada, C.E.; Gil, G.A. Tumor-associated macrophages induce endocrine therapy resistance in ER+ breast cancer cells. *Cancers* **2019**, *11*, 189. [CrossRef] [PubMed]
28. Drits, V.A.; Bookin, A.S. Crystal Structure and X-ray identification of Layered Double hydroxides. In *Layered Double Hydroxides: Present and Future*; Rives, V., Ed.; Nova Science: Hauppauge, NY, USA, 2001; pp. 39–92.
29. Mohanambe, L.; Vasudevan, S. Anionic clays containing anti-inflammatory drug molecules: Comparison of molecular dynamics simulation and measurements. *J. Phys. Chem. B* **2005**, *109*, 15651–15658. [CrossRef]
30. Gu, Z.; Wu, A.; Li, L.; Xu, Z.P. Influence of hydrothermal treatment on physicochemical properties and drug release of anti-inflammatory drugs of intercalated layered double hydroxide nanoparticles. *Pharmaceutics* **2014**, *6*, 235–248. [CrossRef]
31. Figueiredo, M.P.; Cunha, V.R.R.; Cellier, J.; Taviot-Guého, C.; Constantino, V.R.L. Fe(III)-Based Layered Double Hydroxides Carrying Model Naproxenate Anions: Compositional and Structural Aspects. *ChemistrySelect* **2022**, *7*, e202103880. [CrossRef]
32. Kloprogge, J.T.; Frost, R.L. Infrared and Raman Spectroscopic Studies of Layered Double Hydroxides (LDHs). In *Layered Double Hydroxides: Present and Future*; Rives, V., Ed.; Nova Science Publishers: Hauppauge, NY, USA, 2001; pp. 139–192.
33. Gaskell, E.E.; Ha, T.; Hamilton, A.R. Ibuprofen intercalation and release from different layered double hydroxides. *Ther. Deliv.* **2018**, *9*, 653–666. [CrossRef]
34. Wypych, F.; Arízaga, G.G.C.; da Costa Gardolinski, J.E.F. Intercalation and functionalization of zinc hydroxide nitrate with mono- and dicarboxylic acids. *J. Colloid Interface Sci.* **2005**, *283*, 130–138. [CrossRef]
35. Rojas, R.; Palena, M.C.; Jimenez-Kairuz, A.F.; Manzo, R.H.; Giacomelli, C.E. Modeling drug release from a layered double hydroxide-ibuprofen complex. *Appl. Clay Sci.* **2012**, *62–63*, 15–20. [CrossRef]
36. Luengo, C.V.; Crescitelli, M.C.; Lopez, N.A.; Avena, M.J. Synthesis of Layered Double Hydroxides Intercalated With Drugs for Controlled Release: Successful Intercalation of Ibuprofen and Failed Intercalation of Paracetamol. *J. Pharm. Sci.* **2021**, *110*, 1779–1787. [CrossRef]
37. Du, B.-Z.; Wang, R.M. Synthesis and characterizations of naproxen intercalated Mg-Al layered double hydroxides. *J. Chin. Pharm. Sci.* **2010**, *19*, 371–378.
38. Rojas, R.; Linck, Y.G.; Cuffini, S.L.; Monti, G.A.; Giacomelli, C.E. Structural and physicochemical aspects of drug release from layered double hydroxides and layered hydroxide salts. *Appl. Clay Sci.* **2015**, *109–110*, 119–126. [CrossRef]
39. Rojas Delgado, R.; Arandigoyen Vidaurre, M.; de Pauli, C.P.; Ulibarri, M.A.; Avena, M.J. Surface-charging behavior of Zn-Cr layered double hydroxide. *J. Colloid Interface Sci.* **2004**, *280*, 431–441. [CrossRef]
40. Rojas, R.; Barriga, C.; de Pauli, C.P.; Avena, M.J. Influence of carbonate intercalation in the surface-charging behavior of Zn–Cr layered double hydroxides. *Mater. Chem. Phys.* **2010**, *119*, 303–308. [CrossRef]
41. Pavlovic, M.; Rouster, P.; Oncsik, T.; Szilagyi, I. Tuning Colloidal Stability of Layered Double Hydroxides: From Monovalent Ions to Polyelectrolytes. *ChemPlusChem* **2017**, *82*, 121–131. [CrossRef]
42. Vasti, C.; Giacomelli, C.E.; Rojas, R. Pros and cons of coating layered double hydroxide nanoparticles with polyacrylate. *Appl. Clay Sci.* **2019**, *172*, 11–18. [CrossRef]
43. Baino, F.; Yamaguchi, S. The use of simulated body fluid (SBF) for assessing materials bioactivity in the context of tissue engineering: Review and challenges. *Biomimetics* **2020**, *5*, 57. [CrossRef]
44. Lee, K.Y.; Mooney, D.J. Alginate: Properties and biomedical applications. *Prog. Polym. Sci.* **2012**, *37*, 106–126. [CrossRef]
45. Baradaran, T.; Shafiei, S.S.; Mohammadi, S.; Moztarzadeh, F. Poly (ε-caprolactone)/layered double hydroxide microspheres-aggregated nanocomposite scaffold for osteogenic differentiation of mesenchymal stem cell. *Mater. Today Commun.* **2020**, *23*, 100913. [CrossRef]

Article

The Inhibitive Effect of Sebacate-Modified LDH on Concrete Steel Reinforcement Corrosion

David Caballero [1], Ruben Beltrán-Cobos [1], Fabiano Tavares [2], Manuel Cruz-Yusta [1], Luis Sánchez Granados [1], Mercedes Sánchez-Moreno [1,*] and Ivana Pavlovic [1,*]

[1] Departamento de Química Inorgánica, Instituto Universitario de Nanoquímica (IUNAN), Universidad de Córdoba, Campus de Rabanales, E-14014 Cordoba, Spain
[2] Departamento de Mecánica, Universidad de Córdoba, Campus de Rabanales, E-14014 Cordoba, Spain
* Correspondence: msmoreno@uco.es (M.S.-M.); iq2pauli@uco.es (I.P.)

Abstract: In recent decades, layered double hydroxides (LDH) have been proposed as innovative corrosion inhibitors for reinforced concrete. Their protective action is based on the ability to intercalate specific anions in the interlayer and on their ability to exchange the intercalated anion. In the present study, an organically charged LDH, with sebacate anions in the interlayer (LDH-S), is proposed as a water-repellent additive for mortar. The waterproofing efficiency of LDH-S and the associated corrosion inhibition ability has been evaluated in reinforced mortar samples. A 42% decrease in the water capillary absorption coefficient has been estimated when 3% LHD-S is added to a mortar. Both the passivation processes of the steel rebars during the curing period and the initiation of corrosion due to chloride exposure have been studied by electrochemical measurements. Three different mortars have been evaluated: reference mortar (REF), mortar with Mg-Al LDH (LDH), and mortar with LDH-sebacate (LDH-S). The latter has shown an important protective capacity for preventing the initiation of corrosion by chloride penetration, with an inhibitory efficiency of 74%. The presence of LDHs without sebacate in the interlayer also improved the performance of the mortar against rebar corrosion, but with lower efficiency (23% inhibitory efficiency). However, this protection is lost after continued chloride exposure over time, and corrosion initiates similarly to the reference mortar. The low corrosion current density values registered when LDH-S is added to the mortar may be related to the increased electrical resistance recorded in this mortar.

Keywords: LDH-sebacate; corrosion protection; waterproofed reinforced mortar; electrochemical measurements; capillary absorption

Citation: Caballero, D.; Beltrán-Cobos, R.; Tavares, F.; Cruz-Yusta, M.; Granados, L.S.; Sánchez-Moreno, M.; Pavlovic, I. The Inhibitive Effect of Sebacate-Modified LDH on Concrete Steel Reinforcement Corrosion. ChemEngineering 2022, 6, 72. https://doi.org/10.3390/chemengineering6050072

Academic Editors: Miguel A. Vicente, Raquel Trujillano and Francisco Martín Labajos

Received: 22 July 2022
Accepted: 14 September 2022
Published: 20 September 2022

Publisher's Note: MDPI stays neutral with regard to jurisdictional claims in published maps and institutional affiliations.

Copyright: © 2022 by the authors. Licensee MDPI, Basel, Switzerland. This article is an open access article distributed under the terms and conditions of the Creative Commons Attribution (CC BY) license (https://creativecommons.org/licenses/by/4.0/).

1. Introduction

One of the main causes of reinforced concrete deterioration in marine environments is the corrosion of the embedded rebars caused by the action of chloride ions, which reach the steel–concrete interface through the pore network of the concrete [1,2]. Generally, the steel rebars in concrete are passivated due to the high alkaline pH of the aqueous phase. However, if chloride ions reach the steel surface in sufficient concentration, the local rupture of the protective passive layer occurs, and a pitting corrosion process is initiated [3].

A first approach to solve this problem would be to prevent the entry of aggressive species into the cementitious matrix by improving the waterproofing efficiency of concrete [4]. External surface treatments are often suggested for developing waterproof concrete [5]. Different types of inorganic surface treatments have been proposed for improving concrete durability [6], such as surface coatings forming a physical barrier on the concrete surface [7], pore-blocking surface treatments that block capillary pores in concrete surface [8], and hydrophobic impregnations that inhibit water penetration [9].

The performance of surface treatments depends on the initial moisture content and the exposure environments [10]. In addition, surface treatments can degenerate over time, thus decreasing their effectiveness [11]. In this sense, the effectiveness of the concrete waterproofing can be improved by incorporating waterproofing and/or water-repellent agents during the preparation of the mixture, which are distributed in the bulk material. Among the best-known admixtures for waterproofing concrete are crystalline admixtures [12,13], commercial compounds whose composition is not revealed by the supplier.

Layered double hydroxides (LDH) are a class of well-known crystalline compounds, commonly referred to as hydrotalcite-like compounds, because hydrotalcite (HT) is the most studied and representative mineral of this family of compounds. The HT structure can be derived from brucite-like layers, where some of the divalent cations are isomorphically substituted by trivalent ones, and the positive charge generated is balanced by intercalation of hydrated anions between the metal hydroxide layers. LDH can be represented by the general formula $[M^{2+}_{1-x}M^{3+}_{x}(OH)_2]^{x+}A^{n-}_{x/n}\cdot mH_2O$, where M^{2+} and M^{3+} are the divalent and the trivalent metals, and A^{n-} is the anion placed in the LDH interlayer, and is easily substitutable with others present in the aqueous solution [14]. Anions can range from simple inorganic species to complex organic anions, thus changing the interlayer LDH surface from hydrophilic to hydrophobic [14,15].

In the recent years, LDHs have been proposed as corrosion inhibitors of concrete steel reinforcements [16,17]. Both inorganic corrosion inhibitors such as nitrite [18–20] and organic corrosion inhibitors such as benzoate [21], aminobenzoate [22] or phthalate [23] have been intercalated in LDHs of different composition, Mg-Al-LDH [18–20], Ca-Al-LDH [24,25] and Zn-Al-LDH [26]. The corrosion-protective activity of LDHs is based on their ability for ionic exchange: aggressive ions, such as chloride and/or carbonate, can be adsorbed from the aqueous solution in the LDH interlayer, which can then release the corrosion inhibitors [27]. However, efficiency in the long-term of these "smart" corrosion inhibitors could fail, as the chloride trapped in the LDH could be released again [16].

A different way to protect steel rebars from corrosion in concrete is by using waterproofing concrete that prevents the entrance of water, followed by the penetration of the dissolved aggressive ions. Fatty acids are hydrophobic compounds due to the presence of hydrophobic alkyl chains. However, they have been shown to highly affect the properties of cement when added to cement-based materials in a dosage above 0.5% in clinker weight, promoting a significant increase of the setting time [28]. Nevertheless, fatty acids incorporated in the inorganic LDH matrix could be a promising additive for the waterproofing of concrete. A recent study has shown that the presence of ZnAl-LDH intercalated with the sebacate anion (ZnAl-S) in epoxy coatings significantly improves their anticorrosive properties, depending on the degree of its dispersion in the matrix [29].

The present work aimed to explore the waterproofing ability of MgAl-LDH intercalated by sebacate (LDH-S), as well as its protective capacity when added in cementitious mortars as corrosion inhibitors in the presence of chloride ions. The LDH-S was synthetized by the coprecipitation method and characterized by different techniques, such as XRD, FT-IR, TGA and ICP. The water repellent capacity of the mortar containing the LDH-S was assessed by water capillary absorption tests. The corrosion protection ability was monitored by electrochemical measurements considering two stages: the rebar passivation in the alkaline environment of the mortar, and the corrosion initiation due to the chloride penetration through the mortar pores.

2. Materials and Methods

$Mg(NO_3)_2 \cdot 6H_2O$, $Al(NO_3)_3 \cdot 9H_2O$ and sebacic acid $((CH_2)_8(COOH)_2)$ were acquired from Sigma-Aldrich (St. Loui, MO, USA). All of the chemicals were at least 98–99%. Portland CEM II/A-L 42.5 R from Cosmos Cements, siliceous sand with a maximum size of 1.5 mm and tap water were used as raw materials for preparing the mortar samples.

2.1. Synthesis of LDH

Organic LDH, LDH-S, was obtained through the coprecipitation method in a N_2 atmosphere and using CO_2-free water, adding 250 mL of solution containing 0.15 mol of

Mg(NO$_3$)$_2$·6H$_2$O and 0.05 mol of Al(NO$_3$)$_3$·9H$_2$O to 500 mL of the alkaline (0.16 mol of NaOH) 0.125 mol sebacate solution. The LDH suspensions thus obtained were filtered and washed with CO$_2$-free, distilled water and dried at 60 °C. For comparison purposes, a MgAl-CO$_3$ (LDH-CO$_3$) was prepared with the coprecipitation method, dropping a solution containing 0.75 mol of Mg(NO$_3$)$_2$·6H$_2$O and 0.25 mol of Al(NO$_3$)$_3$·9H$_2$O into a solution containing 1.7 mol of NaOH and 0.5 mol of Na$_2$CO$_3$. The resultant suspension was washed with distilled water and dried at 60 °C.

2.2. Characterization of LDH Additives

X-ray diffraction (XRD) patterns were recorded by a Bruker D8 Discover diffractometer (Bruker, Karlsruhe, Germany). The Fourier transform infrared spectra (FT-IR) were recorded by PerkinElmer Frontier MIR using ATR (PerkinElmer España SL, Madrid, Spain).

Elemental chemical analyses were measured by Induced Coupled Plasma mass spectroscopy (ICP-MS) on a PerkinElmer Nexion-X instrument (PerkinElmer España SL, Madrid, Spain).

2.3. Production of Mortar Samples for Water Capillary Absorption Tests

Cylindrical specimens of 70 mm in diameter and 30 mm in height were manufactured for the water capillary absorption test. A cement/sand/water dosage of 1/3/0.5 was used for producing the cementitious mortar. Mortars incorporating additives were mixed with 3% by cement weight of the LDH. Both LDHs (LDH-S and LDH-CO$_3$) were considered.

Mortar samples were demolded 24 h after casting and cured under controlled climate conditions at 21 ± 2 °C and >95% HR. Two samples of each mortar were evaluated for repeatability assessment.

2.4. Production of Reinforced Mortar Samples

Reinforced mortar specimens were prepared by embedding steel rebars 500S (6 mm nominal diameter) in a mortar cover of approximately 1 cm. A mortar with similar composition to the one used for the waterproofing test was used in the reinforced specimens. Mortars incorporating LDH-based additives (M-LDH-S and M-LDH-CO$_3$) in 3% by cement weight were also studied.

The exposed area of the bars was delimited to 6 cm^2 covering the edge of the bar with isolating tape. Before embedding in mortar, steel rebars were cleaned in a HCl:H$_2$O 1:1 with 5 g/L urotropine solution and degreased in acetone. The final aspect of the steel rebar before testing is shown in Figure 1 (left). Two rebars were tested for each mortar for repeatability assessment.

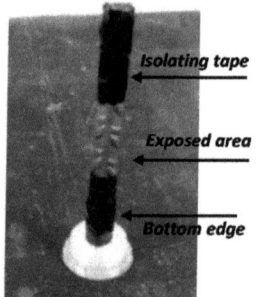 

Figure 1. (**Left**)—Rebar before embedding in mortar. (**Right**)—Electrochemical cell with three reinforced mortar samples and the stainless-steel mesh used as a counter-electrode for the electrochemical tests.

Reinforced samples were demolded 24 h after casting, and immersed in an alkaline solution of Ca(OH)$_2$ for curing. After 28 days curing in the alkaline conditions, the alkaline solution was changed by a 0.5 M NaCl solution for 50 days. In this way, both the passivation and the corrosion stages could be monitored by periodic electrochemical measurements.

2.5. Waterproofing Efficiency of the LDH-Based Additives

The waterproofing efficiency of the studied mortars was assessed after 14 days curing. Measurements were carried out for 3 days according to the Fagerlund method [30]. The mortar specimens were placed 5 mm in contact with water and periodically weighed. From the weight–time curve obtained, two parameters can be deduced: the saturation time, when the weight gain is stabilized, and the water capillary absorption coefficient, estimated according to [30]. For each type of studied mortar, the average value of the two tested samples was considered.

2.6. Electrochemical Tests

A three-electrode electrochemical cell was used for monitoring the corrosion response of the steel rebars, as shown in Figure 1 (right). For the electrochemical measurements, a stainless-steel mesh located around the samples was used as a counter-electrode, an Ag/AgCl electrode in 3 M KCl was used as a reference electrode, and the reinforced samples acted as working electrodes.

An Autolab PGSTAT204 potentiostat/galvanostat with a frequency analyzer module (FRA) using NOVA software was used for the electrochemical tests. The evolution of the corrosion potential (E_{corr}), the electrical resistance of the mortar cover (R_e) and the polarization resistance (R_P) of the steel rebars were periodically measured. R_P values were obtained by linear potential sweep from −10 mV to +10 mV with respect to the measured corrosion potential. R_e was determined by applying an alternating current at a single frequency of 1000 Hz between the rebar and the counter electrode. A potentiostatic method with an amplitude of 10 mV$_{rms}$ around E_{corr} was defined.

Electrochemical measurements were considered only after 4 days curing of the mortars, before non-stable response of the different parameters monitored were registered.

At the end of the test, to accelerate the corrosion process and make the chloride-generated pitting more visible, a constant potential of +250 mV vs. Ag/AgCl reference electrode was applied for 4 h.

The corrosion current density (i_{corr}) was obtained indirectly using the Stern and Geary equation [31], $i_{corr} = \frac{B}{R_P \cdot A}$, where B is a constant equal to 26 mV [32], R_P is the value of polarization resistance, and A is the exposed area. The inhibitory efficiency of the LDH was estimated at the end of the test using the following equation:

$$EI(\%) = \frac{i_{acum,\ REF} - i_{acum,LDH}}{i_{acum,\ REF}} \cdot 100 \qquad (1)$$

where $i_{acum,REF}$ and $i_{acum,LDH}$ are the cumulative values of corrosion current density during the whole test for the reference mortar, and the mortars with LDH, respectively.

3. Results and Discussion

3.1. Characterization of the LDH Additives

The diffraction patterns of both additives, included in Figure 2, were characteristic of LDH compounds, with intense basal reflections (00l) at low angles and low asymmetric reflections at intermediate and high angles [14,15]. An increase in d (003) value for the LDH-S was observed compared to LDH-CO$_3$ (15.6 Å for LDH-S vs. 7.8 Å for LDH-CO$_3$), as expected due to the larger size of sebacate anion and in agreement with [29,33].

Although we carried out a very careful synthesis under inert conditions, the reflection at 24° in 2θ indicates that the LDH-S sample mixed with sebacate also contained a certain amount of carbonate, since this anion has a very high affinity for the intermediate layer. The position of the (012) plane characteristic for LDH compounds, which overlaps with the (009) reflection for LDH-CO$_3$, was maintained unchanged in the HT-S sample pattern [14,34].

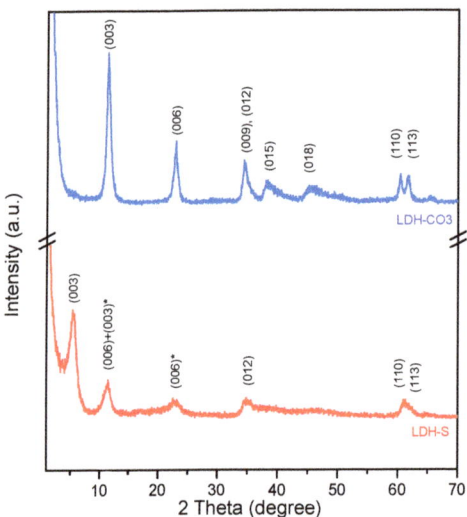

Figure 2. XRD patterns of LDH-CO$_3$ and LDH-S. * Peaks of the carbonate phase in LDH-S.

The FT-IR spectra of LDH-CO$_3$ (Figure 3) shows a characteristic band at around 3500 cm^{-1}, corresponding to the water molecules and free OH$^-$ groups stretching vibrations, the band at 1650 cm^{-1} corresponding to the bending mode vibration of water molecules, and the band at 1375 cm^{-1} of the carbonate antisymmetrical stretching mode. LDH-S spectra showed the most characteristic bands of fatty acid at around 1560 cm^{-1} and 1460 cm^{-1} due to antisymmetrical and symmetrical vibrations of carboxylate groups, respectively, as well as the C-H stretching vibrations at ~2900 cm^{-1} and bending vibrations at ~1408 cm^{-1} of the aliphatic chains. The absence of the protonated carboxylic group (-COOH) band of sebacic acid at ~1700 cm^{-1} confirmed that this compound is intercalated in the interlayer in its anionic form.

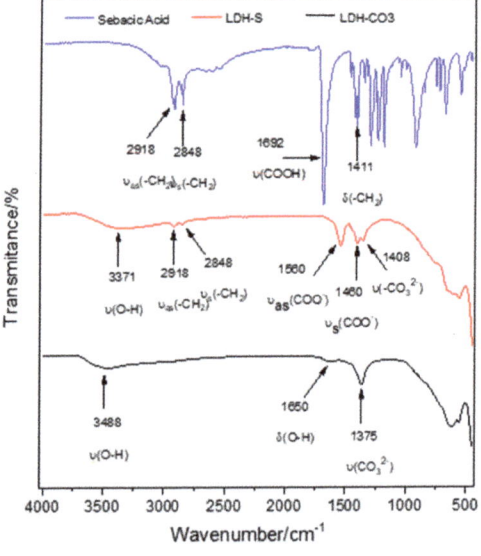

Figure 3. FT-IR spectra patterns of LDH-CO$_3$ and LDH-S.

These findings, in agreement with XRD results (Figure 2), corroborate the intercalation of sebacate anions in the LDH interlayer.

3.2. Water Capillary Absorption Test

In Figure 4, the weight gain as function of the square root of time for the three tested mortars is represented (the average value of two samples per mortar is shown). Typical behavior can be observed in two different stages: during the first hours, a linear weight increase with the time is registered, until the second stage is reached, in which the weight gain is almost stabilized.

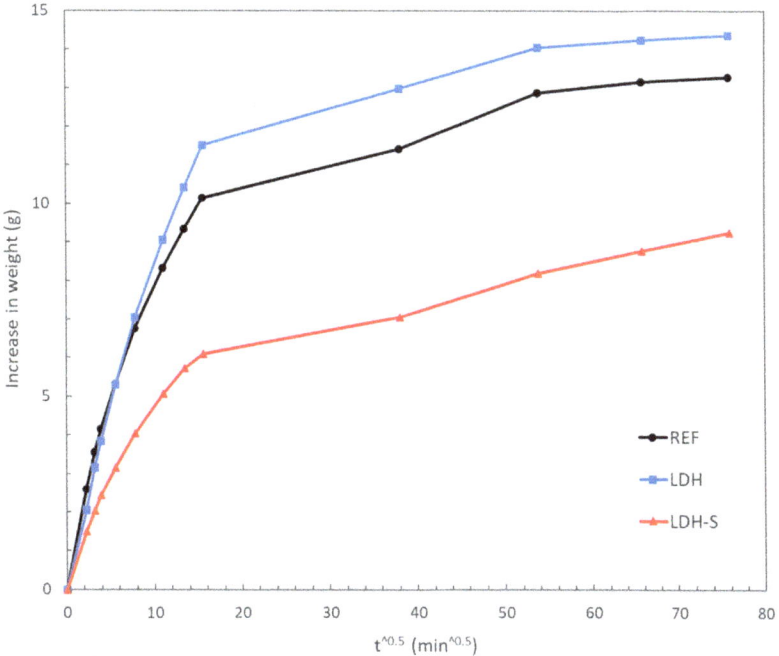

Figure 4. Water capillary absorption tests in mortars after 14 days curing.

In the case of the mortar with waterproofing additive (M-LDH-S), a significant decrease was observed both in the rate of water penetration during the first hours and in the total amount of water absorbed. However, in the case of mortars with LDH-CO_3, similar behavior to the case of the reference was observed. The time of saturation (t_n) estimated of all the studied mortars (between 3–4 h) was as detailed in Table 1. The water capillary absorption coefficient (K), estimated according to [30], is detailed in Table 1 for the different studied mortars.

Table 1. Water capillary absorption coefficients estimated according to the Fagerlund method [30].

Sample	t_n (min)	K (kg/cm^2 min$^{0.5}$)
M-REF	208	0.0184
M-LDH-CO_3	190	0.0220
M-LDH-S	224	0.0104

An increase in the saturation time and a decrease in the water capillary absorption coefficient was obtained in the case of mortar incorporating LDH-S. A waterproofing efficiency of 42% can be estimated from the K values included in Table 1. When LDH-CO_3

was added to the mortar, no improvement in water penetration by capillary absorption was observed. In fact, even higher values of K were obtained in this case.

3.3. Passivation Stage of Reinforced Mortars

The passivation of the steel rebars embedded in the mortars was monitored during the curing period by monitoring the evolution of the corrosion potential, and the corrosion current density estimated by the Stern-Geary equation, as described in the previous section. In each studied mortar, the two rebars showed a similar evolution, and thus, in the results of the present study, the mean value between two rebars is used.

In Figure 5, the evolution of the corrosion potential (E_{corr}) for the three studied mortars is represented. The typical passivation process, with a continuous increase of the potential during the first days of rebar contact with the alkaline phase of the mortar pores, was registered. After about 15 days curing, the three studied mortars showed values of E_{corr} above −0.2 V vs. SCE, that is, the criteria established by Rilem Recommendations [35] as a low probability for active corrosion occuring.

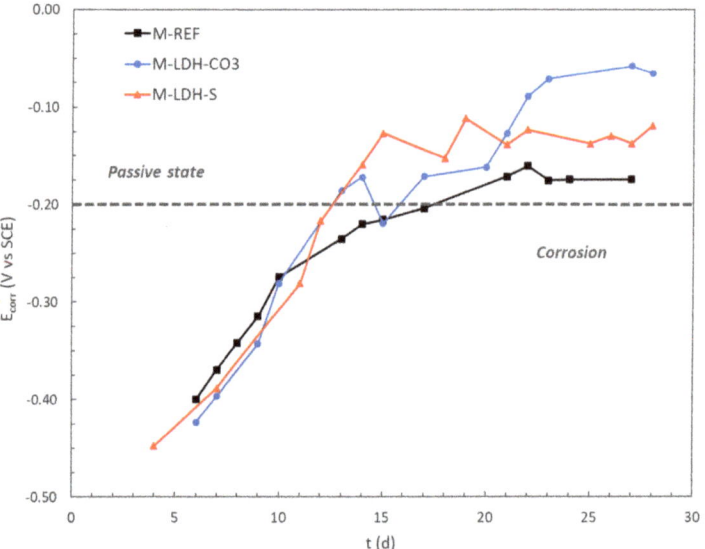

Figure 5. Evolution of the corrosion potential for the three studied reinforced mortars during the passivation stage.

Concerning the evolution of the corrosion current density (i_{corr}), the typical response of a passivation process was also observed for the three studied mortars, as shown in Figure 6. In this case, lower values of i_{corr} during the whole period of curing were registered for the rebars embedded in mortar containing LDH-S. The inhibitive action of the sebacate incorporated in the interlayer of the LDH can be confirmed when compared to the evolution of i_{corr} for the rebars in mortars with LDH-CO_3. In this latter case, higher corrosion rates were registered, as well as a delay in reaching the corrosion values below 0.2 µA/cm^2, which can be considered as the boundary between negligible and active moderate corrosion [36]. The presence of sebacate in the interlayer of the LDH seems to promote the passivation of the rebar, as the passive situation related to a negligible corrosion was reached before the reference case, and lower i_{corr} values were maintained.

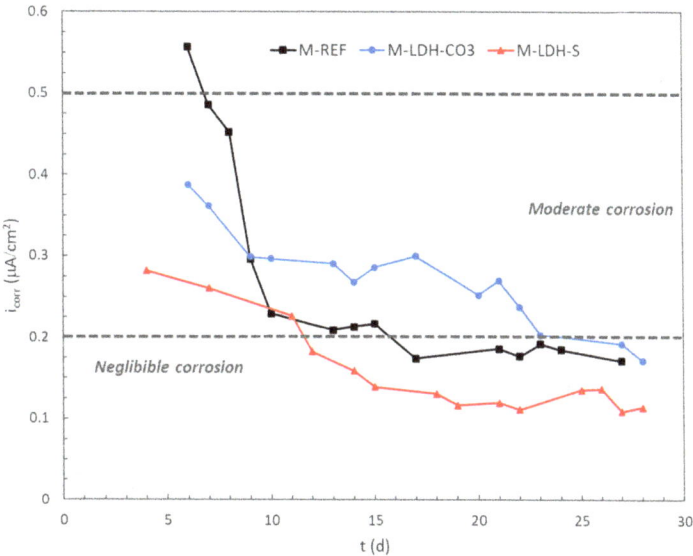

Figure 6. Evolution of the corrosion current density for the three studied reinforced mortars during the passivation stage.

The electrical resistance of the mortar cover was also monitored during the entire passivation period, as shown in Figure 7.

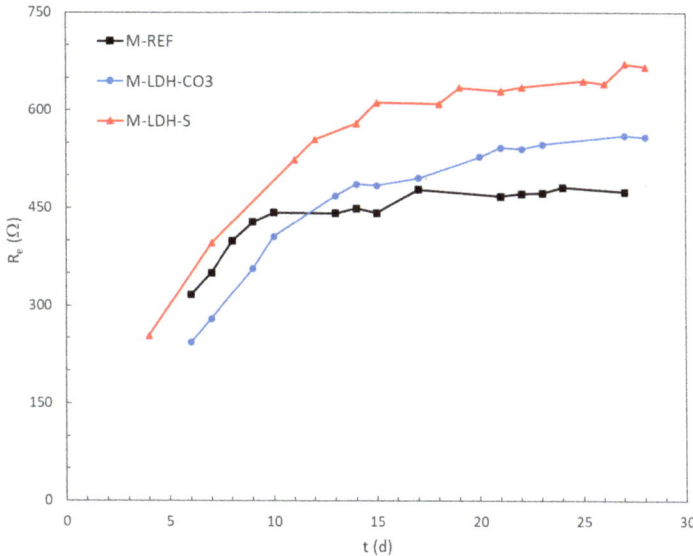

Figure 7. Evolution of the electrical resistance for the three studied reinforced mortars during the passivation period.

A continuous increase during the first days of curing was observed in all the mortars, as expected due to the evolution of the hydrated solid phases in the cementitious matrix. After some time, an almost constant value of the electrical resistance was reached for the different mortars. This steady stage was reached faster in the case of the reference mortar without any additive than in the other mortars, probably due to the continuous release of water from the LDH that promotes later hydration processes of the cementitious matrix [37]. Higher values of the electrical resistance were measured for the M-LDH-S, as expected due to the water-repellent ability of this additive, as has been confirmed in Section 3.2 in the present study. The higher values of the electrical resistance in the mortar incorporating LDH-S seem to be related to a more protective action of this mortar on the passivation processes of the steel rebars (see Figure 6).

3.4. Corrosion Stage of Reinforced Mortars

As detailed in the previous section, after the passivation of the rebars during the curing period in saturated calcium hydroxide, the reinforced samples were exposed to a chloride solution to promote the penetration of the chloride ions through the concrete pores to the rebar. During this stage, the same electrochemical parameters as in the passivation stage were monitored: corrosion potential (E_{corr}) and corrosion current density (i_{corr}).

In Figure 8, the evolution of the corrosion potential during the exposure to chloride penetration is shown. A rapid significant decrease of the corrosion potential in all cases can be observed after just the first day of immersion of the reinforced mortar samples in the solution of sodium chloride.

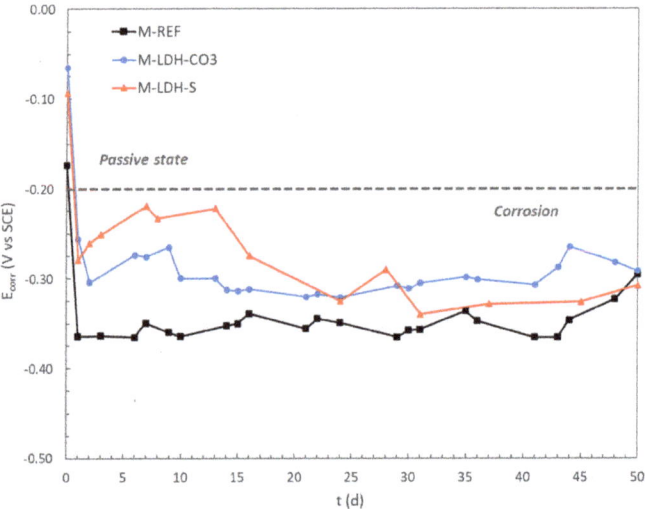

Figure 8. Evolution of the corrosion potential for the three studied reinforced mortars during the corrosion stage (exposure to chloride action).

In the case of the reference mortar, the drop in potential was higher than in the case of the mortar with LDH additives. The potential decrease in the mortar with LDH-S was less significant, and only in this case the E_{corr} seemed to evolve to more anodic values during the first days of exposure to the chloride solution.

When the corrosion current density evolution is considered, the behavior of the three studied reinforced mortars was different during the exposure to chloride action, as can be observed from Figure 9.

In the case of the reference sample, an increase of i_{corr} values to the region considered as moderate corrosion (>0.2 µA/cm^2) was registered after 7 days of exposure to the chloride penetration, indicating the corrosion initiation of the embedded rebars. In the mortar with LDH-CO$_3$, this increase occurred after approximately 25 days of chloride exposure, probably due to the ability of LDH to trap chloride. In fact, the i_{corr} values fluctuated on the limit between passivation and corrosion for a few days, until a new increase of i_{corr} values was produced after 40 days of exposure to chloride action.

When LDH-S was added to the mortar, rebars did not show any sign of corrosion even after 50 days of exposure to chloride, confirming the protective ability of the LDH-S as a corrosion inhibitor. Low values of i_{corr}, below 0.1 µA/cm^2, were maintained during the whole test.

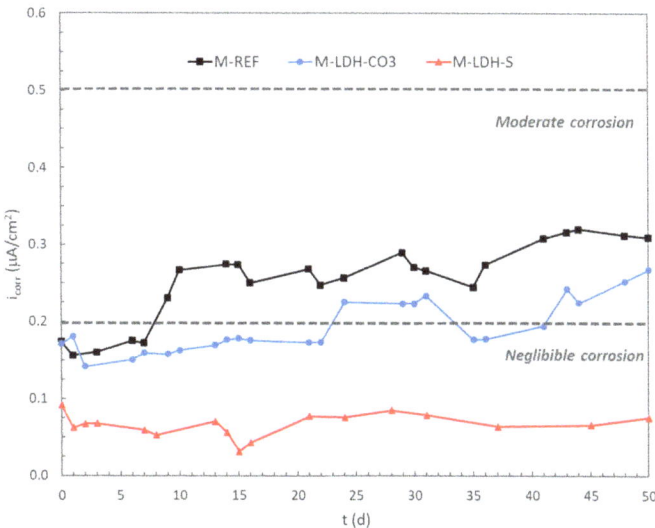

Figure 9. Evolution of the corrosion current density for the three studied reinforced mortars during the corrosion stage (exposure to chloride action).

3.5. Inhibitory Efficiency of LDH-Based Additives

To evaluate the efficiency of the corrosion inhibitors based on LDH, the total charge passed during the entire chloride exposure period (Q_{corr}) has been estimated and represented in Figure 10 for the different studied mortars. In this case, the response of each rebar has been considered, and therefore two points for each mortar are shown in Figure 10.

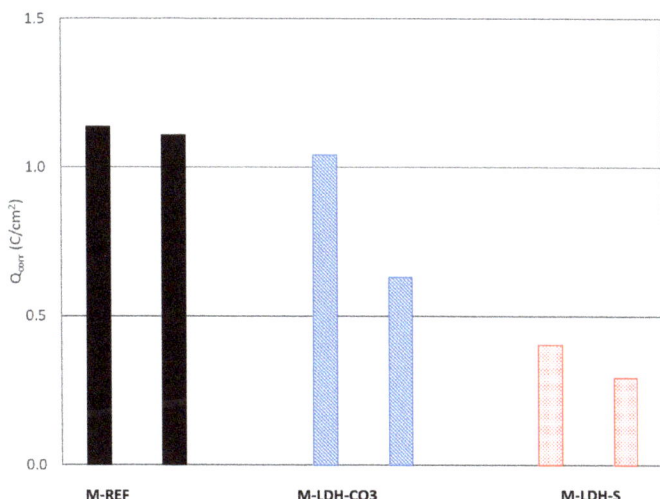

Figure 10. Total charge associated with the corrosion period (exposure to chloride action) for each studied rebar.

While in the case of M-REF and M-LDH-S, a similar response is registered for the two embedded rebars in each case, in the case of M-LDH-CO$_3$, higher variability is observed. Nevertheless, it can be observed that both M-LDH-CO$_3$ and M-LDH-S showed lower values of charge, indicating a certain higher protective ability of these mortars. When LDH-S was added to the mortar, the decrease was significantly higher, confirming the higher efficiency of this additive as a corrosion inhibitor. The inhibitory efficiency of both additives can be estimated using Equation (1), and the mean values of 74% and 23% were obtained for M-LDH-S and M-LDH-CO$_3$, respectively.

At the end of the chloride exposure, one rebar of each mortar was subjected to an anodic current by connecting the rebar to +0.250 V vs. SCE for 4 h to accelerate the corrosive development, making the damage more visible to visual inspection of the rebars. In Figure 11, the photographs of the exposed surface after the anodic polarization are included.

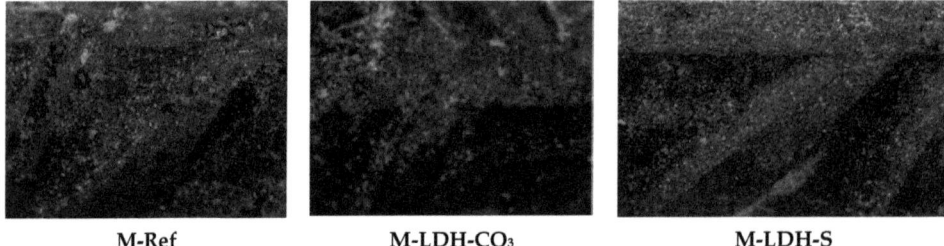

M-Ref   M-LDH-CO$_3$ M-LDH-S

Figure 11. Visual aspect of the rebars at the end of the test.

The visual inspection of the rebar surface confirms the electrochemical results. While the reference mortar and the M-LDH-CO$_3$ show corrosion spots distributed on the exposed surface, in the case of the M-LDH-S mortar, the presence of these spots is not evident.

4. Conclusions

The waterproofing ability of LDH with sebacate (LDH-S) in the interlayer has been confirmed with a decrease of 42% in the water capillary absorption coefficient. This water-repellent capacity resulted in a very effective corrosion inhibition for steel rebars embedded in mortar with 3% LDH-S (74% inhibition). The effect of LDH-S can be observed from the passivation of the rebar due to the contact with the alkaline aqueous phase of mortar pores. The action of LDH-S seems be related to the increase in the electrical resistance values of the mortar cover due to the hydrophobic effect caused by sebacate presence.

A less effective corrosion inhibition, associated with the presence of LDH in the mortar, has been also observed (23% inhibition), probably due to the ability of this additive to trap chloride ions. However, this protection was lost after continued chloride exposure, and corrosion initiated similarly to the reference mortar.

Further studies are underway to analyze the protection mechanism associated to the presence of LDH-S in the mortar.

Author Contributions: Conceptualization, M.C.-Y., L.S.G., M.S.-M. and I.P.; methodology, D.C., R.B.-C., F.T. and M.S.-M.; software, R.B.-C. and F.T.; validation, M.C.-Y., L.S.G., M.S.-M. and I.P.; formal analysis, D.C., M.S.-M. and I.P.; investigation, D.C., L.S.G., M.S.-M. and I.P.; resources, F.T., L.S.G., M.S.-M. and I.P.; data curation, M.S.-M. and I.P.; writing—original draft preparation, M.S.-M. and I.P.; writing—review and editing, M.C.-Y., L.S.G., M.S.-M. and I.P.; visualization, M.C.-Y., L.S.G., M.S.-M. and I.P.; supervision, L.S.G., M.S.-M. and I.P.; project administration, L.S.G. and M.S.-M.; funding acquisition, M.C.-Y., L.S.G., M.S.-M. and I.P. All authors have read and agreed to the published version of the manuscript.

Funding: This research was 80% funded by P.O. FEDER Andalucía (2014–2020)/Consejería de Transformación Económica, Industria, Conocimiento y Universidad/_Proyecto Ref. PY20_00365 "Incorporación de compuestos tipo hidrotalcita (HDL) en tratamientos superficiales multifuncionales para la reparación de infraestructuras de hormigón".

Data Availability Statement: Data can be found on the Zenodo platform (10.5281/zenodo.7092247).

Acknowledgments: The authors thank Cementos Cosmos S.A. for supplying the cement for manufacturing the mortar specimens.

Conflicts of Interest: The authors declare no conflict of interest.

References

1. Alonso, C.; Andrade, C.; Castellote, M.; Castro, P. Chloride threshold values to depassivate reinforcing bars embedded in a standardized OPC mortar. *Cem. Concr. Res.* **2000**, *30*, 1047–1055. [CrossRef]
2. Yuan, Q.; Shi, C.; De Schutter, G.; Audenaert, K.; Deng, D. Chloride binding of cement-based materials subjected to external chloride environment—A review. *Constr. Build. Mater.* **2009**, *23*, 1–13. [CrossRef]
3. Saremi, M.; Mahallati, E. A study on chloride-induced depassivation of mild steel in simulated concrete pore solution. *Cem. Concr. Res.* **2002**, *32*, 1915–1921. [CrossRef]
4. Muhammad, N.Z.; Keyvanfar, A.; Mamjid, M.Z.A.; Shafaghat, A.; Mirza, J. Waterproof performance of concrete: A critical review on implemented approaches. *Constr. Build. Mater.* **2015**, *101*, 80–90. [CrossRef]
5. Sánchez, M.; Faria, P.; Ferrara, L.; Horszczaruk, E.; Jonkers, H.M.; Kwiecien, A.; Mosa, J.; Peled, A.; Pereira, A.S.; Snoeck, D.; et al. External treatments for the preventive repair of existing constructions: A review. *Constr. Build. Mater.* **2018**, *193*, 435–452. [CrossRef]
6. Pan, X.; Shi, Z.; Shi, C.; Ling, T.C.; Li, N. A review on concrete surface treatment Part I: Types and mechanisms. *Constr. Build. Mater.* **2017**, *132*, 578–590. [CrossRef]
7. Almusallam, A.A.; Khan, F.M.; Dulaijan, S.U.; Al-Amoudi, O.S.B. Effectiveness of surface coatings in improving concrete durability. *Cem. Concr. Compos.* **2003**, *25*, 473–481. [CrossRef]
8. Jiang, L.; Xue, X.; Zhang, W.; Yang, J.; Zhang, H.; Li, Y.; Zhang, R.; Zhang, Z.; Xu, L.; Qu, J. The investigation of factors affecting the water impermeability of inorganic sodium silicate-based concrete sealers. *Constr. Build. Mater.* **2015**, *93*, 729–736. [CrossRef]
9. Li, J.; Yi, Z.; Xie, Y. Progress of Silane Impregnating Surface Treatment Technology of Concrete Structure. *Mater. Rev.* **2012**, *26*, 120–125.
10. Bao, J.; Li, S.; Zhang, P.; Xue, S.; Cui, Y.U.; Zhao, T. Influence of exposure environments and moisture content on water repellency of surface impregnation of cement-based materials. *J. Mater. Res. Technol.* **2020**, *9*, 12115–12125. [CrossRef]
11. Allahvirdizadeh, R.; Reshetnia, R.; Dousti, A.; Shekarchi, M. Application of polymer concrete in repair of concrete structures: A literature review. In *Concrete Solutions*, 1st ed.; Grantham, M., Mechtcherine, V., Schneck, U., Eds.; CRC Press: London, UK, 2011; pp. 435–444.
12. Al-Kheetan, M.J.; Rahman, M.M.; Chamberlain, D.A. A novel approach of introducing crystalline protection material and curing agent in fresh concrete for enhancing hudrophobicity. *Constr. Build. Mater.* **2018**, *160*, 644–652. [CrossRef]
13. Al-Rashed, R.; Al-Jabari, M. Concrete protection by combined hygroscopic and hydrophilic crystallization waterproofing applied to fresh concrete. *Case Stud Constr Mater* **2021**, *15*, e00635. [CrossRef]
14. Rives, V. *Layered Double Hydroxides: Present and Future*; Nova Science Publishers, Inc.: New York, NY, USA, 2001.
15. Cavani, F.; Trifirò, F.; Vaccari, A. Hydrotalcite-type anionic clays: Preparation, properties and applications. *Catal. Today* **1991**, *11*, 173–301. [CrossRef]
16. Yang, H.; Xiong, C.; Liu, X.; Liu, A.; Li, T.; Ding, R.; Shah, S.P.; Li, W. Application of layered double hydroxides (LDHs) in corrosion resistance of reinforced concrete—State of the art. *Constr. Build. Mater.* **2021**, *307*, 124991. [CrossRef]
17. Mir, Z.M.; Bastos, A.; Höche, D.; Zheludkevich, M. Recent Advances on the Application of Layered Double Hydroxides in Concrete—A Review. *Materials* **2020**, *13*, 1426. [CrossRef]
18. Cao, Y.; Dong, S.; Zheng, D.; Wang, J.; Zhang, X.; Du, R.; Song, G.; Lin, C. Multifunctional inhibition based on layered double hydroxides to comprehensively control corrosion of carbon steel in concrete. *Corros. Sci.* **2017**, *126*, 166–179. [CrossRef]
19. Zuo, J.; Wu, B.; Luo, C.; Dong, B.; Xing, F. Preparation of MgAl layered double hydroxides intercalated with nitrite ions and corrosion protection of steel bars in simulated carbonated concrete pore solution. *Corros. Sci.* **2019**, *152*, 120–129. [CrossRef]
20. Ma, G.; Xu, J.; Han, L.; Xu, Y. Enhanced inhibition performance of NO_2^- intercalated MgAl-LDH modified with nano-SiO_2 on steel corrosion in simulated concrete pore solution. *Corros. Sci.* **2022**, *204*, 110387. [CrossRef]
21. Wu, B.; Zuo, J.; Dong, B.; Xing, F.; Luo, C. Study on the affinity sequence between inhibitor ions and chloride ions in Mg-Al layer double hydroxides and their effects on corrosion protection for carbon steel. *Appl. Clay Sci.* **2019**, *180*, 105181. [CrossRef]
22. Yang, Z.; Fischer, H.; Cerezo, J.; Mol, J.M.C.; Polder, R. Modified hydrotalcites for improved corrosion protection of reinforcing steel in concrete—Preparation, characterization, and assessment in alkaline chloride solution. *Mater. Corros.* **2016**, *71*, 721–738. [CrossRef]

23. Cao, Y.; Zheng, D.; Dong, S.; Zhang, F.; Lin, J.; Wang, C.; Lin, C. A Composite Corrosion Inhibitor of MgAl Layered Double Hydroxides Co-Intercalated with Hydroxide and Organic anions for Carbon Steel in Simulated Carbonated Concrete Pore Solutions. *J. Electrochem. Soc.* **2019**, *166*, C3106–C3113. [CrossRef]
24. Chen, Y.; Shui, Z.; Chen, W.; Chen, G. Chloride binding of synthetic Ca-Al-NO$_3$ LDHs in hardened cement paste. *Constr. Build. Mater.* **2015**, *93*, 1051–1058. [CrossRef]
25. Yang, Z.; Fischer, H.; Polder, R. Synthesis and characterization of modified hydrotalcites and their ion exchange characteristics in chloride-rich simulated concrete pore solution. *Cem. Concr. Compos.* **2014**, *47*, 87–93. [CrossRef]
26. Tian, Y.; Dong, C.; Wang, G.; Cheng, X.; Li, X. Zn-Al-NO$_2$ layered double hydroxide as a controlled-release corrosion inhibitor for steel reinforcements. *Mater. Lett.* **2019**, *236*, 517–520. [CrossRef]
27. Tian, Y.; Wen, C.; Wang, G.; Deng, P.; Mo, W. Inhibiting property of nitrite intercalated layered double hydroxide for steel reinforcement in contaminated concrete condition. *J. Appl. Electrochem.* **2020**, *50*, 835–849. [CrossRef]
28. Albayrak, A.T.; Yasar, M.; Gurkaynak, M.A.; Gurgey, I. Investigation of the effects of fatty acids on the compressive strength of the concrete and the grindability of the cement. *Cem. Concr. Res.* **2005**, *35*, 400–404. [CrossRef]
29. Nguyen, D.T.; To, H.T.X.; Gervasi, J.; Paint, Y.; Gonon, M.; Olivier, M.G. Corrosion inhibition of carbon steel by hydrotalcites modified with different organic carboxylic acids for organic coatings. *Prog. Org. Coat.* **2018**, *124*, 256–266. [CrossRef]
30. *UNE 83982:2008*; Concrete Durability. Test Methods. Determination of the Capillar Suction in Hardened Concrete; Fagerlund Method. AENOR: Madrid, Spain, 2008.
31. Stern, M.; Geary, A.L. Electrochemical polarization, I. A theoretical analysis of the shape of polarization curves. *J. Electrochem. Soc.* **1957**, *104*, 56–63. [CrossRef]
32. Andrade, C.; Maribona, I.R.; Feliu, S.; González, J.A.; Feliu, S., Jr. The effect of macrocells between active and passive areas of steel reinforcements. *Corr. Sci.* **1992**, *33*, 237–249. [CrossRef]
33. Chaara, D.; Bruna, F.; Ulibarri, M.A.; Draoui, K.; Barriga, C.; Pavlovic, I. Organo/layered double hydroxide nanohybrids used to remove non ionic pesticides. *J. Hazard. Mater.* **2011**, *196*, 350–359. [CrossRef]
34. Dietmann, K.M.; Linke, T.; Trujillano, R.; Rives, V. Effect of Chain Length and Functional Group of Organic Anions on the Retention Ability of MgAl- Layered Double Hydroxides for Chlorinated Organic Solvents. *ChemEngineering* **2019**, *3*, 89. [CrossRef]
35. Elsener, B. RILEM TC 154-EMC: Electrochemical techniques for measuring metallic corrosion—Recommendations—Half-cell potential measurements—Potential mapping on reinforced concrete structures. *Mater. Struct.* **2003**, *36*, 461–471. [CrossRef]
36. Garcés, P.; Andrade, M.C.; Saez, A.; Alonso, M.C. Corrosion of reinforcing steel in neutral and acid solutions simulating the electrolytic environments in the micropores of concrete in the propagation period. *Corros. Sci.* **2005**, *47*, 289–306. [CrossRef]
37. Shui, Z.H.; Yu, R.; Chen, Y.X.; Duan, P.; Ma, J.T.; Wang, X.P. Improvement of concrete carbonation resistance based on a structure modified layered double hydroxides (LDHs): Experiments and mechanism analysis. *Constr. Build. Mater.* **2018**, *176*, 228–240. [CrossRef]

Article

Adsorption Properties and Hemolytic Activity of Porous Aluminosilicates in a Simulated Body Fluid

Olga Yu. Golubeva *, Yulia A. Alikina, Elena Yu. Brazovskaya and Nadezhda M. Vasilenko

Laboratory of Silicate Sorbents Chemistry, Institute of Silicate Chemistry of Russian Academy of Sciences, Adm. Makarova Emb., 2, 199034 St. Petersburg, Russia
* Correspondence: olga_isc@mail.ru or golubeva@iscras.ru

Abstract: A study of the adsorption features of bovine serum albumin (BSA), sodium and potassium cations, and vitamin B1 by porous aluminosilicates with different structures in a medium simulating blood plasma was conducted. The objects of this study were synthetic silicates with a montmorillonite structure $Na_{2x}(Al_{2(1-x)},Mg_{2x})Si_4O_{10}(OH)_2 \cdot nH_2O$ (x = 0.5, 0.9, 1), aluminosilicates of the kaolinite subgroup $Al_2Si_2O_5(OH)_4$ with different particle morphologies (spherical, nanosponge, nanotubular, and platy), as well as framed silicates (Beta zeolite). An assessment of the possibility of using aluminosilicates as hemosorbents for extracorporeal blood purification was carried out. For this purpose, the sorption capacity of the samples both with respect to model medium molecular weight toxicants (BSA) and natural blood components—vitamins and alkaline cations—was investigated. The samples were also studied by X-ray diffraction, electron microscopy, and low-temperature nitrogen adsorption. The zeta potential of the sample's surfaces and the distribution of active centers on their surfaces by the method of adsorption of acid-base indicators were determined. A hemolytic test was used to determine the ability of the studied samples to damage the membranes of eukaryotic cells. Langmuir, Freundlich, and Temkin models were used to describe the experimental BSA adsorption isotherms. To process the kinetic data, pseudo-first-order and pseudo-second-order adsorption models were used. It was found that porous aluminosilicates have a high sorption capacity for medium molecular weight pathogens (up to 12 times that of activated charcoal for some samples) and low toxicity to blood cells. Based on the obtained results, conclusions were made about the prospects for the development of new selective non-toxic hemosorbents based on synthetic aluminosilicates with a given set of properties.

Keywords: aluminosilicates; kaolinite; montmorillonite; zeolites; hemosorbents; adsorption; albumin; vitamins; hemolytic activity; body fluid

Citation: Golubeva, O.Y.; Alikina, Y.A.; Brazovskaya, E.Y.; Vasilenko, N.M. Adsorption Properties and Hemolytic Activity of Porous Aluminosilicates in a Simulated Body Fluid. *ChemEngineering* **2022**, 6, 78. https://doi.org/10.3390/chemengineering6050078

Academic Editors: Miguel A. Vicente, Raquel Trujillano and Francisco Martín Labajos

Received: 7 September 2022
Accepted: 4 October 2022
Published: 6 October 2022

Publisher's Note: MDPI stays neutral with regard to jurisdictional claims in published maps and institutional affiliations.

Copyright: © 2022 by the authors. Licensee MDPI, Basel, Switzerland. This article is an open access article distributed under the terms and conditions of the Creative Commons Attribution (CC BY) license (https://creativecommons.org/licenses/by/4.0/).

1. Introduction

In recent years, more and more attention has been paid to studying the features of the adsorption of protein molecules from model solutions on the surface of sorbents of various natures, in particular, on the surface of clay minerals [1–6]. Interest in such research is related to the possibility of exploring the prospects for using clay minerals to remove proteins from wine, as well as the performance of membranes for protein separation, biosensors, or protein therapy platforms [2,7,8]. In addition, the relevance of these studies is associated with the need to develop biospecific sorbents for the selective adsorption of toxic substances of protein origin that accumulate in the body during oncological, immune, infectious, and other diseases [9,10].

Hemosorption is the most promising method of performing the sorption detoxification of the body [11–13]. Such sorption therapy is based on the adsorption ability of materials to remove toxic substances of various natures from the blood. The first and most common sorbents were materials based on activated carbon. Such materials are capable of removing a variety of toxic molecules-exotoxins (poisons), cytokines, anti-inflammatory mediators, products of a bacterial nature, as well as those arising from cell breakdown [14–17]. However, activated charcoal-based materials have disadvantages, as in the process of hemosorption, there is a partial traumatization and death of blood cells. In addition, in the process of hemosorption on carbon sorbents, along with pathological components, a part of physiologically significant metabolites is removed. In this regard, a promising direction is the development of sorption technologies based on biospecific (selective) hemosorption.

There are a significant number of hemosorbents on the market, but none of them currently fully meet all the requirements for such materials, namely having a high sorption capacity with respect to toxins and metabolites, hemocompatibility, selectivity, and the ability to withstand certain sterilization methods without losing basic properties. The high sorption capacity of a number of inorganic adsorbents has great potential for medical use, however, according to some researchers, inorganic matrices, which usually mean natural porous minerals—clays, zeolites, etc.—are inferior to other adsorbents (activated carbon, synthetic and natural organic polymers) in terms of biocompatibility [18–20]. This problem can be solved by using synthetic inorganic matrices with the following desired characteristics: high sorption characteristics and hemocompatibility due to the absence of impurity phases, controlled chemical and dispersion composition, as well as a certain particle morphology and specified porosity in a wide range (from nano- to macro- and mesopores), which allows the adsorption of biological molecules of different sizes.

Medical sorbents must meet certain requirements—a high degree of chemical purity, a minimum content of impurities, a smooth surface relief, a high sorption capacity for removed substances, and the presence of hemocompatibility [21,22]. Under the conditions of directed hydrothermal synthesis, the porous aluminosilicates of various structures can be obtained with specified characteristics, such as a certain phase and chemical composition, given particle size and morphology, as well as porous textural and sorption characteristics. Preliminary studies of the cytotoxicity and hemolytic activity of synthetic samples of aluminosilicates showed that they do not have the toxicity that is characteristic of natural minerals, which indicates that it is promising to study the possibility of their use as medical sorbents [23,24].

The present work presents the results of a study of porous textural properties, surface properties, hemolytic activity, as well as the features of adsorption by synthetic porous aluminosilicate sorbents with different porosities and particle morphologies from a medium simulating blood plasma, bovine serum albumin, sodium and potassium cations, as well as vitamin B1. Framework aluminosilicates (zeolites), layered silicates with montmorillonite structure, as well as layered silicates of the kaolinite subgroup with spherical, sponge, and platy morphologies were selected as objects for this study.

Bovine serum albumin (BSA) is a water-soluble globular protein (with an approximate molecular size of 9 nm × 8 nm × 6 nm.) [25,26], which is part of the blood serum and blood cytoplasm of animals and plants. Albumin refers to proteins with an average molecular weight of 67–69 kDa. BSA is often used to understand the adsorption mechanism of proteins at solid/liquid interfaces. In this study, BSA acts as a marker of medium molecular weight proteins. It is known that pathogenic compounds formed in the body during oncological, immune, infectious, and other diseases belong to proteins of medium molecular weight [1,27]. The adsorption of albumin by clay minerals has been widely studied [1–6,28], especially with regard to biosensors. Since the value of the isoelectric point of albumin is 5, most studies were carried out with solutions having acidic pH values (4.5) and sometimes at elevated temperatures. At the same time, the requirements for hemosorbents impose certain requirements on the experiments being carried out—the pH values must correspond to the pH of the blood plasma (neutral) and the temperature of the study should not exceed 37 °C. To replicate the conditions of hemosorption as accurately as possible, in this work, studies of the adsorption process were carried out at room temperature at neutral pH, from the medium of a synthetic biological fluid, prepared in accordance with the chemical analysis of human body fluids, with ion concentrations nearly equal to those of the inorganic components of human blood plasma [29,30]. Alkaline cations and vitamins, the adsorption of which was also considered in this work, are essential microelements that are part of the blood and affect the state of the cardiovascular and other human systems. Sodium, potassium, calcium, and magnesium play a central role in the normal regulation of blood pressure [31]. A marked reduction in sodium and potassium intake is effective, even in treating severe hypertension. Thiamin, or vitamin B1, is an essential water-soluble vitamin that acts as a coenzyme in carbohydrate and branched-chain amino acid metabolism [32]. Therefore, the loss of mineral substances during the process of hardware blood purification in the process of hemosorption is extremely undesirable.

The results of the study of the adsorption capacity of synthetic aluminosilicate samples with different porosity (for example, the maximum diameter of zeolite cavities does not exceed 1 nm; and montmorillonites have the ability to change the interlayer distance over a wide range—from 1 Å to complete exfoliation into individual layers) and with different surface properties, this will allow us to evaluate the possibility of developing universal and selective sorbents for carrying out the adsorption of substances of different molecular weights and different molecular sizes.

2. Materials and Methods

2.1. Reagents

The following reagents were used for the synthesis and analysis of the samples: tetraethoxysilane TEOS (($C_2H_5O)_4Si$, special purity grade, $\geq 99.0\%$), aluminum nitrate $Al(NO_3)_3 \cdot 9H_2O$ (reagent grade, $\geq 97.0\%$), magnesium nitrate $Mg(NO_3)_2 \cdot 6H_2O$ (reagent grade), nitric acid HNO_3 (reagent grade, 65 wt%), aqueous ammonia (25 wt% NH_3), ethanol C_2H_5OH (96 wt%), hydrochloric acid HCl (35–38 wt%), sodium hydroxide solution (50 wt% in water), raw halloysite nanotubes (Sigma-Aldrich, Product of Applied Minerals, USA), potassium hydroxide (KOH, 45% aqueous solution), silica sol (LUDOX HS_40, 40%), aluminum sulfate ($Al_2(SO_4)_3 \cdot 18H_2O$, $\geq 98\%$), tetraethylammonium hydroxide (($C_2H_5)_4NOH$, 35% aqueous solution, Sigma), activated charcoal (MW 12.01 g/mol, Fluka Analytical), bovine serum albumin (lyophilized pH~7, Biowest), and vitamin B1 (Thiamine hydrochloride, reagent grade $\geq$ 99%, Hubei Maxpharm Industries).

Simulated body fluid (SBF) was prepared according to the procedure in [25] using the following reagents: NaCl (98%, NevaReactiv), $NaHCO_3$ (99.5%), KCl (NevaReactiv, 99%), $Na_2HPO_4 \cdot 2H_2O$ (98%, Chimmed), $MgCl_2 \cdot 6H_2O$ (98%, NevaReactiv), $CaCl_2 \cdot 2H_2O$ (98%, NavaReactiv), and $(CH_2OH)_3CNH_2$ (Trizma base, Sigma, MW127.14 g/mol). Solutions were prepared in deionized water (Vodolei, NPP Khimelektronika, Russia) with a specific conductivity no higher than 0.2 µS/cm.

2.2. Synthesis of Aluminosilicates

Porous aluminosilicates of various structural types and with different particle morphologies were chosen as the objects of this study. All studied aluminosilicates were synthetic, with the exception of halloysite nanotubes. The main characteristics of the samples, their chemical formulas, and structural types are given in Table 1.

Samples with a montmorillonite structure corresponding to the ideal chemical formula $Na_{2x}(Al_{2(1-x)},Mg_{2x})Si_4O_{10}(OH)_2 \cdot nH_2O$ with various degrees of isomorphic substitution magnesium atoms in octahedral layers were chosen as objects of this study: with x = 1 ($Mg_3Si_4O_{10}(OH)_2 \cdot nH_2O$), x = 0.9 ($Na_{1.8}Al_{0.2}Mg_{1.8}Si_4O_{10}(OH)_2 \cdot H_2O$), and x = 0.5 ($Na_{1.0}Al_{1.0}Mg_{1.0}Si_4O_{10}(OH)_2 \cdot nH_2O$). Samples corresponding to the $Al_2Si_2O_5(OH)_4$ kaolinite formula were synthesized under conditions that made it possible to obtain various particle morphologies—spherical, sponge, and platy. In addition, the sorption and physicochemical properties of the samples were compared with the results of a study of natural halloysite $Al_2Si_2O_5(OH)_4 \cdot nH_2O$ with nanotubular morphology. The zeolite of the structural type Beta was also studied as an object of this study.

The synthesis of all samples was carried out under hydrothermal conditions according to previously developed methods [23,33–37]. The resulting product was washed with water and dried. For zeolite samples, an additional decationization procedure was carried out, that is, the removal of alkaline cations K^+ and Na^+ localized in large cavities. Sample decationization was carried out by the triple treatment of zeolites with an ammonium salt solution followed by drying at 120 °C and the decomposition of the ammonium ion NH_4^+ at 600 °C for 1 h. In addition, the initial zeolite was preliminarily calcined for 2 h at a temperature of 350 °C in order to remove the adsorbed water and residues of organic molecules (tetraethylammonium) from the pores of the zeolite.

Table 1. Main characteristics of the studied samples.

Samples Designation	Mineralogical Name	Structural Type	Chemical Formula (by Synthesis)	Particles Morphology	Synthesis Conditions		Content, wt%				
					T, °C	t, h	SiO_2	Al_2O_3	MgO	Loss on Ignition, %	Additionally.
MT-Al0	Montmorillonite	LS	$Mg_3Si_4O_{10}(OH)_2 \cdot nH_2O$	layers	250	72	59.39	0	28.63	11.45	-
MT-Al0.2	Montmorillonite	LS	$Na_{1.8}Al_{0.2}Mg_{1.8}Si_4O_{10}(OH)_2 \cdot H_2O$	layers	350	72	58.10	5.32	18.31	14.75	Na_2O 3.52
MT-Al1.0	Montmorillonite	LS	$Na_{1.0}Al_{1.0}Mg_{1.0}Si_4O_{10}(OH)_2 \cdot nH_2O$	layers	350	72	53.00	22.82	8.04	13.45	Na_2O 2.69
Kaol-sph	Kaolinite	LS	$Al_2Si_2O_5(OH)_4$	spheres	220	72	44.74	37.22	0	14.74	-
Kaol-sponge	Kaolinite	LS	$Al_2Si_2O_5(OH)_4$	nanosponges	220	72	43.77	36.14	0	15.79	-
Kaol-pl	Kaolinite	LS	$Al_2Si_2O_5(OH)_4$	plates	350	96	45.84	39.48	0	14.05	-
Hal	Halloysite	LS	$Al_2Si_2O_5(OH)_4 \cdot nH_2O$	nanotubes	-	-	46.22	36.38	0	16.04	-
Beta	Zeolite Beta	FS	$H^+{}_7[Al_7Si_{57}O_{128}] \cdot nH_2O$	spheres	135	48	69.38	8.54	0	20.09	Na_2O 0.3, K_2O 0.2

Designations: LS—layered silicate; FS—framed silicate.

2.3. Characterization

The X-ray phase analysis of the samples was carried out using a powder diffractometer Rigaku Corporation, SmartLab 3 (CuKα-radiation, operating mode-40 kV/40 mA; semiconductor point detector (0D)-linear (1D), θ-θ geometry, measurement range 2θ = 5–70° (step 2θ = 0.01°), speed 5°/min).

The samples were chemically analyzed to gravimetrically determine the Si, Mg, and Al contents using a quinolate of the silicon molybdenum complex and by complexometric titration. The sodium and potassium content of the studied samples was determined by atomic absorption spectroscopy (Thermo scientific iCE 3000, Waltham, MA, USA).

The textural parameters of the materials were determined by means of the low-temperature adsorption–desorption of nitrogen. The isotherms were collected using a Quantachrome NOVA 1200e instrument (Quantachrome Instruments, Boynton Beach, FL, USA). Degassing was performed at 300 °C for 12 h. The specific surface area of the sample was calculated by the BET method [38] using NOVAWin (USA) software. The pore size distribution and mean pore diameter were calculated by the Barret-Joyner-Halenda (BJH) method from the desorption curve [39].

The morphology of the samples was studied by scanning electron microscopy (SEM) by using a Carl Zeiss Merlin instrument (Oberkochen, Germany) with a field emission cathode. The beam current and accelerating voltage were 2 nA and 21 kV, respectively. The device was equipped with a two-beam workstation with focused ion and scanning electron beams, a Carl Zeiss Auriga laser with a field emission cathode, a GEMINI electron optics column, and an oil-free vacuum system with a beam current range of 400 pA and an acceleration voltage of 1.5–4 kV. The powders of the samples were directly planted on conductive carbon tape without additional processing.

The electrokinetic (zeta) potential of the samples was determined using the particle size and zeta potential analyzer NaniBrook 90 PlusZeta (Brookhaven Instruments Corporation, USA). The samples were a suspension obtained by dispersing 50 mg of sample in 20 mL of deionized water. Before the measurements, the suspension was subjected to low power (50 W) ultrasonication for two minutes on an ultrasonic processor UP50H.

The functional composition of the surface of the samples was studied by the method of the adsorption of acid-base indicators with different pKa values in the range from −4.4 to

14.2, undergoing a selective adsorption on the surface of active centers with the corresponding pKa values according to the procedure described in [40]. The content of adsorption centers was determined from the change in the optical density of the aqueous solutions of indicators using UV absorption spectroscopy (LEKISS2109UV spectrophotometer).

The adsorption properties of the samples with respect to BSA were studied under static conditions from BSA solutions in SBF with an albumin concentration of 2.4 g/L. The experiments were carried out at room temperature (25 ± 1 °C), which corresponds to the conditions of the hemosorption procedure. To a weighed portion of the sorbent (30 mg), 10 mL of a BSA solution in SBF was added, and the mixture was stirred on a magnetic stirrer for the time necessary to plot the kinetic curves (from 1 to 30 h). After the experiment was completed, the sample was centrifuged. The protein concentration in the supernatant were analyzed with a UV–Vis spectrophotometer (SHIMADZU UV-2600/2700) at 278 nm. Each point of the kinetic curve was taken as the average of three measurements. The BSA concentration was determined using UV–Vis absorption spectroscopy (Shimadzu UV-2600/2700, Shimadzu Europa GmbH) by the optical density at a wavelength of 278 nm.

The capacity of the sorbent, mg/g (the amount of adsorbed substance), was determined by the following Formula (1):

$$X = (C_i - C_f) V_s / m_s, \qquad (1)$$

where C_i is the initial concentration of albumin solution, g/L; C_f is the final concentration after sorption, g/L; V_s is the volume of albumin solution, L; and m_s is the weight of the sorbent sample, g.

To process the kinetic data, pseudo-first-order (PFO) and pseudo-second-order (PSO) adsorption models [41,42] were used. The kinetic expression for PFO, based on the capacitance of a solid, is written in the following form:

$$q_t = q_e \left(1 - e^{-k_1 t}\right), \qquad (2)$$

where q_t and q_e are the sorption capacity at time t and in equilibrium (mg/g), and k_1 is the PFO reaction rate constant, min^{-1}

The mathematical expression for the PSO kinetic model is as follows:

$$q_t = \frac{q_e^2 k_2 t}{1 + q_e k_2 t} \qquad (3)$$

where q_t and q_e are the sorption capacity at time t and in the equilibrium state (mg/g) and k_2 is the PSO rate constant (g/(mg·min)).

The study of the equilibrium adsorption of BSA was carried out at an initial albumin concentration in the range from 100 to 2400 mg/L. For this, 32 mg of a sorbent sample with a weighing accuracy of ±0.0002 g was dispersed in 10 mL of BSA solution in SBF with a given concentration. The experiments were carried out in a static mode in closed glass bottles with a volume of 20 mL with stirring for the time necessary to achieve adsorption equilibrium (from 4 to 24 h depending on the structural type of aluminosilicates). The samples were filtered and the albumin concentration in the filtrate was determined as the arithmetic mean of three measurements. To establish the patterns of sorption, the equations of isotherms were calculated according to the most widely used Langmuir, Freundlich, and Temkin models [43–45]. The parameters of the adsorption equations were calculated by the method of nonlinear regression using the OriginPro 8 program.

The adsorption capacity of the samples for vitamin B1 was determined under static conditions at room temperature (25 ± 1 °C). Vitamin B1 solution in SBF (100 mg/L, 10 mL) was added to 30 mg of the sorbent and stirred on a magnetic stirrer for 1 h. After the experiment was completed, the sample was centrifuged. The vitamin concentration in the supernatant was analyzed with a UV–Vis spectrophotometer (Shimadzu UV-2600/2700, Shimadzu Europa GmbH) at 242 nm. Each concentration was taken as the average of three measurements.

The content of the sodium and potassium cations (in mmol/L) in the SBF solutions after contact with sorbent samples for 1 h was determined by atomic absorption spectroscopy (Thermo scientific iCE 3000, USA). To a weighed portion of the sorbent (30 mg), added 20 mL of SBF was and stirred on a magnetic stirrer for 1 h at room temperature. After the experiment was completed, the sample was centrifuged. The content of the cations in the initial SBF solutions corresponded to the reference values of the content of sodium and potassium cations in human blood plasma and amounted to 142 and 3.43 mmol/L, respectively. The sorption capacity of the samples (C, mg/g) was determined using the following formula (1).

A hemolytic test was used to determine the ability of the studied samples to damage the membranes of eukaryotic cells [46]. Human erythrocytes obtained from the peripheral blood of healthy donors by standard procedure [46,47] were used to determine the hemolytic activity. The studies were carried out according to the previously described method. [23,34]. The final concentration of aluminosilicate preparations in the incubated samples was 10 mg/mL and 0.3 mg/mL. The result of the study was presented as a percentage of hemolysis corresponding to the content of hemoglobin released from destroyed erythrocytes after the incubation of a suspension of erythrocytes with the studied samples of aluminosilicate.

3. Results and Discussion

The X-ray diffraction patterns of samples are shown in Figure 1. The comparison of the diffraction patterns of the samples with the bar charts of the standards allows us to conclude that the single-phase samples of specified structures, montmorillonite, kaolinite, halloysite, and Beta zeolite are used as initial samples. The results of the chemical analysis of the samples (Table 1) confirm that the samples studied are hydrous aluminosilicates with different Si/Al ratios.

Figure 2 shows the SEM images of the samples. It is observed that the aluminosilicate samples are characterized by different particle morphologies. Thus, the main morphology of the samples with the montmorillonite structure are layers self-organized into larger micron size agglomerates (Figure 2a–c). According to previous studies [33], the average particle size of montmorillonite is approximately 20 nm. Samples with a kaolinite structure were obtained with spherical, platy, and sponge morphologies. Particles with a spherical morphology have an average diameter of approximately 200–300 nm (Figure 2g). Samples with a nanosponge morphology are formed by aluminosilicate layers with a thickness of approximately 24–27 nm which are combined into micron-sized agglomerates (Figure 2f). Platy particles have a thickness of approximately 100 nm and an average lateral size of approximately 1 μm (Figure 2d). The raw halloysite has a nanotubular particle shape (Figure 2e). The nanotubes are approximately 700 nm long and 60 nm in diameter. Zeolite Beta particles have a spherical morphology with an average diameter of 300 nm (Figure 2h).

Figure 3 shows low-temperature nitrogen adsorption curves for the studied aluminosilicate samples. All curves can be attributed to a type IV isotherm according to the IUPAC classification. This type of isotherm indicates the presence of both micro- and mesopores [48]. The hysteresis loops are of different shapes, which is associated with the different types and shapes of pores in the samples. The shape of the hysteresis curves for kaolinite samples with a spherical and sponge morphology of particles, tubular halloysite, samples of montmorillonite, and Beta zeolite can be attributed to the *H2* type. This shape of the hysteresis loop points to complicated partly constricted pore network [49]. The shape of the hysteresis loop of the kaolinite sample with a platy morphology can be attributed to the *H3* type and indicates the presence of the aggregates of platy particles that form slit-like pores. The samples differ considerably in their specific surface area (SSA), which increases from 20 to 676 m^2/g depending on the particle morphology (see Table 2). In addition, the samples differ in the average pore diameter. For example, an average pore diameter of MT-Al0 sample is 4.4 nm, while that of MT-Al0.2 and MT-Al1.0 montmorillonites is 1.8 and 3.8 nm, respectively. The average pore size of the samples with the structure of kaolinite and zeolite Beta is 3.7 nm. For the Beta zeolite, this value most likely characterizes the secondary porosity, since the average size of the channels and cavities of this zeolite does not exceed 0.8 nm [50].

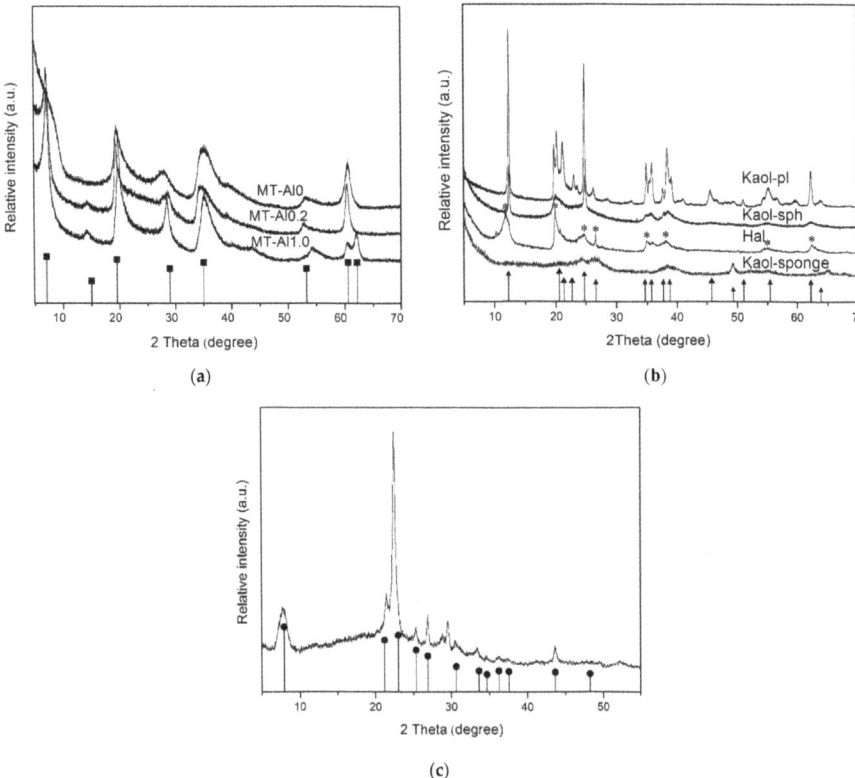

Figure 1. X-ray diffractions patterns of the samples: (**a**) aluminosilicates with montmorillonite structure; (**b**) aluminosilicates with kaolinite structure; and (**c**) zeolite Beta. Bar chart of the standards: ■—raw montmorillonite (PDF No. 48-74); ▲—raw kaolinite (PDF No. 79-1593); *—raw halloysite (PDF No 00–009-0453); ●—Beta (PDF No. 12-204).

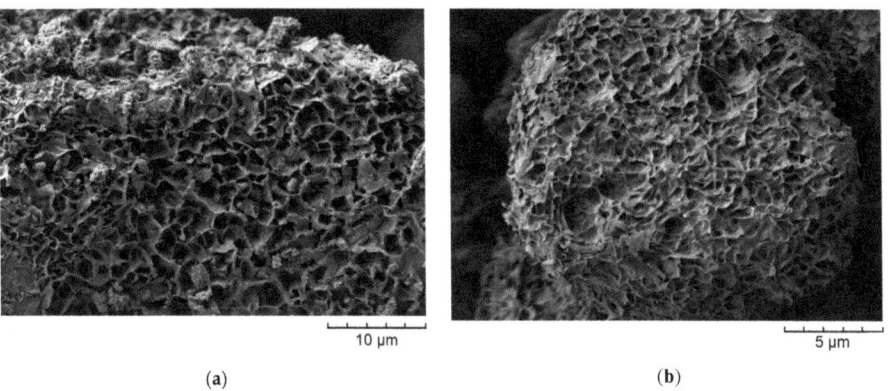

Figure 2. *Cont.*

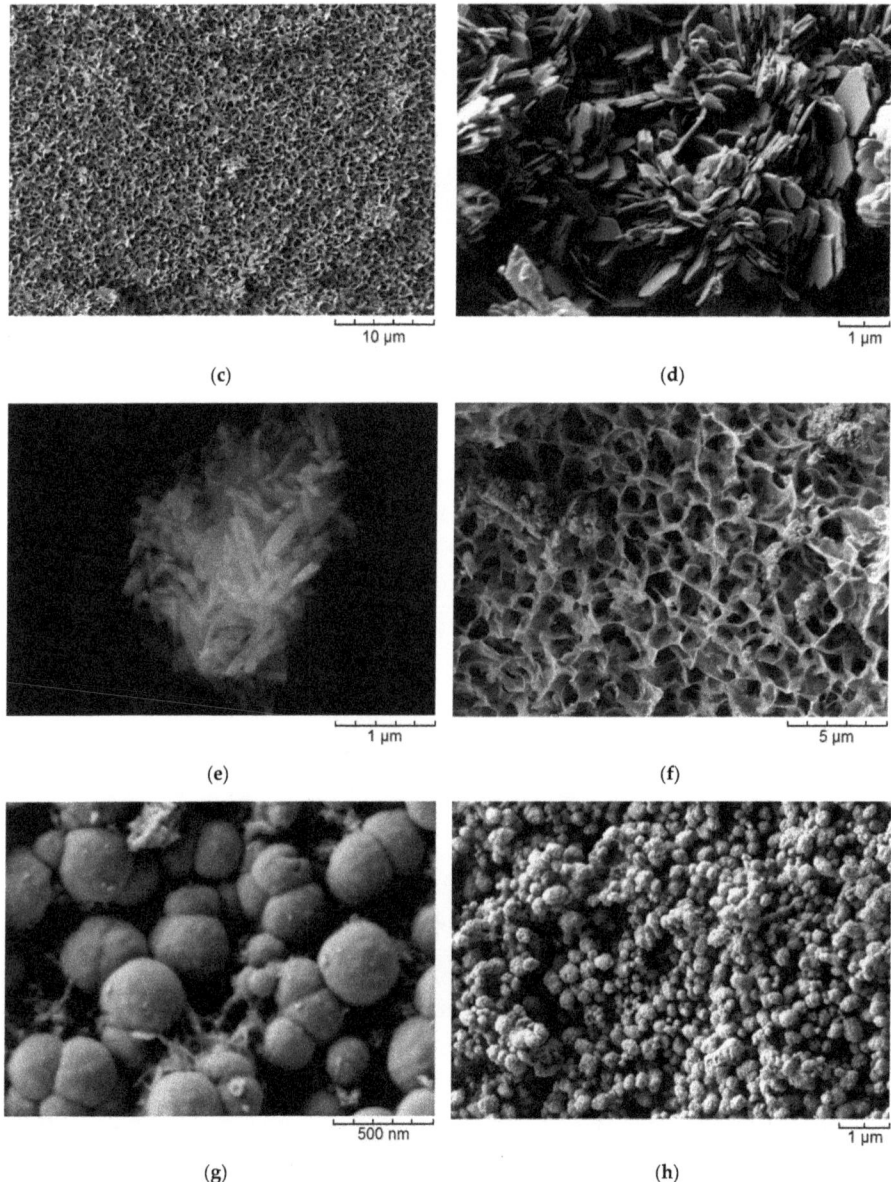

Figure 2. SEM images of the samples: (**a**)—MT-Al0; (**b**)—MT-Al0.2; (**c**)—MT-Al1.0; (**d**)—Kaol-pl; (**e**)—Hal; (**f**)—Kaol-sponge; (**g**)—Kaol-sph; and (**h**)—Beta. Samples are designated in accordance with the designations presented in Table 1.

Along with the porous textural properties of the sorbents, an important role in the choice of sorption materials in medicine is played by the chemical nature of their surface, namely the composition and number of functional groups on the surface. The chemistry of surface compounds determines the course of donor–acceptor interactions, which significantly affects the spectrum of absorbed molecules, and consequently, biochemical parameters ([51]).

The distribution of the adsorption sites on the surface of the studied samples as a function of their pK_a values is shown in Figure 4. These results indicate the presence of different types of adsorption centers on the surface including a Lewis base (pK$_a$ ≤ 0, formed by oxygen atoms) and acidic (pK$_a$ ≥ 14, formed by silicon atoms), Bronsted acidic (0 < pK$_a$ < 6), neutral (pK$_a$~6–8), and basic (8 < pK$_a$ < 14) sites corresponding to hydroxyl groups [40].

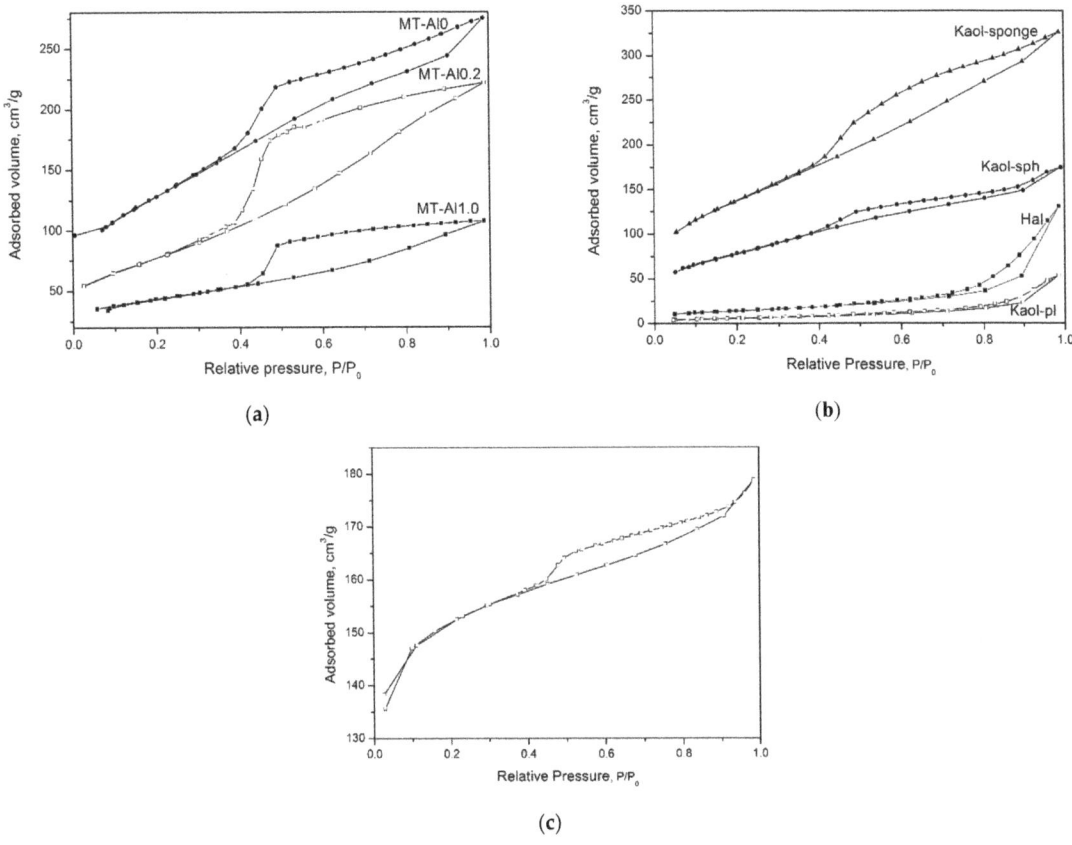

Figure 3. N$_2$ adsorption–desorption isotherms of the samples: (**a**)—aluminosilicates with montmorillonite structure; (**b**)—aluminosilicates with kaolinite structure; and (**c**)—zeolite Beta.

Table 2. Properties of the studied samples of aluminosilicates.

Samples	SSA [a], m²/g	ζ (pH 7), mV	Sorption Capacity for Cations, mg/g		Sorption Capacity for Vitamin B$_1$, mg/g	Hemolysis%, at Sample Concentration	
			Na⁺	K⁺		10 mg/mL	0.3 mg/mL
MT-Al0	549	−15.1 ± 0.2	0	0	22.4 ± 0.9	5.1 ± 0.9	0.6 ± 0.5
MT-Al0.2	320	−33.3 ± 0.3	3.3 ± 0.9	0.20 ± 0.08	39.9 ± 0.1	58.5 ± 5.9	2.0 ± 1.2
MT-Al1.0	190	−34.1 ± 0.9	4.2 ± 0.5	0.14 ± 003	31.6 ± 0.8	86.9 ± 9.0	14.9 ± 4.6
Kaol-sph	240	−18 ± 0.8	2.7 ± 0.3	0.13 ± 0.02	1.37 ± 0.1	23.1 ± 2.6	3.9 ± 4.6
Kaol-sponge	470	−20 ± 0.6	0	0	23.3 ± 0.2	27.0 ± 7.0	2.3 ± 0.8
Kaol-pl	22	−19 ± 0.9	3.1 ± 0.3	0.12 ± 0.03	0.8 ± 0.2	66.1 ± 1.8	3.8 ± 1.6
Hal	41	−28 ± 0.4	4.3 ± 0.2	0.13 ± 0.03	1.1 ± 0.1	97.5 ± 8.6	25.5 ± 9.5
Beta	676	−32.2 ± 0.6	9.1 ± 0.6	0.19 ± 0.05	29.5 ± 0.7	15.7 ± 1.7	0.9 ± 0.2
Carbon	360		0.6 ± 0.4	0	6.7 ± 0.9	8.7 ± 2.3	0.9 ± 0.2

Designations: SSA—specific surface area (m²/g); ζ (pH 7)—zeta potential of the surface as pH 7, mV, [a]—relative error in the specific surface area (SSA) value is 1%.

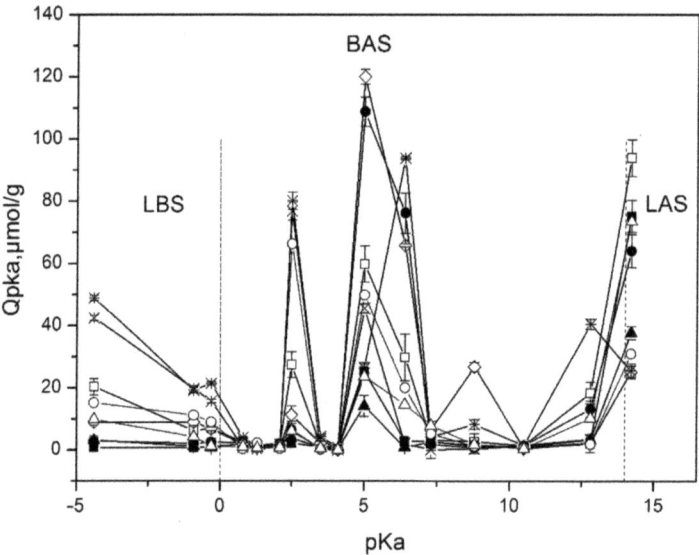

Figure 4. Distribution of the adsorption centers as a function of their pKa values on the surface of aluminosilicate samples and activated carbon: ◇—MT-Al0; ○—MT-Al0.2; △—MT-Al1.0; ●—Kaol-sph; ▲—Kaol-sponge; □—Kaol-pl; ■—Hal; ×—Beta; ✴—activated carbon. LBS—Lewis basic sites; BAS—Brønsted acidic sites; LAS—Lewis acidic sites.

An analysis of the surface of the studied samples by the method of adsorption of acid-base indicators allows us to draw conclusions about the distribution of active centers on the surface, as well as about the change in the strength and ratio between these centers with a change in the chemical composition of aluminosilicates and their morphology. Samples of all compositions contain weakly acidic Brønsted acid sites (BASs) with pK_a 5 and BAS with increased acidity with pK_a 2.5. At the same time, the maximum amount of BAS with pK_a 5 is typical for a sample of MT-Al0 and kaolinite with a spherical particle morphology. Their smallest amount is observed for a kaolinite sample with nanosponge morphology. All samples are characterized by the presence of Lewis basic sites (LBSs) with pK_a -4.4. At the same time, the number of such sites is at its maximum for the zeolite sample, and somewhat less for the kaolinite samples with a platy morphology. In other samples, the content of LBSs with pK_a 4.4 is quite low. The kaolinite sample with platy particle morphology is also characterized by a significant content of LAS with pK_a 14.2. The activated carbon sample is characterized by a high content of active sites with pK_a 6.4, corresponding to Brønsted neutral centers, and a rather low number of active sites with pK_a 5. Most aluminosilicate samples, on the contrary, are characterized by a high number of active centers with pK_a 5 and a small number, or even the complete absence, of active sites with pK_a 6.4. An exception is a sample of kaolinite with a spherical morphology of particles and MT-Al0, which have a large number of active centers with pK_a 5 and with pK_a 6.4. The MT-Al0 sample is also characterized by a large number of active centers with pK_a 8.8. The rest of the samples have practically no active centers in this region.

Comparison of the obtained data with the results of the chemical analysis of the samples and the study of the morphology of their particles allows us to conclude that both the chemical composition and morphology affect the distribution of active centers on the surface of silicate sorbents. Thus, the MT-Al0 sample studied in this work, which does not contain aluminum in its composition, but contains magnesium oxide, is characterized by the largest number of BAS among all samples with pK_a 8.8 and 2.5. Samples of aluminosilicates of the same chemical composition, but with different particle morphologies, such as kaolinite with spherical, platy, and sponge morphologies, as well as nanotubular halloysite, have a different functional composition of active centers on their surface, which may be due to the different availability of these centers, as determined by their morphology. Thus, the largest amount of BAS with pK_a 5 in this series of samples is characteristic of a sample with a spherical particle shape. The sample with a platy morphology has the highest amount of LAS with pK_a 14.2. The sample with nanotubular morphology as a whole has the smallest number of active centers, however, the amount of LAS with pK_a 14.2 in this sample is significant. A comparison of the functional composition of the surface of the studied samples with activated carbon shows that silicates have more active centers both in terms of their number and in terms of their diversity.

The results of the study of the zeta potential of the surface of the samples are presented in Table 2. All the studied samples have a negative surface zeta potential at pH 7, which is typical for aluminosilicates, and ranges from -25 ± 8 mV. Somewhat more negative values are typical for the samples of montmorillonite and zeolite (approximately -30 mV) than for samples of kaolinite—from -18 to -28 mV—depending on the particle morphology. The least negative surface charge among all the samples is characteristic of the MT-Al0 sample (-15 mV), which is associated with the absence of isomorphic substitutions in the octahedral magnesium–oxygen layers. The magnitude of the surface charge can make a significant contribution to the nature of the adsorption of charged molecules. Thus, aluminosilicates, having a negative surface charge, usually sorb positively charged ions from aqueous solutions very well (e.g., methylene blue, vitamin B1) [23,52], and sorb negatively charged ions (e.g., azorubine, 5-fluoracil) [23,53] to a much lesser extent. It is known that the albumin molecule has a net negative charge at a physiological pH [54], which can lead to difficulties in the adsorption of albumin by porous aluminosilicates.

Figure 5 shows the kinetic curves of albumin adsorption by the studied samples of aluminosilicates. The data obtained allow us to conclude that the time to achieve sorption equilibrium, depending on the sorbent, varies in the range from 4 to 24 h. The shortest time to achieve sorption equilibrium (1 and 2 h) is characteristic for activated carbon and Beta zeolite, respectively. It can be seen that the samples of montmorillonite have the highest sorption capacity with respect to albumin, both with isomorphic substitutions (MT-Al1.0 and MT-Al0.2) and without them (Sap), however, the time to reach adsorption equilibrium for them is the longest and reaches 20–24 h. Upon contact with albumin for 24 h, the sorption capacity of the montmorillonite samples reaches 220–250 mg/g. Montmorillonite MT-Al1.0 has the highest sorption capacity. For 24 h of contact, the sorption capacity of MT-Al1.0 reaches 256 mg/g, which is more than 12 times higher than the sorption capacity of activated carbon. The sorption capacity of montmorillonites samples of other compositions is somewhat lower, but they also have rather high values. The obtained values of the sorption capacity of synthetic montmorillonite samples correlate with the previously obtained results of the study of the sorption capacity of raw clays [2,5] and even slightly exceed them, which is due to the absence of impurity phases in the samples under study.

For samples of the kaolinite subgroup with different particle morphologies, the sorption capacity for albumin is significantly lower than that for montmorillonite, and is at the level of 60–80 mg/g for samples with platy, spherical, and tubular morphologies. For samples of kaolinite with a spherical particle morphology, the sorption capacity for albumin is even lower and is at the level of 25–40 mg/g. The sorption capacity of Beta zeolite is 53 mg/g. The sorption capacity for the albumin of all studied aluminosilicate samples exceeds the sorption capacity of activated carbon.

Such adsorption by aluminosilicate samples is associated with the features of their structure and surface properties. No direct relationship between the specific surface area of the samples and their sorption capacity for albumin was found. Thus, the highest values of the specific surface area are typical for MT-Al0 samples (549 m^2/g), kaolinite nanosponges (470 m^2/g), and Beta zeolite (676 m^2/g). However, the sorption capacity of these samples is not the highest. Montmorillonite samples are characterized by the highest sorption capacity for albumin, despite the highest values of the negative zeta potential of the surface. The high sorption capacity of montmorillonite compared to other samples is most likely due to its ability to increase the interlayer distance over a wide range (from 1 Å to complete exfoliation into individual layers) in the process of the adsorption of organic and inorganic molecules and their intercalation in the interlayer space [55,56]. As a result, albumin is located both on the outer (chips, edges, and outer surface of layers) and inner surfaces (interlayer space) of montmorillonite particles. The structures of other aluminosilicate samples do not have this feature, and albumin adsorption mainly occurs on the outer surface of the particles. On the other hand, the ability to intercalate and increase the interlayer distance leads to an increase in the adsorption equilibrium time for samples with the montmorillonite structure.

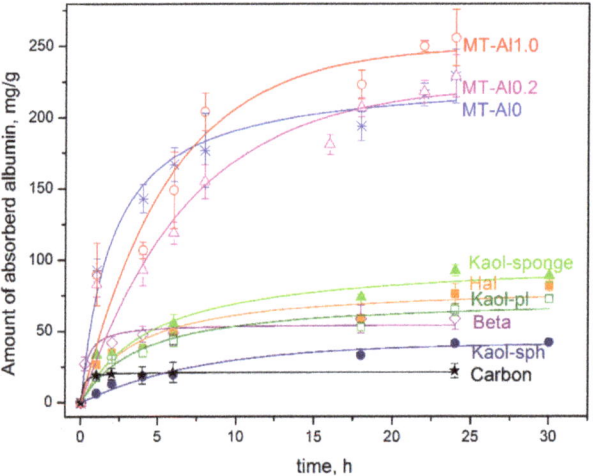

Figure 5. Kinetic curve of albumin by the aluminosilicate samples and activated carbon. Samples are designated in accordance with the designations presented in Table 1.

Based on the results of the graphical processing of the experimental data (Table 3), it was found that the sorption kinetics of all samples, except for MT-Al0.2 and MT-Al1.0 samples, is well described by a pseudo-second-order (PSO) equation: the theoretically calculated values of the sorption capacity q_{calc} are close to those found experimentally, and the high approximation coefficient is 0.92–0.98. The PSO kinetic model is usually associated with the situation where the rate of the direct adsorption/desorption process is rate limiting. Within the framework of kinetic models, the rate constants of the process were calculated (Table 3). The PSO rate constants are the highest for the activated carbon and Beta zeolite samples, which is consistent with the short equilibrium time in the system (1 and 2 h, respectively). The most adequate model for the MT-Al0.2 and MT-Al1.0 samples, taking into account the q_{calc} values, is PFO.

Table 3. Parameters of the kinetic models of sorption of albumin on aluminosilicates with different morphologies.

Samples Morphology	q_{exp}, mg/g	PFO Model			PSO Model		
		q_{calc}	k_1	R^2	q_{calc}	k_2	R^2
MT-Al0	229 ± 19	207 ± 11	0.32 ± 0.07	0.93	229 ± 10	(3 ± 1)·10⁻³	0.97
MT-Al0.2	229 ± 15	222 ± 18	0.14 ± 0.03	0.92	278 ± 32	(5 ± 2)·10⁻⁴	0.93
MT-Al1.0	256 ± 24	250 ± 18	0.17 ± 0.03	0.92	301 ± 30	(7 ± 3)·10⁻⁴	0.93
Kaol-sph	42 ± 4	41 ± 2	0.13 ± 0.02	0.96	52 ± 4	(2 ± 1)·10⁻³	0.98
Kaol-sponge	92 ± 2	86 ± 6	0.18 ± 0.04	0.89	101 ± 9	(2 ± 1)·10⁻³	0.93
Kaol-pl	72 ± 3	64 ± 4	0.22 ± 0.04	0.92	74 ± 5	(3.5 ± 1)·10⁻³	0.95
Hal	82 ± 4	72 ± 6	0.23 ± 0.06	0.87	83 ± 7	(3.4 ± 1)·10⁻³	0.92
Beta	59 ± 7	51 ± 3	2.9 ± 1.1	0.90	55 ± 3	(5 ± 0.2)·10⁻²	0.95
Carbon	22 ± 5	20 ± 1	2.5 ± 0.8	0.98	21 ± 1	0.34 ± 0.18	0.99

Figure 6 shows albumin adsorption isotherms by synthetic aluminosilicates, as well as by activated carbon. The symbols represent the experimental data, and the lines represent the model that best fits the data. Taking into account the high values of the correlation coefficients (R^2) and the close values of the experimental and calculated sorption capacity (Table 4), among the three nonlinear models, the Langmuir isotherm best describes adsorption on all samples, except for Hal and Kaol-pl. This model describes a homogeneous monomolecular adsorption process and assumes that the surface of a solid body contains a finite number of active centers with equal energy. For the Hal and Kaol-pl samples, the Freundlich equation is the most appropriate. According to the Freundlich model, the surface of the studied sorbents contains active centers with different affinity energies for adsorbate molecules. The value of $1/n$ can be considered as an indicator of the inhomogeneity of sorption centers: as the inhomogeneity increases, $1/n \to 0$, and as the homogeneity of centers increases, $1/n \to 1$. At the same time, the data obtained make it possible to characterize aluminosilicates as materials with a high concentration of sorption centers with different degrees of activity, which is consistent with the results of studying the distribution of active centers on the sample surface (Figure 4). The constant K_F has a linear dependence on the adsorption capacity of the adsorbent, i.e., the larger this constant, the greater the adsorption capacity.

Table 2 presents the results of determining the sorption capacity of samples in relation to potassium and sodium cations, as well as vitamin B1 in the SBF medium. The results of the study of the hemolytic activity of the samples are also given there.

Based on the obtained results, it can be concluded that all studied samples absorb sodium and potassium cations from SBF in small amounts, potentially not leading to serious pathological changes. At the same time, two samples—MT-Al0 and kaolinite with nanosponge morphology—do not have a sorption capacity for these cations. The sorption capacity of aluminosilicates with respect to vitamin B1 is relatively high and is the highest for samples with a montmorillonite structure—MT-Al0.2, MT-Al1.0, Sap, and for Beta zeolite. The samples of the kaolinite subgroup with spherical, platy, and tubular particle morphologies are characterized by the lowest sorption capacity. The results of the vitamin B1 adsorption study generally correlate with the results of the BSA adsorption study and can be explained by the structural features of the studied samples.

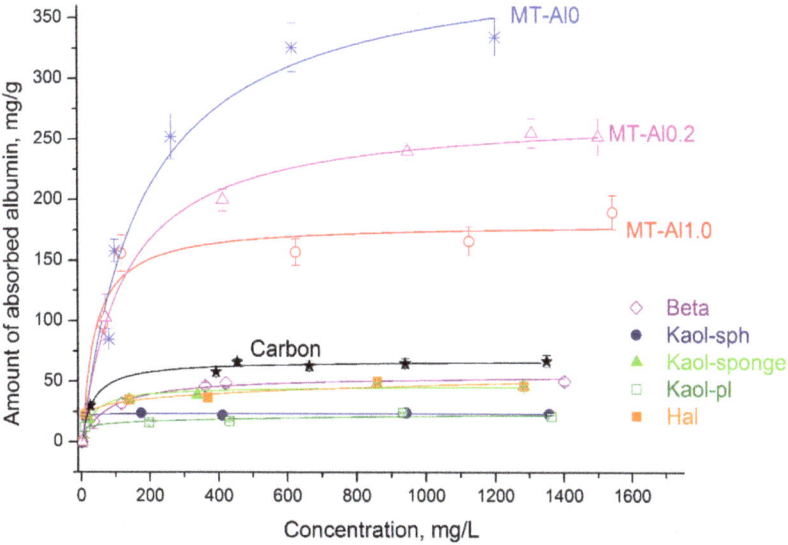

Figure 6. Langmuir and Freundlich adsorption isotherm plots. Langmuir isotherm: MT-Al0; MT-Al1.0; MT-Al0.2; Beta; Kaol-sph; Kaol-sponge; carbon samples. Freundlich isotherm: Hal and Kaol-pl.

Table 4. Equation constants of albumin sorption isotherms.

Sample Morphologies	q_{exp}	Langmuir Equation			Freundlich Equation			Temkin Equation		
		q_m	K_L	R^2	n	K_F	R^2	B_T	A_T	R^2
MT-Al0	334 ± 20	404 ± 34	$(5 ± 1) \cdot 10^{-3}$	0.97	2.9 ± 0.7	30.5 ± 2.9	0.90	89.2 ± 7.5	$(1 ± 0.1) \cdot 10^{-2}$	0.96
MT-Al0.2	255 ± 12	272 ± 5	$(8 ± 1) \cdot 10^{-3}$	0.99	364 ± 0.4	36.0 ± 8.0	0.96	50.0 ± 2.0	$(11 ± 2) \cdot 10^{-2}$	0.94
MT-Al1.0	190 ± 14	180 ± 11	$(25 ± 9) \cdot 10^{-3}$	0.94	7.3 ± 2.5	67.0 ± 6.5	0.93	20.5 ± 6.3	4.6 ± 1.1	0.92
Kaol-sph	24 ± 1	23 ± 1	0.8 ± 0.3	0.99	49.2 ± 5.0	20.4 ± 1.4	0.97	1.0 ± 0.4	3.1 ± 0.2	0.97
Kaol-sponge	48 ± 1	46 ± 2	$(8 ± 3) \cdot 10^{-3}$	0.98	5.2 ± 0.8	12.6 ± 2.3	0.97	6.8 ± 0.6	1.0 ± 0.5	0.96
Kaol-pl	24 ± 1	20 ± 2	$(13 ± 1) \cdot 10^{-2}$	0.87	7.2 ± 2.0	8.1 ± 2.0	0.95	9.7 ± 1.6	2.2 ± 0.9	0.95
Hal	48 ± 3	43 ± 3	$(8 ± 4) \cdot 10^{-2}$	0.91	6.2 ± 1.0	15.3 ± 2.6	0.98	5.4 ± 0.8	5.1 ± 0.3	0.97
Beta	50 ± 6	55 ± 2	$(13 ± 2) \cdot 10^{-2}$	0.99	4.4 ± 1.2	18.0 ± 4.6	0.90	9.7 ± 1.6	2.2 ± 1.5	0.95
Carbon	63 ± 8	66 ± 2	$(31 ± 7) \cdot 10^{-3}$	0.99	4.2 ± 0.8	14.3 ± 3.5	0.98	10.9 ± 1.2	0.6 ± 0.4	0.94

q_m—maximum sorption capacity (mg/g); q_{exp}—experimental value of sorption capacity (mg/g); Langmuir constant related to adsorption free energy (L/mg); B_T—constant related to the heat of adsorption (L/g); K_F—Freundlich constant related to adsorbent capacity (L/g); A_T—dimensionless Temkin isotherm constant.

Blood plasma is an aqueous solution of electrolyte, nutrients, metabolites, proteins, vitamins, trace elements, and signaling substances. The most important characteristic of a selective hemosorbent is the presence of sorption capacity in relation to pathogens, and its absence in relation to other blood components, particularly vitamins and microelements. In this case, the optimal hemosorbent should also not have the ability to destroy blood cells, that is, it should not have hemolytic activity. The results of the study of hemolytic activity, presented in Table 2, indicate that the greatest increase in hemolytic activity (toxicity) at a sample concentration of 10 mg/mL occurs in the series Sap<Carbon<Beta<Kaol-sph<Kaol-sponge<MT-Al0.2<Kaol-pl<MT-Al1.0<Hal. The presence and difference of the hemolytic activity in samples may be associated with differences in their chemical composition, surface properties, and particle shape. Thus, the dependence of the hemolytic activity and cytotoxicity of aluminosilicates of the kaolinite subgroup on the morphology of their particles was shown earlier [57]. It was found that among single-phase samples with the same chemical composition $Al_2Si_2O_5(OH)_2$, samples with a tubular morphology have the highest toxicity, and samples with spherical particles have the lowest toxicity. In addition to this effect, the effect of the influence of the chemical composition on the hemolytic activity was found in the present work. Among the samples with the montmorillonite structure, samples with the highest aluminum content have the highest hemolytic activity. The sample of magnesium silicate montmorillonite ($Mg_3Si_4O_{10}(OH)_2 \cdot nH_2O$) has the lowest hemolytic activity among all the studied samples, including activated carbon.

It should be noted that samples are considered non-toxic if their hemolytic activity does not exceed 5% [58]. With a decrease in the concentration of the studied samples to 0.3 mg/mL, the hemolytic activity of all samples decreases, and for most samples, reaches values not exceeding 5% (with the exception of samples MT-Al1.0 and Hal).

4. Conclusions

This work studied the possibility of using porous aluminosilicates with different structures and particle morphologies as hemosorbents. The sorption capacity of the samples in relation to the model medium molecular weight toxicants (BSA), vitamin B1, and alkaline cations in a simulated body fluid, as well as their hemolytic activity, were studied. It was established that the sorption capacity of aluminosilicate samples is largely determined by their structural features, porous textural characteristics and surface properties (charge and distribution of active centers on the surface). Thus, samples with a montmorillonite structure, which have the ability to increase the interlayer space over a wide range, have the highest sorption capacity with respect to BSA. However, the time to reach sorption equilibrium for such samples is quite long and amounts to approximately 24 h, which is also related to the peculiarities of their structure. The sorption capacity of zeolite samples is several times lower than montmorillonite, however, the sorption equilibrium is reached in 1 h. The Langmuir isotherm best describes adsorption on all samples, except for samples with nanotubular and platy particle morphology. For these samples, the Freundlich equation is the most appropriate. The hemolytic ability of the samples is largely determined by the morphology of the particles and the chemical composition of the samples. Thus, aluminosilicates with tubular particles and samples with the highest aluminum content have the highest hemolytic activity.

The studies of adsorption features and properties of porous aluminosilicates have shown that aluminosilicate sorbents can be considered potential hemosorbents. They have a high sorption capacity for medium molecular weight pathogens (up to 12 times that of activated charcoal for some samples) and low toxicity to blood cells. Directed hydrothermal synthesis makes it possible to obtain aluminosilicates with a given chemical and phase composition, certain surface properties, and porous textural characteristics. It is shown that the chemical composition, surface charge, particle morphology, and structural features determine the adsorption capacity and biological activity of the samples.

Based on the performed study, it can be concluded that the optimal option for the further development of new selective and non-toxic hemosorbents is synthetic magnesium silicate montmorillonite ($Mg_3Si_4O_{10}(OH)_2 \cdot nH_2O$), since it has a significant sorption capacity with respect to BSA, modeling pathogens with an average molecular weight, lack of sorption capacity for potassium and sodium cations from the blood plasma medium, and low hemolytic activity.

Author Contributions: Conceptualization, O.Y.G.; Methodology, O.Y.G.; Validation, O.Y.G., Y.A.A. and E.Y.B.; Formal Analysis, O.Y.G., E.Y.B., N.M.V. and Y.A.A.; Investigation, Y.A.A., E.Y.B. and N.M.V.; Resources, O.Y.G.; Data Curation, O.Y.G., Y.A.A., N.M.V. and E.Y.B.; Writing—Original Draft Preparation, O.Y.G. All authors have read and agreed to the published version of the manuscript.

Funding: This study was funded by the Russian Science Foundation (project No. 22-23-00227, accessed on 21 December 2021, https://rscf.ru/project/22-23-00227/).

Data Availability Statement: Not applicable.

Acknowledgments: The authors are grateful to the employees of the Institute of Experimental Medicine, namely O.V. Shamova and E.V. Vladimirova, for conducting studies on the hemolytic activity of the samples. The authors are grateful to the administration of the St. Petersburg Technological Institute (Technical University) for the opportunity to use the equipment of the Engineering Center (X-ray diffractometer).

Conflicts of Interest: The authors declare no conflict of interest.

References

1. Jaber, M.; Lambert, J.-F.; Balme, S. 8—Protein adsorption on clay minerals. *Dev. Clay Sci.* **2018**, *9*, 255–288. [CrossRef]
2. Kim, H.-M.; Oh, J.-M. Physico–Chemical Interaction between Clay Minerals and Albumin Protein according to the Type of Clay. *Minerals* **2019**, *9*, 396. [CrossRef]
3. Lepoitevin, M.; Jaber, M.; Guégan, R.; Janot, J.-M.; Dejardin, P.; Henn, F.; Balme, S. BSA and lysozyme adsorption on homoionic montmorillonite: Influence of the interlayer cation. *Appl. Clay Sci.* **2014**, *95*, 396–402. [CrossRef]
4. Ralla, K.; Sohling, U.; Riechers, D.; Kasper, C.; Ruf, F.; Scheper, T. Adsorption and separation of proteins by a smectitic clay mineral. *Bioprocess Biosyst. Eng.* **2010**, *33*, 847–861. [CrossRef]
5. Mucha, M.; Maršálek, R.; Bukáčková, M.; Zelenková, G. Interaction among clays and bovine serum albumin. *RSC Adv.* **2020**, *10*, 43927–43939. [CrossRef] [PubMed]
6. Wasilewska, M.; Adamczyk, Z.; Pomorska, A.; Nattich-Rak, M.; Sadowska, M. Human Serum Albumin Adsorption Kinetics on Silica: Influence of Protein Solution Stability. *Langmuir* **2019**, *35*, 2639–2648. [CrossRef]
7. Sommer, S.; Sommer, S.J.; Gutierrez, M. Characterization of Different Bentonites and Their Properties as a Protein-Fining Agent in Wine. *Beverages* **2022**, *8*, 31. [CrossRef]
8. Silva-Barbieri, D.; Salazar, F.N.; López, F.; Brossard, N.; Escalona, N.; Pérez-Correa, J.R. Advances in White Wine Protein Stabilization Technologies. *Molecules* **2022**, *27*, 1251. [CrossRef]
9. Uzhinova, L.D.; Krasovskaja, S.M. Interactions of biospecific sorbents with physiologically active substances. *J. Mater. Sci. Mater. Electron.* **1991**, *2*, 189–192. [CrossRef]
10. Ruiz-Rodríguez, J.C.; Plata-Menchaca, E.P.; Chiscano-Camón, L.; Ruiz-Sanmartin, A.; Ferrer, R. Blood purification in sepsis and COVID-19: What's new in cytokine and endotoxin hemoadsorption. *J. Anesth. Analg. Crit. Care* **2022**, *2*, 15. [CrossRef]
11. Napp, L.C.; Lebreton, G.; De Somer, F.; Supady, A.; Pappalardo, F. Opportunities, controversies, and challenges of extracorporeal hemoadsorption with CytoSorb during ECMO. *Artif. Organs* **2021**, *45*, 1240–1249. [CrossRef] [PubMed]

12. Kovacs, J. Hemoadsorption in Critical Care—It Is a Useful or a Harmful Technique? *J. Crit. Care Med.* **2020**, *6*, 207–209. [CrossRef] [PubMed]
13. Asgharpour, M.; Mehdinezhad, H.; Bayani, M.; Zavareh, M.S.H.; Hamidi, S.H.; Akbari, R.; Ghadimi, R.; Bijani, A.; Mouodi, S. Effectiveness of extracorporeal blood purification (hemoadsorption) in patients with severe coronavirus disease 2019 (COVID-19). *BMC Nephrol.* **2020**, *21*, 356. [CrossRef] [PubMed]
14. Nikolaev, V.G.; Samsonov, V.A. Analysis of medical use of carbon adsorbents in China and additional possibilities in this field achieved in Ukraine. *Artif. Cells Nanomed. Biotechnol.* **2013**, *42*, 1–5. [CrossRef]
15. Mikhalovsky, S.; Nikolaev, V. Chapter 11 Activated carbons as medical adsorbents. *Interface Sci. Technol.* **2006**, *7*, 529–561. [CrossRef]
16. Inoue, S.; Kiriyama, K.; Hatanaka, Y.; Kanoh, H. Adsorption properties of an activated carbon for 18 cytokines and HMGB1 from inflammatory model plasma. *Colloids Surf. B Biointerfaces* **2014**, *126*, 58–62. [CrossRef]
17. Barnes, J.; Cowgill, L.D.; Auñon, J.D. Activated Carbon Hemoperfusion and Plasma Adsorption: Rediscovery and Veterinary Applications of These Abandoned Therapies. *Adv. Small Anim. Care* **2021**, *2*, 131–142. [CrossRef]
18. Maisanaba, S.; Pichardo, S.; Puerto, M.; Gutiérrez-Praena, D.; Cameán, A.M.; Jos, A. Toxicological evaluation of clay minerals and derived nanocomposites: A review. *Environ. Res.* **2015**, *138*, 233–254. [CrossRef]
19. Petushkov, A.; Ndiege, N.; Salem, A.K.; Larsen, S.C. Toxicity of Silica Nanomaterials: Zeolites, Mesoporous Silica, and Amorphous Silica Nanoparticles. *Adv. Mol. Toxicol.* **2010**, *4*, 223–266. [CrossRef]
20. Pavelić, S.K.; Medica, J.S.; Gumbarević, D.; Filošević, A.; Przulj, N.; Pavelić, K. Critical Review on Zeolite Clinoptilolite Safety and Medical Applications in vivo. *Front. Pharmacol.* **2018**, *9*, 1350. [CrossRef]
21. Piskin, E. Potential Sorbents for Medical and Some Related Applications. *Int. J. Artif. Organs* **1986**, *9*, 401–404. [CrossRef] [PubMed]
22. Ankawi, G.; Fan, W.; Pomarè-Montin, D.; Lorenzin, A.; Neri, M.; Caprara, C.; De Cal, M.; Ronco, C. A New Series of Sorbent Devices for Multiple Clinical Purposes: Current Evidence and Future Directions. *Blood Purif.* **2019**, *47*, 94–100. [CrossRef] [PubMed]
23. Golubeva, O.Y.; Alikina, Y.A.; Khamova, T.V.; Vladimirova, E.V.; Shamova, O.V. Aluminosilicate Nanosponges: Synthesis, Properties, and Application Prospects. *Inorg. Chem.* **2021**, *60*, 17008–17018. [CrossRef]
24. Ulyanova, N.Y.; Kurylenko, L.N.; Shamova, O.; Orlov, D.S.; Golubeva, O.Y. Hemolitic Activity and Sorption Ability of Beta Zeolite Nanoparticles. *Glas. Phys. Chem.* **2020**, *46*, 155–161. [CrossRef]
25. Li, Y.; Yang, G.; Mei, Z. Spectroscopic and dynamic light scattering studies of the interaction between pterodontic acid and bovine serum albumin. *Acta Pharm. Sin. B* **2012**, *2*, 53–59. [CrossRef]
26. Yohannes, G.; Wiedmer, S.K.; Elomaa, M.; Jussila, M.; Aseyev, V.; Riekkola, M.-L. Thermal aggregation of bovine serum albumin studied by asymmetrical flow field-flow fractionation. *Anal. Chim. Acta* **2010**, *675*, 191–198. [CrossRef]
27. Lee, S.; Choi, M.C.; Al Adem, K.; Lukman, S.; Kim, T.-Y. Aggregation and Cellular Toxicity of Pathogenic or Non-pathogenic Proteins. *Sci. Rep.* **2020**, *10*, 5120. [CrossRef]
28. Servagent-Noinville, S.; Revault, M.; Quiquampoix, H.; Baron, M.-H. Conformational Changes of Bovine Serum Albumin Induced by Adsorption on Different Clay Surfaces: FTIR Analysis. *J. Colloid Interface Sci.* **2000**, *221*, 273–283. [CrossRef]
29. Ahmed, R.Z.; Patil, G.; Zaheer, Z. Nanosponges—A completely new nano-horizon: Pharmaceutical applications and recent advances. *Drug Dev. Ind. Pharm.* **2012**, *39*, 1263–1272. [CrossRef]
30. Kokubo, T. Surface chemistry of bioactive glass-ceramics. *J. Non-Cryst. Solids* **1990**, *120*, 138–151. [CrossRef]
31. Karppanen, H. Minerals and Blood Pressure. *Ann. Med.* **1991**, *23*, 299–305. [CrossRef] [PubMed]
32. Morris, A.L.; Mohiuddin, S.S. *Biochemistry, Nutrients*; StatPearls Publishing: Treasure Island, FL, USA, 2022.
33. Golubeva, O.Y. Effect of synthesis conditions on hydrothermal crystallization, textural characteristics and morphology of aluminum-magnesium montmorillonite. *Microporous Mesoporous Mater.* **2016**, *224*, 271–276. [CrossRef]
34. Golubeva, O.Y.; Alikina, Y.A.; Kalashnikova, T.A. Influence of hydrothermal synthesis conditions on the morphology and sorption properties of porous aluminosilicates with kaolinite and halloysite structures. *Appl. Clay Sci.* **2020**, *199*, 105879. [CrossRef]
35. Brazovskaya, E.Y.; Golubeva, O.Y. Optimization of the Beta Zeolite Synthesis Method. *Glas. Phys. Chem.* **2021**, *47*, 726–730. [CrossRef]
36. Golubeva, O.Y.; Ul'Yanova, N.Y.; Yakovlev, A.V. Synthesis of zeolite with paulingite structure. *Glas. Phys. Chem.* **2015**, *41*, 413–416. [CrossRef]
37. Golubeva, O.Y.; Ul'Yanova, N.Y. Stabilization of silver nanoparticles and clusters in porous zeolite matrices with Rho, Beta, and paulingite structures. *Glas. Phys. Chem.* **2015**, *41*, 537–544. [CrossRef]
38. Brunauer, S.; Emmett, P.H.; Teller, E. Adsorption of Gases in Multimolecular Layers. *J. Am. Chem. Soc.* **1938**, *60*, 309–319. [CrossRef]
39. Barrett, E.P.; Joyner, L.G.; Halenda, P.P. The Determination of Pore Volume and Area Distributions in Porous Substances. I. Computations from Nitrogen Isotherms. *J. Am. Chem. Soc.* **1951**, *73*, 373–380. [CrossRef]
40. Bardakhanov, S.; Vasiljeva, I.V.; Kuksanov, N.K.; Mjakin, S. Surface Functionality Features of Nanosized Silica Obtained by Electron Beam Evaporation at Ambient Pressure. *Adv. Mater. Sci. Eng.* **2010**, *2010*, 241845. [CrossRef]
41. Revellame, E.D.; Fortela, D.L.; Sharp, W.; Hernandez, R.; Zappi, M.E. Adsorption kinetic modeling using pseudo-first order and pseudo-second order rate laws: A review. *Clean. Eng. Technol.* **2020**, *1*, 100032. [CrossRef]

42. Ho, Y.; McKay, G. Sorption of dye from aqueous solution by peat. *Chem. Eng. J.* **1998**, *70*, 115–124. [CrossRef]
43. Rahimi, M.; Vadi, M. Langmuir, Freundlich and Temkin Adsorption Isotherms of Propranolol on Multi-Wall Carbon Nanotube. *J. Mod. Drug Discov. Drug Deliv. Res.* **2014**, 1–3. Available online: https://zenodo.org/record/893491/files/374178762.pdf?download=1 (accessed on 7 August 2022).
44. Dada, A.O.; Olalekan, A.P.; Olatunya, A.M.; Dada, O.J. Langmuir, Freundlich, Temkin and Dubinin–Radushkevich Isotherms Studies of Equilibrium Sorption of Zn^{2+} Unto Phosphoric Acid Modified Rice Husk. *IOSR J. Appl. Chem.* **2012**, *3*, 38–45. [CrossRef]
45. Ayawei, N.; Ebelegi, A.N.; Wankasi, D. Modelling and Interpretation of Adsorption Isotherms. *J. Chem.* **2017**, *2017*, 3039817. [CrossRef]
46. Staff. Methods in Molecular Biology. Volume 78. Antibacterial Peptide Protocols Edited by William S. Shafer. The Humana Press, Totowa, NJ. 1997. x + 259 pp. 15.5 × 23.5 cm. ISBN 0-89603-408-9. $74.50. *J. Med. Chem.* **1997**, *40*, 4161. [CrossRef]
47. Kopeikin, P.M.; Zharkova, M.S.; Kolobov, A.A.; Smirnova, M.P.; Sukhareva, M.S.; Umnyakova, E.S.; Kokryakov, V.N.; Orlov, D.S.; Milman, B.L.; Balandin, S.V.; et al. Caprine Bactenecins as Promising Tools for Developing New Antimicrobial and Antitumor Drugs. *Front. Cell. Infect. Microbiol.* **2020**, *10*, 552905. [CrossRef]
48. Sing, K.S.W.; Everett, D.H.; Haul, R.A.W.; Moscou, L.; Pierotti, R.A.; Rouquerol, J.; Siemieniewska, T. Reporting Physisorption Data for Gas/Solid Systems. *Handb. Heterog. Catal.* **2008**, 1217–1230. [CrossRef]
49. Niasar, V.; Hassanizadeh, S.M. Analysis of Fundamentals of Two-Phase Flow in Porous Media Using Dynamic Pore-Network Models: A Review. *Crit. Rev. Environ. Sci. Technol.* **2012**, *42*, 1895–1976. [CrossRef]
50. First, E.L.; Gounaris, C.E.; Wei, J.; Floudas, C.A. Computational characterization of zeolite porous networks: An automated approach. *Phys. Chem. Chem. Phys.* **2011**, *13*, 17339–17358. [CrossRef]
51. Boehm, H. Surface Properties of Carbons. *Stud. Surf. Sci. Catal.* **1989**, *48*, 145–157. [CrossRef]
52. Golubeva, O.Y.; Pavlova, S.V.; Yakovlev, A.V. Adsorption and in vitro release of vitamin B1 by synthetic nanoclays with montmorillonite structure. *Appl. Clay Sci.* **2015**, *112–113*, 10–16. [CrossRef]
53. Golubeva, O.Y.; Alikina, Y.; Brazovskaya, E.; Ugolkov, V.V. Peculiarities of the 5-fluorouracil adsorption on porous aluminosilicates with different morphologies. *Appl. Clay Sci.* **2019**, *184*, 105401. [CrossRef]
54. Bert, J.L.; Pearce, R.H.; Mathieson, J.M. Concentration of plasma albumin in its accessible space in postmortem human dermis. *Microvasc. Res.* **1986**, *32*, 211–223. [CrossRef]
55. Block, K.A.; Trusiak, A.; Katz, A.; Alimova, A.; Wei, H.; Gottlieb, P.; Steiner, J.C. Exfoliation and intercalation of montmorillonite by small peptides. *Appl. Clay Sci.* **2015**, *107*, 173–181. [CrossRef] [PubMed]
56. Sainz-Díaz, C.I.; Bernini, F.; Castellini, E.; Malferrari, D.; Borsari, M.; Mucci, A.; Brigatti, M.F. Experimental and Theoretical Investigation of Intercalation and Molecular Structure of Organo-Iron Complexes in Montmorillonite. *J. Phys. Chem. C* **2018**, *122*, 25422–25432. [CrossRef]
57. Golubeva, O.Y.; Alikina, Y.A.; Brazovskaya, E.Y. Particles Morphology Impact on Cytotoxicity, Hemolytic Activity and Sorption Properties of Porous Aluminosilicates of Kaolinite Group. *Nanomaterials* **2022**, *12*, 2559. [CrossRef]
58. Jeswani, G.; Alexander, A.; Saraf, S.; Saraf, S.; Qureshi, A.; Uddin, A. Recent approaches for reducing hemolytic activity of chemotherapeutic agents. *J. Control. Release* **2015**, *211*, 10–21. [CrossRef]

Review

Catalytic Steam Reforming of Biomass-Derived Oxygenates for H$_2$ Production: A Review on Ni-Based Catalysts

Joel Silva [1,2], Cláudio Rocha [2,3], M. A. Soria [1,2,*] and Luís M. Madeira [1,2]

[1] LEPABE—Laboratory for Process Engineering, Environment, Biotechnology and Energy, Department of Chemical Engineering, Faculty of Engineering, University of Porto, Rua Dr. Roberto Frias, 4200-465 Porto, Portugal; joel.alexandre.moreira.silva@gmail.com (J.S.); mmadeira@fe.up.pt (L.M.M.)

[2] ALiCE—Associate Laboratory in Chemical Engineering, Faculty of Engineering, University of Porto, Rua Dr. Roberto Frias, 4200-465 Porto, Portugal; csrocha@fe.up.pt

[3] LSRE-LCM—Laboratory of Separation and Reaction Engineering—Laboratory of Catalysis and Materials, Faculty of Engineering, University of Porto, Rua Dr. Roberto Frias, 4200-465 Porto, Portugal

* Correspondence: masoria@fe.up.pt

Abstract: The steam reforming of ethanol, methanol, and other oxygenates (e.g., bio-oil and olive mill wastewater) using Ni-based catalysts have been studied by the scientific community in the last few years. This process is already well studied over the last years, being the critical point, at this moment, the choice of a suitable catalyst. The utilization of these oxygenates for the production of "green" H$_2$ is an interesting alternative to fuel fossils. For this application, Ni-based catalysts have been extensively studied since they are highly active and cheaper than noble metal-based materials. In this review, a comparison of several Ni-based catalysts reported in the literature for the different above-mentioned reactions is carried out. This study aims to understand if such catalysts demonstrate enough catalytic activity/stability for application in steam reforming of the oxygenated compounds and which preparation methods are most adequate to obtain these materials. In summary, it aims to provide insights into the performances reached and point out the best way to get better and improved catalysts for such applications (which depends on the feedstock used).

Keywords: ethanol; methanol; oxygenates; steam reforming; Ni-based catalysts

1. Introduction

The extensive use of fossil fuels over the last few decades has been contributing significantly to the build-up of greenhouse gases in the atmosphere, which on its hand contributed considerably to global warming. Alternative renewable fuels have been studied, among which hydrogen is seen as a potential candidate to replace fossil fuels as an energy carrier [1–4]. Among other advantages, the following are worth mentioning: hydrogen possesses a gravimetric energy density of 143 MJ·kg^{-1}, more than three times higher than that of gasoline and diesel [4]; it can be produced through several processes, many of which include renewable feedstock [5]; hydrogen can be used either in internal combustion engines or in fuel cells [6–8]; the risk associated to hydrogen handling is equal or lower than that of other fuels [4].

Over the last years, hydrogen has been industrially produced mostly through steam reforming of natural gas, naphtha, heavy oils, and to a lesser extent through coal gasification. Only a very small fraction of all of the hydrogen that has been produced worldwide comes from water and other renewable sources [9–12]. Furthermore, if the electricity used in water electrolysis is generated from fossil fuels, the hydrogen produced through this method cannot be considered renewable either.

Citation: Silva, J.; Rocha, C.; Soria, M.A.; Madeira, L.M. Catalytic Steam Reforming of Biomass-Derived Oxygenates for H$_2$ Production: A Review on Ni-Based Catalysts. *ChemEngineering* **2022**, *6*, 39. https://doi.org/10.3390/chemengineering6030039

Academic Editor: Dmitry Yu. Murzin

Received: 2 March 2022
Accepted: 16 May 2022
Published: 27 May 2022

Publisher's Note: MDPI stays neutral with regard to jurisdictional claims in published maps and institutional affiliations.

Copyright: © 2022 by the authors. Licensee MDPI, Basel, Switzerland. This article is an open access article distributed under the terms and conditions of the Creative Commons Attribution (CC BY) license (https://creativecommons.org/licenses/by/4.0/).

The steam reforming of oxygenated compounds derived from biomass, namely methanol [13–16], ethanol [17–20], and bio-oil [2,21–23], among others, has been largely studied over the last 20 years as an alternative for the well-established fossil-based processes, such as natural gas steam reforming. Furthermore, their contribution to the build-up of greenhouse gases in the atmosphere is significantly lower than that of fossil fuels (e.g., most of the emitted CO_2 belongs to the natural CO_2 cycle) [24]. If these advantages are combined with sustainable biomass exploitation policies that aim to avoid competition between the use of land for the production of bio-fuels, food, animal feed, fiber and ecosystem services [25], big-scale hydrogen production through steam reforming of biomass-derived oxygenates might be a likely scenario in the near future.

The successful development of the steam reforming process applied to these biomasses derived oxygenates is highly dependent on the choice of a suitable catalyst, since the steam reforming process is already well studied. In terms of catalytic performance, a suitable catalyst for hydrogen production through steam reforming should ideally meet the following criteria: be highly active to produce high amounts of hydrogen; be highly selective towards hydrogen so that the production of secondary products is minimized; be capable of maintaining long term activity without suffering from deactivation [26,27]. Furthermore, an easy protocol to activate the catalyst is desired (e.g., activation with the hydrogen-containing feedstock at the reaction temperature, thus avoiding more complex protocols) [27]. A catalyst must be cheap as its cost can contribute significantly to the overall process expenditure [28]. Finally, and in agreement with the current sensitive environmental situation, catalysts must be as environment-friendly as possible.

Different metallic active phases such as based on Ni, Cu, Co, Pt, Pd, Ru, Rh, and/or Ir, among others, have been investigated for the steam reforming of biomass-derived oxygenates for hydrogen production. However, Ni-based catalysts have been extensively studied over the last decades not only due to the fact that they have the potential to be highly active and stable, but also because they are cheaper than, for instance, noble metal-based catalysts. Ni-based catalysts, although widely used in steam reforming processes, are less active than noble metal-based materials and more prone to deactivation; effectively, noble metal-based catalysts perform well—they are stable and exhibit high catalytic activity [29]. However, they are very expensive and need high temperatures to be reduced. In this way, Ni-based materials are more suitable for industrial-scale applications. Moreover, nickel-based catalysts are inexpensive, but under some reaction conditions they suffer from sintering and deactivation by carbon production [30,31]. Nevertheless, the reaction mechanism over Ni-based catalysts follows the same steps as over noble metal-based catalysts for most steam reforming processes [30].

Still, a careful choice of the Ni loading, support, promoter(s) and synthesis method is of uttermost importance, as these factors have a crucial impact on both catalytic performance and price [32].

Since the steam reforming of biomass-derived oxygenates, such as methanol, ethanol and other oxygenates (e.g., bio-oil and glycerol), is still under intensive study, and considering the above reasons that highlight the potential of Ni-based catalysts, a literature review encompassing a wide range of these materials for the different reactions is hereby carried out. Ultimately, this work aims to understand if Ni-based catalysts currently show enough potential for application in steam reforming of oxygenated compounds and which formulations might be more promising for each application. For the first time, and up to the best knowledge of the authors, this review discusses the results obtained with the best Ni-based catalysts developed so far for the steam reforming of different oxygenated compounds, portraying in detail the effect of the preparation method and the deactivation suffered by these materials in long-term tests. Besides that, this work directs future research works about the most critical properties in preparing catalysts with high catalytic performances for the steam reforming of methanol, ethanol and other oxygenates.

2. Methanol Steam Reforming

2.1. Introduction

Methanol is the simplest of all alcohols and can be produced from several types of biomass such as agricultural waste, forestry waste, livestock and poultry waste, fishery waste, sewage sludge [13,21], among others, through pyrolysis, gasification, biosynthesis, electrolysis and photo electrochemical processes [13]—the main route of methanol formation is the through syngas. Even though some of these processes, such as pyrolysis and gasification, allow a significant hydrogen production directly from biomass [33], the production of methanol, later to be converted into hydrogen, is preferred for the following reasons: it is liquid under room conditions and, for that reason, more suitable for use in fuel cells; it is easier to transport; the required infrastructures are already available; and the production and use of methanol (and other biofuels) are considerably more technologically ready [15].

The methanol steam reforming (MSR) is described by the overall reaction shown in Equation (1):

$$CH_3OH + H_2O \rightleftharpoons CO_2 + 3H_2 \left(\Delta H_r^{298\ K} = 49.7\ kJ \cdot mol^{-1} \right) \quad (1)$$

The MSR can be divided into two major reactions: methanol decomposition (Equation (2)) followed by the water-gas shift reaction (WGS) (Equation (3)).

$$CH_3OH \rightleftharpoons CO + 2H_2 \left(\Delta H_r^{298\ K} = 90.2\ kJ \cdot mol^{-1} \right) \quad (2)$$

$$CO + H_2O \rightleftharpoons H_2 + CO_2 \left(\Delta H_r^{298\ K} = -41.2\ kJ \cdot mol^{-1} \right) \quad (3)$$

This process is globally slightly endothermic and, for that reason, it can be carried out at relatively low temperatures (ca. 200–300 °C) [26] when compared to the steam reforming of methane, for example, which is normally carried out at 700–900 °C [29]. Some of the by-products of this process are CO_2 and CO. The production of CO, in particular, should be avoided, especially if the produced hydrogen is to be used in polymer electrolyte fuel cell applications, where the concentration of CO must be lower than 20 ppm to avoid poisoning of the anode catalyst (low temperature fuel cells) [14]. In fact, for low temperature fuel cells in road vehicles applications, the International Organization for Standardization (ISO 14687) recommends a maximum CO concentration of 0.2 ppm in the hydrogen feed stream [34].

Another frequent by-product in MSR is CH_4, whose formation can happen through hydrogenation of either CO and/or CO_2 (Equations (4) and (5), respectively). Besides that, the highly undesired formation of coke deposits on the catalyst surface may also occur (for instance, the Boudouard reaction—Equation (6)).

$$CO + 3H_2 \rightleftharpoons CH_4 + H_2O \left(\Delta H_r^{298\ K} = -206\ kJ \cdot mol^{-1} \right) \quad (4)$$

$$CO_2 + 4H_2 \rightleftharpoons CH_4 + 2H_2O \left(\Delta H_r^{298\ K} = -165\ kJ \cdot mol^{-1} \right) \quad (5)$$

$$2CO \rightleftharpoons CO_2 + C \left(\Delta H_r^{298\ K} = -172\ kJ \cdot mol^{-1} \right) \quad (6)$$

Even though for fuel cell applications the minimization of CO formation is of extreme importance, its conversion into CH_4 is not desired either as it consumes H_2 (3 mol per mol of CO), the target product. Therefore, choosing an appropriate catalyst that maximizes H_2 production (not only due to higher methanol conversions but also due to lower CO and CH_4 production) could solve, at least partially, this problem.

2.2. Nickel-Based Catalysts

Over the years, several materials were studied as catalysts in the MSR reaction. Among them, Cu-based materials are the ones that have been more targeted for this reaction due to the copper's high surface area, high dispersion and small particle size, among other factors [26]. Furthermore, copper is relatively cheap. On the other hand, copper catalysts can be easily deactivated [35]. Group 8–10 metal-based catalysts have also been of significant interest and, even though they were, in general, less active towards MSR than Cu-based—group 11, they have shown better long-term and thermal stability [26]. Among those, Ni-based catalysts, which is one of the cheapest metals of such groups, have been a target of huge interest over the last decades for the steam reforming of oxygenates.

2.2.1. Catalytic Activity and H_2 Selectivity

Monometallic nickel catalysts supported on different materials have been extensively explored. Deshmane et al. [36] tested TiO_2 supported Ni, Co, Cu, Zn, Pd and Sn catalysts for the MSR process. The Ni-based catalyst was the second most active after the Pd-based catalyst in terms of methanol conversion and H_2 selectivity, having converted approximately 86% of methanol with an H_2 selectivity of around 97% (at 350 °C). However, it presented the highest CO selectivity, which would be a problem if the product H_2 stream was to be used in polymer electrolyte fuel cell applications. The authors concluded that the specific metal-support interactions, which controlled both reducibility and metal particles dispersion on the TiO_2 support, had a very significant impact on these results. The low activity of the Ni catalyst in the WGS reaction was the main responsible for the high CO selectivities. In a similar work [37], nickel supported on high surface area mesoporous MCM-41 has also been tested for MSR and an H_2 selectivity of 99.9% was achieved at 350 °C (45.1% of methanol conversion) [37]. However, compared to other metals (Cu, Pd, Sn, Zn and Co), Ni was observed to be one of the less active in terms of methanol conversion, contrary to what was observed for TiO_2 support. The authors suggest that the significantly lower reducibility of 10 wt.% Ni/MCM-41 due to silicates and silicides formation could be the reason for its lower activity. It could also be associated with the lower dispersion observed.

In another work [38], it was observed for a Ni/CeO_2 catalyst that the strong metal-support interactions play a crucial role in the high selectivity towards CO_2 production in the detriment of CO or surface carbon. Specifically, the enhanced water dissociation over reduced CeO_2 and subsequent oxygen transfer from ceria to nickel enhances the surface oxidation ability of the last. Moreover, the methanol conversion was benefitted over this catalyst and full conversion attained at 400 °C. Other monometallic Ni catalysts supported on other simple oxides have also been studied. For example, 12 wt.% Ni/SiO_2 converted 53% of methanol and showed H_2 selectivity as high as 74% at 200 °C only [39], while in another work 10 wt.% Ni/Al_2O_3 exhibited a conversion of only 24% at 400 °C and a H_2 yield of only about 15% [40]. In this last work, a comparison between the Al_2O_3 supported catalyst and Ni catalysts supported on MgO prepared through different methods showed that not only did the Al_2O_3 supported catalyst show higher methanol conversion and H_2 yield (400 °C), but it also showed lower CO selectivity (600 °C). The reason why the MgO-based catalysts showed higher CO selectivity could be due to their higher CO_2 capture ability, which would afterward result in the acceleration of the reverse Boudouard reaction (Equation (6)) and consequent reduction of deposited carbon [40]. This mitigated carbon deposition would certainly be advantageous from a long-term stability standpoint, but the higher CO production would be a problem for polymer electrolyte fuel cell applications. Bobadilla et al. [41] tested Ni nanoparticles supported on a CeO_2 and MgO modified Al_2O_3 support, having reached a methanol conversion of approximately 66% and H_2 yield of around 70% at 350 °C. The reason why the authors chose this modified support was due to the fact that MgO addition favors the gasification of carbon deposits and CeO_2 modification improves the metal dispersion, which results in a more active and stable catalyst [42].

From this preliminary assessment, one might guess that Ni-Cu-based catalysts are probably a good option for MSR that could assure both high activity and stability. Several catalysts combining both nickel and copper active phases have been reported over the years [43–49]. By studying Cu-Ni bimetallic catalysts with different Cu/Ni ratios (and also the corresponding monometallic catalysts), Khzouz et al. [43] observed that while Ni was related to the enhancement of the methanol decomposition reaction, Cu promoted the WGS reaction, especially above 250 °C. Furthermore, the bimetallic Ni-Cu catalysts did not yield any CH_4, thus suggesting that Cu alloying in Ni had an inhibiting effect on methanation reactions (Equations (4) and (5)). However, the formation of CH_4 during MSR is not always inhibited over Cu-Ni-based catalysts, as has been reported in several works [46,50–52]. Besides the operating conditions (e.g., reaction temperature or water/methanol feed molar ratio), factors related to the catalysts are also responsible for such different behaviors, namely the support, metal loading and the preparation method adopted.

Lytkina et al. [53] studied bimetallic Ni-Cu-based catalysts supported on ZrO_2 annealed at different temperatures (350 °C or 400 °C), having observed that the higher annealing temperature led to higher crystallization of the support and, consequently, to the deterioration of its adsorption properties. While methanol molecules adsorb on the metal active sites, water molecules adsorb preferentially on the active sites located in the support [54,55]. This was probably one of the causes of the lower H_2 production for the catalysts whose support was annealed at 400 °C. Simultaneous agglomeration of catalyst particles with resulting decreased active surface area was also probably another cause of such decrease in catalytic activity towards H_2 production.

As already mentioned, the loading of the different species is also of crucial importance for catalytic activity. For bimetallic Cu-Ni catalysts, it has been observed that varying the Cu/Ni ratio (0.25–4), while maintaining the total active metal loading, affected the catalytic activity considerably [43,53]. For both studies, while increasing the amount of Ni enhanced methanol conversion, lower Ni contents yielded higher H_2 yields and lower CO production, due to the WGS reaction. It has been suggested that this happens due to a change in the mechanism of adsorption of methanol molecules on the catalyst surface (see Figure 1) [53]. More specifically, for catalysts with higher copper content, there is a higher tendency for single-site (η^1 in Figure 1) methanol adsorption, while two-sites (η^2 according to Figure 1) adsorption through oxygen and carbon atoms is more likely to occur over group 8–10 metals. While in the first case alcohol and water adsorption on adjacent sites with consequently higher yields of H_2 and CO_2 are more likely, in the second case CO production should be more benefited than carbon dioxide [53,56]. The authors suggest that, alternatively, higher Cu content (lower Ni content) might contribute to deeper oxidation of the carbon atoms of the alcohols (i.e., to CO_2) due to a lower Fermi level [53].

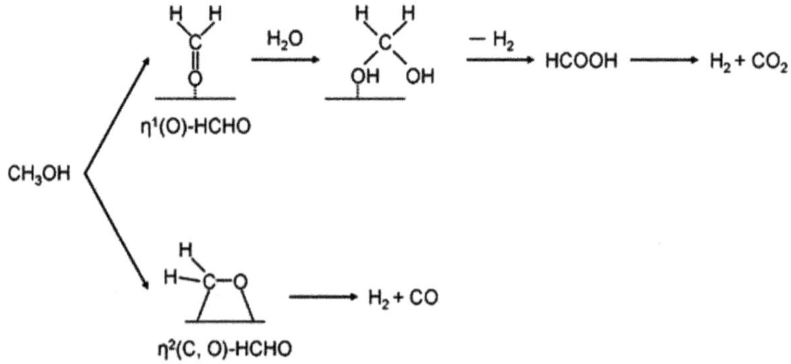

Figure 1. Mechanism of adsorption of methanol molecule over different catalyst surfaces. Reprinted with permission from Ref. [53]. Copyright 2022 Elsevier.

The effect of the total Ni/Cu loading (Ni/Cu of 0.25) on ZrO_2 support has also been analyzed [48]. The increase of the bimetallic loading up to 30 wt.% enhanced the conversion of methanol, having this behavior been correlated to the higher frequency of Ni/Cu core-shell structures observed through transmission electron microscopy (TEM) in the sample with the highest metal content. On the other hand, the sample with a 15 wt.% bimetal loading showed the highest H_2 yield. While only Cu_{core}-Ni_{shell} nanoparticles were identified in the 15 wt.% sample, Cu_{core}-$(NiCu\text{-}alloy)_{shell}$ nanoparticles were also observed in the 30 wt.% sample. The authors suggested that the nanoparticles with only Ni in the shell presented better H_2 production performance than the nanoparticles whose shell consisted of a Ni-Cu alloy.

Ni-Cu bimetallic catalysts supported on metal oxide-stabilized zirconia (ZrO_2) supports doped with Y, La or Ce were analyzed in another work [57]. The results showed that not only the nature but also the composition of the support influence MSR considerably, which supports the above-mentioned bifunctional mechanism. In fact, while doping the ZrO_2 support with La resulted in higher catalytic activity, doping with Y allowed a higher selectivity towards H_2. This suggests that the doped support influences the process selectivity as well. The modification of the ZrO_2 support with Ce led ultimately to a twofold increment in the activity of Cu-Ni-based catalysts comparatively to the ones modified with Y. It was found that 10 wt.% of Ce was the optimum content that allowed to reach the best catalytic performance, increasing the Ce content beyond that point having been shown as detrimental. The authors suggest that this behavior could be either due to the need of the simultaneous presence of Ce anions (Ce^{3+} and Ce^{4+}), being that the presence of Ce^{3+} on the surface of the particles decreases with increasing Ce content, and Zr, or due to the interaction of defects.

In the work of Huang et al. [58], it has been shown that Ni-Cu/Al_2O_4 materials presented high conversion (>99%), high H_2 yield, and high stability in long-term tests, when used as catalysts in the MSR reaction at low temperatures, in the range of 200–300 °C. In addition, this catalyst exhibits high methanol conversion (very close to 100%), high H_2 yield (always close to 90%) and low CO_2 concentration (around 10%) during the long-term experiment performed in this study. Moreover, it was possible to detect the presence of CO in the reactor outlet (around 5%) during 30 h. These results show that the Ni-Cu bimetallic catalysts present suitable properties to be applied in the MSR process and the promotion of these samples with Ce increases the catalytic performance.

Moreover, it was observed [59] that the simultaneous existence of Pt and Ni elements on the surface of CeO_2 support improved the methanol conversion and H_2 production in comparison with the monometallic samples.

Besides Cu and Pt, other metals such as Sn and La [60,61] have been combined with Ni (using Al_2O_3 as support) in other studies, also aiming to optimize the catalytic performance of the MSR process. While La was observed to enhance the catalytic activity towards H_2 production [61], the presence of Sn was reported to favor mainly the stability of the catalyst [60]. This stabilizing effect of tin is further discussed in the next section (Section 2.2.2). Besides assessing the effect of Sn addition, the authors also evaluated the addition of MgO to the Al_2O_3 support, having concluded that the sample whose support incorporated the highest MgO loading (30 wt.%) presented the highest H_2 production. This enhancement was attributed to the decrease of the support acidity and improvement of Ni dispersion. As for the promoting effect of La, it was observed the interaction between La species and NiO and/or Al_2O_3 (support) to form La-Ni oxide and/or La-Ni-Al mixed oxide, which led to the separation of external NiO particles from the Ni-Al interface. The authors hypothesized that this facilitated the formation of smaller separated NiO particles more highly dispersed (Figure 2), which were ultimately responsible for the improved MSR activity [61]. Furthermore, increasing the Ni loading up to 10 wt.% facilitated the reduction of the catalyst at lower temperatures and improved methanol conversion and H_2 selectivity (lower CO selectivity) for similar reasons as for La.

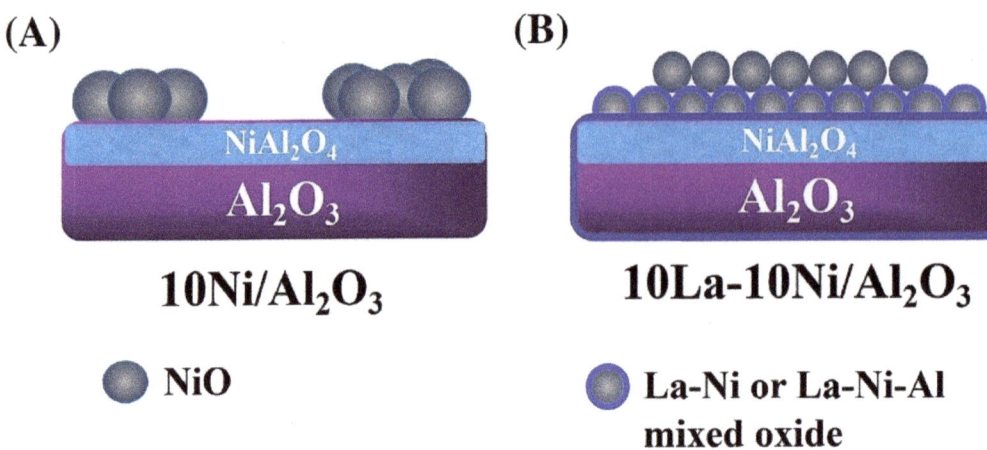

Figure 2. Effect of La addition (**B**) on the surface composition of a NiO supported on Al_2O_3 catalyst (**A**). Reprinted with permission from Ref. [61]. Copyright 2022 Elsevier.

Layered double hydroxides (LDHs) derived catalysts containing nickel have also been a target of interest for MSR over the years. Kim et al. [62] reported a $Cu_{0.55}Ni_{0.10}Zn_{0.10}Al_{0.25}$ LDH which converted about 50% of methanol at 300 °C, being the presence of Ni responsible for a slight improvement of the low temperature activity comparatively to the $Cu_{0.55}Zn_{0.20}Al_{0.25}$ LDH. Qi et al. [63] tested a Ni/Al LDH with a Ni/Al molar ratio of approximately 5.7 which converted 100% of methanol with an H_2 selectivity slightly above 50% and CO selectivity below 10% (at 380 °C). In another work [64], another Ni/Al LDH with a Ni/Al molar ratio of around 4.9 converted 87.9% of methanol, provided an H_2 yield of approximately 77% and produced low levels of CO at 390 °C. Besides the difference in the Ni/Al molar ratio and reaction conditions between both works, the preparation procedures were also different. The effect of this aspect was, in fact, the target of analysis in both works and was certainly decisive in the results obtained. This will be discussed later (cf. Section 2.2.3). The addition of K to LDHs has been found to further enhance its activity in MSR [65]. The use of K_2CO_3, during LDH preparation, as precipitating agent combined with no washing, to keep more K in the final solid, increased the conversion of methanol. Furthermore, post-addition of K_2CO_3 through the incipient wetness method decreased significantly the production of methane. On the other hand, the production of CO suffered a significant increment, which might indicate that the inhibition of methane production could be related to the inhibition of the methanation of CO (Equation (4)). The addition of K also benefited slightly the yield of H_2 and the selectivity towards CO_2 in detriment of CO, especially at temperatures around 400 °C. The catalytic activity of the K-promoted LDH was superior to that of a commercial Cu catalyst (42 wt.% CuO, 47 wt.% ZnO and 10 wt.% Al_2O_3), as shown in Figure 3.

Different studies using Ni-based catalysts suggested the formation of several intermediates on the catalyst surface during the reaction: according to these works, methyl formate and H_2 were produced via dehydrogenation of methanol, followed by the formation of formic acid [66–68]. Then, CO_2 and H_2 were produced by the decomposition of formic acid. The formation of formaldehyde and dioxomethylene during the MSR process was also suggested by other reactions [66].

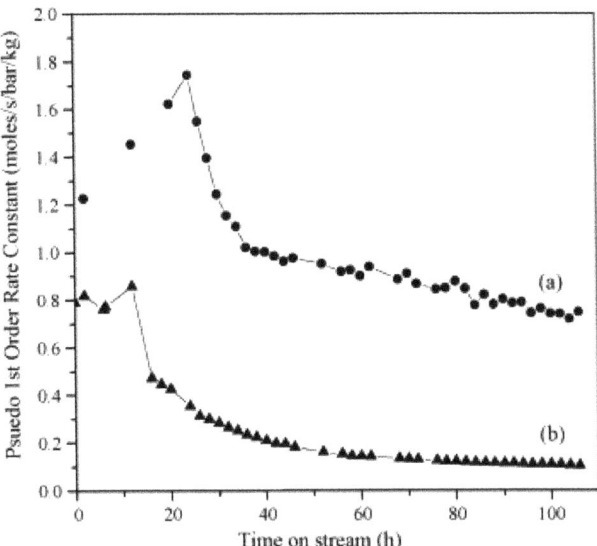

Figure 3. Comparison of the activity in methanol conversion of (**a**) NiAl-LDH with potassium and (**b**) commercial catalyst at 390 °C and steam/carbon molar feed ratio of 1.2. Both catalysts were pre-activated in situ at 240 °C for 6 h on a reactive stream before reaction. Reprinted with permission from Ref. [65]. Copyright 2022 Elsevier.

2.2.2. Deactivation

High catalytic activity and H_2 selectivity alone do not guarantee that a specific material is a good catalyst for a specific reaction. Long-term stability is another crucial criterion that has to be taken into account when choosing a catalyst. To design a catalyst with long-term catalytic stability, it is crucial to understand what phenomena can disturb such stability. Such phenomena are: (i) coke formation due to hydrocarbons decomposition; (ii) sintering and crystallization or segregation of the metal particles caused by thermal effects; (iii) poisoning originated by chemisorption or reaction of certain substrates on the catalyst surface (e.g., H_2S and CO on Pt-based catalyst in hydrogenation reactions and H_2 dissociation in the anode of polymer electrolyte fuel cells, respectively); and (iv) fouling due to solids deposition caused by dusty materials in the feed [69].

Going back to the bimetallic Cu-Ni catalysts, Qing et al. [44] reported a Cu-Ni-Al spinel with Cu/Ni/Al molar ratio of 1/0.05/3 and calcined at 1000 °C that showed stable conversion of methanol during 300 h (Figure 4). Comparatively to the other materials prepared and tested by the authors, the lower particle size, higher specific surface area and pore volume, more hardly-reducible spinel and better sustained release catalytic performance (catalyst used without being previously reduced and active Cu sites gradually generated during the reaction) probably contributed to the enhanced stability of such catalyst.

In another work [48], a 30 wt.% Ni-Cu/ZrO_2 catalyst showed stable methanol conversion (above 90%) and H_2 production (selectivity remained around 60%) for 46 h on-stream at 400 °C, after which the H_2-TPR profile of the catalyst was unaffected. In fact, Lytkina et al. [57] tested several Cu-Ni catalysts supported on metal oxide-stabilized ZrO_2 supports doped with Y, La and Ce and all of them were reported to work for at least 90 h without visible deactivation (constant H_2 selectivity at 300 °C). As already mentioned, even though Cu can be easily deactivated, the presence of a group 8–10 metal, such as Ni, could be partially responsible for this long-term stabilization. In fact, it has been reported that the addition of a second metal improves, in general, the stability of steam reforming catalysts by promoting their hydrogenation activity [69,70] and decreasing the coke deposition [38,40,42,60].

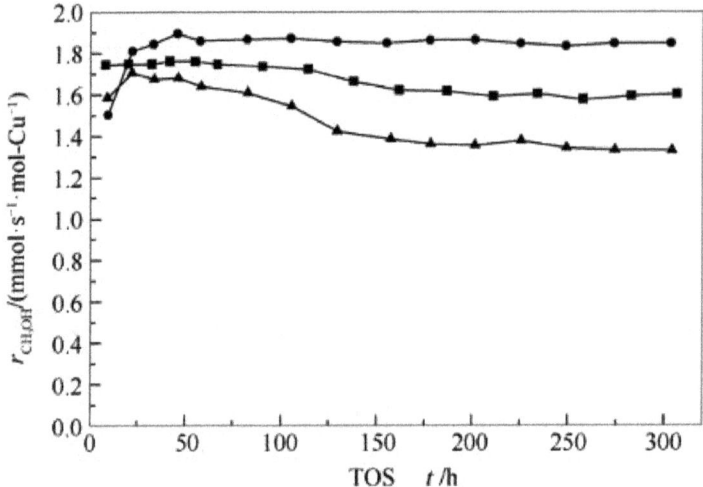

Figure 4. Methanol conversion rate as a function of time-on-stream (TOS) for the catalysts with Cu/Ni/Al molar ratio of 1/0.05/3 calcined at 900 °C (squares), 1000 °C (circles) and 1100 °C (triangles) at 255 °C, 1 bar and WHSV = 2.18 h^{-1}. Reprinted with permission from Ref. [44]. Copyright 2022 Elsevier.

Penkova et al. [60] observed that the simultaneous presence of Ni and Sn results in the formation of an alloy that enhances catalytic stability; this beneficial effect of tin was also corroborated elsewhere [41]. In terms of carbonaceous deposits formation, the addition of Sn resulted in a decrease compared to the monometallic Ni catalysts, indicating that the formation of the NiSn alloy inhibits the formation of NiC and, consequently, coke. It was also observed that the addition of Sn plays an important role in delaying the temperature range of particles agglomeration and, therefore, avoiding low temperature sintering. The authors also concluded that the addition of MgO to Al_2O_3 in the support results in lower coke formation [60]. In fact, after 20 h on-stream, the decrease of the catalytic activity was negligible. This behavior has been attributed to both a decrease in surface acidity of the Al_2O_3 support (confirmed from FTIR of adsorbed pyridine) and an improvement of Ni dispersion caused by the formation of $MgAl_2O_4$ spinel, which inhibits the incorporation of Ni in the Al_2O_3 phase. Consequently, the oxidation of carbonaceous deposits was facilitated. The only carbon deposits detected through temperature-programmed oxidation (TPO) were not very stable. A similar result was observed elsewhere [40] for a Ni catalyst supported on MgO. The authors attributed the stable behavior (20 h) of the catalyst to its capacity to mitigate the agglomeration of Ni particles and to the high basicity of the MgO support. The last enhanced the adsorption of CO_2 and, consequently, promoted the gasification reaction between CO_2 and carbon, as previously discussed.

Concerning bimetallic samples of Ni-Cu, in a recent work by Liu et al. [71] it was observed that the Ni-Cu/Al_2O_3 materials with higher Al content present higher initial catalytic activity, however, show a quick catalyst deactivation, while the catalysts with less Al content show better catalytic stability. The deactivation of the samples was related to the non-spinel CuO particles that are easier to agglomerate and sinter as compared to that from spinel Cu^{2+} species. In this way, the catalysts with more non-spinel CuO show a high initial catalytic activity but a higher deactivation rate.

A comparison between several metals supported on MCM-41 showed that the Ni-based catalyst was less stable than, for instance, the Cu-based catalyst [37]. Such deactivation of the Ni-based catalyst was attributed to the formation of carbonaceous deposits, thermal sintering and changes in the support structure. It is suggested that the metal particles in MCM-41 behave as bulk materials, thus having little metal-support interactions.

This could be one of the reasons why, contrarily to what would be expected, the Ni-based catalyst was less stable than the Cu-based one.

LDHs present good catalytic stability during MSR [64,65]. In both works, the Ni/Al LDHs demonstrated stable MSR activity for approximately 100 h. When the LDH was pre-treated with a diluted H_2 stream [64], stable H_2 production with low levels of CO and no CH_4 occurred. On the other hand, when the catalyst was pre-treated on the reactive stream, it deactivated over time-on-stream with increasing CO formation. The reason for such difference will be further explored in the next section. Besides enhancing the catalytic activity, as already discussed, the addition of K to LDHs also promotes catalytic stability [65]. A K-promoted Ni/Al LDH showed similar stability at 390 °C to that of a commercial Cu catalyst (42 wt.% CuO, 47 wt.% ZnO, and 10 wt.% Al_2O_3).

2.2.3. Effect of the Preparation Method

The preparation method used is also a very important parameter to be considered. The same catalyst formulation prepared through different methods might result in very different performances and costs. The impregnation method is among the most reported for the preparation of heterogeneous catalysts mainly due to its simple execution and low waste streams [72]. From the works already discussed above, a considerable part of them reported the use of such method [38,40,41,43,45,47–49,53,57,60,61], being that both wet impregnation and incipient wetness impregnation were reported. Furthermore, in the case of the bimetallic catalysts, both co-impregnation and sequential impregnation were used.

A comparison between co-impregnation and sequential impregnation, used for the preparation of bimetallic Ni and Cu over CeO_2 catalysts, has been established [73]. The authors observed that the bimetallic catalyst prepared through co-impregnation (Cu-Ni) presented higher methanol conversion and H_2 selectivity during oxidative MSR than the homologous catalyst formulation prepared via sequential impregnation (Ni/Cu and Cu/Ni). However, in a different work [74], the Ni and Cu over ZrO_2 catalysts prepared by sequential impregnation showed higher oxidative MSR activity than the corresponding co-impregnated material. Fukui's theory [75] indicates that higher reactivity occurs when the system's gap-energy is low and its total energy is high. In other words, higher reactivity corresponds to an easier adsorption or desorption of a molecule for a specific reaction. The molecular simulations carried out by López et al. [74] show that the sequentially impregnated materials on ZrO_2 showed lower gap-energy and higher total energy (lower adsorption energy) than the co-impregnated catalyst (Figure 5), thus suggesting that an electron transfer mechanism is benefited at the interface between the support and the bimetallic structures in the first material. This ultimately enhances the redox properties of the catalyst and, consequently, its activity. Furthermore, the observed presence of bimetallic Cu-Ni and core-shell Ni/Cu nanoparticles and crystalline anisotropy of the active phase could have influenced the results greatly. These results show once again that MSR activity is sensitive to the catalyst structure, which on its hand depends on how the bimetallic active phase is impregnated. The opposite behaviors observed in both works could be due to the different supports, having once again in mind the importance of the active phase-support interface in MSR activity.

The precipitation or co-precipitation methods have also been applied to prepare some of the catalysts discussed in this section [40,53,57]. Luo et al. [40] prepared three nano Ni_xMg_yO solid solutions through different methods: (1) incipient wetness impregnation of Ni onto the MgO support prepared via precipitation method; (2) the same method with an added hydrothermal treatment of the support at 100 °C for 24 h after precipitation; (3) co-precipitation of both Mg and Ni salts with the same hydrothermal treatment used in the previous method. Among the three catalysts, the one prepared through the second method showed a superior capacity to convert methanol and produce H_2. By comparing methods (2) and (3), the differences are that while in method (2) Ni was incorporated in the treated support via incipient wetness impregnation, in method (3) Ni was introduced in the catalyst simultaneously with Mg through co-precipitation and went through the same treatment as the support (not Ni) in method (2). Therefore, the lower performance of the catalyst prepared through method (3) is probably associated with these two methodological modifications. TPR analysis showed that the catalyst prepared via method (3) was less reducible than the one prepared through method (2). H_2-TPD results indicated that most Ni species dissolved deeply inside the MgO matrix, being only a minority amount at the subsurface to catalyze the MSR reaction. Furthermore, co-precipitation methods normally involve several washing steps after precipitation, aiming to remove residual nitrates, potassium, sodium and other compounds. This allows avoiding sintering and agglomeration of particles during thermal treatment (higher resistance to sintering was observed for the catalyst prepared through method (3)). Consequently, significant undesired streams contaminated with nitrates are produced. Alternatively, chloride or sulfate-based precursors could be used if they did not poison methanol catalysts. Formate precursors have been researched as possible alternatives that could overcome these limitations [72].

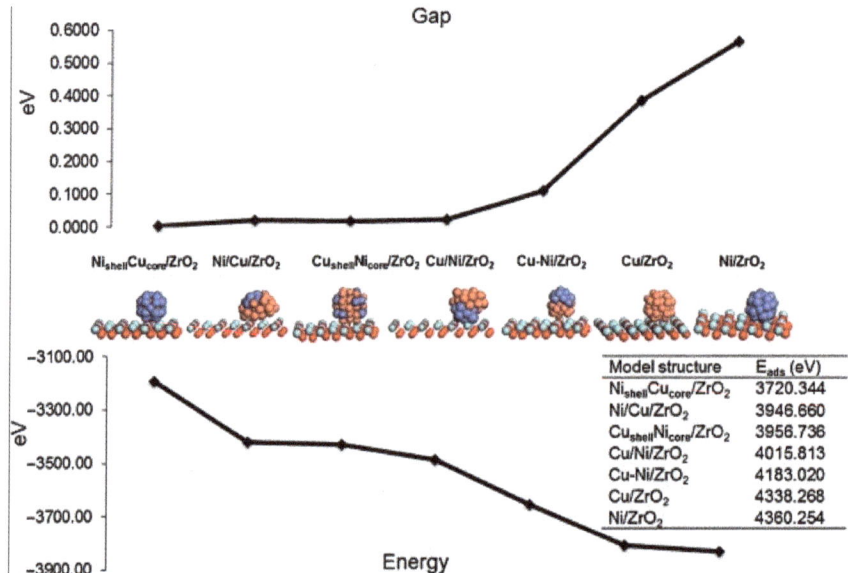

Figure 5. Molecular models of bimetallic and monometallic clusters: the top graph shows the gap-energy, the bottom graph shows the system total energy and the inset shows the adsorption energy of methanol on the surface. Reprinted with permission from Ref. [74]. Copyright 2022 Elsevier.

Other methods such as the sol-gel [39,48,50] and polyol method [41,50] have also been used to prepare Ni-based MSR catalysts. The sol-gel method allows the attainment of good chemical homogeneity, stoichiometry, phase purity, narrow particle size distribution, ultrafine powder and high specific surface area due to molecular and atomic scale mixing and networking of chemical components [76,77]. Regarding the works that used the sol-gel method, none of them established a comparison with other methods. A comparison has, however, been established between Ni/SiO_2 catalysts prepared via wet impregnation and the simple sol-gel method in ethanol steam reforming (ESR) [78]. The catalyst prepared via sol-gel presented good dispersion and considerably higher BET surface area than the catalysts prepared through impregnation. Furthermore, the first produced about twice the amount of H_2 produced by the last. This last result depends not only on the catalyst, but also on the reaction that it catalyzes. In other words, the fact that the sol-gel-prepared catalysts performed better than the catalysts prepared through impregnation in ESR does not guarantee that the same behavior would be observed in MSR. Furthermore, the sol-gel process presents disadvantages such as the high costs of some necessary chemicals and the often large volume shrinkage and cracking due to washing and drying steps mainly [79].

Bobadilla et al. [41] correlated the influence of the synthesis method used to prepare Ni and Ni-Sn supported on CeO_2-MgO-Al_2O_3 with their respective catalytic behavior in MSR. More specifically, the authors established a comparison between the catalysts prepared via deposition of nanoparticles obtained by the polyol method and catalysts prepared through impregnation. The nanoparticles produced via the polyol method presented better activity than the impregnated catalysts, especially the monometallic Ni catalysts. The authors proposed a model of the catalyst surface and phase distribution before and after reduction for the monometallic catalysts prepared through both methods (Figure 6).

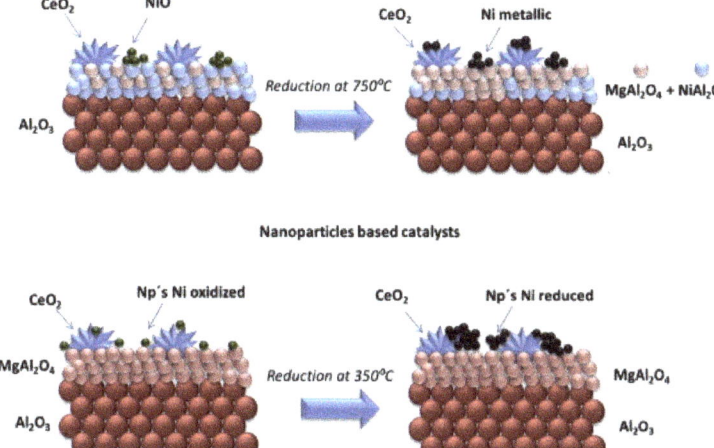

Figure 6. Model of the catalyst surface and phase distribution before and after reduction for the monometallic catalysts prepared through the impregnation method (**above**) and polyol method (**below**). (Np's: Nanoparticles). Reprinted with permission from Ref. [41]. Copyright 2022 Elsevier.

The impregnation method seems to be as good as the polyol method to produce Ni nanoparticles. Even though the catalyst prepared via impregnation required a higher reduction temperature, it also presented higher metal-support interaction and a lower degree of nanoparticles sintering. The monometallic catalyst prepared via impregnation showed CeO_2 fluorite phase and both $MgAl_2O_4$ spinel and $NiAl_2O_4$ spinel, while for the nanoparticles-based catalyst (polyol method) only the CeO_2 fluorite and $MgAl_2O_4$ spinel structures were observed. The $NiAl_2O_4$ spinel in the impregnated sample was transformed into metallic Ni during reduction.

The polyol method was, on the other hand, considered by the authors as the preferable method for the preparation of the bimetallic Ni-Sn catalyst due to its control of composition and the enhanced resistance to sintering and reducibility that it confers to the catalyst. Therefore, polyol method being a better alternative than, for example, impregnation depends significantly on the catalyst being prepared. It has been concluded elsewhere [80] that due to the variety of shapes, compositions and nanostructures that can be produced, significant research has still to be carried out for a better comprehension of the polyol method.

Besides the preparation method, some parameters that are transversal to several methods can also have a significant impact on the catalyst activity and stability. Both drying and calcination conditions are known to be crucial parameters that have to be carefully selected. Going back to the work of Luo et al. [40], the authors observed that the addition of a hydrothermal treatment of the support at 100 °C for 24 h after precipitation increased the methanol conversion from 55.3% to 97.4% and the H_2 yield from 30% up to 58.5% at 600 °C, after 20 h on-stream. This enhancement was attributed to the higher reducibility of the catalysts subjected to the hydrothermal treatment. The catalysts hydrothermally treated also showed superior stability, which was ascribed to their higher surface basicity.

In another work [44] the authors observed that increasing the calcination temperature of Cu-Ni-Al spinel catalysts in the temperature range of 900–1100 °C increased the content of spinel. On the other hand, it also increased the spinel's particle size (sintering) and, consequently, led to a decrease in the specific surface area. Ultimately, the catalysts calcined at 1000 °C presented better catalytic activity and stability (Figure 4). Analysis of the used catalysts showed that the catalysts calcined at 1000 °C presented the smallest size of copper species, which took part in the catalysis of the reaction, in line with the catalytic activity results.

Reduction conditions are also of crucial significance in terms of catalytic performance [46,50]. It has been observed that reduction under H_2 atmosphere at 400 °C or 500 °C yielded different results [46]. More H_2 and CO_2 and less CO were produced over the catalyst reduced at 400 °C compared to the one reduced at 500 °C. Reducing the catalyst at 500 °C resulted in the reduction of Fe from Fe_3O_4, which subsequently reacted with the Ni particles, thus resulting in the formation of a Fe-Ni alloy. It is possible that this alloy covered the pores on the surfaces and lowered the surface area. This benefited the selectivity towards CO at the expense of H_2. In another work [50], besides analyzing the effect of reduction temperature (140, 160 and 180 °C), the authors also analyzed the effect of pretreatment atmosphere (4 h under Ar or 4 h under Ar and then 2 h under H_2 atmosphere) on the MSR activity of Ni-Cu/CaO-SiO_2 prepared via polyol method. Reduction at 160 °C under Ar atmosphere resulted in the highest H_2 yield due to, mainly, the higher dispersion of metal on the support. While increasing the reduction temperature from 160 to 180 °C produced larger Ni and Cu particles on the support and, consequently, poorer dispersion, reduction under H_2 atmosphere, after treatment under Ar atmosphere, led to metal particles agglomeration [81].

The preparation of LDHs is normally carried out via co-precipitation [62–65], and the addition of promoters such as Na or K normally occurs through impregnation [65]. It has been extensively reported that both calcination temperature and pretreatment conditions are decisive regarding LDHs' activity. The results obtained by Qi et al. [63] show that while calcination at 250 °C allowed attaining full methanol conversion only at 380 °C over a NiAl-LDH, calcination at 330 °C and 500 °C resulted in a progressive dislocation of the full conversion temperature towards lower values, with the LDH calcined at 500 °C achieving total conversion at 360 °C. The catalytic performance of the LDH was not only influenced by the pretreatment temperature [63], but also by the pre-treatment atmosphere [64]. While pretreating the LDH under diluted H_2 resulted in high activity and stability, pre-treating it under a reactive stream resulted in higher deactivation, lower H_2 generation and more CO [64].

2.3. Summary

A summary of the MSR catalysts reviewed in this section that showed the most promising results is presented in Table 1. It can be observed that from the list of the 10 most promising catalysts, half of them are bimetallic Ni-and Cu-based materials. All such 5 catalysts showed methanol conversions of at least around 90%, as well as relatively high H_2 yields and selectivities. As already discussed, the catalysts with higher Cu content and lower amounts of Ni tend to be more selective toward H_2 and CO_2 in detriment of CO; almost no CO was produced over $Ni_{0.2}$-$Cu_{0.8}$/ZrO_2 [53]. On the other hand, Ni normally catalyzes mostly the methanol decomposition reaction, thus enhancing methanol conversion. Furthermore, $Ni_{0.2}$-$Cu_{0.8}$/$Ce_{0.1}Zr_{0.9}O_2$-catalyst showed quite promising stable operation [57]. This shows how attractive bimetallic Ni-Cu-based catalysts are for MSR. Moreover, the impregnation method appears to be a critical issue. Finally, the annealing and reduction procedures adopted in both works were carried out at temperatures very similar to those employed during the MSR reaction.

Table 1. Comparison of some of the most promising MSR catalysts reviewed in this section.

Catalyst	Temperature	Feed Flow Rate	Mass of Catalyst	S/C [a]	Conversion of Methanol	H_2 Yield/Selectivity	Stability	Preparation Method	Refs.
5 wt% Ni-5 wt% Cu/Al_2O_3	325 °C	0.06 mL·min^{-1}	3 g	1.7	98.5%	2.2 [b] /n.d.	Result after 3 h	Impregnation	[43]
10.8 wt% Ni-Cu/TiO_2/Monolith	300 °C	1.8 h^{-1} [c]	n.d.	2	92.6%	n.d./92.7% [g]	n.d.	Impregnation	[49]
Ni/Cu/ZnO/Al_2O_3 (22.5/22.5/45/10) [d]	350 °C	150,000 mL·g^{-1}·h^{-1}	0.031 g	1	100%	≈83.3% [f] /n.d.	n.d.	Coating and impregnation	[52]
$Ni_{0.2}$-$Cu_{0.8}$/ZrO_2 (Metals/Carrier = 0.2/1)	325 °C	n.d.	0.3 g	1	≈100%	≈66.7% [f] /n.d.	n.d.	Sequential impregnation over support prepared via precipitation	[53]
$Ni_{0.2}$-$Cu_{0.8}$/$Ce_{0.1}Zr_{0.9}O_2$ (Metals/Carrier = 0.2/1)	350 °C	172 h^{-1} [e]	0.3 g	1	≈86%	n.d./≈99.9% [g]	90 h without deactivation	Sequential impregnation over support prepared via co-precipitation	[57]
10 wt% Ni-10 wt% La/Al_2O_3	350 °C	0.02 mL·min^{-1} 10,920 h^{-1} [e]	0.2 g	3	100%	n.d./≈69% [g]	n.d.	Co-incipient wetness impregnation	[61]
10 wt% Ni/TiO_2	350 °C	2838 h^{-1} [e]	n.d.	3	≈86%	n.d./≈97% [g]	n.d.	Facile one-step synthesis	[36]
10 wt% Ni nanoparticles/15 wt% CeO_2-10 wt% MgO-Al_2O_3	350 °C	8000 mL·g^{-1}·h^{-1}	n.d.	2	≈66%	66.7%/n.d.	Result after 24 h (no significant deactivation)	Polyol and support via co-impregnation	[41]
NiAl-LDH (Ni/Al = 4.9)	390 °C	0.05 mL·min^{-1}	0.15 g	1	94.6%	70% [f] /n.d.	Result after 100 h (no significant deactivation)	Co-precipitation	[64]
3wt% K/$Ni_{0.78}Al_{0.16}(OH)_2(CO_3)_{0.15}$·0.66$H_2O$	390 °C	0.05 mL·min^{-1}	0.15 g	1.2	91%	80% [f] /n.d.	Result after 60 h (no significant deactivation)	Co-precipitation	[65]
(30 wt% Ni)-Cu/Al_2O_4	300 °C	1 mL·min^{-1}	30 g	1.5	≈100%	98% [f] /n.d.	Result after 30 h (no significant deactivation)	Impregnation	[58]

[a] Steam to carbon molar feed ratio. [b] mol_{H_2}·$mol^{-1}_{methanol,in}$. [c] Weight hourly space velocity (mass flow rate of feed/total catalysts weight). [d] Molar ratio. [e] Entering volume flow rate/volume of catalytic layer. [f] Relative to the maximum of 3 mol_{H_2}·$mol^{-1}_{methanol,converted}$. [g] Selectivity (%) = $\frac{H_2}{\sum F_i}$ × 100. n.d.—Not determined.

The bimetallic 10 wt.% Ni-10 wt.% La/Al$_2$O$_3$ catalyst also showed promising results, not only in terms of methanol conversion and H$_2$ production but also regarding its low CO generation [61]. The same cannot be said about the monometallic Ni catalyst reported by Deshmane et al. [36], which showed high CO selectivity despite the high conversion of methanol and selectivity towards H$_2$. Even though the catalyst consisting of nickel nanoparticles supported on CeO$_2$-MgO-Al$_2$O$_3$ mixed oxide [41] showed performance parameters above the average of the catalysts reviewed, it is below all of the others shown in Table 1. More research must be conducted regarding the polyol method. As for the LDH-based materials [64,65], they allowed reaching high conversions of methanol and yields of H$_2$, while showing promisingly stable operation. However, some CO production was observed in both cases. Nevertheless, it has been highlighted in this review that slight changes in the formulation of the LDHs could reduce this problem. However, the drawbacks of the precipitation method (mentioned in more detail in the previous section), namely the undesired streams resulting from washing, must be considered. If such drawbacks are solved, LDH-based materials might be interesting for the MSR. Finally, considering the emergence of CO$_2$ sorption-enhanced reactor concepts and the so extensively reported ability of LDHs to capture CO$_2$ at high temperatures (300–500 °C) [82–84], these materials have great potential to be used as hybrid sorbent-catalyst (so called DFM, or dual function materials).

Finally, along this section, it was observed that the most crucial catalyst properties to achieve high catalytic performance for the MSR process are the metal dispersion and the surface area of the material. In this way, to reach high methanol conversion, high H$_2$ yield and low sub-products production, it is necessary to prepare a catalyst with a high surface area (for instance, using a porous support with a high pore volume) and with a strong interaction between the active phase and the support.

Taking into account the results presented in Section 2.2 (and the main outputs summarized in Table 1), to achieve high methanol conversion and high H$_2$ production in long-term MSR experiments, it is suggested the utilization of a bimetallic Ni-Cu catalyst (for instance, prepared by impregnation) promoted with (or supported on) ZrO$_2$/CeO$_2$.

3. Ethanol Steam Reforming

3.1. Introduction

Renewable ethanol can be produced from several feedstocks, as indicated in Table 2. The production of bio-ethanol from algae, even though possible, is still in an early stage of development [85]. The conversion of the different types of biomass into ethanol varies significantly, mainly in terms of the attainment of sugar solutions. While sugar sources only need an extraction process to attain fermentable sugars, starch sources demand previous hydrolysis to convert starch into glucose. Finally, for lignocellulosic biomass, a pre-treatment is required before hydrolysis so that the cellulose structures are modified for enzyme accessibility [85].

Ethanol is highly available, easy to handle, transport and store and is less toxic than methanol [15,20]. Furthermore, it has higher hydrogen content than methanol (six hydrogen atoms instead of four). These are some of the reasons why it has been extensively investigated for hydrogen production through ethanol steam reforming (ESR). The ESR can be described by the overall reaction shown in Equation (7).

$$C_2H_5OH + 3H_2O \rightarrow 2CO_2 + 6H_2 \left(\Delta H_r^{298\ K} = 174\ \text{kJ·mol}^{-1} \right) \quad (7)$$

This reaction is considerably more endothermic than MSR and for that reason temperatures around 400–600 °C are often adopted, which are still below those typically used in steam reforming of methane.

Table 2. Sources of bio-ethanol [85].

Category	Biomass
Sugar sources	Sugarcane
	Sugar beet
	Sweet sorghum
	Cane
	Molasses
	Beet molasses
	Grape
	Dates
	Watermelon
	Apple
Starch sources	Corn
	Wheat
	Cassava
	Barley
	Canna
	Sorghum grain
	Potato
	Sweet potato
	Yam
	Jerusalem artichoke
	Iles-iles
	Oat
	Banana
Lignocellulosic biomass	Perennial grasses
	Aquatic plants
	Softwood
	Hard wood
	Sawdust
	Pruning
	Bark thinning residues
	Cereal straws
	Stovers
	Bagasse
	Organic municipal solid wastes

Contrarily to MSR, the ESR overall reaction is irreversible since at least one of the reaction steps that ultimately and ideally leads to its conversion into 6 moles of H_2 is irreversible. Regarding the reaction pathways involved in ESR, a consensus has not been reached due to the influence of reaction conditions and a considerable variety of catalysts. Still, there has been some agreement towards the reaction network depicted in Figure 7 [17,19,86].

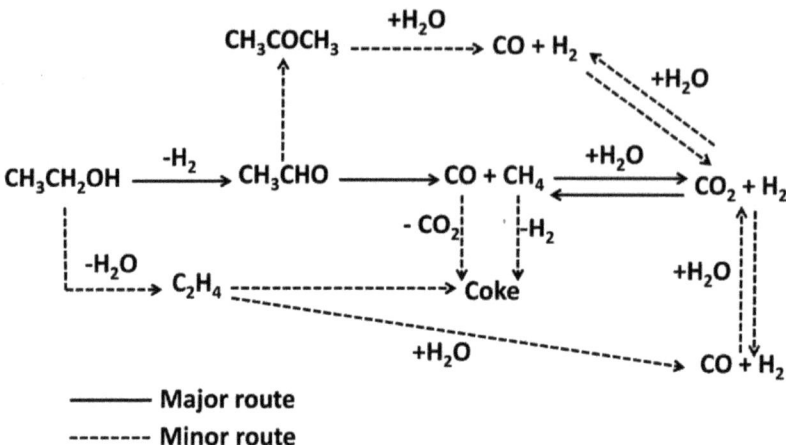

Figure 7. Reaction network proposed for ESR. Reprinted with permission from Ref. [17]. Copyright 2022 Elsevier.

The reactions that have been considered as major and minor routes are presented by order in Table 3. The final major step is the WGS reaction (Equation (3)). Depending on the catalyst that is used, major routes might become minor and vice-versa. Therefore, H_2 production and coke formation in ESR are highly dependent on the catalysts used. Moreover, minimizing CO production is once again crucial if the H_2 that is produced is directed to a polymer electrolyte fuel cell. This will also depend significantly on the choice of an appropriate catalyst. In fact, due to the higher number of possible alternative routes, this choice might be even more crucial here than it was for MSR.

Table 3. Reaction routes considered for the ESR in Figure 7.

Type	Reaction	$\Delta H_r^{298\,K}$ (kJ·mol^{-1})	Eq. Number
Major	$C_2H_5OH \rightarrow CH_3CHO + H_2$ (ethanol dehydrogenation to acetaldehyde)	68	(8)
	$CH_3CHO \rightarrow CH_4 + CO$ (acetaldehyde decomposition)	-	(9)
	$CH_4 + H_2O \rightleftharpoons CO + 3H_2$ (methane steam reforming; the reverse of Equation (4))	206	(10)
Minor	$C_2H_5OH \rightarrow C_2H_4 + H_2O$ (ethanol dehydration to ethylene)	45	(11)
	$C_2H_4 \rightarrow C$ (ethylene polymerization to coke)	-	(12)
	$2CH_3CHO \rightarrow C_3H_6O + CO + H_2$ (acetaldehyde condensation into acetone and subsequent decarboxylation)	-	(13)
	$2CO \rightleftharpoons CO_2 + C$ (Boudouard reaction)	−172	(14)
	$CO + H_2 \rightleftharpoons H_2O + C$ (reduction of carbon monoxide)	−131	(15)
	$CH_4 \rightleftharpoons 2H_2 + C$ (methane cracking)	75	(16)

3.2. Catalysts

On contrary to the MSR, ESR requires catalysts that are active in C-C bond cleavage. Hou et al. [17] observed that Rh-based catalysts are the most active in breaking such bonds. On the other hand, and as already mentioned, nickel-based catalysts are normally cheaper and, therefore, might constitute a potential alternative. Furthermore, these Ni-based materials normally also show good C-C bond breaking activity and are very active in CH_4 reforming (they are used industrially for this reaction). However, challenges related to coke formation and sintering have been identified [17]. In this section, a review of the latest Ni-based catalysts reported in the literature is carried out and a thorough analysis of the best strategies to improve their catalytic activity, selectivity and stability are conducted.

3.2.1. Catalytic Activity and H_2 Selectivity

Once again, the nature of the support that is used is vital for the catalytic activity in ESR and selectivity towards H_2 production in detriment of secondary products such as intermediate liquids, CO, CH_4, and coke. Aiming to better understand this, Dan et al. [87] compared the catalytic performance of monometallic Ni catalysts supported on Al_2O_3 and ZrO_2 and analyzed the effect of modifying both supports with CeO_2 and La_2O_3. The Al_2O_3-supported catalyst presented better catalytic activity than the ZrO_2-supported one. The authors reported that this difference could be mainly due to the higher number of active sites, resulting from the higher surface area, higher Ni dispersion and smaller nanoparticle size. In terms of ethanol conversion, while the addition of both CeO_2 and La_2O_3 to ZrO_2 led to significant improvements, for Al_2O_3 only the modification with La_2O_3 improved the conversion. The authors claim that since these modifications do not improve the intrinsic catalytic activity of the Ni sites (based on turnover frequency calculations), the improved results could be due to the participation of the support in the catalysis of ESR and/or due to the higher amount of Ni active sites available. In terms of H_2 production, while the modifications of the Al_2O_3 support did not result in significant improvements, the addition of Ce and especially La to the ZrO_2 support improved the H_2 yield considerably. Besides the mentioned enhancement of Ni dispersion, La_2O_3 is also known to increase the basicity of the modified catalysts [88]. As for CeO_2, it enhances (besides the Ni dispersion) the reducibility of Ni [89] and the capacity of the support to adsorb water [90]. It has been reported for a bimetallic Ni-Cu catalyst that the reduction ability of CeO_2 promoted the intermediate formation of acetone [91]. Overall, the catalyst supported on La_2O_3-modified ZrO_2 provided the best results with total ethanol conversion and H_2 yield of approximately 60% at 350 °C [87].

As already observed for MSR, the number of modifying agents is also very important. Even though modifying the Al_2O_3 support with La resulted in a slight increment in terms of both ethanol conversion and H_2 production [87], it has been observed in other work [92] that the addition of La_2O_3 to the support could be either beneficial or detrimental depending on the La/Al molar ratio. In this last work, while La/Al ratios in the range of 0.05–0.15 improved the production of H_2 compared to the non-modified catalyst, a La/Al molar ratio of 0.2 was detrimental. Such behavior could be explained by the two opposite trends observed in Figure 8: the Ni surface area and ethanol adsorption capacity. Even though the catalyst with a La/Al ratio of 0.20 presented the highest Ni surface area, it also showed the lowest ethanol adsorption capacity. Although the basicity of La is known to be beneficial to counterbalance the acidity of Al_2O_3, as it reduces the formation of ethylene, too much basicity will negatively affect the dehydrogenation of ethanol [93], thus decreasing H_2 production. For this reason, this catalyst showed not only the lowest H_2 yield, but was also the only catalyst to be unable to fully convert ethanol. The catalyst with a La/Al molar ratio of 0.1 showed the best catalytic performance.

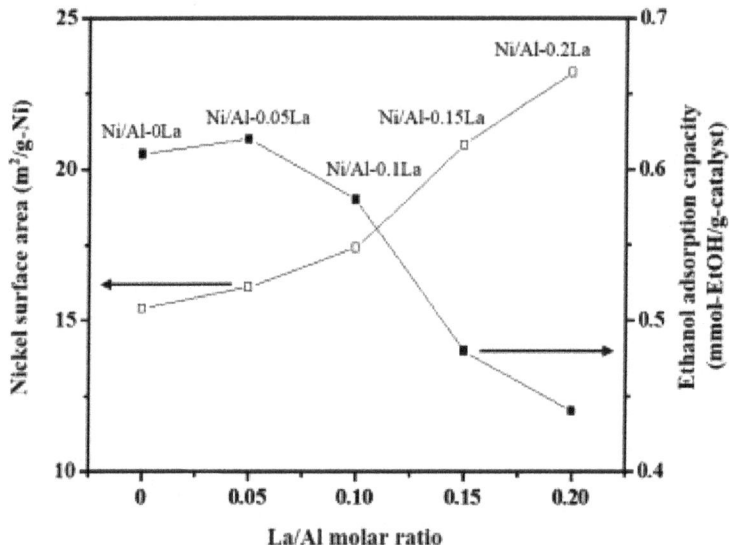

Figure 8. Nickel surface area and ethanol adsorption capacity as a function of the La/Al molar ratio in Ni/xLa$_2$O$_3$-Al$_2$O$_3$ (x = 0, 0.5, 0.1, 0.15 and 0.2) catalysts. Reprinted with permission from Ref. [92]. Copyright 2022 Elsevier.

Trane-Restrup et al. [90] compared the performance of monometallic Ni supported on CeO$_2$, MgAl$_2$O$_4$, Ce$_{0.6}$Zr$_{0.4}$O$_2$, and CeZrO$_4$/MgAl$_2$O$_4$. The catalysts based on the last two supports presented higher catalytic activity, most probably due to their higher water dissociation capacity [94,95]. This would allow higher surface concentrations of OH-species, which would react with carbon species to form H$_2$ and carbon oxides [90]. It has been proposed that reactions with lattice oxygen on these kinds of supports could also contribute to the enhanced activity of the Ni catalysts over these supports [96]. The particle size of Ni obtained over the different supports could not be responsible for the results obtained as the Ce$_{0.6}$Zr$_{0.4}$O$_2$, with the largest Ni particles, presented the highest conversion [90].

Prasongthum et al. [97] prepared a Ni-based catalyst supported on graphene (Ni/CNT-SF) with total ethanol conversion at 300–550 °C. The authors reported that the utilization of this support enhanced the performance of the catalyst due to the tubular structure of the graphene.

In a different work, the effect of modifying Al$_2$O$_3$ support with ZnO was analyzed [98]. While for both Ni/Al$_2$O$_3$ and Ni/ZnAl$_2$O$_4$ small amounts of acetaldehyde and ethylene were formed, indicating the occurrence of both ethanol dehydrogenation and dehydration, over the Ni/ZnO-Al$_2$O$_3$ only H$_2$, CO and CO$_2$ and a small amount of CH$_4$ were observed. This latter case could be a result of the formation of a NiZn alloy with different compositions (NiZn and Ni$_4$Zn) on this catalyst, while on the other two catalysts only metallic Ni was present.

The modification of TiO$_2$ support with montmorillonite (MMT) was also evaluated [99] and an enhancement in the performance during ESR was observed. This modification improved the crystal growth control and produced anatase phase of delaminated MMT/TiO$_2$ nanocomposite. Also, the formation of a Ni-MMT phase at the surface enhanced both Ni reducibility and dispersion, the last having amplified the ability of Ni to break C-C bonds. Elsewhere [100], it was concluded that organically modifying MMT with cetyltrimethylammonium bromide resulted in higher surface area and pore volume and higher Ni dispersion combined with smaller metal particles.

Musso et al. [101] prepared several Ni-based catalysts supported on La and Y via the sol-gel technique. Samples containing the Y element showed higher catalytic performance and lower production of sub-products. One of these materials (containing Y) also showed no catalyst deactivation even in a 50-h long-term experiment at 650 °C. The higher catalytic performance of this material could be related to its structural properties (namely the higher number of oxygen vacancies and better metal-support interaction). Besides that, several Ni-based catalysts supported on ceria with different Ni loadings (10, 13, and 15 wt.%) were prepared by Niazi et al. [102] for the ESR process. The results showed that the Ni content has a positive impact on the catalytic activity of the catalysts (due to the higher number of active sites available). It was verified that, in a general way, the H_2 and CO yields increased with the temperature using all of the prepared materials. In opposition, the production of CH_4 and CO_2 decreased as the temperature increased.

Two works [103,104] have reported monometallic Ni supported on zeolites with considerable ESR activity. In the first case [103], a comparison between hierarchical and non-hierarchical beta zeolite supported catalysts showed that the first presented superior ethanol conversions and H_2 yields over time. Such superiority of the hierarchical beta zeolite is probably associated with the presence of intra-crystalline mesoporous channels, which confine the well dispersed Ni particles, thus improving mass transfer efficiency. Moreover, catalysts having different loadings of Ni (5–15 wt.%) over the hierarchical beta zeolite were tested under the ESR process, being that a loading of 15 wt.% allowed the attainment of approximately complete ethanol conversion and almost 80% of H_2 yield. This was attributed to the strong metal-support interaction and high active Ni surface area. In the second work [104], Ni was impregnated on non-modified and dealuminated BEA zeolite. Even though Ni supported on dealuminated BEA zeolite showed initially lower H_2 selectivity than Ni supported on non-modified zeolite (full conversion in both cases), the second deactivated rapidly while the first maintained the initial H_2 production. Nevertheless, the first catalyst led to higher CO production, which could be a problem in the case of a fuel cell application. On the other hand, it produced less acetaldehyde for the most of the reaction time and significantly less ethylene, being the last species normally an indicator of coke formation.

The effect of different promoting agents on Ni-based catalysts in terms of ESR activity and H_2 selectivity has been widely researched and, in the last years, several interesting findings have been reported. Using Mg to promote Ni-based catalyst was analyzed by Chen et al. [105] and Song et al. [106], being that both report that Mg addition promotes ESR activity towards H_2 production. In the first work [105], the Mg promoter, added to a Ni supported over attapulgite clay catalyst, anchored the Ni species on the support surface, leading to highly dispersed metallic Ni, high metal surface area and the lowest crystal size, ultimately resulting in improved ethanol consumption comparatively to the unpromoted material. In fact, it was observed elsewhere that the increase in the active Ni surface area was accompanied by enhanced ethanol adsorption capacity [106]. Furthermore, modification with Mg also increased H_2 production, as it benefited dehydrogenation of ethanol to form acetaldehyde, and subsequently H_2, in detriment of ethanol dehydration to form ethylene. This is associated with the lowered acidity of the Mg promoted catalysts. The influence of the Mg loading was also analyzed [105], having been concluded that loading of 10 wt.% provided the highest catalytic activity with around 94% of ethanol conversion and H_2 yield of 85% at 500 °C. A Mg loading of 20 wt.%, however, presented similar catalytic activity to the unpromoted catalyst at the same temperature This could be once again associated with the excess of basicity mentioned earlier [93].

A comparison of the Mg promoter with other metal promoters (Ba, Ca and Sr) [106] was also carried out. As can be inferred from Figure 9, all promoted catalysts presented higher H_2 production than the unpromoted material and in the following order: Sr > Mg > Ba > Ca. Furthermore, by establishing a relationship between the catalysts' H_2 yields and active Ni surface areas and ethanol adsorption capacities, the authors concluded that higher H_2 productions occurred over the catalysts with higher Ni surface areas and ethanol adsorption capacities, contrary to the study carried out by Song et al. [92]. In other words, the promotion of the Ni/Al_2O_3-ZrO_2 catalyst with Sr resulted in the highest increase in the surface area of the Ni active phase and ethanol adsorption capacity. The same authors analyzed elsewhere the impact of Sr loading over the same catalyst [107]. Different Sr loadings (0, 2, 4, 6, 8, and 10 wt.%) were used, having been observed that even though complete ethanol conversion was attained, regardless of the Sr loading, the highest H_2 yield was obtained over the catalyst with 6 wt.% of Sr, which presented the highest Ni surface area, lowest particle size, highest dispersion and highest ethanol adsorption capacity. It is also argued that an excess amount of Sr covered the catalyst surface, thus harming H_2 production.

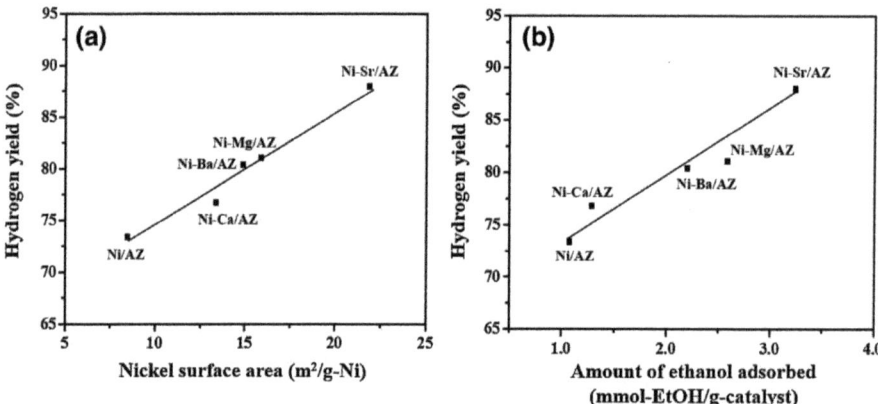

Figure 9. Relationship between the H_2 yield obtained over Ni-X/Al_2O_3-ZrO_2 (X = Ba, Ca, Mg and Sr) and their (**a**) nickel surface area and (**b**) ethanol adsorption capacity. Reprinted with permission from Ref. [106]. Copyright 2022 Elsevier.

A comparison between the promoting effect of K, CeO_2 and ZrO_2 has been established [90]. Enhancement of ethanol conversion was observed for all cases compared to the unpromoted Ni/$MgAl_2O_4$ catalyst, especially when simultaneous modification with K and CeO_2 was carried out. This enhancement in ethanol conversion could be attributed mainly to K modification, as the modification with only CeO_2 did not result in significant enhancement, contrary to what happened when modification with only K was carried out. On the other hand, CeO_2 had a significant role in inhibiting coke formation. Elsewhere [108] it was observed that Ce addition to a Ni/MMT catalyst increased both ethanol conversion and H_2 selectivity by 15% and 24%, respectively. While Ni is responsible for breaking the C-H and C-C bonds of ethanol, CeO_2, with its oxygen vacancies, activates H_2O to produce OH groups, which on their side can react with intermediate products to yield CO_2 and H_2. In fact, the authors observed that increasing the Ce content improved the CO_2/CO ratio, which confirms that CeO_2 benefited the WGS reaction. On the other hand, in a previous work [90], it was observed that when the promotion of the catalyst with $CeZrO_4$ was carried out, higher ethanol conversion was observed. In a different work [109], the addition of Zr to a bimetallic Ni-Co catalyst supported on ordered mesoporous carbon enhanced the ESR catalytic activity in terms of both ethanol conversion and H_2 yield. This was due to the lower crystal size, higher dispersion of the Ni-Co active phase and higher specific surface area of the promoted material.

As for the effect of Co addition, Nejat et al. [110] observed that Co addition benefited both ethanol conversion and H_2 production, regardless of the Co content in the range 1–9 wt.% (constant total metal loading of 10 wt.%). The Co-promoted catalysts showed lower CO and CH_4 production as well. Very similar results were attained elsewhere [111]. Even though neither of these works included a clear explanation of why promotion with Co resulted in enhanced ESR catalytic activity and H_2 selectivity, it has been reported that besides having the ability to break C-C bonds [112], Co also promotes the WGS reaction [113,114]. However, too much Co is detrimental to the desired ESR activity [110] and so a balance between Ni and Co contents is required. The 9 wt.% Ni-1 wt.% Co/MCM-41 catalyst converted 90% of ethanol and produced a H_2 yield of 80%, the best performance among the tested materials in this study [110].

Noble metals have also been combined with Ni aiming to optimize the catalytic activity in ESR towards H_2 production. Combining both Ni and Rh in the active phase has often been observed to be beneficial for ESR activity towards H_2 production [115–117]. The authors in [115] claim that the presence of Rh promotes dehydrogenation of ethanol to acetaldehyde in detriment of ethanol dehydration, as the presence of small quantities of aldehyde was observed for the bimetallic catalysts, but not for the Ni monometallic catalyst. On the other hand, the Rh-promoted catalysts showed higher CH_4 production than the monometallic Ni catalyst, probably due to the higher capacity of hydrogenation of CH_3 species (that were produced during C-C cleavage). However, at higher temperatures (above 500 °C) this trend changed, and the bimetallic catalyst started showing lower CH_4 generation than the monometallic Ni catalyst. This is in agreement with the common report that the promotion of Ni in the active phase with Rh enhances the steam reforming of CH_4 [115–117]. Elsewhere [116] the enhancing effect of Rh promotion was attributed to the easier reduction and smaller crystallite size of Ni on a bimetallic catalyst supported on CeO_2-ZrO_2. Le Valant et al. [117] compared the promoting effect of different metals on a Rh-based catalyst. It was observed that while promotion with Pt had no significant effect on the H_2 yield, both Pd and Ni increased it relatively to the unpromoted catalyst, especially Ni. Furthermore, the Rh-Ni bimetallic catalyst showed higher CO and lower CH_4 production, thus indicating the promotion of CH_4 reforming. Bimetallic catalysts in which Ni was combined with Pt [118,119] and Au [120] have also been studied in the ESR reaction. While significant amounts of acetaldehyde were observed over monometallic Ni catalyst supported on detonation nanodiamond, the same compound was not observed over the bimetallic Pt-Ni on the same support [118]. Furthermore, higher production of H_2 and CO was observed over the bimetallic catalyst than over the monometallic Ni catalyst. It is proposed that the bimetallic catalyst exhibits a synergistic effect and that it could be associated with the influence of both the electronic structure of the catalyst surface and to the sorption properties of the catalyst [119]. The addition of Au to the Ni/SBA-15 catalyst also enhanced the ESR activity [120]. This has been once again attributed to the improved dispersion of the NiO/Ni phase, which resulted in smaller particles, and strengthened NiO/Ni-support interaction. Promotion with B also enhanced ESR activity towards H_2 production due to the formation of a Ni-B alloy [86]. Finally, W was also found to improve catalytic activity due to a synergism between W and Ni and the enhanced steam reforming of methane activity by WO_x [121].

Similar to what was observed for the MSR process, LDH-derived materials containing Ni have been tested in ESR [122–125]. Romero et al. [122] analyzed the effect of changing the Mg content in a Ni-Mg-Al mixed oxide, having found that the Mg/Ni ratio of 0.33 allowed maximum H_2 production and minimum amounts of ethylene and acetaldehyde. On the other hand, more CO and lower ethanol conversion were obtained compared to the materials with a lower Mg/Ni ratio. Such dependence on the Mg/Ni ratio has been attributed to the fact that it influences the interaction of Ni^0 in the oxide matrix, thus also changing the nature of the active sites. In a different work [123] Ce was impregnated into a Ni-Mg-Al LDH derived catalyst, having been determined that 10 wt.% of Ce allows the highest ethanol conversion. The catalysts with more than 10 wt.% of ceria showed lower ESR activity due to ceria particles aggregation, which probably blocked the active sites catalyzing the desired reactions. Elsewhere [125], the introduction of Cu in the LDH structure enhanced Ni reducibility on the derived mixed oxide. Furthermore, the presence of Cu enhanced the production of H_2, as it promoted WGS. A Mg-Al LDH-derived mixed oxide has also been used as support to combine the high basicity of MgO and high activity of Al_2O_3 [124]. Ni supported on this material was able to reach higher ethanol conversion and H_2 production than Ni supported on either MgO or Al_2O_3. Furthermore, Cu-Ni-Co over the same support yielded an even better catalytic performance as the presence of the three metals provided restricted formation of intermediate products and coke, thus leading to higher H_2 yields.

In a general way, the mechanism of the ESR process involves the dehydrogenation of ethanol to yield several intermediates such as ethylene and acetaldehyde [126,127]. These last species can either decompose to CH_4 and CO or transform into ethane and H_2O [127].

3.2.2. Deactivation

Even though there are several possible reasons for catalyst deactivation, coke formation and sintering are the most common ones in the ESR process, as seen previously for the MSR reaction. Regarding coke, there are more coke formation pathways that can happen during ESR than during MSR due to the more complex reaction network of the first (cf. Figure 7).

A comparison in terms of overtime stability has been established between catalysts consisting of Ni over different supports (Al_2O_3, ZrO_2, MgO and CeO_2) [124]. It was observed that while the MgO supported catalyst kept stable activity for several cycles, the other catalysts suffered from practically constant deactivation over cycles. In another work [87] it was observed that while Ni/Al_2O_3 maintained stable activity for 24 h on stream, Ni/ZrO_2 suffered from deactivation. The different stability results regarding Al_2O_3 supported monometallic Ni catalyst between both works [87,124] could be due to the different conditions employed, namely temperature and Water to Ethanol Feed Ratio (WEFR). As for the deactivation suffered by the ZrO_2 supported catalyst, it was related to the formation of large deposits of amorphous carbon, observed by TEM, which could have blocked the access to the active sites. The authors attributed such significant carbon deposition to the large Ni particles and the weak interaction between Ni and the ZrO_2 support [87]. The addition of both CeO_2 and La_2O_3 to the support induced a change in the nature of the formed carbon deposits to filamentous carbon, which, contrarily to amorphous carbon, does not normally envelop the Ni nanoparticles [128]. Therefore, despite the observed formation of coke, no significant blockage of the active sites access took place and so no substantial deactivation was observed after 24 h. Such enhancement in stability was probably a result of the change in the properties of the support's surface and better Ni dispersion. As for the Al_2O_3 supported catalyst and the corresponding catalyst similarly modified, no considerable changes were observed regarding the nature of the deposited carbon (filamentous). Elsewhere [129], Ni supported on CeO_2-MgO converted ethanol completely while showing H_2 selectivity of approximately 70% after 18 h time on stream. Such results are probably associated with the high oxygen storage capacity of the catalyst, which probably allowed the gasification of carbon deposits, thus avoiding deactivation. The presence of Ce in the support could have also promoted water dissociation into –OH and –O species, which would also contribute to coke gasification into CO, CO_2 and H_2 and, consequently, the high H_2 selectivity [130]. Tahir et al. [131] tested a Ni/MMT-TiO_2 catalyst for the steam reforming of ethanol being observed high catalytic performance at 500 °C. After the experimental tests it was determined a significant amount of carbon deposition (graphitic nature), but this had little effect on catalyst activity over the 20 h of the long-term test.

It has been observed by Song et al. [92] that adding too much La_2O_3 to Al_2O_3 can not only affect negatively the catalytic activity, as already discussed in the previous section but also the stability over time. The authors observed that the catalyst with the highest La/Al ratio of 0.20 started deactivating rapidly after 3 h on stream, contrary to what was observed for the other materials with lower La content. The reason for such behavior was not coke formation, since the amount of deposited coke (amorphous and filamentous) decreased with increasing La/Al ratio, as observed by TPO and TEM. The formation of $La_2O_2CO_3$ could have contributed to carbon removal from the catalyst surface [132,133]. Furthermore, the enhanced basicity of the La-rich catalyst could have also played an important role in inhibiting ethylene formation, a main coke precursor. On the other hand, an increase in the size of the Ni particles was observed for the catalyst with the highest La content. This suggests that sintering occurred and could have been the main reason for the observed deactivation. Re-oxidation of metallic Ni did not occur.

In a different work [98] in which a WEFR of only 3 was used, a Ni/Al$_2$O$_3$ catalyst showed deactivation once again. However, it was observed that promoting Al$_2$O$_3$ with ZnO at a Zn/Al ratio of 2 resulted in stable ESR activity for 28 h on stream at 500 °C, probably due to coke formation inhibition, as shown by thermogravimetric analysis and scanning electron microscopy analysis. The calcined Ni/ZnO-Al$_2$O$_3$ precursor is mainly composed of NiO/ZnO crystalline phases dispersed over amorphous Al$_2$O$_3$, which is transformed into a metallic phase over ZnAl$_2$O$_4$ support during reduction (Figure 10). When Zn is in excess (Zn/Al of 2), a mixture of Ni$_x$Zn$_y$ intermetallic phase, containing only small amounts of pure Ni, is observed in the metallic phase of the reduced catalyst. It is possible that this dilution of Ni metal atoms in the NiZn alloy allowed the inhibition of carbon formation, thus resulting in the mentioned increased stability. Furthermore, the alloy was stable and, consequently, not destroyed during the reaction. On the other hand, for Zn/Al of 1/2, the reduced metal consists only of Ni^0, as the limited amount of Zn is only enough to form the ZnAl$_2$O$_4$ support. For this reason, filamentous carbon was formed and resulted in deactivation (Figure 10). Sintering was not significantly observed in either case. Despite this, the catalyst containing an excess of Zn suffered from deactivation at 400 °C and the authors propose that this is due to the re-oxidation of metallic Ni by water during the ESR process.

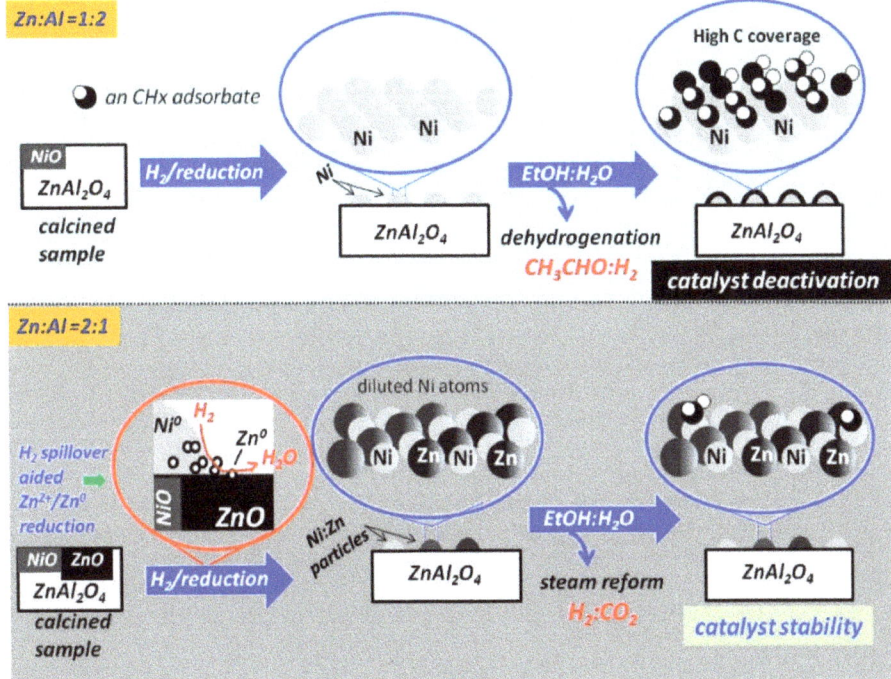

Figure 10. Phase transformation during reduction and reaction steps over Ni/ZnAl having two different Zn/Al molar ratios. Reprinted with permission from Ref. [98]. Copyright 2022 Elsevier.

Two catalysts having MMT in the support have shown promisingly stable behaviors [99,100]. A comparison between Ni/MMT-TiO$_2$ micro-particles and nano-composite at 500 °C showed that while the micro-particles-based catalyst deactivated over time, the nano-composite material did not show signs of deactivation after 20 h [99]. The enhanced stability of the last sample is probably associated with the total inhibition of ethylene formation at 500 °C. This indicates that the ethylene polymerization reaction leading to coke formation did not occur. Also, the presence of MMT in the support of the nano-composite catalyst was probably responsible for the low coke formation observed over this material, as it improved the dispersion of Ni on the surface. On the other hand, Xue-mei et al. [100] observed that Ni supported on MMT presented significant coke deposition accompanied by the formation of acetaldehyde and ethylene at 500 °C, which resulted in deactivation after 10 h of reaction. This indicates that a synergism between MMT and TiO_2 was probably what allowed the positive effect of MMT addition in the work of Mulewa et al. [99]. The modification of MMT with cetrimonium bromide in [100] resulted in enhanced catalytic stability (30 h) due to a reduction of carbon formation, accompanied by a significant reduction of acetaldehyde and ethylene formation. Such reduction was possible due to the immobilization of highly dispersed Ni on the interlayers of the organically modified MMT, which reduced the coating of metallic Ni with carbon.

Besides showing good catalytic activity, the two zeolite-supported monometallic Ni catalysts [103,104], also showed good catalytic stability. The intracrystalline mesoporous structure of the Ni/hierarchical beta zeolite was probably related to the lower deactivation rate compared to the non-hierarchical zeolite. The sol-gel method used also contributed to the observed stability [103]. As for the Ni over dealuminated BEA zeolite tested by Gac et al. [104], it showed superior over time stability than Ni supported on non-modified BEA zeolite. The deactivation of the last catalyst was mainly attributed to the formation of significant amounts of amorphous, graphitic and filamentous carbon deposits, probably associated with the high production of ethylene. On the other hand, much lower carbon deposits were formed over the catalyst whose zeolite support was dealuminated. The enhanced stability of this catalyst is probably related to the structural changes caused by dealumination, which led to enhanced dispersion of nickel nanoparticles (higher active nickel surface area), and a decrease of the acidity of the zeolite support (decreasing the coke formation).

The addition of Mg to Ni-based catalysts has also been reported to allow stable ESR activity [105,106]. Such improvement has been attributed to the suppression of both sintering and carbon deposition [105]. As for coke formation, the addition of Mg decreased the acidic strength of the catalyst, which resulted in lower ethylene formation and, consequently, less carbon deposition. Still, stable ESR activity for 50 h was only reached by the catalyst with the optimal Mg loading of 10 wt.%, which showed the strongest interaction between the Ni active phase and the support and, therefore, the lowest sintering and coke formation rates [105]. The authors attributed this behavior to the lower carbon deposition observed over the Sr promoted catalyst, probably due to its higher basicity (already discussed). As for the effect of Sr loading [107], it was observed that the unpromoted catalyst and the catalyst with the highest Sr loading of 10 wt.% slightly deactivated over a 1000 h period under 450 °C. Besides the presence of filamentous carbon, it was also observed that the 10 wt.% Sr catalyst (weakest metal-support interaction) suffered from sintering. Ultimately, the 6 wt.% Sr catalyst not only presented the best catalytic activity (as already discussed), but also remained stable for 1000 min on stream.

The effect of promoting a $Ni/MgAl_2O_4$ catalyst with K, CeO_2, ZrO_2 and combinations of these promoters (CeO_2-K, $CeZrO_4$, and $K/CeZrO_4$) on the catalytic stability has been studied [90]. As can be seen in Figure 11, the unpromoted catalyst (base) showed rapid deactivation over time. The Ce promoted catalyst showed similar behavior, but with a quicker deactivation during the first hour of reaction and subsequent deactivation (contrary to what was verified in previous works for other catalysts). On the other hand, promotion with K, CeO_2-K, $CeZrO_4$ and $K/CeZrO_4$ resulted in stable activity.

In fact, carbon deposition was lowered by factors of two to four over these promoted catalysts. It is possible that K blocked the active sites for carbon formation, while the redox-active promoters could have contributed to higher OH availability and/or provided lattice oxygen to react with coke precursors.

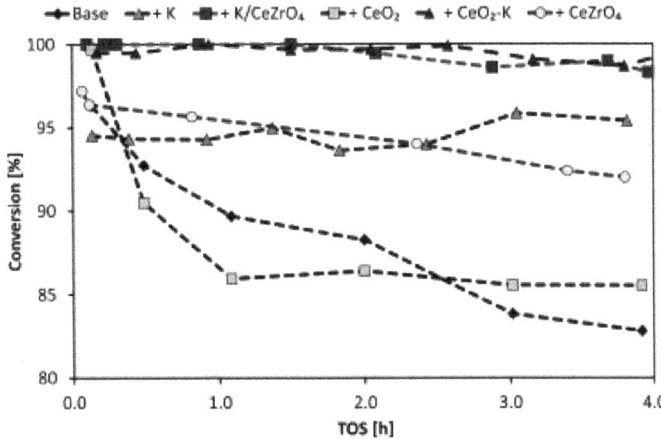

Figure 11. Ethanol conversion as a function of time on stream (TOS) for Ni/MgAl$_2$O$_4$ (base) with different promoters. Reprinted with permission from Ref. [90]. Copyright 2022 Elsevier.

Elsewhere [108], promotion of a Ni-based catalyst with Ce resulted in significantly improved stability over 50 h time on stream compared to the unpromoted material. As already mentioned, such enhancement was related to the suppression of both coke formation and sintering. On one side the high oxygen storage and transport capacity of Ce facilitates carbon species gasification and, on the other side, Ce strengthens the metal-support interaction, thus restraining the growth of Ni particles. Trane-Restrup et al. [90] have shown the formation of carbon whiskers. Such carbonaceous structures have been reported to break catalyst pellets with time, thus leading to increasing pressure drops on the catalyst bed and, consequently, local hot spots in industrial reactors [134]. Moreover, it was observed that Ni sintering contributed to the deactivation of both Ni/CeO$_2$-K/MgAl$_2$O$_4$ and Ni-K/MgAl$_2$O$_4$ catalysts. Finally, the authors tested the promotion with different amounts of sulfur of Ni-CeO$_2$/MgAl$_2$O$_4$, the addition of S having decreased coke deposition. Probably sulfur blocked the step sites on the Ni particles, responsible for carbon deposition. The sulfur loading of 0.03 wt.% allowed the minimum formation of carbon deposits [90]. The positive effect of promotion with Zr addressed by Trane-Restrup et al. [90] was also observed in other work [109], having once again been observed enhanced stability. The authors attributed this enhancement to sintering inhibition, higher dispersion of the Ni-Co active phase on the support and the formation of carbon nanotubes that were unable to deactivate the catalyst. Regarding this last aspect, the catalyst without Zr showed amorphous coke formation which, on contrary to the carbon nanotubes, resulted in ESR active sites deactivation.

The promotion of Ni with Co in the active phase has also been reported to improve the catalytic stability [110]. The authors tested three catalysts for 8 h under ESR reaction, two monometallic Ni and Co over MCM-41 and bimetallic Ni-Co over the same support. The Ni catalyst presented the highest coke formation among the three, followed by the Co catalyst and the bimetallic catalyst presented the lowest carbon deposition. This indicates that Co partially hinders coke formation. Elsewhere [111], a bimetallic Ni-Co/Al$_2$O$_3$ catalyst showed quite stable behavior for 100 h on stream. This improvement, partially caused by the promotion of Ni with Co in the active phase, has been associated with higher surface area, higher metal dispersion and lower particle size. Such enhancements were also caused by specific preparation conditions [111].

Besides improving the catalytic activity of the ESR reaction, promotion with Rh also allows higher long-term stability [115,116]. Campos et al. [115] reported a decrease in coke formation observed over a bimetallic Ni-Rh catalyst of 70 and 560 times relative to monometallic Rh and Ni catalysts, respectively. The authors attributed the enhanced coke resistance of the bimetallic catalyst to i) synergism between Ni and Rh catalytic activities; ii) favored WGS reaction; iii) improved capacity to gasify methyl groups resultant from the decomposition of intermediate products; and iv) no accumulation of CO and/or acetate species. The Rh-Ni/CeO$_2$-La$_2$O$_3$-Al$_2$O$_3$ catalyst was able to still convert more than 80% of ethanol after 144 h on stream at 500 °C. As already discussed for other catalysts, the addition of Rh was also reported to lead to the preferential formation of less encapsulating amorphous coke (which does not deactivate active sites), contrarily to monometallic Ni [116]. Au addition also suppressed, for the reason already mentioned in the previous section, both carbon deposition over Ni particles and their sintering, resulting in stable activity for 25 h on stream [120]. Both W [121] and B [86] promoted Ni catalysts have also shown remarkable stability compared to the respective monometallic Ni catalysts. In fact, the W promoted catalyst maintained an H$_2$ yield of around 80% for 80 h time on stream, to which the inhibition of sintering of Ni particles contributed [121]. As for the effect of B [86], the formation of a Ni-B alloy lowers coke deposition by enhancing the cracking of acetaldehyde and, consequently, avoiding the formation of acetone (coke precursor). Also, the simultaneous presence of Ce in the mixed support led to the formation of CeBO$_3$, which assisted in the removal of carbonaceous deposits. As a result, ethanol conversion was still around 96% after 50 h of reaction.

Finally, Romero et al. [122] observed that the Mg content in a Ni-Mg-Al mixed oxide influenced its stability, besides its catalytic activity (already discussed). More specifically, as shown by the TPO profiles of the used catalysts in Figure 12, higher Mg contents led to lower coke formation (lower gasification to CO$_2$). It has been claimed elsewhere [135,136] that higher Mg content in LDHs results in enhanced oxygen mobility and water adsorption-dissociation capacity, ultimately leading to improved carbon resistance. The presence of Cu in the LDH has also been reported to improve ESR catalytic stability by mitigating the deposition of carbon [125].

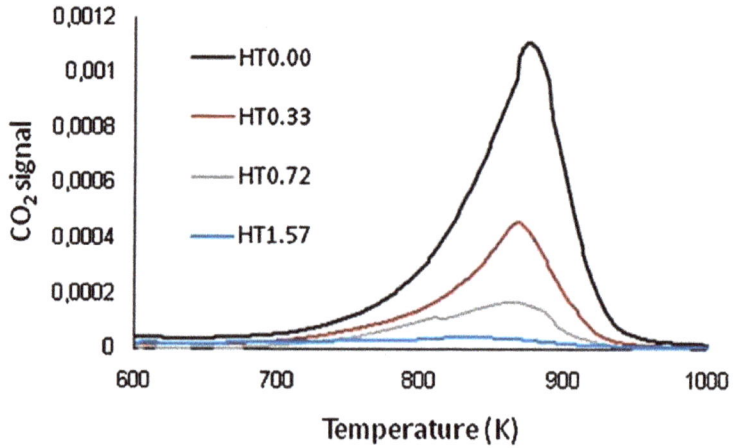

Figure 12. TPO profiles of LDH-derived mixed oxides with different Mg/Ni molar ratios (number in each sample) used in ESR. Reprinted with permission from Ref. [122]. Copyright 2022 Elsevier.

3.2.3. Effect of the Preparation Method

As previously seen for the MSR process, the method through which a catalyst is prepared can have a determining effect on its catalytic activity and stability. The same has been consistently observed for the ESR reaction. Once again, the impregnation method (wet impregnation and incipient wetness impregnation) has been the most commonly used for the preparation of ESR catalysts [87,90–92,98–100,106,107,109–111]. Zhao et al. [111] established a comparison between co-impregnation and sequential impregnation of bimetallic Ni and Co over Al$_2$O$_3$ in terms of ESR performance. As observed for MSR, co-impregnation once again led to higher ethanol conversion and H_2 selectivity (above 350 °C) than sequentially impregnated materials. At 350 °C, the co-impregnated material converted 68.7% of ethanol, while the $Co/Ni/Al_2O_3$ and $Ni/Co/Al_2O_3$ showed conversions of only 50.9% and 36.6%, respectively. Furthermore, the co-impregnated catalyst showed the lowest production of CO and CH_4. These results were mainly due to the higher metal dispersion, lower metal particle size and higher surface area of the co-impregnated material. For the same reasons, while the co-impregnated catalyst showed relatively stable activity for 100 h, both Co/Ni and Ni/Co started deactivating significantly after 60 h and 30 h, respectively. The authors also propose that the co-impregnated material might have benefited from H_2 spillover, which would have been responsible for the observed decrease in carbon deposits, possibly gasified into CH_4 [137].

Elsewhere [138], Ni/SBA-15 was prepared via incipient wetness impregnation using two different Ni precursors: nickel nitrate (commonly used) and nickel citrate. The citrate precursor strengthened the metal-support interaction and improved the dispersion of the smaller nickel particles. Furthermore, analysis of the spent catalysts after 25 h of ESR showed that not only less coke was formed on the catalyst synthesized with the citrate precursor, but it was also easily removable (contrarily to the catalyst prepared with nitrate precursor). This ultimately resulted in higher ESR activity and stability of the catalyst prepared using nickel citrate as a precursor.

The use of chelating agents in the wet impregnation method has also been tested [139]. The authors prepared four different Ni/CeO_2-$MgAl_2O_4$ catalysts, one without a chelating agent (conventional wet impregnation) and three others using different chelating agents: ethylenediaminetetraacetic acid, nitrilotriacetic acid and citric acid. The catalyst prepared with ethylenediaminetetraacetic acid as a chelating agent showed the best catalytic activity among the prepared samples. While the other three catalysts showed production of acetaldehyde and ethylene and a small decay in conversion after 7 h of reaction, the catalyst prepared using ethylenediaminetetraacetic acid did not show either of those. The results for the last catalyst were due to smaller NiO particles, which induced stronger Ni-CeO_2 interaction, and a higher Ce^{3+}/C^{4+} ratio, which means a higher capacity to store and release oxygen, thus leading to the observed higher resistance to coke formation [140]. The other chelating agents also improved such properties comparatively to the catalyst prepared through conventional impregnation, especially nitrilotriacetic acid, but to a lower extent. Contrarily to other methods or method modifications already discussed, this methodology keeps the simplicity of the conventional wet impregnation method. It also allows high synthesis reproducibility [141].

Wu et al. [86] compared the impregnation with the co-precipitation method in terms of their influence on ESR activity and stability. As can be observed in Figure 13, the catalysts prepared via co-precipitation presented higher ethanol conversions and higher stability than the respective B-promoted or unpromoted catalysts prepared through impregnation. First, the materials prepared through co-precipitation presented a higher surface area than the respective impregnated counterparts. This difference was higher between the boron promoted catalysts. The very low surface area obtained for the B promoted catalyst prepared via impregnation was caused by the calcination at 400 °C, which resulted in pore blocking. Despite showing the second higher ethanol conversion, this catalyst showed the lowest H_2 selectivity. Furthermore, co-impregnation improved the dispersion of NiO as a result of allowing Ni^{2+} to interpolate into the $Ce_{0.5}Zr_{0.5}O_2$ solid solution.

A sol-gel iso-volumetric impregnation method was used to prepare Ni/MBeta zeolite [103]. Comparatively to the iso-volumetric impregnation, used to prepare the same catalyst, the sol-gel method improved the same parameters previously discussed (smaller metal particles and improved dispersion), thus resulting in stronger metal-support interaction. This resulted in higher carbon deposition resistance, as the highly dispersed small metal particles make it more difficult for carbon deposits to accumulate on the metal surface and embed it. Therefore, higher conversion of ethanol into H_2 and more stable activity were, once again, obtained via the sol-gel-based method compared to the impregnation-based one.

In a different work, Wu and co-workers [142] analyzed the effect of using different agents (HNO$_3$ or NH$_4$OH) at different ratios with tetraethyl silicate (0.04 or 0.20), which changed the acidity of the preparation solution, to prepare different Ni/SiO$_2$ catalysts. Essentially, changing the preparation solution's pH did not significantly influence the production of H$_2$, but it changed coke formation. The higher basicity of the solution resulted in larger SiO$_2$ particles and lower surface area and porosity and, consequently, higher coke formation.

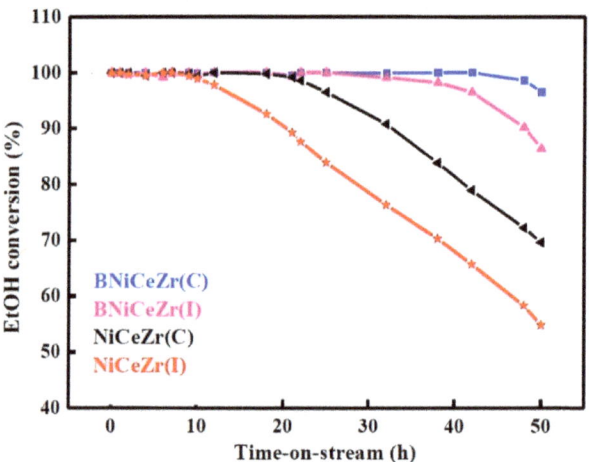

Figure 13. Influence of preparation method of Ni/Ce0.5Zr0.5O2 (C, co-precipitation; I, impregnation) on the catalytic activity of boron promoted and unpromoted catalysts. Reprinted with permission from Ref. [86]. Copyright 2022 Elsevier.

The calcination conditions, such as temperature [143] and atmosphere [142], have also been the target of analysis in terms of impact on ESR activity and catalysts' stability. Nichele et al. [143] prepared Ni/TiO$_2$ catalysts calcined at 500 and 800 °C, being that calcination at 800 °C led to more stable behavior due to stronger metal-support interaction, which ultimately contributed to avoiding sintering. Elsewhere [142], the effect of calcining a Ni/SiO$_2$ catalyst under N$_2$ or air on the catalytic activity and stability in ESR was assessed. The results indicate that calcination under N$_2$ atmosphere resulted in higher H$_2$ yield and lower coke formation comparatively to preparation under air. Calcination under N$_2$ led to the formation of both Ni and NiO phases, while only NiO was formed when air was used. Furthermore, the first showed higher Ni dispersion (smaller Ni particles), probably the main cause of the enhanced activity towards H$_2$ production.

3.3. Summary

A summary of the ESR catalysts reviewed in this section that showed the most promising results is presented in Table 4. These three catalysts showed ethanol conversions of 100%, as well as quite high H$_2$ yields. In these studies, it was verified that the utilization of adequate promoters and supports could be very beneficial for the catalyst performance. For instance, in the work of Song et al. [92], it was observed that the addition of La$_2$O$_3$ in a Ni/Al$_2$O$_3$ catalyst (La/Al between 0.05–0.15) improved significantly the catalytic performance of the material. Moreover, the utilization of graphene and zeolite as supports has shown to be a good solution to achieve high catalytic activities; concerning the study that used zeolite [103], the high performance of the material was attributed to the strong metal-support interaction and high active Ni surface area. In the work of Prasongthum et al. [97], the unique characteristics of the graphene (electron cloud) accelerate the rate of carbon gasification and help the regeneration of the active Ni surface.

Table 4. Comparison of some of the most promising ESR catalysts reviewed in this section.

Catalyst	Temperature	Feed Flow Rate	Mass of Catalyst	S/C [a]	Conversion of Ethanol	H_2 Yield/Selectivity	Stability	Preparation Method	Refs.
15 wt.% Ni/zeolite	550 °C	0.05 mL·min^{-1}	0.1 g	6	≈100%	76% [b]/n.d.	Result after 27 h (no significant deactivation)	Sol-gel + Impregnation	[103]
10 wt.% Ni/CNTs-SF	450 °C	8 g$_{cat}$·h·mol^{-1}	n.d.	9	≈100%	40% [b]/n.d.	Result after 22 h (no significant deactivation)	Sol-gel + Impregnation	[97]
Ni/Al-0.1La	450 °C	23,140 mL·h^{-1} g$_{cat}$$^{-1}$ [c]	0.1 g	6	100%	124% [b]/n.d.	Result after 15 h (no significant deactivation)	Epoxide-initiated sol-gel + Impregnation	[92]
15 wt.% Ni/Y-ZrO$_2$	650 °C	41,000 h^{-1} [c]	0.1 g	4.5	100%	91% [b]/74% [d]	Result after 8 h (no significant deactivation)	Sol-gel	[101]
13 wt.% Ni-4 wt.% Cu/CeO$_2$	600 °C	20,00 mL·h^{-1} g$_{cat}$$^{-1}$ [c]	0.3 g	6	≈100%	n.d./70 [d]	Result after 20 h (no significant deactivation)	Impregnation	[102]

[a] Steam to carbon molar feed ratio. [b] Yield (%) = $\frac{H_2}{H_{2_{Max}}} \times 100$. [c] GHSV: gas hourly space velocity. [d] Selectivity (%) = $\frac{H_2}{\sum F_i} \times 100$. n.d.—Not determined

Besides that, the impregnation method appears to be suitable to impregnate Ni on the support surface, although several studies indicate that co-precipitation increases the catalytic performance of the materials. Finally, the reduction methods adopted in these studies were performed at temperatures very similar to those used during the ESR reaction.

In the previous section, it was also verified that the most important catalyst properties to achieve proper catalytic performance (in terms of catalytic activity and stability) for the ESR process are the number of active sites available and the ability to gasify coke deposits. In this way, to reach high ethanol conversion, high H_2 production and low sub-products production, it is necessary to prepare a catalyst with a proper content of active phase (well dispersed) and promoted with basic oxides (e.g., MgO and CeO_2—these phases inhibited the coke production).

Taking into account the main results shown and discussed in Section 3.2 (and presented in Table 4), it is suggested the preparation of a Ni-based catalyst (for instance, prepared by impregnation) promoted with (or supported on) CeO_2 for the ESR process, to obtain high and stable catalytic performances.

4. Other Oxygenates Steam Reforming

4.1. Introduction

Many recent studies are focused on the sustainable production of fuels and renewable energy sources, to decrease the dependency on fossil fuels and reduce the emissions of CO_2. One of the attractive alternatives for sustainable production of fuels is biomass, which is a renewable and CO_2-neutral emission fuel source [144]. Biomass conversion into different types of fuels can be performed through distinct processes, namely: biological, mechanical, or thermal processes [145–148]. For instance, pyrolysis is the thermal decomposition of the biomass (in the absence of O_2) that produces charcoal, fuel gas and bio-oil. Three different modes of pyrolysis can be differentiated: fast, intermediate, and slow [146,149,150]. The fast pyrolysis, which occurs at high heating rates, @ 450–600 °C and <2 s of residence time, provides a higher yield in bio-oil, while the charcoal and fuel gas can be used to produce heat for the process itself [146]. Compared with biomass, bio-oil has a much higher energy density (ten times higher). Consequently, it is much more suitable for transportation [22,151]. The composition of bio-oil is highly variable depending on several factors such as residence time, heating rate and temperature of the process, and composition of the biomass source itself [152]. A typical composition of bio-oil consists of many different oxygenated species such as alcohols, acids, ketones, phenols, etc. [153–155]. In Table 4 is possible to see an example of bio-oil composition, produced through flash pyrolysis of two different types of biomass: a mix of 85% of pine and 15% of spruce [22].

Another possible attractive alternative of sustainable fuel composed of distinct oxygenated compounds is the olive mill wastewater (OMW, a polluting stream generated from the olive oil producing systems) since the composition of this stream (mainly polyphenols, carbohydrates, fatty acids and water) is very similar to the composition of the bio-oil. Several studies have shown that OMW disposal/discharge causes large environmental impacts due to its high content of organic matter and pollutant load [156–158]. The most referenced compounds are the following: vanillic acid, caffeic acid, tyrosol, p-coumaric acid, cinnamic acid, d-arabinose, d-galactose, d-galacturonic acid, syringic acid, gallic acid, protocatechuic acid, phenol, acetic acid, phenethyl alcohol, guaiacol and benzyl alcohol [159–176]. However, the composition of OMW is highly variable and it suggests that such compounds can be present in major or minor proportion, depending on several factors which include the maturation level of the olive, the region of cultivation, the age of the olive tree, the treatment of the tree, the method of extracting the oil and the weather conditions that the olive was been subjected to in the ripening process [164].

In addition to the polluting effluents already mentioned (bio-oil and OMW), there are other streams with a very similar composition and potential such as palm oil mill effluent (POME) and glycerol. POME is a polluting stream generated from the palm oil producing system, constituted by several oxygenated molecules [177], causing environmental impacts identical to OMW. The composition is also highly variable due to the reasons already mentioned for the composition of the OMW. About the glycerol, it is possible to verify that this oxygenate compound is the main by-product of biodiesel production (100 kg of glycerol/ton of biodiesel), without economic value so far.

These streams could be used directly as combustion fuels, but their poor volatility, high viscosity and coking formation result in problems for equipment. However, experimental studies of oxygenates steam reforming (OSR) showed that this technology is viable to produce hydrogen [178–187]. Additionally, several thermodynamic studies for the steam reforming of several oxygenates have already been performed, which also demonstrate the potential of this technology for the production of hydrogen [152,188–191], though only from the theoretical (thermodynamic) point of view. This technology would enable the production of green H_2, while reducing the pollutant load of these oxygenated streams. The produced biofuel is environmentally attractive since it is renewable [192]. The OSR is described by the overall reaction shown in Equation (17):

$$C_xH_yO_z + (2x - z) H_2O \rightarrow \left(2x - z + \frac{y}{2}\right)H_2 + xCO_2 \quad \left(\Delta H_r^{298 \text{ K}} => 0 \text{ kJ·mol}^{-1}\right) \quad (17)$$

The OSR can be divided into two major reactions: oxygenates decomposition (Equation (18)) to yield syngas (mixture of hydrogen and carbon monoxide) followed by the WGS reaction (Equation (3)).

$$C_xH_yO_z + (x - z)H_2O \rightarrow \left(x - z + \frac{y}{2}\right) H_2 + xCO \quad \left(\Delta H_r^{298 \text{ K}} => 0 \text{ kJ·mol}^{-1}\right) \quad (18)$$

As it was aforementioned for MSR and ESR processes, there are also secondary reactions associated with this process that form some undesired by-products (e.g., methane, coke). For instance, the reactions represented by Equations (4)–(6). Therefore, choosing a proper catalyst, able to maximize conversion and selectivity for H_2 formation, is also required for this reaction.

Several catalysts were extensively studied for the steam reforming of individual model compounds that are present in the bio-oil, OMW, or POME. At the moment, the main challenge is to prepare highly reducible and with high oxygen mobility redox catalysts for OSR. These catalysts must present high performance with high stability. Numerous catalysts have been extensively studied for the steam reforming of individual compounds, which are the main species present in the pollutant effluents considered in this section. The molecules included in this group are acetic acid, phenol and toluene. The performance of such catalysts is affected by the type of support and the promoter agent(s); these topics are discussed in the next sub-section, but only for catalysts with Ni as the active phase. Besides that, some materials already developed for the steam reforming of bio-oil, OMW, glycerol or POME were also discussed.

4.2. Nickel-Based Catalysts

4.2.1. Catalytic Activity and H_2 Selectivity

The Ni-based catalysts are the most used in the steam reforming of oxygenates since they are effective, commercially available and relatively cheap [193–197]. Metal loading is an important parameter in the study of such catalysts' performance. In the study of Zhang et al. [198] (focused on acetic acid steam reforming), it was concluded that the increase of Ni loading from 10 wt.% to higher values (in a Ni/Al_2O_3 catalyst prepared by a wetness impregnation method) did not significantly increase the activity of the catalysts when a steam to carbon feed ratio of two or five was used. However, high content of Ni in the catalyst did prevent coke formation, promoting its stability (due to the higher number of active sites available). In another study of acetic acid steam reforming, Borges et al. [199] verified that the higher is the Ni loading (in LHD-like precursors of Ni-Mg-Al prepared through co-precipitation), the lower will be the temperature and the time necessary for the reduction, since the interactions between Ni and Mg-Al oxides are weaker (TPR peaks shifted to lower temperatures)—c.f. Figure 14.

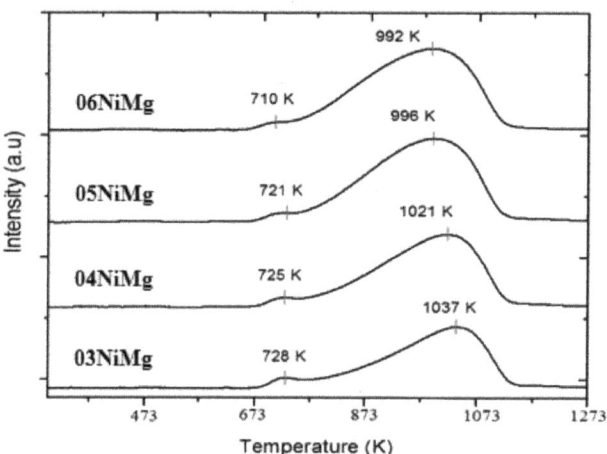

Figure 14. TPR analyses of the calcined hydrotalcite-like precursors (xNiMg) with different Ni loading (x is the molar ratio of Ni^{2+}/Mg^{2+}. Reprinted with permission from Ref. [199]. Copyright 2022 Elsevier.

In several studies, it was found that different supports significantly influenced the catalyst performance in the reforming reaction. Ni-based catalysts supported on various types of Al_2O_3 with different crystalline phases for the steam reforming of acetic acid were prepared and tested by Chen et al. [196]. The crystalline phases of Al_2O_3 support influence the intensity of the interaction between these supports and Ni and, consequently, the formation of metallic Ni. Since the surface of α-Al_2O_3 was mainly formed with bulk NiO, more metallic Ni was present on the Ni/α-Al_2O_3 catalyst after the reduction treatment. In this way, this catalyst presents a higher catalytic activity, as the metallic Ni content caused higher C-C and C-H bonds breaking capability.

In the work of He et al. [200] (toluene steam reforming), it was possible to verify that the high Ni dispersion in Ni/γ-Al_2O_3 increases toluene conversion and H_2 yield. On the other hand, a series of Ni core-shell catalysts with various shell species (i.e., SiO_2, Al_2O_3, CeO_2 and TiO_2) were prepared by Pu et al. [201]—see Figure 15. By comparing the catalytic activities of the catalysts with various shell materials, it was concluded that the improved Ni/Al_2O_3-i catalyst (the nickel precursor was reduced by $NaBH_4$) was the most suitable for the steam reforming of acetic acid, showing much higher catalytic activity than the other materials (Figure 16).

The nature of the support was also studied by Zhang et al. [202], using attapulgite (ATTP) and Al_2O_3 as support of Ni-based catalysts. ATTP has a lower specific surface area and lower thermal stability than Al_2O_3. The low thermal stability negatively affects the catalytic performance for steam reforming of acetic acid. However, the interaction between the Ni species and ATTP is much weaker than that of Ni with Al_2O_3, and, as a consequence, at the low Ni loading, the Ni/ATTP catalyst showed better performance. Chen et al. [203] studied the effect of using biochar as support in a Ni catalyst for the steam reforming of acetic acid. The catalyst characterization showed that after activation, the porosity of the biochar enlarged significantly, and both the surface area and the dispersions of Ni particles increased, increasing the catalytic activity. Kechagiopoulos et al. [204] used a Ni-based catalyst supported in natural material (Ni/olivine) in a spouted bed reactor; it was observed that the catalyst presented a high catalytic activity for the steam reforming of representative model species of the aqueous phase of bio-oil. Besides, Liu et al. [205] stated that the porosity of the support could promote the catalytic performance in bio-oil steam reforming, by comparing the results obtained from a Ni-Mo catalyst supported on natural sepiolite and acidified sepiolite. The authors verified that the acidified support showed higher catalytic performance than the non-acidified sepiolite since the acid treatment changed the internal structure of the support to produce a higher number of pores, leading to an increase of the surface area of the material.

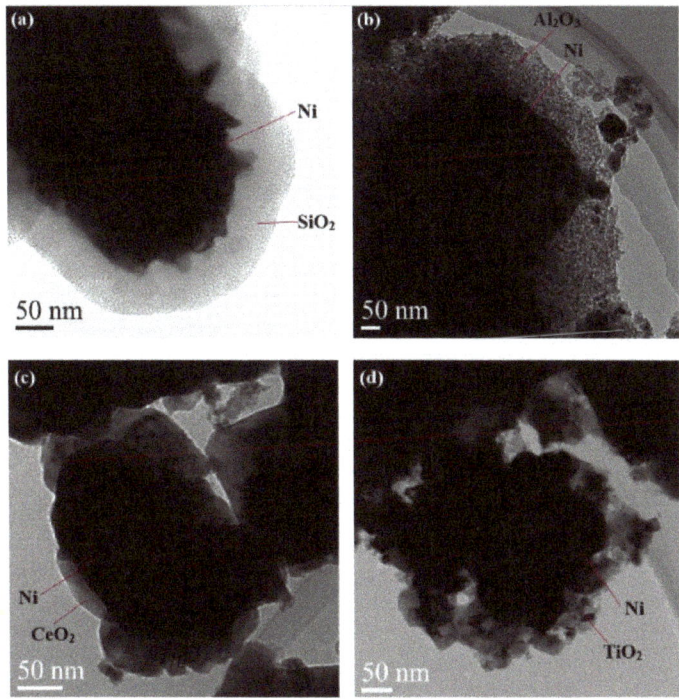

Figure 15. TEM images of the core-shell catalysts reduced at 600 °C: (**a**) Ni/iO$_2$, (**b**) Ni/Al$_2$O$_3$, (**c**) Ni/CeO$_2$, and (**d**) Ni/TiO$_2$. Reprinted with permission from Ref. [201]. Copyright 2022 Elsevier.

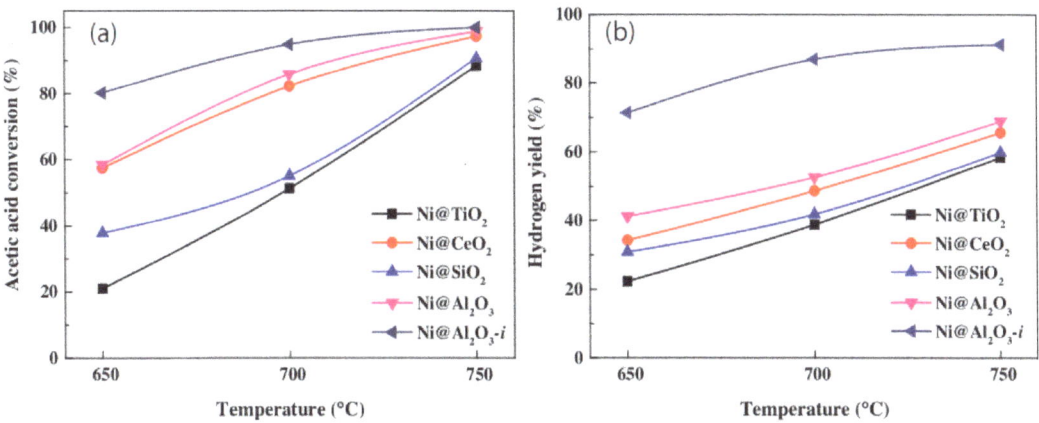

Figure 16. (**a**) Conversions of acetic acid and (**b**) H$_2$ yields in the steam reforming of acetic acid over different catalysts. Reaction conditions: 1 bar, WHSV = 21 h^{-1}, and steam to carbon molar feed ratio = 3.18. Reprinted with permission from Ref. [201]. Copyright 2022 Elsevier.

In several studies dealing with the steam reforming of long-chain oxygenates, the catalysts were often doped with promoters. Galdamez et al. [24], who addressed the steam reforming of bio-oil, concluded that the addition of La_2O_3 in the Ni-based catalyst does not increase the H_2 yield. Nevertheless, Garcia et al. [206] observed that the promotion of a Ni/Al_2O_3 catalyst with MgO and La_2O_3 enhanced the steam adsorption that facilitated the gasification of surface carbon (decreasing coke formation) also during bio-oil steam reforming. The increase in the performance of a Ni/Al_2O_3 catalyst caused by the incorporation of these two promoters was also reported by Bangala et al. [207] (though in naphthalene steam reforming). In the work of Zhang et al. [208] (acetic acid steam reforming), it was verified that the addition of KOH to Ni/Al_2O_3, with the lowest Ni loading, significantly enhances the catalytic activity. Choi et al. [209] studied the steam reforming of acetic acid by using Ni-based catalysts modified by Mg, La, Cu, and K elements; they found that the Ni/Al_2O_3 catalyst modified with Mg showed the best performance at low temperatures.

In the work of Charisiou et al. [210] (steam reforming of glycerol), it was studied the catalytic activity and stability of a Ni-based catalyst promoted with Y and Zr. It was concluded that the addition of Y stabilized the ZrO_2 phase and, for this reason, the utilization of these promoters enhanced the production of H_2 and increased the stability of the catalyst.

Souza et al. [211] prepared Ni-Pt monometallic and bimetallic materials supported on ZrO_2 for the acetic acid steam reforming. The addition of Pt to Ni catalysts caused an increase in the metallic dispersion and a decrease in the nickel reduction temperature; however, the catalytic performance was not improved by Pt addition. Ni monometallic catalyst presented the best catalytic behavior: 100% of conversion and 30% of H_2 yield at 500 °C, without any deactivation during 30 h on stream. The high activity and stability of Ni/ZrO_2 catalyst may be related to its high reduction degree, increasing the availability of metal sites, and its low acidity, reducing coke formation. The fast deactivation of Pt/ZrO_2 catalyst is associated with its highest rate of coke production.

Baamran et al. [212] prepared several Ni-based catalysts supported on TiO_2 for the steam reforming of phenol. As mentioned in previous works, the superior performance of the materials was related to the larger surface area, higher metal dispersion and no internal diffusion inside the pores (these properties were obtained due to the small particle size). The best catalyst (10 wt.% Ni/TiO_2) attained a 98.3% of phenol conversion, 76.9% of H_2 yield and high stability for more than 70 h. Besides that, another work of the research group of Baamran et al. [213] reported a synergistic effect between the TiO_2 and $ZnTiO_3$ phases in Ni-based materials, enhancing the Ni dispersion and, in this way, increasing the catalyst activity and stability. Moreover, Abbas et al. [214] prepared Ni/Co_3O_4-supported TiO_2 catalysts for the phenol steam reforming with continuous H_2 production. Using a feed rate of 10 mL/h, temperature equal to 700 °C, and 0.3 g of catalyst loading, a H_2 yield of 83.5%, a selectivity of 72.8%, and a phenol conversion of 92% were obtained. High stability in terms of production of H_2 after 100 h of reaction was obtained with the best material (no deactivation was verified). This stability was attributed to the strong interaction of the metal-support (improving the metal dispersion and enhancing the reducibility), minimizing the coke formation.

It was reported that the incorporation of Cr in a $Ni/MgO-La_2O_3-Al_2O_3$ catalyst inhibited the formation of Ni_3C [207], modified the metal sites forming alloys with Ni, and reduced the crystallite size (enhancing the crystallinity of the material), increasing the catalyst activity. Besides that, Bangala et al. [207] observed that the presence of TiO_2 decreases the conversion since it reduces the crystallinity and robustness of the material and destroys the Al_2O_3 matrix. It was also observed that the addition of CeO_2 improved the catalyst performance, namely through the inhibition of coke production due to the enhanced catalyst redox properties [195,197,215–218]. So, the addition of basic oxides (e.g., MgO, CeO_2, La_2O_3, etc.) or other promoters to the Al_2O_3 support, in a general way, enhanced the steam reforming catalytic performance.

Besides the monometallic Ni-based catalysts, other elements such as Pt, Co, Rh, Cu, and Fe were used along with Ni in bimetallic catalysts for the steam reforming of oxygenated molecules. A series of Ni-Cu bimetallic catalysts supported on sepiolite (Nix-Cuy/SEP) was prepared by Liang et al. [219] for the steam reforming of phenol. The results showed that the Ni-Cu alloys were successfully synthesized, and the addition of Cu decreased the Ni particle size, improving the redox ability and metal dispersion of bimetallic catalysts (in this case, Cu can be also considered as a promoter of the catalyst). Pant et al. [220] reported that Ni-Co, Ni-Co/CeO_2-ZrO_2, and Ni/La_2O_3-Al_2O_3 catalysts catalyze the acetic acid steam reforming reaction, being that among them, the Ni-Co catalyst was more effective. In this specific case, the unsupported catalyst presents a higher performance due to the combined action of the Ni and Co (Co catalyzes the WGS reaction [221,222]). Besides that, since Al_2O_3 allows the formation of a high quantity of coke, the unsupported catalyst presented higher stability. Mizuno et al. [223] studied the steam reforming of acetic acid over Ni-Co supported in $MgAl_2O_4$. The ketonization reaction occurred on the $MgAl_2O_4$ support and the presence of Co or Ni changed the reaction pathway of adsorbed species, which suppressed the formation of acetone.

Rocha et al. [224,225] tested one commercial catalyst and several prepared materials based on Ni in the OMW steam reforming process (using a synthetic OMW effluent). A catalytic screening was carried out with these materials (at 1 bar and 350/400 °C), and stability tests (at 1 bar and 400 °C) were performed with the most promising samples. The authors reported that the LDH-based catalysts and the Ni-Ru/SiO_2 catalyst prepared in the laboratory showed good catalytic properties (the last one with high deactivation resistance in the long-term test of 24 h) due to a high number of active sites available and high surface area. Using the commercial catalyst (Ni/Al_2O_3-SiO_2), this research group [224] observed that the H_2 and CO_2 production was very high during all of the screening experiments. Still, the CH_4 yield was very close to zero and CO was not detected during the experimental test.

Adhikari et al. [226] compared the catalytic activity of several Ni-based catalysts on different supports for the steam reforming of glycerol: MgO, CeO_2 and TiO_2. At 600 °C, it was found the following order of H_2 selectivity: CeO_2 > MgO > TiO_2. Pant et al. [227] observed that the incorporation of CeO_2 in the Ni-based catalysts affects the reduction of Ni species, enhancing the catalyst activity for the stem reforming of glycerol. In a general way, it was reported that the addition of a CeO_2 promoter enhances the catalytic performance of the Ni-based catalysts for steam reforming processes [228].

For the steam reforming of POME [186], a Ni/Al_2O_3 catalyst was demonstrated to be a good candidate for syngas production and reduction of the organic load of the pollutant stream.

Finally, mixed oxides such as LHD-type oxides (Ni-Mg-Al oxides) have been reported as promising catalyst precursors for the steam reforming of these types of oxygenates since they present high surface areas in comparison with other catalysts [199,229].

In the steam reforming of these oxygenates, it was possible to observe the formation of several surface intermediates. For instance, it was verified the formation of lactic acid, acetaldehyde, propyleneglycol, ethylene glycol, methanol, acetic acid, acetone, hydroxyacetone, acrolein and ethanol during the steam reforming of glycerol [230–232].

4.2.2. Deactivation

The formation of carbon deposits on the catalysts used in the OSR leads to catalyst deactivation (as verified in Sections 2.2.2 and 3.2.2). Therefore, to achieve continuous performance and sustainable H_2 production, it is necessary to study the deactivation of such catalysts.

In the study of Zhang et al. [198], it was verified that the increase of Ni content, from 10 wt.% to higher values, does not significantly affect the catalytic activity of the catalysts. Still, it enhanced the stability, and especially the resistance towards coking in the steam reforming of acetic acid. While the coke formed over the catalyst with the lower Ni loading was mostly amorphous, the coke formed over higher Ni loading was more fibrous [198]. In another work, Zhang et al. [208] also verified that the addition of KOH to Ni/Al$_2$O$_3$ with the low Ni loading not only significantly enhances the catalytic activity but also promotes gasification of the reactive intermediates such as methyl group, carbonyl group, etc. The effect of the support in the stability tests was studied by Zhang et al. [202] (also in acetic acid steam reforming), using attapulgite (ATTP) and Al$_2$O$_3$ as support of a Ni-based catalyst. The stability of the Ni/Al$_2$O$_3$ catalyst was higher than that of Ni/ATTP, due to the higher surface area of Al$_2$O$_3$, and to the nature of the coke formed on the surface of the Ni/Al$_2$O$_3$ catalyst (fibrous) instead of the coke formed on Ni/ATTP (amorphous). Hoang et al. [195] showed that Ni/HT gradually deactivates with the time-on-stream due to coke formation (competitive adsorption in the active sites)—see Figure 17.

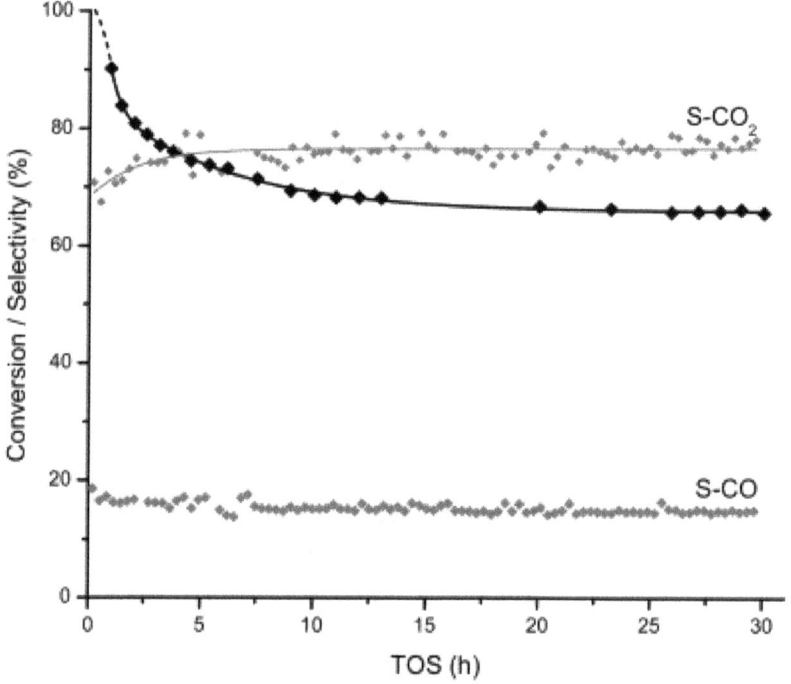

Figure 17. Steam reforming of acetic acid in presence of Ni/HT catalyst with steam to carbon ratio of 5, T = 700 °C, WHSV= 25.2 h^{-1}. Reprinted with permission from Ref. [195]. Copyright 2022 Elsevier.

In the work of Savuto et al. [197] (who studied the steam reforming of tar model compounds), mayenite (Ca$_{12}$Al$_{14}$O$_{33}$) was used as support of Ni, showing excellent oxidation properties that increase the resistance of the catalyst towards coke deposition. On the other hand, Choi et al. [209] verified that a Ni/Al$_2$O$_3$ catalyst with a large amount of weak basic sites and few middle and strong basic sites is required to improve the catalytic performance and minimize coke formation (in acetic acid steam reforming).

It was also reported that, for several Ni-based catalysts, La$_2$O$_3$ led to a decrease in coke formation [194,206,207,217,233,234] and KOH and MgO improved the stability [194,206,207,234–236]. It was also observed in several studies using Ni-based catalysts that the addition of CeO$_2$ improved the catalyst stability, namely through the decrease of coke production [195,197,215–218]. So, the addition of basic oxides to the Al$_2$O$_3$ support, in a general way, enhanced the catalytic performance. The main role of these oxides is to enhance the redox properties of the material [237,238], which increases the oxidation of surface carbon and the stability of the catalyst.

In the study of Rocha et al. [225], it was observed that the increase of the CO production was related to the decreased extent of the WGS reaction (catalyst deactivation), decreasing the production of H_2. Besides that, it was concluded that the production of amorphous coke is the main route for the deactivation of the catalysts, as shown in multiple studies on the steam reforming processes.

Sánchez et al. [239] studied the utilization of a Ni-based catalyst for the steam reforming of glycerol. This work reported a catalyst deactivation after 8 h of a long-term test (at 600 and 650 °C) caused by coke deposition. However, Wen et al. [240] verified that Ni-based supported catalysts suffer quick deactivation for the glycerol steam reforming, but due to the sintering of the Ni particles during the reaction.

Finally, a brief reference to structured mesoporous silicate materials that were reported to be Ni supports less susceptible to deactivation due to coke formation than the conventional microporous supports. Apart from that, it is well known that they also cause less resistance to the diffusion of reactants to the active sites [241].

4.2.3. Effect of the Preparation Method

The preparation method (including the precursors of the species) used is also a very important parameter to be considered since it can affect the activity and stability of the catalysts. For instance, in a recent study by Yu et al. [242], the effect of the type of Ni precursors (used in the incipient wetness impregnation method) on the catalytic behaviors of Ni/Al_2O_3 during steam reforming of acetic acid was studied. It was found that the type of anion in the nickel precursors affects the properties and performance of the Ni/Al_2O_3 catalysts. $NiSO_4/Al_2O_3$ and $Ni(NH_2SO_3)_2/Al_2O_3$ did not show good catalytic activity, while $Ni(NO_3)_2/Al_2O_3$ and $Ni(CH_3COO)_2/Al_2O_3$ showed good and similar activity for the conversion of acetic acid. Nevertheless, $Ni(CH_3COO)_2/Al_2O_3$ was more stable and had a higher resistance toward coke production. Among the nickel precursors investigated, $Ni(CH_3COO)_2$ was more suitable as the CH_3COO^- anion.

Metal salts precursors have significant effects on the properties and catalytic performance of the final catalysts, which should be considered in the preparation of Ni-based catalysts. This topic is not very studied for the steam reforming long-chain oxygenates or even model compounds (apart from methanol and ethanol), and almost all Ni-based catalysts assessed for these reactions are prepared by the traditional impregnation method (see Table 5).

Table 5. Example of the composition of bio-oil [22].

Component	[wt.%]
Water	20–23
Acids	3–22
Sugars	4–9
Phenols	3–4
Lignin	2–18
PAH [a]	8 [ppm]
Others [b]	2–21

[a] Poly aromatic hydrocarbons. [b] Ketones, aldehydes, and alcohols.

4.3. Summary

A summary of the most promising OSR catalysts reviewed in this section is shown in Table 5. From those eleven catalysts, seven are monometallic and the others are bimetallic. Regarding the support, most of the supports are Al_2O_3-based. Besides that, it is worth mentioning that almost all of the eleven catalysts presented in Table 5 were prepared by impregnation. All of these catalysts showed conversions close to 100% and high H_2 yields (always higher than 57%) and selectivities (always higher than 54%). In addition, several catalysts do not present any deactivation, even over long-term experimental tests (30 h).

Analyzing the results of oxygenates conversion, only the catalysts Ni/ABC (ABC—Activated Biochar) and 6.6 wt.% Ni–10 wt.% Fe/$(CeO_2)_{0.4}$-$PG^{0.6}$ (PG, palygorskite) did not convert completely the compounds fed [203,216]. Regarding the H_2 production, it was observed that the monometallic catalysts 15 wt.% Ni/α-Al_2O_3 and 3.5 wt.% Ni/5 wt.% La_2O_3-ZrO_2 demonstrated higher performance (H_2 yield of 90%) in comparison with the remaining materials [194,196]. It is also important to emphasize that two catalysts prepared by co-precipitation [216,233] did not demonstrate loss of activity (during 20 h of operation) through the stability tests (6.6 wt.% Ni–10 wt.% Fe/$(CeO_2)_{0.4}$-$PG_{0.6}$ and 10 wt.% Ni/La_2O_3-Al_2O_3). These results show how attractive Ni-based catalysts are for OSR.

Finally, it was also observed that the most crucial catalyst properties to reach a high catalytic activity/stability for the OSR processes are the number of active sites available (related to metal dispersion), the surface area and the basicity of the material. In this way, to reach high oxygenates conversion, high production of H_2 and low sub-products production, it is necessary to prepare a catalyst promoted with basic oxides, which presents high metal dispersion (defined by the strong interaction between the active phase and the support), and with high surface area (utilization of a proper porous support).

Taking into account the works discussed in Section 4.2 (and observing the data summarized in Table 6), to obtain high catalytic performances during long-term tests in the steam reforming of oxygenates it is suggested the preparation of a Ni-based catalyst by impregnation, supported on Al_2O_3, and promoted with a metal oxide (e.g., MgO or CeO_2—to inhibit the coke production).

Table 6. Comparison of some of the most promising OSR catalysts reviewed in this section.

Catalyst	Temperature	Feed Flow Rate	Mass of Catalyst	S/C [a]	Conversion	H_2 Yield/Selectivity	Stability	Preparation Method	Refs.
3.5 wt.% Ni/5 wt.% La_2O_3-ZrO_2 [1]	700 °C	240,000 h^{-1} [b]	0.050 g	5	100	87 [c]/64 [d]	Lost 7% of H_2 yield in 20 h	Impregnation	[194]
(2.5+2.5) wt.% Ni-Cu/Al_2O_3 [1]	750 °C	28 h^{-1} [e]	0.1 g	1.25	100	67 [c]/57 [d]	Result of 7.5 h (no deactivation)	Impregnation over support prepared by following evaporation-induced-self-assembly	[243]
15 wt.% Ni/α-Al_2O_3 [1]	600 °C	20 h^{-1} [e]	1.5 g	1	≈100	90 [c]/66 [d]	n.d.	Impregnation	[196]
10 wt.% Ni/La_2O_3-Al_2O_3 [1] (La_2O_3/Al_2O_3 = 1:3, weight ratio)	700 °C	10 h^{-1} [f]	0.2 g	2.5	100	73 [c]/59 [d]	Result of 30 h (no deactivation)	Co-precipitation	[233]
15 wt.% Ni/Al_2O_3 [1]	700 °C	7400–10,000 h^{-1} [b]	0.2 g	1	100	57 [c]/54 [d]	n.d.	Incipient wetness impregnation	[244]
5.5 wt.% Cu–2.5 wt.% Ni/$MgAl_2O_4$ [1]	450 °C	9 h^{-1} [e]	0.1 mg	2	100	83 [c]/63 [d]	n.d.	Impregnation	[223]
(6.6+10) wt.% Ni-Fe/$(CeO_2)_{0.4}$-$PG_{0.6}$ [1]	600 °C	14,427 h^{-1} [f]	n.d.	3	≈93	85 [c]/63 [d]	Result of 20 h (no deactivation)	Co-precipitation	[216]
$Ni(NO_3)_2$/Al_2O_3 [1]	600 °C	12.7 h^{-1} [f]	0.5 g	5	≈100	77 [c]/71 [d]	n.d.	Incipient wetness impregnation	[242]
Ni/ABC [1]	700 °C	10 h^{-1} [f]	0.15 g	2.5	91.2	71 [c]/61 [d]	n.d.	Impregnation	[203]
10 wt.% Ni/ATTP [1]	600 °C	7.2 h^{-1} [f]	0.5 g	5	≈100	75 [c]/65 [d]	n.d.	Impregnation	[202]
(10+1) wt.% Ni-Ru/SiO_2 [2]	400 °C	n.d.	0.65 g	694	≈100	84 [c]/68 [d]	Lost 10% of H_2 yield in 20 h	Co-precipitation	[225]

[a] Steam to Carbon Molar Feed Ratio. [b] GHSV: gas hourly space velocity. [c] Yield (%) = $\frac{H_2}{H_{2\,Max.}} \times 100$. [d] Selectivity (%) = $\frac{H_2}{H_2+CH_4+CO_2+CO} \times 100$. [e] WHSV: weight hourly space velocity. [f] LHSV: liquid hourly space velocity. [1] Steam reforming of acetic acid. [2] Steam reforming of OMW. n.d.—Not determined.

5. Conclusions

In the last years, the production of H_2 through the catalytic steam reforming of biomass-derived oxygenates was the target of several studies and such a process has been demonstrated to be a very attractive alternative for green H_2 production. In this way, and to improve the H_2 production, the selection of an appropriate catalyst is fundamental. Nickel-based materials have been widely studied due to their lower price and their performance as compared to noble metal-based materials. This review addressed steam reforming of different biomass-derived oxygenates, namely methanol, ethanol, and other oxygenates (bio-oil, acetic acid, OMW, etc.).

Most of the promising catalysts for the MSR are bimetallic, Ni and Cu-based. The materials demonstrated methanol conversions of around 90%, as well as high H_2 yields and selectivities. The materials with higher Cu content and lower amounts of Ni tend to be more selective towards the production of H_2 and CO_2 (in detriment of CO). The impregnation method for the preparations of these catalysts appears to be an appropriate choice and the annealing/reduction programs adopted were often carried out at temperatures very similar to those used during the MSR. Besides that, the LDH-based materials allowed to reach high methanol conversions and H_2 yields and showed very stable operation. These materials might be a potential option in a MSR process since they present a high potential to be used as hybrid sorbent-catalyst (i.e., dual-functional) materials.

Among the best catalysts for the ESR reaction, half are monometallic and the other half are bimetallic. It was concluded that the presence of CeO_2 in the catalyst composition enhanced the catalytic activity towards H_2 and stability. Several advantages resulting from promoting the catalysts with Rh were identified, despite the high cost of the noble metal (which is attenuated by using low loadings of Rh). The best catalysts used in the ESR process were prepared via simple impregnation techniques.

From the catalysts with the best performances for the OSR, six are monometallic and the others are bimetallic, and almost all of the catalysts presented Al_2O_3-based supports. Most of these catalysts were prepared by impregnation. The catalysts showed practically complete conversions of the oxygenates and moderate to high H_2 yields (>57%) and selectivities (>54%). Still, several catalysts do not present any deactivation in the stability tests, although they have not been tested for extended times on stream (<30 h).

Author Contributions: Conceptualization, J.S., M.A.S. and L.M.M.; investigation, J.S., C.R. and M.A.S.; resources, L.M.M.; writing—original draft preparation, J.S, C.R. and M.A.S.; writing—review and editing, M.A.S. and L.M.M.; supervision, M.A.S. and L.M.M.; funding acquisition, M.A.S. and L.M.M. All authors have read and agreed to the published version of the manuscript.

Funding: This work was financially supported by: LA/P/0045/2020 (ALiCE), UIDB/50020/2020 and UIDP/50020/2020 (LSRE-LCM), and UIDB/00511/2020 and UIDP/00511/2020 (LEPABE) funded by national funds thraough FCT/MCTES (PIDDAC); and project HyGreen&LowEmissions (NORTE-01-0145-FEDER-000077), supported by Norte Portugal Regional Operational Programme (NORTE 2020), under the PORTUGAL 2020 Partnership Agreement, through the European Regional Development Fund (ERDF). This work was also financially supported by the project NORTE-01-0247-FEDER-39789, funded by European Regional Development Fund (ERDF) through the Norte Portugal Regional Operational Programme (NORTE 2020). M.A. Soria thanks the Portuguese Foundation for Science and Technology (FCT) for the financial support of his work contract through the Scientific Employment Support Program (Norma Transitória DL 57/2017).

Conflicts of Interest: The authors declare no conflict of interest.

Notation and Glossary

ABC	Activated Biochar
ATTP	Attapulgite
DFM	Dual Function Materials
ESR	Ethanol Steam Reforming
ISO	International Organization for Standardization
LDH	Layered Double Hydroxides
MSR	Methanol Steam Reforming
MMT	Montmorillonite
OMW	Olive Mill Wastewater
OSE	Oxygenates Steam Reforming
PF	Palygorskite
POME	Palm Oil Mill Effluent
TEM	Transmission Electron Microscopy
TPD	Temperature-Programmed Desorption
TPO	Temperature-Programmed Oxidation
TPR	Temperature-Programmed Reduction
WEFR	Water to Ethanol Feed Ratio
WGS	Water-Gas Shift
WHSV	Weight Hourly Space Velocity
XRD	X-Ray Diffraction

References

1. Dou, B.; Song, Y.; Wang, C.; Chen, H.; Xu, Y. Hydrogen production from catalytic steam reforming of biodiesel byproduct glycerol: Issues and challenges. *Renew. Sustain. Energy Rev.* **2014**, *30*, 950–960. [CrossRef]
2. Ayalur Chattanathan, S.; Adhikari, S.; Abdoulmoumine, N. A review on current status of hydrogen production from bio-oil. *Renew. Sustain. Energy Rev.* **2012**, *16*, 2366–2372. [CrossRef]
3. LeValley, T.L.; Richard, A.R.; Fan, M. The progress in water gas shift and steam reforming hydrogen production technologies—A review. *Int. J. Hydrog. Energy* **2014**, *39*, 16983–17000. [CrossRef]
4. Mazloomi, K.; Gomes, C. Hydrogen as an energy carrier: Prospects and challenges. *Renew. Sustain. Energy Rev.* **2012**, *16*, 3024–3033. [CrossRef]
5. Nikolaidis, P.; Poullikkas, A. A comparative overview of hydrogen production processes. *Renew. Sustain. Energy Rev.* **2017**, *67*, 597–611. [CrossRef]
6. Balat, M. Potential importance of hydrogen as a future solution to environmental and transportation problems. *Int. J. Hydrog. Energy* **2008**, *33*, 4013–4029. [CrossRef]
7. Verhelst, S.; Wallner, T. Hydrogen-fueled internal combustion engines. *Prog. Energy Combust. Sci.* **2009**, *35*, 490–527. [CrossRef]
8. Wang, G.; Yu, Y.; Liu, H.; Gong, C.; Wen, S.; Wang, X.; Tu, Z. Progress on design and development of polymer electrolyte membrane fuel cell systems for vehicle applications: A review. *Fuel Processing Technol.* **2018**, *179*, 203–228. [CrossRef]
9. Kothari, R.; Buddhi, D.; Sawhney, R.L. Comparison of environmental and economic aspects of various hydrogen production methods. *Renew. Sustain. Energy Rev.* **2008**, *12*, 553–563. [CrossRef]
10. Ewan, B.C.R.; Allen, R.W.K. A figure of merit assessment of the routes to hydrogen. *Int. J. Hydrog. Energy* **2005**, *30*, 809–819. [CrossRef]
11. Baharudin, L.; Watson, M.J. Hydrogen applications and research activities in its production routes through catalytic hydrocarbon conversion. *Rev. Chem. Eng.* **2018**, *34*, 43–72. [CrossRef]
12. Uddin, M.N.; Nageshkar, V.V.; Asmatulu, R. Improving water-splitting efficiency of water electrolysis process via highly conductive nanomaterials at lower voltages. *Energy Ecol. Environ.* **2020**, *5*, 108–117. [CrossRef]
13. Shamsul, N.S.; Kamarudin, S.K.; Rahman, N.A.; Kofli, N.T. An overview on the production of bio-methanol as potential renewable energy. *Renew. Sustain. Energy Rev.* **2014**, *33*, 578–588. [CrossRef]
14. Iulianelli, A.; Ribeirinha, P.; Mendes, A.; Basile, A. Methanol steam reforming for hydrogen generation via conventional and membrane reactors: A review. *Renew. Sustain. Energy Rev.* **2014**, *29*, 355–368. [CrossRef]
15. Xuan, J.; Leung, M.K.H.; Leung, D.Y.C.; Ni, M. A review of biomass-derived fuel processors for fuel cell systems. *Renew. Sustain. Energy Rev.* **2009**, *13*, 1301–1313. [CrossRef]
16. Baneshi, J.; Haghighi, M.; Jodeiri, N.; Abdollahifar, M.; Ajamein, H. Urea–nitrate combustion synthesis of ZrO_2 and CeO_2 doped CuO/Al_2O_3 nanocatalyst used in steam reforming of biomethanol for hydrogen production. *Ceram. Int.* **2014**, *40*, 14177–14184. [CrossRef]
17. Hou, T.; Zhang, S.; Chen, Y.; Wang, D.; Cai, W. Hydrogen production from ethanol reforming: Catalysts and reaction mechanism. *Renew. Sustain. Energy Rev.* **2015**, *44*, 132–148. [CrossRef]

18. Badwal, S.P.S.; Giddey, S.; Kulkarni, A.; Goel, J.; Basu, S. Direct ethanol fuel cells for transport and stationary applications—A comprehensive review. *Appl. Energy* **2015**, *145*, 80–103. [CrossRef]
19. Frusteri, F.; Freni, S. Bio-ethanol, a suitable fuel to produce hydrogen for a molten carbonate fuel cell. *J. Power Sources* **2007**, *173*, 200–209. [CrossRef]
20. Ni, M.; Leung, D.Y.C.; Leung, M.K.H. A review on reforming bio-ethanol for hydrogen production. *Int. J. Hydrog. Energy* **2007**, *32*, 3238–3247. [CrossRef]
21. Nabgan, W.; Tuan Abdullah, T.A.; Mat, R.; Nabgan, B.; Gambo, Y.; Ibrahim, M.; Ahmad, A.; Jalil, A.A.; Triwahyono, S.; Saeh, I. Renewable hydrogen production from bio-oil derivative via catalytic steam reforming: An overview. *Renew. Sustain. Energy Rev.* **2017**, *79*, 347–357. [CrossRef]
22. Trane, R.; Dahl, S.; Skjøth-Rasmussen, M.S.; Jensen, A.D. Catalytic steam reforming of bio-oil. *Int. J. Hydrog. Energy* **2012**, *37*, 6447–6472. [CrossRef]
23. Arregi, A.; Amutio, M.; Lopez, G.; Bilbao, J.; Olazar, M. Evaluation of thermochemical routes for hydrogen production from biomass: A review. *Energy Convers. Manag.* **2018**, *165*, 696–719. [CrossRef]
24. Galdámez, J.R.; García, L.; Bilbao, R. Hydrogen Production by Steam Reforming of Bio-Oil Using Coprecipitated Ni−Al Catalysts. Acetic Acid as a Model Compound. *Energy Fuels* **2005**, *19*, 1133–1142. [CrossRef]
25. Somerville, C.; Youngs, H.; Taylor, C.; Davis, S.C.; Long, S.P. Feedstocks for Lignocellulosic Biofuels. *Science* **2010**, *329*, 790–792. [CrossRef]
26. Sá, S.; Silva, H.; Brandão, L.; Sousa, J.M.; Mendes, A. Catalysts for methanol steam reforming—A review. *Appl. Catal. B Environ.* **2010**, *99*, 43–57. [CrossRef]
27. Li, D.; Nakagawa, Y.; Tomishige, K. Development of Ni-Based Catalysts for Steam Reforming of Tar Derived from Biomass Pyrolysis. *Chin. J. Catal.* **2012**, *33*, 583–594. [CrossRef]
28. Baddour, F.G.; Snowden-Swan, L.; Super, J.D.; Van Allsburg, K.M. Estimating Precommercial Heterogeneous Catalyst Price: A Simple Step-Based Method. *Org. Process Res. Dev.* **2018**, *22*, 1599–1605. [CrossRef]
29. Angeli, S.D.; Monteleone, G.; Giaconia, A.; Lemonidou, A.A. State-of-the-art catalysts for CH_4 steam reforming at low temperature. *Int. J. Hydrog. Energy* **2014**, *39*, 1979–1997. [CrossRef]
30. Llorca, J.; Corberán, V.C.; Divins, N.J.; Fraile, R.O.; Taboada, E. Chapter 7-Hydrogen from Bioethanol. In *Renewable Hydrogen Technologies*; Gandía, L.M., Arzamendi, G., Diéguez, P.M., Eds.; Elsevier: Amsterdam, The Netherlands, 2013; pp. 135–169. [CrossRef]
31. Bao, Z.; Yu, F. Chapter Two-Catalytic Conversion of Biogas to Syngas via Dry Reforming Process. In *Advances in Bioenergy*; Li, Y., Ge, X., Eds.; Elsevier: Amsterdam, The Netherlands, 2018; Volume 3, pp. 43–76.
32. Silva, J.M.; Soria, M.A.; Madeira, L.M. Challenges and strategies for optimization of glycerol steam reforming process. *Renew. Sustain. Energy Rev.* **2015**, *42*, 1187–1213. [CrossRef]
33. Molino, A.; Chianese, S.; Musmarra, D. Biomass gasification technology: The state of the art overview. *J. Energy Chem.* **2016**, *25*, 10–25. [CrossRef]
34. International Organization for Standardization. *Hydrogen Fuel-Product Specification-Part 2: Proton Exchange Membrane Fuel Cell Applications for Road Vehicles*; International Organization for Standardization: Geneva, Switzerland, 2012.
35. Kurtz, M.; Wilmer, H.; Genger, T.; Hinrichsen, O.; Muhler, M. Deactivation of Supported Copper Catalysts for Methanol Synthesis. *Catal. Lett.* **2003**, *86*, 77–80. [CrossRef]
36. Deshmane, V.G.; Owen, S.L.; Abrokwah, R.Y.; Kuila, D. Mesoporous nanocrystalline TiO_2 supported metal (Cu, Co, Ni, Pd, Zn, and Sn) catalysts: Effect of metal-support interactions on steam reforming of methanol. *J. Mol. Catal. A Chem.* **2015**, *408*, 202–213. [CrossRef]
37. Abrokwah, R.Y.; Deshmane, V.G.; Kuila, D. Comparative performance of M-MCM-41 (M: Cu, Co, Ni, Pd, Zn and Sn) catalysts for steam reforming of methanol. *J. Mol. Catal. A Chem.* **2016**, *425*, 10–20. [CrossRef]
38. Liu, Z.; Yao, S.; Johnston-Peck, A.; Xu, W.; Rodriguez, J.A.; Senanayake, S.D. Methanol steam reforming over Ni-CeO_2 model and powder catalysts: Pathways to high stability and selectivity for H_2/CO_2 production. *Catal. Today* **2018**, *311*, 74–80. [CrossRef]
39. Shetty, K.; Zhao, S.; Cao, W.; Siriwardane, U.; Seetala, N.V.; Kuila, D. Synthesis and characterization of non-noble nanocatalysts for hydrogen production in microreactors. *J. Power Sources* **2007**, *163*, 630–636. [CrossRef]
40. Luo, X.; Hong, Y.; Wang, F.; Hao, S.; Pang, C.; Lester, E.; Wu, T. Development of nano Ni_xMgyO solid solutions with outstanding anti-carbon deposition capability for the steam reforming of methanol. *Appl. Catal. B Environ.* **2016**, *194*, 84–97. [CrossRef]
41. Bobadilla, L.F.; Palma, S.; Ivanova, S.; Domínguez, M.I.; Romero-Sarria, F.; Centeno, M.A.; Odriozola, J.A. Steam reforming of methanol over supported Ni and Ni-Sn nanoparticles. *Int. J. Hydrog. Energy* **2013**, *38*, 6646–6656. [CrossRef]
42. Xu, C.; Koel, B.E. Influence of alloyed Sn atoms on the chemisorption properties of Ni(111) as probed by RAIRS and TPD studies of CO adsorption. *Surf. Sci.* **1995**, *327*, 38–46. [CrossRef]
43. Khzouz, M.; Gkanas, E.I.; Du, S.; Wood, J. Catalytic performance of Ni-Cu/Al_2O_3 for effective syngas production by methanol steam reforming. *Fuel* **2018**, *232*, 672–683. [CrossRef]
44. Qing, S.; Hou, X.; Liu, Y.; Wang, L.; Li, L.; Gao, Z. Catalytic performance of Cu-Ni-Al spinel for methanol steam reforming to hydrogen. *J. Fuel Chem. Technol.* **2018**, *46*, 1210–1217. [CrossRef]
45. Lytkina, A.A.; Orekhova, N.V.; Ermilova, M.M.; Belenov, S.V.; Guterman, V.E.; Efimov, M.N.; Yaroslavtsev, A.B. Bimetallic carbon nanocatalysts for methanol steam reforming in conventional and membrane reactors. *Catal. Today* **2016**, *268*, 60–67. [CrossRef]
46. Huang, Y.-H.; Wang, S.-F.; Tsai, A.-P.; Kameoka, S. Catalysts prepared from copper–nickel ferrites for the steam reforming of methanol. *J. Power Sources* **2015**, *281*, 138–145. [CrossRef]

47. Khouz, M.; Wood, J.; Pollet, B.; Bujalski, W. Characterization and activity test of commercial Ni/Al$_2$O$_3$, Cu/ZnO/Al$_2$O$_3$ and prepared Ni–Cu/Al$_2$O$_3$ catalysts for hydrogen production from methane and methanol fuels. *Int. J. Hydrog. Energy* **2013**, *38*, 1664–1675. [CrossRef]
48. Pérez-Hernández, R.; Gutiérrez-Martínez, A.; Espinosa-Pesqueira, M.E.; Estanislao, M.L.; Palacios, J. Effect of the bimetallic Ni/Cu loading on the ZrO2 support for H$_2$ production in the autothermal steam reforming of methanol. *Catal. Today* **2015**, *250*, 166–172. [CrossRef]
49. Tahay, P.; Khani, Y.; Jabari, M.; Bahadoran, F.; Safari, N. Highly porous monolith/TiO$_2$ supported Cu, Cu-Ni, Ru, and Pt catalysts in methanol steam reforming process for H$_2$ generation. *Appl. Catal. A Gen.* **2018**, *554*, 44–53. [CrossRef]
50. Yang, R.-X.; Chuang, K.-H.; Wey, M.-Y. Hydrogen production through methanol steam reforming: Effect of synthesis parameters on Ni–Cu/CaO–SiO$_2$ catalysts activity. *Int. J. Hydrog. Energy* **2014**, *39*, 19494–19501. [CrossRef]
51. Suetsuna, T.; Suenaga, S.; Fukasawa, T. Monolithic Cu–Ni-based catalyst for reforming hydrocarbon fuel sources. *Appl. Catal. A Gen.* **2004**, *276*, 275–279. [CrossRef]
52. Lorenzut, B.; Montini, T.; De Rogatis, L.; Canton, P.; Benedetti, A.; Fornasiero, P. Hydrogen production through alcohol steam reforming on Cu/ZnO-based catalysts. *Appl. Catal. B Environ.* **2011**, *101*, 397–408. [CrossRef]
53. Lytkina, A.A.; Zhilyaeva, N.A.; Ermilova, M.M.; Orekhova, N.V.; Yaroslavtsev, A.B. Influence of the support structure and composition of Ni–Cu-based catalysts on hydrogen production by methanol steam reforming. *Int. J. Hydrog. Energy* **2015**, *40*, 9677–9684. [CrossRef]
54. Duprez, D.; Pereira, P.; Miloudi, A.; Maurel, R. Steam dealkylation of aromatic hydrocarbons: II. Role of the support and kinetic pathway of oxygenated species in toluene steam dealkylation over group VIII metal catalysts. *J. Catal.* **1982**, *75*, 151–163. [CrossRef]
55. Duprez, D. Selective steam reforming of aromatic compounds on metal catalysts. *Appl. Catal. A Gen.* **1992**, *82*, 111–157. [CrossRef]
56. Takezawa, N.; Iwasa, N. Steam reforming and dehydrogenation of methanol: Difference in the catalytic functions of copper and group VIII metals. *Catal. Today* **1997**, *36*, 45–56. [CrossRef]
57. Lytkina, A.A.; Orekhova, N.V.; Ermilova, M.M.; Yaroslavtsev, A.B. The influence of the support composition and structure (MXZr1-XO2-δ) of bimetallic catalysts on the activity in methanol steam reforming. *Int. J. Hydrog. Energy* **2018**, *43*, 198–207. [CrossRef]
58. Huang, H.-K.; Chih, Y.-K.; Chen, W.-H.; Hsu, C.-Y.; Lin, K.-J.; Lin, H.-P.; Hsu, C.-H. Synthesis and regeneration of mesoporous Ni–Cu/Al$_2$O$_4$ catalyst in sub-kilogram-scale for methanol steam reforming reaction. *Int. J. Hydrog. Energy* **2021**. [CrossRef]
59. Pérez-Hernández, R. Reactivity of Pt/Ni supported on CeO$_2$-nanorods on methanol steam reforming for H$_2$ production: Steady state and DRIFTS studies. *Int. J. Hydrog. Energy* **2021**, *46*, 25954–25964. [CrossRef]
60. Penkova, A.; Bobadilla, L.; Ivanova, S.; Domínguez, M.I.; Romero-Sarria, F.; Roger, A.C.; Centeno, M.A.; Odriozola, J.A. Hydrogen production by methanol steam reforming on NiSn/MgO–Al$_2$O$_3$ catalysts: The role of MgO addition. *Appl. Catal. A Gen.* **2011**, *392*, 184–191. [CrossRef]
61. Lu, J.; Li, X.; He, S.; Han, C.; Wan, G.; Lei, Y.; Chen, R.; Liu, P.; Chen, K.; Zhang, L.; et al. Hydrogen production via methanol steam reforming over Ni-based catalysts: Influences of Lanthanum (La) addition and supports. *Int. J. Hydrog. Energy* **2017**, *42*, 3647–3657. [CrossRef]
62. Kim, W.; Mohaideen, K.K.; Seo, D.J.; Yoon, W.L. Methanol-steam reforming reaction over Cu-Al-based catalysts derived from layered double hydroxides. *Int. J. Hydrog. Energy* **2017**, *42*, 2081–2087. [CrossRef]
63. Qi, C.; Amphlett, J.C.; Peppley, B.A. Product composition as a function of temperature over NiAl-layered double hydroxide derived catalysts in steam reforming of methanol. *Appl. Catal. A Gen.* **2006**, *302*, 237–243. [CrossRef]
64. Qi, C.; Amphlett, J.C.; Peppley, B.A. Hydrogen production by methanol reforming on NiAl layered double hydroxide derived catalyst: Effect of the pretreatment of the catalyst. *Int. J. Hydrog. Energy* **2007**, *32*, 5098–5102. [CrossRef]
65. Qi, C.; Amphlett, J.C.; Peppley, B.A. K (Na)-promoted Ni, Al layered double hydroxide catalysts for the steam reforming of methanol. *J. Power Sources* **2007**, *171*, 842–849. [CrossRef]
66. Frank, B.; Jentoft, F.C.; Soerijanto, H.; Kröhnert, J.; Schlögl, R.; Schomäcker, R. Steam reforming of methanol over copper-containing catalysts: Influence of support material on microkinetics. *J. Catal.* **2007**, *246*, 177–192. [CrossRef]
67. Bepari, S.; Kuila, D. Steam reforming of methanol, ethanol and glycerol over nickel-based catalysts-A review. *Int. J. Hydrog. Energy* **2020**, *45*, 18090–18113. [CrossRef]
68. Mosińska, M.; Szynkowska-Jóźwik, M.I.; Mierczyński, P. Catalysts for Hydrogen Generation via Oxy–Steam Reforming of Methanol Process. *Materials* **2020**, *13*, 5601. [CrossRef]
69. Boskovic, G.; Baerns, M. Catalyst Deactivation. In *Basic Principles in Applied Catalysis*; Baerns, M., Ed.; Springer: Berlin/Heidelberg, Germany, 2004; pp. 477–503. [CrossRef]
70. Figueiredo, J.L.; Ramôa Ribeiro, F. *Catálise Heterogénea*, 2nd ed.; Gulbenkian, F.C., Ed.; Fundação Calouste Gulbenkian: Lisbon, Portugal, 2007.
71. Liu, Y.; Kang, H.; Hou, X.; Zhang, L.; Qing, S.; Gao, Z.; Xiang, H. Cu-Ni-Al spinel catalyzed methanol steam reforming for hydrogen production: Effect of Al content. *J. Fuel Chem. Technol.* **2020**, *48*, 1112–1121. [CrossRef]
72. Munnik, P.; de Jongh, P.E.; de Jong, K.P. Recent Developments in the Synthesis of Supported Catalysts. *Chem. Rev.* **2015**, *115*, 6687–6718. [CrossRef]
73. Pérez-Hernández, R.; Mendoza-Anaya, D.; Martínez, A.G.; Gómez-Cortés, A. Catalytic steam reforming of methanol to produce hydrogen on supported metal catalysts. In *Hydrogen Energy-Challenges and Perspectives*; Minic, D., Ed.; InTech: London, UK, 2012; pp. 149–174.

74. López, P.; Mondragón-Galicia, G.; Espinosa-Pesqueira, M.E.; Mendoza-Anaya, D.; Fernández, M.E.; Gómez-Cortés, A.; Bonifacio, J.; Martínez-Barrera, G.; Pérez-Hernández, R. Hydrogen production from oxidative steam reforming of methanol: Effect of the Cu and Ni impregnation on ZrO2 and their molecular simulation studies. *Int. J. Hydrog. Energy* **2012**, *37*, 9018–9027. [CrossRef]
75. Yang, W.; Parr, R.G. Hardness, softness, and the fukui function in the electronic theory of metals and catalysis. *Proc. Natl. Acad. Sci. USA* **1985**, *82*, 6723–6726. [CrossRef]
76. Chen, D.-H.; He, X.-R. Synthesis of nickel ferrite nanoparticles by sol-gel method. *Mater. Res. Bull.* **2001**, *36*, 1369–1377. [CrossRef]
77. Bhosale, R.; Shende, R.; Puszynski, J. H_2 Generation From Thermochemical Water-Splitting Using Sol-Gel Synthesized Zn/Sn/Mn-doped Ni-Ferrite. *I.RE.CH.E.* **2010**, *2*, 852–862.
78. Wu, C.; Williams, P.T. A Novel Nano-Ni/SiO_2 Catalyst for Hydrogen Production from Steam Reforming of Ethanol. *Environ. Sci. Technol.* **2010**, *44*, 5993–5998. [CrossRef] [PubMed]
79. Carter, C.B.; Norton, M.G. (Eds.) Sols, Gels, and Organic Chemistry. In *Ceramic Materials: Science and Engineering*; Springer New York: New York, NY, USA, 2007; pp. 400–411. [CrossRef]
80. Fiévet, F.; Ammar-Merah, S.; Brayner, R.; Chau, F.; Giraud, M.; Mammeri, F.; Peron, J.; Piquemal, J.Y.; Sicard, L.; Viau, G. The polyol process: A unique method for easy access to metal nanoparticles with tailored sizes, shapes and compositions. *Chem. Soc. Rev.* **2018**, *47*, 5187–5233. [CrossRef] [PubMed]
81. Wang, G.; Takeguchi, T.; Yamanaka, T.; Muhamad, E.N.; Mastuda, M.; Ueda, W. Effect of preparation atmosphere of Pt–SnOx/C catalysts on the catalytic activity for H_2/CO electro-oxidation. *Appl. Catal. B Environ.* **2010**, *98*, 86–93. [CrossRef]
82. Silva, J.M.; Trujillano, R.; Rives, V.; Soria, M.A.; Madeira, L.M. High temperature CO_2 sorption over modified hydrotalcites. *Chem. Eng. J.* **2017**, *325*, 25–34. [CrossRef]
83. Silva, J.M.; Trujillano, R.; Rives, V.; Soria, M.A.; Madeira, L.M. Dynamic behaviour of a K-doped Ga substituted and microwave aged hydrotalcite-derived mixed oxide during CO_2 sorption experiments. *J. Ind. Eng. Chem.* **2019**, *72*, 491–503. [CrossRef]
84. Rocha, C.; Soria, M.A.; Madeira, L.M. Effect of interlayer anion on the CO_2 capture capacity of hydrotalcite-based sorbents. *Sep. Purif. Technol.* **2019**, *219*, 290–302. [CrossRef]
85. Zabed, H.; Sahu, J.N.; Suely, A.; Boyce, A.N.; Faruq, G. Bioethanol production from renewable sources: Current perspectives and technological progress. *Renew. Sustain. Energy Rev.* **2017**, *71*, 475–501. [CrossRef]
86. Wu, R.-C.; Tang, C.-W.; Huang, H.-H.; Wang, C.-C.; Chang, M.-B.; Wang, C.-B. Effect of boron doping and preparation method of Ni/Ce0.5Zr0.5O2 catalysts on the performance for steam reforming of ethanol. *Int. J. Hydrog. Energy* **2019**, *44*, 14279–14289. [CrossRef]
87. Dan, M.; Mihet, M.; Tasnadi-Asztalos, Z.; Imre-Lucaci, A.; Katona, G.; Lazar, M.D. Hydrogen production by ethanol steam reforming on nickel catalysts: Effect of support modification by CeO_2 and La_2O_3. *Fuel* **2015**, *147*, 260–268. [CrossRef]
88. Bussi, J.; Musso, M.; Veiga, S.; Bespalko, N.; Faccio, R.; Roger, A.-C. Ethanol steam reforming over NiLaZr and NiCuLaZr mixed metal oxide catalysts. *Catal. Today* **2013**, *213*, 42–49. [CrossRef]
89. Biswas, P.; Kunzru, D. Steam reforming of ethanol for production of hydrogen over Ni/CeO_2–ZrO2 catalyst: Effect of support and metal loading. *Int. J. Hydrog. Energy* **2007**, *32*, 969–980. [CrossRef]
90. Trane-Restrup, R.; Dahl, S.; Jensen, A.D. Steam reforming of ethanol: Effects of support and additives on Ni-based catalysts. *Int. J. Hydrog. Energy* **2013**, *38*, 15105–15118. [CrossRef]
91. Dancini-Pontes, I.; DeSouza, M.; Silva, F.A.; Scaliante, M.H.N.O.; Alonso, C.G.; Bianchi, G.S.; Medina Neto, A.; Pereira, G.M.; Fernandes-Machado, N.R.C. Influence of the CeO_2 and Nb2O5 supports and the inert gas in ethanol steam reforming for H_2 production. *Chem. Eng. J.* **2015**, *273*, 66–74. [CrossRef]
92. Song, J.H.; Yoo, S.; Yoo, J.; Park, S.; Gim, M.Y.; Kim, T.H.; Song, I.K. Hydrogen production by steam reforming of ethanol over Ni/Al_2O_3-La_2O_3 xerogel catalysts. *Mol. Catal.* **2017**, *434*, 123–133. [CrossRef]
93. Di Cosimo, J.I.; Díez, V.K.; Xu, M.; Iglesia, E.; Apesteguía, C.R. Structure and Surface and Catalytic Properties of Mg-Al Basic Oxides. *J. Catal.* **1998**, *178*, 499–510. [CrossRef]
94. Rioche, C.; Kulkarni, S.; Meunier, F.C.; Breen, J.P.; Burch, R. Steam reforming of model compounds and fast pyrolysis bio-oil on supported noble metal catalysts. *Appl. Catal. B Environ.* **2005**, *61*, 130–139. [CrossRef]
95. Takanabe, K.; Aika, K.; Seshan, K.; Lefferts, L. Sustainable hydrogen from bio-oil—Steam reforming of acetic acid as a model oxygenate. *J. Catal.* **2004**, *227*, 101–108. [CrossRef]
96. Matas Güell, B.; Babich, I.; Nichols, K.P.; Gardeniers, J.G.E.; Lefferts, L.; Seshan, K. Design of a stable steam reforming catalyst—A promising route to sustainable hydrogen from biomass oxygenates. *Appl. Catal. B Environ.* **2009**, *90*, 38–44. [CrossRef]
97. Prasongthum, N.; Xiao, R.; Zhang, H.; Tsubaki, N.; Natewong, P.; Reubroycharoen, P. Highly active and stable Ni supported on CNTs-SiO2 fiber catalysts for steam reforming of ethanol. *Fuel Processing Technol.* **2017**, *160*, 185–195. [CrossRef]
98. Anjaneyulu, C.; Costa, L.O.O.d.; Ribeiro, M.C.; Rabelo-Neto, R.C.; Mattos, L.V.; Venugopal, A.; Noronha, F.B. Effect of Zn addition on the performance of Ni/Al_2O_3 catalyst for steam reforming of ethanol. *Appl. Catal. A Gen.* **2016**, *519*, 85–98. [CrossRef]
99. Mulewa, W.; Tahir, M.; Amin, N.A.S. MMT-supported Ni/TiO_2 nanocomposite for low temperature ethanol steam reforming toward hydrogen production. *Chem. Eng. J.* **2017**, *326*, 956–969. [CrossRef]
100. Yin, X.-m.; Xie, X.-m.; Wu, X.; An, X. Catalytic performance of nickel immobilized on organically modified montmorillonite in the steam reforming of ethanol for hydrogen production. *J. Fuel Chem. Technol.* **2016**, *44*, 689–697. [CrossRef]

101. Musso, M.; Cardozo, A.; Romero, M.; Faccio, R.; Segobia, D.; Apesteguía, C.; Bussi, J. High performance Ni-catalysts supported on rare-earth zirconates (La and Y) for hydrogen production through ethanol steam reforming. Characterization and assay. *Catal. Today* **2021**. [CrossRef]
102. Niazi, Z.; Irankhah, A.; Wang, Y.; Arandiyan, H. Cu, Mg and Co effect on nickel-ceria supported catalysts for ethanol steam reforming reaction. *Int. J. Hydrog. Energy* **2020**, *45*, 21512–21522. [CrossRef]
103. Wang, S.; He, B.; Tian, R.; Sun, C.; Dai, R.; Li, X.; Wu, X.; An, X.; Xie, X. Ni-hierarchical Beta zeolite catalysts were applied to ethanol steam reforming: Effect of sol gel method on loading Ni and the role of hierarchical structure. *Mol. Catal.* **2018**, *453*, 64–73. [CrossRef]
104. Gac, W.; Greluk, M.; Słowik, G.; Millot, Y.; Valentin, L.; Dzwigaj, S. Effects of dealumination on the performance of Ni-containing BEA catalysts in bioethanol steam reforming. *Appl. Catal. B Environ.* **2018**, *237*, 94–109. [CrossRef]
105. Chen, M.; Wang, Y.; Yang, Z.; Liang, T.; Liu, S.; Zhou, Z.; Li, X. Effect of Mg-modified mesoporous Ni/Attapulgite catalysts on catalytic performance and resistance to carbon deposition for ethanol steam reforming. *Fuel* **2018**, *220*, 32–46. [CrossRef]
106. Song, J.H.; Han, S.J.; Yoo, J.; Park, S.; Kim, D.H.; Song, I.K. Hydrogen production by steam reforming of ethanol over Ni–X/Al_2O_3–ZrO_2 (X=Mg, Ca, Sr, and Ba) xerogel catalysts: Effect of alkaline earth metal addition. *J. Mol. Catal. A Chem.* **2016**, *415*, 151–159. [CrossRef]
107. Song, J.H.; Han, S.J.; Yoo, J.; Park, S.; Kim, D.H.; Song, I.K. Effect of Sr content on hydrogen production by steam reforming of ethanol over Ni-Sr/Al_2O_3-ZrO_2 xerogel catalysts. *J. Mol. Catal. A Chem.* **2016**, *418–419*, 68–77. [CrossRef]
108. Li, L.; Tang, D.; Song, Y.; Jiang, B.; Zhang, Q. Hydrogen production from ethanol steam reforming on Ni-Ce/MMT catalysts. *Energy* **2018**, *149*, 937–943. [CrossRef]
109. Gharahshiran, V.S.; Yousefpour, M. Synthesis and characterization of Zr-promoted Ni-Co bimetallic catalyst supported OMC and investigation of its catalytic performance in steam reforming of ethanol. *Int. J. Hydrog. Energy* **2018**, *43*, 7020–7037. [CrossRef]
110. Nejat, T.; Jalalinezhad, P.; Hormozi, F.; Bahrami, Z. Hydrogen production from steam reforming of ethanol over Ni-Co bimetallic catalysts and MCM-41 as support. *J. Taiwan Inst. Chem. Eng.* **2019**, *97*, 216–226. [CrossRef]
111. Zhao, X.; Lu, G. Modulating and controlling active species dispersion over Ni–Co bimetallic catalysts for enhancement of hydrogen production of ethanol steam reforming. *Int. J. Hydrog. Energy* **2016**, *41*, 3349–3362. [CrossRef]
112. Llorca, J.; Homs, N.; Ramirez de la Piscina, P. In situ DRIFT-mass spectrometry study of the ethanol steam-reforming reaction over carbonyl-derived Co/ZnO catalysts. *J. Catal.* **2004**, *227*, 556–560. [CrossRef]
113. Sutton, D.; Kelleher, B.; Ross, J.R.H. Review of literature on catalysts for biomass gasification. *Fuel Processing Technol.* **2001**, *73*, 155–173. [CrossRef]
114. Davda, R.R.; Shabaker, J.W.; Huber, G.W.; Cortright, R.D.; Dumesic, J.A. A review of catalytic issues and process conditions for renewable hydrogen and alkanes by aqueous-phase reforming of oxygenated hydrocarbons over supported metal catalysts. *Appl. Catal. B Environ.* **2005**, *56*, 171–186. [CrossRef]
115. Campos, C.H.; Pecchi, G.; Fierro, J.L.G.; Osorio-Vargas, P. Enhanced bimetallic Rh-Ni supported catalysts on alumina doped with mixed lanthanum-cerium oxides for ethanol steam reforming. *Mol. Catal.* **2019**, *469*, 87–97. [CrossRef]
116. Mondal, T.; Pant, K.K.; Dalai, A.K. Oxidative and non-oxidative steam reforming of crude bio-ethanol for hydrogen production over Rh promoted Ni/CeO_2-ZrO_2 catalyst. *Appl. Catal. A Gen.* **2015**, *499*, 19–31. [CrossRef]
117. Le Valant, A.; Can, F.; Bion, N.; Duprez, D.; Epron, F. Hydrogen production from raw bioethanol steam reforming: Optimization of catalyst composition with improved stability against various impurities. *Int. J. Hydrog. Energy* **2010**, *35*, 5015–5020. [CrossRef]
118. Mironova, E.Y.; Lytkina, A.A.; Ermilova, M.M.; Efimov, M.N.; Zemtsov, L.M.; Orekhova, N.V.; Karpacheva, G.P.; Bondarenko, G.N.; Muraviev, D.N.; Yaroslavtsev, A.B. Ethanol and methanol steam reforming on transition metal catalysts supported on detonation synthesis nanodiamonds for hydrogen production. *Int. J. Hydrog. Energy* **2015**, *40*, 3557–3565. [CrossRef]
119. Mironova, E.Y.; Ermilova, M.M.; Orekhova, N.V.; Muraviev, D.N.; Yaroslavtsev, A.B. Production of high purity hydrogen by ethanol steam reforming in membrane reactor. *Catal. Today* **2014**, *236*, 64–69. [CrossRef]
120. He, S.; He, S.; Zhang, L.; Li, X.; Wang, J.; He, D.; Lu, J.; Luo, Y. Hydrogen production by ethanol steam reforming over Ni/SBA-15 mesoporous catalysts: Effect of Au addition. *Catal. Today* **2015**, *258*, 162–168. [CrossRef]
121. Kim, D.; Kwak, B.S.; Min, B.-K.; Kang, M. Characterization of Ni and W co-loaded SBA-15 catalyst and its hydrogen production catalytic ability on ethanol steam reforming reaction. *Appl. Surf. Sci.* **2015**, *332*, 736–746. [CrossRef]
122. Romero, A.; Jobbágy, M.; Laborde, M.; Baronetti, G.; Amadeo, N. Ni(II)–Mg(II)–Al(III) catalysts for hydrogen production from ethanol steam reforming: Influence of the Mg content. *Appl. Catal. A Gen.* **2014**, *470*, 398–404. [CrossRef]
123. Bepari, S.; Basu, S.; Pradhan, N.C.; Dalai, A.K. Steam reforming of ethanol over cerium-promoted Ni-Mg-Al hydrotalcite catalysts. *Catal. Today* **2017**, *291*, 47–57. [CrossRef]
124. Shejale, A.D.; Yadav, G.D. Cu promoted Ni-Co/hydrotalcite catalyst for improved hydrogen production in comparison with several modified Ni-based catalysts via steam reforming of ethanol. *Int. J. Hydrog. Energy* **2017**, *42*, 11321–11332. [CrossRef]
125. Passos, A.R.; Pulcinelli, S.H.; Santilli, C.V.; Briois, V. Operando monitoring of metal sites and coke evolution during non-oxidative and oxidative ethanol steam reforming over Ni and NiCu ex-hydrotalcite catalysts. *Catal. Today* **2019**, *336*, 122–130. [CrossRef]
126. Wang, F. *Hydrogen Production from Steam Reforming of Ethanol Over an Ir/Ceria-Based Catalyst: Catalyst Ageing Analysis and Performance Improvement upon Ceria Doping*; Université Claude Bernard-Lyon I: Villeurbanne, France, 2012.
127. Mhadmhan, S.; Natewong, P.; Prasongthum, N.; Samart, C.; Reubroycharoen, P. Investigation of Ni/SiO2 Fiber Catalysts Prepared by Different Methods on Hydrogen production from Ethanol Steam Reforming. *Catalysts* **2018**, *8*, 319. [CrossRef]

128. Xu, W.; Liu, Z.; Johnston-Peck, A.C.; Senanayake, S.D.; Zhou, G.; Stacchiola, D.; Stach, E.A.; Rodriguez, J.A. Steam Reforming of Ethanol on Ni/CeO$_2$: Reaction Pathway and Interaction between Ni and the CeO$_2$ Support. *ACS Catal.* **2013**, *3*, 975–984. [CrossRef]
129. Santander, J.A.; Tonetto, G.M.; Pedernera, M.N.; López, E. Ni/CeO$_2$–MgO catalysts supported on stainless steel plates for ethanol steam reforming. *Int. J. Hydrog. Energy* **2017**, *42*, 9482–9492. [CrossRef]
130. Zhuang, Q.; Qin, Y.; Chang, L. Promoting effect of cerium oxide in supported nickel catalyst for hydrocarbon steam-reforming. *Appl. Catal.* **1991**, *70*, 1–8. [CrossRef]
131. Tahir, M.; Mulewa, W.; Amin, N.A.S.; Zakaria, Z.Y. Thermodynamic and experimental analysis on ethanol steam reforming for hydrogen production over Ni-modified TiO$_2$/MMT nanoclay catalyst. *Energy Convers. Manag.* **2017**, *154*, 25–37. [CrossRef]
132. Chen, X.; Liu, Y.; Niu, G.; Yang, Z.; Bian, M.; He, A. High temperature thermal stabilization of alumina modified by lanthanum species. *Appl. Catal. A Gen.* **2001**, *205*, 159–172. [CrossRef]
133. Carrera Cerritos, R.; Fuentes Ramírez, R.; Aguilera Alvarado, A.F.; Martínez Rosales, J.M.; Viveros García, T.; Galindo Esquivel, I.R. Steam Reforming of Ethanol over Ni/Al$_2$O$_3$–La$_2$O$_3$ Catalysts Synthesized by Sol−Gel. *Ind. Eng. Chem. Res.* **2011**, *50*, 2576–2584. [CrossRef]
134. Lee, S. Concepts in Syngas Manufacture. By Jens Rostrup-Nielsen and Lars J. Christiansen. *Energy Technol.* **2013**, *1*, 419–420. [CrossRef]
135. Melo, F.; Morlanés, N. Synthesis, characterization and catalytic behaviour of NiMgAl mixed oxides as catalysts for hydrogen production by naphtha steam reforming. *Catal. Today* **2008**, *133–135*, 383–393. [CrossRef]
136. Melo, F.; Morlanés, N. Naphtha steam reforming for hydrogen production. *Catal. Today* **2005**, *107-108*, 458–466. [CrossRef]
137. Conner, W.C.; Falconer, J.L. Spillover in Heterogeneous Catalysis. *Chem. Rev.* **1995**, *95*, 759–788. [CrossRef]
138. He, S.; Mei, Z.; Liu, N.; Zhang, L.; Lu, J.; Li, X.; Wang, J.; He, D.; Luo, Y. Ni/SBA-15 catalysts for hydrogen production by ethanol steam reforming: Effect of nickel precursor. *Int. J. Hydrog. Energy* **2017**, *42*, 14429–14438. [CrossRef]
139. Villagrán-Olivares, A.C.; Gomez, M.F.; Barroso, M.N.; Abello, M.C. Hydrogen production from ethanol: Synthesis of Ni catalysts assisted by chelating agents. *Mol. Catal.* **2020**, *481*, 110164. [CrossRef]
140. Tarditi, A.M.; Barroso, N.; Galetti, A.E.; Arrúa, L.A.; Cornaglia, L.; Abello, M.C. XPS study of the surface properties and Ni particle size determination of Ni-supported catalysts. *Surf. Interface Anal.* **2014**, *46*, 521–529. [CrossRef]
141. van Dillen, A.J.; Terörde, R.J.A.M.; Lensveld, D.J.; Geus, J.W.; de Jong, K.P. Synthesis of supported catalysts by impregnation and drying using aqueous chelated metal complexes. *J. Catal.* **2003**, *216*, 257–264. [CrossRef]
142. Wu, C.; Dupont, V.; Nahil, M.A.; Dou, B.; Chen, H.; Williams, P.T. Investigation of Ni/SiO2 catalysts prepared at different conditions for hydrogen production from ethanol steam reforming. *J. Energy Inst.* **2017**, *90*, 276–284. [CrossRef]
143. Nichele, V.; Signoretto, M.; Menegazzo, F.; Rossetti, I.; Cruciani, G. Hydrogen production by ethanol steam reforming: Effect of the synthesis parameters on the activity of Ni/TiO$_2$ catalysts. *Int. J. Hydrog. Energy* **2014**, *39*, 4252–4258. [CrossRef]
144. Bridgwater, A.V. Review of fast pyrolysis of biomass and product upgrading. *Biomass Bioenergy* **2012**, *38*, 68–94. [CrossRef]
145. Akhtar, J.; Amin, N.A.S. A review on process conditions for optimum bio-oil yield in hydrothermal liquefaction of biomass. *Renew. Sustain. Energy Rev.* **2011**, *15*, 1615–1624. [CrossRef]
146. Bridgwater, T. Biomass for energy. *J. Sci. Food Agric.* **2006**, *86*, 1755–1768. [CrossRef]
147. Faba, L.; Díaz, E.; Ordóñez, S. Recent developments on the catalytic technologies for the transformation of biomass into biofuels: A patent survey. *Renew. Sustain. Energy Rev.* **2015**, *51*, 273–287. [CrossRef]
148. Sadhukhan, J.; Martinez-Hernandez, E.; Murphy, R.J.; Ng, D.K.S.; Hassim, M.H.; Siew Ng, K.; Yoke Kin, W.; Jaye, I.F.M.; Leung Pah Hang, M.Y.; Andiappan, V. Role of bioenergy, biorefinery and bioeconomy in sustainable development: Strategic pathways for Malaysia. *Renew. Sustain. Energy Rev.* **2018**, *81*, 1966–1987. [CrossRef]
149. Dhyani, V.; Bhaskar, T. A comprehensive review on the pyrolysis of lignocellulosic biomass. *Renew. Energy* **2018**, *129*, 695–716. [CrossRef]
150. Mondal, P.; Dang, G.S.; Garg, M.O. Syngas production through gasification and cleanup for downstream applications—Recent developments. *Fuel Processing Technol.* **2011**, *92*, 1395–1410. [CrossRef]
151. Raffelt, K.; Henrich, E.; Koegel, A.; Stahl, R.; Steinhardt, J.; Weirich, F. The BTL2 process of biomass utilization entrained-flow gasification of pyrolyzed biomass slurries. *Appl. Biochem. Biotechnol.* **2006**, *129*, 153–164. [CrossRef]
152. Soria, M.A.; Barros, D.; Madeira, L.M. Hydrogen production through steam reforming of bio-oils derived from biomass pyrolysis: Thermodynamic analysis including in situ CO2 and/or H$_2$ separation. *Fuel* **2019**, *244*, 184–195. [CrossRef]
153. Basagiannis, A.C.; Verykios, X.E. Reforming reactions of acetic acid on nickel catalysts over a wide temperature range. *Appl. Catal. A Gen.* **2006**, *308*, 182–193. [CrossRef]
154. Marquevich, M.; Czernik, S.; Chornet, E.; Montané, D. Hydrogen from Biomass: Steam Reforming of Model Compounds of Fast-Pyrolysis Oil. *Energy Fuels* **1999**, *13*, 1160–1166. [CrossRef]
155. Radlein, D.; Piskorz, J.; Scott, D.S. Fast pyrolysis of natural polysaccharides as a potential industrial process. *J. Anal. Appl. Pyrolysis* **1991**, *19*, 41–63. [CrossRef]
156. Paredes, M.J.; Moreno, E.; Ramos-Cormenzana, A.; Martinez, J. Characteristics of soil after pollution with wastewaters from olive oil extraction plants. *Chemosphere* **1987**, *16*, 1557–1564. [CrossRef]
157. DellaGreca, M.; Monaco, P.; Pinto, G.; Pollio, A.; Previtera, L.; Temussi, F. Phytotoxicity of low-molecular-weight phenols from olive mill wastewaters. *Bull. Environ. Contam. Toxicol.* **2001**, *67*, 352–359. [CrossRef]
158. Rana, G.; Rinaldi, M.; Introna, M. Volatilisation of substances alter spreading olive oil waste water on the soil in a Mediterranean environment. *Agric. Ecosyst. Environ.* **2003**, *96*, 49–58. [CrossRef]

159. Montero, C.; Oar-Arteta, L.; Remiro, A.; Arandia, A.; Bilbao, J.; Gayubo, A.G. Thermodynamic comparison between bio-oil and ethanol steam reforming. *Int. J. Hydrog. Energy* **2015**, *40*, 15963–15971. [CrossRef]
160. Casanovas, A.; Galvis, A.; Llorca, J. Catalytic steam reforming of olive mill wastewater for hydrogen production. *Int. J. Hydrog. Energy* **2015**, *40*, 7539–7545. [CrossRef]
161. Tosti, S.; Fabbricino, M.; Pontoni, L.; Palma, V.; Ruocco, C. Catalytic reforming of olive mill wastewater and methane in a Pd-membrane reactor. *Int. J. Hydrog. Energy* **2016**, *41*, 5465–5474. [CrossRef]
162. Tosti, S.; Accetta, C.; Fabbricino, M.; Sansovini, M.; Pontoni, L. Reforming of olive mill wastewater through a Pd-membrane reactor. *Int. J. Hydrog. Energy* **2013**, *38*, 10252–10259. [CrossRef]
163. Gebreyohannes, A.Y.; Mazzei, R.; Giorno, L. Trends and current practices of olive mill wastewater treatment: Application of integrated membrane process and its future perspective. *Sep. Purif. Technol.* **2016**, *162*, 45–60. [CrossRef]
164. Aggoun, M.; Arhab, R.; Cornu, A.; Portelli, J.; Barkat, M.; Graulet, B. Olive mill wastewater microconstituents composition according to olive variety and extraction process. *Food Chem.* **2016**, *209*, 72–80. [CrossRef]
165. El-Abbassi, A.; Kiai, H.; Hafidi, A. Phenolic profile and antioxidant activities of olive mill wastewater. *Food Chem.* **2012**, *132*, 406–412. [CrossRef]
166. Daâssi, D.; Lozano-Sánchez, J.; Borrás-Linares, I.; Belbahri, L.; Woodward, S.; Zouari-Mechichi, H.; Mechichi, T.; Nasri, M.; Segura-Carretero, A. Olive oil mill wastewaters: Phenolic content characterization during degradation by Coriolopsis gallica. *Chemosphere* **2014**, *113*, 62–70. [CrossRef]
167. Fki, I.; Allouche, N.; Sayadi, S. The use of polyphenolic extract, purified hydroxytyrosol and 3,4-dihydroxyphenyl acetic acid from olive mill wastewater for the stabilization of refined oils: A potential alternative to synthetic antioxidants. *Food Chem.* **2005**, *93*, 197–204. [CrossRef]
168. Dermeche, S.; Nadour, M.; Larroche, C.; Moulti-Mati, F.; Michaud, P. Olive mill wastes: Biochemical characterizations and valorization strategies. *Process Biochem.* **2013**, *48*, 1532–1552. [CrossRef]
169. Kyriacou, A.; Lasaridi, K.E.; Kotsou, M.; Balis, C.; Pilidis, G. Combined bioremediation and advanced oxidation of green table olive processing wastewater. *Process Biochem.* **2005**, *40*, 1401–1408. [CrossRef]
170. Araújo, M.; Pimentel, F.B.; Alves, R.C.; Oliveira, M.B.P.P. Phenolic compounds from olive mill wastes: Health effects, analytical approach and application as food antioxidants. *Trends Food Sci. Technol.* **2015**, *45*, 200–211. [CrossRef]
171. Kaleh, Z.; Geißen, S.U. Selective isolation of valuable biophenols from olive mill wastewater. *J. Environ. Chem. Eng.* **2016**, *4*, 373–384. [CrossRef]
172. Caputo, A.C.; Scacchia, F.; Pelagagge, P.M. Disposal of by-products in olive oil industry: Waste-to-energy solutions. *Appl. Therm. Eng.* **2003**, *23*, 197–214. [CrossRef]
173. Vlyssides, A.G.; Loizides, M.; Karlis, P.K. Integrated strategic approach for reusing olive oil extraction by-products. *J. Clean. Prod.* **2004**, *12*, 603–611. [CrossRef]
174. Paredes, C.; Cegarra, J.; Roig, A.; Sánchez-Monedero, M.A.; Bernal, M.P. Characterization of olive mill wastewater (alpechin) and its sludge for agricultural purposes. *Bioresour. Technol.* **1999**, *67*, 111–115. [CrossRef]
175. Feki, M.; Allouche, N.; Bouaziz, M.; Gargoubi, A.; Sayadi, S. Effect of storage of olive mill wastewaters on hydroxytyrosol concentration. *Eur. J. Lipid Sci. Technol.* **2006**, *108*, 1021–1027. [CrossRef]
176. Hamden, K.; Allouche, N.; Damak, M.; Elfeki, A. Hypoglycemic and antioxidant effects of phenolic extracts and purified hydroxytyrosol from olive mill waste in vitro and in rats. *Chem.-Biol. Interact.* **2009**, *180*, 421–432. [CrossRef]
177. O-Thong, S.; Suksong, W.; Promnuan, K.; Thipmunee, M.; Mamimin, C.; Prasertsan, P. Two-stage thermophilic fermentation and mesophilic methanogenic process for biohythane production from palm oil mill effluent with methanogenic effluent recirculation for pH control. *Int. J. Hydrog. Energy* **2016**, *41*, 21702–21712. [CrossRef]
178. Bizkarra, K.; Bermudez, J.M.; Arcelus-Arrillaga, P.; Barrio, V.L.; Cambra, J.F.; Millan, M. Nickel based monometallic and bimetallic catalysts for synthetic and real bio-oil steam reforming. *Int. J. Hydrog. Energy* **2018**, *43*, 11706–11718. [CrossRef]
179. Kechagiopoulos, P.N.; Voutetakis, S.S.; Lemonidou, A.A.; Vasalos, I.A. Hydrogen Production via Steam Reforming of the Aqueous Phase of Bio-Oil in a Fixed Bed Reactor. *Energy Fuels* **2006**, *20*, 2155–2163. [CrossRef]
180. Remiro, A.; Arandia, A.; Oar-Arteta, L.; Bilbao, J.; Gayubo, A.G. Regeneration of NiAl2O4 spinel type catalysts used in the reforming of raw bio-oil. *Appl. Catal. B Environ.* **2018**, *237*, 353–365. [CrossRef]
181. Remiro, A.; Valle, B.; Aguayo, A.T.; Bilbao, J.; Gayubo, A.G. Operating conditions for attenuating Ni/La2O3–αAl2O3 catalyst deactivation in the steam reforming of bio-oil aqueous fraction. *Fuel Processing Technol.* **2013**, *115*, 222–232. [CrossRef]
182. Valle, B.; Aramburu, B.; Olazar, M.; Bilbao, J.; Gayubo, A.G. Steam reforming of raw bio-oil over Ni/La2O3-αAl2O3: Influence of temperature on product yields and catalyst deactivation. *Fuel* **2018**, *216*, 463–474. [CrossRef]
183. Xie, H.; Yu, Q.; Wei, M.; Duan, W.; Yao, X.; Qin, Q.; Zuo, Z. Hydrogen production from steam reforming of simulated bio-oil over Ce–Ni/Co catalyst with in continuous CO2 capture. *Int. J. Hydrog. Energy* **2015**, *40*, 1420–1428. [CrossRef]
184. Silva, J.M.; Ribeiro, L.S.; Órfão, J.J.M.; Soria, M.A.; Madeira, L.M. Low temperature glycerol steam reforming over a Rh-based catalyst combined with oxidative regeneration. *Int. J. Hydrog. Energy* **2019**, *44*, 2461–2473. [CrossRef]
185. Rocha, C.; Soria, M.A.; Madeira, L.M. Olive mill wastewater valorization through steam reforming using hybrid multifunctional reactors for high-purity H_2 production. *Chem. Eng. J.* **2022**, *430*, 132651. [CrossRef]
186. Ng, K.H.; Cheng, Y.W.; Lee, Z.S.; Cheng, C.K. A study into syngas production from catalytic steam reforming of palm oil mill effluent (POME): A new treatment approach. *Int. J. Hydrog. Energy* **2019**, *44*, 20900–20913. [CrossRef]

187. Cheng, Y.W.; Ng, K.H.; Lam, S.S.; Lim, J.W.; Wongsakulphasatch, S.; Witoon, T.; Cheng, C.K. Syngas from catalytic steam reforming of palm oil mill effluent: An optimization study. *Int. J. Hydrog. Energy* **2019**, *44*, 9220–9236. [CrossRef]
188. Ighalo, J.O.; Adeniyi, A.G. Factor effects and interactions in steam reforming of biomass bio-oil. *Chem. Pap.* **2020**, *74*, 1459–1470. [CrossRef]
189. Rocha, C.; Soria, M.A.; Madeira, L.M. Steam reforming of olive oil mill wastewater with in situ hydrogen and carbon dioxide separation–Thermodynamic analysis. *Fuel* **2017**, *207*, 449–460. [CrossRef]
190. Rocha, C.; Soria, M.A.; Madeira, L.M. Thermodynamic analysis of olive oil mill wastewater steam reforming. *J. Energy Inst.* **2019**, *92*, 1599–1609. [CrossRef]
191. Macedo, M.S.; Soria, M.A.; Madeira, L.M. Glycerol steam reforming for hydrogen production: Traditional versus membrane reactor. *Int. J. Hydrog. Energy* **2019**, *44*, 24719–24732. [CrossRef]
192. Leal, A.L.; Soria, M.A.; Madeira, L.M. Autothermal reforming of impure glycerol for H_2 production: Thermodynamic study including in situ CO_2 and/or H_2 separation. *Int. J. Hydrog. Energy* **2016**, *41*, 2607–2620. [CrossRef]
193. Medrano, J.A.; Oliva, M.; Ruiz, J.; Garcia, L.; Arauzo, J. Catalytic steam reforming of acetic acid in a fluidized bed reactor with oxygen addition. *Int. J. Hydrog. Energy* **2008**, *33*, 4387–4396. [CrossRef]
194. Matas Güell, B.; Silva, I.M.T.d.; Seshan, K.; Lefferts, L. Sustainable route to hydrogen–Design of stable catalysts for the steam gasification of biomass related oxygenates. *Appl. Catal. B Environ.* **2009**, *88*, 59–65. [CrossRef]
195. Hoang, T.M.C.; Geerdink, B.; Sturm, J.M.; Lefferts, L.; Seshan, K. Steam reforming of acetic acid–A major component in the volatiles formed during gasification of humin. *Appl. Catal. B Environ.* **2015**, *163*, 74–82. [CrossRef]
196. Chen, G.; Tao, J.; Liu, C.; Yan, B.; Li, W.; Li, X. Steam reforming of acetic acid using Ni/Al_2O_3 catalyst: Influence of crystalline phase of Al_2O_3 support. *Int. J. Hydrog. Energy* **2017**, *42*, 20729–20738. [CrossRef]
197. Savuto, E.; Navarro, R.M.; Mota, N.; Di Carlo, A.; Bocci, E.; Carlini, M.; Fierro, J.L.G. Steam reforming of tar model compounds over Ni/Mayenite catalysts: Effect of Ce addition. *Fuel* **2018**, *224*, 676–686. [CrossRef]
198. Zhang, Z.; Hu, X.; Li, J.; Gao, G.; Dong, D.; Westerhof, R.; Hu, S.; Xiang, J.; Wang, Y. Steam reforming of acetic acid over Ni/Al_2O_3 catalysts: Correlation of nickel loading with properties and catalytic behaviors of the catalysts. *Fuel* **2018**, *217*, 389–403. [CrossRef]
199. Borges, R.P.; Ferreira, R.A.R.; Rabelo-Neto, R.C.; Noronha, F.B.; Hori, C.E. Hydrogen production by steam reforming of acetic acid using hydrotalcite type precursors. *Int. J. Hydrog. Energy* **2018**, *43*, 7881–7892. [CrossRef]
200. He, L.; Hu, S.; Jiang, L.; Liao, G.; Chen, X.; Han, H.; Xiao, L.; Ren, Q.; Wang, Y.; Su, S.; et al. Carbon nanotubes formation and its influence on steam reforming of toluene over Ni/Al_2O_3 catalysts: Roles of catalyst supports. *Fuel Processing Technol.* **2018**, *176*, 7–14. [CrossRef]
201. Pu, J.; Nishikado, K.; Wang, N.; Nguyen, T.T.; Maki, T.; Qian, E.W. Core-shell nickel catalysts for the steam reforming of acetic acid. *Appl. Catal. B Environ.* **2018**, *224*, 69–79. [CrossRef]
202. Zhang, C.; Hu, X.; Yu, Z.; Zhang, Z.; Chen, G.; Li, C.; Liu, Q.; Xiang, J.; Wang, Y.; Hu, S. Steam reforming of acetic acid for hydrogen production over attapulgite and alumina supported Ni catalysts: Impacts of properties of supports on catalytic behaviors. *Int. J. Hydrog. Energy* **2019**, *44*, 5230–5244. [CrossRef]
203. Chen, J.; Wang, M.; Wang, S.; Li, X. Hydrogen production via steam reforming of acetic acid over biochar-supported nickel catalysts. *Int. J. Hydrog. Energy* **2018**, *43*, 18160–18168. [CrossRef]
204. Kechagiopoulos, P.N.; Voutetakis, S.S.; Lemonidou, A.A.; Vasalos, I.A. Hydrogen Production via Reforming of the Aqueous Phase of Bio-Oil over Ni/Olivine Catalysts in a Spouted Bed Reactor. *Ind. Eng. Chem. Res.* **2009**, *48*, 1400–1408. [CrossRef]
205. Liu, S.; Chen, M.; Chu, L.; Yang, Z.; Zhu, C.; Wang, J.; Chen, M. Catalytic steam reforming of bio-oil aqueous fraction for hydrogen production over Ni–Mo supported on modified sepiolite catalysts. *Int. J. Hydrog. Energy* **2013**, *38*, 3948–3955. [CrossRef]
206. Garcia, L.a.; French, R.; Czernik, S.; Chornet, E. Catalytic steam reforming of bio-oils for the production of hydrogen: Effects of catalyst composition. *Appl. Catal. A Gen.* **2000**, *201*, 225–239. [CrossRef]
207. Bangala, D.N.; Abatzoglou, N.; Chornet, E. Steam reforming of naphthalene on $Ni–Cr/Al_2O_3$ catalysts doped with MgO, TiO_2, and La_2O_3. *AIChE J.* **1998**, *44*, 927–936. [CrossRef]
208. Zhang, Z.; Hu, X.; Gao, G.; Wei, T.; Dong, D.; Wang, Y.; Hu, S.; Xiang, J.; Liu, Q.; Geng, D. Steam reforming of acetic acid over $NiKOH/Al_2O_3$ catalyst with low nickel loading: The remarkable promotional effects of KOH on activity. *Int. J. Hydrog. Energy* **2019**, *44*, 729–747. [CrossRef]
209. Choi, I.-H.; Hwang, K.-R.; Lee, K.-Y.; Lee, I.-G. Catalytic steam reforming of biomass-derived acetic acid over modified Ni/γ-Al_2O_3 for sustainable hydrogen production. *Int. J. Hydrog. Energy* **2019**, *44*, 180–190. [CrossRef]
210. Charisiou, N.D.; Siakavelas, G.; Tzounis, L.; Dou, B.; Sebastian, V.; Hinder, S.J.; Baker, M.A.; Polychronopoulou, K.; Goula, M.A. Ni/Y_2O_3–ZrO_2 catalyst for hydrogen production through the glycerol steam reforming reaction. *Int. J. Hydrog. Energy* **2020**, *45*, 10442–10460. [CrossRef]
211. Souza, I.C.A.; Manfro, R.L.; Souza, M.M.V.M. Hydrogen production from steam reforming of acetic acid over Pt–Ni bimetallic catalysts supported on ZrO_2. *Biomass Bioenergy* **2022**, *156*, 106317. [CrossRef]
212. Baamran, K.S.; Tahir, M.; Mohamed, M.; Hussain Khoja, A. Effect of support size for stimulating hydrogen production in phenol steam reforming using Ni-embedded TiO_2 nanocatalyst. *J. Environ. Chem. Eng.* **2020**, *8*, 103604. [CrossRef]
213. Baamran, K.S.; Tahir, M. Ni-embedded TiO_2-$ZnTiO_3$ reducible perovskite composite with synergistic effect of metal/support towards enhanced H_2 production via phenol steam reforming. *Energy Convers. Manag.* **2019**, *200*, 112064. [CrossRef]

214. Abbas, T.; Tahir, M.; Saidina Amin, N.A. Enhanced Metal–Support Interaction in Ni/CO_3O_4/TiO_2 Nanorods toward Stable and Dynamic Hydrogen Production from Phenol Steam Reforming. *Ind. Eng. Chem. Res.* **2019**, *58*, 517–530. [CrossRef]
215. Pu, J.; Ikegami, F.; Nishikado, K.; Qian, E.W. Effect of ceria addition on NiRu/CeO2Al2O3 catalysts in steam reforming of acetic acid. *Int. J. Hydrog. Energy* **2017**, *42*, 19733–19743. [CrossRef]
216. Wang, Y.; Chen, M.; Yang, J.; Liu, S.; Yang, Z.; Wang, J.; Liang, T. Hydrogen Production from Steam Reforming of Acetic Acid over Ni-Fe/Palygorskite Modified with Cerium. *BioResources* **2017**, *12*, 4830–4853. [CrossRef]
217. Li, L.; Jiang, B.; Tang, D.; Zhang, Q.; Zheng, Z. Hydrogen generation by acetic acid steam reforming over Ni-based catalysts derived from La1−xCexNiO3 perovskite. *Int. J. Hydrog. Energy* **2018**, *43*, 6795–6803. [CrossRef]
218. Zhao, X.; Xue, Y.; Lu, Z.; Huang, Y.; Guo, C.; Yan, C. Encapsulating Ni/CeO_2-ZrO_2 with SiO_2 layer to improve it catalytic activity for steam reforming of toluene. *Catal. Commun.* **2017**, *101*, 138–141. [CrossRef]
219. Liang, T.; Wang, Y.; Chen, M.; Yang, Z.; Liu, S.; Zhou, Z.; Li, X. Steam reforming of phenol-ethanol to produce hydrogen over bimetallic NiCu catalysts supported on sepiolite. *Int. J. Hydrog. Energy* **2017**, *42*, 28233–28246. [CrossRef]
220. Pant, K.K.; Mohanty, P.; Agarwal, S.; Dalai, A.K. Steam reforming of acetic acid for hydrogen production over bifunctional Ni–Co catalysts. *Catal. Today* **2013**, *207*, 36–43. [CrossRef]
221. Lee, Y.-L.; Jha, A.; Jang, W.-J.; Shim, J.-O.; Jeon, K.-W.; Na, H.-S.; Kim, H.-M.; Lee, D.-W.; Yoo, S.-Y.; Jeon, B.-H.; et al. Optimization of Cobalt Loading in Co–CeO_2 Catalyst for the High Temperature Water–Gas Shift Reaction. *Top. Catal.* **2017**, *60*, 721–726. [CrossRef]
222. Sabnis, K.D. *Structure-Activity Relationships for the Water-Gas Shift Reaction over Supported Metal Catalysts*; Purdue University: West Lafayette, IN, USA, 2015.
223. Mizuno, S.C.M.; Braga, A.H.; Hori, C.E.; Santos, J.B.O.; Bueno, J.M.C. Steam reforming of acetic acid over $MgAl_2O_4$-supported Co and Ni catalysts: Effect of the composition of Ni/Co and reactants on reaction pathways. *Catal. Today* **2017**, *296*, 144–153. [CrossRef]
224. Rocha, C.; Soria, M.A.; Madeira, L.M. Screening of commercial catalysts for steam reforming of olive mill wastewater. *Renew. Energy* **2021**, *169*, 765–779. [CrossRef]
225. Rocha, C.; Soria, M.A.; Madeira, L.M. Use of Ni-containing catalysts for synthetic olive mill wastewater steam reforming. *Renew. Energy* **2022**, *185*, 1329–1342. [CrossRef]
226. Adhikari, S.; Fernando, S.D.; To, S.D.F.; Bricka, R.M.; Steele, P.H.; Haryanto, A. Conversion of Glycerol to Hydrogen via a Steam Reforming Process over Nickel Catalysts. *Energy Fuels* **2008**, *22*, 1220–1226. [CrossRef]
227. Pant, K.K.; Jain, R.; Jain, S. Renewable hydrogen production by steam reforming of glycerol over Ni/CeO_2 catalyst prepared by precipitation deposition method. *Korean J. Chem. Eng.* **2011**, *28*, 1859. [CrossRef]
228. Buffoni, I.N.; Pompeo, F.; Santori, G.F.; Nichio, N.N. Nickel catalysts applied in steam reforming of glycerol for hydrogen production. *Catal. Commun.* **2009**, *10*, 1656–1660. [CrossRef]
229. Silva, O.C.V.; Silveira, E.B.; Rabelo-Neto, R.C.; Borges, L.E.P.; Noronha, F.B. Hydrogen Production Through Steam Reforming of Toluene Over Ni Supported on MgAl Mixed Oxides Derived from Hydrotalcite-Like Compounds. *Catal. Lett.* **2018**, *148*, 1622–1633. [CrossRef]
230. Manfro, R.L.; Ribeiro, N.F.P.; Souza, M.M.V.M. Production of hydrogen from steam reforming of glycerol using nickel catalysts supported on Al_2O_3, CeO_2 and ZrO_2. *Catal. Sustain. Energy* **2013**, *1*, 60–70. [CrossRef]
231. Carrero, A.; Calles, J.A.; García-Moreno, L.; Vizcaíno, A.J. Production of Renewable Hydrogen from Glycerol Steam Reforming over Bimetallic Ni-(Cu,Co,Cr) Catalysts Supported on SBA-15 Silica. *Catalysts* **2017**, *7*, 55. [CrossRef]
232. Cheng, F.; Dupont, V. Nickel catalyst auto-reduction during steam reforming of bio-oil model compound acetic acid. *Int. J. Hydrog. Energy* **2013**, *38*, 15160–15172. [CrossRef]
233. Wang, M.; Zhang, F.; Wang, S. Effect of La_2O_3 replacement on γ-Al_2O_3 supported nickel catalysts for acetic acid steam reforming. *Int. J. Hydrog. Energy* **2017**, *42*, 20540–20548. [CrossRef]
234. Matas Güell, B.; Babich, I.V.; Lefferts, L.; Seshan, K. Steam reforming of phenol over Ni-based catalysts–A comparative study. *Appl. Catal. B Environ.* **2011**, *106*, 280–286. [CrossRef]
235. Frusteri, F.; Freni, S.; Chiodo, V.; Spadaro, L.; Di Blasi, O.; Bonura, G.; Cavallaro, S. Steam reforming of bio-ethanol on alkali-doped Ni/MgO catalysts: Hydrogen production for MC fuel cell. *Appl. Catal. A Gen.* **2004**, *270*, 1–7. [CrossRef]
236. Ahmed, T.; Xiu, S.; Wang, L.; Shahbazi, A. Investigation of Ni/Fe/Mg zeolite-supported catalysts in steam reforming of tar using simulated-toluene as model compound. *Fuel* **2018**, *211*, 566–571. [CrossRef]
237. Fally, F.; Perrichon, V.; Vidal, H.; Kaspar, J.; Blanco, G.; Pintado, J.M.; Bernal, S.; Colon, G.; Daturi, M.; Lavalley, J.C. Modification of the oxygen storage capacity of CeO_2–ZrO_2 mixed oxides after redox cycling aging. *Catal. Today* **2000**, *59*, 373–386. [CrossRef]
238. Vidal, H.; Kašpar, J.; Pijolat, M.; Colon, G.; Bernal, S.; Cordón, A.; Perrichon, V.; Fally, F. Redox behavior of CeO_2–ZrO2 mixed oxides: II. Influence of redox treatments on low surface area catalysts. *Appl. Catal. B Environ.* **2001**, *30*, 75–85. [CrossRef]
239. Sánchez, E.A.; D'Angelo, M.A.; Comelli, R.A. Hydrogen production from glycerol on Ni/Al_2O_3 catalyst. *Int. J. Hydrog. Energy* **2010**, *35*, 5902–5907. [CrossRef]
240. Wen, G.; Xu, Y.; Ma, H.; Xu, Z.; Tian, Z. Production of hydrogen by aqueous-phase reforming of glycerol. *Int. J. Hydrog. Energy* **2008**, *33*, 6657–6666. [CrossRef]
241. Cakiryilmaz, N.; Arbag, H.; Oktar, N.; Dogu, G.; Dogu, T. Effect of W incorporation on the product distribution in steam reforming of bio-oil derived acetic acid over Ni based Zr-SBA-15 catalyst. *Int. J. Hydrog. Energy* **2018**, *43*, 3629–3642. [CrossRef]

242. Yu, Z.; Hu, X.; Jia, P.; Zhang, Z.; Dong, D.; Hu, G.; Hu, S.; Wang, Y.; Xiang, J. Steam reforming of acetic acid over nickel-based catalysts: The intrinsic effects of nickel precursors on behaviors of nickel catalysts. *Appl. Catal. B Environ.* **2018**, *237*, 538–553. [CrossRef]
243. Pekmezci Karaman, B.; Cakiryilmaz, N.; Arbag, H.; Oktar, N.; Dogu, G.; Dogu, T. Performance comparison of mesoporous alumina supported Cu & Ni based catalysts in acetic acid reforming. *Int. J. Hydrog. Energy* **2017**, *42*, 26257–26269. [CrossRef]
244. Goicoechea, S.; Kraleva, E.; Sokolov, S.; Schneider, M.; Pohl, M.-M.; Kockmann, N.; Ehrich, H. Support effect on structure and performance of Co and Ni catalysts for steam reforming of acetic acid. *Appl. Catal. A Gen.* **2016**, *514*, 182–191. [CrossRef]

Review

Layered Double Hydroxide/Nanocarbon Composites as Heterogeneous Catalysts: A Review

Didier Tichit [1,*] and Mayra G. Álvarez [2,*]

1. ICGM, Université de Montpellier, CNRS, ENSCM, 34296 Montpellier, France
2. GIR-QUESCAT, Departamento de Química Inorgánica, Facultad de Ciencias Químicas, Universidad de Salamanca, 37008 Salamanca, Spain
* Correspondence: didier.tichit@enscm.fr (D.T.); mgalvarez@usal.es (M.G.Á.)

Abstract: The synthesis and applications of composites based on layered double hydroxides (LDHs) and nanocarbons have recently seen great development. On the one hand, LDHs are versatile 2D compounds that present a plethora of applications, from medicine to energy conversion, environmental remediation, and heterogeneous catalysis. On the other, nanocarbons present unique physical and chemical properties owing to their low-dimensional structure and sp^2 hybridization of carbon atoms, which endows them with excellent charge carrier mobility, outstanding mechanical strength, and high thermal conductivity. Many reviews described the applications of LDH/nanocarbon composites in the areas of energy and photo- and electro-catalysis, but there is still scarce literature on their latest applications as heterogeneous catalysts in chemical synthesis and conversion, which is the object of this review. First, the properties of the LDHs and of the different types of carbon materials involved as building blocks of the composites are summarized. Then, the synthesis methods of the composites are described, emphasizing the parameters allowing their properties to be controlled. This highlights their great adaptability and easier implementation. Afterwards, the application of LDH/carbon composites as catalysts for C–C bond formation, higher alcohol synthesis (HAS), oxidation, and hydrogenation reactions is reported and discussed in depth.

Keywords: layered double hydroxides; carbon materials; composites; catalysts; C–C bond formation; oxidation; hydrogenation

1. Introduction

Nanocomposites combining LDHs and carbon-based materials as building blocks have attracted a great deal of interest lately, particularly as electrocatalysts and photocatalysts. Several excellent reviews related to the synthesis, characterization, and application of LDH/carbon nanocomposites highlight their remarkable properties and performance for energy storage and conversion, environmental protection, and pollution abatement [1–7]. Although less developed, the applications of LDH/carbon nanocomposites as nanofillers, non-enzymatic sensors, adsorbents for water remediation, and drug delivery systems have been also reported in several reviews [8–11]. Some of them point out the use of LDH/carbon nanocomposites as catalysts in acid–base and redox reactions [3,8,11,12]. However, there is now a remarkable number of publications reporting the high efficiency of LDH/carbon nanocomposites in these reactions.

This type of application was reported for the first time in 2005 [13,14]. MgAl-LDH/carbon nanofiber (CNF) composites were used in the base-catalyzed condensation of acetone, reaching specific activity four times higher than that of the unsupported MgAl-LDH. Moreover, impregnation of a MgAl-LDH/CNF composite with Pd led to a bifunctional catalyst achieving the single-stage synthesis of methylisobutylketone (MIBK) from acetone, with initial activity five times higher than the mechanical mixture of activated MgAl-LDH and Pd/CNF catalysts. Afterwards, various LDH/carbon nanocatalysts have been implemented, displaying highly successful performance compared to unsupported LDH-based catalysts for base-catalyzed C–C bond-forming reactions, as well as for oxidation and reduction reactions. Their higher efficiency than the single MgAl-LDH mainly results from the higher specific surface areas and porosity. Moreover, interactions between LDH nanosheets or metal nanoparticles (NPs) obtained from LDH precursors and nanocarbons improve electron transfer and avoid aggregation of the LDH or metal NPs. The LDH/carbon composites also exhibit high thermal and chemical stabilities. They generally retain the intrinsic properties of the individual LDH and carbon components with additional synergistic effects.

The hierarchical nanocomposites derived from a multitude of LDH compositions and 0D carbon dots (CD), 1D nanofibers (CF), single-walled carbon nanotubes (SWCNT) and multi-walled carbon nanotubes (MWCNT), and 2D graphene-like compounds give rise to a large family of materials.

The main characteristics of each type of carbon material have been extensively described [15–19]. Graphene, particularly its oxidized or reduced forms, i.e., graphene oxide (GO) or reduced graphene oxide (rGO), is currently the most widely used nanocarbon component in LDH/carbon nanocatalysts. The use of nitrogen-doped graphene is also now emerging. The success of the graphene-like supports for the design of LDH-containing composites is partly based on their 2D structural compatibility, which allows intercalation. The resulting sandwich-type structures are rarely used as catalysts, where disordered arrangements with more accessible and dispersed active sites are desirable [20,21]. GO and rGO are more likely chosen for their electronic and thermal conductivities, their ability to enhance the dispersion and avoid aggregation of the LDH, and to enhance the mechanical and chemical stability of the composites. CNF and CNT appear as the second type of nanocarbons used. More recently, carbon dots (CD) and nitrogen-doped carbon dots (NCD) have been considered promising to develop composites with improved basic properties, fast electron transfer through strong metal–CD interaction, and high mechanical resistance [20–22].

This review highlights the research implemented on the LDH/nanocarbon composites for base-catalyzed C–C bond formation, oxidation, hydrogenation reactions, and HAS. Firstly, the properties of the different types of carbon materials involved as building blocks of the composites will be summarized. Then, the synthesis methods of the LDH/carbon nanocomposites will be described. Afterwards, application of LDH/carbon composites as catalysts in the previously cited reactions will be reported. Finally, the main features emerging from the large survey of the literature considered in the review will be summarized and discussed. In addition, several possible future search directions will be indicated.

2. LDHs and Carbon Materials

2.1. LDHs

LDHs are built up of brucite ($Mg(OH)_2$)-like layers, which consist of magnesium ions surrounded by six hydroxyl groups in an octahedral geometry where the divalent metal (M^{2+}) is isomorphically substituted by a trivalent one (M^{3+}). The excess of positive charge is balanced by intercalated anions coexisting with water molecules. The general formula of LDH can be written as $[M^{2+}_{1-x}M^{3+}_x(OH)_2]^{x+} (A^{m-}_{x/m})^{x-} \cdot nH_2O$, where M^{2+} and M^{3+} are di- and trivalent cations, respectively, and A^{m-} are the interlayer anions. The molar ratio $M^{3+}/(M^{2+} + M^{3+})$ generally ranges from 0.2 to 0.33, although some different M^{2+}/M^{3+} molar ratios have also been reported [23]. LDHs exhibit remarkable versatility for the preparation of base- and metal-supported catalysts due to their variety of compositions and their activity either in the lamellar form or as layered double oxides (LDO) obtained by thermal decomposition. Moreover, they are highly suitable as precursors of metal-supported or multifunctional catalysts, with large specific surface areas and peculiar metal–support interactions [23–26]. However, LDHs are prone to particle aggregation, dissolution in liquid media, and sheet stacking, which causes low dispersion of metal NPs and hinders reactants' accessibility to the active sites. This contributes to reducing the efficiency of the LDH-derived catalysts. Many of these drawbacks have been prevented by the dispersion of LDH-based catalysts on various supports. Among them, nanocarbons are increasingly used due to their complementary properties with LDHs.

2.2. Carbon Materials

Despite exhibiting the same general properties, i.e., electron conductivity, high mechanical strength, high thermal conductivity, and chemical inertness, each nanocarbon has specific properties. For catalytic applications, the nanocarbons should have a high surface area, suitable pore size, high graphitization degree, and strong interfacial coupling. In particular, 2D graphene-like materials, 1D carbon nanotubes (CNT), carbon fibers (CF) and nanofibers (CNF), and 0D carbon dots (CD) have been considered as components of the LDH/nanocarbons [15]. These nanocarbon components are very versatile for surface modification and functionalization, which is necessary to ensure interfacial coupling with LDHs.

Graphene-like materials, i.e., graphene (G), graphene oxide (GO), and reduced graphene oxide (rGO), are the most reported materials involved in LDH/nanocarbon composites. G consists of a single layer of sp^2 and sp^3-hybridized carbon atoms organized in a 2D hexagonal lattice. The synthesis of graphene-like materials involves either bottom-up methods starting from carbon molecules or top-down methods using a carbon source, generally graphite. For catalytic applications, chemical synthesis is most suitable since it provides a reactive surface with high density of functional groups. Hummers' method and its modified methods are the most common routes for graphite oxidation [27]. The obtained GO can be easily exfoliated. Through a subsequent reduction step with a chemical agent or by ultrasonication or thermal or hydrothermal treatment, GO sheets are transformed into reduced graphene oxide (rGO).

A CNT can be considered as a rolled-up graphene sheet to form cylindrical molecules. They are classified as single-walled carbon nanotubes (SWCNT) and multi-walled carbon nanotubes (MWCNT) according to the number of rolled-up graphene layers forming the tubular nanostructure. CNTs present high strength, electrical and thermal conductivity and stability, and a high surface area, which makes them very attractive for catalytic applications.

CFs and CNFs exhibit structures and properties closely related and similar to those of CNTs. However, the geometry of CNFs, formed by regularly stacked truncated conical graphene layers, is different to that of CNTs [19]. CNFs can be defined as linear filaments formed of sp^2 carbon atoms, giving a flexible, highly graphitic structure with lengths from several nanometers to microns. Similar to CNTs, CNFs present a high surface area and chemically active end planes, which facilitate its functionalization.

Carbon dots (CD), including carbon quantum dots CQD, graphene quantum dots (GQD) and N-doped carbon dots (NCD), and the CD/inorganic nanocomponents, are emerging as cutting-edge materials for the development of advanced catalysts. CD exhibit a size below 10 nm, whose structure is composed of sp^2 and sp^3 hybridized carbon atoms in the core and the outer part, respectively [28]. A variety of top-down and bottom-up syntheses have been developed for producing CD with different characteristics.

The typical synthesis, structure, and properties of the nanocarbons used in the LDH/nanocarbon-derived catalysts are summarized in Table 1. More extensive descriptions can be found in several recent reviews [15–17,19,28–31].

Table 1. Properties and synthesis methods of different types of nanocarbons.

Nanocarbon	Properties	Synthesis Methods
Graphene-like materials	sp2-sp3 electronic configuration with free π-electrons Semiconductor, fast electron transfer Highly functionalizable Strong hybridization with electronic state of catalyst species, strong interfacial coupling π-π conjugation interaction of reactants Extremely high theoretical surface area	Confined self-assembly Chemical vapor deposition Arc discharge Epitaxial growth on SiC layer Unzipping of carbon nanotubes Mechanical exfoliation of graphite Sonication of graphite Electrochemical exfoliation/functionalization of graphene Chemical synthesis/exfoliation
CNT	sp2 electronic configuration with free π-electrons Highly graphitic Enhanced charge transport Tailorable acid/basicity and easy functionalization High surface area Good thermal stability	Arc discharge Laser ablation Chemical vapor deposition
CNF	Facile and eco-friendly preparation Semiconductive, electronic structure similar to graphite Chemically active edges, easy functionalization Excellent thermal resistance High surface area	Chemical vapor deposition Floating catalyst method Electrospinning/carbonization
CD	sp2 hybridization Water solubility, easy functionalization Low cost and toxicity Quantum confinement properties, semiconductor	Microwave-assisted Combustion/hydrothermal Supporting synthesis method Arc discharge Laser ablation Electrochemical synthesis Chemical oxidation

3. Preparation Methods

Among the various preparation methods of LDH/carbon nanocatalysts, two stand out: self-assembly of LDHs and carbon components and coprecipitation of LDHs on the carbon component acting as a growth substrate [1,3,8,10,11]. The self-assembly method is scarcely utilized, and an overwhelming majority of LDH/carbon nanocomposites are prepared by coprecipitation, which can be easily adapted to the nature of the carbon material and the design of various structures.

3.1. Self-Assembly Method

The self-assembly method, particularly employed in the case of GO, is based on the electrostatic interaction between the negatively and positively charged surfaces of the GO and LDH, respectively. It leads to different nanocomposite structures depending on the previous treatment of the LDH mixed with the GO nanosheets exfoliated by ultrasonication. When the LDH is exfoliated in water, the resulting LDH/GO composite is poorly homogeneous, exhibiting LDH and GO domains both separated and in intimate contact [20]. On the contrary, LDH nanosheets exfoliated in dimethyl formamide (DMF), when restacking is avoided during drying, leads to a nanocomposite with ultrathin LDH nanosheets highly dispersed on GO, boosting the catalytic performance [32].

A nanostructured Fe_3O_4@GO@Zn-Ni-Fe-LDH magnetic composite was developed by Zeynizadeh et al. combining three components: magnetic Fe_3O_4 nanoparticles (MNP), GO, and ZnNiFe-LDH [33]. Fe_3O_4 MNP allow the recovery and reusability of the nanocatalyst, while their immobilization on GO prevents aggregation. The overall synthetic process of the Fe_3O_4@GO@Zn-Ni-Fe-LDH composite started with the oxidation of graphite to GO (Hummers' procedure), followed by the coprecipitation of Fe_3O_4 on GO exfoliated by sonication to give the Fe_3O_4@GO component. Then, the Fe_3O_4@GO constituent was assembled with a previously prepared ZnNiFe-LDH, which led to a nanocomposite where LDH was disposed in plane with the GO constituent, generating a highly porous material.

CD/LDH catalysts have also been generally prepared by self-assembly, mixing LDH and CD aqueous solutions without particular pretreatments in the preformed constituents [34]. In other cases, the LDH nanosheets are stabilized with an organic polymer to avoid particle aggregation and favor water dispersibility and CD dispersion on the LDH sheets. For instance, Yang et al. developed a method for the synthesis of an LDH@PDA@PNIPAM@Pd/CD nanocatalyst where polydopamine (PDA) was deposited on the MgAl-LDH nanosheets, dispersed by ultrasonic treatment through the mussel chemical method [22]. Afterwards, sulfhydryl-terminated poly(N-isopropylacrylamide) (PNIPAM) polymer was grafted via Michael addition. Finally, the LDH@PDA@PNIPAM@Pd/CD nanocatalyst was obtained by loading Pd and CD on the polymer brush-modified MgAl-LDH nanosheets dispersed by sonication (Figure 1a). This preparation favored both the formation of small and uniform CD and Ag NPs. The different components (LDH and CD) in the composite promoted electron transfer and migration but stabilized and improved the dispersion of Pd NPs.

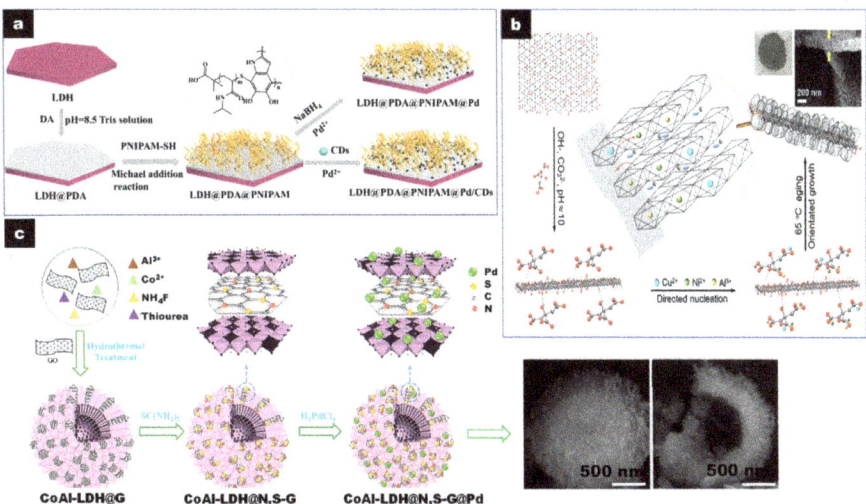

Figure 1. (**a**) Schematic synthesis of LDH@PDA@PNIPAM@Pd/CDs and LDH@PDA@PNIPAM@Pd catalysts. Reproduced with permission [22]. Copyright 2021, Elsevier. (**b**) Schematic synthesis strategy for nanoarray-like xCu-LDH/rGO nanohybrids. Reproduced with permission [35]. Copyright 2019, ACS. (**c**) Formation process of hollow flower-like and SEM images of CoAl-LDH@N,S-G@Pd. Reproduced with permission [36]. Copyright 2019, Elsevier.

3.2. Coprecipitation Method

Coprecipitation of LDH on the carbon support is facilitated by the strong adsorption ability of the metal cations in the solution with the negatively charged carbon surface. This gives rise to highly dispersed LDH nucleation sites.

Except for CD and amorphous carbon, other nanocarbons, such as G, CF, and CNT, need to be oxidized before the reaction due to their chemical inertness and hydrophobicity. Oxygenated functional groups are introduced through different procedures. GO is produced via oxidative exfoliation of graphite using the Hummers or modified Hummers method [27]. The functionalized surface of GO allows the nucleation and growth of LDH by electrostatic forces between GO and the LDH cation precursors [20,21,32,33,37–40]. Treatment of GO with citric acid (CA) leads to hierarchical LDH/GO composites containing completely exfoliated GO, supporting small LDH particles vertically oriented, due to the complexation effect of citrate on the LDH nucleation sites [35,39,41]. The preparation of nanocomposites based on CA-assisted synthesis is schematized in Figure 1b, with several key features at each main step favoring the formation of nanosheet array-like structures. Addition of the mixed base solution up to pH 10 to the CA-GO suspension induced deprotonation of the –COOH groups of CA (pK_{a1} = 3.13, pK_{a2} = 4.76, pK_{a3} = 5.40) and a decrease in the negative surface charge of GO. This led to hydrogen bonding between the carboxyl groups of the citrate species and the OH groups of GO. The dispersibility of GO was enhanced due to electrostatic repulsion among the negatively charged CA-functionalized sheets. When the mixed metal cations and the alkaline solutions were simultaneously added into the CA-GO suspension, the metal cations were electrostatically bonded to the citrate species, leading to highly dispersed nucleation sites and LDH sheets. The LDH nuclei presented faster growth in the a,b than in the c direction, resulting in plate-like LDH crystals, oriented vertically and covering both sides of the GO surface, preventing their restacking.

Functionalization of CNT goes from oxidation by sonication or reflux in a strong acid solution to the adsorption of organic molecules or coating with a polymer, e.g., polyacrylic acid (PAA), introducing carboxyl groups [42–47]. CNT functionalization followed by the addition of an organic linker efficiently immobilized the LDH cation precursors on the surface. For instance, L-cysteine (L-cys) has been added to functionalized PAA-CNT. The LDH was then coprecipitated on the L-cys-PAA-CNT materials, leading to the nanocomposite. L-cys contained on the surface of PAA-CNTs can interact with the cations, forming the LDH, acting as a bridging linker between LDH and CNT [47].

The presence of L-cyst led to the coprecipitation of a highly dispersed LDH phase on the CNT and inhibited the crystal growth of the LDH compared to single-functionalized PAA-CNT. This strategy can be applied to obtain highly dispersed metallic NPs derived from the LDH over CNT. In contrast, studies involving either CNF or CF report simple oxidation by refluxing in concentrated nitric acid [14,48].

Coprecipitation of LDH on the nanocarbon has been performed following different approaches. The metal precursor salts of the LDH can be added to the carbon material (GO, functionalized CNT, CD) with subsequent addition of the alkaline solution [20,37,38,47,49]. When LDH is precipitated on CNT, the particles nucleate and grow within the open framework of CNT, as well as separated domains. Due to the different geometry of the two components (1D vs. 2D), the contact area between the particles is relatively small.

Intimate contact has been found between LDH and GO particles in LDH/rGO nanocatalysts, where the LDH nanoplatelets are generally disposed in plane with the rGO surface, except for high LDH loadings, where the nanoplates grow perpendicularly. Separate aggregated LDH domains can be also observed depending on the loading and preparation route.

Previously oxidized CNF have also been used as a substrate for the precipitation and growth of LDH through incipient impregnation of the Mg^{2+} and Al^{3+} salt precursors and an alkaline solution, followed by hydrothermal treatment (50 °C for 24 h). The LDH platelets presented a lateral size of ca. 20 nm supported on the CNF, with a loading of 11 wt% [50–52].

In many other cases, the nanocarbon was first dispersed in the alkaline solution and the metal salt solution was subsequently added [39,40,42,44,53]. It led to well-dispersed and non-aggregated LDH nanosheets over the nanocarbons, particularly when GO contained citrate complexing groups, where nanoarray-like structures were observed. The different hierarchical structures induced by the order of addition of the metal cation salts and the alkaline solutions account for different supersaturation levels during synthesis.

The simultaneous addition of the metal-containing salt solution and the alkaline solution at constant pH into a suspension containing the well-dispersed nanocarbon (GO, CF, CNT) has been also reported [33,43,48]. In such cases, the carbon support was completely wrapped by well-dispersed LDH nanoflakes oriented vertically. Contrarily, a single-drop method, where the base solution was added dropwise to a mixture of GO and Ru, Co, and Al chloride salt solution, led to nanocomposites with parallel orientation of the LDH platelets on GO (Ru/LDH-GO-P) [54]. The high local supersaturation level in the former case favored the nucleation and perpendicular growth of the LDH platelets. Contrarily, in the single-drop method, the low supersaturation favored growth and led to parallel orientation of the LDH platelets onto GO with higher interaction.

LDH can also be coprecipitated on GO or N- and S-doped GO using urea or thiourea as a hydrolysis agent under hydrothermal conditions. Rohani et al. reported a catalyst with hollow inner and mesoporous hierarchically flower-like outer structures using a template-free method (Figure 1c) [36]. The preparation process was based on the synthesis of CoAl-LDH@GO hollow spheres previously reported involving Ostwald ripening by coprecipitation of the Co and Al metal salts with urea and ammonium fluoride within a GO suspension in ethanol [55]. The obtained hollow CoAl-LDH@GO was then dispersed in thiourea as source of N and S dopants and hydrothermally treated (180 °C for 12 h) to obtain CoAl-LDH@N,S-G hollow spheres. Finally, the Pd NPs were introduced to obtain CoAl-LDH@N,S-G@Pd hollow spheres. During the process, most of the Pd (II) was reduced to Pd (0) due both to the presence of reductive Co (II) in the LDH and of the electron rich N,S–G. SEM images show a spherical flower-like architecture of the assembly of CoAl-LDH and G with a dense outer part and a hollow inner part (Figure 1c) of the hierarchical CoAl-LDH@N,S-G@Pd. HRTEM confirmed the presence of (012) planes of CoAl-LDH and a good combination between Pd NP and G sheets of ~2 nm thickness. These structural features, creating high surface areas, also improved the dispersion of Pd NPs, electronically enriched due to the interaction with the basic LDH support.

The slow urea hydrolysis maintains low supersaturation during the coprecipitation and a lower nucleation than growth rate, leading to LDH materials with large crystallites and a homogeneous particle size distribution [56–58]. Urea hydrolysis also gives rise to GO reduction to G [32].

The influence of the synthesis route on the final structure and features of the composites was noticed in the work of Álvarez et al., which studied two series of MgAl-LDH/rGO with $0.5 \leq$ LDH/rGO ≤ 10 mass ratio by the direct coprecipitation or self-assembly of MgAl-LDH and rGO. The LDH/rGO samples were dried either in static air at 80 °C or by freeze drying (Table 2, entry 7) [20].

Table 2. Composition and preparation method of the LDH/nanocarbon composites, and catalytic reactions performed.

	Comp.	Preparation Method	Reaction (Substrate)	Reaction Conditions	Ref.
1	CNF/MgAl-LDH	Pore precip^{ion} by incipient imp^{ion}	Condens^{ion} (Acetone)	1.8 mol acetone; m_{cat}: 1 g; activated 450 °C; rehyd. gas phase; $T = 0$ °C	[13]
2	CNF/MgAl-LDH	Pore precip^{ion} by incipient imp^{ion}	MIBK synthesis (Acetone)	1.8 mol acetone; m_{cat}: 1 g; calcined 450 °C; red. 250 °C (H_2); rehyd. gas phase. $T = 60$ °C, P $H_2 = 1.2$ bar	[14]

Table 2. Cont.

	Comp.	Preparation Method	Reaction (Substrate)	Reaction Conditions	Ref.
3	CNF/MgAl-LDH	Pyrolysis/hydration of electrospun PVA/PEO/MgAl–nitrate	Transest^{ion} (DEC, glycerol)	DEC: glycerol: 16 mol/mol; m_{cat}: 0.4 g; calcined 450 °C; rehyd. gas phase; T = 130 °C	[59]
4	MWCNT/MgAl-LDH	Coprecip^{ion}	Condens^{ion} (Acetone)	0.8 mol acetone; m_{cat} adjusted to contain 50 mg LDH; calcined 450 °C; rehyd. liq. phase; T = 0 °C	[44]
5	CNF/MgAl-LDH	Pore precip^{ion} by incipient imp^{ion}	Transest^{ion} (DEC, glycerol)	DEC: glycerol: 17 mol/mol; m_{cat}: 0.3 g; calcined 500 °C; rehyd. liq. or gas phase; T = 130 °C	[60]
6	CNF/MgAl-LDH	Pore precip^{ion} by incipient imp^{ion}	Condens^{ion} (Acetone)	100 g acetone; 1 g supported catalyst (0.3 g bulk catalyst); calcined 500 °C; rehyd. gas phase	[60]
7	rGO/MgAl-LDH	Cop^{ion} or self-assembly; Freeze drying or drying in air (80 °C)	Condens^{ion} (Acetone)	27.5 mmol acetone; 1.5 wt% LDH/rGO respect to acetone; activated 450 °C; T = 0 °C	[20]
8	rGO/MgAl-LDH	Cop^{ion} LDH on GO	Claisen-Schmidt condens^{ion} (BALD, acetophenone)	BALD $^{(a)}$/acetophenone = 1.05; m_{cat}: 25 wt%; calcined 450 °C; T = 40 °C	[37]
9	rGO/CeMgAl-LDH	Cop^{ion} LDH on GO	Knoev.^{el} (BALD, diethylmalonate)	BALD: diethylmalonate 2:3; m_{cat}: 1 wt%; non-activated; T = 160 °C.	[61]
10	rGO/CeMgAl-LDH	Cop^{ion} LDH on GO	Cascade oxidation-Knoev.^{el} (BA/benzoyl acetonitrile)	BA $^{(b)}$/benzoyl acetonitrile 0.83; m_{cat}: 1 g; non-calcined; P O$_2$ = 1 atm: T = 80 °C	[61]
11	GO/CuAl-LDH GO/CoAl-LDH	Cop^{ion} LDH from metal nitrates with NaOH + Na$_2$CO$_3$ on GO) Cop^{ion} LDH from metal chlorides by urea on GO (97 °C; 48 h)	Ullmann (Iodobenzene)	Iodobenzene (2 mmol), DMSO (4 mmol), m_{cat}: 0.25 g; non-activated; T = 110 °C	[21]
12	N,S-G/CoAl-LDH/Pd	Mixture Al^{3+}, Co^{2+}, NH$_4$F, and thiourea hydrothermally treated. Then + H$_2$PdCl$_4$	Sonogashira (Aryl halides/Phenylacetylene)	Aryhalide (1 mmol), phenylacetylene (1.2 mmol), 0.06 mol% Pd; cat. non-activated; T = 100 °C.	[36]
13	rGO/CoAl-LDH/Pd	Cop^{ion} Co and Al nitrates on the CA-modified GO (pH 10). Aging 65 °C; 4 h	Heck (Iodobenz./Styrene)	DMF (12 mL), H$_2$O (4 mL), K$_2$CO$_3$ (3 mmol), aryl halide (1 mmol), styrene (1.2 mmol); cata non-activated	[53]
14	GO/ZnNiFe-LDH/Fe$_3$O$_4$	Fe$_3$O$_4$ on GO. Then LDH immobilized on Fe$_3$O$_4$@GO	One-pot Knoev.^{el}-Michael (BALD/4-hydroxycoumarin)	4-hydroxycoumarin/BALD = 1: 0.5; m_{cat}: 30 mg; non-activated; reflux H$_2$O (2 mL)	[33]

Table 2. Cont.

	Comp.	Preparation Method	Reaction (Substrate)	Reaction Conditions	Ref.
15	rGO/RuCoAl-LDH	Ru, Co, and Al chlorides added into GO susp.^{ion}. NaOH + Na$_2$CO$_3$ then added until pH 10. Aging 90 °C; 6 h Ru, Co, and Al chlorides and NaOH + Na$_2$CO$_3$ sol.^{ion} simultaneously added to GO susp.^{ion} until pH 10. Aging 90 °C; 6 h	One-pot oxidation + Knoev.^{el} cond^{ion} (Cinnamyl alcohol/ethyl cyanoacetate (substituted BA/ethyl cyanoacetate or malonitrile))	BA (1 mmol) and toluene (7 mL) in 50 mL flask kept 60 °C (O$_2$, 2 h, 0.10 MPa). After ethyl cyanoacetate (1.5 mmol) added at 60 °C: cata non-activated	[54]
16	rGO/NiAl-LDH/Au	LDH/rGO: Cop^{ion} metal nitrates and exfol. GO dispersed into NaOH + Na$_2$CO$_3$ Au/NiAl-LDH/rGO: (1) PVA aq. sol. (PVA/Au = 1.2) + aq. HAuCl$_4$ +NaBH$_4$ (NaBH$_4$/Au = 5); (2) LDH/rGO added in the Au-containing colloid	Oxidation (BA)	m_{cata} = 0.4 g; non-activated; 40 mL BA, no solvent; T = 140 °C; $P\,O_2$ = 2 bar	[40]
17	GO/CoAl-LDH	Assembly of GO and LDH exfoliated in formamide	Oxidation (BA)	m_{cata} = 0.1 g; non-activated; 1 mmol BA, 5 mL DMF; T = 120 °C; $P\,O_2$ = 1 bar	[32]
18	NCD/MgAl-LDH/Au	Cop^{ion} LDH in presence NCD (pH 10.5). Then addition of HAuCl$_4$ into aq. susp^{ion} NCD/LDH and poly(N-vinyl-2-pyrrolidone) + NaBH$_4$	Oxidation (BA)	m_{cata} = 0.1 g; non-activated; 10 mL BA, no solvent; T = 120 °C; $P\,O_2$ = 0.4 MPa	[49]
19	CNT/MgAl-LDH/Ru	LDH-CNT: Acid-treated CNT dispersed in NaOH + Na$_2$CO$_3$ then addition of metal nitrates (pH 10.5). Treatment 100 °C; 16 h. Wetness impr. LDH-CNT with RuCl$_3$·3H$_2$O (Ru = 1 wt%)	Oxidation (BA)	m_{cata} = 0.2 g (0.02 mmol Ru); reduced 400 °C (H$_2$); 2 mmol BA, 105 mL toluene, 6 mL H$_2$O; T = 85 °C; $P\,O_2$ = 1 bar; O$_2$ flow rate 25 mL min^{-1}	[42]
20	GO/CoCuAl-LDH	Cop^{ion} LDH from metal nitrates on GO with NaOH + Na$_2$CO$_3$ (pH 10). GO/LDH = 0.2–0.8 wt/wt	Oxidation (EB)$^{(c)}$	EB 10 mmol; TBHP 40 mmol; m_{cata} = 0.1 g; non-activated; T = 120 °C	[38]

Table 2. Cont.

	Comp.	Preparation Method	Reaction (Substrate)	Reaction Conditions	Ref.
21	CNT/ZnCr-LDH	Cop^{ion} LDH from nitrate salts on acid-treated CNT and NaOH + Na$_2$CO$_3$	Oxidation (EB)	EB: 10 mL (81.7 mmol); no solvent; P$_{O2}$ =1 MPa; m$_{cata}$ = 0.1 g; non-activated; T = 130 °C	[62]
22	rGO/NiCo-LDH	GO colloid added in mixed sol. (V$_{C2H5OH}$:V$_{H2O}$ = 1:1) and sonicated. Metal chloride sol^{ion} (Ni/Co = 1) added dropwise to GO susp^{ion} with NH$_4$Cl (pH 9). Treatment 120 °C; 12 h.	Oxidation (Sty)$^{(d)}$	1.1 mL Sty; m$_{cata}$ = 0.03 g; non activated; 10 mL acetonitrile, 1.8 mL TBHP (70%, aq. sol.); T = 80 °C	[63]
23	rGO/CuMgAl-LDH	Cop^{ion} LDH into CA $^{(e)}$-GO suspension	Reduction (4-NP)	1 mM 4-NP (200 μL) + 10 mM NaBH$_4$ (2.5 mL) + 10 μL cata. susp. (2.5 mg mL^{-1}) non-activated	[39]
24	rGO/CuMgAl-LDH	Cop^{ion} LDH into CA-GO suspension	Reduction (4-NP)	200 μL 4-NP (200 μL, 1 mM) + NaBH$_4$ (2.5 mL, 10 mM) + cata. (20 μL, 1 mg mL^{-1}) calcined 600 °C	[41]
25	rGO/CuNiAl-LDH	Cop^{ion} LDH into CA-GO suspension	Reduction (4-NP)	4-NP (200 μL, 1 mM) + NaBH$_4$ sol (2.5 mL, 10 mM) + cata. (10 μL, 2.5 mg mL^{-1}) non-activated	[35]
26	CDs/MgAlCe-LDH/Ag	LDH Cop^{ion} and dispersed into AgNO$_3$ aq. sol. Then addition of CDs aq. sol. and UV irradiation	Reduction (4-NP, MB, MO, CR, RhB, R6G)	4-NP (200 μL, 10 mM) + NaBH$_4$ sol (200 μL, 0.1 M) + cata. (20 μL, 1 mg mL^{-1}) non-activated	[34]
27	CDs/MgAl-LDH/PNIPAM/Pd	LDH funct^{ed} with PNIPAM (Mussel). Then addition of Pd and CDs	Reduction; Knoev.^{el}; One-pot Knoev.^{el}-reduction (MB, RhB, CR, R6G, MO. 4-NP, o-NO, m-NP, BALD, malonitrile)	MB (2 mL, 0.013 mM) + NaBH$_4$ sol (1 mL, 0.5 M) + cata. (20 μL, 0.05 mg mL^{-1}) non-activated	[22]
28	rGO/NiAl-LDH/Pt	Cop^{ion} LDH on GO. Then impregnation with H$_2$PtCl$_6$	Reduction (p-nitrophenol, P-NA)	P-NA (3 mL, 0.1 M) + NaBH$_4$ sol (0.1 mL, 0.1 M) + m$_{cat}$: 1 mg activated 600 °C (N$_2$)	[64]
29	PAA-CNT/NiAl-LDH	LDH cop^{ion} in presence PAA functionalized CNT (P-CNT) and L-cysteine	Reduction o-(CNB)	T = 140 °C, P H$_2$ = 2 MPa; cata activated 500 °C (10% v/v H$_2$/Ar)	[47]
30	rGO/NiAl-LDH	LDH cop^{ion} in presence GO	Reduction (CALD)	T = 120 °C, P H$_2$ = 1 MPa; cata activated 600 °C (N$_2$)	[65]
31	MWCNT/CuMgAl-LDH	LDH cop^{ion} in presence pretreated MWCNT	Reduction (Glycerol)	Glycerol (5 g, 60 wt%); P H$_2$ 2 Mpa; T = 180 °C; m$_{cat}$: 0.2 g activated 400 °C (air)	[66]

Table 2. Cont.

	Comp.	Preparation Method	Reaction (Substrate)	Reaction Conditions	Ref.
32	CFs/CuCoAl-LDH	Cop^{ion} of Cu, Co, and Al nitrates on CF (pH 9.5–10). Aging 60 °C	HAS (H$_2$/CO/N$_2$)	3 MPa syngas mixture H$_2$/CO/N$_2$ = 8:4:1 (GHSV: 3900 mL (g$_{cat}$ h)$^{-1}$); T = 220 °C; cata reduced 450 °C (H$_2$)	[48]
33	CNT/CuCoAl-LDH	Cu, Co, Al nitrates dissolved into CNT sol^{ion} then + (NaOH + Na$_2$CO$_3$) at pH 9.5. Aging RT	HAS (H$_2$/CO/N$_2$)	3 MPa syngas mixture H$_2$/CO/N$_2$ = 8:4:1 (GHSV: 3900 mL (g$_{cat}$ h)$^{-1}$). T = 230 °C; cata reduced 450 °C (H$_2$)	[43]
34	CFs/CuFeMg-LDH	Cop^{ion} Cu, Fe and Mg nitrates and on CF at pH 9.5–10. Aging 65 °C	HAS (H$_2$/CO/N$_2$)	3 MPa syngas mixture H$_2$/CO/N$_2$ = 8:4:1 (GHSV: 3900 mL (g$_{cat}$ h)$^{-1}$). T = 280 °C; cata reduced 500 °C (H$_2$)	[67]
35	CFs/CuZnAl-LDH/K	Cop^{ion} Zn and Cu nitrates on CF and Al nitrate at pH 5.5–6.2. Aging 110 °C. K$_2$CO$_3$ imp^{ion}	Isobutanol synthesis (H$_2$/CO)	H$_2$/CO = 2 (GHSV: 3900 h^{-1}), m$_{cat}$: 560 mg; T = 320 °C, P = 4 MPa; cata activated 400 °C (N$_2$) + 320 °C (N$_2$/H$_2$ = 4:1)	[68,69]

(a) Benzaldehyde; (b) Benzyl alcohol; (c) Ethylbenzene; (d) Styrene; (e) Citric acid.

The LDH:rGO ratio, the preparation, and the drying methods influenced the crystallinity of the composites, which was higher in the self-assembled than in the coprecipitated and freeze-dried samples. Intimate contact between LDH and rGO components occurred in all samples. However, coprecipitation led to more homogeneous LDH/rGO composites, with LDH nanoplatelets markedly smaller than those obtained by self-assembly. Moreover, drying at 80 °C developed more ordered structures by hydrogen bonding between both LDH and rGO building blocks than freeze drying, where restacking of LDH and rGO layers was avoided. The self-assembled samples exhibited, in general, a heterogeneous structure regardless of the drying method.

Differently from coprecipitation or self-assembly methods, LDH/CF composites have been prepared through the reconstruction of an LDO dispersed on pyrolyzed poly(vinyl alcohol) (PVA)/poly(ethylene oxide) (PEO) fibers obtained by electrospinning. In this case, MgAl-LDH precursors were directly introduced throughout the microstructured fiber matrix, whose confined space limited the LDH crystal growth, although this can also reduce their accessibility. The PVA/PEO/MgAl fibers were obtained from a PVA/PEO/MgAl precursor aqueous solution following three consecutive steps: (i) PVA/PEO/MgAl fiber templating by electrospinning of the precursor solution; (ii) carbonization of the PVA/PEO/MgAl fibers at 450 °C, leading to Mg(Al)O mixed oxide supported on carbon fiber; (iii) hydration in the gas phase leading to the reconstruction of Mg(Al)O into MgAl-LDH and the formation of the LDH/CF composite (Table 2, entry 3) [59]. Pyrolysis and rehydration resulted in the formation of agglomerated LDH particles both over the surface and within the network formed by the fused CNF. HRTEM and XRD analyses confirmed the formation of a crystalline layer of approximately 10–20 nm thickness containing LDH crystallites of approximately 4–8 nm (5–10 layers).

4. Applications of LDH/Nanocarbon Composites in Heterogeneous Catalysis
4.1. Base and Multi-Step Reactions

A series of publications reported the use of MgAl-LDH NPs deposited on various nanocarbons (CNF, MWCNT, or GO) as solid base catalysts for the self-condensation of acetone, the single-stage synthesis of MIBK, the Claisen–Schmidt condensation of acetophenone and benzaldehyde, the Knoevenagel reaction, and the transesterification of glycerol with diethyl carbonate [13,14,20,37,44,59–61].

The preparation method, activation procedure, and experimental conditions used in the different reactions considered in this review are summarized in Table 2.

4.1.1. Self-Condensation of Acetone and Single-Stage Synthesis of MIBK

The pioneering work of Winter et al. reports catalysts based on MgAl-LDH particles of controlled size supported on CNF to improve the number of accessible active sites and the LDH's mechanical strength [13,14]. The MgAl-LDH/CNF composite was used as a base catalyst for the self-condensation of acetone to diacetone alcohol (DAA), the single-stage synthesis of MIBK, and the condensation of citral and acetone to pseudoionone. The MgAl-LDH/CNF nanocatalysts were activated at 500 °C, followed by rehydration under decarbonated water-saturated N_2 flow, providing highly efficient base sites [50–52,70–72].

The small lateral dimension (21 nm) of the LDH platelets in the LDH/CNF catalysts increased the number of active sites situated at the edges, improving by approximately four times the catalytic activity in the self-condensation of acetone if compared with the unsupported LDH. The number of basic sites correlated well with the initial specific activity, confirming the influence of reducing the size of the LDH platelets. Selectivity to DAA was higher than 98%. Reactivation of the used catalyst by heat treatment and rehydration removed the adsorbed side products and restored the activity without leaching of the LDH phase.

Winter et al. also reported the preparation of a multifunctional Pd-MgAl-LDH/CNF catalyst, containing acid, base, and hydrogenating sites required for the single-stage synthesis of MIBK from acetone [14]. The as-synthesized LDH/CNF impregnated with Pd(acac)$_2$ was then heated at 500 °C and reduced at 250 °C, and was rehydrated to obtain Brønsted basic sites, favoring the condensation of acetone with respect to Lewis basic sites (Table 2, entry 2).

The initial activity of the activated Pd-LDH/CNF was five times higher than that of a physical mixture of LDH and Pd/CNF. The formation of DAA was faster than the dehydration of DAA to mesityl oxide (MO), which was almost not observed, being rapidly hydrogenated to MIBK. The sum of selectivities into DAA, MO, and MIBK was higher than 99%.

These were the first reports on LDH/CNF-supported catalysts that confirmed that the catalytically active sites are situated at the edges of the LDH platelets and their amount greatly increases when the platelets are well-dispersed on a support.

Commercially available MWCNT were used by Celaya-Sanfiz et al. as supports for LDHs [44]. MWCNT have several advantages over CNF, particularly a higher aspect ratio, smaller diameter, and enhanced mechanical strength.

Coprecipitation of MgAl-LDH into a dispersion of previously oxidized MWCNT led to the MWCNT/MgAl-LDH nanocomposite with LDH loading ranging from 33 to 83 wt%. Activation was performed by calcination and rehydration in the liquid phase. The catalysts presented LDH particles disposed within the open structure of the MWCNT, decreasing the specific surface area from 230 m^2 g^{-1} of bare MWCNT to 119–85 m^2 g^{-1} depending on the LDH loading. The LDH crystallite sizes of the composites were slightly smaller than in the bulk LDH, and, apparently, they decreased with the increasing MWCNT content. However, larger LDH platelets were observed after hydration (ca. 62 nm) than for the as-synthesized MWCNT/LDH (ca. 30 nm). Álvarez et al. reported a similar trend for hydrated MgAl-LDH/CNF, assigned to dissolution/recrystallization during hydration in the liquid phase [60]. The basicity of the MWCNT/LDH ranged from 1.6 to 2.4 mmol g$_{LDH}^{-1}$ of CO_2 desorbed for the composites with LDH content of 33–83 wt%, which is higher than the values reported previously in LDH/CNF samples, obtained by CO_2 adsorption, with lower LDH loadings (11 to 20 wt%) [14,59,60].

The self-condensation of acetone at 0 °C with the hydrated MWCNT/LDH composites led to DAA only. There was an almost threefold increase in the initial reaction rate (V_0)

from ca. 71 to ca. 196 mmol$_{DAA}$ g$_{LDH}$ h^{-1} when going from the mere activated LDH to the MWCNT/LDH catalyst containing 83 wt% LDH. V_0 decreased greatly with the MWCNT loading up to 9.8 mmol$_{DAA}$ g$_{LDH}$ h^{-1} at 33 wt% LDH. This value can be compared to V_0 of 542 mmol$_{DAA}$ g$_{LDH}$ h^{-1} reported by Winter et al. for an LDH loading of 11 wt% [14]. Notably, in this latter case, the V_0 of the LDH/CNF catalyst was fourfold higher than that of bulk LDH. Compared to LDH/CNF, the catalytic improvement of MWCNT/LDH is less important despite its higher LDH content. This accounts for the presence of carboxylated carbonaceous fragments generated during the MWCNT oxidation, which inhibits the active sites.

Álvarez et al. were the first to investigate the LDH/graphene-like composites as acid–base catalysts for C–C bond reactions, particularly the self-condensation of acetone [20]. Two series of MgAl-LDH/rGO with $0.5 \leq$ LDH/rGO ≤ 10 mass ratio were prepared by either direct coprecipitation or self-assembly. In addition, both series of LDH/rGO samples were dried either in static air at 80 °C or by freeze drying (Table 2 entry 7).

After activation at 450 °C, all composites exhibited the XRD pattern of rGO at low LDH loading and of Mg(Al)O mixed oxide at high LDH loading; meanwhile, HRTEM images revealed similar morphologies and structural features to the non-calcined parent samples. The surface areas of the activated composites with LDH:rGO = 10 were in the range of 170–240 m^2 g^{-1}.

Only DAA was obtained with all the activated LDH/rGO catalysts, with a clear influence of the LDH/rGO mass ratio on the activity, from 19.8 to 104.7 mmol g$_{LDH}$$^{-1}$ for LDH/rGO mass ratios of 0.5 and 10, respectively. Only when LDH/rGO ≥ 5, the activity was higher than that of the bulk LDH (51.2 mmol g$_{LDH}$$^{-1}$), with a maximum at LDH/rGO = 10, achieving the best compromise between the number and accessibility of sites. The composite prepared by coprecipitation and freeze drying exhibited remarkable activity, leading to a DAA amount of 104.7 mmol g$_{LDH}$$^{-1}$ after 8 h, compared to around 70–80 mmol g$_{LDH}$$^{-1}$ for LDH/rGO prepared by self-assembly and either dried in air or freeze-dried. This can be related to the reduced lateral dimension of the LDH platelets (30–40 nm) and the disordered structure of the composite, with poorly stacked rGO and LDH sheets in weak interaction, which enhanced the accessibility to the active sites. Contrarily, those samples with lower LDH content exhibited poor activity significantly lower than that of the bulk calcined LDH. This was attributed to their low number of active sites and to the layer-by-layer hybridization between the LDH layers and the rGO sheets, hindering accessibility to the active sites.

Adsorption of CO_2 on both series of samples with amounts from 63 to 157 µmol g$_{cat}$$^{-1}$ for LDH/rGO-0.5 and from 361 to 407 µmol g$_{cat}$$^{-1}$ for LDH/rGO-10 confirmed that the improved catalytic activity accounted for the LDH loading. Moreover, a clear correlation was found between the global rate of reaction and the number of stronger basic sites of the LDH/rGO series.

The basicity, determined by the adsorption of CO_2, and the catalytic activity in the self-condensation reaction of acetone for the different MgAl-LDH/nanocarbon catalysts are compared in Table 3. All the reactions were performed at 0 °C with a similar LDH/acetone weight ratio of ~1.10 for CNF and MWCNT supports, and of 0.36 for rGO.

The number of basic sites varied in a large range, pointing to the influence of the loading and particle size of the LDH component and the structure of the carbonaceous support, which led to different accessibilities to the active sites in the composites. The unexpectedly different number of basic sites found between the two LDH/CNF samples with the same LDH loading and particle size probably resulted from the different conditions of CO_2 adsorption performed at 0 and 25 °C, respectively. The higher number of basic sites of MWCNT/LDH compared to LDH/rGO having similar LDH loading (80–90 wt%) and particle size (~50 nm) typically accounts for the lower accessibility to the sites in the case of rGO due to its intimate contact with the LDH sheets. The LDH particles are, on the contrary, well exposed within the open framework of the MWCNT, acting as a supporting scaffold. The reaction rates are difficult to confirm as the values were calculated over very different times, but a global tendency can be underlined regarding the efficiency of the supports as CNF > MWCNT > rGO. However, the absence of a correlation between the basic properties and the catalytic activity among the different materials confirms that accessibility to the sites is a determining parameter.

Table 3. Basicity and catalytic activity in the self-condensation of acetone at 0 °C for different MgAl-LDH/carbon nanocomposites.

Carbon	Activation (Rehyd.^{ion})	m_{HT} [mg]	m_{acet} [mol]	m_{HT}/m_{acet} × 10^3 [g/g]	CO_2 ads. [mmol g_{HT}^{-1}]	React. Rate [mmol$_{DAA}$ g_{HT}^{-1} h^{-1}]	Ref.
CNF	Calc. 450 °C (gas phase)	110	1.8	1.05	0.75 [a]	542 [b]	[14]
CNF	Calc. 450 °C (gas phase)	120	1.72	1.20	0.37 [c]	nd	[60]
MWCNT	Calc. 450 °C (liq. phase)	50	0.8	1.08	1.8 [c]	196 [d]	[44]
rGO	Calc. 450 °C	0.57	2.75 × 10^{-2}	0.36	0.41 [c]	0.052 [e]	[20]

[a] CO_2 adsorption at 0 °C; [b] over the first 15 min; [c] CO_2 adsorption at 25 °C; [d] from the amount of DAA produced over 100 h; [e] from the amount of DAA produced over 8 h.

4.1.2. Claisen–Schmidt Reaction (Synthesis of Chalcone)

LDH/rGO composites were investigated in the Claisen–Schmidt condensation of acetophenone and benzaldehyde to chalcone [37]. A wide range of basic catalysts were used for this reaction, but selectivity to chalcone is generally low because of the side reactions, such as Michael addition [73–79]. A series of catalysts with $0.5 \leq$ MgAl-LDH/rGO ≤ 20 mass ratio was prepared by coprecipitation and drying in air at 80 °C, as previously reported (Table 2, entry 7) [20].

The number of basic sites of the activated catalysts, evaluated by the amount of adsorbed CO_2, increased from 52 to 265 µmol g_{cat}^{-1} when the LDH content increased. For LDH loadings higher than 80 wt%, the amount adsorbed was similar to or higher than in the bulk activated LDH (128 to 265 µmol g_{cat}^{-1} compared to 129 µmol g_{cat}^{-1}, respectively), all exhibiting similar specific surface areas (~280 m^2 g^{-1}). Thus, dispersion of the LDH nanoparticles on the rGO surface increased the number and accessibility toward highly basic O^{2-} sites. Accordingly, conversion reached 100% after 4 h with activated LDH/rGO catalysts when the condensation of acetophenone and benzaldehyde was carried out at 40 °C in the presence of either polar protic (methanol, MeOH), polar aprotic (acetonitrile, ACN), or non-polar (toluene) solvents or neat conditions. Bare activated LDH needed 8 h to complete the reaction.

The LDH/rGO ratio and the nature of the solvent greatly influenced the distribution of products. Chalcone, with yields in the range from 65 to 100%, was the main product using ACN, toluene, and in neat conditions. The weak acidity of ACN poisoned the stronger basic sites of the catalyst, inhibiting the Michael addition. Meanwhile, in solventless conditions, a Michael addition product was formed with ca. 20% selectivity along with c,t-chalcone (ca. 80% selectivity) due to the stronger basic sites, which initiated a side reaction.

With MeOH only, c,t-chalcone with yields in the range from 20% to 75% and an aldol product were formed. The Michael addition was totally inhibited due to the acidity of MeOH being slightly higher than that of ACN. Toluene led to the higher yield of c-chalcone and the Michael addition product, which was likely related to its very weak acidity and the different adsorption of the reactants on the catalyst surface.

4.1.3. Knoevenagel and One-Pot Oxidative Knoevenagel Reactions

Recently, Ce-containing MgAl-LDH-GO composites bearing different GO loading (5–25 wt%) were tested as catalysts in the Knoevenagel condensation of benzaldehyde with dimethyl malonate to diethyl benzylidene malonate (DBM) and subsequently to cinnamic acid (CNA), and the one-pot cascade oxidation–Knoevenagel condensation of benzyl alcohol (BA) and benzoyl acetonitrile to yield 2-benzoyl-3-phenylacrylonitrile (BPA) [61]. The Ce-MgAl-LDH/GO composite prepared by coprecipitation (Table 2, entry 9) presented the typical structural features corresponding to both LDH and GO. The acidity and basicity measurements showed a synergistic interaction between the Ce-containing MgAl-LDH and GO rather than a simple additive effect. Thus, while Ce-MgAl-LDH solid showed a mainly basic and GO a mainly acidic character, the composite presented both basic and acidic sites. Their numbers passed through a maximum for 15 wt% of GO (2.98 mmol$_{\text{acrylic acid}}$ g^{-1} and 0.46 mmol$_{\text{pyridine}}$ g^{-1}, basic and acid sites, respectively).

The conversion of the mere solids Ce-MgAl-LDH and GO (8% and 5% after 5 h, respectively) was rather low compared to the composites. The catalytic activity, in terms of aldehyde conversion, of the composites increased with the GO content up to an optimum for 15 wt% (24% and 76% after 5 h and 24 h, respectively), in line with the acidity and basicity results. Then, it decreased for higher GO content. The yield to CNA, with a maximum of ca. 45% after 24 h of reaction for the catalyst containing 15 wt% of GO, followed the same trend.

The ratio between the basic and acid sites, associated with the LDH and GO components, respectively, was a key factor determining the product distribution, since the basic sites are involved in the proton abstraction from the α-position of diethyl malonate to undergo nucleophilic addition to the carbonyl group of the benzaldehyde yielding DBM. However, selectivity to DBM decreased with the basicity. This is assigned to the favored decarbethoxylation of DBM by the basic sites, first leading to the ethyl cinnamate ester intermediate. At the same time, conversion of the latter into CNA occurs by the acid sites. Indeed, it was observed that the pure GO phase (essentially acid) led to high selectivity to CNA (99%), indicating a key role in its production in the composites. Again, the catalyst with 15 wt% GO presented the highest values of both total acidic and basic sites, but also showed the highest basic-to-acidic site ratio.

In the case of the cascade reaction, the conversion of BA was higher for the Ce-MgAl LDH-GO composites compared to the LDH and GO materials alone, the latter being completely inactive. BA conversion decreased with the increase in the GO concentration, which was attributed to a decrease in the number of accessible redox active sites needed in the first step of the reaction due to GO restacking at higher content. However, with lower GO content, the accessibility to the redox sites was increased by separation of the LDH particles, which favored the oxidation of BA.

4.1.4. Transesterification of Glycerol

The Mg(Al)O mixed oxide and meixnerite-like compound (OH-MgAl-LDH), obtained upon rehydration in the liquid or gas phase of Mg(Al)O, presented high efficiency in the base-catalyzed transesterification of glycerol with diethyl carbonate (DEC), leading to glycerol carbonate [80,81]. This reaction was also studied using MgAl-LDH/CNF with LDH nanoparticles of controlled size (20 nm), prepared following the protocol of Winter et al., with LDH content of ca. 12 wt% and activated either by calcination at 500 °C (LDH-CNFc) or by calcination followed by rehydration in the liquid phase (LDH-CNFrl) or gas phase (LDH-CNFrg) (Table 2, entries 5 and 6) [60].

The LDH-CNFc was the most efficient catalyst, showing a V_0 of 53 mmol Gly g$^{-1}_{\text{HT}}$ h^{-1} better than LDH-CNFrl and LDH-CNFrg (3 and 7 mmol Gly g$^{-1}_{\text{HT}}$ h^{-1}, respectively) and by far higher than bare calcined LDH (0.18 mmol Gly g$^{-1}_{\text{HT}}$ h^{-1}). Accordingly, total glycerol conversion was obtained after 1.5 h for LDH-CNFc but after 3 h for LDH-CNFrg and 8 h for LDH-CNFrl; meanwhile, the calcined LDH (LDHc) needed 22 h for total conversion. The nearly four times higher number of basic sites in LDH-CNFc than in the bulk catalyst (LDHc) cannot account for its 300 times higher initial reaction rate. This indicates different adsorption of the reactants depending on the different polarity of the catalysts. The heat-treated CNF is less polar than the bulk LDH and this improves DEC adsorption and the reaction rate. Likewise, the different behavior of the rehydrated samples can be explained by their different content of physisorbed and interlayer water.

The LDH-CNFc catalysts exhibited good stability, with slight decrease in glycerol conversion after three runs, without leaching of the active phase.

The transesterification of glycerol with DEC was also studied with catalysts formed by MgAl-LDH NP supported on CNF prepared by electrospinning [59]. The loading of LDH in the composite was ca. 20 wt%.

Glycerol transesterification with MgAl-LDH/CF was threefold higher than that of bare MgAl-LDH (5.6×10^{-3} vs. 1.7×10^{-3} mmol Gly g^{-1}_{HT} s^{-1}, respectively). This is consistent with their different numbers of basic sites located at the edges of the platelets. However, the activity of MgAl-LDH/CF (5.6×10^{-3} mmol Gly g^{-1}_{HT} h^{-1} after 10 h) was by far lower than that of the MgAl-LDH/CNFrg catalyst (7 mmol Gly g^{-1}_{HT} h^{-1} after 1 h) reported by Álvarez et al., both studies being performed in similar conditions [60]. The different behavior of MgAl-LDH/CF and MgAl-LDH/CNFrg catalysts both rehydrated in the gas phase can be related to their physico-chemical properties. LDH loading in MgAl-LDH/CF was slightly higher than in MgAl-LDH/CNFrg (20 vs. 12 wt%) and the mean crystallite size was also smaller in the former catalyst, two features able to improve the catalytic efficiency. However, the specific surface area was significantly higher for MgAl-LDH/CNFrg (164 m^2 g^{-1}) than for MgAl-LDH/CF (15 m^2 g^{-1}) due to the different textures of the CNF, being of lower porosity when prepared by electrospinning. Consistently, the amount of adsorbed CO_2 was higher by a factor of 1.8 in MgAl-LDH/CNFrg (0.37 mmol g^{-1}_{LDH}) than in MgAl-LDH/CF (0.21 mmol g^{-1}_{LDH}) and in both cases higher than in the bulk rehydrated MgAl-LDH (0.13 mmol g^{-1}_{LDH}). Therefore, despite the potentially higher number of active sites located at the edges of its smaller particles, MgAl-LDH/CF was found to be less active due to the lower accessibility of the particles embedded in the carbon fibers prepared by electrospinning. Moreover, covering of the LDH particles by coke provided by the carbonization process cannot be ruled out.

4.1.5. Ullman, Sonogashira, and Heck Reactions

Ahmed et al. investigated CuAl-LDH and CoAl-LDH supported on GO as catalysts for the classical Ullmann reaction [21]. The CuAl-LDH/GO composite (Cu:Al = 2:1; LDH:GO = 20:1) was prepared by coprecipitation of the CuAl-LDH onto an aqueous GO dispersion. Meanwhile, the CoAl-LDH/GO composite (Co:Al = 2:1; LDH:GO = 20:1) was prepared via a urea-mediated coprecipitation method (Table 2 entry 11). The LDH content was 97 wt% in both composites and the XRD patterns were typical of LDH, while GO was considered well exfoliated. Crystallite sizes (c direction) were around 19 and 43 nm for CuAl-LDH/GO and CoAl-LDH/GO, respectively. SEM images showed lateral particle sizes of 100 nm for CuAl-LDH. Consistently, they were larger (up to 8 μm) with a hexagonal shape for CoAl-LDH synthesized by the urea method. The specific surface area was higher for CuAl-LDH/GO (44 m^2 g^{-1}) than for CoAl-LDH/GO (17 m^2 g^{-1}). It increased by around 60–70% compared to the bare LDHs upon introduction of 3 wt% GO, with the creation of mesopores of 2–3 nm.

The catalysts' activity in the condensation of iodobenzene to biphenyl was CuAl-LDH > CoAl-LDH > CuAl-LDH/GO > CoAl-LDH/GO. Remarkably, these results were obtained without the addition of a reducing agent, contrarily to usual operating conditions. The slightly lower activity of the LDH/GO catalysts compared to their bulk counterparts was attributed to a restacking of the GO layers in DMSO. In contrast, GO in the composite catalysts greatly improved their stability over reusability cycles. Thereby, the biphenyl yield dropped by around 30% after five reaction cycles with the LDH/GO catalysts, whereas it dropped by around 55% and 72% with CuAl-LDH and CoAl-LDH samples, respectively.

Wang et al. designed hierarchically structured nanoarray-like catalysts Pdx/rGO@CoAl-LDH for the Heck reaction using a lattice atomic-confined in situ reduction strategy of Pd^{2+} induced by the well-dispersed Co^{2+} atoms in the LDH layers [53]. CoAl-LDH provides both reductive Co^{2+} sites able to form highly dispersed Pd^0 NP and basic sites able to increase their electron density. Moreover, graphene enhances electron conductivity at the same time, which provides high mechanical strength and a large surface area, inducing high dispersion of LDH nanoplates and improved adsorption of aromatic reactants via π–π interactions.

The catalysts presented a hierarchically structured nanoarray-like morphology with LDH nanoplates of ~65 × 7.5 nm grown perpendicular to both sides of the rGO layers. Pd NPs (<2 nm) were highly dispersed and preferentially located at the edges of the LDH nanoplatelets due to the interaction of the Pd precursor and the pending OH groups, suggesting strong metal–support interaction (SMSI).

Based on the time (t) required to reach the maximum conversion (C) and TOF values, hierarchical Pd0.6/rGO@CoAl-LDH (t = 20 min, C = 98.1%, TOF = 981 h-1) presented the highest catalytic activity in the Heck reaction between iodobenzene and styrene compared to non-hierarchical Pd0.6/rGO@CoAl-LDH-h (t = 50 min, C = 98.2%, TOF = 393 h-1) and control catalysts Pd0.92/CoAl-LDH (t = 60 min, C = 95.0%, TOF = 317 h-1) and Pd2 + 0.78/GO (t = 270 min, C = 98.1%, TOF = 73 h-1) (Figure 2a). The rGO layers largely contributed to the activity, and the hierarchical composite improved it due to its structured nanoarray-like morphology, which allowed a reduction in the Pd NP size and enhancement in their dispersion. It also provided more accessible active sites (Figure 2b).

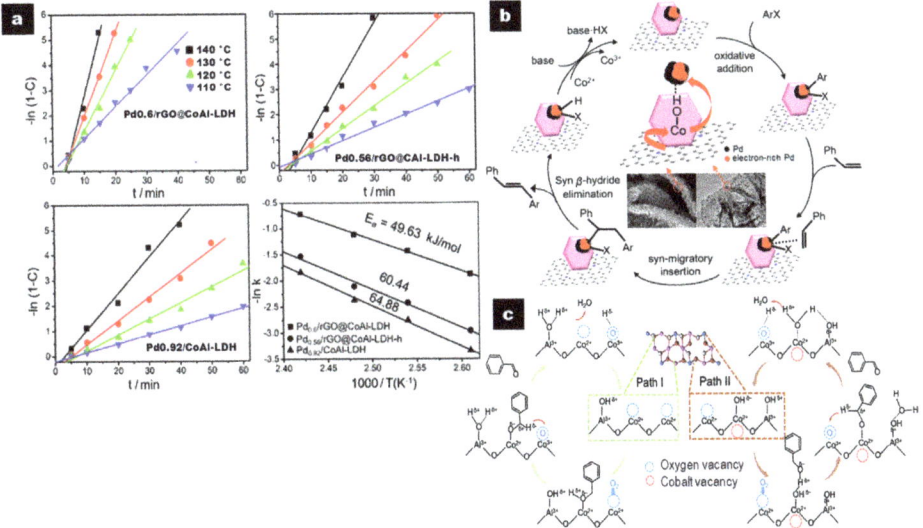

Figure 2. (a) −ln(1−C) against time (t) and Arrhenius plots for the Heck reaction of iodobenzene and styrene on Pdx/rGO@CoAl-LDH at varied temperatures; (b) plausible mechanism for the Heck coupling reaction of aryl halides with styrene on Pdx/rGO@CoAl-LDH catalysts. Reproduced with permission [53]. Copyright 2017, ACS. (c) The possible reaction pathways for the oxidation of BA over the CoAl-ELDH/GO catalyst. Reproduced with permission [32]. Copyright 2018, ACS.

When the Heck reaction was performed with the series of hierarchical Pdx/rGO@CoAl-LDH (x = 1.9, 1.2, 0.6, 0.33, 0.0098), the TOF increased from 846 h-1 to 2982 h-1 when Pd loading decreased from 1.9 to 0.33 wt%, in line with a concurrent increase in dispersion. The hierarchical Pd0.0098/rGO@CoAl-LDH catalyst with ultrasmall Pd NPs (1.3 nm) greatly dispersed exhibited a TOF value of 160,000 h-1 with 16.0% of iodobenzene conversion after 1 h, which is considerably higher than the values previously reported with Pd-based heterogeneous catalysts [53].

Moreover, the hierarchical composites showed a broad range of substrate applicability, having conversions higher than 94.3% within 40 min in reactions between varied substituted aryl halides with styrene and its derivatives. The catalysts could be reused in more than five runs without noticeable loss of activity.

Hierarchical flower-like CoAl-LDH@N,S-G@Pd catalysts were evaluated in the Sonogashira alkynylation reaction between a variety of aryl halides and phenylacetylene with optimized conditions (solvent, temperature, and base) [36]. The Pd and N,S-G loading was ~1.20 wt% and ~5 wt%, respectively. XPS analysis identified Co, Co^{2+} in CoAl-LDH structure, and Pd^0, Pd^{2+} species in the CoAl-LDH@N,S-G@Pd composite.

Product yields higher than 90% were obtained in the coupling reaction of both electron-deficient and electron-rich aryl iodides with phenylacetylene, although completion of the reaction was reached after 3 h with the methyl substituent. The coupling was also effective with substituted aryl bromides and aryl chlorides, but longer reaction times were needed to achieve completion. This hierarchical catalyst presented activities comparable with other Pd-containing catalysts reported, with the advantages of a shorter reaction time, mild reaction conditions, high product yield, and smaller amount of Pd [36].

The efficiency of the CoAl-LDH@N,S-G@Pd composite can be ascribed to its mesoporous spherical flower-like architecture (Figure 1c) with a hollow structure, which facilitates the diffusion of the reactants to the Pd sites, also favored by the high dispersion and stabilization of the Pd NPs by the well-ordered Co(II) in the LDH sheets. Moreover, the π-π stacking of G with aromatic substrates makes the reactants more accessible to Pd sites, and the electron enrichment of Pd by the basic LDH and the N,S-G increases the rate of the oxidative addition step.

The absence of Pd leaching from CoAl-LDH@N,S-G@Pd showed that the reaction was purely heterogeneous. Moreover, the catalyst was highly recyclable. A decrease of only ~12% in the product yield in the model reaction was observed after six catalytic runs.

4.1.6. One-Pot and Cascade Reactions

A magnetic Fe_3O_4@GO@Zn-Ni-Fe-LDH system showed high efficiency in the one-pot Knoevenagel–Michael reaction between 4-hydroxycoumarin and a variety of benzaldehydes substituted with different electron-donating and electron-withdrawing groups [33]. The biscoumarin compounds were obtained in 87–95% yield within 30–40 min, showing the good activity of the catalyst. The condensation of 4-hydroxycoumarin and benzaldehyde with the Fe_3O_4@GO@Zn-Ni-Fe-LDH catalyst gave 95% yield of 3-3′-benzylidenebis(4-hydroxycoumarin) in 3 min, the highest performance compared with other catalytic systems such as CuO-CeO_2 [82], SiO_2-OSO_3H NP [83], or phosphotungstic acid [84], which led to 93–94% yields in 12–30 min. GO provides a substrate for the formation of well-dispersed ZnNiFe-LDH platelets with acid–base properties and highly accessible sites required for the targeted reaction. After magnetic separation of Fe_3O_4@GO@Zn-Ni-Fe-LDH from the reaction media and washing, the catalyst can be reused up to five times, with a small loss in the yield of 3-3′-benzylidenebis(4-hydroxycoumarin) from 95% to 87%.

Zhang et al. have prepared nanocomposites based on RuCoAl-LDH and GO following either single-drop addition or simultaneous addition at constant pH for the cascade reaction between cinnamyl alcohol and ethyl cyanoacetate to produce cynnamilidene ethyl cyanoacetate [54].

The single-drop method led to the parallel orientation of the LDH platelets on the surface of GO (Ru/LDH-GO-P). Meanwhile, simultaneous addition led to the perpendicular orientation of the LDH platelets on GO (Ru/LDH-GO-V).

Ru/LDH-GO-P, with 99% yield of cinnamylidene ethyl cyanoacetate, was significantly more active than Ru/LDH-GO-V, with 48% yield (60 °C under O_2; 3 h). Ru/LDH-GO-P presented the highest amount of weak and medium–strong basic sites. The higher catalytic efficiency of Ru/LDH-GO-P was also in line with its higher abundance of O_2^{2-} and O_2^- species detected by O_2-TPD, whose formation was promoted by the higher amount of Co^{3+} species and the higher specific surface area compared to Ru/LDH-GO-V. Two control catalysts, Ru/LDH-sd and Ru/LDH-cp, obtained by single-drop and coprecipitation methods, respectively, were poorly active (30% and 14% yield, respectively), showing that the presence of GO in the nanocomposites improves the dispersion of the active sites of the LDH nanosheets. The physical mixtures of the three components also were less active (<36% yield) than the corresponding nanocomposites. Ru/LDH-GO-P could be reused at least five times without an obvious decrease in the activity.

Interestingly, the Ru/LDH-GO-P catalyst showed high efficiency in the one-pot oxidation–Knoevenagel condensation reactions involving benzyl alcohols substituted with either an electron-withdrawing (NO_2, Cl, Br) or an electron-donating group (CH_3, CH_3O) and active methylene groups (ethyl cyanoactate or malonitrile). Product yields were in the range of 91 to 99% within 2–3 h.

The different LDH/carbon nanocatalysts with hierarchical structures involved in the Ullmann, Sonogashira, Heck, and one-pot reactions are summarized in Table 2, entries 11–15.

It is noteworthy that the hierarchical nanocomposites combining transition metal-containing LDH, particularly CoAl-LDH, and Pd, with GO or rGO show remarkable efficiency for the achievement of a wide range of C–C coupling, such as Ullman, Sonogashira, and Heck reactions. Varying the LDH composition, e.g., RuCoAl-LDH, ZnNiFe-LDH, allows multifunctional catalysts that are able to perform one-pot Knoevenagel–Michael and oxidation–Knoevenagel cascade reactions. This behavior results from several main characteristics of these nanocomposites. The large surface area of the GO or rGO support favors the dispersion, and the number, strength, and accessibility of the most active sites of low coordination located at the edges of the LDH nanoplatelets. Moreover, it is possible to adjust the supersaturation rate during coprecipitation to obtain LDH nanosheets vertically or horizontally oriented on GO. Vertical orientation gives rise to the most efficient catalysts. Defects created by the reduction of GO to rGO during preparation of the nanocomposites are the preferential nucleation sites of the LDH crystallites and concur with their high dispersion. For CoAl-LDH and Pd systems, the presence of Co^{2+} species in the LDH facilitates the reduction of Pd^{2+}, giving rise to highly dispersed Pd NPs of small size. Moreover, electron transfer occurs from both the LDH nanosheets with basic character and rGO to the Pd NPs. All these features contribute to enhancing the reactivity of Pd for oxidative addition. The specific surface area of the nanocomposites reaches up to 180 $m^2\,g^{-1}$, accounting for the dispersion of the GO nanosheets due to the presence of LDH decreasing π-π interactions, generally responsible for their stacking. These properties make the Pd-containing nanocomposites more active than the classical catalysts in the C–C coupling reactions. The LDH-based nanocomposites can also lead to bi- and tri-functional catalysts required to achieve one-pot reactions and magnetic separation of the catalyst.

It is important to highlight the high robustness and stability of the LDH/carbon nanocomposites, which can be subjected to five or six cycles of regeneration with a decrease in activity not exceeding 30%.

4.2. Oxidation Reactions

LDH/nanocarbon composites are very attractive for a variety of oxidation reactions because the tunable composition of LDH allows the required catalytic active sites to be obtained, particularly efficient electron-deficient metal species. Moreover, the intrinsic LDH basicity can induce multifunctional composites. LDH/carbon nanocomposites are also designed for the oxidation of aromatic compounds due to their highly favorable adsorption on the carbon surface through π-π interaction closely to the dispersed active sites. The reactivity will also benefit from the balance between the hydrophobic and hydrophilic character induced by the carbon and LDH components, respectively, the large specific surface areas, and the strong interaction between the components. The preparation methods of the LDH/carbon nanocomposites and the conditions of the oxidation reactions considered are summarized in Table 2, entries 16–22.

4.2.1. Oxidation of Primary Alcohols

The development of highly active and selective catalysts for the solvent-free oxidation of primary alcohols is still challenging. Miao et al. reported for the first time the use of an LDH/graphene composite as a support for Au NPs for the selective oxidation of benzyl alcohol (BA) [40]. Its structure consisted of thin graphene sheets decorated by nanosized LDH particles grown parallel or perpendicular to the surface with an average size ~62 nm and Au NPs with an average particle size of 2.63 nm. Both the oxygenic functional groups and the defect sites on the surface of rGO acted as anchoring sites for the nucleation of LDH and Au NPs, with strong SMSI, high dispersion, and small size.

The BA conversion of Au/NiAl-LDH/rGO achieved in solvent-free conditions (140 °C; P O_2: 2 bar) reached ca. 62% after 10 h, which was higher than that of Au/GO (7.1%), Au/NiAl-LDH (51.8%), and the physical mixture of Au/GO and Au/NiAl-LDH (38.2%). Selectivity toward the targeted benzylaldehyde of the Au/NiAl-LDH/rGO and Au/NiAl-LDH catalysts after 10 h reaching 65.2% and 63.3%, respectively, was higher than that of Au/GO (60.6%), showing that the over-oxidation of BA was prevented on less oxygenated rGO than GO. The main by-products were toluene, benzoic acid, and benzyl benzoate, the latter reaching up to 35% selectivity at higher BA conversion through the reaction of benzaldehyde with BA to form hemiacetyl, which was then oxidized. The optimum GO/NiAl-LDH mass ratio of 1:2.8 led to a 40% benzaldehyde yield.

Upon reusing Au/NiAl-LDH/rGO, the conversion of BA dropped by only 10%, with selectivity to benzaldehyde still reaching 68% after the third recycling step. This stability probably results from the strong anchoring of the Au NPs, which prevents their agglomeration, and from the presence of NiAl-LDH, preventing the agglomeration of rGO.

Wang et al. developed an approach aiming to replace noble metal-containing catalysts for the oxidation of BA using CoAl-LDH [32]. Co-based catalysts have demonstrated high efficiency in the oxidation of alcohols. However, the high Co loading required to reach high conversion leads to poorly dispersed active species and low TOF values. Furthermore, the addition of basic promotors is common to improve the activity. Co-containing LDH can provide both highly dispersed Co species and high basicity. Exfoliation of LDH is also an interesting method to improve the accessibility to the active sites. Wang et al. prepared CoAl-ELDH/GO composites achieving first the coprecipitation of CoAl-LDH using the urea method. An aqueous suspension of GO exfoliated by ultrasonication was then added into the suspension of CoAl-LDH nanosheets previously exfoliated in formamide (ELDH) (Table 2, entry 17). Both ultrathin ELDH nanosheets and GO nanosheets with apparent thickness in the range of 2.4–3.3 nm and 1.1–1.2 nm, respectively, were observed in the CoAl-ELDH/GO composite, with Co/Al = 1.6 and GO content of 15.3 wt%.

CoAl-ELDH/GO was significantly more active (92.2% conversion) than CoAl-LDH (37.3%), GO (10.7%), and their physical mixture (51.9%) in the BA oxidation (DMF; 120 °C; 4 h). The conversion reached with the CoAl-LDH and GO physical mixture corresponds to the sum of the values of the two components, showing that there is no synergetic effect.

Benzaldehyde was the main product of the reaction, with similar selectivity of 99.2% for CoAl-ELDH/GO and the bulk CoAl-LDH, while it decreased to 91.5% for GO with the formation of significant amounts of benzoic acid (5.9%) and benzyl benzoate (2.5%). These results revealed the predominant role of CoAl-LDH in the catalytic performance, while GO acted as a poorly active support. The positive effect of the dispersion of CoAl-LDH on GO is evidenced by the TOF value, being five times higher for CoAl-ELDH/GO (1.14 h^{-1}) than for the bulk CoAl-LDH (0.23 h^{-1}).

Investigation of the surface defects, local atomic arrangement, and electronic structure revealed that the Co-O_{OH} and Co$\cdots$Co distances remained unchanged but that the coordination number decreased significantly in CoAl-ELDH/GO in comparison to the bulk CoAl-LDH. This suggests that coordinatively unsaturated CoO_{6-x} octahedra were formed in the ultrathin nanosheets of the composite with the generation of Co and O vacancies. The positron annihilation spectra (PAS) of CoAl-ELDH/GO showed the presence of larger amounts of Co vacancies and of negatively charged V_{Co}-Co-$OH^{\delta-}$ sites than in the bulk CoAl-LDH, with more lattice oxygen atoms exposed. Accordingly, both the strength and density of the basic sites of CoAl-ELDH/GO were improved with respect to bulk CoAl-LDH. The mobility of oxygen species was higher in the CoAl-ELDH/GO than in the bulk CoAl-LDH. This was consistent with the higher number of oxygen vacancies and surface O_2^- in the composite. The oxygen vacancies together with Co-OH$^{\delta-}$ adjacent to Co vacancies (V_{Co}-Co-OH$^{\delta-}$) were found by DFT calculation to be the sites of stronger BA adsorption via its OH group.

The CoAl-ELDH/GO catalyst showed good stability over six runs. Furthermore, it was also highly active and selective in the oxidation of a wide range of other benzylic alcohols.

The authors proposed a mechanism with two possible pathways for the oxidation of BA under molecular oxygen on CoAl-ELDH/GO (Figure 2c). O_2 initially adsorbed on the oxygen vacancies captured electrons from adjacent Co^{2+}, giving activated O^- and Co^{3+} species. The O-H group of the BA molecule simultaneously adsorbed on an oxygen vacancy (path I) or a V_{Co}-Co-OH$^{\delta-}$ site (path II) was activated, leading to H abstraction and the formation of an unstable metal-alkoxide species. Further, the α-C$^{\delta+}$-H$^{\delta-}$ bond cleavage on an activated O^- site accepting H$^{\delta-}$ led to the formation of benzaldehyde. The catalytic cycle was completed by the oxidation of the hydride by activated O^-, with the concurrent reduction of Co^{3+}-O^- to Co^{2+}-O^- followed by the desorption of H_2O.

CD-containing LDH-based nanocomposites are scarcely investigated in the literature. Notwithstanding, several recent papers indicate exciting outlooks for such composites, such as an N-doped CD/CoAl-LDH/g-C_3N_4 heterojunction photocatalyst and GdDy-LDH assembled with doxorubicin and folate–carbon dots, designed as cancer-targeted therapeutic agents or CD/LDH phosphors [85–87].

Interesting papers report nanocomposites with LDH and CD or N-doped CD (NCD) as supports of Au NPs and Ag NPs for heterogeneous catalysts. CD or NCD provide stabilizing and reducing ability, electro-donating capacity improving their basicity, and strong metal–carbon interaction favorable to the reduction of organic water pollutants and oxidation of alcohols [34,49].

Supported Au NPs on a NCD/MgAl-LDH composite (Au/NCD/MgAl-LDH) prepared by the coprecipitation of MgAl-LDH and NCD, and subsequent Au introduction by the deposition–reduction approach, gave rise to catalysts with improved basicity and metal–support interaction (Table 2, entry 18) [49]. Nitrogen atoms incorporated into carbon materials provided basic species inducing a Lewis basicity to the neighboring carbon atoms [88,89].

The Au/NCD/MgAl-LDH composite contained Au^0 NPs of approximately 3.46 nm average size, uniformly dispersed and poorly aggregated, with a loading of 0.3 wt%. The highly dispersed NCD component on the surface of MgAl-LDH increased the surface density of stronger basic sites compared to the Au/MgAl-LDH sample. In addition, SMSI was produced in the Au/NCD/MgAl-LDH composite, indicated by BE values of Au^0 species lower than for the Au/MgAl-LDH. This is due to the strong coordination of electron-donating N atoms in NCD and Au^0 NPs.

The BA oxidation with the Au/NCD/MgAl-LDH composite was conducted without a solvent and the addition of bases in the reaction media, as previously reported for an Au/NiAl-LDH/rGO composite [40].

MgAl-LDH and NCD/MgAl-LDH were poorly active in BA oxidation, leading to conversion of 4.2% and 5.8%, respectively, after 4 h. The conversion increased upon introduction of Au on MgAl-LDH (38.2%) and further with the introduction of NCD (47.3%) with the Au/NCD/MgAl-LDH composite at similar Au loading compared to Au/MgAl-LDH (0.82% and 0.91%, respectively). This suggests that the improvement of basicity promotes the activity of the composite. Consistently, the initial TOF based on surface Au atoms after 0.5 h of reaction greatly increased from 8591 h^{-1} with Au/MgAl-LDH up to 20175 h^{-1} with Au/NCD/MgAl-LDH. The role of the basicity was also confirmed when the catalytic results were compared with those previously obtained with the Au/NiAl-

LDH/rGO composite [40]. Conversion of BA was indeed 49 and 35% with Au/NCD/MgAl-LDH and Au/NiAl-LDH/RGO, respectively, after 5 h.

Benzaldehyde was the main reaction product, with benzoic acid and benzyl benzoate as by-products. The conversion increased with the Au loading, with a concurrent slight decrease in benzaldehyde selectivity.

Similar selectivity to benzaldehyde of approximately 80%, with benzyl benzoate as the main by-product, was obtained with Au/NiAl-LDH/rGO [40]. However, it must be underlined that Au/NiAl-LDH/rGO exhibited a specific surface area of 172.5 $m^2\ g^{-1}$ and Au NP size of 2.63 nm, while they were 61 $m^2\ g^{-1}$ and 3.46 nm, respectively, for Au/NCD/MgAl-LDH. These structural features were more favorable to the former catalyst. Moreover, Au/NCD/MgAl-LDH was more active despite the softer reaction conditions used, i.e., 120 °C reaction temperature vs. 140 °C for Au/NiAl-LDH/rGO. The presence of NCD and of MgAl-LDH instead of NiAl-LDH contributes to improving the basicity of Au/NCD/MgAl-LDH in comparison to Au/NiAl-LDH/rGO and, therefore, the catalytic activity in the oxidation of BA.

BA conversion with Au/NCD/MgAl-LDH decreased only by 6.0% after five consecutive runs, instead of 12.4% with Au/MgAl-LDH, showing the high stability of the former composite, whose Au leaching was around 1.0%.

Shan et al. developed a method aiming to obtain highly thermodynamically stable Pickering emulsion using an amphiphilic nanocomposite based on LDH and CNT components. The selective oxidation of BA was studied using an Ru-based LDH-CNT catalyst as a solid emulsifier [42].

LDH-CNT composites were prepared by the coprecipitation of MgAl-LDH on the acid-treated CNT (Table 2 entry 19). The Ru/MgAl-LDH-CNT catalyst was obtained by wet impregnation of the MgAl-LDH-CNT support with $RuCl_3 \cdot 3H_2O$ and reduced at 400 °C under H_2 flow. The collapsed LDH structure obtained after reduction was recovered after redispersion for 3 h in a water–oil interface (H_2O:toluene = 1:2), giving a homogeneous emulsion.

The Ru content was ca. 0.7 wt% in the Ru/LDH-CNT composite, Ru/LDH, and Ru/CNT, with particle sizes of 2–3 nm uniformly dispersed on the surfaces. They also showed similar specific surface areas of ~90 $m^2\ g^{-1}$. However, the catalytic activities of the three materials revealed different behaviors for the selective oxidation of BA to benzaldehyde. The selectivity was in all cases 99.9%, but conversion of 92% was reached by Ru/LDH-CNT, compared to 52% and less than 5% for Ru/CNT and Ru/LDH, respectively, after 5 h of reaction. These results can be attributed to the different capacities of LDH, CNT, and LDH-CNT to stabilize water–toluene emulsions. LDH-CNT presented a 100% stabilized volume fraction with the smallest emulsion droplets (30–150 μm) much lower than those of CNT and unstable hydrophilic LDH nanosheets at the water–oil interface (100–300 μm). This resulted in an increased emulsion interfacial surface area, where the Pickering interfacial catalysis (PIC) process took place, leading to the high catalytic activity observed with Ru/LDH-CNT in the aerobic oxidation of BA. The advantage of the PIC process is obvious when one considers that 54% conversion of BA is obtained over Ru/LDH-CNT with toluene as the solvent.

With LDH at 85 °C, coalescence of the droplets at the water–oil interface occurred and the LDH was transferred to the aqueous phase. Therefore, Ru/LDH did not catalyze the oxidation of BA likely dissolved in toluene. CNT dispersed at the water–oil interface allowed PIC emulsion catalysis to occur, although partial coalescence of the droplets occurred. Ru/CNT (conversion 52%) was then found more active than Ru/LDH (conversion < 5%). LDH-CNT exhibited the higher thermostability at the water–oil interface due to the smaller size of the droplets and the restricted rotation of the emulsifier induced by the nanosheet-shaped structure, which led to higher conversion on Ru/LDH-CNT. The positive effect of LDH nanosheets for the PIC process was confirmed using a Ru/LDH-CNT catalyst whose LDH-CNT support was hydrothermally treated for 4 h instead of 16 h, leading to a smaller LDH particle size. The conversion of BA on this Ru/LDH-CNT reached 84% at 5 h, higher than that over Ru/CNT despite a lower emulsion volume (55% versus 77%). The three-phase contact angle of the LDH-CNT at the water–oil interface of 97°, instead of 110° for CNT, showed that the wettability of CNT is modulated in the presence of LDH, leading to a more stable emulsion, accounting for the higher stability of the Pickering emulsion with LDH-CNT.

Table 4 summarizes the reaction conditions and catalytic results of the considered nanocomposite catalysts. Benzaldehyde is always the main product of the reaction. Higher conversion of BA and selectivity to benzaldehyde are obtained in the presence of a solvent, DMF, or water–toluene, than in solvent-free conditions. In the latter, higher catalytic activity is obtained when the basicity of the catalyst is improved. The CoAl-ELDH/GO composite allows similar conversion and selectivity to the Ru-containing catalyst to be obtained in the PIC process, showing that a non-noble-metal-containing catalyst can be efficient in the oxidation reaction. It can be noted that the PIC process allows the performance of the reaction at a lower temperature. In the case of Au-containing catalysts, the high selectivity to benzyl benzoate probably results from the reaction conditions, particularly the high oxygen pressure and temperature, promoting the reaction of benzaldehyde and BA.

Table 4. Conditions of reaction and results of aerobic oxidation of BA over different nanocomposite catalysts.

Catalyst	BA [mmol]	Solv	m_{cat} [g]	P_{O_2} [bar]	T [°C]	t [h]	SS [m^2 g^{-1}]	Conv. [%]	BAL [%]	BB [%]	Ref
Au/NiAl-LDH/rGO	4.3	no	0.4	2	140	5	172.5	35	80	20	[40]
CoAl-ELDH/GO	1	DMF	0.1	1	120	4		92.2	99.3	0	[42]
Au/NCD/MgAl-LDH	89	no	0.1	4	120	5	61	49	83.4	16	[32]
Ru/LDH-CNT (Pickering)	2	H$_2$O/Tol^{ne}	0.2	1	85	5	89	92	99.9	6	[49]

BAL: Benzaldehyde; BB: Benzyl benzoate.

4.2.2. Oxidation of Alkylaromatics

Nanocomposites based on non-noble-metal-containing LDH and carbon compounds offer wide possibilities to obtain highly efficient catalysts for the selective oxidation of alkylaromatics able to fulfil sustainable chemistry requirements. Two main model reactions have been particularly investigated: the selective oxidation of ethylbenzene (EB) to acetophenone (AP) and the oxidation of styrene to styrene oxide (SO).

Among the selective oxidation of alkylaromatic compounds, that of EB to produce AP is relevant because AP is an important intermediate to produce esters, aldehydes, and pharmaceuticals. Cobalt-based metal oxides have been reported as efficient catalysts for the oxidation of EB. For example, hierarchical flower-like core–shell-structured CoZnAl-MMO supported on amorphous alumina microspheres or a flower-like Al$_2$O$_3$@CoCuAl-MMO catalyst exhibited high activity and selectivity toward AP [90,91].

CoCuAl-LDH on a graphene sheet nanocomposite prepared by coprecipitation of the LDH on GO (Table 2 entry 20) was also tested in EB oxidation [38].

The main products of the oxidation of EB at 120 °C using tert-butyl hydroperoxide (TBHP) as an oxidant were AP, BA, and 1-phenylethanol (1-PA). The catalytic activity of the CoCuAl-LDH/graphene nanocomposites depended on the graphene/LDH mass ratio. A mass ratio of 0.4 led to a maximum EB conversion of 96.8% and AP selectivity of 95.4%. GO and graphene were poorly active, with conversion not exceeding 13.5%, while pristine CoCuAl-LDH exhibited significant conversion (68.8%) and selectivity to targeted

AP (88.1%). Moreover, the catalyst could be recycled at least four times without significant loss of activity.

CoCuAl-LDH/graphene nanocomposites presented improved catalytic performance if compared to other types of catalysts previously reported, despite the different reaction conditions used in each case [38].

The higher activity shown by CoCuAl-LDH/graphene nanocomposites is assignable to: (i) the adsorption of EB on the graphene through π-π interaction in close proximity to Co^{2+} and Cu^{2+} active sites; (ii) the high dispersion of CoCuAl-LDH nanoplatelets of small size; (iii) promotion of THBP activation due to the strong interaction of graphene with Co^{2+} and Cu^{2+} sites; and (iv) the preferential adsorption of EB than water on the hydrophobic graphene surface.

ZnCr-LDH/CNT was the best-performing of the different carbonaceous composites studied by Zhao et al. in solvent-free aerobic EB oxidation with O_2 at 130 °C, with conversion of 54.2% and AP selectivity of 93.7% [62]. The ZnCr-LDH/CNT nanocomposite, prepared as shown in Table 2, entry 21, presented LDH platelets with an average size of ~10 nm, highly dispersed on the surface of CNT (Figure 3a). Larger ZnCr-LDH particles (30–40 nm) were obtained over the graphene (G) surface, while aggregation occurred on active carbon (AC). These ZnCr-LDH/G and ZnCr-LDH/C composites yielded EB conversion of 45.1% and 40.0%, respectively, with AP selectivity of ca. 87% in both cases. The physical mixture of CNT and ZnCr-LDH (20 wt% CNT) leading to EB conversion of 28.5% was less active than ZnCr-LDH/CNT and the pristine ZnCr-LDH (37.9%), revealing a synergistic effect in the composite. The strong interaction between ZnCr-LDH particles and CNT was revealed by the largest positive shift in the BE of the Zn 2p and Cr 2p regions in ZnCr-LDH/CNT in comparison to the pristine ZnCr-LDH.

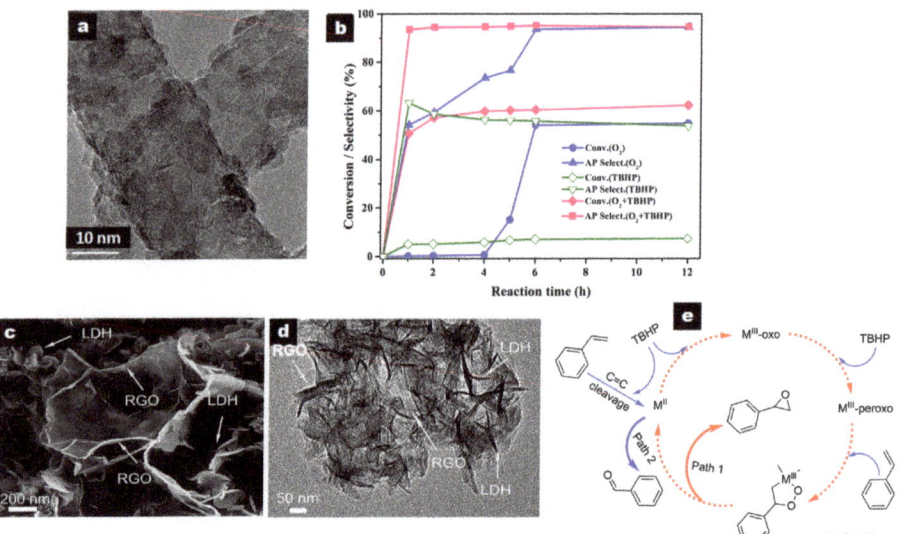

Figure 3. (**a**) HRTEM image of ZnCr-LDH/CNT composite; (**b**) ZnCr-LDH/CNT performance with single molecular oxygen, single TBHP (0.5 mL, 70% aqueous solution), and mixed oxidants of O_2 and TBHP, respectively (reaction conditions: EB, 10 mL; P_{O2}, 1.0 MPa; 130 °C; catalyst, 0.1 g). Reproduced with permission [62]. Copyright 2018, Elsevier. (**c**,**d**) SEM and TEM images of NiCo-LDH/RGO$_5$ and (**e**) proposed reaction pathway for the epoxidation for styrene. Reproduced with permission [63]. Copyright 2019, Elsevier.

Interestingly, oxidation of EB over ZnCr-LDH/CNT using either O_2, a small amount of TBHP (0.5 mL, 70% aqueous solution), or both oxidants (Figure 3b) led to contrasting results.

An induction period of around 4 h was observed using O_2 that was, on the contrary, absent using TBHP, but EB conversion was very low, reaching only 7.1% with AP selectivity of 55.8% after 6 h. Using both O_2 and TBHP, EB conversion of 50.8% with high AP selectivity of 93.6% was achieved after only 1 h. The addition of butylated hydroxytoluene, as a radical scavenger, inhibited the reaction performed under O_2, suggesting that it obeyed a free radical process because free radicals generated from TBHP greatly enhanced the reactivity. The ZnCr-LDH/CNT catalyst was also found to be active for the oxidation of other alkylaromatics.

The high efficiency with a clear synergistic effect between the components of the ZnCr-LDH/CNT catalyst can be mainly related to the dispersion of the ZnCr-LDH nanoplatelets on the CNT surface, which increases the accessibility, the strength of the active sites, and their strong electronic interaction. Such features promote EB adsorption on CNT through π-π interaction and its contact with the adjacent Cr^{3+} sites interacting with O_2.

The CoCuAl-LDH/graphene catalyst (G/LDH mass ratio of 0.4), with an AP yield of 92.35% using TBHP, is significantly more efficient than ZnCr-LDH/CNT as reported by Xie et al., leading to an AP yield of ca. 4% or 50.8% with TBHP or O_2, respectively, which emphasizes the higher intrinsic efficiency of the CoCuAl-LDH/graphene catalyst [38,90]. This was also confirmed in the oxidation of cumene, 1,2,3,4-tetrahydronaphtalene, and diphenylmethane, with conversions of 84.8, 98.1, and 97.3%, respectively, with the CoCuAl-LDH/graphene nanocatalyst, and of 49.0, 53.6, and 63.8%, respectively, with ZnCr-LDH/CNT.

The different behavior of the CoCuAl-LDH/graphene and ZnCr-LDH/CNT catalysts mainly accounts for their different numbers of active sites. The total loading of Co^{2+} and Cu^{2+} indeed reached 11.2 wt% in CoCuAl-LDH/graphene, while Cr^{3+} loading was 0.87 wt% in ZnCr-LDH/CNT. Notably, the performance of the CoCuAl-LDH/graphene and ZnCr-LDH/CNT catalysts was better than that of a range of previously reported catalysts. The literature data suggest that higher EB conversions are generally obtained when using TBHP operating at a lower temperature (70–130 °C) but a larger reaction time (8–24 h) than with O_2 (120–160 °C and 1–9 h) [38,90].

The epoxidation of styrene is of outstanding importance. For cleaner processes, oxidation of styrene with H_2O_2 or TBHP rather than with organic peracids is preferred, with higher selectivity to styrene oxide generally obtained with TBHP.

Shen et al. have reported a series of NiCo-LDH/rGO nanocomposites (NiCo-LDH/RGO$_x$, x = 1–10 wt%), synthesized by a one-pot hydrothermal method, as catalysts using TBHP as an oxidant [63]. A GO colloid was dispersed in a C_2H_5OH/H_2O solution, where a Ni and Co chloride salt solution (Ni/Co = 1) was added dropwise along with NH_4Cl under stirring (pH adjusted to 9). The suspension was hydrothermally treated (120 °C for 12 h), producing both LDH crystallization and GO reduction.

LDH nanoplatelets were dispersed around rGO as nanoflower clusters (Figure 3c,d) and presented a smaller size (50–100 nm) than in the pristine NiCo-LDH (ca. 200 nm).

There was a significant influence of the rGO content on the catalytic activity of the reaction performed at 80 °C for 8 h. Styrene conversion increased from 88.6% to 95.1% when the rGO content increased from 0% to 5 wt%. Such behavior accounted for an increase in the number and strength of accessible Ni^{2+} and Co^{2+} active sites, leading to improved THBP binding (Figure 3e). This gave rise to M^{III}-oxo and M^{III}-peroxo species, able to attack the C=C bond of styrene, producing a peroxo metallocycle that was then decomposed to form styrene oxide (path 1). Benzaldehyde was also generated due to the cleavage of C=C bond of the styrene adsorbed on M^{II} species.

The best SO yield (70.2%) was obtained with 5 wt% of rGO and decreased when the rGO content increased; meanwhile, selectivity to BAL remained almost similar around 27%, regardless of the rGO content. These catalytic results improve those ones previously reported with catalysts such as Fe-sal-CMK-3, NiO NP and Au/L-Fe$_3$O$_4$, giving SO yields of 32.4%, 10.8%, and 53.4%, respectively [92–94]. NiCo-LDH/RGO catalysts showed remarkable stability along five catalytic runs, with styrene conversion decreasing only by around 2% and SO selectivity remaining constant.

In summary, the previous results illustrate the great potential of the LDH/nanocarbon composites to perform the oxidation of various substrates through the tailoring of their composition. The LDH composition and/or the introduction of nitrogen-doped carbon dots (NCD) allow the adjustment of the basicity of the final nanocomposite. The ability to obtain Pickering emulsions highly stabilized with LDH/nanocarbon composites offers interesting outlooks that have not yet been explored in the hydrogenation of benzaldehyde. The LDH/nanocarbon catalysts are particularly suitable for the oxidation of molecules containing aromatic cycles due to the π-π interactions with the carbon component and the enhanced electron transfer.

4.3. Hydrogenation Reactions

Since 2012, Mg, Co, Ni, or Cu-based LDHs have been combined with carbon materials, i.e., amorphous carbon, GO, CNT, or CD, to obtain hierarchical structures acting as precursors of catalysts for a wide variety of hydrogenation reactions. They were involved in the reduction of organic dyes and nitroarenes, the selective hydrogenation of C=O or C=C conjugated bonds and of 5-hydroxymethylfurfural (HMF) to 2,5-dimethylfuran (DMF), the hydrodechlorination (HDC) of chlorobenzene, and the hydrogenolysis of glycerol [22,34,35,39,41,47,64–66,95–99].

The nature of the components and the preparation methods of the LDH/nanocarbon nanocomposites and the hydrogenation reactions considered are summarized in Table 2, entries 23–30.

4.3.1. Hydrogenation of Organic Dyes and Nitroarenes

Nitro derivatives have a highly pollutant nature, especially 4-nitrophenol (4-NP) (or *p*-nitrophenol), which is anthropogenic, carcinogenic, and toxic, with, in addition, high solubility and stability in water. Reduction of 4-NP to produce 4-aminophenol (4-AP) is an important issue because this latter is a valuable industrial raw material and pharmaceutical intermediate. Several transition metals, mainly Cu, Co, and Ni and their oxides, have been widely used for the reduction of 4-NP [99–102]. A wide variety of LDH/nanocarbon composites have been also explored as catalysts or precursors for the reduction of 4-NP and organic dyes.

$Cu_xMg_{3-x}Al$-LDH/rGO (x = 0.5, 1, 1.5) and $Cu_{3-x}Ni_xAl$-LDH/rGO (x = 2.5, 2.0, 1.5) nanocomposites with nanoarray-like structures (Figure 1) were prepared according to the CA-assisted aqueous-phase coprecipitation method and studied for the reduction of 4-NP with an excess amount of $NaBH_4$ [35,39,41].

The reduction of GO and the presence of abundant and smaller sp^2 graphitic regions, with the creation of defects acting as nucleation sites, account for the vertical and uniform growth of the LDH in the nanoarray-like $Cu_xMg_{3-x}Al$-LDH/rGO (x = 0.5, 1, 1.5) (named xCu-LDH/rGO) nanocomposites, reducing the strong interparticle interaction among LDH nanoplates (Table 2, entry 23). Thus, the hierarchical nanoarray-like morphology is in line with the higher specific surface area of the composites (204–215 $m^2\ g^{-1}$) compared to Cu_1Mg_2Al-LDH (named CuAl-LDH) (102 $m^2\ g^{-1}$).

Complete conversion in the reduction of 4-NP was achieved within 2.5 min with the xCu-LDH/rGO composites, in comparison to 11 min with CuAl-LDH, while GO and Mg_3Al-LDH were inactive [39]. The active Cu-related species were more efficient in the hierarchical nanoarray-like structure. Both the apparent reaction rate constant (pseudo-first-order kinetics) and the TOF number increased with the Cu content and were higher than with pure Cu-LDH or previously reported catalysts [103–105]. Cu 2p XPS, XRD, and HRTEM analyses revealed that the active sites were Cu^{2+} ions belonging to LDH nanosheets and Cu^+ formed by reduction upon dissociation of $NaBH_4$ in water, which led to the in situ formation of Cu_2O NPs on the LDH nanosheets. Indeed, the presence of spherical Cu_2O NPs (~7 nm) was ascertained by HRTEM, with also the presence of interfaces between LDH domains supporting Cu_2O NPs and single rGO layers. A synergistic Cu_2O-Cu-LDH-rGO three-phase interface is then suggested, which can be considered as the most active catalytic domain due to the enhanced electron transfer. The xCu-LDH/rGO composites act as efficient Cu_2O reservoirs without needing a pre-reduction step. These composites were also active for the reduction of various nitroarenes (2-NP, 3-NP, 2,4-dinitrotoluene, and 4-nitrobenzaldehyde).

The high catalytic efficiency of the composites accounts for: (i) the electronic mobility of rGO in the Cu_2O-Cu-LDH-rGO three-phase interface, which facilitates the electron transfer from BH_4^- to 4-NP mediated by small Cu_2O NPs; (ii) the enhanced contact between the reactant and the active Cu_2O centers due to the π–π stacking interaction between rGO and adsorbed 4-NP; (iii) the improved diffusion and transfer of 4-NP and BH_4^- to the catalytic sites due to the large specific surface area of the composites.

Considering the well-known influence of calcination to improve metal–support interaction, Dou et al. investigated the catalytic activity of the $xCu@Cu_2O/MgAlO$-rGO hierarchical nanocomposites obtained by calcination (600 °C under N_2 flow) of the $Cu_xMg_{3-x}Al$-LDH/rGO precursors (x = 0.5, 1, 1.5) (Table 2, entry 24) [41].

The $xCu@Cu_2O/MgAlO$-rGO nanocomposites contained Cu, Cu_2O, and CuO phases according to the reduction of Cu^{2+} to Cu^+ and Cu^0 by rGO, and poorly crystallized MgAlO mixed oxide with highly dispersed Cu-based spherical-like particles. Their average size increased (12.8 to 40.3 nm) with the copper content, and they were located near the border between MgAlO and rGO layers. These structural features conferred upon the composite catalysts a mesoporous structure with high specific surface areas, decreasing from 200 to 157 $m^2 g^{-1}$ when the copper content increased.

FTIR suggested charge transfer between π electrons of rGO and the copper atom, weakening the Cu-O bonds and stabilizing Cu^+. Catalysts $1.0Cu@Cu_2O/MgAlO$-rGO and $1.5Cu@Cu_2O/MgAlO$-rGO exhibited core–shell-like $Cu@CuO$ NPs with Cu^0 in the core and Cu_2O in the outer shell, while $0.5Cu@Cu_2O/MgAlO$ showed a mixture of Cu nanocrystallites and Cu_2O phases. These results confirmed the in situ self-reduction of Cu^{2+} into Cu^0 by rGO and Cu_2O NPs located on the border between MgAlO and rGO, creating a Cu-Cu_2O-MgAlO-rGO four-phase interface. These latter must greatly improve the catalytic activity, particularly due to the presence of well-dispersed and poorly aggregated core–shell-like $Cu@Cu_2O$ NPs in the MgAlO matrix, in close contact with rGO layers, and electron transfer from Cu or Cu_2O to MgAlO and rGO (XPS).

Complete reduction of 4-NP to 4-AP occurred within 3, 1, and 1.2 min with $0.5Cu@Cu_2O/MgAlO$-rGO, $1.0Cu@Cu_2O/MgAlO$-rGO, and $1.5Cu@Cu_2O/MgAlO$-rGO catalyst, respectively. The 1.0CuMgAlO with CuO phases supported on MgAlO led to the reduction of 88% of 4-NP within 17 min. The dramatically higher activity of the $xCu@Cu_2O/MgAlO$-rGO catalysts may be related to the four-phase synergistic effect and the presence of core–shell-like $Cu@Cu_2O$ NPs. The apparent rate constant (pseudo-first-order kinetics) was 2.03-fold higher for $1.0Cu@Cu_2O/MgAlO$-rGO and $1.5Cu@Cu_2O/MgAlO$-rGO than for $0.5Cu@Cu_2O/MgAlO$-rGO. This clearly showed the higher catalytic efficiency of the core–shell-like $Cu@Cu_2O$ NPs than the mixture of Cu and Cu_2O NPs, in agreement with the core–shell metal–metal oxide interaction. Moreover, $1.0Cu@Cu_2O/MgAlO$-rGO, with the smaller core–shell-like $Cu@Cu_2O$ NPs, showed 1.2-fold higher activity than its non-calcined precursor. Therefore, calcination, which allows a shift from the three-phase to the four-phase interface system and the formation of core–shell-like $Cu@Cu_2O$ NPs interacting with MgAlO and rGO, creates more active species. In addition, $1.0Cu@Cu_2O/MgAlO$-rGO exhibited remarkable activity in 4-NP reduction comparable to that of several noble-metal-containing catalysts [103,106–108].

The peculiarity of the four-phase interface system is that the core–shell-like $Cu@Cu_2O$ NPs play a key role as efficient electron transport entities instead of Cu_2O alone in the three-phase interface system [39]. Indeed, 4-NP anions and BH_4^- are both adsorbed on the surface of the catalyst, supporting the electrophilic $Cu@Cu_2O$ entities, while BH_4^- reacts with H_2O to produce H_2. Dissociation of H_2 on $Cu@Cu_2O$ generates Cu-H species, reducing $-NO_2$ groups to $-NH_2$ groups through several sequential steps, including the formation of nitroso groups and hydroxylamine intermediates and the generation of 4-AP.

The high dispersion of the active Cu@Cu$_2$O species and the strong four-phase synergistic effect also serve to improve the catalytic activity. rGO facilitates both the electron transfer from BH$_4^-$ to 4-NP and the adsorption of 4-nitrophenate anions through π-π interaction with the surface-active Cu@Cu$_2$O NPs.

A very interesting proof-of-concept for the use of the −1.0Cu@Cu$_2$O/MgAlO-rGO catalyst in a fixed bed reactor for the treatment of industrial effluents has been given by the reduction of 4-NP, methyl orange (MO), and a mixture of 4-NP and MO. Complete reduction of 4-NP to 4-AP and complete degradation of MO as an individual or mixture of substrates were obtained.

Bimetallic Cu-based catalysts, especially with Ni as a second transition metal, have been found more active than monometallic ones for the reduction of nitroarenes, particularly 4-NP [109–112]. Wei et al. have considered that bi-transition-metal-based Cu$_{3-x}$Ni$_x$Al-LDH/rGO nanocomposites may improve the catalytic activity compared to the previous Cu$_x$Mg$_{3-x}$Al-LDH/rGO [35]. Therefore, a series of Cu$_{3-x}$Ni$_x$Al-LDH/rGO (x = 2.5, 2, 1.5) composites was prepared via an AC-assisted aqueous-phase coprecipitation method (Table 2, entry 25). The samples exhibited the same structural and morphological features as the Cu$_x$Mg$_{3-x}$Al-LDH/rGO series, with highly uniform, dispersed nanosheets forming an array-like structure, high accessibility and surface area, and strong interaction between LDH and GO, which led to electron transfer. The BE values for Cu 2p$_{3/2}$ and Ni 2p$_{3/2}$ varied with the Cu^{2+} on Ni^{2+} molar ratio, suggesting an electronic interaction and synergistic effect.

The time to reach the entire conversion of 4-NP to 4-AP ranged as follows: Cu$_1$Ni$_2$Al-LDH/rGO (1.5 min) < Cu$_{1.5}$Ni$_{1.5}$Al-LDH/rGO (2 min) < Cu$_{0.5}$Ni$_{2.5}$Al-LDH/rGO (2.5 min). This shows that the activity depended on the composition. Cu$_1$Ni$_2$Al-LDH/rGO was more active than the single transition metal sample Cu$_1$Mg$_2$Al-LDH/rGO, reaching complete conversion in 2 min [39]. Because complete 4-NP conversion occurred in 3 min with Cu$_1$Ni$_2$Al-LDH and both Ni$_3$Al-LDH and rGO were inactive, it was concluded that the Cu-related species were the active sites, whose efficiency was improved due to the synergistic effect with Ni^{2+}. Moreover, the positive influence of the dispersed Cu species in the LDH nanosheets strongly interacting with rGO was reflected by the 1.3-fold higher k$_{app}$ values (pseudo-first-order kinetics) obtained with the composites than with their corresponding Cu$_{3-x}$Ni$_x$Al-LDH.

The excess of NaBH$_4$ benefited the formation of small Cu$_2$O NPs, also directed by the isolation and stabilization effects of the Ni-OH groups of the LDH layers. XPS peaks in Cu$_1$Ni$_2$Al-LDH/rGO upon reduction with NaBH$_4$ showed an increase in electron density around Ni and Cu core level compared to untreated Cu$_1$Ni$_2$Al-LDH/rGO. Moreover, 15.9% of total Cu in the LDH was in situ reduced to Cu$_2$O species. Therefore, Cu^{2+} ions can be considered as a reservoir continuously providing highly active Cu$_2$O NPs when an excess of NaBH$_4$ is used in the reduction reaction, as already reported for the Cu$_x$Mg$_{3-x}$Al-LDH/rGO composites [39]. Moreover, in the reaction medium, the electron transfer from Cu to Ni (slightly more electronegative) improved the hydrophilicity of the surface of Cu$_2$O in the Cu$_1$Ni$_2$Al-LDH/rGO composite when compared to the single transition metal Cu$_1$Mg$_2$Al-LDH/rGO. Cu$_2$O NPs were more stabilized in the former hybrid due to the stronger interaction with the OH groups of LDH. All these features suggest that the three-phase interface Cu$_2$O-Ni-OH-rGO between the Cu$_1$Ni$_2$Al-LDH domains and the single-layer rGO induces a strong synergistic effect, enhancing the catalytic activity.

Furthermore, 4-NP conversion of 90.3% after 10 catalytic cycles demonstrated the remarkable reusability of the Cu$_1$Ni$_2$Al-LDH/rGO catalyst, whose slight deactivation was mainly due to the increase in the average Cu$_2$O NP size (from ~3.8 to ~9.8 nm).

The previous results emphasize the influence of the composites' composition (Cu$_x$Mg$_{1-x}$Al-LDH/rGO and Cu$_{3-x}$Ni$_x$Al-LDH/rGO) and of their topotactic decomposition (xCu@Cu$_2$O/MgAlO-rGO). The main physico-chemical characteristics, the performance in the reduction of 4-NP, and the recyclability of the most active catalysts in each series are compared in Table 5.

Table 5. Physico-chemical characteristics, catalytic performance in the reduction of 4-NP, and reusability of the nanocomposites containing Cu-based LDH and rGO.

Catalyst	SA [m^{2}g^{-1}]	Active Interf. Domain	Size Cu [nm]	t [min]	k$_{app}$ [×10^{-3} s^{-1}]	TOF [h^{-1}]	Rec.[a] [nb]	Ref
Cu$_1$Mg$_2$Al-LDH/rGO (1.0Cu-LDH/rGO)	210.5	Cu$_2$O-Cu(LDH)-rGO	Cu$_2$O (~6.8)	2	25.11	161.9	20	[39]
1.0Cu@Cu$_2$O/MgAlO-rGO	160	Cu-Cu$_2$O-MgAlO-rGO	Cu@Cu$_2$O (24.3/core 12.2)	1	55.35	199.6	25	[41]
Cu$_1$Ni$_2$Al-LDH/rGO	151	Cu$_2$O-Ni-OH (CuNiAl-LDH)-rGO	Cu$_2$O (~3.8)	1.5	34.37	197.6	10	[35]

[a] Number of reuse cycles.

All the as-prepared composites exhibited ultrathin LDH nanosheets vertically grown on both sides of the rGO substrate, giving rise to large surface areas (>150 m^2 g^{-1}) and mesopores (2–15 nm). Their catalytic efficiency arose from the existence of three-phase interface domains formed upon in situ reduction with NaBH$_4$ during the reaction process. They contained well-dispersed Cu$_2$O NPs of small size (3–7 nm), continuously provided by the reduction of Cu^{2+} in the LDH layers, which acted as a reservoir of the active species. The Cu$_2$O NPs supported on the LDH layers in close contact with the rGO layers benefited from significant electron transfer. These features and the π-π stacking effect between the catalytic surface and the aromatic cycle of the nitroarenes led to a synergistic effect. This was enhanced in the bi-transition-metal composites due to the higher isolation and stabilization effect induced by Ni-OH rather than Mg-OH species and to the electron transfer from Cu to Ni.

The topotactic decomposition of the Cu$_x$Mg$_{1-x}$Al-LDH/rGO precursors by calcination led to four-phase interfaces in the xCu@Cu$_2$O/MgAlO-rGO nanocomposites with core–shell-like Cu@Cu$_2$O NPs instead of Cu$_2$O NPs in the non-calcined precursor. The former Cu species improved the synergistic effect and the catalytic performance due to faster electron transfer, with also the contribution of the larger surface area and mesoporosity. It is noteworthy that the reduction of the Cu$_2$O species into well-dispersed Cu crystallites was observed by HRTEM in the Cu$_x$Mg$_{1-x}$Al-LDH/rGO precursors recycled 20 times, with only a slight decrease in activity.

Remarkably, calcination of the single-transition-metal-containing composites gave rise to catalysts (xCu@Cu$_2$O/MgAlO-rGO) exhibiting rather similar performance to the non-calcined bi-transition-metal Cu$_{3-x}$Ni$_x$Al-LDH/rGO composites. Then, core–shell-like Cu@Cu$_2$O NPs presented the same efficiency as regards electron transfer and the adsorption of reactants as materials with the presence of a second transition metal. This provides many possibilities to tune the catalytic properties.

The LDH/rGO composites presented significantly higher catalytic reduction efficiency for 4-NP than several recently reported catalysts [64,103–105,113,114].

Iqbal et al. have designed a multicomponent CD/Ag@MgAlCe-LDH catalyst combining Ag NPs and CD NPs with Ce-doped MgAl-LDH for the reduction of 4-NP and several organic dyes [34]. Ag NPs act as electron relays between electron donors and electron acceptor species, making them efficient redox catalysts [115]. CD acted as stabilizers and reducing agents of Ag NPs, preventing their aggregation. LDHs doped with rare-earth elements behave as highly basic structures or precursors of mixed oxides with tunable redox properties.

The CD/Ag@MgAlCe-LDH nanocomposite was obtained by dispersion of a previously coprecipitated Ce-doped MgAl-LDH (MgAlCe-LDH) into an aqueous solution of

AgNO$_3$ followed by the addition of CD. The obtained mixed solution was exposed to UV light (254 nm) to induce a photoreaction (Table 2, entry 26).

The introduction of Ce^{3+} into the brucite-like layers in CD/Ag@MgAlCe-LDH was confirmed. Additionally, a cerium phase and Ag NPs were identified. Photoluminescence experiments revealed electron transfer from CD to Ag NPs, and transient photocurrent measurements showed that the introduction of CD enhanced the photocurrent value. Moreover, CD/Ag@MgAlCe-LDH exhibited a photocurrent intensity two-fold higher than Ag@MgAlCe-LDH which evidenced a synergistic effect of CD and Ag NPs and a more efficient electron–hole pair separation.

The complete reduction of 4-NP into 4-AP in the presence of NaBH$_4$ (4-NP/NaBH$_4$ = 3:200) occurred at approximately 120 s upon addition of the CD/Ag@MgAlCe-LDH catalyst. On the contrary, the Ag@MgAlCe-LDH, CD@MgAlCe-LDH, and MgAlCe-LDH catalysts were practically inactive after reaction times above 840 s. CD/Ag@MgAlCe-LDH was also very active for the reduction of various organic dyes, with complete degradation achieved for rhodamine 6G (R6G) (90 s), MB, MO, and RhB (120 s), and Congo red (CR) (440 s). Based on the values of the apparent kinetic rate constants (k_{app}) (pseudo-first-order reaction) for the degradation of the different substrates, the activity of the catalysts ranged as follows: CD/Ag@MgAlCe-LDH > Ag@MgAlCe-LDH > CD@MgAlCe-LDH > MgAlCe-LDH. The k_{app} for the reduction of 4-NP (38×10^{-3} s^{-1}) is comparable to that obtained with the best-performing CuMgAl-LDH/rGO and CuNiAl-LDH/rGO catalysts, leading to k_{app} values in the range from 25.11×10^{-3} to 55.35×10^{-3} s^{-1} (Table 5) [39,41]. The CD/Ag@MgAlCe-LDH catalyst can be recycled and reused eight times without loss of activity.

In this catalyst, CD acted as a reducing agent of Ag$^+$ and coordinatively bonded to the Ag NPs, preventing their aggregation. Moreover, the small size of the CD and Ag NPs increased the number of accessible active sites and provided an effective interface, improving the electron transfer. Moreover, the MgAlCe-LDH improved the catalytic activity through its high specific surface area and behaved as a basic co-catalyst. The LDH's hydrophilic surface also facilitated the adsorption of BH$_4^-$ on Ag NPs and the adsorption of the 4-NP or organic dye substrates. Ce ions were involved in the redox reaction through the reversible Ce^{3+}/Ce^{4+} oxidation state. The CD NPs on the Ag NPs and the Ce ions on the MgAlCe-LDH support both promoted the electron transfer and migration.

Yang et al. achieved the combination of Pd NPs as an electron transfer system and CD as a reducing and stabilizing agent in a nanocomposite, where the originality is the use of a polymer (poly(N-isopropyl acrylamide) surface-modified LDH as a support (Figure 1; Table 2, entry 27) [22]. This multicomponent nanocatalyst displayed good water dispersibility.

The content and average diameter of the Pd NPs were 5.86 wt% and 3.7 nm, respectively, and the average diameter of CD was 4.25 nm in LDH@PDA@PNIPAM@Pd/CD. Pd NPs exhibited a more uniform size and distribution on the surface than on a reference LDH@PDA@PNIPAM@Pd composite reduced with NaBH$_4$, due to the stabilizing and reducing effects of CD. The sequential increase in intensity of the N 1s XPS peak after PDA modification, PNIPAM grafting, and CD loading in the LDH@PDA@PNIPAM@Pd/CD material confirmed that these components were supported on the LDH surface. Moreover, the characteristic Pd 3d$_{5/2}$ and Pd 3d$_{3/2}$ XPS peaks of Pd NPs were attributed to three different Pd^{2+} species: unreduced Pd^{2+}; PdO in contact with air, and ultra-small Pd clusters.

Reduction of MB into leucomethylene blue in the presence of an excess of NaBH$_4$ was completed after 32 s for LDH@PDA@PNIPAM@Pd/CDs and 64, 32, 80, and 80 s for LDH@PDA@PNIPAM@Pd, LDH@Pd/CD, LDH@Pd, and Pd/CD control catalysts, respectively. Meanwhile, LDH, LDH@PDA@PNIPAM, and CD were almost inactive. LDH@PDA@PNIPAM@Pd/CD was also active for the reduction of CR, MO, R6G, and RhB and the reduction of 4-NP, o-NP, and m-NP.

These results showed that both the grafting of PNIPAM and the presence of CD improve the catalytic performance. The polymer grafting increases the dispersibility in water and the surface area of the LDH@PDA. This favors the dispersion of Pd NPs of small size immobilized on the polymer brushes. Meanwhile, the high water dispersibility improves the contacting surface area between the reactants and the catalytically active sites. The reactants and BH_4^- are strongly adsorbed on the LDH support around the Pd NPs. These latter are stabilized and poorly aggregated due to the presence of CD, which improves the electronic transfer between the species. All these synergistic effects serve to enhance the catalytic activity and reusability.

Interestingly, LDH@PDA@PNIPAM@Pd/CD was also investigated as a bifunctional catalyst for the one-pot Knoevenagel condensation–reduction tandem reaction of 4-nitrobenzaldehyde and malonitrile. The Knoevenagel reaction was first performed (60 °C for 6 h) and the resulting mixture subjected to reduction with $NaBH_4$. Conversion reached 78% in the first condensation step and 64% after the second step with reduction of the alkenyl and nitro groups to alkane and amine groups. These results need to be optimized but represent a promising approach for the development of other types of tandem reactions.

NiO has been scarcely investigated for the reduction of 4-NP, although CuO, Co_3O_4, Fe_2O_3, and NiO exhibit almost similar activity [116]. Akbarzadeh and Gholami prepared a Pt-modified NiO-Al_2O_3 nanocomposite derived from GO-supported NiAl-LDH as a catalyst for 4-NP reduction (Table 2, entry 28) [64].

The characteristic XRD reflections of NiAl-LDH in the NiAl-LDH/G precursor moved to those of NiO after calcination at 600 °C in Pt-NiO/G. The highly dispersed Pt^0 NPs in low amounts (2.3 wt%) were not detected. TEM images showed that they were highly dispersed on the mixed oxide.

The performance of the Pt-NiO/G catalyst was compared to that of NiO, NiO/G, and Pt-NiO catalysts obtained by calcination of NiAl-LDH, NiAl-LDH/G, and Pt-NiAl-LDH precursors, respectively. The latter was obtained by impregnation of H_2PtCl_6 on NiAl-LDH. The reduction of 4-NP was achieved with $NaBH_4$:4-NP = 1:1000.

Both NiO (k_{app} = 1.37 × 10^{-3} s^{-1}) and NiO/G (k_{app} = 2.07 × 10^{-3} s^{-1}) were poorly active (pseudo-first-order kinetics). The activity was greatly improved upon introduction of Pt in Pt-NiO (k_{app} = 14.87 × 10^{-3} s^{-1}) due to the synergistic effect between NiO and well-dispersed Pt^0 NPs. Supporting Pt-NiO on the G support in the Pt-NiO/G composite led then to a two-fold increase in the k_{app} value, reaching 33.9 × 10^{-3} s^{-1}. However, the Pt-NiO/G was poorly stable because the reaction time for reduction increased by approximately 50% after five consecutive cycles.

The mechanism of the reduction was explored by varying the BH_4^- and 4-NP concentrations. The non-linear dependence of the k_{app} values with the concentration of substrate and 4-NP suggested a Langmuir–Hinshelwood mechanism. The hydrogen transfer from the adsorbed BH_4^- to 4-phenolate anions was the rate-limiting step of the reaction. The higher catalytic efficiency of Pt-NiO/G is likely due to the electron transfer from Pt to NiO, leading to an electron-rich area. The active hydrogen species and electrons formed from BH_4^- reacting with the Pt-NiO surface are transferred to the adsorbed 4-NP, thus generating 4-AP. The authors did not consider the promoting effect of rGO, which likely also improved the electron transfer as Pt-NiO was less active than Pt-NiO/G.

4.3.2. Hydrogenation of Nitro Compounds and α-β-Unsaturated Aldehydes

Previous works have shown that Ni-supported or Ni alloys efficiently replace noble-metal-containing catalysts in the selective hydrogenation of o-chloronitrobenzene (o-CNB) to o-chloroaniline (o-CAN) [117,118]. Wang et al. prepared dispersion-enhanced Ni-supported catalysts derived from nanocomposites based on NiAl-LDH and PAA (polyacrylic acid) and L-cysteine-functionalized CNT components (Ni-L/P-CNT) (Table 2, entry 29) [47]. In these catalysts, Ni NPs with a narrow size distribution of ca. 6.0 nm highly dispersed on the CNT support exhibited thin and faceted aspects accounting for SMSI. Ni^0 NPs were larger and highly aggregated in Ni/P-CNT. Dispersion of Ni^0 NPs was ~11.8% and ~20.5% in Ni/P-CNT and Ni-L/P-CNT, respectively. It was improved in Ni-L/P-CNT due to the optimized dispersion of the precursor NiAl-LDH on CNT.

The small LDH crystallite size in LDH-L/P-CNT made the reduction of Ni^{2+} in Ni(Al)O mixed oxide easier if compared to the LDH/P-CNT catalyst (without L-cysteine), where a mixture of NiO and Ni^0 was found after reduction.

Ni-L/P-CNT exhibited superior catalytic performance in the liquid-phase hydrogenation of o-CNB (140 °C, P H_2 2 MPa, 150 min), compared with the LDH/P-CNT catalyst and that prepared by conventional impregnation. For similar Ni content (~21 wt%), Ni-L/P-CNT presented o-CNB conversion of 99.3% and selectivity to o-CAN of 98.8%. In contrast, Ni/P-CNT presented conversion of 2.4% and selectivity of 98.5% due to the higher specific surface area (206 vs. 109 m^2 g^{-1}), reduction degree (95 vs. 46%), and dispersion (20.5 vs. 11.8%) of Ni^0 species in the former catalyst, as well as the existence of an electronic interaction between metal and support. This latter modified the mode of the metal–reactant interaction and thus enhanced the selectivity towards the desired products. The SMSI increased the high electronic density for π-electrons of the CNT support, inducing the enhanced oriented adsorption of o-CNB with the nitro group and repulsion of the chlorine group, whose hydrogenolysis was inhibited. This accounted for the higher o-CAN selectivity of the Ni/P-CNT and Ni-L/P-CNT catalysts compared to single Ni-impregnated CNT. Ni-L/P-CNT could be recycled up to five times, maintaining an o-CAN yield of 95% without Ni leaching.

Xie et al. used a NiAl-LDH/G composite as a precursor for catalysts to perform the liquid-phase hydrogenation of cinnamaldehyde (CALD) into hydrocinnamaldehyde [65]. NiAl-LDH/G was synthesized via the classical coprecipitation method of LDH in the presence of GO (Table 2, entry 30). Several comparative samples, i.e., NiAl-LDH, NiAl-LDH/C composite (C: active carbon), coprecipitated $Ni(OH)_2$/G, and impregnated $Ni(NO_3)_2$/G, were also prepared. The catalysts (Ni-L/G, Ni-L/C, Ni-Co/G, and Ni-Im/G) were obtained by the heating of the precursors at 600 °C under N_2 flow.

In the NiAl-LDH/G composite, GO was reduced into graphene, supporting highly dispersed LDH nanosheets (average size 22 nm) homogeneously dispersed and anchored on both sides of the exfoliated graphene sheets. Subsequently, small, highly homogeneous supported Ni NPs were obtained upon thermal treatment.

The activity of the Ni-L/G catalyst was evaluated in the hydrogenation reaction of CALD (120 °C, 1 MPa H_2) and compared to that of Ni-Co/G, Ni-Im/G, and Ni-L/C catalysts.

The products of the reaction are hydrocinnamaldehyde (HCAL), cinnamyl alcohol (COL), and hydrocinnamyl alcohol (HCOL). Ni-L/G exhibited the lowest particle size (12.6 nm), highest dispersion (36.5%), and largest specific surface area (182.6 m^2 g^{-1}) and it was the most active (100% conversion) and selective catalyst toward HCAL (94.8%). The strong interaction between NiAl-LDH and graphene, which inhibited both the aggregation of the Ni^{2+}-containing LDH sheets and the restacking of the graphene sheets, accounted for the high dispersion of the Ni^0 NPs upon in situ reduction and the improvement in the catalytic activity. On the contrary, weak interaction between the metal precursor species and the support in $Ni(OH)_2$/G and $Ni(NO_3)_2$/G samples led to poorly dispersed and large Ni NPs in Ni-Co/G (dispersion 19.8%; particle size 38.6 nm) and Ni-Im/G (dispersion 10.7%; particle size 68.4 nm), giving rise to lower conversion of 62% and 37.9%, respectively. Despite the similar specific surface areas (175.2 vs. 182.6 m^2 g^{-1}), Ni particle sizes (13.4 vs. 12.6 nm), and dispersion (34.6 vs. 36.5%) of Ni-L/C and Ni-L/G, the former catalyst led to lower conversion (56.5%). This shows that the 2D structure and the electronic properties of graphene, favoring the accessibility of the reactant to the active sites and electronic transfer toward the Ni NPs, improve the catalytic efficiency. Ni-L/G could be recycled up to five times while maintaining its catalytic performance.

4.3.3. Hydrogenolysis of Glycerol

The rising production of biodiesel gives rise to huge amounts of glycerol. Several means of revalorizing glycerol have been implemented using LDH-based materials, such as base-catalyzed transesterification with diethyl carbonate, dimethyl carbonate, or methyl stearate; carboxylation with urea; selective etherification to short-chain polyglycerols;

acetalization with acetone; oxidation into commodity chemicals, and steam reforming into H_2 [59,60,119–126]. It has been already reported that LDH/CNF and LDH/CF composites can act as efficient catalysts for the transesterification of glycerol with diethylcarbonate [59,60].

Xia et al. have used a nanocomposite obtained from CuMgAl-LDH and MWCNT components to valorize glycerol through hydrogenolysis to propanediols (PDOs) [66].

A series of MWCNT-CuMgAl-LDH composites were prepared by coprecipitation of CuMgAl-LDH into an aqueous suspension of pretreated MWCNT. The MWCNT's weight varied between 1.5 and 6 wt%. The precursors were heated at 400 °C and subsequently reduced at 300 °C to obtain the MWCNT-Cu/MgAlO catalysts (Table 2, entry 31). The MWCNT enabled the ordered accumulation of lamellae, bringing narrow-sized, doublet pore channels and high surface area of the catalysts. Indeed, the specific surface areas of the MWCNT-Cu/MgAlO catalysts, in the range of 130–218 $m^2 \ g^{-1}$, were higher than that of the LDH Cu-supported catalyst Cu/MgAlO (128 $m^2 \ g^{-1}$). The MWCNT-Cu/MgAlO catalysts showed a first group of mesopores whose average size decreased from 12.8 to 4.7 nm when the MWCNT content increased from 1.5 to 6 wt%. Other mesopores of 40–50 nm were attributed to the gap of MWCNT-pillared Cu/MgAlO lamellae. The close contact of Cu/MgAlO with MWCNT also enhanced the reducibility of CuO due to H_2 dissociation on MWCNT and spillover to Cu^{2+}, and improved Cu^0 dispersion.

The conversion of glycerol (26%) and the selectivity to 1,2-PDO (97.8%) both increased with the incorporation of MWCNT in the catalysts, having a maximum for (3%) MWCNT-Cu/MgAlO, with conversion and selectivity reaching 64.8 and 99.3%, respectively. At the same time, the activity per exposed Cu atom increased from 8.6 to 17.3 h^{-1}. Remarkably, the conversion and selectivity to 1,2-PDO were higher than those obtained with $Pd_{0.04}Cu/MgAlO$ (conversion 35.5% and selectivity 97.2%) and $Rh_{0.02}Cu/MgAlO$ (conversion 32.1% and selectivity 96.3%) catalysts, which is of practical interest due to the lower price of MWCNT than noble metals.

The 1,2-PDO yield decreased from 64.3% to 41.3% after five catalytic cycles performed with (3%)MWCNT-Cu/MgAlO, but this was mainly due to the loss of catalyst upon separation, because the activity of surface Cu atoms decreased only slightly (17.2 to 16.6 h^{-1}) and no leaching of Cu occurred.

The results obtained in hydrogenation reactions deserve several general comments regarding the properties of the LDH/nanocarbon catalysts. Assemblies of Ni- or Cu-containing LDH and carbon supports succeeded to obtain active and selective catalysts, mainly resulting from the ability to concurrently obtain highly dispersed metal NPs at high loading, SMSI, and tailored acid–base and textural properties, leading to synergistic effects. The large majority of the LDH/carbon-derived supported catalysts combine NiAl-LDH and amorphous carbon, CNT, or rGO. A comparison of the mean size and dispersion of Ni NPs of the most efficient catalysts at high Ni loading (22–42 wt% normalized to the Al content) and almost similar activation conditions allows us to highlight the influence of the nature of the carbon support. For an Ni particle size of ca. 13 nm, GO led to higher dispersion. This suggests higher metal–support interaction with the former support, which is consistent with the uniform size of the particles observed by TEM [65]. The high dispersion (20.5%) observed at Ni loading of ~40 wt% (based on the same Al content in the catalyst of ~6.5 wt%) on the PAA and L-cysteine-functionalized CNT support confirms that the surface groups of L-cysteine interacting with the Ni^{2+} and Al^{3+} cations allowed a highly dispersed NiAl-LDH phase to be obtained. It led subsequently to highly dispersed Ni^0 NPs of small size after reduction. It is noteworthy that the reduction of NiO depended on the nature of the carbon support. Treatment under nitrogen of composites composed of LDH graphene induced the complete reduction of NiO, where the supports acted as reducing agents. Meanwhile, although reduction was performed under hydrogen, Ni^{2+} was not totally reduced in the case of LDH/CNT composites, indicating that CNT is a less efficient reducing agent.

The works dealing with the series of hierarchical nanosheet array-like CuMgAl-LDH/rGO composites highlight the influence of their composition and activation mode [35,39,41]. With Cu and Ni (CuNiAl-LDH), the active species in the 4-NP hydrogenation is Cu_2O, even though Ni is the major species. There is electron transfer from Cu to the more electronegative Ni, and the more electrophilic surface of Cu_2O is more stabilized by the Ni-OH groups of the LDH. There is a strong synergistic effect among the in situ reduced Cu_2O species, Ni-OH (of LDH layers), and rGO, with an enhanced electron transfer ability [35].

It can be pointed out that in the CD/Ag@MgAlCe-LDH nanocomposite, the carbon compound did not act as a support but as NPs, decorating the active Ag NPs, thus improving electron transfer. It also produced the reduction of Ag^+. The Ce-doped MgAl-LDH was not only a support of high specific surface area but also a co-catalyst, providing basic sites and the redox ability of the cerium ions. The CD/Ag@MgAlCe-LDH nanocomposite represents a remarkable example of a synergistic effect and electronic interactions between the components.

4.4. HAS and Direct Synthesis of Isobutanol from Syngas

Catalysts based on Co, Fe, or Ni (e.g., CuFe- and CuCo- catalysts) have attracted particular interest for syngas (CO + H_2) conversion to higher alcohols, albeit suffering from several drawbacks such as poor stability, low alcohol productivity, and low selectivity. LDHs are very promising to obtain efficient catalysts for HAS. CO conversion from 17 to 57% and alcohol selectivity from 45 to 60% have been reported using catalysts prepared from CuFeMg-LDH, CuFeMgAl-LDH, CoZnGaAl-LDH, CoMn-LDH, and CuCoAl-LDH precursors [127–132].

Dispersion of the LDH on a carbonaceous support greatly improved the performance of the obtained composite catalysts by avoiding hot spots during the reaction and the agglomeration of LDH particles, which decrease selectivity to higher alcohols.

Several works deal with catalysts combining CuCoAl-LDH with CF or CNT and CuFeMg-LDH with CF highly selective to C_{2+}OH alcohols, and combining CuZnAl-LDH with CF for the direct synthesis of isobutanol from syngas [43,48,67–69]. The different nanocomposites were prepared by coprecipitation of LDH on acid-treated CF or CNT. The LDH crystallite size decreased with the CF or CNT content, in agreement with the increasing dispersion of the nucleation centers, generating more sheets and preventing their stacking. All nanocomposites presented a mesoporous structure, where the CF and CNT supports were completely wrapped by the highly dispersed LDH nanosheets. Metal NPs or alloys were formed upon reduction. The reduced composites exhibited the same lamellar morphology as the precursors without agglomeration of the particles and metal NPs with a uniform spherical-like shape.

The preparation of the LDH/carbon nanocatalysts and the conditions of activation are summarized in Table 2, entries 32–35.

CO conversions in the range of 32.5%–38.5% for CuCo/Al_2O_3/CFs (220 °C) and of 39.1–44.6% for CuCo-Red/CNT catalysts (230 °C), with selectivities to C_{+2}OH in the range from 40.6% to 61.2% in the HAS from syngas, were higher than that found for reduced CuCoAl-LDH (CO conversion 30.1 and 34.7% at 220 and 230°C, respectively, and selectivity ~24.5%). Increasing the CF or CNT content in the nanocomposite favored the CO conversion and C_{+2}OH selectivity at the expense of hydrocarbon, CO_2, and methanol. The rise in CO conversion can be related to the smaller size of the CuCo-alloy NPs, their higher dispersion, and the H_2 adsorption/activation on the supports. Indeed, CF or CNT was found to promote the reduction of Cu^{2+} and Co^{2+} due to the activation and spillover of H_2 and the lowering of the crystallite size of the LDH. The Cu and Co interaction in the alloy is highly favorable to C_{+2}OH selectivity. Cu is the active species responsible for CO molecular activation and insertion, whereas Co is responsible for CO dissociative activation and chain propagation.

The improved selectivity to C_{+2}OH also accounted for the good thermal conductivity of CF and CNT supports, which suppress hot spots responsible for the hydrocarbons' formation. Simultaneously, water produced during HAS can react with CO via the water–gas shift reaction (WGSR) to produce CO_2. This reaction, faster at higher temperatures, is lowered by the suppression of hot spots and by the higher diffusion of water in the porous structure of the supported catalysts. The lower hydrocarbon selectivity observed with CNT-containing nanocomposites suggests that thermal conductivity is probably better with CNT than CF.

In the CuFeMg-LDH/CF composite developed by Cao et al., Cu acts as the active site for CO molecular activation and insertion, and Fe for CO dissociation and chain propagation, with a cooperative effect between these sites [67]. Moreover, dispersion of the CuFe dual sites on the CF support improved the catalytic performance in comparison to an unsupported catalyst.

The CuFeMg-LDH/CF (Cu/Fe/Mg molar ratio 1:1:1) composite was prepared as CuCo-LDH/CF composites, leading to LDH nanosheets (200–230 nm) uniformly attached to the CF support and not agglomerated (Table 2, entry 32) [48]. The CuFe-Red/CF catalyst was obtained by direct reduction at 450 °C of the CuFeMg-LDH/CF precursor without intermediate calcination to avoid the reduction of Cu^{2+} to Cu^0 by CF. It contained Cu^0 and Fe^0 particles supported on MgO and CF (30 wt%), with Cu^0 particles size (5.2 nm) smaller than in the unsupported catalyst (6.1 nm).

Both CuFe-Red/CF and CuFe-Red presented higher CO conversion and $C_{+2}OH$ selectivity than Cu-Fe catalysts previously reported in the literature for HAS, according to the strong synergistic effect between the Cu and Fe species obtained from the CuFe-LDH precursor [133,134]. CuFe-Red/CF performed better than CuFe-Red, with a significant decrease in hydrocarbon and CO_2 selectivity due to the improvement in the thermal conductivity upon introduction of CF. Only a slight deactivation in CuFe-Red/CF was observed during 500 h of reaction, with CO conversion decreasing from 35.4% to 30.1% and selectivity to alcohols from 41.1% to 33.9%. The spent CuFe-Red/CF catalyst contained Cu^0, MgO, and CF, as in the fresh catalyst, but Fe^0 NPs converted into iron carbide (Fe_2C) species. Cu and Fe_2C were active species for HAS. After 500 h of reaction, there was no sintering of the particles, but separation of the Cu and Fe species was the key reason for the deactivation and the enhancement of hydrocarbon selectivity.

With CO conversion of 35.4% at 280 °C, CuFe-Red/CF showed lower activity than $CuCo/Al_2O_3/30\%CF$, whose conversion reached 38.5% at 230 °C, both catalysts having the same CF loading of 30%. Selectivity to CO_2 was enhanced with CuFe-Red/CF compared to $CuCo/Al_2O_3/30\%CF$ due to the higher reaction temperature, causing the WGSR to occur more rapidly, generating CO_2.

Isobutanol is a valuable platform chemical obtained from the HAS from syngas. $Cu/ZnO/Al_2O_3$-based catalysts are the most interesting for the industrial synthesis of isobutanol from syngas [135]. Alkali-doped Cu/Zn catalysts have been largely investigated for HAS in order to increase the yield of higher alcohols [136–140].

Huang et al. have prepared CuZnAl-LDH/CF composites by coprecipitation for the direct synthesis of isobutanol from syngas, focusing on the CF content and on the addition of K (Table 2, entry 35) [68,69]. Well-crystallized CuO and weakly crystallized ZnO were the only phases detected in the composites. Reduction at 320 °C led to Cu^0 NPs and to a CuO/ZnO solid solution whose formation was promoted upon the addition of ACF or K. The content of CuO/ZnO solid solution increased from 0.7 to 3.7 and 5.1 wt% when moving from $CuO/ZnO/Al_2O_3$ to $CuO/ZnO/Al_2O_3/30\%ACF$ and $CuO/ZnO/Al_2O_3/30\%ACF/K$. The reducibility of Cu^{2+} into the $CuO/ZnO/Al_2O_3/ACF$ catalysts was improved with the ACF content (0 to 30 wt%). Interestingly, EDS analysis revealed similar amounts of Cu and Zn on $CuO/ZnO/Al_2O_3/30\%ACF$ and $CuO/ZnO/Al_2O_3/30\%ACF/K$ samples, while a Cu enrichment was produced on the $Cu/ZnO/Al_2O_3$ surface (without ACF), showing that K and/or ACF promoted the formation of the CuO/ZnO solid solution.

Moderate CO adsorption is beneficial to alcohol synthesis and particularly for isobutanol selectivity resulting from CO insertion by reversal aldol condensation at the β-carbon of n-propanol [141]. CO-TPD analysis showed that the peak area of the moderately CO adsorbed species increased as $CuO/ZnO/Al_2O_3/30\%ACF/K > CuO/ZnO/Al_2O_3/30\%ACF > CuO/$

ZnO/Al$_2$O$_3$. Then, both ACF and K promoted the adsorption of CO species with moderate strength.

The CuO/ZnO/Al$_2$O$_3$/ACF composites with CO conversion in the range from 24.94% to 47.27% were more active than the non-supported CuO/ZnO/Al$_2$O$_3$ mixed oxide with conversion of 24.52%. This agreed with an increase in the specific surface area and dispersion of the LDH nanoflakes in the composites, leading to more accessible Cu^{2+} active sites. Conversion decreased at 40% ACF content (42.61%) due to the low amount of CuO/ZnO/Al$_2$O$_3$ active phase. Selectivity to alcohols decreased upon addition of ACF (>10%) at the expense of alkanes. However, methanol and isobutanol were the two main products. Addition of ACF increased significantly the isobutanol selectivity, reaching a maximum of 19.88% for 30% ACF. CO conversion and isobutanol selectivity further increased upon addition of K, although to a lower extent than upon addition of ACF. The best-performing catalyst was CuO/ZnO/Al$_2$O$_3$/30%ACF/K, with CO conversion of 49.77% and isobutanol selectivity of 22.13% owing to the synergistic effect between ACF and K. Both promoted electron transfer to Cu and adsorption of CO with moderate strength, favorable to CO insertion and carbon chain propagation. Otherwise, they enhanced the content of active CuO/ZnO solid solution.

A comparison of the results obtained from the CuCoAl-LDH/CF, CuCoAl-LDH/CNT, and CuFeMg-LDH/CF composites, on one hand, and from the CuZnAl-LDH/CF composite, on the other hand, can hardly be achieved due to the different reaction temperatures in the range of 220–280 °C in the former case and of 320 °C in the latter. However, the results deserve several comments. A different product distribution was obtained at almost similar CO conversion when CuCo alloy and Cu with Fe$_2$C NPs or CuO/ZnO solid solution were the main active species. The former species gave rise to higher hydrocarbon selectivities despite the lower reaction temperature, which suggests better thermal conductivity in the presence of ZnO. Ethanol and C$_{4+}$OH were the main alcohols formed. The CuO/ZnO solid solution as the active species led to higher alcohol selectivity, with methanol and isobutanol as the main products. The higher alcohol yield was accompanied by a higher CO$_2$ yield according to the formation of a large amount of water and then WGSR with CO.

5. Concluding Remarks and Perspectives

This review evidences that the catalysts derived from LDH/nanocarbon composites deserve significant attention due to their huge potentialities apart from electro- and photocatalysis. They are yet employed with remarkable efficiency in a large range of reactions, including C–C cross-coupling (e.g., aldolization, Michael, Knoevenagel, Ullman, Sonogashira, Heck, synthesis of chalcone, etc.), oxidation, hydrogenation, HAS, and cascade reactions (e.g., Knoevenagel–Michael and oxidation–Knoevenagel). These hierarchical catalysts will be undeniably developed in the near future due to their unique tailoring for targeted reactions. A dramatic variety of assemblies can be obtained, as illustrated through several remarkable examples. Assembling hollow flower-like LDH and N, S-doped graphene acting as a support of Pd NPs led to composites with hollow inner and mesoporous hierarchically flower-like outer structures. Sandwich-like superstructures resulted from the intercalation of LDH and rGO and of MMO and MWCNT. Magnetic properties can be obtained combining Fe$_3$O$_4$ NP, GO, and LDH. Colloidal suspensions of LDH/GO composites resulted from the self-assembly of previously exfoliated LDH and GO. A uniform nanoarray-like framework was obtained by the coprecipitation of CuMgAl-LDH on exfoliated and citrate-functionalized GO.

Several specific properties account for the various morphologies and the structural and textural features. LDH nucleates from smaller sp^2 graphitic domains and some unrepaired defect sites on the nanocarbons. The resulting growth of highly dispersed and poorly aggregated LDH nanosheets in turn impedes the restacking and agglomeration of the graphene sheets. Both features give rise to highly accessible and dispersed active sites. Dispersion of the LDH nanosheets can be also greatly improved through pre-functionalization of the carbon support, as achieved with PAA-CNT and subsequent immobilization bridging linkers, e.g., L-cysteine. Moreover, the supersaturation rate of the LDH during coprecipitation allows control of the orientation of the nanosheets on the nanocarbon.

Other specific properties of the LDH/nanocarbon composites, such as the hydrophobic/hydrophilic balance and the amphiphilic property exhibited by LDH/CNT nanocomposites used as highly thermodynamically stable Pickering emulsifiers, are emerging. Adsorption of aromatic reactants via π-π stacking with graphene-like and CNT supports is highly favorable to catalytic reactivity.

One of the main specificities of the LDH/carbon composites is the interfacial electron transfer. It is facilitated by the considerable inherent electronic mobility in the carbon components, leading to charge redistribution upon interaction with the less conductive LDH phases. Differences have been previously reported among the variety of nanocarbons with, for instance, higher charge transfer ability of the planar π-electrons of the highly graphitized structure of graphene, compared to CNT with a more distorted sp^2 structure [142–145].

Evidence of the in situ self-reduction of cations such as Ni^{2+} induced by different carbon supports, i.e., rGO or CNT with SMSI, has been given. It occurs upon heating in the absence of H_2 in LDH/rGO composites or under hydrogen in LDH/CNT composites. Reduction of the cations can also occur during the reaction processes using chemical agents ($NaBH_4$) thus avoiding the pre-reduction step.

The textures that improve the mass transport and accessibility of the reactants to the active sites, the electron transfer leading to SMSI and the improved adsorption of the reactants, and the dispersion of the active sites are the main features responsible for the synergistic effects observed in the nanocomposites in comparison to physical mixtures of the components. However, a specificity of the LDH/nanocarbon catalysts is the multi-phase synergistic effect illustrated through outstanding examples.

It must be underlined for future developments that basic sites can be introduced through several routes in LDH/nanocarbon composites. LDH or LDO components can indeed provide Lewis- or Brønsted-type basic sites. Another approach is the doping of the LDH/nanocarbon with an alkaline cation (e.g., K^+) or the introduction of N atoms in GO (N,S-GO), amorphous carbon, or CD (NCD).

Some strong tendencies can be drawn from the reported results. The large majority of the composites are prepared by coprecipitation of the LDH phase in the presence of the nanocarbon. Otherwise, well-known methods for controlling the size and avoiding aggregation of the LDH particles have been employed. We can underline the separated nucleation and aging step process, the exfoliation of the nanosheets, and the use of urea as an alkaline controlling agent. Regarding the composition of the LDH, MgAl-LDH is the most widely used for base-catalyzed reactions, and the Cu- or Ni-containing LDH for redox reactions.

GO is undeniably the most significant nanocarbon in all the reported works. It is generally exfoliated by sonication, leading to almost individual nanosheets. Its reduction in the alkaline media creates defect sites very useful as nucleation sites to improve the dispersion of the coprecipitated LDH phase.

It is important to highlight the good durability and cyclability of the LDH/nanocarbon catalysts, which is a critical property for practical applications. It is consistent with the high mechanical strength of the nanocarbons and the robustness of the assemblies with LDH, the SMSI leading to very weak leaching. Moreover, regeneration of the active sites is facilitated by the high dispersion and accessibility.

The above overview suggests several perspectives for the larger development of LDH/nanocarbon catalysts:

1. Those based on N-doped nanocarbons (N-GO, N-CD, N-graphene quantum dots) would lead to hierarchical structures exhibiting original basic properties. They will efficiently perform a larger range of base-catalyzed reactions, particularly those requiring strong basic strength (e.g., isomerization of olefines).
2. The ability to introduce basic sites by different routes (LDH and/or carbon compound) and metal NPs through transition metal cations in the LDH or by impregnation

offers unlimited possibilities to design multifunctional catalysts. This will allow large implementation of one-pot and cascade reactions. It is perhaps the most promising application.

3. LDHs intercalated with catalytically active anionic species by direct exchange or reconstruction have not been yet exploited in LDH/carbon compounds. This would be a new, original route to synthesize multifunctional catalysts.
4. There is a great need to optimize the interface between the nanocarbons and the LDHs, determining to a large extent the charge transfer and cyclability of the obtained catalysts. This will result from control of the synthesis conditions. Because direct growth of LDH on the nanocarbon by coprecipitation appears the most convenient and largely used method, great effort must be devoted to control the supersaturation level during synthesis. Moreover, coprecipitation of LDH through separated nucleation and aging steps will be more extensively exploited and, when large LDH particle size is needed, the urea synthesis method will be preferred. Furthermore, exfoliation of LDH could be more largely developed, particularly using organic solvents, leading to exceptional results.
5. CD-based LDH/nanocarbons offer possibilities to improve the electron transfer and to decorate the metal NPs. These composites have been scarcely studied, although they are potential multifunctional catalysts for cascade reactions.
6. A major challenge is the large scale and the reproducibility of the synthesis of the LDH/nanocarbon hierarchical structures. Synthesis in continuous flow of rGO/ZnAl-LDH has been reported [146,147]. This approach must be extended.

Therefore, the first concern in the new directions of research involving LDH/nanocarbon catalysts will be the tailored combination of components chosen to provide the specific sites involved in the targeted reactions. Original LDH compositions including rare earth and noble metal cations must be considered. Besides the most adapted nanocarbon, other oxides and/or dopants (e.g., Na^+, K^+, F^-) able to induce, for instance, magnetic properties or improve basicity could be added. A preparation method exploiting several intrinsic properties of the LDHs (exfoliation, reconstruction, anionic exchange) will largely contribute to inducing synergistic effects and porous hierarchical structures. At present, one of the most promising catalytic applications of these materials concerns the valorization of biomass. The LDH/nanocarbon catalysts can be adapted to one-pot and cascade reactions in liquid media, including aqueous media, which has gradually led to their use in this application field. The noticeable advantage of the LDH/nanocarbon composites to obtain good-performing supported metal catalysts at low noble metal loading or avoiding the use of highly dispersed transition metals (Ni, Co, Cu, etc.) without leaching and good recyclability will be largely exploited to develop industrial processes of biomass valorization at low cost. Another promising direction is the synthesis of fine chemicals for pharmaceutical applications, where cascade reactions are particularly suitable to reduce costs and improve the atom economy of the processes.

Author Contributions: Conceptualization, D.T. and M.G.Á.; resources, D.T. and M.G.Á.; writing—original draft preparation, D.T. and M.G.Á.; writing—review and editing D.T. and M.G.Á.; supervision, D.T. and M.G.Á. All authors have read and agreed to the published version of the manuscript.

Funding: This research received no external funding.

Informed Consent Statement: Not applicable.

Conflicts of Interest: The authors declare no conflict of interest.

References

1. Varadwaj, G.B.B.; Nyamori, V.O. Layered double hydroxide- and graphene-based hierarchical nanocomposites: Synthetic strategies and promising applications in energy conversion and conservation. *Nano Res.* **2016**, *9*, 3598–3621. [CrossRef]
2. Zhao, M.; Zhao, Q.; Li, B.; Xue, H.; Pang, H.; Chen, C. Recent progress in layered double hydroxide based materials for electrochemical capacitors: Design, synthesis and performance. *Nanoscale* **2017**, *9*, 15206–15225. [CrossRef] [PubMed]

3. Zhao, M.Q.; Zhang, Q.; Huang, J.Q.; Wei, F. Hierarchical Nanocomposites Derived from Nanocarbons and Layered Double Hydroxides–Properties, Synthesis, and Applications. *Adv. Func. Mater.* **2012**, *22*, 675. [CrossRef]
4. Kulandaivalu, S.; Azman, N.H.N.; Sulaiman, Y. Advances in Layered Double Hydroxide/Carbon Nanocomposites Containing Ni^{2+} and $Co^{2+}/^{3+}$ for Supercapacitors. *Front. Mater.* **2020**, *7*, 147. [CrossRef]
5. Tang, C.; Titirici, M.M.; Zhang, Q. A review of nanocarbons in energy electrocatalysis: Multifunctional substrates and highly active sites. *J. Energy Chem.* **2017**, *26*, 1077–1093. [CrossRef]
6. Tang, C.; Wang, H.F.; Zhu, X.L.; Li, B.Q.; Zhang, Q. Advances in Hybrid Electrocatalysts for Oxygen Evolution Reactions: Rational Integration of NiFe Layered Double Hydroxides and Nanocarbon. *Part. Part. Syst. Charact.* **2016**, *33*, 473–486. [CrossRef]
7. Song, B.; Zeng, Z.; Zeng, G.; Gong, J.; Xiao, R.; Ye, S.; Chen, M.; Lai, C.; Xu, P.; Tang, X. Powerful combination of g-C_3N_4 and LDHs for enhanced photocatalytic performance: A review of strategy, synthesis, and applications. *Adv. Colloid Interface Sci.* **2019**, *272*, 101999. [CrossRef]
8. Daud, M.; Kamal, M.S.; Shehzad, F.; Al Harthi, M. Graphene/Layered Double Hydroxides Nanocomposites: A Review of Recent Progress in Synthesis and Applications. *Carbon* **2016**, *104*, 241–252. [CrossRef]
9. Gu, P.; Zhang, S.; Li, X.; Wang, X.; Wen, T.; Jehan, R.; Alsaedi, A.; Hayat, T.; Wang, X. Recent advances in layered double hydroxide-based nanomaterials for the removal of radionuclides from aqueous solution. *Environ. Pollut.* **2018**, *240*, 493–505. [CrossRef]
10. Pang, H.; Wu, Y.; Wang, X.; Hu, B.; Wang, X. Recent Advances in Composites of Graphene and Layered Double Hydroxides for Water Remediation: A Review. *Chem. Asian J.* **2019**, *14*, 2542–2552. [CrossRef] [PubMed]
11. Cao, Y.; Li, G.; Li, X. Graphene/layered double hydroxide nanocomposite: Properties, synthesis, and applications. *Chem. Eng. J.* **2016**, *292*, 207–223. [CrossRef]
12. Fan, G.; Li, F.; Evans, D.G.; Duan, X. Catalytic applications of layered double hydroxides: Recent advances and perspectives. *Chem. Soc. Rev.* **2014**, *43*, 7040–7066. [CrossRef] [PubMed]
13. Winter, F.; Van Dillen, A.J.; de Jong, K.P. Supported hydrotalcites as highly active solid base catalysts. *Chem. Commun.* **2005**, *31*, 3977–3979. [CrossRef] [PubMed]
14. Winter, F.; Koot, V.; van Dillen, A.J.; Geus, J.W.; de Jong, K.P. Hydrotalcites supported on carbon nanofibers as solid base catalysts for the synthesis of MIBK. *J. Catal.* **2005**, *236*, 91–100. [CrossRef]
15. Liang, Y.N.; Oh, W.D.; Li, Y.; Hu, X. Nanocarbons as platforms for developing novel catalytic composites: Overview and prospects. *Appl. Catal. A Gen.* **2018**, *562*, 94–105. [CrossRef]
16. Bhuyan, M.S.A.; Uddin, M.N.M.; Islam, M.; Bipasha, F.A.; Hossain, S.S. Synthesis of graphene. *Int. Nano Lett.* **2016**, *6*, 65–83. [CrossRef]
17. Mallakpour, S.; Khadem, E. Carbon nanotube–metal oxide nanocomposites: Fabrication, properties and applications. *Chem. Eng. J.* **2016**, *302*, 344–367. [CrossRef]
18. Sousa, H.B.A.; Martins, C.S.M.; Prior, J.A.V. You Don't Learn That in School: An Updated Practical Guide to Carbon Quantum Dots. *Nanomaterials* **2021**, *11*, 611. [CrossRef]
19. Feng, L.; Xie, N.; Zhong, J. Carbon Nanofibers and Their Composites: A Review of Synthesizing, Properties and Applications. *Materials* **2014**, *7*, 3919–3945. [CrossRef]
20. Álvarez, M.G.; Tichit, D.; Medina, F.; Llorca, J. Role of the synthesis route on the properties of hybrid LDH-graphene as basic catalysts. *Appl. Surf. Sci.* **2017**, *396*, 821–831. [CrossRef]
21. Ahmed, N.S.; Menzel, R.; Wang, Y.; Garcia-Gallastegui, A.; Bawaked, S.M.; Obaid, A.Y.; Basahel, S.N.; Mokhtar, M. Graphene-oxide-supported CuAl and CoAl layered double hydroxides as enhanced catalysts for carbon-carbon coupling via Ullmann reaction. *J. Solid State Chem.* **2017**, *246*, 130–137. [CrossRef]
22. Yang, Y.; Zhu, W.; Cui, D.; Lü, C. Mussel-inspired preparation of temperature-responsive polymer brushes modified layered double hydroxides@Pd/carbon dots hybrid for catalytic applications. *Appl. Clay Sci.* **2021**, *200*, 105958. [CrossRef]
23. Tichit, D.; Coq, B. Catalysis by Hydrotalcites and Related Materials. *Cattech* **2003**, *7*, 206–217. [CrossRef]
24. Takehira, K. Recent development of layered double hydroxide-derived catalysts − Rehydration, reconstitution, and supporting, aiming at commercial application. *Appl. Clay Sci.* **2017**, *136*, 112–141. [CrossRef]
25. Debecker, D.P.; Gaigneaux, E.M.; Busca, G. Exploring, Tuning, and Exploiting the Basicity of Hydrotalcites for Applications in Heterogeneous Catalysis. *Chem. Eur. J.* **2009**, *15*, 3920–3935. [CrossRef]
26. Li, C.; Wei, M.; Evans, D.G.; Duan, X. Layered Double Hydroxide-based Nanomaterials as Highly Efficient Catalysts and Adsorbents. *Small* **2014**, *10*, 4469. [CrossRef]
27. Hummers, W.S.; Offeman, R.E. Preparation of Graphitic Oxide. *J. Am. Chem. Soc.* **1958**, *80*, 1339. [CrossRef]
28. Lim, S.Y.; Shen, W.; Gao, Z. Carbon quantum dots and their applications. *Chem. Soc. Rev.* **2015**, *44*, 362–391. [CrossRef]
29. Inagaki, M.; Tsumura, T.; Kinumoto, T.; Toyoda, M. Graphitic carbon nitrides (g-C_3N_4) with comparative discussion to carbon materials. *Carbon* **2019**, *141*, 580–607. [CrossRef]
30. Álvarez, M.G.; Marcu, I.C.; Tichit, D. *Progress in Layered Double Hydroxides—From Synthesis to New Applications*; Nocchetti, M., Costantino, U., Eds.; World Scientific Publishing Ltd.: Singapore, 2022; pp. 189–362.
31. Zheng, Y.; Liu, J.; Liang, J.; Jaroniec, M.; Qiao, S.Z. Graphitic Carbon Nitride Materials: Controllable Synthesis and Applications in Fuel Cells and Photocatalysis. *Energy Environ. Sci.* **2012**, *5*, 6717–6731. [CrossRef]

32. Wang, Q.; Chen, L.; Guan, S.; Zhang, X.; Wang, B.; Cao, X.; Yu, Z.; He, Y.; Evans, D.G.; Feng, J.; et al. Ultrathin and Vacancy-Rich CoAl-Layered Double Hydroxide/Graphite Oxide Catalysts: Promotional Effect of Cobalt Vacancies and Oxygen Vacancies in Alcohol Oxidation. *ACS Catal.* **2018**, *8*, 3104–3115. [CrossRef]
33. Zeynizadeh, B.; Gilanizadeh, M. Synthesis and characterization of a magnetic graphene oxide/Zn–Ni–Fe layered double hydroxide nanocomposite: An efficient mesoporous catalyst for the green preparation of biscoumarins. *New J. Chem.* **2019**, *43*, 18794–18804. [CrossRef]
34. Iqbal, K.; Iqbal, A.; Kirillov, A.M.; Shan, C.; Liu, W.; Tang, Y. A new multicomponent CDs/Ag@Mg–Al–Ce-LDH nanocatalyst for highly efficient degradation of organic water pollutants. *J. Mater Chem. A* **2018**, *6*, 4515–4524. [CrossRef]
35. Wei, Z.; Li, Y.; Dou, L.; Ahmad, M.; Zhang, H. $Cu_{3-x}Ni_x$Al-Layered Double Hydroxide-Reduced Graphene Oxide Nanosheet Array for the Reduction of 4-Nitrophenol. *ACS Appl. Nano Mater.* **2019**, *2*, 2383. [CrossRef]
36. Rohani, S.; Ziarani, G.M.; Ziarati, A.; Badiei, A. Designer 3D CoAl-layered double hydroxide@N, S doped graphene hollow architecture decorated with Pd nanoparticles for Sonogashira couplings. *Appl. Surf. Sci.* **2019**, *496*, 143599. [CrossRef]
37. Álvarez, M.G.; Crivoi, D.G.; Medina, F.; Tichit, D. Synthesis of Chalcone Using LDH/Graphene Nanocatalysts of Different Compositions. *ChemEngineering* **2019**, *3*, 29. [CrossRef]
38. Xie, R.; Fan, G.; Yang, L.; Li, F. Highly Efficient Hybrid Cobalt–Copper–Aluminum Layered Double Hydroxide/Graphene Nanocomposites as Catalysts for the Oxidation of Alkylaromatics. *ChemCatChem* **2016**, *8*, 363–371. [CrossRef]
39. Dou, L.; Zhang, H. Facile assembly of nanosheet array-like CuMgAl-layered double hydroxide/rGO nanohybrids for highly efficient reduction of 4-nitrophenol. *J. Mater. Chem. A* **2016**, *4*, 18990–19002. [CrossRef]
40. Miao, M.Y.; Feng, J.T.; Jin, Q.; He, Y.F.; Liu, Y.N.; Du, Y.Y.; Zhang, N.; Li, D.Q. Hybrid Ni–Al layered double hydroxide/graphene composite supported gold nanoparticles for aerobic selective oxidation of benzyl alcohol. *RSC Adv.* **2015**, *5*, 36066–36074. [CrossRef]
41. Dou, L.; Wang, Y.; Li, Y.; Zhang, H. Novel core–shell-like nanocomposites $xCu@Cu_2O$/MgAlO-rGO through an in situ self-reduction strategy for highly efficient reduction of 4-nitrophenol. *Dalton Trans.* **2017**, *46*, 15836–15847. [CrossRef]
42. Shan, Y.; Yu, C.; Yang, J.; Dong, Q.; Fan, X.; Qiu, J. Thermodynamically Stable Pickering Emulsion Configured with Carbon-Nanotube-Bridged Nanosheet-Shaped Layered Double Hydroxide for Selective Oxidation of Benzyl Alcohol. *ACS Appl. Mater. Interfaces* **2015**, *7*, 12203–12209. [CrossRef] [PubMed]
43. Cao, A.; Liu, G.; Wang, L.; Liu, J.; Yue, Y.; Zhang, L.; Liu, Y. Growing layered double hydroxides on CNTs and their catalytic performance for higher alcohol synthesis from syngas. *J. Mater. Sci.* **2016**, *51*, 5216–5231. [CrossRef]
44. Celaya-Sanfiz, A.; Morales-Vega, N.; De Marco, M.; Iruretagoyena, D.; Mokhtar, M.; Bawaked, S.M.; Basahel, S.N.; Al Thabaiti, S.A.; Alyoubi, A.O.; Shaffer, M.S.P. Self-condensation of acetone over Mg–Al layered double hydroxide supported on multi-walled carbon nanotube catalysts. *J. Mol. Catal. A Chem.* **2015**, *398*, 50–57. [CrossRef]
45. Lan, M.; Fan, G.; Sun, W.; Li, F. Synthesis of hybrid Zn–Al–In mixed metal oxides/carbon nanotubes composite and enhanced visible-light-induced photocatalytic performance. *Appl. Surf. Sci.* **2013**, *282*, 937–946. [CrossRef]
46. Wang, H.; Xiang, X.; Li, F. Hybrid ZnAl-LDH/CNTs nanocomposites: Noncovalent assembly and enhanced photodegradation performance. *AIChE J.* **2010**, *56*, 768–778. [CrossRef]
47. Wang, J.; Fan, G.; Li, F. A hybrid nanocomposite precursor route to synthesize dispersion-enhanced Ni catalysts for the selective hydrogenation of o-chloronitrobenzene. *Catal. Sci. Technol.* **2013**, *3*, 982–991. [CrossRef]
48. Wang, L.; Cao, A.; Liu, G.; Zhang, L.; Liu, Y. Bimetallic CuCo nanoparticles derived from hydrotalcite supported on carbon fibers for higher alcohols synthesis from syngas. *Appl. Surf. Sci.* **2016**, *360*, 77–85. [CrossRef]
49. Guo, Y.; Fan, L.; Liu, M.; Yang, L.; Fan, G.; Li, F. Nitrogen-Doped Carbon Quantum Dots-Decorated Mg-Al Layered Double Hydroxide-Supported Gold Nanocatalysts for Efficient Base-Free Oxidation of Benzyl Alcohol. *Ind. Eng. Chem. Res.* **2020**, *59*, 636–646. [CrossRef]
50. Roelofs, J.C.A.A.; Lensveld, D.J.; van Dillen, A.J.; de Jong, K.P. On the Structure of Activated Hydrotalcites as Solid Base Catalysts for Liquid-Phase Aldol Condensation. *J. Catal.* **2001**, *203*, 184–191. [CrossRef]
51. Roelofs, J.C.A.A.; van Dillen, A.J.; de Jong, K.P. Base-catalyzed condensation of citral and acetone at low temperature using modified hydrotalcite catalysts. *Catal. Today* **2000**, *60*, 297–303. [CrossRef]
52. Roelofs, J.C.A.A.; van Dillen, A.J.; de Jong, K.P. Condensation of citral and ketones using activated hydrotalcite catalysts. *Catal. Lett.* **2001**, *74*, 91–94. [CrossRef]
53. Wang, Y.; Dou, L.; Zhang, H. Nanosheet Array-Like Palladium-Catalysts Pd_x/rGO@CoAl-LDH via Lattice Atomic-Confined in Situ Reduction for Highly Efficient Heck Coupling Reaction. *ACS Appl. Mater. Interfaces* **2017**, *9*, 38784–38795. [CrossRef] [PubMed]
54. Zhang, W.; Wang, Z.; Zhao, Y.; Miras, H.N.; Song, Y.F. Precise Control of the Oriented Layered Double Hydroxide Nanosheets Growth on Graphene Oxides Leading to Efficient Catalysts for Cascade Reactions. *ChemCatChem* **2019**, *11*, 5466–5474. [CrossRef]
55. Huang, P.; Liu, J.; Wei, F.; Zhu, Y.; Wang, X.; Cao, C.; Song, W. Size-selective adsorption of anionic dyes induced by the layer space in layered double hydroxide hollow microspheres. *Mater. Chem. Front.* **2017**, *1*, 1550–1555. [CrossRef]
56. Costantino, U.; Marmottini, F.; Nocchetti, M.; Vivani, R. New Synthetic Routes to Hydrotalcite-Like Compounds—Characterisation and Properties of the Obtained Materials. *Eur. J. Inorg. Chem.* **1998**, *10*, 1439–1446. [CrossRef]
57. Adachi-Pagano, M.; Forano, C.; Besse, J.P. Synthesis of Al-rich hydrotalcite-like compounds by using the urea hydrolysis reaction—control of size and morphology. *J. Mater. Chem.* **2003**, *13*, 1988–1993. [CrossRef]

58. Ogawa, M.; Kaiho, H. Homogeneous Precipitation of Uniform Hydrotalcite Particles. *Langmuir* **2002**, *18*, 4240–4242. [CrossRef]
59. Modesto-Lopez, L.B.; Chimentao, R.J.; Álvarez, M.G.; Rosell-Llompart, J.; Medina, F.; Llorca, J. Direct growth of hydrotalcite nanolayers on carbon fibers by electrospinning. *Appl. Clay Sci.* **2014**, *101*, 461–467. [CrossRef]
60. Álvarez, M.G.; Frey, A.M.; Bitter, J.H.; Segarra, A.M.; de Jong, K.P.; Medina, F. On the role of the activation procedure of supported hydrotalcites for base catalyzed reactions: Glycerol to glycerol carbonate and self-condensation of acetone. *Appl. Catal. B Environ.* **2013**, *134*, 231–237. [CrossRef]
61. Stamate, A.E.; Pavel, O.D.; Zavoianu, R.; Brezestean, I.; Ciorita, A.; Birjega, R.; Neubauer, K.; Koeckritz, A.; Marcu, I.C. Ce-Containing MgAl-Layered Double Hydroxide-Graphene Oxide Hybrid Materials as Multifunctional Catalysts for Organic Transformations. *Materials* **2021**, *14*, 7457. [CrossRef]
62. Zhao, Y.; Xie, R.; Lin, Y.; Fan, G.; Li, F. Highly efficient solvent-free aerobic oxidation of ethylbenzene over hybrid Zn–Cr layered double hydroxide/carbon nanotubes nanocomposite. *Catal. Commun.* **2018**, *114*, 65–69. [CrossRef]
63. Shen, J.; Ye, S.; Xu, X.; Liang, J.; He, G.; Chen, H. Reduced graphene oxide based NiCo layered double hydroxide nanocomposites: An efficient catalyst for epoxidation of styrene. *Inorg. Chem. Commun.* **2019**, *104*, 219–222. [CrossRef]
64. Akbarzadeh, E.; Gholami, M.R. Pt–NiO–Al_2O_3/G derived from graphene-supported layered double hydroxide as efficient catalyst for *p*-nitrophenol reduction. *Res. Chem. Intermed.* **2017**, *43*, 5829–5839. [CrossRef]
65. Xie, R.; Fan, G.; Ma, Q.; Yang, L.; Li, F. Facile synthesis and enhanced catalytic performance of graphene-supported Ni nanocatalyst from a layered double hydroxide-based composite precursor. *J. Mater. Chem. A* **2014**, *2*, 7880–7889. [CrossRef]
66. Xia, S.; Zheng, L.; Ning, W.; Wang, L.; Chen, P.; Hou, Z. Multiwall carbon nanotube-pillared layered $Cu_{0.4}/Mg_{5.6}Al_2O_{8.6}$: An efficient catalyst for hydrogenolysis of glycerol. *J. Mater. Chem. A* **2013**, *1*, 11548–11552. [CrossRef]
67. Cao, A.; Yang, Q.; Wei, Y.; Zhang, L.; Liu, Y. Synthesis of higher alcohols from syngas over CuFeMg-LDHs/CFs composites. *Int. J. Hydrogen Energy* **2017**, *42*, 17425–17434. [CrossRef]
68. Cheng, S.Y.; Gao, Z.H.; Kou, W.; Liu, Y.; Huang, W. Direct Synthesis of Isobutanol from Syngas over Nanosized Cu/ZnO/Al_2O_3 Catalysts Derived from Hydrotalcite-like Materials Supported on Carbon Fibers. *Energy Fuels* **2017**, *31*, 8572–8579. [CrossRef]
69. Kou, J.W.; Cheng, S.Y.; Gao, Z.H.; Cheng, F.Q.; Huang, H. Synergistic effects of potassium promoter and carbon fibers on direct synthesis of isobutanol from syngas over Cu/ZnO/Al_2O_3 catalysts obtained from hydrotalcite-like compounds. *Solid State Sci.* **2019**, *87*, 138–145. [CrossRef]
70. Tichit, D.; Naciri Bennani, M.; Figueras, F.; Tessier, R.; Kervennal, J. Aldol condensation of acetone over layered double hydroxides of the meixnerite type. *Appl. Clay Sci.* **1998**, *13*, 401–415. [CrossRef]
71. Abello, S.; Medina, F.; Tichit, D.; Perez Ramirez, J.; Cesteros, Y.; Salagre, P.; Sueiras, J.E. Nanoplatelet-based reconstructed hydrotalcites: Towards more efficient solid base catalysts in aldol condensations. *Chem. Commun.* **2005**, *11*, 1453–1455. [CrossRef]
72. Abello, S.; Medina, F.; Tichit, D.; Perez Ramirez, J.; Groen, J.C.; Sueiras, S.E.; Salagre, P.; Cesteros, Y. Aldol Condensations over Reconstructed Mg–Al Hydrotalcites: Structure–Activity Relationships Related to the Rehydration Method. *Chem. Eur. J.—A Eur. J.* **2005**, *11*, 728–739. [CrossRef]
73. Evranos Aksoz, B.; Ertan, R. Chemical and Structural Properties of Chalcones. *IFABAD J. Pharm. Sci.* **2011**, *36*, 223–242.
74. Chimenti, F.; Fioravanti, R.; Bolasco, A.; Chimenti, P.; Secci, D.; Rossi, F.; Yáñez, M.; Orallo, F.; Ortuso, F.; Alcaro, S. Chalcones: A Valid Scaffold for Monoamine Oxidases Inhibitors. *J. Med. Chem.* **2009**, *52*, 2818–2824. [CrossRef]
75. Mahapatra, D.K.; Bharti, S.K.; Asati, V. Anti-cancer chalcones: Structural and molecular target perspectives. *Eur. J. Med. Chem.* **2015**, *98*, 69–114. [CrossRef]
76. Hargrove-Leak, S.C.; Amiridis, M.D. Substitution effects in the heterogeneous catalytic synthesis of flavanones over MgO. *Catal. Commun.* **2002**, *3*, 557–563. [CrossRef]
77. Climent, M.J.; Corma, A.; Iborra, S.; Velty, A. Activated hydrotalcites as catalysts for the synthesis of chalcones of pharmaceutical interest. *J. Catal.* **2004**, *221*, 474–482. [CrossRef]
78. Solhy, A.; Tahir, R.; Sebti, S.; Skouta, R.; Bousmina, M.; Zahouily, M.; Lazrek, M. Efficient synthesis of chalcone derivatives catalyzed by re-usable hydroxyapatite. *Appl. Catal. A Gen.* **2010**, *374*, 189–193. [CrossRef]
79. Jioui, I.; Danoun, K.; Solhy, A.; Jouiad, M.; Zahouily, M.; Essaid, B.; Len, C.; Fihri, A. Modified fluorapatite as highly efficient catalyst for the synthesis of chalcones via Claisen–Schmidt condensation reaction. *J. Ind. Eng. Chem.* **2016**, *39*, 218–225. [CrossRef]
80. Álvarez, M.G.; Segarra, A.M.; Contreras, S.; Sueiras, J.E.; Medina, F.; Figueras, F. Enhanced use of renewable resources: Transesterification of glycerol catalyzed by hydrotalcite-like compounds. *Chem. Eng. J.* **2010**, *161*, 340–345. [CrossRef]
81. Álvarez, M.G.; Chimentão, R.J.; Figueras, F.; Medina, F. Tunable basic and textural properties of hydrotalcite derived materials for transesterification of glycerol. *Appl. Clay Sci.* **2012**, *58*, 16–24. [CrossRef]
82. Albadi, J.; Mansournezhad, A.; Salehnasab, S. Green synthesis of biscoumarin derivatives catalyzed by recyclable CuO–CeO_2 nanocomposite catalyst in water. *Res. Chem. Intermed.* **2015**, *41*, 5713–5721. [CrossRef]
83. Sadeghi, B.; Tayebe, Z. A Fast, Highly Efficient, and Green Protocol for Synthesis of Biscoumarins Catalyzed by Silica Sulfuric Acid Nanoparticles as a Reusable Catalyst. *J. Chem.* **2013**, *2013*, 179013. [CrossRef]
84. Singh, P.; Kumar, P.; Katyal, A.; Kalra, R.; Dass, S.K.; Prakash, S.; Chandra, R. Phosphotungstic Acid: An Efficient Catalyst for the Aqueous Phase Synthesis of Bis-(4-hydroxycoumarin-3-yl)methanes. *Catal. Lett.* **2010**, *134*, 303–308. [CrossRef]
85. Jo, J.W.K.; Kumar, S.; Tonda, S. N-doped C dot/CoAl-layered double hydroxide/g-C_3N_4 hybrid composites for efficient and selective solar-driven conversion of CO_2 into CH_4. *Compos. Part B* **2019**, *176*, 107212. [CrossRef]

86. Nava Andrade, K.; Knauth, P.; López, Z.; Hirata, G.A.; Guevara Martinez, S.J.; Carbajal Arízaga, G.G. Assembly of folate-carbon dots in GdDy-doped layered double hydroxides for targeted delivery of doxorubicin. *Appl. Clay Sci.* **2020**, *192*, 105661. [CrossRef]
87. Liu, W.; Liang, R.; Lin, Y. Confined synthesis of carbon dots with tunable long-wavelength emission in a 2-dimensional layered double hydroxide matrix. *Nanoscale* **2020**, *12*, 7888–7894. [CrossRef]
88. Wang, Y.; Wang, X.; Antonietti, M. Polymeric Graphitic Carbon Nitride as a Heterogeneous Organocatalyst: From Photochemistry to Multipurpose Catalysis to Sustainable Chemistry. *Angew. Chem. Int. Ed.* **2012**, *51*, 68–89. [CrossRef]
89. Watanabe, H.; Asano, S.; Fujita, S.I.; Yoshida, H.; Arai, M. Nitrogen-Doped, Metal-Free Activated Carbon Catalysts for Aerobic Oxidation of Alcohols. *ACS Catal.* **2015**, *5*, 2886–2894. [CrossRef]
90. Xie, R.; Fan, G.; Yang, L.; Li, F. Solvent-free oxidation of ethylbenzene over hierarchical flower-like core–shell structured Co-based mixed metal oxides with significantly enhanced catalytic performance. *Catal. Sci. Technol.* **2015**, *5*, 540–548. [CrossRef]
91. Xie, R.; Fan, G.; Yang, L.; Li, F. Hierarchical flower-like Co–Cu mixed metal oxide microspheres as highly efficient catalysts for selective oxidation of ethylbenzene. *Chem. Eng. J.* **2016**, *288*, 169–178. [CrossRef]
92. Yang, F.; Zhou, S.J.; Gao, S.Y.; Liu, X.F.; Long, S.F.; Kong, Y. In situ embedding of ultra-fine nickel oxide nanoparticles in HMS with enhanced catalytic activities of styrene epoxidation. *Microporous Mesoporous Mater.* **2017**, *238*, 69–77. [CrossRef]
93. Liu, J.Y.; Chen, T.T.; Jian, P.M.; Wang, L.X. Hierarchical 0D/2D Co_3O_4 hybrids rich in oxygen vacancies as catalysts towards styrene epoxidation reaction. *Chin. J. Catal.* **2018**, *39*, 1942–1950. [CrossRef]
94. Huang, C.L.; Zhang, H.Y.; Sun, Z.Y.; Zhao, Y.F.; Chen, S.; Tao, R.T.; Liu, Z.M. Porous Fe_3O_4 nanoparticles: Synthesis and application in catalyzing epoxidation of styrene. *J. Colloid Interface Sci.* **2011**, *364*, 298–303. [CrossRef]
95. Yang, L.; Jiang, Z.; Fan, G.; Li, F. The promotional effect of ZnO addition to supported Ni nanocatalysts from layered double hydroxide precursors on selective hydrogenation of citral. *Catal. Sci. Technol.* **2014**, *4*, 1123–1131. [CrossRef]
96. Ma, N.; Song, Y.; Han, F.; Waterhouse, G.I.N.; Li, Y.; Ai, S. Highly selective hydrogenation of 5-hydroxymethylfurfural to 2,5-dimethylfuran at low temperature over a Co–N–C/NiAl-MMO catalyst. *Catal. Sci. Technol.* **2020**, *10*, 4010–4018. [CrossRef]
97. Wang, Y.; Wang, J.; Fan, G.; Li, F. Synthesis of a novel Ni/C catalyst derived from a composite precursor for hydrodechlorination. *Catal. Commun.* **2012**, *19*, 56–60. [CrossRef]
98. Wang, J.; Fan, G.; Li, F. Carbon-supported Ni catalysts with enhanced metal dispersion and catalytic performance for hydrodechlorination of chlorobenzene. *RSC Adv.* **2012**, *2*, 9976–9985. [CrossRef]
99. Jiang, J.; Lim, Y.S.; Park, S.; Kim, S.H.; Yoon, S.; Piao, L. Hollow porous Cu particles from silica-encapsulated Cu_2O nanoparticle aggregates effectively catalyze 4-nitrophenol reduction. *Nanoscale* **2017**, *9*, 3873–3880. [CrossRef]
100. Bai, S.; Shen, X.P.; Zhu, G.X.; Li, M.Z.; Xi, H.T.; Chen, K.M. In situ Growth of Ni_xCo_{100-x} Nanoparticles on Reduced Graphene Oxide Nanosheets and Their Magnetic and Catalytic Properties. *ACS Appl. Mater. Interf.* **2012**, *4*, 2378–2386. [CrossRef]
101. Bordbar, M.; Negahdar, N.; Nasrollahzadeh, M.M. *Melissa Officinalis* L. leaf extract assisted green synthesis of CuO/ZnO nanocomposite for the reduction of 4-nitrophenol and Rhodamine B. *Sep. Purif. Technol.* **2018**, *191*, 295–300. [CrossRef]
102. Pang, J.J.; Li, W.T.; Cao, Z.H.; Xu, J.J.; Li, X.; Zhang, X.K. Mesoporous Cu_2O–CeO_2 composite nanospheres with enhanced catalytic activity for 4-nitrophenol reduction. *Appl. Surf. Sci.* **2018**, *439*, 420–429. [CrossRef]
103. Ye, W.C.; Yu, J.; Zhou, X.X.; Gao, D.Q.; Wang, D.A.; Wang, C.M.; Xue, D.S. Green synthesis of Pt–Au dendrimer-like nanoparticles supported on polydopamine-functionalized graphene and their high performance toward 4-nitrophenol reduction. *Appl. Catal. B* **2016**, *181*, 371–378. [CrossRef]
104. Konar, S.; Kalita, H.; Puvvada, N.; Tantubay, S.; Mahto, M.K.; Biswas, S.; Pathak, A. Shape-dependent catalytic activity of CuO nanostructures. *J. Catal.* **2016**, *336*, 11–22. [CrossRef]
105. Sasmal, A.K.; Dutta, S.; Pal, T. A ternary Cu_2O–Cu–CuO nanocomposite: A catalyst with intriguing activity. *Dalton Trans.* **2016**, *45*, 3139–3150. [CrossRef] [PubMed]
106. Liu, L.J.; Chen, R.F.; Liu, W.K.; Wu, J.M.; Gao, D. Catalytic reduction of 4-nitrophenol over Ni-Pd nanodimers supported on nitrogen-doped reduced graphene oxide. *J. Hazard. Mater.* **2016**, *320*, 96–104. [CrossRef]
107. Wang, X.; Liu, D.P.; Song, S.Y.; Zhang, H.J. Pt@CeO_2 Multicore@Shell Self-Assembled Nanospheres: Clean Synthesis, Structure Optimization, and Catalytic Applications. *J. Am. Chem. Soc.* **2013**, *135*, 15864–15872. [CrossRef]
108. Jiang, Y.F.; Yuan, C.Z.; Xie, X.; Zhou, X.; Jiang, N.; Wang, X.; Imran, M.; Xu, A.W. A Novel Magnetically Recoverable Ni-CeO_{2-x}/Pd Nanocatalyst with Superior Catalytic Performance for Hydrogenation of Styrene and 4-Nitrophenol. *ACS Appl. Mater. Interfaces* **2017**, *9*, 9756–9762. [CrossRef]
109. Wu, G.Q.; Liang, X.Y.; Zhang, L.J.; Tang, Z.Y.; Al-Mamun, M.; Zhao, H.J.; Su, X.T. Fabrication of Highly Stable Metal Oxide Hollow Nanospheres and Their Catalytic Activity toward 4-Nitrophenol Reduction. *ACS Appl. Mater. Interfaces* **2017**, *9*, 18207–18214. [CrossRef]
110. Yu, C.; Fu, J.J.; Muzzio, M.; Shen, T.; Su, D.; Zhu, J.J.; Sun, S.H. CuNi Nanoparticles Assembled on Graphene for Catalytic Methanolysis of Ammonia Borane and Hydrogenation of Nitro/Nitrile Compounds. *Chem. Mater.* **2017**, *29*, 1413–1418. [CrossRef]
111. Fang, H.; Wen, M.; Chen, H.X.; Wu, Q.S.; Li, W.Y. Graphene stabilized ultra-small CuNi nanocomposite with high activity and recyclability toward catalysing the reduction of aromatic nitro-compounds. *Nanoscale* **2016**, *8*, 536–542. [CrossRef]
112. Kohantorabi, M.; Gholami, M.R. Kinetic Analysis of the Reduction of 4-Nitrophenol Catalyzed by CeO_2 Nanorods-Supported CuNi Nanoparticles. *Ind. Eng. Chem. Res.* **2017**, *56*, 1159–1167. [CrossRef]
113. Krishna, R.; Fernandes, D.M.; Ventura, J.; Freire, C.; Titus, E. Novel synthesis of highly catalytic active Cu@Ni/RGO nanocomposite for efficient hydrogenation of 4-nitrophenol organic pollutant. *Int. J. Hydrogen Energy* **2016**, *41*, 11608–11615. [CrossRef]

114. Qi, H.T.; Yu, P.; Wang, Y.X.; Han, G.C.; Liu, H.B.; Yi, Y.P.; Li, Y.L.; Mao, L.Q. Graphdiyne Oxides as Excellent Substrate for Electroless Deposition of Pd Clusters with High Catalytic Activity. *J. Am. Chem. Soc.* **2015**, *137*, 5260–5263. [CrossRef]
115. Mallick, K.; Witcomb, M.; Scurrell, M. Silver nanoparticle catalysed redox reaction: An electron relay effect. *Mater. Chem. Phys.* **2006**, *97*, 283–287. [CrossRef]
116. Mandlimath, T.R.; Gopal, B. Catalytic activity of first row transition metal oxides in the conversion of p-nitrophenol to p-aminophenol. *J. Mol. Catal. A Chem.* **2011**, *350*, 9–15. [CrossRef]
117. Chen, J.; Yao, N.; Wang, R.; Zhang, J. Hydrogenation of chloronitrobenzene to chloroaniline over Ni/TiO$_2$ catalysts prepared by sol–gel method. *Chem. Eng. J.* **2009**, *148*, 164–172. [CrossRef]
118. Yan, X.; Sun, J.; Wang, Y.; Yang, J. A Fe-promoted Ni–P amorphous alloy catalyst (Ni–Fe–P) for liquid phase hydrogenation of *m*- and *p*-chloronitrobenzene. *J. Mol. Catal. A Chem.* **2006**, *252*, 17–22. [CrossRef]
119. Sun, Y.; Gao, X.; Yang, N.; Tantai, X.; Xiao, X.; Jiang, B.; Zhang, L. Morphology-Controlled Synthesis of Three-Dimensional Hierarchical Flowerlike Mg–Al Layered Double Hydroxides with Enhanced Catalytic Activity for Transesterification. *Ind. Eng. Chem. Res.* **2019**, *58*, 7937–7947. [CrossRef]
120. Balsamo Mendieta, S.; Heredia, A.; Crivello, M. Nanoclays as dispersing precursors of La and Ce oxide catalysts to produce high-valued derivatives of biodiesel by-product. *Mol. Catal.* **2020**, *481*, 110290. [CrossRef]
121. Álvarez, M.G.; Chimentão, R.; Barrabés, N.; Föttinger, K.; Gispert-Guirado, F.; Kleymenov, E.; Tichit, D.; Medina, F. Structure evolution of layered double hydroxides activated by ultrasound induced reconstruction. *Appl. Clay Sci.* **2013**, *83*, 1–11. [CrossRef]
122. Wang, D.; Zhang, X.; Cong, X.; Liu, S.; Zhou, D. Influence of Zr on the performance of Mg-Al catalysts via hydrotalcite-like precursors for the synthesis of glycerol carbonate from urea and glycerol. *Appl. Catal. A Gen.* **2018**, *555*, 36–46. [CrossRef]
123. Sangkhum, P.; Yanamphorn, J.; Wangriya, A.; Ngamcharussrivichai, C. Ca–Mg–Al ternary mixed oxides derived from layered double hydroxide for selective etherification of glycerol to short-chain polyglycerols. *Appl. Clay Sci.* **2019**, *173*, 79–87. [CrossRef]
124. Li, X.; Zheng, L.; Hou, Z. Acetalization of glycerol with acetone over Co[II](Co[III]$_x$Al$_{2-x}$)O$_4$ derived from layered double hydroxide. *Fuel* **2018**, *233*, 565–571. [CrossRef]
125. Talebian-Kiakalaieh, A.; Amin, N.A.S.; Rajaei, K.; Tarighi, S. Oxidation of bio-renewable glycerol to value-added chemicals through catalytic and electro-chemical processes. *Appl. Energy* **2018**, *230*, 1347–1379. [CrossRef]
126. Jing, F.; Liu, S.; Wang, R.; Li, X.; Yan, Z.; Luo, S.; Chu, W. Hydrogen production through glycerol steam reforming over the NiCe$_x$Al catalysts. *Renew. Energy* **2020**, *158*, 192–201. [CrossRef]
127. Gao, W.; Zhao, Y.; Liu, J.; Huang, Q.; He, S.; Li, C.; Zhao, J.; Wei, M. Catalytic conversion of syngas to mixed alcohols over CuFe-based catalysts derived from layered double hydroxides. *Catal. Sci. Technol.* **2013**, *3*, 1324–1332. [CrossRef]
128. Han, X.; Fang, K.; Sun, Y. Effects of metal promotion on CuMgFe catalysts derived from layered double hydroxides for higher alcohol synthesis via syngas. *RSC Adv.* **2015**, *5*, 51868–51874. [CrossRef]
129. Han, X.; Fang, K.; Zhou, J.; Zhao, L.; Sun, Y. Synthesis of higher alcohols over highly dispersed Cu–Fe based catalysts derived from layered double hydroxides. *J. Colloid Interface Sci.* **2016**, *470*, 162–171. [CrossRef]
130. Ning, X.; An, Z.; He, J. Remarkably efficient CoGa catalyst with uniformly dispersed and trapped structure for ethanol and higher alcohol synthesis from syngas. *J. Catal.* **2016**, *340*, 236–247. [CrossRef]
131. Liao, P.; Zhang, C.; Zhang, L.; Yang, Y.; Zhong, L.; Wang, H.; Sun, Y. Higher alcohol synthesis via syngas over CoMn catalysts derived from hydrotalcite-like precursors. *Catal. Today* **2018**, *311*, 56–64. [CrossRef]
132. Cao, A.; Liu, G.; Yue, Y.; Zhang, L.; Liu, Y. Nanoparticles of Cu–Co alloy derived from layered double hydroxides and their catalytic performance for higher alcohol synthesis from syngas. *RSC Adv.* **2015**, *5*, 58804–58812. [CrossRef]
133. Xiao, K.; Bao, Z.; Qi, X.; Wang, X.; Zhong, L.; Lin, M. Unsupported CuFe bimetallic nanoparticles for higher alcohol synthesis via syngas. *Catal. Commun.* **2013**, *40*, 154–157. [CrossRef]
134. Ding, M.; Qiu, M.; Liu, J.; Li, Y.; Wang, T.; Ma, L.; Wu, C. Influence of manganese promoter on co-precipitated Fe–Cu based catalysts for higher alcohols synthesis. *Fuel* **2013**, *109*, 21–27. [CrossRef]
135. Kattel, S.; Ramirez, P.J.; Chen, J.G.; Rodriguez, J.A.; Liu, P. Active sites for CO$_2$ hydrogenation to methanol on Cu/ZnO catalysts. *Science* **2017**, *355*, 1296–1299. [CrossRef]
136. Smith, K.J.; Anderson, R.B. A chain growth scheme for the higher alcohols synthesis. *J. Catal.* **1984**, *85*, 428–436. [CrossRef]
137. Nunan, J.G.; Bogdan, C.E.; Klier, K.; Smith, K.J.; Young, C.W.; Herman, R.G. Methanol and C$_2$ oxygenate synthesis over cesium doped CuZnO and Cu/ZnO/Al$_2$O$_3$ catalysts: A study of selectivity and ^{13}C incorporation patterns. *J. Catal.* **1988**, *113*, 410–433. [CrossRef]
138. Nunan, J.G.; Herman, R.G.; Klier, K. Higher alcohol and oxygenate synthesis over Cs/Cu/ZnO/M$_2$O$_3$ (M = Al, Cr) catalysts. *J. Catal.* **1989**, *116*, 222–229. [CrossRef]
139. Spivey, J.J.; Egbebi, A. Heterogeneous catalytic synthesis of ethanol from biomass-derived syngas. *Chem. Soc. Rev.* **2007**, *36*, 1514–1528. [CrossRef]
140. Subramani, V.; Gangwal, S.K. A Review of Recent Literature to Search for an Efficient Catalytic Process for the Conversion of Syngas to Ethanol. *Energy Fuels* **2008**, *22*, 814–839. [CrossRef]
141. Herman, R.G. Advances in catalytic synthesis and utilization of higher alcohols. *Catal. Today* **2000**, *55*, 233–245. [CrossRef]
142. Li, L.; Zhu, Z.H.; Yan, Z.F.; Lu, G.Q.; Rintoul, L. Catalytic ammonia decomposition over Ru/carbon catalysts: The importance of the structure of carbon support. *Appl. Catal. A Gen.* **2007**, *320*, 166–172. [CrossRef]

143. Nie, R.; Wang, J.; Wang, L.; Qin, Y.; Chen, P.; Hou, Z. Platinum supported on reduced graphene oxide as a catalyst for hydrogenation of nitroarenes. *Carbon* **2012**, *50*, 586–596. [CrossRef]
144. Ye, A.; Fan, W.; Zhang, Q.; Deng, W.; Wang, Y. CdS–graphene and CdS–CNT nanocomposites as visible-light photocatalysts for hydrogen evolution and organic dye degradation. *Catal. Sci. Technol.* **2012**, *2*, 969–978. [CrossRef]
145. Li, Y.; Gao, W.; Ci, L.; Wang, C.; Ajayan, P.M. Catalytic performance of Pt nanoparticles on reduced graphene oxide for methanol electro-oxidation. *Carbon* **2010**, *48*, 1124–1130. [CrossRef]
146. Luo, X.; Yuan, S.; Pan, X.; Zhang, C.; Du, S.; Liu, Y. Synthesis and Enhanced Corrosion Protection Performance of Reduced Graphene Oxide Nanosheet/ZnAl Layered Double Hydroxide Composite Films by Hydrothermal Continuous Flow Method. *ACS Appl. Mater. Interfaces* **2017**, *9*, 18263–18275. [CrossRef]
147. Tichit, D.; Layrac, G.; Gérardin, C. Synthesis of layered double hydroxides through continuous flow processes: A review. *Chem. Eng. J.* **2019**, *369*, 302–332. [CrossRef]

Review

Review of the Application of Hydrotalcite as CO_2 Sinks for Climate Change Mitigation

David Suescum-Morales [1], José Ramón Jiménez [1,*] and José María Fernández-Rodríguez [2,*]

1   Área de Ingeniería de la Construcción, Universidad de Córdoba, E.P.S. de Belmez. Avenida de la Universidad s/n, Belmez, E-14240 Córdoba, Spain; p02sumod@uco.es
2   Área de Química Inorgánica, Universidad de Córdoba, E.P.S. de Belmez. Avenida de la Universidad s/n, Belmez, E-14240 Córdoba, Spain
*   Correspondence: jrjimenez@uco.es (J.R.J.); um1feroj@uco.es (J.M.F.-R.)

Abstract: In recent decades, the environmental impact caused by greenhouse gases, especially CO_2, has driven many countries to reduce the concentration of these gases. The study and development of new designs that maximise the efficiency of CO_2 capture continue to be topical. This paper presents a review of the application of hydrotalcites as CO_2 sinks. There are several parameters that can make hydrotalcites suitable for use as CO_2 sinks. The first question is the use of calcined or uncalcined hydrotalcite as well as the temperature at which it is calcined, since the calcination conditions (temperature, rate and duration) are important parameters determining structure recovery. Other aspects were also analysed: (i) the influence of the pH of the synthesis; (ii) the molar ratio of its main elements; (iii) ways to increase the specific area of hydrotalcites; (iv) pressure, temperature, humidity and time in CO_2 absorption; and (v) combined use of hydrotalcites and cement-based materials. A summary of the results obtained so far in terms of CO_2 capture with the parameters described above is presented. This work can be used as a guide to address CO_2 capture with hydrotalcites by showing where the information gaps are and where researchers should apply their efforts.

Keywords: CO_2 sinks; calcined hydrotalcite; one-coat mortar; CO_2 curing

1. Introduction

The incessant consumption of energy, which goes hand in hand with modern life in developed countries, has negative effects on the quality of the environment and ecosystems. These impacts, caused by greenhouse gases (GHGs), are leading many countries to adopt responsible environmental policies. Carbon dioxide (CO_2) is the dominant anthropogenic greenhouse gas (76%), responsible for global warming [1]. Before the industrial revolution (1760s), the CO_2 level was about 280 ppm [2], while today it could be considered at an average level of 400 ppm [3,4]. According to the International Energy Agency (IEA, 2017), the global mean temperature has increased by 1 °C above the preindustrial level due to anthropogenic greenhouse gas emissions [5]. The increase in global average temperature is expected to reach 1.5 °C by the end of 2040 [6,7], and it is therefore necessary to take measures to reduce these CO_2 levels.

To reduce these levels, two main carbon capture (CC) technologies are being presented [6,8]: (i) carbon capture, transport and storage technologies (CCS); and (ii) carbon capture and utilisation technologies (CCU). CCSs are primarily aimed at mitigating GHGs when fossil fuels are used for energy generation. CCS technologies are classified into three types: pre-combustion, post-combustion, and oxy-fuel combustion capture [6]. CCS technologies would remove around 20% of GHG emission by 2050 [9]. Captured CO_2 can be a source of recycled carbon, and CCU can provide more services and greater climate change savings than capturing and storing CO_2 underground [10]. The use of CO_2 gives an added value to these GHGs, which, together with the circular economy concept, can mitigate climate change [11,12]

Therefore, CO_2 capture is expected to play an important role in the commercialisation of future CCS technologies [13]. Many countries and research teams are considering various candidate processes and materials [14], such as absorption [15], adsorption [16,17]; membranes [18] and cryogenic distillation systems [19]. Solid sorbents (carbon, silica, calcium oxide, among others) are expanding as an emerging alternative for CO_2 capture, due to their great characteristics for such a task [20–22]. According to their temperature of use, solid sorbents can be classified as follows [23]: (a) low temperature (<200 °C) [24–27]; (b) intermediate temperature (between 200–400 °C) [28,29]; and (c) high-temperature (> 400 °C).

Hydrotalcites are brucite-like layered materials, that have been known for over 150 years with a general formula of $[M^{2+}_{1-x}M^{3+}_{x}(OH)_2]^{x+}(A^{n-})_{x/n} \cdot mH_2O$, where M^{2+} and M^{3+} represent divalent and trivalent cations, respectively, and A represents the interlayered anion [30–35]. Hydrotalcites are known in the bibliography as LDHs (Layered double hydroxides). The layer charge is determined by the molar ratio, that is $x = M^{3+}/(M^{3+} + M^{2+})$, and it varies between 0.2 and 0.4 [36–38]. The Mg_3AlCO_3 hydrotalcite is a type of LDH commonly found in nature. Reviewing the literature, a wide range of applications of these materials can be found [3]: catalytic applications [36,39–42], medical applications [43–46] as additives for polymers [47,48], for adsorption of pollutants [49–51], water decontamination [52,53], waste barriers [54–58], among various other applications. Several chemical companies (e.g., BASF, SASOL, Clariant, Kisuma Chemicals, Sakai Chemicals, etc.) produce several thousands of tonnes yearly, so it is an easily-available product [19]. Hydrotalcites, as such, are not good CO_2 absorbents due to poor basic properties and presence of entities that hinder CO_2 adsorption and are therefore subjected to thermal treatment (around 500 °C) to obtain nearly amorphous metastable mixed solid solutions known as calcined layered double hydroxides (CLDHs) [19]. In CLDHs, there is a loss of mass and a breakdown of its structure, forming an oxide, according to Equation (1) [59]:

$$Mg_6Al_2(OH)_{16}CO_3 \cdot 4H_2O \rightarrow Mg_6Al_2O_9 + 12H_2O + CO_2 \quad (1)$$

The CO_2 emitted during calcination (Equation (1)) is identical to the CO_2 captured during the synthesis of hydrotalcite (Equation(2)):

$$Mg_6Al_2O_9 + 12H_2O + CO_2 \rightarrow Mg_6Al_2(OH)_{16}CO_3 \cdot 4H_2O \quad (2)$$

Consequently, the CO_2 balance of the calcined hydrotalcite is 0 (Equations (1) and (2)). In this sense, all the CO_2 captured by the calcined hydrotalcite represents a negative CO_2 balance. This will reduce the carbon footprint of those materials to which calcined hydrotalcite is added.

Another characteristic of hydrotalcite is the memory effect, which allows the reconstruction of the original shape of hydrotalcite when it is in a humid environment and in the presence of CO_2. CLDHs in a CO_2 environment return to its initial state of LDHs [60–63]. The CO_2 capture balance by the hydrotalcite in its reconstruction is positive and hence the interest of using this material as a CO_2 capture material is very great (in the last decade) [64]. Nowadays, the challenge is to develop new types of hydrotalcites with higher CO_2 sorption capacities, higher sorption/desorption kinetics, and good stability throughout consecutive reutilisation cycles in similar operation conditions as those applied in a sorption-enhanced steam reforming process [65].

These hydrotalcites (in their LDH or CLDHs form) can also be found as additives to cement-based materials to improve resistance to chloride attack [66], durability [67–70] and even the use of LDH as additives to improve the thermal insulation of the intumescent fire retardant (IFR) coating [71]. Wu et al. [72] indicated that the structure regeneration of CLDHs in a cement paste environment had also been revealed [69]. After calcination, a large number of active sites produced what in favour of the improvement effect of CLDH on cement [72]. Since hydrotalcites may be incorporated into various building material mixtures, mortars, concretes and backfills, their application as accessible and affordable materials is prospective [73–75]. However, research in the field of hydrotalcites and cement-based materials is still insufficient [72,76]. It is also difficult to find studies that add LDH or CLDH to alkaline-activated materials [77,78].

It is even more difficult to find studies in which CLDHs are used as additives in order to increase the CO_2 capture capacity of cement-based materials. Ma et al. [79] reported the adsorption of CO_3^{2-} by CLDH seven times faster than LDH due to the release of anions after calcination, which can be very beneficial for CO_2 capture. Suescum-Morales et al. [3,59] added different percentages of CLDH (calcined Mg_3AlCO_3) in a one-coat mortar in order to increase the CO_2 capture capacity. Adding 5% of calcined hydrotalcite increased the CO_2 capture capacity by 8.52% with respect to the reference mortar.

This paper presents a review of the application of hydrotalcites as CO_2 sinks. Different aspects were analysed: pH of the synthesis, the molar ratio (Mg/Al), the specific area, pressure, temperature and time in CO_2 absorption. A summary of the results obtained so far in terms of CO_2 capture with the parameters described above is presented. This work can be used as a guide to address CO_2 capture with hydrotalcites by showing where the information gaps are and where researchers should apply their efforts.

2. Calcined or Uncalcined Hydrotalcite to Capture CO_2?

The answer is immediate: calcined hydrotalcite or its use under high temperatures (around 400 °C so that the hydrotalcite becomes oxide and can be rebuilt in contact with CO_2). LDH are poor CO_2 adsorbents in their natural or unburned form, which is due to a poor basic property and the presence of entities that hinder CO_2 adsorption. Hence, they are subjected to thermal treatment to obtain nearly amorphous metastable mixed solid solutions (CLDHs) [19]. Figure 1 shows two diagrams representing what happens during the calcination of hydrotalcite of Mg_3AlCO_3 as shown (A) by W.J. Long et al. [77] and (B) by Lauermannová et al. [73]. Both schemes attempt to represent the collapse of the structure due to the loss of interlayer anions and moisture.

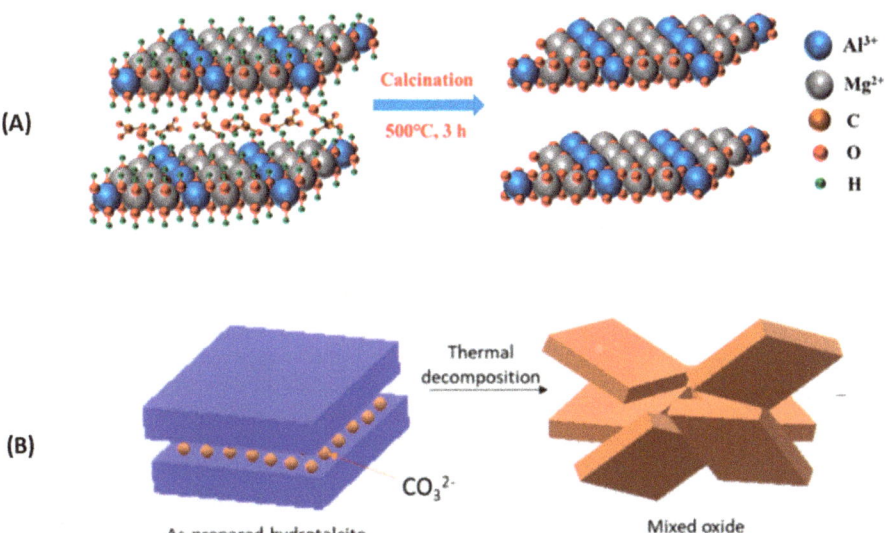

Figure 1. Structural changes of LDH after calcination according to adapted from W.J. Long et al. [77] (**A**) and Lauermannová et al. [73] (**B**) (open Access).

2.1. Thermal Behaviour of Hydrotalcite by TGA/DTA

LDH undergoes several stages until it reaches CLDHs. There is even research that attempts to explain this process in great detail and focuses on it alone [80,81]. It is very important to choose a suitable calcination temperature, avoiding it being too high (to avoid higher energy consumption), or being too low (not producing the collapse of the structure and hindering a higher CO_2 capture). To distinguish the different stages, it is useful to rely on thermogravimetric analysis (TGA) and differential thermal analysis (DTA) carried out by Suescum-Morales et al. [33], shown in Figure 2. It should be noted that different variations in temperature ranges may be encountered, approximately as shown below.

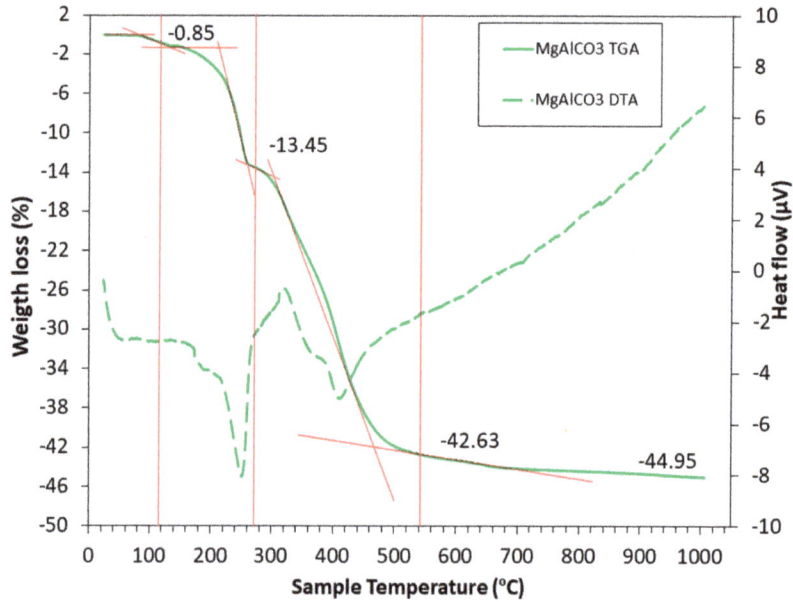

Figure 2. TGA (solid lines) and DTA (dotted lines) for commercial hydrotalcite of MgAlCO$_3$ [33].

First of all, a loss of humidity is observed, up to a temperature of about 105 °C. The second stage occurs from 105 to 270 °C, where the water hydration of the hydrotalcite structure is lost (leading to a decrease in basal spacing) [82–84]. There are authors who indicate that between 200 and 300 °C, the OH^- groups attached to Al^{3+} are lost [19]. The third stage, from 270 to 540 °C, is where the dehydroxilation of the hydrotalcite takes place and the loss of the carbonate anion of the interlayer in the form of CO_2 occurs. The layer structure collapses (Figure 1), and the LDH converts to a mixed-oxide MgO-like phase [85]. In the last stage, from 540 to 1000 °C, very small weight losses are observed, attributed to the loss of residual OH^- groups.

From the above, it can be seen that the calcination temperature affects the capacity to capture CO_2 in CLDHs. Different structural characteristics are presented in the different stages of thermal decomposition. From this analysis, the ideal calcination temperature of the hydrotalcite under study can be determined, which is characteristic and unique depending on the type of hydrotalcite. Most research indicates that the calcination temperature of a Mg-Al LDH is around 400 °C [86–88]. However, the performance of a TGA/DTA for each specific case would allow observing the exact calcination temperature. Therefore, specific experimental tests are necessary to further verify the influence of the calcination temperature on adsorption. This is discussed in the following sections of this review, using XRD, SEM/TEM, pH measurements in the synthesis, influence of molar ratio, and specific surface area measurements.

2.2. Thermal Behaviour of Hydrotalcite by XRD

The appropriate calcination temperature can also be determined by XRD temperature variation analysis. The thermal decomposition sequence of Mg-Al-CO$_3$ hydrotalcite is well documented. Miyata, 1980 and Hibino et al., 1995 [82,89] studied the XRD variation at different temperatures, shown in Figure 4 (non-calcined, 150, 250, 350, 500, 850 and 1000 °C). The diffraction patterns of MgO can be identified between 400 °C and 850 °C, given the amorphous nature of Al_2O_3 at this temperature. For 900 °C the spinal phase ($MgAl_2O_4$) was formed. Given these experiences, the following chemical equations can be posed, with the different temperature ranges, to help us understand the process (Equations (3)–(5)):

$$Mg_6Al_2(OH)_{16}CO_3 \cdot 4H_2O \rightarrow Mg_6Al_2(OH)_{16}CO_3 + 4H_2O \quad 100 < T < 250\,°C \quad (3)$$

$$Mg_6Al_2(OH)_{16}CO_3 \rightarrow 6Mg(Al)O + 8H_2O + CO_2 \quad 400 < T < 850\,°C \quad (4)$$

$$6MgO + Al_2O_3 \rightarrow MgAl_2O_4 + 5MgO \quad 900 < T < 1000\,°C \quad (5)$$

Figure 3 shows the similar XRD obtained by several authors by calcination at 500 °C for different periods: firstly, the XRD obtained by W.J. Long et al. [77] (Figure 3a) shows how at a temperature of 500 °C for 3 h the layered structure collapses and also shows the production of mixed oxides. Figure 3b shows a similar result, but in this case using a calcination time of 2 h, and the same temperature (500 °C) [33]. This leads to a large saving of energy in the calcination of LDH, with a consequent lowering of the carbon footprint. Already, Z. Yang et al. [67] calcined at 500 °C for 3 h, obtaining a similar result. Even Q.Tao et al. [90] heated at 500 °C for 4 h, obtaining similar results (Figure 3c). Similar results in XRD were obtained by S.I. Garcés Polo et al. [91] for CLDH. None of the previous authors [33,67,77,91] indicated the amount of sample used in the calcination, which may have led to these observed differences. Although different calcination temperatures have been used with similar results, there are no economic studies of the cost of calcination; studies of this type, with times, temperatures and quantities of LDH to be calcined, would be very useful for these materials in industrial applications. It would also be very important to carry out a real carbon footprint calculation (UNE EN 15804:2012), which determines the real CO_2 sink capacity in each specific case (amount of hydrotalcite, type of furnace, etc.). Only two studies have been found that indicate the amount of hydrotalcite calcined (100 and 1 g respectively) [92,93]. Annotations of this type, i.e., what quantity is fed into the LDH kiln, are very important in order to maximise the efficiency of the calcination process.

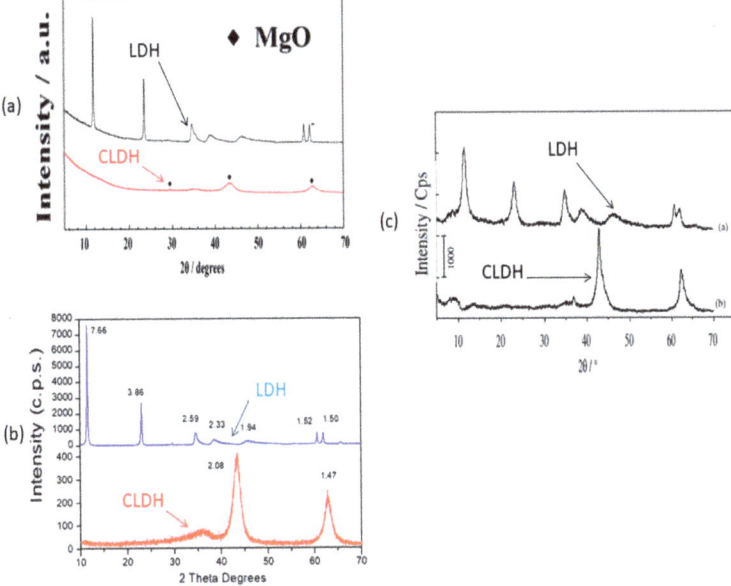

Figure 3. XRD patterns of (**a**) LDH and CLDH adapted from W.J. Long et al. [77]; (**b**) LDH and CLDH of Suescum-Morales et al. [33]; (**c**) LDH and CLDH adapted from Q.Tao et al. [90].

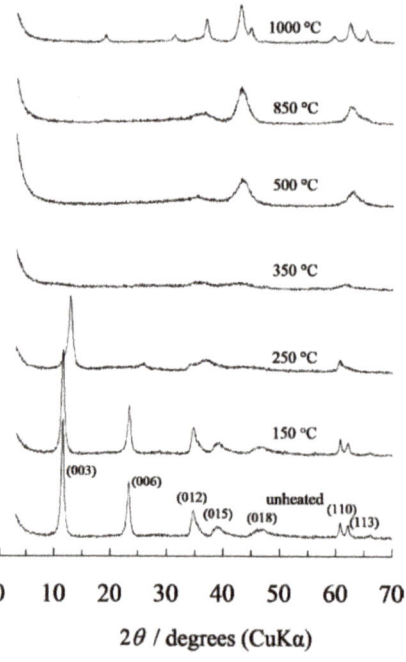

Figure 4. XRD of MgAlCO$_3$ formed between room temperature and 1000 °C [94].

2.3. Thermal Behaviour of Hydrotalcite by SEM/TEM

Figure 5a shows SEM images of commercial LDH and CLDH Mg-Al from W.J. Long et al. [77]. After 3 h at 500 °C, the structure collapses. A decrease in size was also observed. P. Cai et al. [95] obtained similar results (Figure 5b) with the same temperature and time of calcination. After 4 h at 450 °C on Mg-Al LDH, C. Geng et al. [96] observed a decrease in particle size, and the hexagonal shape was hardly noticeable (Figure 5c). No other studies using different temperatures (different at 500 °C) and calcination times have been found that show SEM images of LDH and CLDH. Studies along these lines could fill this information gap.

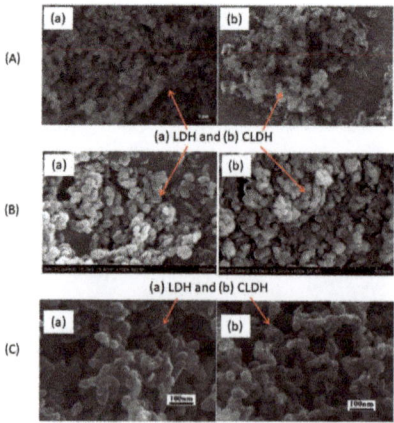

Figure 5. SEM images of LDH and CLDH Mg-Al of (**A**) adapted from W.J. Long et al. [77] and (**B**) adapted from P. Cai et al. [95] and (**C**) adapted from C. Geng et al. [96].

Figure 6A shows the TEM images of LDH and CLDH, after 2 h at 500 °C, obtained by Suescum-Morales et al. [33]. In CLDH, small pores were formed, attributable to the dehydration process, dehydroxilation of the OH^- groups, and to the decomposition of the interlayer carbonate. C Hobbs et al. [97] studied the evolution of a Mg-Al LDH under different temperatures using a rate of 10 °C/min (Figure 6B). At a temperature of 20 °C (LDH), they have a well-defined platelet shape; the porous structure is clearly visible at 850 °C. S Luo et al. [98] also obtained the same porous structure in CLDH, as shown in Figure 6C.

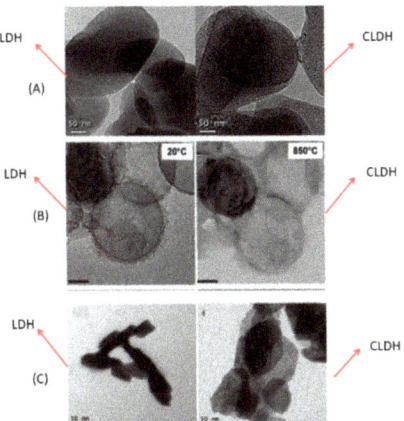

Figure 6. TEM images of LDH and CLDH Mg-Al of (**A**) Suescum-Morales et al. [33]; (**B**) adapted from C Hobbs et al. [97] (open access) and (**C**) adapted from S Luo et al. [98].

Different heating rates have been used in LDH calcination; a heating rate of 10 °C/min was used in Mg/Al LDH by several authors [99–102] and a heating rate of 5 °C/min was used by [103]. It is known that if the heating rate is slower, the thermal effects are better observed (better porous structure). However, if the ramp is slower, it takes more time and wastes more energy; if the heating is too fast, the porous structure will be worse. These differences should be extensively studied, both from an energy point of view (higher consumption and, therefore, higher carbon footprint produced by the kiln when the larger ramp is used), and from the point of view of the efficiency of the CLDH itself. Another very important aspect is the kinetics of LDH, which has been extensively discussed in other research [104,105].

3. Influence of the pH Used in the Synthesis in CO_2 Capture

Generally, LDH used for CO_2 capture has been synthesised by the coprecipitation method [33,34]. In this respect, Wang et al. [106] studied the effect of pH variation on the synthesis (coprecipitation method) in the range of 6.5–14. Subsequently, they studied the effect of the variation of this parameter on CO_2 capture. The pH that produced the best adsorption was 12 (23.76 mg/g). In other previous research, Wang et al. [88] indicated that the best pH value used for the synthesis of hydrotalcite oriented for CO_2 capture was between 10 and 12.

The crystallinity of the HT samples increases with the pH value of the synthesis, while the BET surface area decreases with increasing synthesis pH (in the range of 6.5–9 pH values). However, from a pH value of 10, the BET surface area increases suddenly, which can be seen in Figure 7, taken from Wang et al. [88].

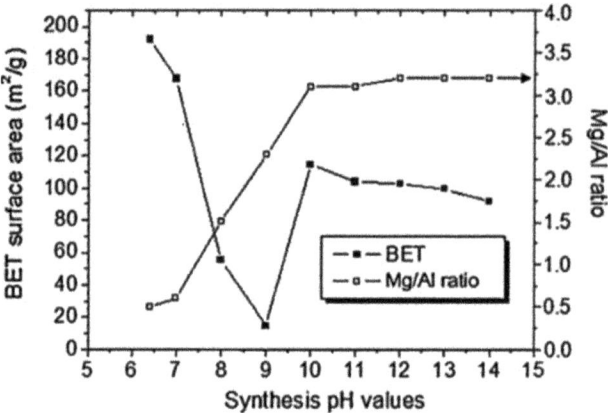

Figure 7. BET surface area and Mg/Al ratio as function of synthesis pH values adapted from Wang et al [88].

In other research studies, a pH of 7 was used by León et al. [25] and Rossi et al. [27], a pH of 9 was used by Torres-Rodriguez et al. [26], and a pH of 8 by Suescum-Morales et al. [33]. In summary, we must take into account that the pH value in the synthesis is fundamental in the development of the morphology, porous structure, as well as in the chemical composition (Mg/Al ratio).

4. Influence of the Molar Ratio (Mg/Al) of Its Main Elements in CO_2 Capture

The properties of LDH are also strongly influenced by the M^{2+}/M^{3+} ratio, cation type and anion type. For a higher capture capacity of CO_2, large interlayer spacing, high layer charge density and a greater number of basic sites are desirable. A high Al content increases the density, but decreases the layer spacing. A high Mg content increases the number of basic sites. In the consulted bibliography, optimal Mg/Al ratios are considered to vary from 1:1 to 3:1 [19,106–109].

Kim et al. [110] studied the effect of high Mg/Al ratios on the CO_2 sorption with hydrotalcites prepared with the coprecipitation method using ratios between 3 and 30. The highest CO_2 capture capacity was obtained (407.9 mg/g) for an Mg/Al ratio of 20, using temperatures of 240 °C.

M. Salomé Macedo et al. [111] studied the influence of different Mg/Al ratios (from 2 to 20) to be used as CO_2 sorbents at high temperature. The best results were obtained for an Mg/Al ratio of 7 at 1 bar of pressure and 300 °C (71.3 mg/g). These authors indicated that, in general, one observes a gradual increase in the sorption capacity for the same synthesis pH with the increase in the Mg/Al ratio.

The relationship between the pH of the synthesis, the molar ratio and the specific surface area obtained is very important. For example, Kim et al. [110] obtained the highest surface area with a molar ratio of 3 (256 m^2/g$^-$). M. Salomé Macedo et al. [111] showed that the BET specific surface area and total pore volume of samples clearly depend on the Mg/Al molar ratio. Suescum-Morales et al. [33] showed that the higher the specific surface area, the higher the CO_2 capture capacity. Therefore, the specific surface area seems to be a very important factor in determining the CO_2 capture.

As a summary, it can be said that although the molar ratio is very important in terms of CO_2 capture, there are also many other factors that affect this parameter: the pH of the synthesis, the nature of the anion, pressure used, and absorption temperature, among others, are the most important.

5. Ways to Increase the Specific Area of Hydrotalcites

The extensive literature presented in this paper highlights the significant interest of researchers in using LDH and CLDH as CO_2 adsorbents. Several strategies have been implemented in order to increase the CO_2 capture capacities: replacing cations or anions [88], different preparation methods [112], calcination, working temperatures and pressures [113] and alkali doping [29,114], among others.

A good strategy would be to increase the specific surface area, as a larger specific surface area would lead to a higher CO_2 capture capacity. Wang et al. [115] intercalated long carbon-chain organic anions, and it increased the CO_2 capture capacity. This was due to the increase in the interlayer distance from 0.78 to 3.54 nm (Figure 8). A similar strategy was followed by Li et al. [116], except that in this case they used K_2CO_3 in the LDH precursor of Mg_3Al-stearate, again increasing the interlayer distance, and achieving a higher CO_2 capture capacity. A similar strategy was also followed by A. Hanif et al. [117].

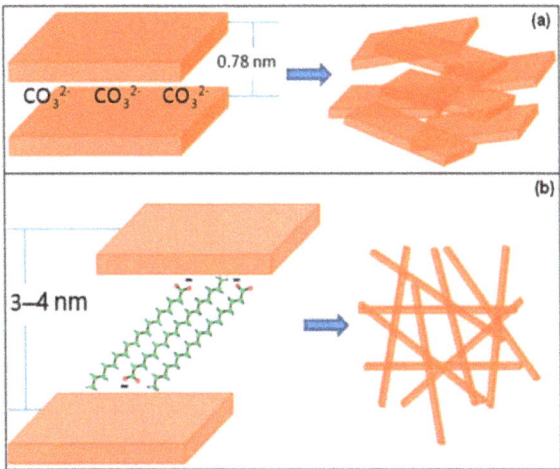

Figure 8. General schemes of the structural changes of (**a**) Mg_3Al-CO_3 and (**b**) with organic anions adopted from abstract of Wang et al [115].

Another way to increase the specific surface area would be to directly exfoliate the LDH in layers using bottom-up or top-down methods [118], although some authors indicate that the layers could be restacked after drying [119]. To avoid this, García-Gallastegui et al. [86] used graphene oxide as a support, where the negatively charged graphene oxide flakes were well dispersed in the positively charged LDH layers. Another strategy was followed by Othman et al. [120], who coated a $MgAlCO_3$ hydrotalcite with zeolites in order to increase the CO_2 capture capacity. K. Wu et al. [121] were the first to use mesoporous alumina to load LDH using the coprecipitation method, obtaining a large BET surface area (278–378 m^2/g).

A new method, called aqueous miscible organic solvent treatment (AMOST), uses solvents such as acetone and methanol to wash the LDH wet slurry and remove the intercalated water [119,122]. This method achieves larger surface areas and larger pore volumes [119]. Figure 9 shows the diagram of the synthetic process used for AMOST using acetone and K_2CO_3.

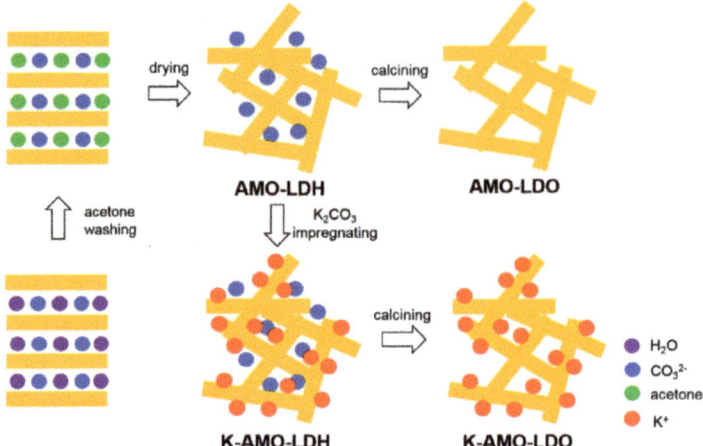

Figure 9. General scheme of synthetic process of AMOST method using by adopted from Zhu et al [122].

6. Pressure, Temperature, and Capacity in CO_2 Absorption and Use of CO_2 Captured

There are several parameters that can vary the CO_2 capture capacity of a hydrotalcite; among the most important are temperature and pressure [33]. Table 1 shows a comparison of the capture capacity of different types of hydrotalcite under different conditions.

Table 1. Maximum adsorption capacities of CO_2 for hydrotalcites reported in the literature.

Refs.	Type	LDH to CLDH?	Mg/Al Molar Ratio	Pressure (Atm.)	Temperature Isotherm (°C)	Capacity Adsorption (mg/g)
[29]	Alkali-modified (K and CS)	300 °C for 3 h	-	2	400	25.52
[38]	K promoted * Mg_3AlCO_3	LDH	-	16.50	400	28.60
[122]	* Mg_3AlCO_3 with treatment AMOST	450 °C for 3 h	3	1	400	30.58
[22]	K promoted * Mg_3AlCO_3	400 °C for 4 h	-	3.5	400	41.80
[64]	K promoted * Mg_3AlCO_3	400 °C for 3 h	-	10	400	25.68
[2]	K promoted commercial hydrotalcite	400 °C for 6 h	-	30	400	21.18
[117]	* Mg_3AlCO_3	450 °C for 10 h	2	13	350	44.95
[114]	K promoted * Mg_3AlCO_3	450 °C for 3 h	2.9	20	350	44.94
[28]	* Mg_3AlCO_3	400 °C for 4h	3	1	300	26.4
[7]	* Mg_3AlCO_3	400 °C for 2h	2	1	300	46.21
[91]	* Mg_3AlCO_3	500 °C for 4 h	3	43.42	300	144.32
[92]	Hydrotalcite of K-Na	650 °C for 6 h (100 g)	3 (K/Na ratio)	1.34	300	34.03
[112]	* Mg_3AlCO_3	400 °C for 2h	2	1	300	41.53
[116]	* K-Mg-Al	400 °C for 6 h	3	-	300	54.57
[86]	* Mg-Al with graphene oxide	400 °C for 4 h	-	-	300	12.84
[35]	Alkali metal (Na, Cs and K) with * Mg_3AlCO_3	300 °C for -h	2	0.15	300	21.12
[123]	K-loaded CNF supported hydrotalcite	500 °C for 4 h	-	1.1	250	62.27
[110]	* (Mg/Al = 20)	No information	20	1	240	407.97

Table 1. Cont.

Refs.	Type	LDH to CLDH?	Mg/Al Molar Ratio	Pressure (Atm.)	Temperature Isotherm (°C)	Capacity Adsorption (mg/g)
[88]	* Mg_3AlCO_3	-	3	1	200	23.32
[106]	* Mg_3AlCO_3	400 °C for 1 h	3.1	1	200	23.76
[23]	* Mg_3AlCO_3	400 °C for 4 h		1	200	39.60
[106]	* Mg_3AlCO_3	400 °C for 6 h	3	1	200	36.52
[124]	K promoted Gallium substituted hydrotalcite	400 °C	-	-	200	39.80
[121]	Mesoporous alumina with Mg-Al LDH	No	4	1	200	68.64
[26]	* Mg_3AlCO_3	550 °C for 1 h	3	1	80	93.72
[25]	* Mg_3AlCO_3	450 °C for 10 h	3	1	50	45.76
[27]	* Mg_3AlCO_3	400 °C for 1 h	3	1	50	40.04
[120]	* Mg-Al with coated zeolites	400 °C for 15 h	3	1	30	197.73
[93]	* Ni-Mg-Al	650 °C for 7 h (1 g)	-	1	20	70.62
[125]	* Cu-Al	600 °C for 75 min	3 (Cu/Al ratio)	1	20	20.54
[33]	* Mg_3AlCO_3	500 °C for 2 h	3	34.28	0	142.02
[34]	Organohydrotalcites TDD [1]	500 °C for 2 h	3	35	0	176.66

* Hidrotalcite of. [1] Tetradecanedioate anions.

As can be seen in Table 1, it is uncommon to use pressures higher than atmospheric pressure to measure CO_2 capture capacity, except in some research [2,22,29,38,64,91,92,114,117]. It is much more unusual to find research using high pressures and low temperatures (between 0 and 40 °C) [33,34]. At a pressure of 35 atm, an amount of 1.34 g of CLDH reduce CO_2 in 1 m³ of air to preindustrial level [34]. Therefore, researchers have to pool their efforts to study CO_2 capture at high pressures and low application temperatures with calcined hydrotalcites.

Another very important factor to take into account will be the analysis of the reversibility conditions of CO_2 adsorption–desorption. For those materials in which it is reversible, these could be used to purify different gas streams in cyclic adsorption–desorption processes, as in the case of the use of CaO [21]. CaO could be used to capture CO_2 from thermal power plants or cement plants, to be stored and subsequently sold for use in different industrial processes. There is research in which CO_2 is used as a curing gas [11], capturing CO_2 (5.55 kg CO_2/t mix), improving mechanical properties and decreasing curing times (1 day in a CO_2 chamber is similar to 7 days in a conventional chamber), which can lead to an improvement in productivity. In addition, in order to avoid the inconvenience of using an accelerated carbonation chamber, an unprecedented strategy has been introduced, where an aqueous solution with injected CO_2 is used as kneading water, with very promising results [12]. In the case of irreversible processes, they could be used as CO_2 sinks, either alone or incorporated into other materials.

7. Combined Use of Hydrotalcites and Cement-Based Materials

The good and promising results obtained by the scientific community as adsorbents of hydrotalcites suggest the idea of incorporating them into construction materials, such as cement-based materials [66–70]. The ion exchange is the key feature that makes the use of LDH/CLDH in building materials attractive. These, together with the memory effect of CLDH, are the two suitable factors for the ion exchange and capture mechanism shown in Figure 10 [75].

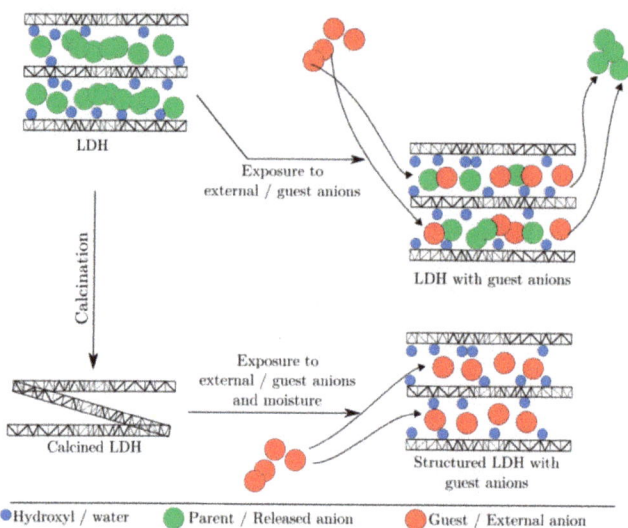

Figure 10. Schematic representation of the LDH/CLDH structure and ion exchange/captured mechanism represented by Zahid M. Mir [75] (open access).

In recent years, LDH/CLDH have emerged as a new class of engineering materials [75], which can aid in the corrosion control of concrete structures and potentially prolong their service life. Calcined hydrotalcite improves the durability of cement-base materials. One of the first findings mixing hydrotalcites with concrete dates back to 2004 [126]. In 2013, there were two very important contributions in this field [69,126].

This includes the use of LDH as an additive to improve the thermal insulation of the intumescent fire retardant (IFR) coating [71]. Wu et al. [72] indicated that the structure regeneration of CLDHs in a cement paste environment was also revealed [69]. After calcination, a large number of active sites produced were improving the effect of CLDH on cement [72]. Since hydrotalcites may be incorporated into various building materials mixtures, mortars, concretes and backfills, their applications as accessible and affordable materials is prospective [73–75]. However, research in the field of hydrotalcites and cement-based materials is still insufficient [72,76].

It is also difficult to find studies that add LDH or CLDH to alkaline activated materials [77,78]. Mixing LDH/CLDH in alkali-activated materials to improve the durability of these materials may be a very interesting field for the scientific community.

It is even more difficult to find studies in which CLDHs are used as additives in order to increase the CO_2 capture capacity of cement-based materials. Ma et al. [79] reported the adsorption of CO_3^{2-} by CLDH seven times faster than LDH due to the release of anions after calcination, which can be very beneficial for CO_2 capture. Suescum-Morales et al. [3,59] added different percentages of CLDH (Mg_3AlCO_3) in a one-coat mortar in order to increase the CO_2 capture capacity. Adding 5% of calcined hydrotalcite increased the CO_2 capture capacity by 8.52% with respect to the reference mortar. One m^2 of one-coat mortar with 5% of calcined hydrotalcites cleans 5540 m^3 of air [3]. The use of these one-coat mortars in building facades is a very promising strategy due to the large surface area exposed to the atmosphere. Studies along these lines should be reinforced to improve this information gap.

8. Conclusions

Application of hydrotalcites as CO_2 sinks and climate change mitigation is an emerging line of research. In this review, a guide on CO_2 capture by hydrotalcites is presented.

Firstly, the behaviour of hydrotalcites and the structural change from LDH to CLDH and their properties with the calcination of LDH are presented. After a review of the literature, although the calcination temperatures are roughly similar, the times required to produce these changes, as well as the amount of hydrotalcite to be calcined, are somewhat diverse. Researchers must work together to clarify these factors so that they are not ambiguous data. These data would be very important since, for industrial use and a real application of these materials as CO_2 sinks, technical and economic calculations are necessary, which indicate the feasibility of using these products by companies. On the other hand, hydrotalcites can also be excessively calcined, with the consequent waste of energy (in time or temperature).

Secondly, the pH value in the synthesis is very important in the development of the morphology, porous structure, as well as in the chemical composition. The molar ratio (Mg/Al) is very important in terms of CO_2 capture, although there are also many other factors that affect this parameter: pH of the synthesis, nature of the anion, pressure used and absorption temperature.

It is also important to develop strategies that increase the specific surface area of hydrotalcites, as this is one of the most important factors in CO_2 absorption capacity. After a comparison of different studies (32 papers) on the capture capacity of LDH and CLDH, a gap in information on CO_2 capture capacities at high pressure and low temperature has been observed. Therefore, researchers have to pool their efforts to study CO_2 capture at high pressures and low application temperatures using calcined hydrotalcites.

It is very difficult to find studies in which CLDHs are used as additives in order to increase the CO_2 capture capacity of cement-based materials. The ion exchange and the memory effect of CLDH are key to the use of hydrotalcites as CO_2 capture additives in cement-based materials. Studies along these lines should be reinforced to improve this information gap and mitigate climate change.

Author Contributions: Conceptualization, D.S.-M. and J.M.F.-R.; Methodology, J.R.J.; writing—original draft preparation, D.S.-M.; writing—review and editing, J.M.F.-R. and J.R.J.; supervision, J.M.F.-R.; project administration, J.M.F.-R. and J.R.J. All authors have read and agreed to the published version of the manuscript.

Funding: This research has been supported from the Andalusian Regional Government, Spain (UCO-FEDER 20 REF. 1381172-R).

Data Availability Statement: Not applicable.

Acknowledgments: D. Suescum-Morales also acknowledges funding from MECD-Spain (http://www.mecd.gob.es/educacion-mecd/) FPU 17/04329 (accessed on 1 April 2022). The authors wish to thank the IE 57164 project for the implementation and improvement of scientific and technological infrastructures and equipment supported by the Andalusian Regional Government (FEDER 2011).

Conflicts of Interest: The authors declare no conflict of interest.

References

1. USEP Agency. Global Greenhouse Gas Emissions Data. 2022. Available online: https://www.epa.gov/ghgemissions/global-greenhouse-gas-emissions-data (accessed on 24 March 2022).
2. Van Selow, E.R.; Cobden, P.D.; Verbraeken, P.A.; Hufton, J.R.; Brink, R.W.V.D. Carbon Capture by Sorption-Enhanced Water−Gas Shift Reaction Process using Hydrotalcite-Based Material. *Ind. Eng. Chem. Res.* **2009**, *48*, 4184–4193. [CrossRef]
3. Suescum-Morales, D.; Cantador-Fernández, D.; Jiménez, J.R.; Fernández, J.M. Potential CO_2 capture in one-coat limestone mortar modified with Mg_3Al–CO_3 calcined hydrotalcites using ultrafast testing technique. *Chem. Eng. J.* **2021**, *415*, 129077. [CrossRef]
4. Martins, V.F.D.; Miguel, C.V.; Gonçalves, J.C.; Rodrigues, A.E.; Madeira, L.M. Modeling of a cyclic sorption–desorption unit for continuous high temperature CO_2 capture from flue gas. *Chem. Eng. J.* **2022**, *434*, 134704. [CrossRef]
5. IEA. Global Energy and CO_2 Status Report 2017. Glob. Energy CO_2 Status Rep. 2017. Available online: https://www.iea.org/publications/freepublications/publication/GECO2017.pdf (accessed on 1 April 2022).
6. Yadav, S.; Mondal, S. A review on the progress and prospects of oxy-fuel carbon capture and sequestration (CCS) technology. *Fuel* **2021**, *308*, 122057. [CrossRef]
7. Rocha, C.; Soria, M.; Madeira, L.M. Effect of interlayer anion on the CO_2 capture capacity of hydrotalcite-based sorbents. *Sep. Purif. Technol.* **2019**, *219*, 290–302. [CrossRef]
8. Bui, M.; Adjiman, C.S.; Bardow, A.; Anthony, E.J.; Boston, A.; Brown, S.; Fennell, P.S.; Fuss, S.; Galindo, A.; Hackett, L.A.; et al. Carbon capture and storage (CCS): The way forward. *Energy Environ. Sci.* **2018**, *11*, 1062–1176. [CrossRef]

9. Yadav, S.; Mondal, S.S. A complete review based on various aspects of pulverized coal combustion. *Int. J. Energy Res.* **2019**, *43*, 3134–3165. [CrossRef]
10. Christensen, T.H.; Bisinella, V. Climate change impacts of introducing carbon capture and utilisation (CCU) in waste incineration. *Waste Manag.* **2021**, *126*, 754–770. [CrossRef]
11. Suescum-Morales, D.; Kalinowska-Wichrowska, K.; Fernández, J.M.; Jiménez, J.R. Accelerated carbonation of fresh cement-based products containing recycled masonry aggregates for CO_2 sequestration. *J. CO2 Util.* **2021**, *46*, 101461. [CrossRef]
12. Suescum-Morales, D.; Fernández-Rodríguez, J.M.; Jiménez, J.R. Use of carbonated water to improve the mechanical properties and reduce the carbon footprint of cement-based materials with recycled aggregates. *J. CO2 Util.* **2022**, *57*, 101886. [CrossRef]
13. Boot-Handford, M.E.; Abanades, J.C.; Anthony, E.J.; Blunt, M.J.; Brandani, S.; Mac Dowell, N.; Fernández, J.R.; Ferrari, M.-C.; Gross, R.; Hallett, J.P.; et al. Carbon capture and storage update. *Energy Environ. Sci.* **2013**, *7*, 130–189. [CrossRef]
14. Jung, W.; Lee, J. Economic evaluation for four different solid sorbent processes with heat integration for energy-efficient CO_2 capture based on PEI-silica sorbent. *Energy* **2021**, *238*, 121864. [CrossRef]
15. Jung, W.; Lee, M.; Hwang, G.S.; Kim, E.; Lee, K.S. Thermodynamic modeling and energy analysis of a polyamine-based water-lean solvent for CO_2 capture. *Chem. Eng. J.* **2020**, *399*, 125714. [CrossRef]
16. Jung, W.; Park, J.; Won, W.; Lee, K.S. Simulated moving bed adsorption process based on a polyethylenimine-silica sorbent for CO_2 capture with sensible heat recovery. *Energy* **2018**, *150*, 950–964. [CrossRef]
17. Jung, W.; Park, S.; Lee, K.S.; Jeon, J.-D.; Lee, H.K.; Kim, J.-H.; Lee, J.S. Rapid thermal swing adsorption process in multi-beds scale with sensible heat recovery for continuous energy-efficient CO_2 capture. *Chem. Eng. J.* **2019**, *392*, 123656. [CrossRef]
18. Lee, S.; Kim, J.-K. Process-integrated design of a sub-ambient membrane process for CO_2 removal from natural gas power plants. *Appl. Energy* **2019**, *260*, 114255. [CrossRef]
19. Bhatta, L.K.G.; Subramanyam, S.; Chengala, M.D.; Olivera, S.; Venkatesh, K. Progress in hydrotalcite like compounds and metal-based oxides for CO_2 capture: A review. *J. Clean. Prod.* **2015**, *103*, 171–196. [CrossRef]
20. Wang, J.; Huang, L.; Yang, R.; Zhang, Z.; Wu, J.; Gao, Y.; Wang, Q.; O'Hare, D.; Zhong, Z. Recent advances in solid sorbents for CO_2 capture and new development trends. *Energy Environ. Sci.* **2014**, *7*, 3478–3518. [CrossRef]
21. Quesada Carballo, L.; Perez Perez, M.; Cantador Fernández, D.; Caballero Amores, A.; Fernández Rodríguez, J.M. Optimum Particle Size of Treated Calcites for CO_2 Capture in a Power Plant. *Materials* **2019**, *12*, 1284. [CrossRef]
22. Halabi, M.; de Croon, M.; van der Schaaf, J.; Cobden, P.; Schouten, J. High capacity potassium-promoted hydrotalcite for CO_2 capture in H_2 production. *Int. J. Hydrog. Energy* **2012**, *37*, 4516–4525. [CrossRef]
23. Kou, X.; Guo, H.; Ayele, E.G.; Li, S.; Zhao, Y.; Wang, S.; Ma, X. Adsorption of CO_2 on $MgAl-CO_3$ LDHs-Derived Sorbents with 3D Nanoflower-like Structure. *Energy Fuels* **2018**, *32*, 5313–5320. [CrossRef]
24. Williams, G.R.; O'Hare, D. Towards understanding, control and application of layered double hydroxide chemistry. *J. Mater. Chem.* **2006**, *16*, 3065–3074. [CrossRef]
25. León, M.; Díaz, E.; Bennici, S.; Vega, A.; Ordóñez, S.; Auroux, A. Adsorption of CO_2 on Hydrotalcite-Derived Mixed Oxides: Sorption Mechanisms and Consequences for Adsorption Irreversibility. *Ind. Eng. Chem. Res.* **2010**, *49*, 3663–3671. [CrossRef]
26. Torres-Rodríguez, D.A.; Lima, E.; Valente, J.S.; Pfeiffer, H. CO_2 Capture at Low Temperatures (30–80 °C) and in the Presence of Water Vapor over a Thermally Activated Mg–Al Layered Double Hydroxide. *J. Phys. Chem. A* **2011**, *115*, 12243–12250. [CrossRef] [PubMed]
27. Rossi, T.M.; Campos, J.; Souza, M.M.V.M. CO_2 capture by Mg–Al and Zn–Al hydrotalcite-like compounds. *Adsorption* **2015**, *22*, 151–158. [CrossRef]
28. Hutson, N.D.; Attwood, B.C. High temperature adsorption of CO_2 on various hydrotalcite-like compounds. *Adsorption* **2008**, *14*, 781–789. [CrossRef]
29. Oliveira, E.L.; Grande, C.A.; Rodrigues, A.E. CO_2 sorption on hydrotalcite and alkali-modified (K and Cs) hydrotalcites at high temperatures. *Sep. Purif. Technol.* **2008**, *62*, 137–147. [CrossRef]
30. Allmann, R. The crystal structure of pyroaurite. *Acta Crystallogr. Sect. B Struct. Crystallogr. Cryst. Chem.* **1968**, *24*, 972–977. [CrossRef]
31. Taylor, H.F.W. Crystal structures of some double hydroxide minerals. *Miner. Mag.* **1973**, *39*, 377–389. [CrossRef]
32. Miyata, S. The syntheses of hydrotalcite-like compounds and their structure and physico-chemical properties—I: The systems $Mg^{2+}-Al^{3+}-NO_3^-$, $Mg^{2+}-Al^{3+}-Cl^-$, $Mg^{2+}-Al^{3+}-ClO_4^-$, $Ni^{2+}-Al^{3+}-Cl^-$ and $Zn^{2+}-Al^{3+}-Cl^-$. *Clays Clay Miner.* **1975**, *23*, 369–375. [CrossRef]
33. Suescum-Morales, D.; Cantador-Fernández, D.; Jiménez, J.; Fernández, J. Mitigation of CO_2 emissions by hydrotalcites of Mg_3Al-CO_3 at 0 °C and high pressure. *Appl. Clay Sci.* **2020**, *202*, 105950. [CrossRef]
34. Fernandez, D.C.; Morales, D.S.; Jiménez, J.R.; Fernández-Rodriguez, J.M. CO_2 adsorption by organohydrotalcites at low temperatures and high pressure. *Chem. Eng. J.* **2021**, *431*, 134324. [CrossRef]
35. Faria, A.C.; Trujillano, R.; Rives, V.; Miguel, C.; Rodrigues, A.; Madeira, L.M. Alkali metal (Na, Cs and K) promoted hydrotalcites for high temperature CO_2 capture from flue gas in cyclic adsorption processes. *Chem. Eng. J.* **2021**, *427*, 131502. [CrossRef]
36. Cavani, F.; Trifirò, F.; Vaccari, A. Hydrotalcite-type anionic clays: Preparation, properties and applications. *Catal. Today* **1991**, *11*, 173–301. [CrossRef]
37. Costantino, U.; Nocchetti, M.; Sisani, M.; Vivani, R. Recent progress in the synthesis and application of organically modified hydrotalcites. *Zeitschrift für Kristallographie* **2009**, *224*, 273–281. [CrossRef]

38. Ding, Y.; Alpay, E. Equilibria and kinetics of CO_2 adsorption on hydrotalcite adsorbent. *Chem. Eng. Sci.* **2000**, *55*, 3461–3474. [CrossRef]
39. Schulze, K.; Makowski, W.; Chyzy, R. Nickel doped hydrotalcites as catalyst precursors for the partial oxidation of light paraffins. *Appl. Clay Sci.* **2001**, *18*, 59–69. [CrossRef]
40. Reichle, W. Catalytic reactions by thermally activated, synthetic, anionic clay minerals. *J. Catal.* **1985**, *94*, 547–557. [CrossRef]
41. Vaccari, A. Preparation and catalytic properties of cationic and anionic clays. *Catal. Today* **1998**, *41*, 53–71. [CrossRef]
42. Rives, V.; Ulibarri, M.A. Layered double hydroxides (LDH) intercalated with metal coordination compounds and oxometalates. *Coord. Chem. Rev.* **1999**, *181*, 61–120. [CrossRef]
43. Ishihara, Y.; Okabe, S. Effects of cholestyramine and synthetic hydrotalcite on acute gastric or intestinal lesion formation in rats and dogs. *Am. J. Dig. Dis.* **1981**, *26*, 553–560. [CrossRef] [PubMed]
44. Barlattani, M.; Mantera, G.; Fasani, R.; Carosi, M. Efficacy of antacid treatment in peptic ulcerative patients: Therapeutic value of synthetic hydrotalcite (Talcid). *Clin. Trials J.* **1982**, *19*, 359–367.
45. Ookubo, A.; Ooi, K.; Hayashi, H. Hydrotalcites as Potential Adsorbents of Intestinal Phosphate. *J. Pharm. Sci.* **1992**, *81*, 1139–1140. [CrossRef] [PubMed]
46. Rives, V.; Carriazo, D.; Martín, C. Heterogeneous Catalysis by Polyoxometalate-Intercalated Layered Double Hydroxides. In *Pillared Clays and Related Catalysts*; Springer: New York, NY, USA, 2010.
47. Mori, K.; Nakamura, Y.; Kikuchi, I. Modification of poly(vinyl chloride). XLI. Effect of hydrotalcite on the stabilization of poly(vinyl chloride) by 6-anilino-1,3,5-triazine-2,4-dithiol and zinc stearate. *J. Polym. Sci. Part C Polym. Lett.* **1981**, *19*, 623–628. [CrossRef]
48. Van der Ven, L.; van Gemert, M.; Batenburg, L.; Keern, J.; Gielgens, L.; Koster, T.; Fischer, H. On the action of hydrotalcite-like clay materials as stabilizers in polyvinylchloride. *Appl. Clay Sci.* **2000**, *17*, 25–34. [CrossRef]
49. Hermosín, M.; Pavlovic, I.; Ulibarri, M.; Cornejo, J. Hydrotalcite as sorbent for trinitrophenol: Sorption capacity and mechanism. *Water Res.* **1996**, *30*, 171–177. [CrossRef]
50. Cheng, W.; Wan, T.; Wang, X.; Wu, W.; Hu, B. Plasma-grafted polyamine/hydrotalcite as high efficient adsorbents for retention of uranium (VI) from aqueous solutions. *Chem. Eng. J.* **2018**, *342*, 103–111. [CrossRef]
51. Ogata, F.; Ueta, E.; Kawasaki, N. Characteristics of a novel adsorbent Fe–Mg-type hydrotalcite and its adsorption capability of As(III) and Cr(VI) from aqueous solution. *J. Ind. Eng. Chem.* **2018**, *59*, 56–63. [CrossRef]
52. Miyata, S. Anion-Exchange Properties of Hydrotalcite-Like Compounds. *Clays Clay Miner.* **1983**, *31*, 305–311. [CrossRef]
53. Goh, K.-H.; Lim, T.-T.; Dong, Z. Application of layered double hydroxides for removal of oxyanions: A review. *Water Res.* **2008**, *42*, 1343–1368. [CrossRef]
54. Abdelouas, A. Formation of Hydrotalcite-like Compounds During R7T7 Nuclear Waste Glass and Basaltic Glass Alteration. *Clays Clay Miner.* **1994**, *42*, 526–533. [CrossRef]
55. Wang, S.-D.; Scrivener, K.L. Hydration products of alkali activated slag cement. *Cem. Concr. Res.* **1995**, *25*, 561–571. [CrossRef]
56. Faucon, P.; Le Bescop, P.; Adenot, F.; Bonville, P.; Jacquinot, J.; Pineau, F.; Felix, B. Leaching of cement: Study of the surface layer. *Cem. Concr. Res.* **1996**, *26*, 1707–1715. [CrossRef]
57. Paul, M.; Glasser, F. Impact of prolonged warm (85 °C) moist cure on Portland cement paste. *Cem. Concr. Res.* **2000**, *30*, 1869–1877. [CrossRef]
58. Scheidegger, A.M.; Wieland, E.; Scheinost, A.; Dähn, R.; Tits, J.; Spieler, P. Ni phases formed in cement and cement systems under highly alkaline conditions: An XAFS study. *J. Synchrotron Radiat.* **2001**, *8*, 916–918. [CrossRef] [PubMed]
59. Suescum-Morales, D.; Fernández, D.C.; Fernández, J.M.; Jiménez, J.R. The combined effect of CO_2 and calcined hydrotalcite on one-coat limestone mortar properties. *Constr. Build. Mater.* **2021**, *280*, 122532. [CrossRef]
60. He, P.; Shi, C.; Tu, Z.; Poon, C.S.; Zhang, J. Effect of further water curing on compressive strength and microstructure of CO_2-cured concrete. *Cem. Concr. Compos.* **2016**, *72*, 80–88. [CrossRef]
61. Erans, M.; Jeremias, M.; Zheng, L.; Yao, J.G.; Blamey, J.; Manovic, V.; Fennell, P.S.; Anthony, E.J. Pilot testing of enhanced sorbents for calcium looping with cement production. *Appl. Energy* **2018**, *225*, 392–401. [CrossRef]
62. Takehira, K.; Shishido, T.; Shoro, D.; Murakami, K.; Honda, M.; Kawabata, T.; Takaki, K. Preparation of egg-shell type Ni-loaded catalyst by adopting "Memory Effect" of Mg–Al hydrotalcite and its application for CH_4 reforming. *Catal. Commun.* **2004**, *5*, 209–213. [CrossRef]
63. Gao, Z.; Sasaki, K.; Qiu, X. Structural Memory Effect of Mg–Al and Zn–Al layered Double Hydroxides in the Presence of Different Natural Humic Acids: Process and Mechanism. *Langmuir* **2018**, *34*, 5386–5395. [CrossRef]
64. Yong, Z.; Rodrigues, E. Hydrotalcite like compounds as adsorbents for carbon dioxide. *Energy Convers. Manag.* **2002**, *43*, 1865–1876. [CrossRef]
65. Rocha, C.; Soria, M.; Madeira, L.M. Doping of hydrotalcite-based sorbents with different interlayer anions for CO_2 capture. *Sep. Purif. Technol.* **2019**, *235*, 116140. [CrossRef]
66. Qu, Z.; Yu, Q.; Brouwers, H. Relationship between the particle size and dosage of LDHs and concrete resistance against chloride ingress. *Cem. Concr. Res.* **2018**, *105*, 81–90. [CrossRef]
67. Yang, Z.; Fischer, H.; Cerezo, J.; Mol, J.M.C.; Polder, R. Aminobenzoate modified MgAAl hydrotalcites as a novel smart additive of reinforced concrete for anticorrosion applications. *Constr. Build. Mater.* **2013**, *47*, 1436–1443. [CrossRef]

68. Yang, Z.; Fischer, H.; Polder, R. Synthesis and characterization of modified hydrotalcites and their ion exchange characteristics in chloride-rich simulated concrete pore solution. *Cem. Concr. Compos.* **2014**, *47*, 87–93. [CrossRef]
69. Yang, Z.; Fischer, H.; Polder, R. Modified hydrotalcites as a new emerging class of smart additive of reinforced concrete for anticorrosion applications: A literature review. *Mater. Corros.* **2013**, *64*, 1066–1074. [CrossRef]
70. Lozano-Lunar, A.; Álvarez, J.I.; Navarro-Blasco, Í.; Jiménez, J.R.; Fernández-Rodriguez, J.M. Optimisation of mortar with Mg-Al-Hydrotalcite as sustainable management strategy lead waste. *Appl. Clay Sci.* **2021**, *212*, 106218. [CrossRef]
71. Hu, X.; Zhu, X.; Sun, Z. Fireproof performance of the intumescent fire retardant coatings with layered double hydroxides additives. *Constr. Build. Mater.* **2020**, *256*, 119445. [CrossRef]
72. Wu, Y.; Duan, P.; Yan, C. Role of layered double hydroxides in setting, hydration degree, microstructure and compressive strength of cement paste. *Appl. Clay Sci.* **2018**, *158*, 123–131. [CrossRef]
73. Lauermannová, A.-M.; Paterová, I.; Patera, J.; Skrbek, K.; Jankovský, O.; Bartůněk, V. Hydrotalcites in Construction Materials. *Appl. Sci.* **2020**, *10*, 7989. [CrossRef]
74. Gomes, C.; Mir, Z.; Sampaio, R.; Bastos, A.; Tedim, J.; Maia, F.; Rocha, C.; Ferreira, M. Use of ZnAl-Layered Double Hydroxide (LDH) to Extend the Service Life of Reinforced Concrete. *Materials* **2020**, *13*, 1769. [CrossRef] [PubMed]
75. Mir, Z.M.; Bastos, A.; Höche, D.; Zheludkevich, M.L. Recent Advances on the Application of Layered Double Hydroxides in Concrete—A Review. *Materials* **2020**, *13*, 1426. [CrossRef] [PubMed]
76. Cao, L.; Guo, J.; Tian, J.; Xu, Y.; Hu, M.; Wang, M.; Fan, J. Preparation of Ca/Al-Layered Double Hydroxide and the influence of their structure on early strength of cement. *Constr. Build. Mater.* **2018**, *184*, 203–214. [CrossRef]
77. Long, W.-J.; Xie, J.; Zhang, X.; Fang, Y.; Khayat, K.H. Hydration and microstructure of calcined hydrotalcite activated high-volume fly ash cementitious composite. *Cem. Concr. Compos.* **2021**, *123*, 104213. [CrossRef]
78. Long, W.; Xie, J.; Zhang, X.; Kou, S.; Xing, F.; He, C. Accelerating effect of calcined hydrotalcite-Na_2SO_4 binary system on hydration of high volume fly ash cement. *Constr. Build. Mater.* **2022**, *328*, 127068. [CrossRef]
79. Ma, J.; Duan, P.; Ren, D.; Zhou, W. Effects of layered double hydroxides incorporation on carbonation resistance of cementitious materials. *J. Mater. Res. Technol.* **2019**, *8*, 292–298. [CrossRef]
80. Vágvölgyi, V.; Palmer, S.J.; Kristóf, J.; Frost, R.L.; Horváth, E. Mechanism for hydrotalcite decomposition: A controlled rate thermal analysis study. *J. Colloid Interface Sci.* **2007**, *318*, 302–308. [CrossRef]
81. Yahyaoui, R.; Jimenez, P.E.S.; Maqueda, L.A.P.; Nahdi, K.; Luque, J.M.C. Synthesis, characterization and combined kinetic analysis of thermal decomposition of hydrotalcite ($Mg_6Al_2(OH)_{16}CO_3 \cdot 4H_2O$). *Thermochim. Acta* **2018**, *667*, 177–184. [CrossRef]
82. Miyata, S. Physico-Chemical Properties of Synthetic Hydrotalcites in Relation to Composition. *Clays Clay Miner.* **1980**, *28*, 50–56. [CrossRef]
83. Kannan, V.R.S.; Velu, S. Synthesis and physicochemical properties of cobalt aluminium hydrotalcites. *J. Mater. Sci.* **1995**, *30*, 1462–1468. [CrossRef]
84. Cocheci, L.; Barvinschi, P.; Pode, R.; Popovici, E.; Seftel, E.M. Structural Characterization of Some Mg/Zn-Al Type Hy-drotalcites Prepared for Chromate Sorption from Wastewater. *Chem. Bull.* **2010**, *55*, 40–45.
85. Palmer, S.J.; Spratt, H.J.; Frost, R.L. Thermal decompostition of hydrotalcites with variable cationic ratios. *J. Therm. Anal. Calorim.* **2009**, *95*, 123–129. [CrossRef]
86. Garcia-Gallastegui, A.; Iruretagoyena, D.; Gouvea, V.; Mokhtar, M.; Asiri, A.M.; Basahel, S.N.; Al-Thabaiti, S.A.; Alyoubi, A.O.; Chadwick, D.; Shaffer, M.S.P. Graphene Oxide as Support for Layered Double Hydroxides: Enhancing the CO_2 Adsorption Capacity. *Chem. Mater.* **2012**, *24*, 4531–4539. [CrossRef]
87. Ram Reddy, M.K.; Xu, Z.P.; Lu, G.Q.; Diniz da Costa, J.C. Layered Double Hydroxides for CO_2 Capture: Structure Evolution and Regeneration. *Ind. Eng. Chem. Res.* **2006**, *45*, 7504–7509. [CrossRef]
88. Wang, Q.; Wu, Z.; Tay, H.H.; Chen, L.; Liu, Y.; Chang, J.; Zhong, Z.; Luo, J.; Borgna, A. High temperature adsorption of CO_2 on Mg–Al hydrotalcite: Effect of the charge compensating anions and the synthesis pH. *Catal. Today* **2011**, *164*, 198–203. [CrossRef]
89. Hibino, T. Decarbonation Behavior of Mg-Al-CO_3 Hydrotalcite-like Compounds during Heat Treatment. *Clays Clay Miner.* **1995**, *43*, 427–432. [CrossRef]
90. Tao, Q.; Zhang, Y.; Zhang, X.; Yuan, P.; He, H. Synthesis and characterization of layered double hydroxides with a high aspect ratio. *J. Solid State Chem.* **2006**, *179*, 708–715. [CrossRef]
91. Garcés-Polo, S.; Villarroel-Rocha, J.; Sapag, K.; Korili, S.; Gil, A. Adsorption of CO_2 on mixed oxides derived from hydrotalcites at several temperatures and high pressures. *Chem. Eng. J.* **2018**, *332*, 24–32. [CrossRef]
92. Martunus; Othman, M.R.; Fernando, W.J.N. Elevated temperature carbon dioxide capture via reinforced metal hydrotalcite. *Microporous Mesoporous Mater.* **2011**, *138*, 110–117. [CrossRef]
93. Aschenbrenner, O.; McGuire, P.; Alsamaq, S.; Wang, J.; Supasitmongkol, S.; Al-Duri, B.; Styring, P.; Wood, J. Adsorption of carbon dioxide on hydrotalcite-like compounds of different compositions. *Chem. Eng. Res. Des.* **2011**, *89*, 1711–1721. [CrossRef]
94. Forano, C.T.-G.C.; Hibino, T.; Leroux, F. Layered double hydroxides. In *Hand-Book of Clay Science*; Bergaya, F., Theng, B.K.G., Lagaly, G., Eds.; Elsevier: Newnes, Australia, 2006; ISBN 978-0-08-044183-2.
95. Cai, P.; Zheng, H.; Wang, C.; Ma, H.; Hu, J.; Pu, Y.; Liang, P. Competitive adsorption characteristics of fluoride and phosphate on calcined Mg–Al–CO3 layered double hydroxides. *J. Hazard. Mater.* **2012**, *213–214*, 100–108. [CrossRef] [PubMed]
96. Geng, C.; Xu, T.; Li, Y.; Chang, Z.; Sun, X.; Lei, X. Effect of synthesis method on selective adsorption of thiosulfate by calcined MgAl-layered double hydroxides. *Chem. Eng. J.* **2013**, *232*, 510–518. [CrossRef]

97. Hobbs, C.; Jaskaniec, S.; McCarthy, E.K.; Downing, C.; Opelt, K.; Güth, K.; Shmeliov, A.; Mourad, M.C.D.; Mandel, K.; Nicolosi, V. Structural transformation of layered double hydroxides: An in situ TEM analysis. *npj 2D Mater. Appl.* **2018**, *2*, 4. [CrossRef]
98. Luo, S.; Guo, Y.; Yang, Y.; Zhou, X.; Peng, L.; Wu, X.; Zeng, Q. Synthesis of calcined La-doped layered double hydroxides and application on simultaneously removal of arsenate and fluoride. *J. Solid State Chem.* **2019**, *275*, 197–205. [CrossRef]
99. Elhalil, A.; Qourzal, S.; Mahjoubi, F.; Elmoubarki, R.; Farnane, M.; Tounsadi, H.; Sadiq, M.; Abdennouri, M.; Barka, N. Defluoridation of groundwater by calcined Mg/Al layered double hydroxide. *Emerg. Contam.* **2016**, *2*, 42–48. [CrossRef]
100. Harizi, I.; Chebli, D.; Bouguettoucha, A.; Rohani, S.; Amrane, A. A New Mg–Al–Cu–Fe-LDH Composite to Enhance the Adsorption of Acid Red 66 Dye: Characterization, Kinetics and Isotherm Analysis. *Arab. J. Sci. Eng.* **2018**, *44*, 5245–5261. [CrossRef]
101. Chebli, D.; Bouguettoucha, A.; Reffas, A.; Tiar, C.; Boutahala, M.; Gulyas, H.; Amrane, A. Removal of the anionic dye Biebrich scarlet from water by adsorption to calcined and non-calcined Mg–Al layered double hydroxides. *Desalin. Water Treat.* **2016**, *57*, 22061–22073. [CrossRef]
102. Li, R.; Zhan, W.; Song, Y.; Lan, J.; Guo, L.; Zhang, T.C.; Du, D. Template-free synthesis of an eco-friendly flower-like Mg/Al/Fe-CLDH for efficient arsenate removal from aqueous solutions. *Sep. Purif. Technol.* **2021**, *282*, 120011. [CrossRef]
103. Liu, T.; Chen, Y.; Yu, Q.; Fan, J.; Brouwers, H. Effect of MgO, Mg-Al-NO3 LDH and calcined LDH-CO3 on chloride resistance of alkali activated fly ash and slag blends. *Constr. Build. Mater.* **2020**, *250*, 118765. [CrossRef]
104. Singh, R.; Reddy, M.R.; Wilson, S.; Joshi, K.; da Costa, J.C.D.; Webley, P. High temperature materials for CO_2 capture. *Energy Procedia* **2009**, *1*, 623–630. [CrossRef]
105. Ebner, A.D.; Reynolds, S.P.; Ritter, J.A. Nonequilibrium Kinetic Model That Describes the Reversible Adsorption and Desorption Behavior of CO_2 in a K-Promoted Hydrotalcite-like Compound. *Ind. Eng. Chem. Res.* **2007**, *46*, 1737–1744. [CrossRef]
106. Wang, Q.; Tay, H.H.; Guo, Z.; Chen, L.; Liu, Y.; Chang, J.; Zhong, Z.; Luo, J.; Borgna, A. Morphology and composition controllable synthesis of Mg–Al–CO_3 hydrotalcites by tuning the synthesis pH and the CO_2 capture capacity. *Appl. Clay Sci.* **2012**, *55*, 18–26. [CrossRef]
107. Yong, Z.; Mata, V.; Rodrigues, E. Adsorption of Carbon Dioxide onto Hydrotalcite-like Compounds (HTlcs) at High Temperatures. *Ind. Eng. Chem. Res.* **2001**, *40*, 204–209. [CrossRef]
108. Yang, W.; Kim, Y.; Liu, P.K.T.; Sahimi, M.; Tsotsis, T.T. A study by in situ techniques of the thermal evolution of the structure of a Mg–Al–CO_3 layered double hydroxide. *Chem. Eng. Sci.* **2002**, *57*, 2945–2953. [CrossRef]
109. Peng, J.; Iruretagoyena, D.; Chadwick, D. Hydrotalcite/SBA15 composites for pre-combustion CO_2 capture: CO_2 adsorption characteristics. *J. CO2 Util.* **2018**, *24*, 73–80. [CrossRef]
110. Kim, S.; Jeon, S.G.; Lee, K.B. High-Temperature CO_2 Sorption on Hydrotalcite Having a High Mg/Al Molar Ratio. *ACS Appl. Mater. Interfaces* **2016**, *8*, 5763–5767. [CrossRef]
111. Macedo, M.S.; Soria, M.; Madeira, L.M. High temperature CO_2 sorption using mixed oxides with different Mg/Al molar ratios and synthesis pH. *Chem. Eng. J.* **2021**, *420*, 129731. [CrossRef]
112. Silva, J.; Trujillano, R.; Rives, V.; Soria, M.; Madeira, L.M. High temperature CO_2 sorption over modified hydrotalcites. *Chem. Eng. J.* **2017**, *325*, 25–34. [CrossRef]
113. Gao, Y.; Zhang, Z.; Wu, J.; Yi, X.; Zheng, A.; Umar, A.; O'Hare, D.; Wang, Q. Comprehensive investigation of CO_2 adsorption on Mg–Al–CO_3 LDH-derived mixed metal oxides. *J. Mater. Chem. A* **2013**, *1*, 12782–12790. [CrossRef]
114. Walspurger, S.; de Munck, S.; Cobden, P.; Haije, W.; Brink, R.V.D.; Safonova, O. Correlation between structural rearrangement of hydrotalcite-type materials and CO_2 sorption processes under pre-combustion decarbonisation conditions. *Energy Procedia* **2011**, *4*, 1162–1167. [CrossRef]
115. Wang, Q.; Tay, H.H.; Zhong, Z.; Luo, J.; Borgna, A. Synthesis of high-temperature CO_2 adsorbents from organo-layered double hydroxides with markedly improved CO_2 capture capacity. *Energy Environ. Sci.* **2012**, *5*, 7526–7530. [CrossRef]
116. Li, S.; Shi, Y.; Yang, Y.; Zheng, Y.; Cai, N. High-Performance CO_2 Adsorbent from Interlayer Potassium-Promoted Stearate-Pillared Hydrotalcite Precursors. *Energy Fuels* **2013**, *27*, 5352–5358. [CrossRef]
117. Hanif, A.; Dasgupta, S.; Divekar, S.; Arya, A.; Garg, M.O.; Nanoti, A. A study on high temperature CO_2 capture by improved hydrotalcite sorbents. *Chem. Eng. J.* **2014**, *236*, 91–99. [CrossRef]
118. Wang, Q.; O'Hare, D. Recent Advances in the Synthesis and Application of Layered Double Hydroxide (LDH) Nanosheets. *Chem. Rev.* **2012**, *112*, 4124–4155. [CrossRef] [PubMed]
119. Wang, Q.; O'Hare, D. Large-scale synthesis of highly dispersed layered double hydroxide powders containing delaminated single layer nanosheets. *Chem. Commun.* **2013**, *49*, 6301–6303. [CrossRef]
120. Othman, M.; Rasid, N.; Fernando, W. Mg–Al hydrotalcite coating on zeolites for improved carbon dioxide adsorption. *Chem. Eng. Sci.* **2006**, *61*, 1555–1560. [CrossRef]
121. Wu, K.; Ye, Q.; Wang, L.; Meng, F.; Dai, H. Mesoporous alumina-supported layered double hydroxides for efficient CO_2 capture. *J. CO2 Util.* **2022**, *60*, 101982. [CrossRef]
122. Zhu, X.; Chen, C.; Suo, H.; Wang, Q.; Shi, Y.; O'Hare, D.; Cai, N. Synthesis of elevated temperature CO_2 adsorbents from aqueous miscible organic-layered double hydroxides. *Energy* **2018**, *167*, 960–969. [CrossRef]
123. Meis, N.N.A.H.; Bitter, J.H.; de Jong, K.P. On the Influence and Role of Alkali Metals on Supported and Unsupported Activated Hydrotalcites for CO_2 Sorption. *Ind. Eng. Chem. Res.* **2010**, *49*, 8086–8093. [CrossRef]

124. Yavuz, C.T.; Shinall, B.D.; Iretskii, A.V.; White, M.G.; Golden, T.; Atilhan, M.; Ford, P.C.; Stucky, G.D. Markedly Improved CO_2 Capture Efficiency and Stability of Gallium Substituted Hydrotalcites at Elevated Temperatures. *Chem. Mater.* **2009**, *21*, 3473–3475. [CrossRef]
125. Lwin, Y.; Abdullah, F. High temperature adsorption of carbon dioxide on Cu–Al hydrotalcite-derived mixed oxides: Kinetics and equilibria by thermogravimetry. *J. Therm. Anal.* **2009**, *97*, 885–889. [CrossRef]
126. Raki, L.; Beaudoin, J.; Mitchell, L. Layered double hydroxide-like materials: Nanocomposites for use in concrete. *Cem. Concr. Res.* **2004**, *34*, 1717–1724. [CrossRef]

Article

30 Years of Vicente Rives' Contribution to Hydrotalcites, Synthesis, Characterization, Applications, and Innovation

Raquel Trujillano

Departamento de Química Inorgánica, Universidad de Salamanca, 37008 Salamanca, Spain; rakel@usal.es

Abstract: Hydrotalcite is the name of a mineral discovered in Sweden in 1842 whose formula is $Mg_6Al_2(OH)_{16}CO_3 \cdot 4H_2O$ and presents a layered crystal structure that consists of positively charged hydroxide layers neutralized by interlayer anions as carbonate, also containing water molecules. The ease of their synthesis and the possibility of incorporating other layer cations and interlayer anions have made this type of layered double hydroxides (LDH) a group of very interesting materials for industry. In addition to LDH and due to the name of the most representative mineral, this group of compounds is commonly called hydrotalcite-like materials, or simply hydrotalcites. Another way of referring to them is as anionic clays because of their layered structure but, unlike classical clays, their layers are positive and their interlayers are anionic. The main fields of application of these solids comprise catalysis, catalyst support, anion scavengers, polymer stabilizers, drug carriers, or adsorbents. This paper briefly summarizes some of the work carried out by Professor Rives over more than thirty years, focused, among other topics, on the study of the synthesis, characterization, and applications of hydrotalcites. This research has led him to train many researchers, to collaborate with research groups around the world and to publish reference papers and books in this field. This contribution, written to be included in the Special Issue "A Themed Issue in Honor of Prof. Dr. Vicente Rives", edited on the occasion of his retirement, only shows a small part of his scientific research and intends to value and recognize his cleverness and his enormous scientific and human quality.

Keywords: hydrotalcite; LDH (layered double hydroxide); synthesis of hydrotalcites; hydrotalcites as catalysts; hydrotalcites as adsorbents

Citation: Trujillano, R. 30 Years of Vicente Rives' Contribution to Hydrotalcites, Synthesis, Characterization, Applications, and Innovation. *ChemEngineering* **2022**, *6*, 60. https://doi.org/10.3390/chemengineering6040060

Academic Editor: Isabella Nova

Received: 8 June 2022
Accepted: 30 June 2022
Published: 1 August 2022

Publisher's Note: MDPI stays neutral with regard to jurisdictional claims in published maps and institutional affiliations.

Copyright: © 2022 by the author. Licensee MDPI, Basel, Switzerland. This article is an open access article distributed under the terms and conditions of the Creative Commons Attribution (CC BY) license (https://creativecommons.org/licenses/by/4.0/).

1. Introduction

Hydrotalcite-like compounds (HT), so-called "anionic clays", are layered double hydroxides with general formula $[M_{1-x}M'_x(OH)_2]^{x+}(X)_{x/m} \cdot nH_2O$, where M = Mg(II), Zn(II), Ni(II), …, M' = Al(III), Fe(III), … and X = CO_3^{2-}, NO_3^{-}, SO_4^{2-}, … [1]. This is the sentence that, in 1991, started a new line of research for Professor Rives' research group at the Universidad de Salamanca. It can be said that it was "the principle of a beautiful friendship" with these solids for Prof. Rives and for many people that were a part of his team during this time. The study of these layered double hydroxides (LDH) has led to the collaboration of Prof. Rives with researchers from all over the planet. These collaborations implied a great advance in the knowledge of this group of compounds.

As indicated, hydrotalcite-like compounds or hydrotalcites (HT), layered double hydroxides (LDH), or anionic clays, are the common names for these kinds of materials. The first name is an extension of that of the mineral found in Snarum, Norway [2,3], extended for designating the large family of solids, mainly synthetic, with its structure. As these solids have, contrary to the natural clays, their layers positively charged and so, their interlayer space occupied by exchangeable anions, they are also known as anionic clays. Finally, attending to their layered structure and chemical composition, they belong to the family of the layered double hydroxides (Figure 1).

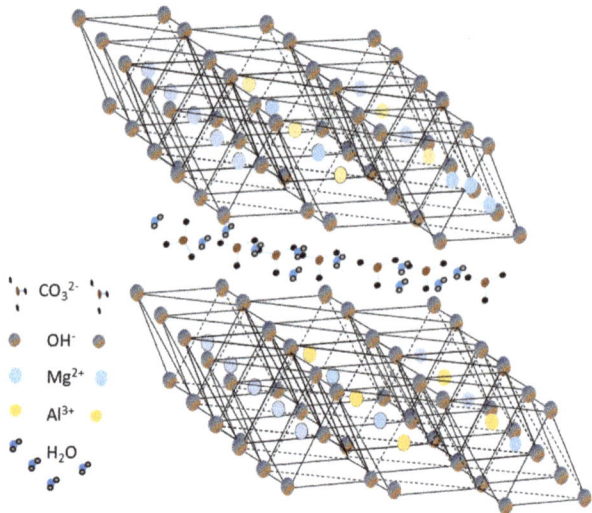

Figure 1. Structure of Mg–Al–carbonate hydrotalcite.

The formula of this mineral was first given by the Italian mineralogist Ernesto Manasse in 1915 [4], and so manasseite is another name used for this solid, although it slightly differs from hydrotalcite in its chemical composition. In 1942, two Swiss scientists, Feitknecht and Gerber, described for the first time the synthesis of this compound [5]. During the second half of the 20th century, a great number of researchers have reported on its crystal structure [6–8]. All these aspects, together with the synthesis methods and the large number of possible applications, were largely commented on in the first book entirely dedicated to these compounds, edited by Prof. Rives in 2001 [9]. The entire book has been cited more than 1400 times, according to GoogleScholar, even more, if considering the separate chapters. The high interest in hydrotalcites has given rise to the publication of thousands of scientific articles and bibliographic review papers (9300 documents when looking for "hydrotalcite" at Web of Science^{TM}). One of the most important revision papers is that published by Rives and Ulibarri in 1999, which reviewed the scientific articles published up to that date on the synthesis, properties, and applications of hydrotalcite-like materials containing intercalated anions constituted by metal complexes or oxometalates [10] (cited 770 times according to Web of Science and 1000 times according to GoogleScholar). This review described the role of catalysts, sensors, electrodes, etc., that these materials can play, as obtained or after thermal treatments.

The aim of this work is to summarize the scientific works dedicated to the knowledge of LDH that Professor Rives has carried out and shared, together with his collaborators, with the entire scientific community. Another no less important purpose of this paper is to acknowledge the time dedicated and the will, kindness, good work, and brilliance with which Professor Rives has been able to teach and supervise all the research students that have had the luck of working with him.

2. Synthesis and Characterization

Rives et al. reported for the first time, in 1991, the synthesis and characterization of the solid obtained after thermal treatments of Co,Al-hydroxycarbonate compound containing Co(II) cations with a hydrotalcite structure [1]. In the same year, they described the synthesis by the coprecipitation method of layered double hydroxides, also with the hydrotalcite-like structure and containing carbonate as interlayer anion and Ni(II) and Al(III) cations in the layers [11]. Several solids with different Ni/Al ratios were prepared and studied by means of FT-IR and V-UV/DR spectroscopies. The local environment of the Ni(II) cations was found to be octahedral in all cases, but the orientation of the interlayer carbonate anions varied depending on the Ni/Al ratio. Although the synthesis of this type of hydrotalcite had been previously reported by Reichle [12], its behavior upon aging treatments was not clear. Rives et al. studied in-depth for the first time the Co(II)/Co(III) oxidation in these compounds, linked to the synthesis and aging treatment experimental conditions. Aging was carried out by thermal treatments, and so the same authors, in 1992, continued their study by analyzing the effect of hydrothermal treatments on the nature of the hydrotalcites when varying the time and temperature [1]. Consequently, they also examined the physicochemical properties of MgAl hydrotalcites synthesized by the coprecipitation method and they established the bases that explained the variation of the crystallinity with the temperature, time, and type of aging treatment [13,14]. The Mg/Al ratio increased, and the specific surface area and the water content decreased, when the hydrothermal treatment was prolonged.

To explain the importance of these materials for their use in catalysis, a study of their acid/base properties was conducted in 1993 on a MgAl-HT and on their calcination products [15]. The adsorption of pyridine to measure the acidity, and formic acid to study the basicity on the surface of the HT was carried out by using FT-IR spectroscopy. The acid and basic sites of these solids were described and the authors confirmed that Lewis acid sites were linked to Al(III) ions and the basic ones were correlated to isolated surface hydroxyl groups. More recently a wide study of the control of the synthesis conditions to obtain hydrotalcite-like materials with tuned properties has been done on MgAl hydroxycarbonates [16]. In this work, the authors tried to find the optimal conditions to prepare samples with pre-designed structural and textural properties. They found that a sample obtained after 48 h of conventional hydrothermal treatment can be synthesized by aging only 10 h under microwave irradiation. Hydrothermal treatment under microwave irradiation needed shorter treatment times and, thus, lower energy consumption, leading to hydrotalcites with higher crystallinity than conventional hydrothermal treatments. This last work was based on the studies previously carried out on the incidence of microwave thermal treatments, in which the changes in the crystallinity, the textural and structural characteristics, or the porosity of different hydrotalcites were explored [17–19].

The research in HT containing oxidizable cations continued with the report for the first time of the synthesis of an LDH containing V(III) cations in the layers [20]. To avoid vanadium (III) oxidation, the synthesis was carefully designed; the preparation of an Mg-V-carbonate HT was carried out under a nitrogen atmosphere using Schlenk techniques. Characterization results using powder X-ray diffraction (PXRD) infrared spectroscopy with Fourier Transform (FT-IR) and extended X-ray absorption fine structure (EXAFS), among other techniques, confirmed the formation of hydrotalcites with Mg(II) and V(III) cations in the layers, CO_3^{2-} as an interlayer anion, and with a well-ordered layered structure. Subsequently, Labajos et al. described different ways of obtaining and aging Mg(II)-V(III)-HT with carbonate as interlayer anion, widely studying the effect of drying under air or under vacuum and the incidence of the hydrothermal treatment in the as-synthesized samples and its calcination products [21] was tested. The use of vacuum for drying and hydrothermal treatment after synthesis gave rise to better-crystallized samples. The first description in the literature of the synthesis and characterization of Ni(II)-V(III) hydrotalcites with different Ni/V molar ratios and the study of their calcination products was carried out by Rives's research group in 1999 [22]. The synthesis of this system was similar to the previously reported to avoid the V(III) oxidation [21,22], obtaining well-crystallized hydrotalcite with Ni(II) and V(III) in the brucite-like layers, and carbonate with the plane parallel to the brucite-like layers as interlayer anion [22].

The characterization of this type of layered solids by different thermal techniques, namely thermogravimetry (TG) Differential Thermal Analysis (DTA), and Temperature Programmed Reduction (TPR), has been largely implemented along to improve the knowledge of the properties of the HT [9,23]. TG and DTA analysis were demonstrated to be essential to analyzing the nature and the strength of the bond of interlayer anions and molecules, and also the number of volatile species. Thus, Labajos and Rives investigated the thermal evolution of the layer Cr(III) cations of HT compounds [23]. The presence of chromate species after calcination at medium temperatures in MgCr and NiCr of HT was recognized by TPR analysis of the samples [23]. TPR analysis of a series of carbonate-HT with reducible cations such as Co(II)-Fe(III), Ni(II)-Cr(III), Mg(II)-Mn(III), and MgAl-vanadate were carried out. The results proved the ability of these techniques to complete the characterization of these solids, mainly focusing on their applicability as catalysts and/or catalyst precursors. [24].

The interest in HT containing cations with redox properties increased because of their potential as oxidation or photooxidation catalysts. Thus, a comparative study of vanadate exchanged MgAl, MgCr, NiAl, and NiCr-HT obtained by means of different synthesis methods was reported in 1994–1995 [25–27]. The synthesis procedures tested to introduce the vanadate anions in the interlayer were direct coprecipitation, direct or indirect reconstruction, and pre-swelling with glycerol, among others. These preparations were also done to compare the results with those of hydrotalcites previously synthesized containing vanadium (III) within the layers [24]. The different synthesis procedures gave solids with similar composition and layered structure but with differences in the crystallinity degree and specific surface area and porosity. These kinds of differences are interesting to obtain final materials with specific properties in catalytic reactions [28–33]. The use of alternative routes to synthesize hydrotalcite-like materials with decavanadate in the interlayer was also assessed. For this, the application of an ultrasonic method was used to obtain MgAl-decavanadate by means of anionic exchange between carbonate and decavanadate. This anionic exchange was achieved by submitting a suspension of MgAl-carbonate LDH precursor in a vanadate solution to ultrasonic treatment. The exchange was easily achieved, and the crystallinity of the final solid was higher than that of the precursor [34]. The synthesis of polyoxovanadate pillared systems of ZnAl-HT was also tested and in this case, the structural characterization was implemented by an X-ray absorption and diffraction study [35]. The inclusion of anions other than vanadate in the interlayer space was successfully achieved as Zn,Al-HT interlayered with hexacyanoferrate(II) and hexacyanoferrate(III) anions were synthesized by reconstruction of the solid obtained by calcination of the carbonate form at 500 °C, anionic exchange of nitrate- and terephthalate-containing HT, and also direct synthesis [36].

Anions other than carbonate or decavanadate, such as oxalate, borate, or silicate anions, have also been used as interlayer anions that compensate for the layer charge, and the synthesis procedure gave perfectly characterized solids having the hydrotalcite-like structure. The layered solids were synthesized by using two different procedures [37,38]. The first one was the anion exchange method by changing the nitrate anion of a MgAl-nitrate HT for the different anions, borate, silicate, or $[Cr(C_2O_4)_3]^{3-}$. The reconstruction method of Mg-Al-carbonate calcined HT was also used for obtaining the Mg,Al layered double hydroxide with $[Cr(C_2O_4)_3]^{3-}$ anions in the interlayer. All solids obtained showed PXRD patterns typical of layered solids with a hydrotalcite-like structure (Figure 2).

The effect of the Mg/Al molar ratio on the disposition and number of anions in the interlayer spacing was thoroughly studied. The interlayer spacing value depended on this ratio, and thus also depended on the precise orientation of the interlayer anions [37,38]. In this sense, the "memory effect" is a very interesting property of this kind of compound that is based on the ability of the oxides derived from the calcination of HT in reconstructing the layer structure when rehydrated in the presence of anions. This effect was also widely depicted in 1999 by submitting to rehydration the oxides obtained after calcining the original solids at increasing temperatures [39]. The crystalline phases formed were assessed by means of PXRD and the Al cations environment by 27Al MAS NMR. It was observed that the reconstruction could be complete when the calcination temperature was below 550 °C and the time of rehydration 24 h. When samples were calcined at 750 °C, a rehydration time of 3 days was necessary. Finally, after calcination at 1000 °C, only a partial reconstruction was possible. However, the NMR results indicated that some of the Al(III) cations remained in tetrahedral sites [39]. This work aroused the interest of many researchers and has been widely cited (more than 190 times according to Web of Science™, and more than 250 times according to GoogleScholar) in different HT research publications. Recently, it was cited in a review that summarizes the effect of aluminum content on optical, electrical, and dielectric properties of mixed metal oxides obtained by calcination of Mg-Al-nitrate HT [40].

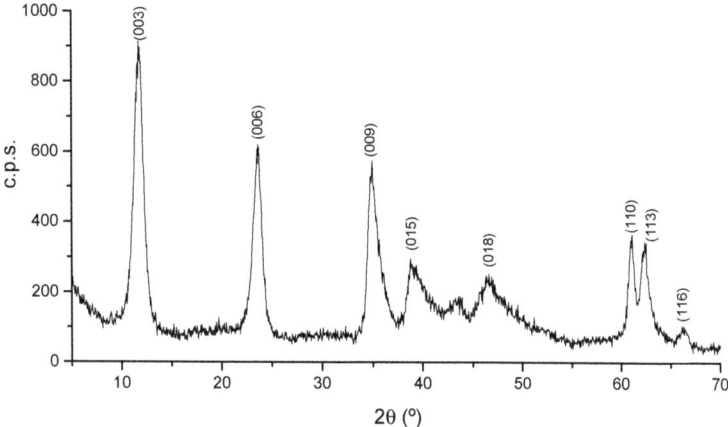

Figure 2. PXRD of a Mg−Al−carbonate hydrotalcite obtained by the coprecipitation method. The Miller indices are assigned according to the JCPDS 22-700.

The calcination products of all these LDH have always shown a high interest because of the possibility of obtaining double, ternary, or quaternary mixed oxides with high specific surface area and dispersion degree, and so suitable for catalysts reactions. Countless LDH have been prepared to improve the synthesis methods and also the characterization procedures of the solids derived from their calcination [41–43]. Ulibarri et al. studied the synthesis of hydrotalcites with oxidable cations in the layers, such as Ni and Mn, and their incidence in the mixed oxides obtained from their calcination at different temperatures [43]. Fernández et al. synthesized Mg-Al-Fe hydrotalcites with different Al/Fe molar ratios to test the effect of iron content on the crystalline phases formed upon thermal decomposition [44]. Trujillano et al. described the synthesis and characterization of Co-Fe-carbonate HT and their calcination products. The formation of $Co_x(II)$-$Co_y(III)$-$Fe_z(III)$ oxides with spinel-like structure and different cation molar ratios, depending on the temperature and time of calcination, was widely studied [45,46]. A Ni-Fe-carbonate HT was prepared by testing two different aging treatments and calcining temperatures to each original sample [47]. X-ray absorption results for calcined samples showed that Ni(II) cations were located in octahedral sites, while Fe(III) cations were in octahedral and tetrahedral sites in the same proportion. Coordination parameters at the first and second shells of Ni(II) and Fe(III) cations for the sample calcined at 750 °C coincided with those expected for a mixture of NiO and the $NiFe_2O_4$ spinel, phases detected by X-ray diffraction [47]. For the sample calcined at 450 °C, although PXRD only detected crystalline NiO, X-ray absorption measurements also indicated that Fe(III) ions were forming an amorphous phase. This study also allowed us to know that the reducibility of cations in the sample calcined at 450 °C depended on the synthesis method and the formation of the spinel phase. This article has also received large attention (109 Web of Science citations, and 148 GoogleScholar citations up to May 2022). Recently, it has been used as a reference in a review dealing with the use of hydrotalcite as adsorbents for the removal of heavy metal anions as water contaminants [48].

To study systems with more than two cations in the layer and to test the nature of derived mixed oxides containing yttrium obtained by its calcination, a new Mg,Al,Y-LDH with the hydrotalcite-like structure as a precursor of mixed oxides was prepared. This study was based on industry interest in yttrium garnets as gemstones and as magnetic materials. The synthesis was carried out by coprecipitation of Mg(II), Al(III), and Y(III) cations in alkaline solutions. The nature of mixed oxides obtained by calcination at increasing temperatures was assessed meanly by X-ray diffraction [49].

Other type of LDH is the so-called hydrocalumite, which contains Ca (II) and Al (III) as layer cations and chloride as interlayer anion. For the synthesis of this compound, Al was recovered from a saline slag generated during the recycling of this element, trying to minimize the contamination and valorize this residue [50]. The synthesis of hydrocalumite was favored by using microwave radiation. The optimum aging temperature under MW irradiation was 125 °C, as it allowed to obtain pure hydrocalumite, with high crystallinity and composed of LDH-type aggregates of regular hexagonal particles. The high crystallinity of the solids justified their low S_{BET} values. Globally, solids with textural properties comparable to those prepared from pure commercial reagents, can be prepared by using the aluminum slag, which supposes a high added value for this residue. [50]. Depending on the conditions, hydrocalumite was accompanied by impurities of calcite and katoite, the thermal behavior of this system has been studied in detail by a combination of TG, DTA, and analysis of the evolved gases (EGA) by means of mass spectrometry (MS) [51].

3. Applications

The catalytic activities for acetaldehyde self-condensation and 2-methylbut-3-yn-2-ol conversion of both Mg-Al and Li-Al HT with carbonate, chloride, nitrate, or vanadate as anions were studied by Rives et al. [52]. The solids were tested as-synthesized and after calcination at different temperatures. The authors concluded that the surface acidity that was measured using Hammett indicators depended on the nature of the anions. Another conclusion was that the host matrix had no major effect on the acidity or the catalytic activity, although the MgAl materials showed higher thermal stability. They also confirmed that the activity was related mostly to the nature of the interlayer anion, and higher catalytic activity was found for the solids containing vanadate.

As hydrotalcite-like solids act as catalysts precursors, some catalytic reactions have been tested using the mixed oxides obtained upon calcination of synthetic LDH. In 1997, MgCr-, NiCr- and NiAl-Vanadate hydrotalcites with decavanadate as interlayer anion were prepared. The original solids were calcined and then the surface acidity was studied by monitoring the adsorption of pyridine by FT-IR spectroscopy and tested for 2-propanol oxidative dehydrogenation. The samples NiAl-vanadate presented the highest acidity while the selectivity to the oxidative dehydrogenation was larger on the MgCr- Vanadate, and NiCr- Vanadate samples calcined at 300 °C [53]. The isopropanol oxidation reaction on calcined LDH with different interlayer anions (carbonate, nitrate, silicate, and borate), was studied in 2001 [53]. The original solids were submitted at 600 °C and their surface acidity was determined by ammonia adsorption; surface Lewis acid sites and Brönsted basic sites were studied. The samples derived from carbonate and nitrate hydrotalcites showed the highest activity for acetone formation due to their surface acidity and basicity. In contrast, samples derived from silicate- or borate-HT showed lower activity due to the lack of surface basic sites [54]. MgAl HT was also used in a new and green methodology of synthesis of two chiral building blocks. This synthesis implied a regioselective Baeyer-Villiger reaction and can be considered a green procedure because hydrogen peroxide was the only oxidant used and the catalyst was a non-contaminant MgAl-carbonate HT [55]. Furthermore, the use of carvone, a cheap commercial compound, enhances the value of this methodology (Figure 3).

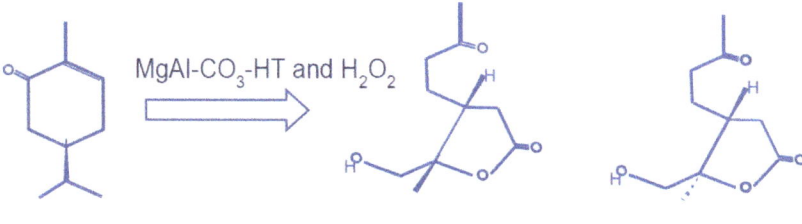

Figure 3. Green synthesis of two chiral building blocks by using hydrotalcites as catalysts.

A wide catalytic study was centered on the influence of Zn content of Ni–Cr–Al mixed-oxide catalysts with a spinel-like structure on the catalytic performance in acetylene hydrogenation. The physicochemical properties of these oxides were also analyzed. The catalysts derived from the calcination of several HT with different cationic molar ratios and formula $(Zn_xNi_yAl_zCr_t (OH)_2) (CO_3)_{z+t/2}$ mH_2O, and the joint action of ZnO and Cr_2O_3, as activity modifier and support modifier, respectively, was tested [56–58].

Continuing with the use of calcined HT for the acetylene hydrogenation reaction, Monzón et al. compared the previous catalysts but doped with Fe(III) instead of Cr(III) [59]. The researchers tested the effect of the Ni/Zn molar ratio on the activity, selectivity, and coke formation of NiO-ZnO-Al_2O_3 catalysts depending on the Fe/Al cationic molar ratio. They concluded that Ni was not the only one responsible for the performance of the catalyst and that the presence of ZnO gave rise to a significant decrease in the coke formation (Figure 4). The same effect was observed when the catalysts contained Cr(III) instead of Fe(III).

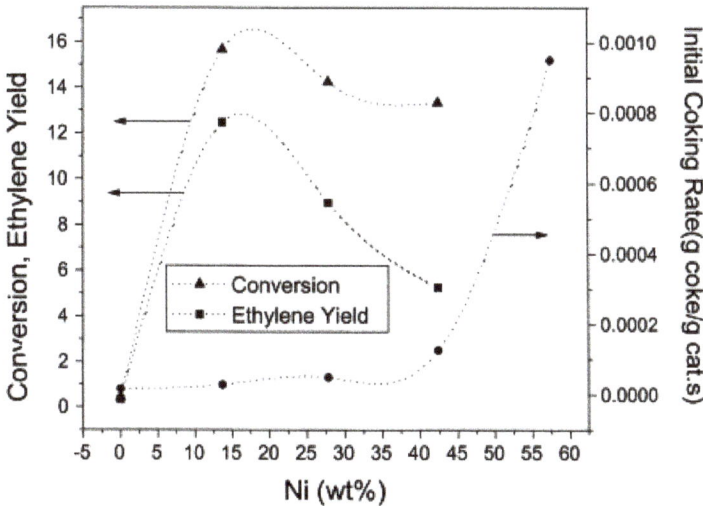

Figure 4. Conversion and ethylene yield (left) and initial rate of coke formation (right), versus the Ni loading in the different catalysts derived from calcined HT. Reprinted from Appl. Clay Sci. 1998, 13, 363–379. Copyright (1998) with permission from Elsevier.

In these works, the authors assessed that the calcination of HT was a successful method to obtain highly dispersed mixed oxides. On the other hand, they concluded that the selectivity in the hydrogenation of acetylene, a structure-sensitive reaction, and the coke formation, can be easily modified by varying the chemical composition of the catalysts [59]. Otherwise, the influence of the Ni/Mg cationic molar ratio on the precursor HT was linked with the variation of the selectivity, the activity, and coke formation of NiO-MgO-Al_2O_3 catalysts modified with Cr(III) for acetylene hydrogenation [60]. An optimal

Ni/Mg molar ratio was found, also observing that the addition of MgO can modulate the catalytic properties of Ni because the conversion, selectivity, and yield to ethylene increased when the Mg/Ni ratio increased, and coke formation simultaneously decreased.

Another catalytic reaction tested using hydrotalcite-derived catalysts was hydroxylation of phenol. For this purpose, Rives et al. synthesized CuCoAl ternary HT with (Cu + Co)/Al atomic ratio of 3.0 and different Cu/Co atomic ratios from 80:20, to 20:80 [61]. The original solids were obtained by the coprecipitation method under low supersaturation. These solids showed a layered hydrotalcite-like structure and their crystallinity increased with the copper concentration. Another remarkable aspect was the increase in the total pore volume and the specific surface area with cobalt content. Calcination of the original hydrotalcites at 500 °C gave rise to the solids whose PXRD patterns corresponded with those of the tenorite and spinel phases. These oxides were tested in the liquid phase hydroxylation of phenol using H_2O_2 as an oxidant. The higher activity was presented by the calcined solid with a Cu/Co atomic ratio of 75:25 It was also concluded that an increase in substrate/catalyst ratio enhanced the conversion of phenol. Using oxidants other than H_2O_2 and solvents other than water did not show measurable conversion of phenol. The authors confirmed the influence of the presence of cobalt and of the surface properties on the activity of copper in the control of the development of the reaction [61].

Mg-V-hydrotalcites were synthesized with different Mg/V molar ratios and then calcined at 800 °C to be tested for oxidative dehydrogenation of propane and n-butane. The relative amounts of $Mg_3(VO_4)$ and MgO determined the formation of different vanadates, which influenced the performance of the oxides obtained in oxidative dehydrogenation of propane and n-butane [62].

Thanks to their high specific surface area, another application of the HT and their calcination products is the adsorption and removal of pollutant compounds. Thus, Trujillano et al. used HT solids to eliminate the salts responsible for the degradation of the stones of the cultural heritage as well as for their cleaning [63]. In this work, the authors used the MgAl calcined hydrotalcite as an absorbent of anionic contaminants of the stone as a complement of the conventional clays (sorbents of cationic contaminants) that were used by the same research group as sorbents of cationic pollutants and salts [64]. Both surveys demonstrated the possibility of using effective, cheap, non-aggressive, and green procedures to remove pollutants and salts. The authors proved that hydrotalcites and conventional clays are highly effective agents for cleaning stone material in an easy way. This investigation was part of a European scientific and multidisciplinary collaboration project between researchers from a consolidated research unit formed by the Instituto de Recursos Naturales y Agrobiología at Salamanca (IRNASA-CSIC) and the University of Salamanca, dedicated to the cleaning and protection of the artistic heritage in stone, and the Spanish research team was headed by Professor Rives.

Owing to their wide range of chemical composition, HT can show a wide range of colors after calcination. It is well known that on heating the hydrotalcites decompose forming mostly amorphous species when calcining at low temperatures. When the calcination temperature increases, the crystallization of the divalent oxide is first observed, and at higher temperatures segregation of spinel takes place. If the mixed oxides obtained contain transition metal cations, they can be applied as ceramic pigments [65]. With the purpose of starting a study on the use of HT as pigments, a series of MgAl-carbonate HT with Mg/Al molar ratios of 4/1, 2/1, $\frac{1}{2}$, and 1/4 were prepared by the coprecipitation method to ascertain the different steps of the degradation of this kind of solids under calcination depending on their composition. The solids obtained had as formula $[Mg_xAl_{1-x}(OH)_2]$ $(CO_3{}^{2-})_{(x/2)} \cdot nH_2O$ and the optimum x value for hydrotalcite formation was between 0.20 and 0.35. The thermal decomposition of the solids obtained was systematically studied by following the amount of Al in tetrahedral positions and the inversion degree of the spinel obtained at each temperature was determined to be higher as the temperature of calcination increased [65].

Other hydrotalcites prepared by precipitation at constant pH to be used as pigments were those containing Mg (II) and Al(III), doped with 0.5%, 1%, 2%, or 3% of Cr(III) and Y(III) [66]. The solids were calcined at 1200 °C for 5 h in air to give solids with a mixed structure (spinel and rock salt for MgO). Their color was pink, which made these solids suitable for being used as ceramic pigments. The incorporation of Cr in the structure of the spinel modified somewhat the chromaticity coordinates, but quantitative changes follow different trends. Nevertheless, these changes were qualitatively like those observed for the original LDH, so the authors found it easy to predict the expected color of the calcined solids by monitoring the color of the uncalcined precursors [66].

To test the incidence of the preparation method in the final color of the pigments obtained by calcination of HT, nickel-aluminum HT was prepared by conventional coprecipitation and by coprecipitation in the presence of a surfactant. Larger and better-ordered crystals were obtained when the synthesis method was the inverse micelles route [67]. Calcination of the original samples gave rise to the formation of homogeneously dispersed mixed oxides. The lightness and chromaticity coordinates of the solids (HT precursors and calcined solids) were analyzed to ascertain if the preparation procedure influenced the final color. Both the preparation method and the calcination treatment had a little effect on the precise chromaticity coordinates (green/red and blue/yellow) and an important effect on the luminosity (whiteness/darkness) of the solids. The color coordinates of these solids showed that an enhancement of the green color was observed when the HT was prepared in an organic medium. After calcination, this solid also showed a larger luminosity than the solid derived from the HT prepared by conventional coprecipitation [67].

As the thermal decomposition of HT leads to highly dispersed mixed oxides, the precise color of the final oxides can be tuned by controlling the precise chemical composition of LDH precursors containing transition metal cations in the layers. Rives et al. synthesized hydrotalcites containing Co(II) and Cr(III) cations with Co/Cr molar ratios ranging from 2 to 0.5, which were tested as pigments upon calcination. The authors developed the synthesis of some selected samples whose colors were of better quality than two commercial black pigments [68].

The preparation of HT-containing lanthanide cations has also been reported. A series of Mg-Al-carbonate HT with 4% loading of different lanthanide cations in the brucite-like layers was reported [69]. All the original mixtures had an M(II)/M(III) molar ratio of 3, and among the trivalent cations, different Ln/Al molar ratios were considered: 4/96 for Tb-Al, Er-Al, and Yb-Al HT, and 2/2/96 for Er-Yb-Al HT. The as-synthesized solids showed the hydrotalcite-type structure, without contaminating phases and they were calcined at 1000 °C forming perfectly dispersed mixed oxides, with the MgO and the $MgAl_2O_4$ spinel structure and without any differentiated phase containing lanthanide ions. All samples prepared and their calcined products were white, except the Er (III)-containing samples, which revealed a weak pink color. The analysis of the color of all the samples was made by determining the colorimetric coordinates L*a*b* of the CIE system (International Commission on Illumination). All the samples containing Tb(III) were submitted to luminescence studies, exhibiting green fluorescence under irradiation of 254 and 365 nm. The emission spectra showed a series of narrow lines ascribed to Tb(III) transitions. The decay curves monitored at 543 nm indicated the presence of a single local Tb(III) environment in the parent and calcined samples. For the calcined samples, the photoluminescence evidence supported the insertion of Tb(III) in $MgAl_2O_4$ rather than in MgO [69].

The synthesis, structural characteristics, and applications of some Cu(II)-containing HT were also depicted. Originally, due to the difficulty in the obtaining of hydrotalcites with Cu (II) as the only divalent cation, ethylene glycol was used to reduce the divalent cations of a CuAl-LDH and a NiAl-LDH by means of the so-called polyol process. In this test, it was found that the degree of reduction of both cations to their metallic forms depended on the nature of the interlayer anions as well as on the time and temperature of treatment of the reaction mix [70]. The synthesis, structural and textural characteristics, and the catalytic performance on the hydroxylation of phenol of CuNiAl and CuCoAl hydrotalcites were described by Kannan and Rives, together with other researchers [71–73]. These authors also studied the transformation of Cu-rich hydrotalcites (Cu + M(II))/Al (with M(II) = Co, Ni, or Mg) when calcined at high temperatures, observing considerable differences in the thermal change temperatures of these samples depending on the co-divalent metal cations [61].

The synthesis under several procedures of NiGa, NiAl, CuAl, or CuAlFe hydrotalcites with carbonate or surfactants as interlayer anions was reported trying to establish a relationship between their properties and the preparation method. The surfactants used were alkyl sulfonates of organic chains with different lengths or alkyl-benzene sulfonates. These materials were submitted to calcination and the thermal effects at increasing temperatures and the solids formed were defined [74–79]. The magnetic properties of CuAl LDH with carbonate or anionic surfactant were also investigated [78], and low oxidation states of copper (Cu_2O or Cu) were obtained over a mixed oxide phase when Cu(II)Al (III)-LDH containing organic sulfonates in the interlayer were calcined. This reduction of Cu(II) was not possible when the interlayer anion was carbonate, and this process was not observed for cations other than Cu(II) [79]. Ternary hydrotalcite containing Ni, Cu, and Al cations with different cationic molar ratios were also synthesized to measure the effect of copper on the reaction of isomerization of eugenol, observing a decrease in the activity when increasing the amount of copper [80].

The use of hydrotalcites as drug carriers for controlled release is a research line that V. Rives began to study at the beginning of this century. In 2004, MgAl LDH with naproxen in the interlayer was prepared by using two synthetic procedures: reconstruction and coprecipitation. From the X-ray results discussion, the authors concluded that the drug was incorporated in the interlayer of the solids with a tilted bilayer orientation with the carboxylate groups linked to the brucite-like sheets [81].

The intercalation of naproxen in LDHs by coprecipitation and anion-exchange was also studied on Mg-Al-Fe HT [82]. The drug release rate was tested in vitro, to ascertain if the LDH can act as an additive or as a matrix. The intercalation of the drug forming bilayers and the possibility of its exchange in the biological medium led to a release much slower than when the LDH was only physically mixed with the drug [82]. Later, Rives et al. collected the existing literature on hydrotalcites–drugs interactions in two bibliographic reviews, published in 2013 and 2014, respectively [83,84].

The studies on CO_2 sorption on hydrotalcites started by using calcined HT as sorbents, research done in collaboration with researchers from the Universade de Porto (Portugal). In this case, the preoccupation with the production of huge amounts of CO_2 from the combustion of fossil fuels, and the greenhouse effect of this gas, justified the use of calcined hydrotalcites as CO_2 adsorbents, finding high efficiency. This suitability was due to the ability of these solids to operate in the temperature interval found in both post- and pre-combustion streams (i.e., 200–400 °C) [85–88].

So as to ascertain the incidence of water in the CO_2 capture, a calcined Mg-Al-Ga HT impregnated with K, aged under microwave irradiation was checked (Figure 5). It was observed that the presence of water significantly increased the CO_2 capture. So did higher total pressures and temperatures. Finally, Energy-dispersive X-ray spectroscopy results showed that under wet conditions potassium and gallium were mobilized towards the sorbent surface and this fact contributed to enhancing the sorbent behavior of the HT.

Different procedures of synthesis, the characterization results, and the thermal properties of composites having hydrotalcites as nanofillers can be found in the literature of this research group [89–94]. The preparation by in situ polymerization was tested with the synthesis of a polyethylenterephthalate-hydrotalcite by using polyethylenterephthalate and different amounts of a hydrotalcite with dodecylsulphate as interlayer anion. The characterization results permitted conclude that the maximum amount of inorganic additive that could be well dispersed was between 2–10%, while higher loadings did not allow a complete dispersion and formed agglomerates [90]. The in-situ polymerization method was proved to be the best one to obtain PET-LDH composites with low contents of

LDHs. The PET-LDH composites thus prepared were thermally more stable than similar composites prepared by blending [90].

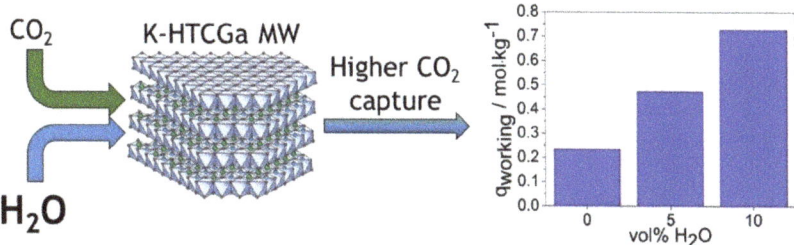

Figure 5. CO2 sorption experiments over Mg–Al–Ga HT. Reprinted from J. Ind. Eng. Chem. 2019, 72, 491–503. Copyright (2019) with permission from Elsevier.

To evaluate the influence of the hydrotalcite as nanofillers in a polyester resin, two LDH with organic interlayer anions, adipate-LDH and 2-methyl-2-propene-1-sulfonate-LDH were synthesized and characterized. Flexural tests showed that incorporation of organo-LDH in the resin reduced the flexural strength of the resin. The study of fire reaction properties indicated up to 46% reduction in the polyester flammability [90]. Polyamide6.6/Mg,Al/adipate-LDH nanocomposites were prepared by solid state polymerization [92] and the structural characterization and thermal measurements results led to the conclusion that the best dispersion and the best thermal stability of the nanocomposites were achieved for the composite with 0.1% LDH content, and it was higher than that of pure polyamide [92].

The interest in the synthesis of hydrotalcites for the remediation of soils and groundwater contamination was evidenced by the research in the synthesis of MgAl and CaAl HT labeled with fluorescein by intercalation of this molecule [95,96]. These fluorescent solids can act as tracer particles for in situ remediation strategies. Their particle size was tuned by varying the synthesis conditions and aging time with an easy and economic synthesis method. If co-injected with reactive particles showing similar properties, the reactants could easily be detected, thanks to the fluorescent particles, in the subsurface, and their potential movement and spreading rate, caused by groundwater flow, could be followed [95,96].

Hydrotalcites with intercalated oxidizing agents have been proved to act as reactants for remediation strategies because HT serves as supports for the oxidizing agents during their injection to penetrate aquifers. So, the efficiency of LDH intercalated with permanganate and peroxydisulphate anions was tested in batch experiments using trichloroethene or 1,1,2-trichloroethane as the target contaminants. The oxidation processes produced by the intercalated oxidizing agents gave rise to the degradation of the contaminants by the HT-based solids [97].

Hydrotalcite-like compounds treated with acetone to modify their surface area and particle aggregate can be used as sorbents of ecosystems pollutants as chlorinated hydrocarbons [98] or other toxic chemicals such as perfluorooctane sulfonate and perfluorooctanoate [99]. These tuned solids have proved to be powerful and interesting toxic sorbents because of their high anion exchange capacity and high specific surface area [100].

Different ways of preparing hydrotalcite and the phenomena occurring during their synthesis continue being studied [100–102]. So, the synthesis of ZnAl-nitrate LDH has recently been carried out in the presence of methylamine, dimethylamine, and trimethylamine, and under microwave hydrothermal treatment. Solids with high crystal sizes and low particle size distributions were obtained [100]. ZnAl-carbonate HT and the oxides obtained from their calcination have also been proved to be efficient in the photodegradation of 4-nitrophenol, one of the most common contaminants in industrial waters [102]. The catalysts were tested as prepared and after calcination at 650 °C, a treatment that gave rise to the formation of a ZnO phase dispersed on an Al_2O_3 amorphous phase. The best performance was found for the calcined solid, which removed the contaminant by adsorption-degradation. The high specific surface area of the solid led to a high adsorption capacity and the high dispersion of its active form allowed a higher degradation power than commercial ZnO that was tested as reference.

4. Conclusions

The work on hydrotalcites done by Professor Rives and the research team that has worked with him is not easy to summarize. Innovation in hydrotalcite synthesis methods has resulted in the knowledge of procedures that lead to obtaining hydrotalcites tailored to the final application. This fact is demonstrated by the extensive and detailed characterization work carried out for more than thirty years. Likewise, the usefulness of these solids as catalysts, catalyst supports, contaminant adsorbents, etc., has been constantly verified. The contribution to the improvement of synthesis procedures of solids that allow for reducing pollution and energy expenditure has been constant in the research projects on hydrotalcites. Such work has been published in several scientific journals and the publications have been widely studied and cited by researchers around the world.

Funding: This research received no external funding.

Acknowledgments: The editors of this Special Issue, M.A. Vicente, F.M. Labajos and R. Trujillano, we are extremely grateful to have had Vicente Rives as head of our Group. Vicente, we acknowledge you for everything, for every moment, for every idea shared, for every project, for every correction, for every one of the talks, courses, coffees, laughs, classes, and for your guidance in our careers. Thank you for being a boss, a teacher, and a colleague.

Conflicts of Interest: The authors declare no conflict of interest.

References

1. Ulibarri, M.A.; Fernández, J.M.; Labajos, F.M.; Rives, V. Synthesis and characterization of $[Co_{1-x}Al_x(OH)2] (CO_3)_{x/2} \cdot n H_20$]. *Chem. Mater.* **1991**, *3*, 626–630. [CrossRef]
2. Hochstetter, C. Untersuchung Über Die Zusammensetzung einiger Mineralien. *J. Prakt. Chem.* **1842**, *27*, 375–378. [CrossRef]
3. Caillère, S. Sur l'hydrotalcite de Snarum (Norvège). *Bull. Soc. Miner.* **1944**, *67*, 411–419. [CrossRef]
4. Manasse, E. Idrotalcite e piroaurite. *Atti Soc. Toscana Sc. Nat. Proc. Verb.* **1915**, *24*, 92–97.
5. Feitknecht, W.; Gerber, M. Zur Kenntnis der Doppelhydroxyde und basischen Doppelsalze III. Über Magnesium-Aluminiumdoppelhydroxyd. *Helv. Chim. Acta* **1942**, *25*, 131–137. [CrossRef]
6. Allmann, R. The crystal structure of pyroaurite. *Acta Cryst. B* **1968**, *24*, 972–977. [CrossRef]
7. Taylor, H.F.W. Segregation and Cation-Ordering in Sjögrenite and Pyroaurite. *Mineral Mag.* **1969**, *37*, 338–342. [CrossRef]
8. Ross, G.J.; Kodama, H. Properties of a Synthetic Magnesium-Aluminum Carbonate Hydroxide and its Relationship to Magnesium-Aluminum Double Hydroxide, Manasseite and Hydrotalcite. *Am. Mineral.* **1967**, *752*, 1036–1047.
9. Rives, V. (Ed.) *Layered Double Hydroxides: Present and Future*; Nova Science Publishers: New York, NY, USA, 2001.
10. Rives, V.; Ulibarri, M.A. Layered double hydroxides (LDH) intercalated with metal coordination compounds and oxometalates. *Coord. Chem. Rev.* **1999**, *181*, 61–120. [CrossRef]
11. Labajos, F.M.; Rives, V.; Ulibarri, M.A. A FT-IR and V-UV Spectroscopic Study of Nickel-Containing Hydrotalcite-Like Compounds, $[Ni_{1-x}Al_x(OH)_2](CO_3)_{x/2} \cdot nH_2O$. *Spectrosc. Lett.* **1991**, *24*, 499–505. [CrossRef]
12. Reichle, W.T. Synthesis of anionic clay minerals (mixed metal hydroxides, hydrotalcite). *Solid State Ion.* **1986**, *22*, 135. [CrossRef]
13. Rives, V.; Labajos, F.M.; Ulibarri, M.A. Effect of hydrothermal and thermal treatments on the physicochemical properties of Mg-Al hydrotalcite-like materials. *J. Mat. Sci.* **1992**, *27*, 1546–1552.
14. Arco, M.; Rives, V.; Trujillano, R. Surface and textural properties of hydrotalcite-like materials and their decomposition products. *Stud. Surf. Sci. Catal.* **1994**, *87*, 507–515.
15. Arco, M.C.; Martín, C.; Martín, I.; Rives, V.V.; Trujillano, R. A FTIR spectroscopic study of surface-acidity and basicity of mixed Mg,Al-oxides obtained by thermal-decomposition of hydrotalcite *Spectrochim. Acta* **1993**, *49*, 1575–1582.
16. Trujillano, R.; González-García, I.; Morato, A.; Rives, V. Controlling the synthesis conditions for tuning the properties of hydrotalcite-like materials at the nano scale. *Chem. Eng.* **2018**, *2*, 31. [CrossRef]
17. Benito, P.; Labajos, F.M.; Rocha, J.; Rives, V. Influence of microwave radiation on the textural properties of layered double hydroxides. *Micropor. Mesopor. Mat.* **2006**, *94*, 148–158. [CrossRef]
18. Benito, P.; Labajos, F.M.; Rives, V. Microwave-assisted synthesis of layered double hydroxides. In *Solid State Chemistry Research Trends*; Buckley, R.W., Ed.; NOVA Sci. Pub. Inc.: New York, NY, USA, 2007; pp. 173–225.

19. Herrero, M.; Benito, P.; Labajos, F.M.; Rives, V. Change in microporosity of granitic building stones upon stabilisation of Co^{2+} in LDH by microwave assisted ageing. *J. Solid State Chem.* **2007**, *180*, 873–884. [CrossRef]
20. Rives, V.; Labajos, F.M.; Ulibarri, M.A.; Malet, P. A New Hydrotalcite-like Compound Containing V(III) Ions in the Layers. *Inorg. Chem.* **1993**, *32*, 5000–5001. [CrossRef]
21. Labajos, F.M.; Rives, V.; Malet, P.; Centeno, M.A.; Ulibarri, M.A. Synthesis and characterisation of hydrotalcite-like compounds containing V(III) in the layers, and of their calcination products. *Inorg. Chem.* **1996**, *35*, 1154–1160. [CrossRef]
22. Labajos, F.M.; Sastre, M.-D.; Trujillano, R.; Rives, V. New layered double hydroxides with the hydrotalcite structure containing Ni(II) and V(III). *J. Mater. Chem.* **1999**, *9*, 1033–1039. [CrossRef]
23. Labajos, F.M.; Rives, V. Thermal evolution of Cr(III) ions in hydrotalcite-like compounds. *Inorg. Chem.* **1996**, *35*, 5313–5318. [CrossRef]
24. Rives, V.; Ulibarri, M.A.; Montero, A. Application of temperature-programmed reduction to the characterization of anionic clays. *Appl. Clay Sci.* **1995**, *10*, 83–93. [CrossRef]
25. Ulibarri, M.A.; Labajos, F.M.; Rives, V.; Trujillano, R.; Kagunya, W.; Jones, W. Comparative Study of the Synthesis and Properties of Vanadate-Exchanged Layered Double Hydroxides. *Inorg. Chem.* **1994**, *33*, 2592–2599. [CrossRef]
26. Kooli, F.; Rives, V.; Ulibarri, M.A. Preparation and Study of Decavanadate-Pillared Hydrotalcite-like Anionic Clays Containing Transition Metal Cations in the Layers. 1. Samples Containing Nickel-Aluminum Prepared by Anionic Exchange and Reconstruction. *Inorg. Chem.* **1995**, *34*, 5114–5121. [CrossRef]
27. Ulibarri, M.A.; Labajos, F.M.; Rives, V.; Trujillano, R.; Kagunya, W.; Jones, W. Effect of intermediates on the nature of polyvanadate-intercalated layered double hydroxides. *Mol. Cryst. Liq. Cryst.* **1994**, *244*, 167–172. [CrossRef]
28. Kooli, F.; Rives, V.; Ulibarri, M.A. Preparation and Study of Decavanadate-Pillared Hydrotalcite-like Anionic Clays Containing Transition Metal Cations in the Layers. 2. Samples containing Magnesium-Chromium and Nickel-Chromium. *Inorg. Chem.* **1995**, *34*, 5122–5128. [CrossRef]
29. Crespo, I.; Barriga, C.; Ulibarri, M.A.; González-Bandera, G.; Malet, P.; Rives, V. An X-ray diffraction and absorption study of the phases formed upon calcination of Zn-Al-Fe hydrotalcites. *Chem. Mat.* **2001**, *15*, 1518–1527. [CrossRef]
30. Kooli, F.; Rives, V.; Ulibarri, M.A.; Jones, W. Pillaring of layered double hydroxides possessing variable layer charge with vanadate polyoxoanions. In *Advances in Porous Materials*; Materials Research Society, Sympisum Proceedings; Komarneni, S., Smith, D.M., Beck, J.S., Eds.; Cambridge University Press: Cambridge, UK, 1995; Volume 371, pp. 143–149.
31. Arco, M.; Galiano, M.V.G.; Rives, V.; Trujillano, R.; Malet, P. Preparation and study of decavanadate-pillared hydrotalcite-like anionic clays containing cobalt and chromium. *Inorg. Chem.* **1996**, *35*, 6362–6372. [CrossRef]
32. Arco, M.; Rives, V.; Trujillano, R.; Malet, P. Thermal behaviour of Zn, Cr layered double hydroxides with the hydrotalcite-like structure containing carbonate or decavanadate. *J. Mat. Chem.* **1996**, *6*, 1419–1428. [CrossRef]
33. Kooli, F.; Holgado, M.; Rives, V.; San Román, M.; Ulibarri, M.A. A simple conductivity study of decavanadate intercalation in hydrotalcite. *Mat. Res. Bull.* **1997**, *32*, 977–982. [CrossRef]
34. Kooli, F.; Jones, W.; Rives, V.; Ulibarri, M.A. An alternative route to polyoxometalate-exchanged layered double hydroxides: The use of ultrasounds. *J. Mat. Sci. Lett.* **1997**, *16*, 27–29. [CrossRef]
35. Barriga, C.; W Jones, W.; Malet, P.; Rives, V.; Ulibarri, M.A. Synthesis and characterisation of polyoxovanadate-pillared Zn, Al layered double hydroxides: An X-ray absorption and diffraction study. *Inorg. Chem.* **1998**, *37*, 1812–1820. [CrossRef]
36. Crespo, I.; Barriga, C.; Rives, V.; Ulibarri, M.A. Intercalation of iron hexacyano complexes in Zn,Al hydrotalcite. *Solid State Ion.* **1997**, *101*, 729–735. [CrossRef]
37. Arco, M.; Gutierrez, S.; Martin, C.; Rives, V.; Rocha, J. Effect of the Mg/Al ratio on borate (or silicate)/nitrate exchange in hydrotalcite. *J. Solid State Chem.* **2000**, *151*, 272–280. [CrossRef]
38. Arco, M.; Gutierrez, S.; Martin, C.; Rives, V. Intercalation of $[Cr(C_2O_4)_3]^{3-}$ complex in Mg,Al layered double hydroxides. *Inorg. Chem.* **2003**, *42*, 4232–4240. [CrossRef] [PubMed]
39. Rocha, J.; Arco, M.; Rives, V.; Ulibarri, M.A. Reconstruction of layered double hydroxides from calcined precursors: A powder XRD and ^{27}Al MAS NMR study. *J. Mater. Chem.* **1999**, *9*, 2499–2503. [CrossRef]
40. Lahkale, R.; Sadik, R.; Elhatimi, W.; Bouragba, F.Z.; Assekouri, A.; Chouni, K.; Rhalmi, O.; Sabbar, E. Optical, electrical and dielectric properties of mixed metal oxides derived from Mg-Al Layered Double Hydroxides based solid solution series. *Phys. B Cond. Matt.* **2022**, *626*, 413367. [CrossRef]
41. Barriga, C.; Kooli, F.; Rives, V.; Ulibarri, M.A. Layered hydroxycarbonates with the hydrotalcite structure containing Zn, Al and Fe. In *Synthesis of Porous Materials*; Occelli, M.L., Kessler, H., Eds.; Marcel Dekker Inc.: New York, NY, USA, 1996; pp. 661–674.
42. Fernández, J.M.; Barriga, C.; Ulibarri, M.A.; Labajos, F.M.; Rives, V. Preparation and thermal-stability of manganese-containing hydrotalcite, $[Mg_{0.75}Mn(II)_{0.04}Mn(III)_{0.21}(OH)_2] (CO_3)_{0.11} \cdot NH_2O$. *J. Mat. Chem.* **1994**, *4*, 1117–1121. [CrossRef]
43. Barriga, C.; Fernández, J.M.; Ulibarri, M.A.; Labajos, F.M.; Rives, V. Synthesis and characterisation of new hydrotalcite-like compounds containing Ni and Mn in the layers and of their calcination products. *J. Solid State Chem.* **1996**, *124*, 205–213. [CrossRef]
44. Fernández, J.M.; Ulibarri, M.A.; Labajos, F.M.; Rives, V. The effect of iron on the crystalline phases formed upon thermal decomposition of Mg-Al-Fe hydrotalcites. *J. Mat. Chem.* **1998**, *8*, 2507–2517. [CrossRef]
45. Arco, M.; Trujillano, R.; Kassabov, S.; Rives, V. Spectroscopic properties of Co-Fe hydrotalcites. *Spectrosc. Lett.* **1998**, *31*, 859–869. [CrossRef]

46. Arco, M.; Rives, V.; Trujillano, R. Cobalt-iron hydroxycarbonates and their evolution to mixed oxides with the spinel structure. *J. Mater. Chem.* **1998**, *8*, 761–767. [CrossRef]
47. Arco, M.; Malet, P.; Trujillano, R.; Rives, V. Synthesis and Characterization of Hydrotalcites Containing Ni(II) and Fe(III) and Their Calcination Products. *Chem. Mater.* **1999**, *11*, 624–633. [CrossRef]
48. Dong, Y.; Kong, X.; Luo, X.; Wang, H. Adsorptive removal of heavy metal anions from water by layered double hydroxide: A review. *Chemosphere* **2022**, *162*, 134685. [CrossRef]
49. Fernández, J.M.; Barriga, C.; Ulibarri, M.A.; Labajos, F.M.; Rives, V. New hydrotalcite-like compounds containing yttrium. *Chem. Mat.* **1997**, *9*, 312–318. [CrossRef]
50. Jiménez, A.; Misol, A.; Morato, A.; Rives, V.; Vicente, M.A.; Gil, A. Optimization of hydrocalumite preparation under microwave irradiation for recovering aluminium from a saline slag. *Appl. Clay Sci.* **2021**, *212*, 10621. [CrossRef]
51. Jiménez, A.; Rives, V.; Vicente, M.A. Thermal study of the hydrocalumite–katoite–calcite system. *Thermochim. Acta* **2022**, *713*, 179242. [CrossRef]
52. Chisem, I.; Jones, W.; Martín, C.; Martín, C.; Rives, V. Probing the surface acidity of lithium aluminum and magnesium aluminum layered double hydroxides. *J. Mater. Chem.* **1998**, *8*, 1917–1926. [CrossRef]
53. Kooli, F.; Martín, C.; Rives, V. FT-IR spectroscopy study of surface acidity and iso-propanol decomposition on mixed oxides obtained upon calcination of layered double hydroxides. *Langmuir* **1997**, *13*, 2303–2306. [CrossRef]
54. del Arco, M.; Gutiérrez, S.; Martín, C.; Rives, V. FTIR study of isopropanol reactivity on calcined layered double hydroxides. *Phys. Chem. Chem. Phys.* **2001**, *3*, 119–126. [CrossRef]
55. Rodilla, J.M.; Neves, P.P.; Pombal, S.; Rives, V.; Trujillano, R.; Díez, D. Hydrotalcite catalysis for the synthesis of new chiral building blocks. *Nat. Prod. Res.* **2016**, *30*, 834–840. [CrossRef] [PubMed]
56. Rives, V.; Labajos, F.M.; Trujillano, R.; Romeo, E.; Royo, C.; Monzón, A. Acetylene hydrogenation on Ni–Al–Cr oxide catalysts: The role of added Zn. *Appl. Clay Sci.* **1998**, *13*, 363–379. [CrossRef]
57. Romeo, E.; Royo, C.; Monzón, A.; Trujillano, R.; Labajos, F.M.; Rives, V. Study of mixed oxides prepared from hydrotalcite-type precursors as hydrogenation catalysts. In *Actas del 16 Simposio Iberoamericano de Catálisis*; Centeno, A., Giraldo, S.A., Páez Mozo, E.A., Eds.; Universidad Industrial De Santander (UIS): Bucaramanga, Colombia, 1998; pp. 567–572.
58. Monzón, A.; Romeo, E.; Royo, C.; Trujillano, R.; Labajos, F.M.; Rives, V. Desarrollo de óxidos mixtos de ni como catalizadores de hidrogenación selectiva (Development of Ni mixed oxides as catalysts for selective hydrogenation). *Av. Ing. Química* **1998**, *8*, 24–27.
59. Monzón, A.; Romeo, E.; Royo, C.; Trujillano, R.; Labajos, F.M.; Rives, V. Use of hydrotalcites as catalytic precursors of multimetallic mixed oxides. Application in the hydrogenation of acetylene. *Appl. Catal. A Gen.* **1999**, *185*, 53–63. [CrossRef]
60. Romeo, E.; Royo, C.; Monzón, A.; Trujillano, R.; Labajos, F.M.; Rives, V. Preparation and characterisation of Ni-Mg-Al hydrotalcites as hydrogenation catalysts. *Stud. Surf. Sci. Catal.* **2000**, *130*, 2099–2104.
61. Rives, V.L.; Dubey, A.; Kannan, S. Synthesis, Characterization and catalytic hydroxylation of phenol over CuCoAl ternary hydrotalcites. *Phys. Chem. Chem. Phys.* **2001**, *3*, 4826–4836. [CrossRef]
62. Holgado, M.J.; Labajos, F.M.; Montero, M.J.S.; Rives, V. Thermal decomposition of Mg/V hydrotalcites and catalytic performance of the products in oxidative dehydrogenation reactions. *Mat. Res. Bull.* **2003**, *38*, 1879–1891. [CrossRef]
63. Trujillano, R.; Vicente, M.A.; Rives, V. Utilización de arcillas catiónicas y aniónicas en la limpieza y desalado del patrimonio artístico (The use of cationic and anionic clays for cleaning and salt removal of cultural heritage). In *Actas del IV Congreso de Rehabilitación del Patrimonio Arquitectónico y Edificación*; Fernández Matrán, M.-A., Castro Castellano, J., Eds.; Centro Internacional para la Conservación del Patrimonio (CICOP España): La Habana, Cuba, 1998; pp. 192–193.
64. Trujillano, R.; García-Talegón, J.; Iñigo, A.C.; Vicente, M.A.; Rives, V.; Molina, E. Removal of salts from granite by sepiolite. *Appl. Clay Sci.* **1995**, *9*, 459–463. [CrossRef]
65. Nebot-Diaz, I.; Rives, V.; Rocha, J.; Carda, J.B. Thermal decomposition study of hydrotalcite-like compounds. *Bol. Soc. Esp. Ceram.* **2002**, *41*, 411–414.
66. García-García, J.M.; Pérez-Bernal, M.E.; Ruano-Casero, R.J.; Rives, V. Chromium and yttrium-doped magnesium aluminum oxides prepared from layered double hydroxides. *Solid State Sci.* **2007**, *9*, 1115–1125. [CrossRef]
67. Pérez-Bernal, M.E.; Ruano-Casero, R.J.; Benito, F.; Rives, V. Nickel–aluminum layered double hydroxides prepared via inverse micelles formation. *J. Solid State Chem.* **2009**, *182*, 1593–1601. [CrossRef]
68. Rives, V.; Pérez-Bernal, M.E.; Ruano-Casero, R.J.; Nebot-Díaz, I. Development of a black ceramic pigment from non-stoichiometric hydrotalcites. *J. Eur. Ceram. Soc.* **2012**, *32*, 975–987. [CrossRef]
69. Vicente, P.; Pérez-Bernal, M.E.; Ruano-Casero, R.J.; Ananias, D.; Almeida-Paz, F.A.; Rocha, J.; Rives, V. Luminescence properties of lanthanide-containing layered double hydroxides. *Micropor. Mesopor. Mat.* **2016**, *226*, 209–220. [CrossRef]
70. Kooli, F.; Rives, V.; Jones, W. Reduction of Ni(II)-Al(III) and Cu(II)-Al(III) layered double hydroxides to metallic Ni(0) and Cu(0) via polyol treatment. *Chem. Mat.* **1997**, *9*, 2231–2235. [CrossRef]
71. Rives, V.; Kannan, S. Layered double hydroxides with the hydrotalcite-type structure containing Cu^{2+}, Ni^{2+} and Al^{3+}. *J. Mat. Chem.* **2000**, *10*, 489–495. [CrossRef]
72. Dubey, A.; Rives, V.; Kannan, S. Catalytic hydroxylation of phenol over ternary hydrotalcites containing Cu, Ni and Al. *J. Molec. Catal. A Gen.* **2002**, *181*, 151–160. [CrossRef]

73. Kannan, S.; Rives, V.; Knözinger, H. High-temperature transformations of cu-rich hydrotalcites. *J. Solid State Chem.* **2004**, *177*, 319–331. [CrossRef]
74. Alvarez, A.; Trujillano, R.; Rives, V. Differently aged gallium-containing layered double hydroxides. *Appl. Clay Sci.* **2013**, *80*, 326–333. [CrossRef]
75. Trujillano, R.; Holgado, M.J.; Rives, V. Alternative synthetic routes for NiAl layered double hydroxides with alkyl and alkylbenzene, sulfonates. *Stud. Surf. Sci. Catal.* **2002**, *142*, 1387–1394.
76. Kloprogge, J.T.; Hickey, H.; Trujillano, R.; Holgado, M.J.; San Roman, M.S.; Rives, V.; Martens, W.N.; Frost, R.L. Characterization of intercalated Ni/Al hydrotalcites prepared by the partial decomposition of urea. *Cryst. Growth Des.* **2006**, *6*, 1533–1536. [CrossRef]
77. Trujillano, R.; Holgado, M.J.; González, J.L.; Rives, V. Cu-Al-Fe layered double hydroxides with CO_3^{2-} and anionic surfactants with different alkyl chains in the interlayer. *Solid State Sci.* **2005**, *7*, 931–935. [CrossRef]
78. Trujillano, R.; Holgado, M.J.; Pigazo, F.; Rives, V. Preparation, physicochemical characterisation and magnetic properties of Cu-Al layered double hydroxides with CO_3^{2-} and anionic surfactants with different alkyl chains in the interlayer. *Physica B* **2006**, *373*, 267–273. [CrossRef]
79. Trujillano, R.; Holgado, M.J.; Rives, V. Obtention of low oxidation states of copper from Cu^{2+}-Al^0 layered double hydroxides containing organic sulfonates in the interlayer. *Solid State Sci.* **2009**, *11*, 688–693. [CrossRef]
80. Jinesh, C.M.; Rives, V.; Carriazo, D.; Antonyraj, C.A.; Kannan, S. Influence of copper on the isomerization of eugenol for as-synthesized NiCuAl ternary hydrotalcites: An understanding through physicochemical study. *Catal. Lett.* **2010**, *134*, 337–342. [CrossRef]
81. Arco, M.; Gutiérrez, S.; Martín, C.; Rives, V.; Rocha, J. Synthesis and characterization of layered double hydroxides (LDH) intercalated with non-steroidal anti-inflammatory drugs (NSAID). *J. Solid State Chem.* **2004**, *177*, 3954–3962. [CrossRef]
82. Del Arco, M.; Fernández, A.; Martín, C.; Rives, V. Release studies of different NSAIDs encapsulated in Mg, Al, Fe-hydrotalcites. *Appl. Clay Sci.* **2009**, *42*, 538–544. [CrossRef]
83. Rives, V.; Arco, M.; Martín, C. Layered double hydroxides as drug carriers and for controlled release of non-steroidal antiinflammatory drugs (NSAIDs): A review. *J. Controll. Rel.* **2013**, *169*, 28–39. [CrossRef]
84. Rives, V.; Arco, M.; Martín, C. Intercalation of drugs in layered double hydroxides and their controlled release: A review. *Appl. Clay Sci.* **2014**, *88–89*, 239–269. [CrossRef]
85. Silva, J.M.; Trujillano, R.; Rives, V.; Soria, M.A.; Madeira, L.M. High temperature CO_2 sorption over modified hydrotalcites. *Chem. Eng. J.* **2017**, *325*, 25–34. [CrossRef]
86. Miguel, C.V.; Trujillano, R.; Rives, V.; Vicente, M.A.; Ferreira, A.F.P.; Rodrigues, A.E.; Mendes, A.; Madeira, L.M. High temperature CO_2 sorption with gallium-substituted and promoted hydrotalcites. *Sep. Purif. Technol.* **2014**, *127*, 202–211. [CrossRef]
87. Silva, J.M.; Trujillano, R.; Rives, V.; Soria, M.A.; Madeira, L.M. Dynamic behaviour of a K-doped Ga substituted and microwave aged hydrotalcite-derived mixed oxide during CO_2 sorption experiments. *J. Ind. Eng. Chem.* **2019**, *72*, 491–503. [CrossRef]
88. Faria, A.C.; Trujillano, R.; Rives, V.; Miguel, C.V.; Rodrigues, A.E.; Madeira, L.M. Alkali metal (Na, Cs and K) promoted hydrotalcites for high temperature CO_2 capture from flue gas in cyclic adsorption processes. *Chem. Eng. J.* **2022**, *427*, 131502. [CrossRef]
89. Martínez-Gallegos, S.; Herrero, M.; Rives, V. Preparation of composites by in situ polymerisation of pet-hydrotalcite using dodecylsulphate. *Mat. Sci. For.* **2008**, *587*, 568–571.
90. Martínez-Gallegos, S.; Herrero, M.; Labajos, F.M.; Barriga, C.; Rives, V. Dispersion of layered double hydroxides in poly(ethylene terephthalate) by in situ polymerization and mechanical grinding. *Appl. Clay Sci.* **2009**, *45*, 44–49. [CrossRef]
91. Pereira, C.M.; Herrero, M.; Labajos, F.M.; Marques, A.V. Rives Preparation and properties of new flame retardant unsaturated polyester nanocomposites based on layered double hydroxides. *Polym. Degrad. Stab.* **2009**, *94*, 939–946. [CrossRef]
92. Herrero, M.; Benito, P.; Labajos, F.M.; Rives, V.; Zhu, Y.D.; Allen, G.C.; Adams, J.M. Structural characterization and thermal properties of polyamide 6.6/Mg,Al/Adipate-LDH nanocomposites obtained by solid state polymerization. *J. Solid State Chem.* **2010**, *183*, 1645–1651. [CrossRef]
93. Rives, V.; Labajos, F.M.; Herrero, M. Effect of preparation procedures on the properties of LDH/organo nanocomposites. In *Nanocomposites: Synthesis, Characterization and Applications*; Wang, X., Ed.; Nova Sci. Pub. Inc.: New York, NY, USA, 2013; pp. 169–202.
94. Rives, V.; Labajos, F.M.; Herrero, M. Layered double hydroxides as nanofillers of composites and nanocomposute materials based on polyethylene. In *Polyethylene Based Blends, Composites and Nanocomposites*; Visakh, P.M., Martínez Morlanes, M.J., Eds.; Wiley Inc.: Hoboken, NJ, USA; Scrivener Pub.: New York, NY, USA, 2015; pp. 163–199.
95. Dietmann, K.M.; Linke, T.; Reischer, M.; Rives, V. Fluorescing Layered Double Hydroxides as Tracer Materials for Particle Injection during Subsurface Water Remediation. *ChemEngineering* **2020**, *4*, 53. [CrossRef]
96. Dietmann, K.M.; Linke, T.; Trujillano, R.; Rives, V. Effect of Chain Length and Functional Group of Organic Anions on the Retention Ability of MgAl-Layered Double Hydroxides for Chlorinated Organic Solvents. *ChemEngineering* **2019**, *3*, 89. [CrossRef]
97. Dietmann, K.M.; Linke, T.; Nogal Sánchez, M.d.; Pérez Pavón, J.L.; Rives, V. Layered Double Hydroxides with Intercalated Permanganate and Peroxydisulphate Anions for Oxidative Removal of Chlorinated Organic Solvents Contaminated Water. *Minerals* **2020**, *10*, 462. [CrossRef]

98. Alonso-de-Linaje, V.; Mangayayam, M.C.; Tobler, D.J.; Dietmann, K.M.; Espinosa, R.; Rives, V.; Dalby, K.N. Sorption of chlorinated hydrocarbons from synthetic and natural groundwater by organo-hydrotalcites: Towards their applications as remediation nanoparticles. *Chemosphere* **2019**, *236*, 124369. [CrossRef]
99. Alonso-de-Linaje, V.; Mangayayam, M.C.; Tobler, D.; Rives, V.; Espinosa, R.; Kim, K.N. Enhanced sorption of perfluorooctane sulfonate and perfluorooctanoate by hydrotalcites. *Environm. Techol. Inn.* **2021**, *21*, 101231. [CrossRef]
100. Misol, A.; Labajos, F.M.; Morato, A.; Rives, V. Synthesis of Zn, Al layered double hydroxides in the presence of amines. *Appl. Clay Sci.* **2020**, *189*, 105539. [CrossRef]
101. Misol, A.; Jiménez, A.; Morato, A.; Labajos, F.M.; Rives, V. Quantification by Powder X-ray Diffraction of Metal Oxides Segregation During Formation of Layered Double Hydroxides. *Eur. J. Eng. Technol. Res.* **2020**, *5*, 1243–1248.
102. Trujillano, R.; Nájera, C.; Rives, V. Activity in the photodegradation of 4-nitrophenol of a Zn, Al hydrotalcite-like solid and the derived alumina-supported ZnO. *Catalysts* **2020**, *10*, 702. [CrossRef]

Review

Overview on Photoreforming of Biomass Aqueous Solutions to Generate H₂ in the Presence of g-C₃N₄-Based Materials

E. I. García-López [1,*], L. Palmisano [2] and G. Marcì [2]

[1] Department of Biological, Chemical and Pharmaceutical Sciences and Technologies (STEBICEF), University of Palermo, Viale delle Scienze, 90128 Palermo, Italy
[2] "Schiavello-Grillone" Photocatalysis Group, Department of Engineering, University of Palermo, Viale delle Scienze, 90128 Palermo, Italy
* Correspondence: elisaisabel.garcialopez@unipa.it

Abstract: Photoreforming (PR) of biomass can be considered a viable technology under mild experimental conditions to produce hydrogen with a high reaction rate using compounds from renewable resources and waste materials. The application of biomass PR gives rise to both hydrogen generation and biomass waste valorization. The process could be scaled up to obtain hydrogen under natural sunlight irradiation, and research on polymeric carbon nitride (g-C_3N_4)-based photocatalysts has been widely carried out in recent years. The non-metallic-based carbon nitride materials are economical and (photo)stable polymer semiconductors, and their physicochemical surface and electronic properties are optimal for obtaining H_2, which can be considered a gas that does not cause major environmental problems. Some hindrances related to their structure, such as the low absorption of visible light and the relatively high recombination rate of electron-hole pairs, restrict the performance; therefore, it is necessary to improve their activity and the yield of the reaction by modifying them in various ways. Various types of solutions have been proposed in this regard, such as, for example, their coupling with other semiconductors to form composite materials. The current mini-review aims to overview the PR field, reporting some of the most interesting papers devoted to understanding the role of g-C_3N_4 in biomass PR. Information on many physico-chemical aspects related to the performance of the process and possible ways to obtain better results than those present up to now in the literature will be reported.

Keywords: polymeric carbon nitride; C_3N_4; H_2; hydrogen production; photoreforming; biomass; photocatalysis

1. Introduction

The significant rise of the Earth's inhabitants and industrial progress has led to increasingly greater use of fossil fuels, with consequent depletion of energy resources and built up of environmental pollution.

Hydrogen is a fuel with a high calorific value (142 MJ·kg^{-1}) which is storable (although some precautions are necessary), clean and ecological, and its combustion leads to the formation of water without coproduction of other gases such as CO_2 or particulate matter [1]. It is predicted to be one of the most important energy carriers of the future and has been produced worldwide in recent years, mainly from steam reforming of natural gas, coal, or crude oil, and only a very small proportion has derived from biofuel reforming and electrolysis [2]. Thermal catalytic reforming of fossil fuels, mainly hydrocarbons, represents the most important source of hydrogen (about 95%), while the alternative use of biomass can be conceived as a second-generation technology. However, a large amount of heat required for the thermal catalytic reforming processes of aqueous solutions of organic compounds is a major obstacle to environmental sustainability; therefore, the development of technologies capable of exploiting solar energy and renewable raw materials to obtain green hydrogen from renewable resources such as water and biomass are welcome [3].

Heterogeneous photocatalysis can play an essential role in the progress of sustainable energy, and it is considered one of the best strategies for transforming solar energy into chemical energy. It is a green and low-cost technology that has demonstrated its potential use for environmental remediation, even if no large-scale applications have been realized and only niche applications can be hypothesized [4]. Since 1972, when Fujishima and Honda reported the possibility of obtaining "green" H_2 by the water-splitting process using a TiO_2 electrode under simulated solar irradiation [5], interest in this technology has increased tremendously. Green hydrogen can be obtained not only by photocatalytic water splitting but also by the photoreforming (PR) of an organic or inorganic substrate.

The photocatalytic reactions are started by the excitation of a semiconductor, which, when exposed to light having an energy equal to or greater than its band gap, causes the promotion of electrons (e^-) from the valence band (VB) to the conduction band (CB). The electron-hole pairs formed can recombine, release energy, or migrate and be trapped by active sites present on the semiconductor surface or by adsorbed species that are reduced or oxidized.

Figure 1 reports three important heterogeneous photocatalytic processes. Their application to water and air remediation has been largely studied [4]. The figure shows the classical application of the photocatalytic technology where the photoproduced electron-hole pairs must give rise to oxidation and a reduction reaction, respectively. The positive holes can react with donor species, often water molecules, which form very oxidant hydroxyl radicals ($\cdot OH$), and contemporaneous electrons can react with adsorbed molecules, usually O_2, because the organic photo-oxidation processes are generally carried out in the air. Consequently, O_2 present in the air can react with electron-producing $\cdot O_2^-$ which in turn reacts with water-affording hydroxyl and hydroperoxide ($\cdot OOH$) radicals. The oxidant species formed unselectively react, giving rise to intermediates and, ultimately, the final products, which are mainly CO_2 and H_2O when the organic molecules in the suspension of the semiconductor only contain C and H. Notably, sulfates or nitrates can be formed from the mineralization of the organic substrate. Under anaerobic conditions, the splitting of water can produce H_2 and O_2 in a process called water photosplitting. This occurs when the energy of the electrons is sufficient to reduce protons to H_2, i.e., when $E_{CB} < E_{(H2/H+)}$, and holes can oxidize water to O_2, i.e., when $E_{VB} > E_{(O2/H2O)}$. Instead of H_2O, an organic molecule can act as a hole trap, and, therefore, the photogenerated holes, which act directly or indirectly through the production of hydroxide radicals, can lead to the production of a partially oxidized derivative of the organic molecule or to its complete decomposition to CO_2 and H_2O. This process, which uses organic molecules as traps for holes, is called photoreforming (PR). In this case, the organic species can be a biomass derivative, which is represented in Figure 1 simply as an organic molecule. Figure 1 also shows merely the photocatalytic oxidation of an organic molecule in the presence of oxygen, and this type of reaction has been studied extensively with the aim of breaking down polluting organic molecules. In the last case, the difference with respect to the photoreforming process is that in the organic photo-oxidation process, the molecular oxygen acts as an electron trap instead of water or protons (as shown in Figure 1), and, consequently, H_2 is not formed.

Indeed, the reduction of water or H^+ takes place only under anaerobic conditions and always gives rise to the formation of H_2, as shown in Figure 1. Many articles claim to perform "water splitting" despite the fact that an organic species has been used as a hole scavenger; however, when O_2 (from water oxidation) is not obtained, it is more correct to refer to these reactions as PR instead of water splitting, as often reported in the literature [7]. Despite its attractiveness, water photo-splitting is a more difficult reaction than PR, both thermodynamically and kinetically. The H_2 formation rate is much higher in the PR process than in the photo-splitting one because the oxidation reaction of the organic species with photogenerated holes is irreversible. Conversely, H_2 and O_2 from water splitting can easily reform H_2O, decreasing the efficiency of converting light to hydrogen. The suppression of electron-hole recombination is also more difficult during the photo-splittingsplitting of water.

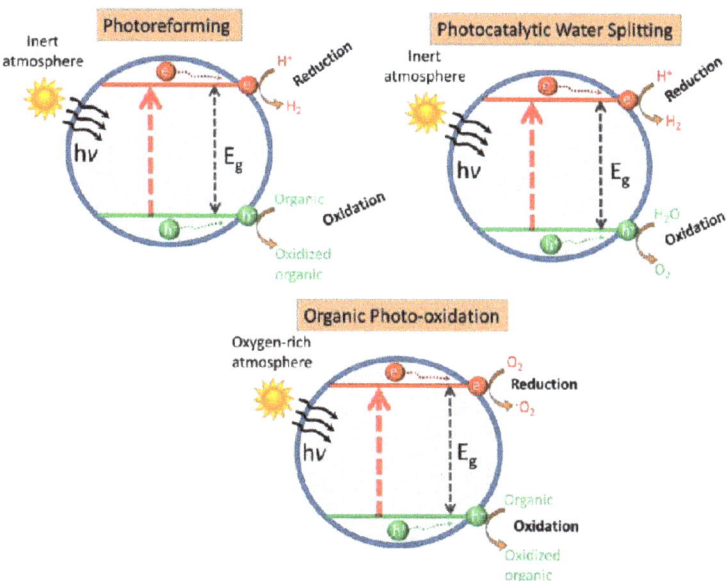

Figure 1. Reaction pathways of water photo-splitting and photoreforming under anaerobic conditions along with the oxidation of an organic molecule in the presence of oxygen. Reprinted with permission from Ref. [6].

The overall photocatalytic process to form H_2 can be considered for both water splitting and biomass PR. The splitting of water involves the concurrent **O**xygen **E**volution **R**eaction (OER) shown in Equation (1) and the H_2 **E**volution **R**eaction (HER), as reported in Equations (2) and (3):

$$2\,H_2O \rightarrow O_2 + 4\,H^+ + 4\,e^- \qquad (1)$$

$$2\,H^+ + 2\,e^- \rightarrow H_2 \qquad (2)$$

or

$$2\,H_2O + 2\,e^- \rightarrow H_2 + 2\,OH^- \qquad (3)$$

with an overall process represented by Equation (4):

$$H_2O \rightarrow H_2 + O_2 \quad \Delta E^\circ = -1.23\ \text{V} \qquad (4)$$

The highly reducing electrons can give rise to the hydrogen evolution reaction in Equations (2) and/or (3). As mentioned before, H_2 generation from water splitting possesses a large thermodynamic barrier due to the difficult oxygen evolution reaction (Equation (1)) with $E^\circ = +1.23$ V vs. **N**ormal **H**ydrogen **E**lectrode (NHE) at pH = 0.

The PR process is considered non-selective and involves the splitting of water to generate H_2 through a reduction reaction and the simultaneous partial oxidation of an organic molecule to other species with higher added value or its mineralization to CO_2 and H_2O, all in one process. In PR, which must take place under anaerobic conditions, the electrons in the photocatalyst migrate to the (CB) and reduce the protons of H_2O to H_2. The photogenerated holes in the (VB) of the photocatalyst (see Figure 1), on the other hand, oxidize the organic substrate.

The oxidizing species are the holes (h+) in the VB, which can also form radical oxidizing species (reactive oxygen species, ROS), which in turn attack the biomass by oxidizing it as in the reaction shown in (Equation (5)) for glucose, where the latter molecule can obviously be considered only a model molecule representative of biomass. The overall PR process (Equation (6)) is, however, nearly energetically neutral ($\Delta E^\circ = +0.001$ V), which means that energy is only needed to overcome the activation barriers. Thus, low-energy photons present in visible light that are very abundant in the solar spectrum can be used for biomass PR.

$$C_6H_{12}O_6 + 6\,H_2O \rightarrow 6\,CO_2 + 24\,H^+ + 24\,e^- \quad E° = -0.01\text{ V vs. RHE} \quad (5)$$

$$C_6H_{12}O_6 + 6\,H_2O \rightarrow 12\,H_2 + 6\,CO_2 \quad \Delta E° = +0.001\text{ V} \quad (6)$$

The effect of sacrificial reagents used as hole scavengers to boost hydrogen production was first reported in the 1980s by Kawai and Sakata [8]. Organic or inorganic species, including biomass compounds, may be used for the production of H_2. Photoreforming has not received as much attention as water photo-splitting despite its interest and the high efficiencies reported. It is worth noting that its selectivity exceeds that of thermocatalytic processes due to the milder (environmental) conditions under which it proceeds, with comparable activity.

The idea is to obtain the desired products, possibly by increasing the selectivity toward the most interesting chemical, preventing the obtaining of dangerous or worthless compounds such as CO_2. A mixture of various organic compounds is, therefore, undesirable. Toe et al. attribute the poor selectivity during PR to (i) unwanted adsorption/desorption of reacting species on the photocatalyst surface; (ii) uncontrolled formation of radical species (e.g., OH radicals) with a strong oxidant power; and (iii) saturation of the surface with products that gives rise to over-oxidation [9].

2. Organic Molecules as Hole Scavengers in Photoreforming: From Model Molecules to Biomass

Research on how to enhance selectivity toward photocatalytic organic transformation has long been studied, but mainly under aerobic conditions, where O_2 traps the electron from the conduction band. One of the most used organic substrates used for selective PR is methanol [10], whose partial oxidation gives rise to formaldehyde, formic acid, methyl formate, or ethylene glycol. Ethanol is also often used as a hole scavenger [11], forming mainly acetaldehyde, acetic acid, 1,1-diethoxyethane, and 2,3-butanediol. As shown in Scheme 1, when glycerol acts as a hole scavenger [12], dihydroxyacetone (DHA), glyceraldehyde, and glyceric acid are formed, while among the aromatics, benzyl alcohol gives rise to benzaldehyde, benzoic acid, hydrobenzoin, 2-phenylacetophenone, and benzoin.

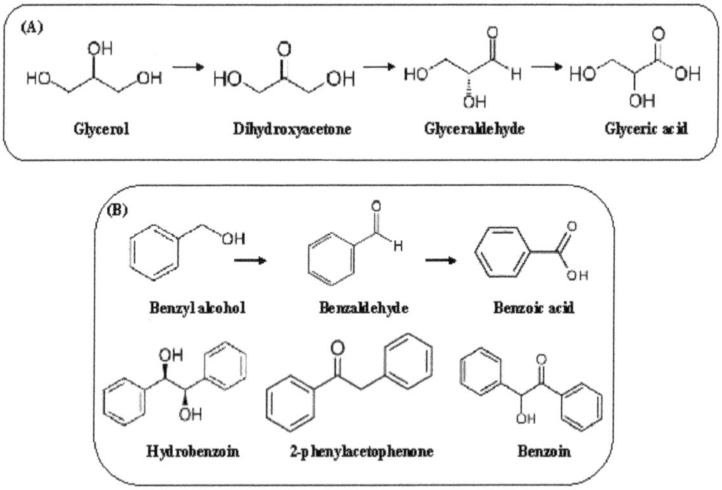

Scheme 1. Partial oxidation products obtained from glycerol (**A**) and benzyl alcohol (**B**) as hole scavengers in the photoreforming reaction.

In Scheme 2, it can be seen, among the furans (they are the most used hole scavengers), that (A) 5-hydroxymethyl-2-furfural (HMF) gives rise to 2,5-furandicarboxaldehyde, also called 2,5-diformylfuran (FDC), and 2,5-furandicarboxylic acid (FDCA), while (B) furfuryl alcohol is partially oxidized to furfural and furoic acid.

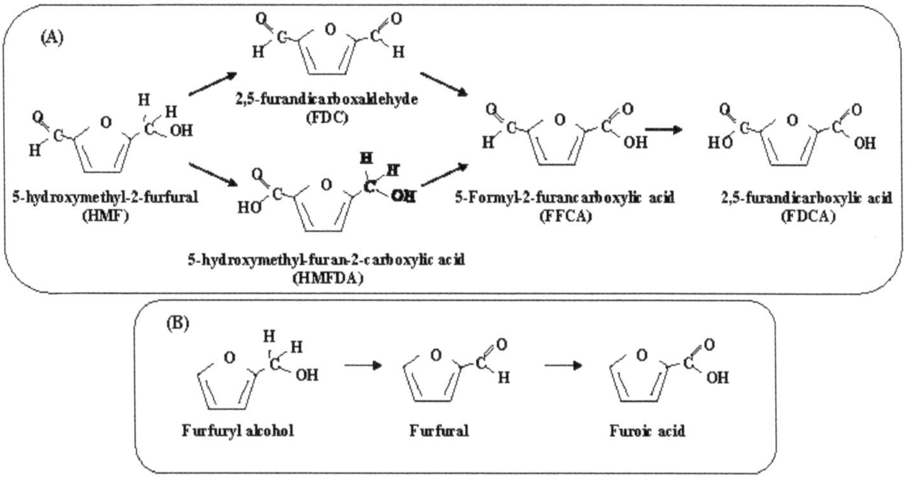

Scheme 2. Oxidation pathways of 5-hydroxymethyl-2-furfural (HMF) (**A**) and furfuryl alcohol (**B**) to their corresponding aldehydes and acids.

Furfural and 5-hydroxymethyl-2-furfural (HMF) are extensively accessible biomass-derived renewable chemical feedstocks, and their oxidation to 2,5 furandicarboxylic acid (FDCA) and furoic acid, respectively, is a research area with great possibilities for application in cosmetics, food, optics, and polymer industries. Water-based oxidation of furfural/HMF is a cost-effective approach to generating, at the same time, H_2 and furoic acid/FDCA. Nevertheless, this process is today limited to (photo)electrochemical methods that can be difficult to improve and scale up.

A wide range of biomass-derived compounds bearing a large variety of oxygen-containing functionalities can be used as a hole scavenger in the PR reaction, the most widely used being methanol, ethanol, and glycerol, but also aldehydes and alcohols of various kinds [13,14], saccharides [15] and others, such as amines [15] as triethanolamine (TEOA), or polysaccharides, including cellulose [16]. Monosaccharides, such as pentoses (ribose, arabinose) and hexoses (glucose, galactose, fructose, mannose), and organic acids (acetic acid, formic acid) have also been extensively used as hole scavengers [17].

Based on the species analyzed in an aqueous solution during the glucose PR, we hypothesized the reaction sequence reported in Scheme 3. Glucose was transformed into arabinose by α-cleavage in the presence of H_2O producing equimolar amounts of formic acid and H_2, and subsequently, arabinose into erythrose by the same mechanism. The greater quantity of H_2 compared to formic acid can be explained by taking into account that its formation comes not only from glucose degradation but also from the splitting of water [18].

Di- or polysaccharides show that the hydrogen production rate is one or two orders of magnitude higher with respect to those reached in the presence of pure water [19].

The oxidation of the organic species has been considered unselective; however, some reports conclude that the organic molecules can be partially oxidized instead of mineralized, so PR for H_2 production can be more widely considered as a synthesis process where, for instance, alcohol is selectively oxidized to an aldehyde [9,20,21].

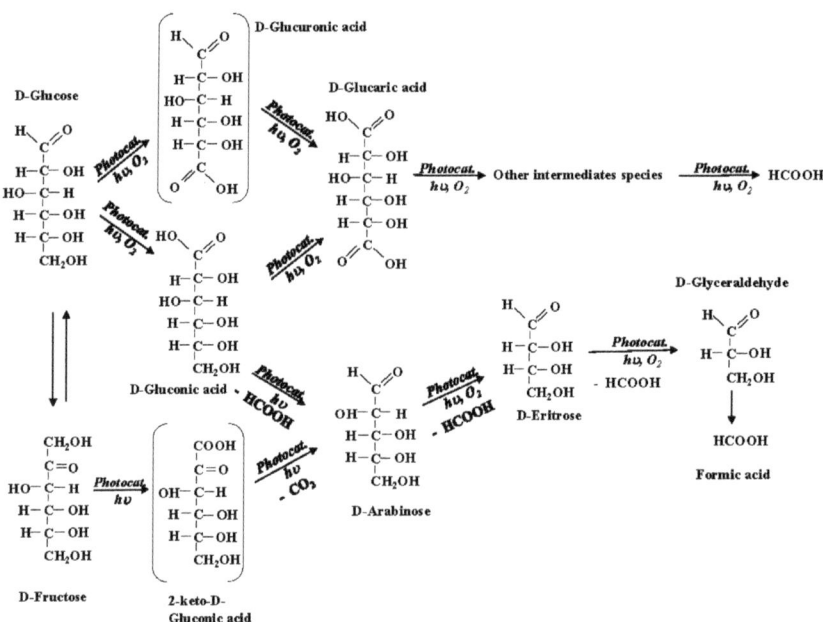

Scheme 3. Reaction pathway by using glucose as a hole scavenger in H_2 generation in the presence of Pt-TiO$_2$ photocatalysts in the anaerobic system [18].

Di- or polysaccharides show that the hydrogen production rate is one or two orders of magnitude higher with respect to those reached in the presence of pure water [19].

The oxidation of the organic species has been considered unselective; however, some reports conclude that the organic molecules can be partially oxidized instead of mineralized, so PR for H_2 production can be more widely considered as a synthesis process where, for instance, alcohol is selectively oxidized to an aldehyde [9,20,21].

Importantly, not all organic species are equally valid for this role, as Mills et al. pointed out. Organic species need to possess a suitable functional group (e.g., alcohol, carbonyl) and a hydrogen atom in the "α" position with respect to them [22]; organics without the alpha hydrogen, such as ketones and carboxylic acids, may result ineffective in PR.

Catalysis is among the principles of green chemistry [23], and together with the paradigm of obtaining a high reaction efficiency, assessed by the chemical yield, waste minimization is also one of the principles considered to support environmental sustainability [24], along with the energy expended in a process and the type of the catalysts that must be non-polluting. In this context, not only materials derived from waste are gaining more attention in recent years, but also the use of waste itself with a view to promoting a circular economy.

In a remarkable process, real biomass feedstocks can be used as hole scavengers in the PR reaction; raw feedstocks, such as wood, rice husks, sawdust, and algae, have been used in PR [25]. Lignocellulosic and agro-industrial waste and residues represent important feedstock for modern biorefineries aimed at the sustainable production of renewable energy and chemicals [26]. The most important biomass feedstocks are of three types: (i) lignocelluloses, (ii) starch-and sugar-based crops, (iii) vegetable oil crops [27], being the lignocellulosic substrates those present in greater quantities in the biosphere.

Lignocellulose is the biomass most abundant, and as illustrated in Figure 2, it is a combination of carbohydrate polymers (cellulose and hemicellulose) and aromatic polymers (lignin), including all dry waste derived from plants (biomass); moreover, it has an evolved structure to provide mechanical and chemical stability [28].

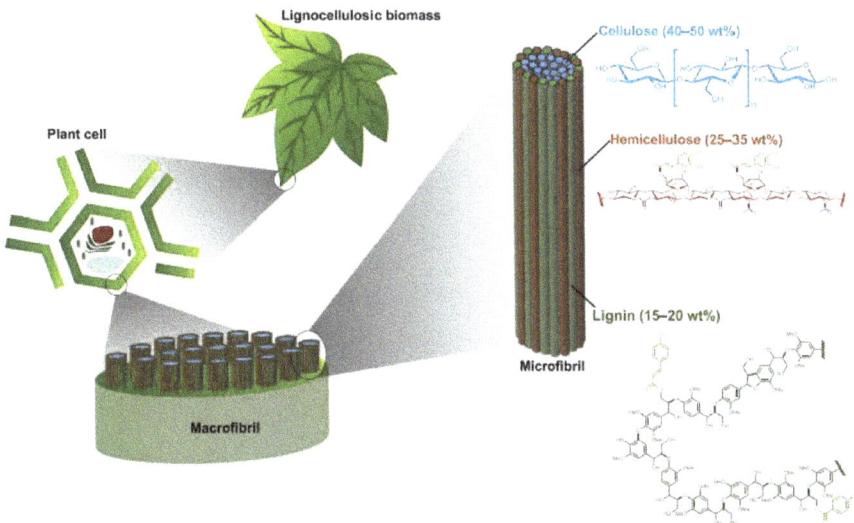

Figure 2. Schematic illustration of the chemical and spatial structure of lignocellulose. Reprinted with permission from Ref. [28]; © 2020 reprinted with permission from Elsevier.

Cellulose and hemicellulose are formed by long chains of hexose and pentose sugars such as $C_6H_{12}O_6$ or $C_5H_{10}O_5$, which can be easily photo-reformed by virtue of their polarity and the high content of hydroxyl groups (-OH), as shown in Figure 2. Lignin (also called lignocellulose) is a heavy and complex organic polymer consisting mostly of phenolic compounds, mainly found in the cell wall of plants. It is a polymer chain composed of phenylpropane molecules. Cellulose and lignin represent about 70% of the total biomass.

The content of chemical species varies significantly depending on the type of biomass selected. For example, considering biomass from food waste, cereals contain 70–80% carbohydrates (they represent very suitable biomass for photoreforming), while meat consists mainly of proteins and fats, both of which are unsuitable to be used in H_2O-based PR because of their structural complexity and hydrophobicity. Sugars were extensively studied as probe substrates for biomass PR since, as mentioned above, most of the biomass is based on saccharide chains (cellulose and hemicellulose). The most consolidated PR of glucose is carried out in the presence of TiO_2 [18]. An increase in temperature from 30 to 60 °C improved the H_2 production.

Chemisorbed biomass consumes all of the reactive oxygen species (ROS) produced in the protocatalytic process (h^+, ˙OH, ˙O_2^- and $^1O_2$), generating the corresponding oxidized species in the PR reaction. In general, as an example, a monosaccharide as glucose generates small acidic species as lactic acid by isomerization, retro-aldol reaction, and dehydration [29]. Moreover, the cleavage of one C-C bond could give rise to a succession of arabinose, erythrose, and formic acid products in photocatalytic glucose partial oxidation and hence in reforming [30].

Disaccharides (maltose, sucrose, lactose) generally provided worse results than monosaccharides. The PR of soluble polysaccharides occurred at smaller H_2 yields, probably because of their high molecular weights and the presence of hydrogen bonds in the structures. The use of lignocellulose in PR is energy-demanding, because it requires the breaking of its structure. Depending on the kind of biomass chosen, the accessible chemical content for the PR is 55–95% by weight [31]. This percentage could increase with the incessant progress in photocatalytic materials able to photoreforming lignin, which, however, being a complex polymer, remains difficult to photo-reformate. Lignocellulose PR technology is still not suitable for the industrial expectations of the difficulty in the deconstruction of lignocellulosic feedstocks. In fact, the complex structure is difficult to be degraded and protect against microbial attack. Moreover, the strong hydrogen bonds between the chains within the cellulose microfibrils make lignocellulose recalcitrant to chemical transformation.

As pointed out by Uekert, before using it for PR, the biomass feedstocks in general must suffer a pre-treatment. In the first stage, a mechanical treatment would produce small pieces; hence, a chemical pre-treatment would solubilize the organics to facilitate the contact between the hole scavenger and the photocatalyst during the PR, increasing the H_2 evolution rate. Some reports are devoted to the hydrolysis and solubilization at acidic or alkaline pHs, exposition of the feedstock to high-pressure saturated steam, enzymatic processes (mainly to hydrolyze cellulose) or others; however, no systematic study has been devoted to this important stage as far as the research has been focused on synthetic aqueous solutions of model molecules [32].

3. Some Considerations and Details on g-C_3N_4-Based Photocatalysts for H_2 Production

3.1. g-C_3N_4 as Photocatalyst

The ideal photocatalyst in a PR reaction should be selective versus the most value-added species deriving from organics oxidation, stable and cheap in order to be used on an industrial scale. Important parameters are band edges, optical absorbance, and carrier mobility. The biggest challenge for a photocatalyst is to possess suitable visible light absorption in order to harvest solar light and high carrier mobility, i.e., reduced recombination. Many semiconductors have been studied as photocatalysts, but certainly, TiO_2 has been the most studied due to its activity along with cost-effectiveness, safety, and (photo)chemical stability.

In the last decades, polymeric carbon nitride, known as melon, C_3N_4 polymer, or g-C_3N_4, has successfully been employed in photocatalysis due to its low cost, non-toxicity, thermal stability, appropriate band gap, and suitable photocatalytic performance. g-C_3N_4 consists of a conjugated polymeric system. Indeed, it is constituted by s-triazine or tri-s-triazine units interconnected via tertiary amines. The atoms in the layers are arranged in honeycomb configurations with strong covalent bonds. Interactions between the two-dimensional sheets (2D) are weak van der Waals forces. This material is highly stable in various solvents, including H_2O, diethylether, acetic acid, alcohols, N,N-dimethylformamide (DMF), toluene, tetrahydrofuran (THF), and NaOH aqueous solution. The g-C_3N_4 can be prepared by thermal condensation of nitrogenous precursors such as melamine, urea, thiourea, cyanamide, dicyandiamide, and ammonium thiocyanate. From these precursors, the organic moiety is transformed stepwise by slow calcination into the yellow-brown melon structure based on tri-s-triazine (heptazine) rings. Poly-addition and poly-condensation reactions occur together in a continuous manner to build the melon sheets that constitute the g-C_3N_4 structure [33]. As shown in Figure 3A, melon is structurally related to the more condensed (fully dehydrogenated) graphitic carbon nitride, g-C_3N_4, a visible-absorbing semiconductor. Melon, a highly ordered polymer, is the first formed polymeric g-C_3N_4 structure. Further reaction leads to more condensed and less defective g-C_3N_4 species, based on tri-s-triazine (C_6N_7) units as elementary building blocks. Melamine, melem, and melon are triazine- and heptazine (tri-s-triazine unit)-based molecular compounds to prepare g-C_3N_4. As illustrated in Figure 3B, triazine (C_3N_3) and heptazine (C_6N_7) rings are the basic tectonic units of g-C_3N_4.

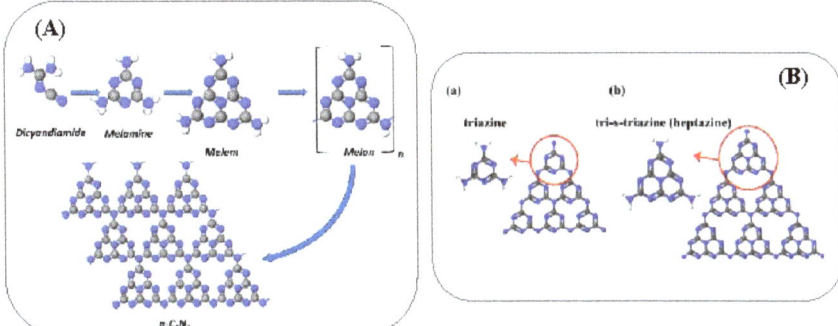

Figure 3. (**A**) Structures of melem, melon, and g-C_3N_4 obtained from the thermal condensation of dicyandiamide to form g-C_3N_4. Reprinted with permission from Ref. [34]; © 2023 MDPI. (**B**) (**a**) Triazine and (**b**) tri-s-triazine (heptazine) structures of g-C_3N_4 (gray, blue, and white balls are carbon, nitrogen, and hydrogen, respectively). Reprinted with permission from Ref. [35]; © 2022 MDPI.

Antonietti's group introduced this metal-free semiconductor in 2006 as a catalyst [36] and then, in 2009, as a heterogeneous photocatalyst for H_2 evolution [37]. It has been used as a photocatalyst for selective redox transformations [9,38]; indeed, the potential of the valence band (VB) and the absence of hydroxyl groups on the surface hinders the direct formation of ·OH radicals, species responsible for unselective oxidation of substrates. The HOMO-LUMO gaps of melem, polymeric melon (the building unit of g-C_3N_4), and an infinite sheet of a hypothetically fully condensed g-C_3N_4 were 3.5, 2.6, and 2.1 eV, respectively [37]. The calculated band gap of polymeric melon is very close to the experimentally measured medium-band gap of 2.7 eV, as reported by Antonietti et al., with the edges of the conduction band and valence band lying at −1.13 V and +1.57 V (vs. NHE at pH = 7) [39].

The photocatalytic activity of C_3N_4 has been hampered by several important drawbacks that are challenging to overcome, as evidenced in Figure 4; in particular, low charge carrier mobility and the fast recombination of the charge carriers restrict its practical use very significantly.

Several methods have been explored to modify/improve/optimize the structure of C_3N_4 by top-down strategies such as acid treatment, exfoliation, or etching, as well as bottom-up approaches [40]. Some synthetic strategies, such as nanostructure design, for instance, by using soft or rigid templates, electronic structure alteration via incorporation of dopant atoms, generation of point defects via vacancies, or the supramolecular pre-organization, allow to increase their specific surface area and to modify the value of the band gap. Furthermore, the deposition of noble metals and the construction of composites can improve the electron-hole separation, as will be better described below [41].

3.2. g-C_3N_4 in Anaerobic Conditions: Strategies to Improve Its Photocatalytic Activity

g-C_3N_4 has been used as a photocatalyst for water photo-splitting for the first time to obtain H_2 and O_2 under visible light irradiation by Antonietti's group [37]. The production of H_2 from an aqueous solution of triethanolamine (10 vol%) after 72 h of reaction passed from 7 to 770 μmol when 3 wt% of Pt nanoparticles were added to the bare semiconductor under irradiation with a wavelength longer than 420 nm. On the other hand, under UV irradiation at λ > 300 nm, the total H_2 production in 19 h was ca. 4.5 mmol in the presence of Pt/g-C_3N_4 photocatalyst.

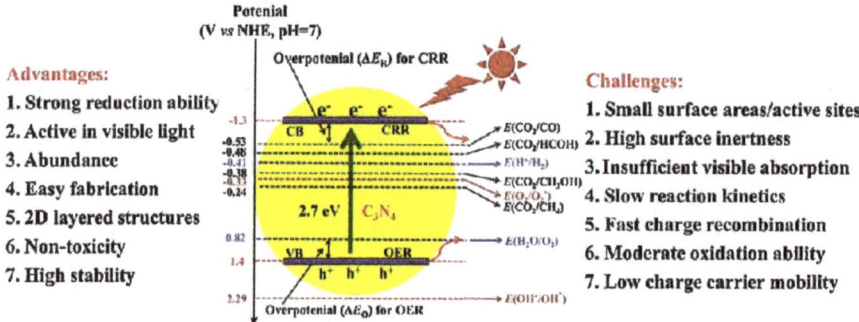

Figure 4. Advantages and disadvantages of g-C_3N_4 in photocatalysis along with the estimated position of the g-C_3N_4 band edges at pH 7 and reduction potentials of the relevant reactions related to water splitting and CO_2 reduction. Reprinted with permission from Ref. [41]; © 2017 Elsevier.

Very few papers report the evolution of H_2 from H_2O without the use of a hole scavenger. Ong et al. report only one example in the presence of a composite of g-C_3N_4/polypyrrole [42]. Polypyrrole injects electrons into the g-C_3N_4 conduction band, and the holes in the valence band of g-C_3N_4 react with water giving rise to H_2O_2. The authors do not propose a trap for polypyrrole valence band holes, which could hardly be transferred to solution species; hence, a self-oxidation of the polymer (sacrificial agent) would likely occur [43]. Alternatively, Liu et al. report O_2 and H_2 evolution in the absence of scavengers using as photocatalyst metal-free carbon nanodot/g-C_3N_4 nanocomposite. They calculated quantum efficiencies of 16% (at λ = 420 nm), 6.3% (λ = 580 nm), and 4.4% (λ = 600 nm) and obtained an overall efficiency in the solar energy conversion equal to ca. 2.0%. The mechanism of water oxidation is a process with two 2-electron steps yielding firstly H_2O_2 that subsequently decomposes to provide O_2 and H_2O. The rate increases with carbon nanodot loading because they catalyze the rate-limiting step, i.e., the H_2O_2 decomposition [44]. The results reported by Liu et al. are surprisingly high.

Generally, a hole scavenger is employed to minimize electron-hole recombination, often with suitable quantum yields, as reviewed by Naseri et al. [45], but as claimed before, the reaction cannot be considered water photo-splitting, but it regards the PR of the scavenger employed. The oxidation reaction is the rate-determining step of the overall reaction, so the use of electron donors (hole scavengers) clearly improves the activity [46].

In order to increase the catalytic activity of the pristine material, other additional strategies aimed at improving the absorption of visible light, reducing the electron-hole recombination rate, and improving the reaction kinetics have been extensively investigated. Such approaches include nanostructure architecture, doping, heterojunction, and the use of plasmonic metals and co-catalysts.

Noble metals such as Rh, Au, Ag, and particularly Pt are often used as co-catalysts because the metals trap photogenerated electrons as well as decrease the overpotential facilitating multielectron transfer reactions. Often, RuO_2 is also present to sink the holes, avoiding the photoproduced couples (h^+/e) recombination [47]. The large work function and low activation energy of Pt allow this metal to be the most effective cocatalyst for H_2 evolution [48]. The optimal amount of Pt follows a bell shape because the photocatalytic activity increases with increasing the amounts of metal. However, further increasing could lead to a decrease in the number of surface active sites on the semiconductor photocatalyst and even to a shielding effect on the incident light, as well as can favor electron/hole recombination, acting the metal as a recombination center. Furthermore, the presence of noble metal nanoparticles on the semiconductor surface can generate visible activated surface plasmon resonance (SPR) to activate the photocatalytic process to visible light irradiation.

Battula et al. report that Pt-C_3N_4 exhibited high selectivity to 2,5-diformylfuran (DFF) during the partial oxidation of 10 mL of an aqueous solution of 5-hydroxymethyl furfural

(10 mM) [49]. The obtained results are modest in terms of photoreforming, and a H_2 production rate of 12 µmol h^{-1} m^{-2} (corresponding to 2040 µmol h^{-1} g_{cat}^{-1}) was obtained with a DFF yield of 13.8% with >99% selectivity after 6 h under simulated solar light (light intensity of 100 mW cm^{-2}) light. The selectivity was maintained even after 48 h of experiment, with an improved DFF yield of 38.4% and H_2 production rate of 36 µmol h^{-1} m^{-2} (corresponding to 6120 µmol h^{-1} g_{cat}^{-1}).

Transition metals such as Fe, Co, and Ni or inorganic compounds such as $(OH)_2$, MoS_2, WS_2, NiS, NiO, $Ni(OH)_2$, or CoP have also been used as co-catalysts for the PR. The analogous layered structures of inorganic semiconductors, such as MoS_2 and g-C_3N_4, gave rise to a composite that remarkably increased the photocatalytic H_2 evolution. This performance has been attributed to the similar layered geometries of the solid photocatalysts, which allow the composite to improve the mobility of charge carriers at the interfaces and hence their lifetime [50]. Interestingly, it was observed that by increasing from 0D to 1D, to 2D, and to 3D, the dimensions of conjugated polymers, the mobility of electrons rises, and, at the same time, the binding energies of the bound electron/hole pairs are reduced. For instance, it has been observed a decrease in the electron/hole recombination rate in the presence of graphene nanosheets/g-C_3N_4 composite for the photocatalytic H_2 evolution. Xiang et al. observed that an amount of 1 wt% of graphene on g-C_3N_4 was an optimum during the photoreforming of an aqueous solution of methanol carried out under visible light irradiation (λ > 400 nm). In fact, the quantity of H_2 resulted ca. three-fold that found by using the pristine g-C_3N_4. The heterojunction between g-C_3N_4 and graphene increases the electrical conductivity and improves the carrier separation. It is also capable of storing and transporting electrons to the reaction sites [51].

Improved performance for photocatalytic H_2 production has been reported by Sun et al., who highlighted the synergistic effect of rGO nanosheets and Pt nanoparticles. In fact, the rGO nanosheets, which act as electron transfer mediators, capture the electrons photogenerated by g-C_3N_4 and then transfer them to the Pt cocatalyst, while the nanoparticles of Pt act as reduction active sites to promote the H_2 evolution reaction [52].

In addition, Yan et al. report that in the composites g-C_3N_4/rGO, the carbon nitride plays the role of the photocatalyst, whereas reduced graphene oxide can collect and transport electrons to reaction sites improving in this way the activity [53]. Organic polymers as poly(3-hexylthiophene) can exhibit semiconducting properties (band gap ca. 2 eV), and its composite with g-C_3N_4, using Pt as cocatalyst, has been used for H_2 production from aqueous ascorbic acid under visible light irradiation [54]. The poly(3-hexylthiophene)/g-C_3N_4 composite enabled outstanding activities (>300 mmol h^{-1} g_{cat}^{-1}) for irradiation with λ > 500 nm. Although deactivation leading to 30% lower H_2 production rates after several days of operation has been observed, these results encourage further investigation of this inexpensive carbon-based material.

Interestingly, sunlight can also be used as an irradiation source to obtain H_2 from water by triethanolamine PR using rGO nanosheets/C_3N_4 composites. The highest H_2 production observed on Ag-loaded samples (the wt% of Ag in the catalyst was in the range of 1–5%) was 525 µmol h^{-1} g_{cat}^{-1}, which increased to ca. 88 mmol h^{-1} g_{cat}^{-1} when 1 wt% of Pt was also present. The apparent quantum efficiency (AQE) reported for the best material was ca. 9% using, as an irradiation source, indifferently visible LED or natural sunlight [55].

Spinel ferrite-g-C_3N_4 systems also take advantage of sunlight for H_2 generation through PR [56]. For instance, the evolved H_2 rate by a ferrite-g-C_3N_4 composite gave rise to 10 times more H_2 than with the bare g-C_3N_4. The efficiency of the composite was justified by claiming its optimized light absorption ability [57].

As far as composite semiconductor materials are concerned, an interesting idea is the Z-scheme. Bard suggested the first mechanism proposal of this system in 1979 [58]. It involves the transfer of electrons from the LUMO of the CB of WO_3 to the HOMO of the VB of C_3N_4, thus gaining the capability to react for both electrons and holes compared to heterojunctions [59]. Few Z-scheme has been published in PR [60–62]. In the Z-scheme g-C$_3$N$_4$/WO$_3$ published by Yu et al., a host-guest concept is conceived. Authors introduce WO$_3$ nanocuboids in the host g-C$_3$N$_4$ achieving an intimate interfacial contact with an injection of electrons from the WO$_3$ conduction band (CB) to the valence band (VB) of g-C$_3$N$_4$, as reported in Figure 5 [62]. The photocatalytic H$_2$ production was evaluated by using an aqueous solution of triethanolamine, which acted as a hole scavenger under simulated solar irradiation. The presence of 1 wt% Pt on g-C$_3$N$_4$ gave rise to an H$_2$ amount of 0.44 mmol h^{-1} g$_{cat}$$^{-1}$. Indeed, the electrons from the CB of g-C$_3$N$_4$ (ca.−1.1 V vs. NHE) possess enough reducing power for the H$_2$ production, whereas pristine WO$_3$ resulted in virtual inactivity. The use of g-C$_3$N$_4$/WO$_3$ as a photocatalyst significantly enhanced the H$_2$ production to 3.12 mmol h^{-1} g$_{cat}$$^{-1}$, i.e., it increased seven times [62].

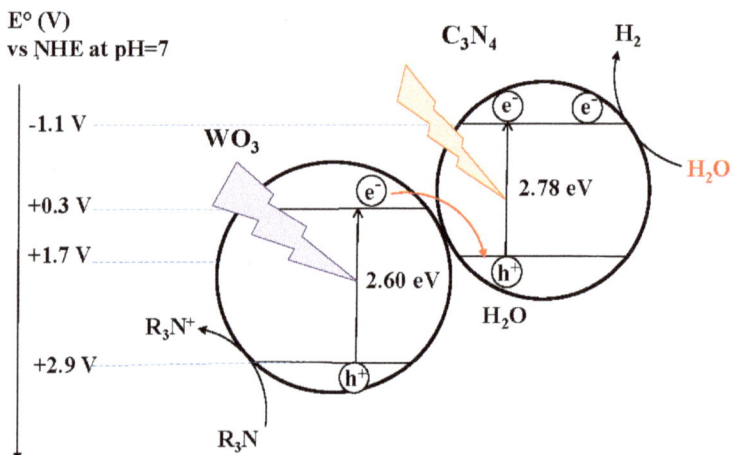

Figure 5. Z-scheme of visible light harvesting g-C$_3$N$_4$/WO$_3$ nanocomposite. CB and VB energy levels have been taken from [62].

Composites of Nb$_2$O$_5$/g-C$_3$N$_4$ exhibited high visible light absorption resulting in a notable photocatalytic activity under simulated sunlight irradiation using triethanolamine or methanol as hole scavengers. An amount of 110 mmol h^{-1} g$_{cat}$$^{-1}$ of H$_2$ has been produced using a sample containing 10 wt% of g-C$_3$N$_4$ over niobium oxide, more than double that obtained with the pristine semiconductor [63]. The enhanced photocatalytic activity has been attributed to the energy and the fast separation of photogeneration of electron-hole pairs at the Nb$_2$O$_5$/g-C$_3$N$_4$ interface through a direct Z-scheme, as suggested in Figure 6.

3.3. Real Wastes as Hole Scavengers in the Photoreforming Process in the Presence of g-C$_3$N$_4$-Based Semiconductors

The use of real biomass has been very rarely reported in the context of photoreforming technology, particularly in the presence of g-C$_3$N$_4$-based materials as photocatalysts. Uekert et al. have been the most active group, and pioneer, as far as we know, in using real waste in this process. It is interesting to mention that they have used not only biomass but real food wastes and synthetic polymers as hole scavengers. As well known, polymers are of particular concern due to their non-biodegradability and accumulation in the environment. In addition, plastics can be used in PR even if its oxidation is not a simple task because non-biodegradable polymers are composed of long hydrocarbon chains such as polyethylene or polystyrene, which are difficult to reform due to the stability of their C-C bonds [9,64]. Uekert et al. have studied this problem and proposed the use of carbon nitride/nickel phosphide composite as a photocatalyst to obtain H$_2$ by the PR of poly(ethylene-terephthalate) (PET) and poly(lactic acid) (PLA) under alkaline aqueous conditions [65]. They used as photocatalysts a cyanamide-functionalized carbon nitride

(CN_x) coupled with a nickel phosphide (Ni_2P). The same group has previously used bare C_3N_4 for lignocellulose PR [66]. They compared the activity of this material with that of CdS/CdO_x quantum dots in an alkaline aqueous solution via solar illumination at room temperature producing H_2 [67]. These authors have proven that glucose, fructose, galactose, sucrose, but also raw biomass such as starch, casein, bovine serum albumin (BSA), glycerol, castor oil, and soybean oil are suitable hole scavengers in PR by using photocatalysts both CdS/CdO_x in alkaline medium (KOH 10M) and the composite containing C_3N_4 in aqueous suspension, i.e., H_2NCN_x/Ni_2P. They observed that simple soluble molecules such as sugars, glutamic acid, and glycerol gave rise to the highest yields of H_2, though the activity decreased when the molecule to be oxidized became more complex. The same research group has also focused on the interesting aim of developing the use of food waste in PR. They argue that food waste PR can be applied to small off-grid systems to simultaneously handle food waste and generate H_2. Both CdS/CdO_x in an alkaline solution and the C_3N_4-based composite H_2NCN_x/Ni_2P at neutral pH have been active in this task [32]. Uekert et al. have selected casein, fructose, and starch as case studies because they are present in commonly discarded foods (bread, cheese, and apples). After 5 days of irradiation, conversions in H_2 with CdS/CdO_x in KOH were ca. 16–27% measured with respect to the theoretical yield of hydrogen, whereas they were 3–7% in the presence of the C_3N_4-based photocatalyst in KOH and 1–4% in water [32].

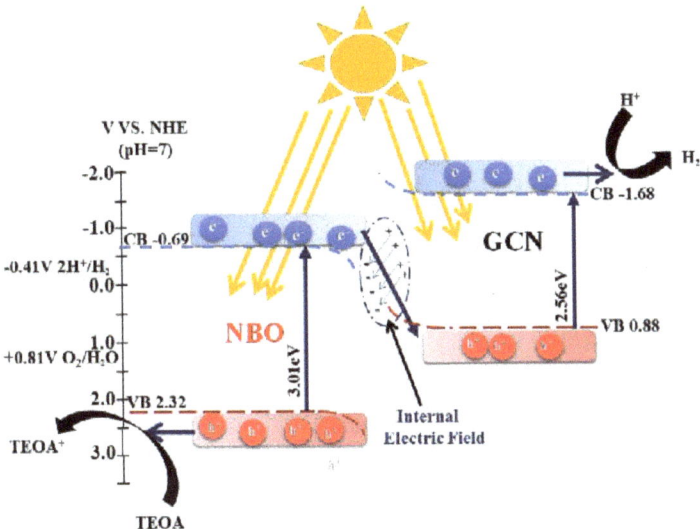

Figure 6. Scheme of the possible photocatalytic mechanism in the presence of Nb_2O_5/g-C_3N_4 photocatalysts. Reprinted with permission from Ref. [63]; © 2019 MDPI.

In any case, the most abundant renewable resource for the production of aromatic chemicals remains lignin, the natural amorphous polymer that acts as the essential glue that gives plants their structural integrity, which constitutes 15–30% by weight of the biomass [68]. Lignin mainly contains C-O bonds and C-C bonds, wherein the β-O-4 bond represents the main content [69,70]. The latter bond can be broken into more chemically useful fragments. Lignin can be converted into different aromatic compounds, which are in turn used to synthesize other value-added chemicals [21,31,70] and is able to produce H_2 in the order of mmol h^{-1} g_{cat}^{-1} [9,19,55].

4. Conclusions

Hydrogen production by photocatalysis in the presence of renewable solar energy is one of the most versatile and environmentally benign paths for research to pursue.

Photoreforming offers a simple, sunlight-driven method for transforming biomass waste into valuable chemicals and clean H_2 fuel. In this manner, H_2 can be produced at room temperature and atmospheric pressure by a simple, efficient, low-cost, and sustainable process, with the use of a heterogeneous photocatalyst using biomass, solar light, and water. Photoreforming involves the splitting of water to generate H_2 through a reduction reaction and the simultaneous oxidation of an organic species to obtain other molecules with higher added value or, simply, to completely oxidize (mineralize) organics to CO_2 and H_2O. The current manuscript overviews some of the most relevant studies focused on photoreforming using g-C_3N_4-based materials. Almost all of them report the use of model organic molecules as hole scavengers, considering them as biomass prototype species to investigate the photoreforming process at bench scale in order to understand the physical-chemical features of the process. The published studies have demonstrated that the PR of biomass is a promising approach to the sustainable generation of H_2 and feedstock chemicals. The simplicity of this process, which is capable of producing clean H_2 also at room temperature, is of considerable advantage if compared to thermochemical methods, but efficiencies are still lower than those of conventional processes. The use of g-C_3N_4 has been explored due to the several advantages in terms of cost, atoxicity, and thermodynamic constraints. Indeed, it has been demonstrated that g-C_3N_4 is able to work as a photocatalyst also under visible light irradiation. However, numerous disadvantages have also emerged, as concerns about its use as a pristine solid due to the high recombination rate of e^-/h^+ couples and the low oxidant capability of its valence band. The use of heterojunctions has been proposed as the best strategy to take advantage of the g-C_3N_4 potentialities, particularly in the presence of semiconductor oxides. The latter types of materials are particularly promising for PR reactions. In the future, a more transversal and interdisciplinary approach will be necessary in order to jump into the view of the use of real waste, currently pioneered faced by very few groups, as Uekert et al. In perspective, future studies should focus on the development of narrow band-gap materials to enhance solar energy conversion efficiency and to lower the required driving force, improving the selectivity toward high-value products. Furthermore, the use of real waste as hole scavengers would be approached. Photoreforming offers a unique sunlight-driven platform for transforming biomass waste resources, even when combined with other types of waste, into both valuable H_2 and organic chemicals. To this aim, an intense dialogue among chemists, materials science scientists, engineers, and technologists is mandatory.

Author Contributions: Conceptualization, E.I.G.-L., L.P. and G.M.; methodology, E.I.G.-L. and G.M.; software, E.I.G.-L. and G.M.; investigation, E.I.G.-L. and G.M.; resources, E.I.G.-L., L.P. and G.M.; data curation, E.I.G.-L., L.P. and G.M.; writing—original draft preparation, E.I.G.-L.; writing—review and editing, E.I.G.-L., L.P. and G.M.; visualization, E.I.G.-L., L.P. and G.M.; supervision, L.P. and G.M.; project administration, G.M.; funding acquisition, L.P. All authors have read and agreed to the published version of the manuscript.

Funding: This research received no external funding.

Data Availability Statement: No further data available.

Conflicts of Interest: The authors declare no conflict of interest.

References

1. Ng, C.H.; Teo, S.H.; Mansir, N.; Islam, A.; Joseph, C.G.; Hayase, S.; Taufiq-Yap, Y.H.; Yap, T. Recent advancements and opportunities of decorated graphitic carbon nitride toward solar fuel production and beyond. *Sustain. Energy Fuels* **2021**, *5*, 4457–4511. [CrossRef]
2. *DOE Hydrogen and Fuel Cells Program: 2014 Annual Progress Report*; US Department of Energy: Washington, DC, USA, 2014.
3. Turner, J.A. Sustainable hydrogen production. *Science* **2004**, *305*, 972–974. [CrossRef] [PubMed]
4. García-López, E.I.; Palmisano, L. Chapter 1: Fundamentals of photocatalysis: The role of the photocatalysts in heterogeneous photo-assisted reactions. In *Materials Science in Photocatalysis*, 1st ed.; García-López, E.I., Palmisano, L., Eds.; Elsevier: Amsterdam, The Netherlands, 2021; pp. 3–9.

5. Fujishima, A.; Honda, K. Electrochemical Photolysis of Water at a Semiconductor Electrode. *Nature* **1972**, *238*, 37–38. [CrossRef] [PubMed]
6. Samage, A.; Gupta, P.; Halakarni, M.A.; Nataraj, S.K.; Sinhamahapatra, A. Progress in the Photoreforming of Carboxylic Acids for Hydrogen Production. *Photochemistry* **2022**, *2*, 40. [CrossRef]
7. Jitputti, J.; Pavasupree, S.; Suzuki, Y.; Yoshikawa, S. Synthesis and photocatalytic activity for water-splitting reaction of nanocrystalline mesoporous titania prepared by hydrothermal method. *J. Solid State Chem.* **2007**, *180*, 1743–1749. [CrossRef]
8. Kawai, T.; Sakata, T. Conversion of carbohydrate into hydrogen fuel by a photocatalytic process. *Nature* **1980**, *286*, 474–476. [CrossRef]
9. Toe, C.Y.; Tsounis, C.; Zhang, J.; Masood, H.; Gunawan, D.; Scott, J.; Amal, R. Advancing photoreforming of organics: Highlights on photocatalyst and system designs for selective oxidation reactions. *Energy Environ. Sci.* **2021**, *14*, 1140–1175. [CrossRef]
10. Al-Mazroai, L.S.; Bowker, M.; Davies, P.; Dickinson, A.; Greaves, J.; James, D.; Millard, L. The photocatalytic reforming of methanol. *Catal. Today* **2007**, *122*, 46–50. [CrossRef]
11. Romero Ocana, I.; Beltram, A.; Delgado Jaen, J.J.; Adami, G.; Montini, T.; Fornasiero, P. Photocatalytic H_2 production by ethanol photodehydrogenation: Effect of anatase/brookite nanocomposites composition. *Inorg. Chim. Acta* **2015**, *431*, 197–205. [CrossRef]
12. Daskalaki, V.M.; Kondarides, D.I. Efficient production of hydrogen by photo-induced reforming of glycerol at ambient conditions. *Catal. Today* **2009**, *144*, 75–80. [CrossRef]
13. Bahruji, H.; Bowker, M.; Davies, P.R.; Pedrono, F. New insights into the mechanism of photocatalytic reforming on Pd/TiO_2. *Appl. Catal. B* **2011**, *107*, 205–209. [CrossRef]
14. Bowker, M.; Morton, C.; Kennedy, J.; Bahruji, H.; Greaves, J.; Jones, W.; Davies, P.R.; Brookes, C.; Wells, P.P.; Dimitratos, N. Hydrogen production by photoreforming of biofuels using Au, Pd and Au-Pd/TiO_2 photocatalysts. *J. Catal.* **2014**, *10*, 10–15. [CrossRef]
15. Bowker, M.; Bahruji, H.; Kennedy, J.; Jones, W.; Hartley, G.; Morton, C. Photocatalytic Window: Photo-reforming of organics and water splitting for sustainable hydrogen production. *Catal. Lett.* **2015**, *145*, 214–219. [CrossRef]
16. Caravaca, A.; Jones, W.; Hardacre, C.; Bowker, M. H_2 production by the photocatalytic reforming of cellulose and raw biomass using Ni, Pd, Pt and Au on titania. *Proc. R. Soc. A* **2016**, *472*, 20160054. [CrossRef] [PubMed]
17. Ma, J.; Liu, K.; Yang, X.; Jin, D.; Li, Y.; Jiao, G.; Zhou, J.; Sun, R. Recent Advances and Challenges in Photoreforming of Biomass-Derived Feedstocks into Hydrogen, Biofuels, or Chemicals by Using Functional Carbon Nitride Photocatalysts. *ChemSusChem* **2021**, *14*, 4903–4922. [CrossRef]
18. Bellardita, M.; García-López, E.I.; Marcì, G.; Palmisano, L. Photocatalytic formation of H_2 and value-added chemicals in aqueous glucose (Pt)-TiO_2 suspension. *Int. J. Hydrogen Energy* **2016**, *41*, 5934–5947. [CrossRef]
19. Huang, C.W.; Nguyen, B.S.; Wu, J.C.S.; Nguyen, V.H. A current perspective for photocatalysis towards the hydrogen production from biomass-derived organic substances and water. *Int. J. Hydrogen Energy* **2020**, *45*, 18144–18159. [CrossRef]
20. Wang, J.; Zhao, H.; Lui, P.; Yasri, N.; Zhong, N.; Kibria, M.G.; Hu, J. Selective superoxide radical generation for glucose photoreforming into arabinose. *J. Energy Chem.* **2022**, *74*, 324–331. [CrossRef]
21. Marcì, G.; García-López, E.I.; Palmisano, L. Polymeric carbon nitride (C_3N_4) as heterogeneous photocatalyst for selective oxidation of alcohols to aldehydes. *Catal. Today* **2018**, *315*, 126–137. [CrossRef]
22. Bowker, M.; O'Rourke, C.; Mills, A. The Role of Metal Nanoparticles in Promoting Photocatalysis by TiO_2. *Top. Curr. Chem.* **2022**, *380*, 17. [CrossRef]
23. Anastas, P.T.; Warner, J.C. *Green Chemistry: Theory and Practice*; Oxford University Press: Oxford, UK, 1998.
24. Sheldon, R.A. Fundamentals of green chemistry: Efficiency in reaction design. *Chem Soc. Rev.* **2012**, *41*, 1437–1451. [CrossRef] [PubMed]
25. Butburee, T.; Chakthranont, P.; Phawa, C.; Faungnawakij, K. Beyond Artificial Photosynthesis: Prospects on Photobiorefinery. *ChemCatChem* **2020**, *12*, 1873–1890. [CrossRef]
26. Rodríguez-Padrón, D.; Puente-Santiago, A.R.; Balu, A.M. Environmental catalysis: Present and future. *ChemCatChem* **2019**, *11*, 18–38. [CrossRef]
27. Huber, G.W.; Iborra, S.; Corma, A. Synthesis of transportation fuels from biomass: Chemistry, catalysts, and engineering. *Chem. Rev.* **2006**, *106*, 4044–4098. [CrossRef] [PubMed]
28. Bertella, S.; Luterbacher, J.S. Lignin Functionalization for the Production of Novel Materials. *Trends Chem.* **2020**, *2*, 440–453. [CrossRef]
29. Zhao, H.; Li, C.; Yong, X.; Kumar, P.; Palma, B.; Hu, Z.; Tendeloo, G.; Siahrostami, S.; Larter, S.; Zheng, D.; et al. Coproduction of hydrogen and lactic acid from glucose photocatalysis on band-engineered $Zn_{1-x}Cd_xS$ homojunction. *iScience* **2021**, *24*, 102109. [CrossRef]
30. Bellardita, M.; García-López, E.I.; Marcì, G.; Megna, B.; Pomilla, F.R.; Palmisano, L. Photocatalytic conversion of glucose in aqueous suspensions of heteropolyacid-TiO_2 composites. *RSC Adv.* **2015**, *5*, 59037. [CrossRef]
31. Kuehnel, M.F.; Reisner, E. Solar Hydrogen Generation from Lignocellulose. *Angew. Chem. Int. Ed.* **2018**, *57*, 3290–3296. [CrossRef]
32. Uekert, T.; Pichler, C.M.; Schubert, T.; Reisner, E. Solar-driven reforming of solid waste for a sustainable future. *Nat. Sustain.* **2021**, *4*, 383–391. [CrossRef]
33. Muhmood, T.; Xia, M.; Lei, W.; Wang, F. Erection of duct-like graphitic carbon nitride with enhanced photocatalytic activity for ACB photodegradation. *J. Phys. D Appl. Phys.* **2018**, *51*, 065501. [CrossRef]

34. Zhurenok, A.V.; Vasilchenko, D.B.; Kozlova, E.A. Comprehensive Review on g-C_3N_4-Based Photocatalysts for the Photocatalytic Hydrogen Production under Visible Light. *Int. J. Mol. Sci.* **2023**, *24*, 346. [CrossRef] [PubMed]
35. Alaghmandfard, A.; Ghandi, K. A Comprehensive Review of Graphitic Carbon Nitride (g-C_3N_4)-Metal Oxide-Based Nanocomposites: Potential for Photocatalysis and Sensing. *Nanomaterials* **2022**, *12*, 294. [CrossRef] [PubMed]
36. Goettmann, F.; Fischer, A.; Antonietti, M.; Thomas, A. Metal-free catalysis of sustainable Friedel-Crafts reactions: Direct activation of benzene by carbon nitrides to avoid the use of metal chlorides and halogenated compounds. *Chem. Commun.* **2006**, *5*, 4530–4532. [CrossRef]
37. Wang, X.C.; Maeda, K.; Thomas, A.; Takanabe, K.; Xin, G.; Domen, K.; Antonietti, M. A metal-free polymeric photocatalyst for hydrogen production from water under visible light. *Nat. Mater.* **2009**, *8*, 76–80. [CrossRef]
38. Hollmann, D.; Karnahl, M.; Tschierlei, S.; Kailasam, K.; Schneider, M.; Radnik, J.; Grabow, K.; Bentrup, U.; Junge, H.; Beller, M.; et al. Structure-Activity Relationships in Bulk Polymeric and Sol–Gel-Derived Carbon Nitrides during Photocatalytic Hydrogen Production. *Chem. Mater.* **2014**, *26*, 1727–1733. [CrossRef]
39. Cui, Y.J.; Ding, Z.X.; Liu, P.; Antonietti, M.; Fu, X.Z.; Wang, X.C. Metal-free activation of H_2O_2 by g-C_3N_4 under visible light irradiation for the degradation of organic pollutants. *Phys. Chem. Chem. Phys.* **2012**, *14*, 1455–1462. [CrossRef] [PubMed]
40. Niu, P.; Zhang, L.; Liu, G.; Cheng, H. Graphene-like carbon nitride nanosheets for improved photocatalytic activities. *Adv. Funct. Mater.* **2012**, *22*, 4763–4770. [CrossRef]
41. Wen, J.; Xie, J.; Chen, X.; Li, X. A review on g-C_3N_4-based photocatalysts. *Appl. Surf. Sci.* **2017**, *391*, 72–123. [CrossRef]
42. Ong, W.J.; Tan, L.L.; Ng, Y.H.; Yong, S.T.; Chai, S.P. Graphitic Carbon Nitride (g-C_3N_4)-Based Photocatalysts for Artificial Photosynthesis and Environmental Remediation: Are We a Step Closer To Achieving Sustainability? *Chem. Rev.* **2016**, *116*, 7159–7329. [CrossRef]
43. Sui, Y.; Liu, J.; Zhang, Y.; Tian, X.; Chen, W. Dispersed conductive polymer nanoparticles on graphitic carbon nitride for enhanced solar-driven hydrogen evolution from pure water. *Nanoscale* **2013**, *5*, 9150–9155. [CrossRef]
44. Liu, J.; Liu, Y.; Liu, N.; Han, Y.; Zhang, X.; Huang, H.; Lifshitz, Y.; Lee, S.; Zhong, J.; Kang, Z. Metal-free efficient photocatalyst for stable visible water splitting via a two-electron pathway. *Science* **2015**, *347*, 970–974. [CrossRef]
45. Naseri, A.; Samadi, M.; Pourjavadi, A.; Moshfegh, A.Z.; Ramakrishna, S. Graphitic carbon nitride (g-C_3N_4)-based photocatalysts for solar hydrogen generation: Recent advances and future development directions. *J. Mater. Chem. A* **2017**, *5*, 23406–23433. [CrossRef]
46. Cao, S.; Low, J.; Yu, J.; Jaroniec, M. Polymeric photocatalysts based on graphitic carbon nitride. *Adv. Mater.* **2015**, *27*, 2150–2176. [CrossRef] [PubMed]
47. Yang, J.; Wang, D.; Han, H.; Li, C. Roles of cocatalysts in photocatalysis and photoelectrocatalysis. *Acc. Chem. Res.* **2013**, *46*, 1900–1909. [CrossRef] [PubMed]
48. Maeda, K.; Wang, X.; Nishihara, Y.; Lu, D.; Antonietti, M.; Domen, K. Photocatalytic activities of graphitic carbon nitride powder for water reduction and oxidation under visible light. *J. Phys. Chem. C* **2009**, *113*, 4940–4947. [CrossRef]
49. Battula, V.R.; Jaryal, A.; Kailasam, K. Visible light-driven simultaneous H_2 production by water splitting coupled with selective oxidation of HMF to DFF catalyzed by porous carbon nitride. *J. Mater. Chem. A* **2019**, *7*, 5643–5649. [CrossRef]
50. Hou, Y.; Laursen, A.B.; Zhang, J.; Zhang, G.; Zhu, Y.; Wang, X.; Dahl, S.; Chorkendorff, I. Layered nanojunctions for hydrogen-evolution catalysis. *Angew. Chem. Int. Ed.* **2013**, *52*, 3621–3625. [CrossRef]
51. Xiang, Q.; Yu, J.; Jaroniec, M. Preparation and Enhanced Visible-Light Photocatalytic H_2-Production Activity of Graphene/C_3N_4 Composites. *J. Phys. Chem. C* **2011**, *115*, 7355–7363. [CrossRef]
52. Sun, Q.; Wang, P.; Yu, H.G.; Wang, X.F. In situ hydrothermal synthesis and enhanced photocatalytic H_2-evolution performance of suspended rGO/g-C_3N_4 photocatalysts. *J. Mol. Catal. A* **2016**, *424*, 369–376. [CrossRef]
53. Yan, J.Q.; Peng, W.; Zhang, S.S.; Lei, D.P.; Huang, J.H. Ternary Ni_2P/reduced graphene oxide/g-C_3N_4 nanotubes for visible light-driven photocatalytic H_2 production. *Int. J. Hydrogen Energy* **2020**, *45*, 16094–16104. [CrossRef]
54. Zhang, X.H.; Peng, B.S.; Zhang, S.; Peng, T.Y. Robust Wide Visible-Light-Responsive Photoactivity for H_2 Production over a Polymer/Polymer Heterojunction Photocatalyst: The Significance of Sacrificial Reagent. *ACS Sustain. Chem. Eng.* **2015**, *3*, 1501–1509. [CrossRef]
55. García-López, E.I.; Lo Meo, P.; Megna, B.; Palmisano, L.; Marcì, G. C_3N_4/reduced graphene oxide based photocatalysts for H_2 evolution from aqueous solutions of oxygenated organic molecules. *Catal. Today* **2022**, in press. [CrossRef]
56. Acharya, R.; Pati, S.; Parida, K. A review on visible light driven spinel ferrite-g-C_3N_4 photocatalytic systems with enhanced solar light utilization. *J. Mol. Liq.* **2022**, *357*, 119105. [CrossRef]
57. Jo, W.K.; Moru, S.; Tonda, S. Magnetically responsive $SnFe_2O_4$/g-C_3N_4 hybrid photocatalysts with remarkable visible-light-induced performance for degradation of environmentally hazardous substances and sustainable hydrogen production. *Appl. Surf. Sci.* **2020**, *506*, 144939. [CrossRef]
58. Bard, A.J. Photoelectrochemistry and heterogeneous photocatalysis at semiconductors. *J. Photochem.* **1979**, *10*, 59–75. [CrossRef]
59. Zhou, P.; Yu, J.; Jaroniec, M. All-Solid-State Z-Scheme Photocatalytic Systems. *Adv. Mater.* **2014**, *26*, 4920–4935. [CrossRef]
60. Yu, Z.B.; Xie, Y.P.; Liu, G.; Lu, G.Q.; Ma, X.L.; Cheng, H.M. Self-assembled CdS/Au/ZnO heterostructure induced by surface polar charges for efficient photocatalytic hydrogen evolution. *J. Mater. Chem. A* **2013**, *1*, 2773–2776. [CrossRef]
61. Fu, N.; Jin, Z.; Wu, Y.; Lu, G.; Li, D. Z-Scheme Photocatalytic System Utilizing Separate Reaction Centers by Directional Movement of Electrons. *J. Phys. Chem. C* **2011**, *115*, 8586–8593. [CrossRef]

62. Yu, W.; Chen, J.; Shang, T.; Chen, L.; Gu, L.; Peng, T. Direct Z-scheme g-C_3N_4/WO_3 photocatalyst with atomically defined junction for H_2 production. *Appl. Catal. B* **2017**, *219*, 693–704. [CrossRef]
63. Idrees, F.; Dillert, R.; Bahnemann, D.; Butt, F.K.; Tahir, M. In-Situ Synthesis of Nb_2O_5/gC_3N_4 Heterostructures as Highly Efficient Photocatalysts for Molecular H_2 Evolution under Solar Illumination. *Catalysts* **2019**, *9*, 169. [CrossRef]
64. Han, M.; Zhu, S.; Xia, C.; Yang, B. Photocatalytic upcycling of poly(ethylene terephthalate) plastic to high-value chemicals. *Applied Catal. B* **2022**, *316*, 121662. [CrossRef]
65. Uekert, T.; Kasap, H.; Reisner, E. Photoreforming of nonrecyclable plastic waste over a carbon nitride/nickel phosphide catalyst. *J. Am. Chem. Soc.* **2019**, *141*, 15201–15210. [CrossRef]
66. Kasap, H.; Achilleos, D.S.; Huang, A.; Reisner, E. Photoreforming of Lignocellulose into H_2 Using Nanoengineered Carbon Nitride under Benign Conditions. *J. Am. Chem. Soc.* **2018**, *140*, 11604–11607. [CrossRef] [PubMed]
67. Uekert, T.; Kuehnel, M.F.; Wakerley, D.W.; Reisner, E. Plastic waste as a feedstock for solar-driven H_2 generation. *Energy Environ. Sci.* **2018**, *11*, 2853–2857. [CrossRef]
68. Zakzeski, J.; Bruijnincx, P.C.A.; Jongerius, A.L.; Weckhuysen, B.M. The Catalytic Valorization of Lignin for the Production of Renewable Chemicals. *Chem. Rev.* **2010**, *110*, 3552–3599. [CrossRef] [PubMed]
69. Shi, C.; Kang, F.; Zhu, Y.; Teng, M.; Shi, J.; Qi, H.; Huang, Z.; Si, C.; Jiang, F.; Hu, J. Photoreforming lignocellulosic biomass for hydrogen production: Optimized design of photocatalyst and photocatalytic system. *Chem. Eng. J.* **2023**, *452*, 138980. [CrossRef]
70. Liu, X.; Duan, X.; Wei, W.; Wang, S.; Ni, B.J. Photocatalytic conversion of lignocellulosic biomass to valuable products. *Green Chem.* **2019**, *21*, 4266–4289. [CrossRef]

Disclaimer/Publisher's Note: The statements, opinions and data contained in all publications are solely those of the individual author(s) and contributor(s) and not of MDPI and/or the editor(s). MDPI and/or the editor(s) disclaim responsibility for any injury to people or property resulting from any ideas, methods, instructions or products referred to in the content.

Review

Mesoporous Silica-Based Catalysts for Biodiesel Production: A Review

Is Fatimah [1,*], Ganjar Fadillah [1], Suresh Sagadevan [1,2], Won-Chun Oh [3] and Keshav Lalit Ameta [4]

1. Department of Chemistry, Faculty of Mathematics and Natural Sciences, Kampus Terpadu UII, Universitas Islam Indonesia, Jl. Kaliurang Km 14, Sleman, Yogyakarta 55584, Indonesia; ganjar.fadillah@uii.ac.id (G.F.); drsureshnano@gmail.com (S.S.)
2. Nanotechnology & Catalysis Research Centre, University of Malaya, Kuala Lumpur 50603, Malaysia
3. Department of Advanced Materials, Engineering at Hanseo University, Seosan-si 356-706, Republic of Korea; wc_oh@hanseo.ac.kr
4. Centre for Applied Chemistry, School of Applied Material Sciences, Central University of Gujarat, Gandhinagar 382030, India; klameta@cug.ac.in
* Correspondence: isfatimah@uii.ac.id

Abstract: High demand for energy consumption forced the exploration of renewable energy resources, and in this context, biodiesel has received intensive attention. The process of biodiesel production itself needs to be optimized in order to make it an eco-friendly and high-performance energy resource. Within this scheme, development of low-cost and reusable heterogeneous catalysts has received much attention. Mesoporous silica materials with the characteristics of having a high surface area and being modifiable, tunable, and chemical/thermally stable have emerged as potential solid support of powerful catalysts in biodiesel production. This review highlights the latest updates on mesoporous silica modifications including acidic, basic, enzyme, and bifunctional catalysts derived from varied functionalization. In addition, the future outlook for progression is also discussed in detail.

Keywords: biodiesel; mesoporous silica; catalyst; heterogeneous catalyst

1. Introduction

Renewable and clean energy is still a very important issue in the world due to declining fossil fuel supplies. In response to this, producing sustainable energy with less CO_2 emission by using biomass feedstock is one of the strategies. In addition, by applying the circular economy principle, biomass feedstock derived from agricultural or forestry cycles can help and enhance food security, environmental security, etc. [1]. As well as reducing global CO_2 emission. Within this scheme, biodiesel is a good alternative due to its huge potential raw material and simple production. In terms of chemical composition and applicability, biodiesel can be a replacement for diesel fuel, which is the petroleum oil fraction containing 8–21 carbon atoms. Originating from different raw material, biodiesel is composed of mono-alkyl esters of long-chain fatty acids derived from various feedstocks such as plant oil, animal fats, or other lipids that are also known as triglycerides. Its composition leads biodiesel to have better performance with respect to diesel fuel, especially in terms of less harmful compounds such as sulfur, less toxicity, and better biodegradability. A life cycle assessment showed a lower CO_2 footprint of by biodiesel than diesel fuel. For example, the reduction in life-cycle greenhouse gas (GHG) is valued as ranging from 40% to 69% compared with petroleum diesel [2,3]. Higher free oxygen in biodiesel results in lower emission and full combustion. However, the corrosion and stability viscosity of biodiesel are still drawbacks of its use. The oxygen-containing compounds such as fatty acids can easily adsorb humidity or react with rubbing surfaces, leading to reduced adhesion between contacting asperities and thereby limiting friction, wear, and seizure. In practical terms, biodiesel harms the rubber parts of machinery [4]. This leads us to recommend the use of biodiesel in a blend with petroleum diesel.

There are numerous techniques that could be used to improve the quantity and quality of biodiesel production from biomass sources such as vegetable oils. Essentially, these depend on the synthesis and quality monitoring of the production, in which esterification and transesterification are the most popular and inexpensive methods. Both esterification and transesterification are alcoholysis reactions, in which, in principle, the ester of triglyceride is converted into methyl ester or ethyl ester by methanol or ethanol as the reactant, and glycerol is produced as a by-product. In addition, esterification converts free fatty acid.

Transesterification is a reversible reaction that can mostly be initiated by the presence of a catalyst. This can theoretically be acid or base catalysis, and in such a situation, simultaneous esterification and transesterification can be performed in a production system. The catalysts can either be in a different phase to the reactant, called a heterogeneous catalyst, or in the same phase as the reactant, called a homogeneous catalyst. The traditional method of transesterification is normally catalyzed using diluted sodium hydroxide as a homogeneous catalyst, which produces a fast reaction but tends to be corrosive. Moreover, recycled waste oils and greases as well as many other low-cost feedstocks usually contain water and are about 10–25% FFA. The existing FFAs could be reacted through saponification, which leads to reduced reaction rates and yields. This is a problem in terms of purification, and to overcome this, an extra pre-treatment step is required in order to meet ASTM biodiesel standards. Apart from these, homogeneous catalysts cannot be recycled and reused, and, moreover, vast amounts of water are required for the washing step. Similar effects are expressed by other homogeneous catalysts such as sulfuric acid or phosphoric acid.

Furthermore, progress towards overcoming these drawbacks has been demonstrated by using solid catalysts as in a heterogenous catalysis system. The appropriateness of the solid catalyst depends mainly on the composition of triglyceride and free fatty acids (FFAs) in oil or fat, and the reaction condition/system [5]. Although only a small amount in the production system, a catalyst could determine the total cost of production due to its effectivity and reusability in the process. By taking the basic principle of heterogeneous catalysis, a solid catalyst should accelerate a reaction via its capability to adsorb the reactant, for more conducting surface reaction, followed by desorption of the product.

This means that not only is the basicity or acidity of the catalyst surface important, but the solid catalyst should have chemical stability, thermal stability, and high adsorption capacity, mainly for organic compounds. Inorganic solids having porosity and a high specific surface area fit this purpose. Considering the molecular size of triglyceride, alcohol, and the possible products of the reaction, the modifiable pore size, and the possible functionalization of the surface to make it more hydrophobic make mesoporous solids fit the application [6]. On the other hand, the development of catalysts for use in biodiesel production itself has progressed, for example, the use of various types of acid/base catalysts that could be active at room temperature. These were developed and adapted to consider the operational conditions of the reaction with the lowest possible energy. With a similar purpose to the other mechanism, enzyme-based catalysts are also an advancement that is considered fundamental and prospective for the future. This forced modifiable solid supports to be adaptive in terms of catalyst sustainability [5,7].

Inorganic synthetic and natural solids such as metal oxide, mixed noble metal oxide composites, zeolite, clay, carbon, silica, silica alumina, and their modified forms have been reported. Among these materials, mesoporous silica-based catalysts demonstrated a superiority for biodiesel production. Based on the progress of published papers, an increasing amount of research on mesoporous silica-based catalyst is demonstrated, as can be seen from the comparison presented in Figure 1.

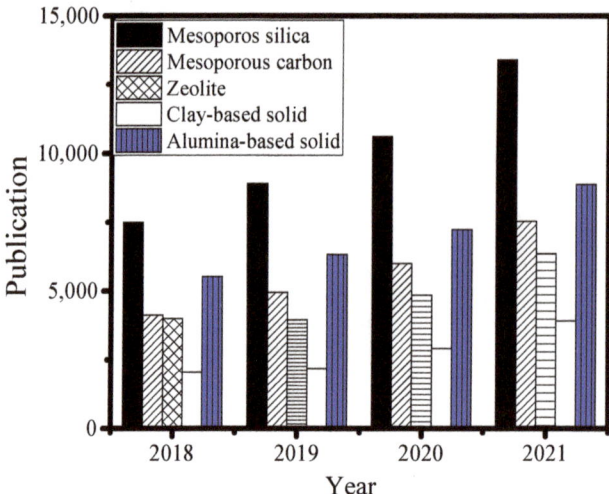

Figure 1. Popularity of mesoporous silica-based catalysts for biodiesel production [Source: SCOPUS database, February, 2023].

The intensive interest in mesoporous silica-based catalysts is expressed by the increasing number of published papers ranging from 2018 to 2021. Individually, the objectives are higher compared to the published papers on mesoporous carbon, zeolite, clay, and alumina-based solids catalysts [8–11]. These indicate the superiority, potency, and possible intensification of mesoporous silica-based catalysts. Some highlighted features of mesoporous silica-based catalysts in biodiesel production are correlated with the tunable porous size and structure and easy embedding of functional groups consisting of either acid/base sites or the hydrophobic/enzymatic sites. Referring to these objectives and publication progress, this review studied the bigger picture of the development of mesoporous silica-based catalysts for biodiesel production, potency, and future challenges in depth [12–14]. The study will begin with the kinds and classification of mesoporous silica materials, modified mesoporous silica in terms of biodiesel production mechanisms, and, furthermore, some possible challenges that need to be overcome and explored for future industrialization.

2. Various Types of Mesoporous Catalysts

Mesoporous catalyst materials have a well-defined pore network with sizes ranging from 2 to 50 nm. This mesoporous material exhibits a unique pore structure with interconnected voids and channels, giving it porosity and a large surface area. Various mesoporous catalyst materials have been developed, including mesoporous carbon materials, metal oxides, metal organic frameworks (MOFs), and mesoporous silica (MCM-41 and SBA-15). These materials have characteristics depending on the raw materials and synthesis methods used. Several previous studies have reported various metal oxides as fillers for mesoporous materials such as Al_2O_3, ZrO_2, TiO_2, and ZnO_2, showing good performance as catalysts due to their high firmness and low temperature deposition capacity [15]. Apart from metal oxides, a mesoporous material that has been widely developed and used as a catalyst in the biodiesel process is mesoporous carbon material. Mesoporous carbon provides a large surface area and uniform porosity, so it has a high absorption of long FFA chains during the conversion process [16]. However, the catalytic performance of this material is affected by the modification or functionalization of the functional groups on its surface. Several methods, such as chemical and electrochemical reduction, can be used to achieve this modification. Apart from these two types of mesoporous materials, mesoporous MOFs are promising in the catalytic process of biodiesel production. MOFs are a class of crystalline materials consisting of metal ions or clusters coordinated with organic ligands. These materials have a highly ordered structure with an orderly arrangement of metal knots or groups interconnected by organic linkages, forming a three-dimensional framework. MOF fabric has a high surface area and porosity, so it is very suitable as a catalyst. However, this material still has disadvantages such as low stability, a complex synthesis process, and, in some cases, the larger pores in the mesoporous MOFs can be blocked or inaccessible to molecules, reducing mass transport efficiency and inhibiting the material's catalytic activity or adsorption capacity [17]. The latter is a mesoporous silica material with a hexagonal structure with several prominent properties, such as high porosity, thick pore walls, and high thermal stability. Compared to metal oxide catalysts and others, using mesoporous silica as a catalyst in biodiesel production offers several advantages, such as high surface area, modifiable pore size and structure, good thermal stability, acid–base versatility, and low cost or availability of raw materials. For example, mesoporous silica catalysts can be modified to have different acidic or basic properties depending on the specific reaction requirements. This versatility makes it possible to adapt the trigger's properties to the particular conditions of the biodiesel production process, such as esterification or transesterification reactions. Metal oxide catalysts usually have a fixed acidity or basicity, limiting their flexibility [18]. Silica-based catalysts are generally more cost-effective due to the abundance of raw materials, making them easier to obtain for synthesis and commercialization than metal oxide-based catalysts.

3. Mesoporous Silica Materials and Modified Forms

Mesoporous silica is an inorganic polymer nanomaterial having a high specific surface area and specifically with a pore size ranging from 2 to 50 nm. In principle, mesoporous silica is synthesized through a controllable polymerization of silica precursors, mainly based on the Stöber method [19]. Typically, the synthesis mechanism involves the hydrolysis and condensation of silanes in basic, acidic, or neutral aqueous solution, called sol–gel processes. Such techniques with various methods of polymerization, the use of surfactant or template, and other specific conditions of sol–gel mechanisms were applied to achieve a uniformly adjustable pore size and highly ordered channel structure [20]. From these synthesis conditions, an impressive diversity of synthesis approaches have been developed and patented and now well-established for specified mesoporous silica materials. In 1992, a new family of mesoporous material was synthesized by Mobil Oil Company, and then called MCM-41. The material has a homogeneous ordered pore size distribution ranging between 2 nm and 10 nm, and went on to form the basis of developing cubic-MCM-48 and lamellar-MCM-50. These pioneering findings were followed by various kinds of mesoporous silica materials such as Michigan State University (MSU-1), hexagonal mesoporous silica (HMS), Technische Universiteit Delft (TUD), Santa Barbara Amorphous (SBA), Meso Cellular Form (MCF), etc. [21–23].

Based on the synthesis mechanism, it can be concluded that there are some crucial factors that determine the pore size distribution and ordered structure, i.e., the rate and mechanism of polymerization, gelation mechanism, and the presence of surfactant, co-surfactant, solvent, and co-solvent types. The surfactant influences the pore size distribution due to their molecular size, hydrophobicity, and ionic/non-ionic state [19].

The highly specific surface area of mesoporous silica materials provides potential for various heterogeneous catalysis applications. From a green chemistry perspective, some mesoporous silica materials can be prepared using low-cost raw materials such as clay (bentonite, kaolinite), fly ash, and bottom ash [13,24,25]. Regarding the surface catalysis mechanism, specific features such as pore size can be adjusted according to the type of material. Moreover, special characteristics required in catalysis such as surface acidity, basicity, or the strength of certain molecular interactions can be achieved by modifying the surface of the material. The attachment of certain functional groups to the surface can be carried out based on the surface properties of the material, which is rich in hydroxyl groups. The modification can be physical or chemical modification. Physical modification is mainly performed through adsorption, wrapping, and other physical effects. Meanwhile, chemical modification commonly utilizes the Lewis acidity–basicity interaction between the coupling agent containing organic functional groups. Through this mechanism, functional groups such as -SH, -NH$_2$, -Cl, or -CN could be tethered onto the surface via a grafting functional group [8,26,27]. Figure 2 represents the simple mechanism of sulfonic acid tethering on mesoporous silica's surface to enhance the surface acidity for the transesterification reaction.

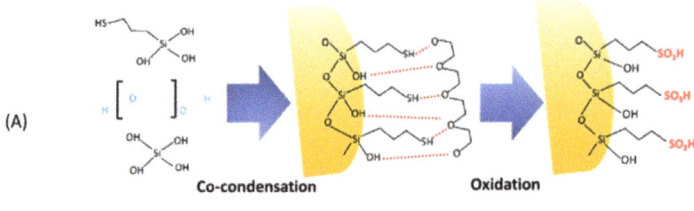

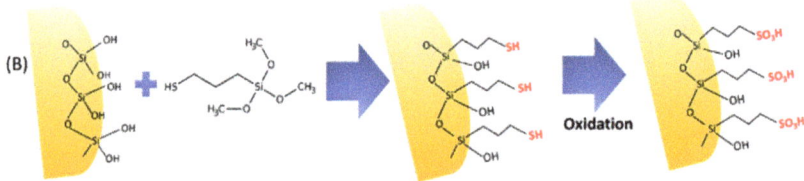

Figure 2. Scheme of sulfonic acid tethering onto mesoporous silica through (**A**) co-condensation and (**B**) post-synthesis methods.

Generally, there are two main mechanisms for surface modifications: the simultaneous/co-condensation sol–gel process and post-synthesis methods. Co-condensation is a one-step synthesis method in which the organic, acid/base, or enzyme precursor is added into the reactant of silica polymerization. With different steps, in the post-synthesis, the functional groups are tethered or anchored onto the silica surface after the mesoporous structure is created. Figure 2 describes the mesoporous silica functionalization by an SO$_3$H functional group through the co-condensation and post-synthesis methods [28,29]. According to several works [29–32], both methods result in SO$_3$-functionalized mesoporous silica with enhanced surface acidity, acid catalysis capability, and reusability for biodiesel production without any significant difference. However, somehow, the features could be remarkably different for other functionalizations. In the next part, a description of the kind, method, and specificity of functionalized mesoporous silica as a catalyst for biodiesel production is discussed. The discussed functionalization is classified into: (i) acid or base surface functionalization; (ii) metal or metal oxide supporting; (iii) ionic liquid functionalization; and (iv) anchoring enzymes on the surface of mesoporous silica.

3.1. Acid- or Base-Functionalized Mesoporous Silica

For the transesterification and esterification mechanisms in biodiesel production, both acid and base catalysts could be employed. Acid surface functionalization onto mesoporous silica is performed through the attachment of sulfonic acid, organo-sulfonic acid, phosphoric acid, tungstophosphoric acid, etc. Meanwhile, basicity enhancement by attaching an alkaline earth base and alkylamine functionalizations are the strategies that have been reported. In many kinds of catalysis, sulfonic acid and amine modifications are the most common functional groups introduced onto mesoporous silica. Sulfonic acid-functionalized MCM-41, SBA, MCM-48, and KIT are some examples in this scheme, and, specifically for biodiesel production application, Table 1 lists the details of the acid functionalization method and the impact of the functionalization on the catalytic activity in biodiesel production.

A series of organosulfonic acid-functionalized SBA for the transesterification of soybean oil was reported in [33]. The organic sulfonic acids are propyl-sulfonic acid ($Pr-SO_3H$), arene-SO_3H ($Ar-SO_3H$), fluoro-sulfonic acid ($F-SO_3H$), and modified arene-SO_3H. Except for $F-SO_3H$ and modified arene SO_3-H functionalization, the organosulfonic acid-SBA 15 samples were prepared using co-condensation methods. Tetraethyl orthosilicate (TEOS, 98%, Aldrich) was used as silica precursor in the synthesis, and for the $Pr-SO_3H$ and $Ar-SO_3H$ functionalizations, (3-Mercaptopropyl) trimethoxy silane (MPTMS) and 2-(4-chlorosulfonyl phenyl) ethyltrimethoxy silane (CSPTMS) were utilized as the anchoring agent before oxidization into organo-sulfonic acid conversion. The post-synthesis of $F-SO_3H$-SBA-15 was performed by interacting perfluorosulfonic acid precursor with calcined SBA-15 (refer to previous investigation [34]). Meanwhile, the $Ar-SO_3H$ was then furthermore modified through a reaction with methoxytrimethyl and methylation to obtain modified $Ar-SO_3H$-SBA-15. The surface acidity of the materials was assayed through potentiometric titration using TMA as the basic standard solution. Details from the surface acidity measurement revealed that there is a significantly increased exchange of TMA with the modification. The surface acidity is relevant with the increased yield of soybean transesterification by using butanol over a microwave-assisted reaction. Among the organosulfonic acids, $F-SO_3H$-SBA-15 was the most superior catalyst as it produced about 80% of butyl ester in a short time. However, the recyclability of the catalyst is still an important issue to be resolved as the catalysts lost their activity after first usage. This is demonstrated by $F-SO_3H$-SBA-15, in which the catalyst could not be regenerated. The surface acidity seems to be diluted in the reaction system. A similar phenomenon of the superiority of $F-SO_3H$-SBA-15's catalytic activity was demonstrated in the methanolysis of palm oil, in comparison with the activity of $Pr-SO_3H$-SBA-15 and $Ar-SO_3H$-SBA-15. In addition, the $Ar-SO_3H$-SBA-15 showed recyclability until the third use without any significant yield reduction [34]. Similarly, SO_3H-HM-ZSM-5 exhibited excellent improvement of oleic acid conversion until 100% at a low temperature of reaction (88 °C) [35].

From several studies presented in Table 2, it can be summarized that the surface acidity and pore distribution of functionalized mesoporous silica are more important for determining catalyst activity and yield than the hydrophobicity character [36,37]. Although, in general, sulfonation tends to decrease the specific surface area of mesoporous silica materials, a significant enhancement of surface acidity was expressed by post-synthesis sulfonic acid functionalization onto MCM-48, MCM-41, KIT-6, and SBA-15 [38,39]. The significant role of surface acidity is also reflected by the reduction in activation energy (E_a) caused by aril sulfonic acid functionalization, which referred to acid stabilization rather than the hydrophobicity and specific surface area [40]. However, sometimes, the compared physicochemical characterization highlighted that the specific surface area data are in line with the order of surface acidity. Both characteristics are proportional with the catalytic activity, so the conversion and yield are in following order: SO_3H-MCM-48 > SO_3H-MCM-41 > SO_3H-SBA-15. According to a more detailed assay using NH_3 temperature program desorption (NH_3-TPD), the materials are rich in low acidity, as identified by the release of protons at temperatures ranging from 100 to 350 °C. The acidity fits with the optimum reaction for palmitic acid, which lay at around 160 °C. A comparative study on the type of organosulfonic acid was performed by evaluating propyl sulfonic acid ($Pr-SO_3H$), phenol sulfonic acid (-$PhSO_3H$), and the methyl- and hexyl-grafted phenol sulfonic acid SBA and MCF [41]. Remarkably, it is noted that the grafting for hydrophobicity enhancement does not effectively increase the conversion and selectivity, but it tends to maintain the stability of the catalyst, at least for a second catalytic run [34,42].

In the co-condensation preparation of propylsulfonic acid ($PrSO_3H$)/KIT-6, the $PrSO_3H$ content affects the pore expanding. As the pore accessibility becomes rate-limiting in the esterification of long-chain fatty acids such as lauric and palmitic acids, the increasing pore size by sulfonic acid attachment improved the turnover frequency until 70% [43]. A more simple procedure to enhance surface acidity was expressed by the heteropolyacid ($H_3PW_{12}O_{40}$)/(HPA) impregnation of mesoporous silica structure. An increased conversion of palmitic acid was demonstrated by HPA/MCM-41 [44]. The catalyst showed 100% conversion toward palmitic acid and a turnover number of 1992. It is also important to note that the catalyst is easily recyclable, without any activity change until the fourth cycle.

A study on the preparation of HPA-impregnated KIT-6 (HPA/KIT-6) with varied content (10, 20, and 40 wt%.) as a catalyst for neem oil transesterification revealed the significant improvement in yield, conversion, and reusability of the catalyst, which was related to the presence of surface acidity in the composite. With a similar objective, a study on HPA immobilization in MCM-48 was also performed with the content ranging from 0 to 50% for palmitic acid transesterification using cetyl alcohol [45]. The combination of Lewis and Brønsted acidity influencing the conversion of oleic acid was reported for zirconium-doped MCM-41-supported WO_3. This depends on the WO_3 loading, ranging from 15 to 25 wt%. The better WO_3 dispersion supports the activity and stability at high temperatures [46]. High conversion was maintained at 97% even in the condition of the presence of 5.5 wt% of water, suggesting that water is not adsorbed on the active centers of the catalyst and oleic acid molecules.

However, it is not only surface acidity that influences the optimum conversion, but the fine dispersion of small clusters of HPWA on the catalyst surface also plays a significant role. The surface area and pore distribution do not linearly respond to the HPA content, but an optimum is reached at about 15 wt%. [47]. A similar effect on the increased pore distribution, where the increasing pore accessibility of the fatty acid led to optimum surface reaction of canola oil transesterification, was represented by HPA immobilization on aluminophosphate [47]. A similar trend was reported for the comparison of HPA dispersion onto MCM-48, which showed the order of catalytic activity was as follows: MCM-48 > SBA-15 > MCM-41. The higher activity is correlated with the structural geometry of the pores [48]. Furthermore, the combination of sulfonic acid and HPA expressed a synergistic effect, enhancing the activity and hydrothermal stability of the mesoporous silica [49].

Table 1. Various acid/base-functionalized mesoporous silica for transesterifications.

Catalyst	Preparation Method	Transesterification Reaction	Remark	Reference
Ar-SO_3H/SBA-15	Aryl-sulfonic acid-functionalized SBA-15 material was syn-thesized by following the one-step co-condensation procedure.	Crude soybean oil transesterification	Catalyst showed recyclability until third use without any significant yield reduction. Arene structure gave higher surface acidity and tends to give better catalyst stability.	[34]
SO_3H/SBA-15	SO_3H/SBA-15 was prepared through the co-condensation method.	Olive pomace oil transesterification	Increased conversion of oleic acid was reported. The catalyst has reusability properties.	[36]
Propyl sulfonic-KIT-6	KIT-6 silica functionalized with sulfonic acid through the co-condensation method.	Cashew nut oil transesterification by butanol	The KIT-6 propylsulfonic acid catalyst was able to produce a 70% butyl ester yield.	[39]

Table 1. Cont.

Catalyst	Preparation Method	Transesterification Reaction	Remark	Reference
SO$_3$H/MCM-41	Material was prepared by using polystyrene as a template and p-toluenesulfonic acid (TsOH) as a carbon precursor and —SO3H source.	Oleic oil transesterification	Catalyst showed recyclability until fifth use without any significant conversion reduction.	[37]
Propyl sulfinic-KIT 6	KIT was prepared by using pluronic acid P123: TEOS: BuOH:HCl:H$_2$O = 0.017:1:0.31:1.83:195. After calcination, silicas were functionalized with sulfonicacid groups by post-grafting using mercaptopropyl trimethoxysilane (MPTS 95%) and the thiol was converted by oxidation using H$_2$O$_2$.	Propanoic and hexanoic esterification	The enhancements in turnover frequency (TOF) toward propanoic and hexanoic acid esterification were 40 and 70%, respectively.	[43]
12-Tungstophosphoric acid anchored to MCM-41	MCM-41 was synthesized through the sol–gel method using surfactant cetyl trimethyl ammonium bromide (CTAB), NaOH, and TEOS. 12 Tungstophosphoric acid (12-TPA) was impregnated by stirring at 100 °C for 10 h.	Transesterification of palmitic acid	The catalyst shows high activity in terms of 100% conversion toward palmitic acid and a high turnover number of 1992.	[44]
12-TPA/MCM-48	MCM-48 was prepared through the sol–gel method with composition of 1 M TEOS: 12.5 M NH$_4$OH:54 M EtOH: 0.4 M CTAB: 174 M H$_2$O. 12-TPA was impregnated by incipient impregnation.	Transesterification of jatropha oil (JO)	The uniform dispersion of HPA inside the 3D channels of MCM-48 influenced the increasing activity for the esterification of oleic acid under mild conditions. The catalyst could be used for biodiesel production from WCO and JO with very high conversion: 95% and 93%, respectively.	[48]
HPA/KIT-6	KIT-6 was synthesized through hydrothermal condensation using precursor at a molar ratio of 1 TEOS: 0.017 P123:1.83 HCl (35%):1.3 n-BuOH: 195 H$_2$O. HPA functionalization to KIT-6 was conducted by impregnation.	Transesterification of neem oil	The conversion of neem oil depends on Brønsted acid sites, large surface area, pore size, and the fine dispersion of HPA in the composite. The optimum HPA content in the composite is 20%.	[50]
1,5,7-triazabicyclo [4.4.0]dec-5-ene (TBD)/SBA-15	SBA-15 was prepared using a P123 templating agent. 1,5,7-triazabicyclo [4.4.0]dec-5-ene [TBD] was functionalized through the adsorption method in a nitrogen environment.	Transesterification of soybean oil	The higher the grafted base amount, the higher the FAME yield.	[51]

Table 1. *Cont.*

Catalyst	Preparation Method	Transesterification Reaction	Remark	Reference
TBD/MCM-41	Material was prepared through the post-synthesis method. TBD was anchored by immersing MCM-41 in TBD using tetrahydrofuran (THF) as a solvent, followed by filtration.	Transesterification of soybean oil	The TON was 57, higher than that of MCM-41 (48). The catalyst is reusable.	[52]
Piperazine/MCM-41	Material was prepared through the post-synthesis method. Piperazine was anchored by using the reflux method in dry toluene and propylamine in a N_2 atmosphere.	Transesterification of soybean oil	The TON was 1270, higher than that of MCM-41 (48). The catalyst is reusable without any activity loss in the second cycle.	[52]
Amine-functionalized SBA-15 and MCM-41	Material was prepared through the post-synthesis method. Amine functionalization was conducted by grafting in anhydrous toluene under argon.	Transesterification of glyceryl tributyrate	The aniline-functionalized OMS materials display the highest conversion in transesterification.	[53]
Diphenylamine(DPA)/SBA-15 and DPA/MCM-48	Material was prepared through the co-condensation method.	Transesterification of oleic acid	Diphenylammonium salts were immobilized onto meso-porous silicas using either the co-condensation or grafting technique. The resulting catalysts were highly effective at esterifying the FFA in greases (12–40 wt% FFA) to FAME but displayed only minimal activity in transesterifying glycerides.	[54]
Sulfonated phosphotungstic acid-modified ordered mesoporous silica (HPW/OMS-SO$_3$H)	HPW/OMS-SO$_3$H was prepared through the co-condensation method in non-hydrochloric acid solution.	Transesterification of oleic acid	The catalyst showed very high hydrothermal stability and recycling performance. The reaction catalyzed by 0.3HPW/OMS-SO$_3$H-5 followed pseudo-first-order kinetics, and Ea was found to be 22.46 kJ/mol.	[49]

The surface basicity modification of mesoporous silica is usually performed by anchoring alkylamine and other nitrogen ligands, and the immobilization of alkali/alkali earth oxide such as CaO, MgO, and Li$_2$O [45,52,53,55]. The modification of mesoporous silica with 3-aminopropyltriethoxysilane (RNH$_2$), 2,3-aminoethylamino)propyltrimethoxysilane (NN), and 3-diethylaminopropyltrimethoxysilane (DN) showed increased basicity, which related to the increasing turn-over number. The compared kinds of amines confirmed the superior performance of the tertiary amine [56]. The presence of tertiary amines can also facilitate an increase in nucleophilicity due to the presence of three alkyl groups, making the catalyst more reactive and able to form more robust interactions with the reactants during the transesterification process. This process causes faster kinetics and increases efficiency compared to secondary and primary amines.

SBA-15 immobilized with 1,3-dicyclohexyl-2-octylguanidine (DCOG) was prepared due to the fact that guanidine compounds are strong bases with higher basicities comparable to carbonates and alkali hydroxide [51,57,58]. The preparation involved the covalent attachment of DCOG onto the SBA15 surface, which increased the solid basicity and recyclability [58,59]. With the similar reason of the strength of basicity, 1,5,7-triazabicycloij[4.4.0]dec-5-ene (TBD) and 1,1,3,3-tetramethylguanidine (TMG) were used to functionalize SBA-15 and MCF [41,51]. A study on TBD functionalization to SBA-15 showed that the higher the amount of TBD grafted to SBA-15, the higher the FAME yield obtained, even though the SBA-15 support does not express the activity. Conversely, the free TBD base is extremely reactive, as a complete trans-esterification into FAMEs was observed after just 1 h of reaction.

From the study, the functionalized materials became active at lower reaction temperatures, and the activity is greatly influenced by the support. The activity is closely related to the surface basicity, which results from the stability of the cations playing a role in the catalysis mechanism. Similar to the result of [56], the catalytic activity is related to the stability of cations caused by the delocalization of the positive charge on the electronegative nitrogen atoms, and also the possible resonance. In line with this finding, the superiority of guanidine-modified MCM-41 with respect to the piperazine-modified sample was reported. The turnover number (TON) of guanidine-modified MCM-41 in the soybean oil transesterification was 1270, while that of piperazine-MCM-41 was 57. A comparison of the modifiers, presented in Figure 3, shows the superiority of aniline-functionalized ordered mesoporous silica (OMS) [53]. However, when comparing mesoporous silica functionalized with aniline and tertiary amine, the sample functionalized with aniline has a more substantial base site and has higher basicity than the tertiary amine. Basicity is very important in the transesterification reaction. After all, it helps activate the alcohol in the conversion to biodiesel. In addition, mesoporous silica functionalized with aniline has a large pore size, which allows for increased accessibility of the reactants into the active site of the catalyst [60].

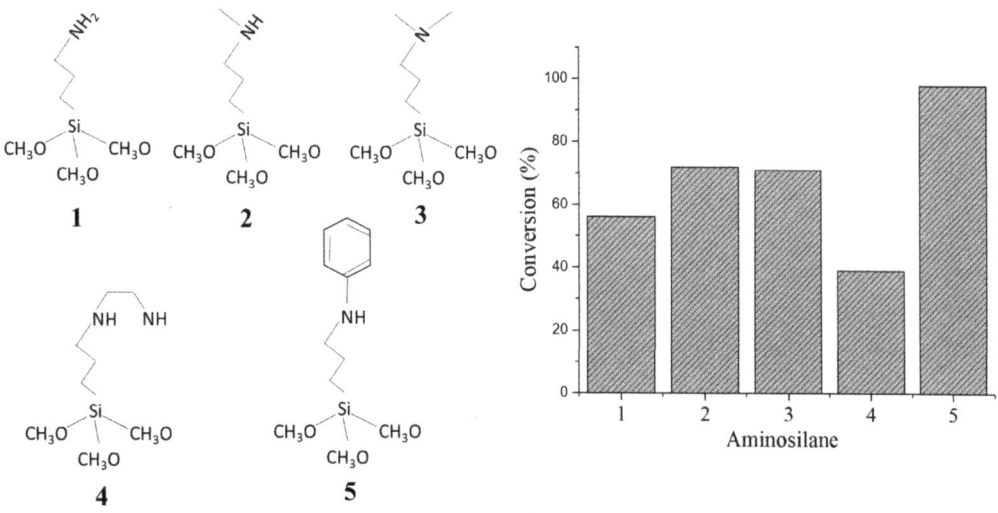

Figure 3. Comparison on the amine modifier n transesterification conversion.

The loss of activity becomes an important issue of acid/base-functionalized mesoporous silica. According to some papers, the reduced activity is mainly caused by the blocking solid porosity and the leaching of the active phase. One interesting case was expressed by guanidine-modified MCM-41, in which the conversion decreased significantly from 99% in the first use to 86% and 26% in second and third usage, respectively. An investigation on the possibility of guanidine leaching indicated that the anchored functional group is still maintained. 13C CPMAS NMR analysis of the catalysts demonstrated that the decreased catalytic activity was caused by the neutralization of the basic sites by the free fatty acids present in the oil. The binding causes catalyst poisoning [52]. In addition, the leaching of non-chemically bonded amine groups is the main reason for the loss of activity after recycling. Other factors include the non-intensive interaction between bulky surface and triglyceride, which leads to the tendency to be more active for converting free fatty acids (FFAs) [54]. Recyclability is achievable by the combination of acid/base-functionalized magnetic mesoporous silica [61] and is similarly reported for other modifications such as enzymes and metal/metal oxide [61,62].

3.2. Metal- or Metal Oxide-Impregnated Mesoporous Silica

Adding metal species or metal oxides to mesoporous silica can produce new properties and functions, especially for catalyst applications. The synergistic combination of mesoporous silica's high surface area and pore structure with the catalytic properties of metals or metal oxide species opens opportunities for advanced materials with tailored functionality and enhanced performance. The choice of metal modification or metal oxide for mesoporous silica is application-dependent. As a catalyst for biodiesel production, the catalytic activity and stability of the material will be affected by the type of metal or metal oxide modified on the surface of the mesoporous silica. Therefore, further studies and literature reviews related to comparing the use of metal or metal oxides need to be conducted. Other than amine functional groups, impregnating CaO, MgO, and Li_2O into mesoporous silica materials are also popular methods for improving total basicity and the yield of biodiesel. Impregnation is a simple method to disperse the alkaline earth oxide. Table 2 presents some metal/metal oxide-modified mesoporous silicas and their performance in biodiesel production. In general, the dispersion does not affect the structure of the mesoporous silica much but generally reduces the specific surface area of the material. The high total basicity, homogeneous dispersion of metal oxide, particle size, and strong interaction between support and metal oxide determine the conversion. This is reflected by the higher activity of CaO/SBA-15 compared to CaO/MCM-41 [63]. In order to achieve a homogeneous distribution of Ca in the structure of the nanocomposite, a synthesis route can be taken by including calcium acetate or calcium citrate in the silica gel hydrolysis [64]. From the range of studies, it can be concluded that the modification of basicity or acidity in mesoporous silica significantly influences the catalytic activity and reusability of the catalyst, but the main factor in increasing the features is the character of the support. The specific surface area of the mesoporous silica determines the possibility of functional organic molecules blocking the solid pores, but, on the other hand, changes the surface hydrophobicity.

Even though the study on MgO dispersion into various mesoporous silicas such as MCM-41, KIT-6, and SBA-15 revealed that the characteristics of the host such as surface area, basicity, and porosity are important for the conversion enhancement, as well as the surface Mg concentration, the improvement in the catalyst activity seems to be from the combined effects of the multiple attributes of its host material and metal oxide distribution [65,66]. This shows that the activity of a base catalyst does not predominantly depend upon its basic properties. As an example, the stability of MgO/ZSM-5 is recognized as an important parameter to ensure the reusability of the catalyst, as the yield of biodiesel produced from *Spirulina* oil was still above 85% until the fifth cycle. The conversion obeys pseudo-first-order kinetics with the activation energy of 49.67 kJ/mol [67]. It is also important to note that mesoporous silica itself does not show any catalytic activity for the reaction. Within the scheme of enhancing surface acidity/basicity, supporting metals and metal oxides added onto mesoporous silica materials are well-known active catalysts for various reactions. Similar to the functional group modification, in principle, the supporting procedure could be the incorporation of metal into the framework of mesoporous aluminosilicates to improve the acidity and, in the other method, the impregnation of metal or metal oxide.

Various metal and metal oxides such as W, Ce, Zr, Ca, Mo, and Zn were reported as the active site for biodiesel conversion from many plant oils [68,69]. In addition, the combination of both methods for bimetal-modified mesoporous silica has also been attempted. For example, Ce/Al-MCM-41 and Zr/Al-MCM-41 were prepared through the impregnation of Ce and Zr metal precursor into Al-MCM-41, which was directly synthesized through the sol–gel method of Si and Al precursors [70]. In the case of Al-MCM-41, Al-MCM-48, or Al-SBA-15, for example, the precursors are usually transformed via alkali fusion in the hydrothermal condition with the presence of surfactant as a templating agent [71–75]. The isomorphic substitution of silica with alumina improves the Brønsted acidity of the mesoporous sample, and the presence of the ionic charges of the structure means there is potential to be replaced with metal or metal oxide precursors. As-synthesized Ce/MCM-41 shows extreme improvement for sunflower conversion into biodiesel with respect to the homogeneous distribution of Ce in the structure, as the identified agglomeration leads to the reduced conversion.

In addition, the reusability of Ce/MCM-41 is an important feature with respect to the pristine MCM-41, in which loss of the activity is easy [76]. The loading of active metals such as Ce, Ca, Ti, and Zn onto Al-MCM-41 or other mesoporous materials support could be performed various intensification methods. One of these is ultrasound irradiation [76,77]. The Ca dispersed into Al-MCM-41 demonstrated the significant effect of ultrasound-assisted dispersion on the uniformly distributed Ca in the nanocomposite [70]. A similar result was reported for ultrasound-assisted Zr dispersion and the sulfation of Zr/MCM-41. The irradiation of ultrasonic waves during sulfuric acid impregnation over Zr/MCM-41 improves the morphology, particle size, surface area, and particle distribution compared to non-sonicated sample. More intense irradiation leads to the formation of smaller particle sizes, and highly dispersed particles were found in intensely irradiated Zr/MCM-41 catalysts [78]. The higher dispersion of Zr with respect to the higher specific surface area determines the accessibility of the reactant into solid surface and, furthermore, gives more stability until the fifth cycle. The adoption of green synthesis of nanoparticles for nanocomposite preparation is also a challenging topic. A homogeneous dispersed ZnO/MCM-41 catalyst was successfully prepared by using orange peel extract [79].

Table 2. Some metal/metal oxide-modified mesoporous silica catalysts and their performance in biodiesel production.

Catalyst	Transesterification Reaction	Remark	Reference
Zirconium-doped MCM-41-supported WO_3.	Transesterification of oleic oil	High conversion was maintained at 97% even at in condition of the presence of 5.5 wt% of water, suggesting that water is not adsorbed on the active centers of the catalyst and oleic acid molecules.	[46]
CaO/SBA-15	Transesterification of sunflower oil and castor oil by using methanol	The conversion was 65.7 and 95% for sunflower oil and castor oil, respectively.	[63]
MgO/SBA-15	Transesterification of lauric acid with butanol	Incorporation of Mg into mesoporous silica does not affect the structure. The catalysts were able to promote the esterification of lauric acid with 1-butanol, giving good yields at ambient pressure.	[80]
MgO/ZSM-5	Transesterification of *Spirulina* oil	Catalyst is reusable until the fifth cycle.	[67]

Table 2. Cont.

Catalyst	Transesterification Reaction	Remark	Reference
MgO/KIT 6	Transesterification of vegetable oil	Reaction conversion of 96%.	[66]
Cs/SBA-15	Transesterification of canola oil	A conversion of 99% was achieved with the pressure of 3 MPa and reaction temperature of 260 °C.	[81]
Ce/MCM-41	Transesterification of sunflower oil	Catalyst shows stability, which is related to the homogeneous distribution of Ce in the nanocomposite.	[76]
Zr/MCM-41	Transesterification of sunflower oil	Catalyst was prepared through ultrasound irradiation. It was found that the frequency of ultrasound influenced the Zr distribution and specific surface area, thus affecting the catalyst stability.	[78]
ZnO/MCM-41	Jatropha oil transesterification	Catalyst was prepared using orange peel extract as a green reductor of ZnO nanoparticles. Catalyst showed high activity (97% conversion).	[79]
TiO_2/MCM-48	Palmitic acid photocatalytic transesterification	The prepared material shows photocatalytic activity for the photocatalytic esterification of palmitic acid, and the material is recyclable until the 10th cycle.	[77]
Cr/SiO_2	Palmitic acid photocatalytic transesterification	The prepared material shows photocatalytic activity for the photocatalytic esterification of palmitic acid under solar irradiation, and the material is recyclable until the 10th cycle.	[82]

Photocatalytic transesterification is another green chemistry approach for biodiesel production, and the utilization of TiO_2 nanoparticles and TiO_2/MCM-48 was successfully recorded [77,83]. In principle, as the photocatalyst is impinged by the UV light, there will be electron excitation from the valence band (VB) into the conductance band (CB), at the same time, generating the same number of photogenerated holes (h^+) in the VB. The combination of the generated hole with adsorbed methanol as the reactant leads to the formation of free radicals ($CH_3O \cdot$) and, similarly, with the adsorbed free fatty acids (FFAs), R-COOH generates R-COO·. The collisions between methanol radicals ($CH_3O \cdot$) and the carbonyl carbons on R-COOH· form intermediates. In a further third step, the dehydration of the intermediate rearrangement produces biodiesel. Technically, vigorous stirring in the process of reactant adsorption and product desorption is conducive to the reaction since it facilitates the transfer rate of reactants and products at the interface and in the liquid phase [77,84,85]. TiO_2 is a well-known photoactive material for this mechanism, as the band gap energy effectively catches the photon source for creating radicals for the mechanism. As a non-photocatalytic mechanism, the study on palmitic oil transesterification expressed that the conversion depends on several factors of the reaction such as the length of the carbon chain of methanol, methanol:oil ratio, and stirring speed. For the various alcohols, the conversion was increased in the following order: butanol < isopropanol < ethanol < methanol [77]. By a different photocatalytic mechanism, Cr/SiO_2 demonstrated photoactivity in transesterification under solar radiation. The prepared Cr/SiO_2 composite photocatalysts expressed strong photo-absorption both in the UV and visible electromagnetic regions (230, 380, 440 nm, and between 500 and 700 nm). The photo-illumination of the photocatalyst caused an electronic reduction of Cr^{6+} and Cr^{3+}, which generated H^+, $CH_3O \cdot$, and R-COOH· radicals in high concentrations at the photocatalyst surface. A further surface mechanism is similar to the mechanism of the TiO_2 photoactive material [82]. In order to enhance the stability of the catalyst for low-quality oil to biodiesel production, the biofunctionalization of metals/metal oxides with acid/base functional groups was attempted [86]. Bifunctional acid–base catalyst of iron (II)-impregnated double-shelled hollow mesoporous silica (Fe/DS-HMS-NH) and amine-functionalized Ni/Mo-mesoporous silica are the examples for this scheme [87,88]. Based on the studies that have been explored, metal oxide modifications, such as supported metal oxide nanoparticles, can offer better longevity and stability for biodiesel production catalysts. However, metal modification can still be effective, especially if the reaction conditions are mild or cost considerations are critical.

3.3. Mesoporous Silica-Immobilized Lipase

The use of lipase (triacylglycerol acylhydrolase; EC 3.1.1.3) in biodiesel production was reported in 1990 by Mittelbach for the alcoholysis of sunflower oil. The relatively high conversion of nonaqueous alcoholysis in the mild reaction conditions, easy isolation of biodiesel and glycerol of glycerin without further purification, and production without the formation of chemical waste are general attractive features of lipase-catalyzed conversion. As with many other enzymes, the specificity of lipases has great importance in the conversion mechanism. Specifically, 1,3-specific lipases are capable of releasing fatty acids from positions 1 and 3 of a glyceride and hydrolyzing the ester bonds of a plant oil. This generally takes place in two main steps; hydrolysis of the ester bond and esterification with the second substrate.

For industrial applications including biodiesel conversion, microbial-generated lipases are preferred because of their short generation time. Among many different bacterial and fungal sources such as *Candida antarctica*, *Candida rugosa*, *Aspergillus niger*, *Chromobacterium Viscosum*, *Mucor miehei*, and *Rhizopus oryzae*, *Candida rugosa* is the most-used microorganism for lipase production. Although there are several advantages, the use of lipase enzyme comes with problems regarding its expense and instability. Lipase enzymes could lose their activity due to inhibition by the reaction condition and the presence of by-products. The inhibition by insoluble glycerol as a by-product occurs naturally.

Recycling, reusability, and the stability of enzymes in the face of environmental changes are important factors, and as an attempt to achieve these feature, immobilizing lipase into supporting solids is an intelligent strategy. From a technical perspective, lipase immobilization could be conducted to maintain operational stability and catalyst recovery. Immobilized enzymes can be separated from the reaction mixture more easily, and at the same time allow the enzyme to be studied under harsher environmental conditions. Immobilizing lipase is a mid-stream processing strategy along with intensification, optimization, and process design. Many solid supports have also been reported as host of lipase, including polymeric materials, silica-based material, and silica-alumina-based materials. With the consideration of chemical stability and capability of hosting the enzyme effectively, mesoporous silica materials have also been reported. Lipase immobilization can be performed through various mechanisms such as adsorption, cross-linking, encapsulation, entrapment, and hydrophobic interaction [89].

Adsorption is one of the simplest methods, particularly for mesoporous silica support. Generally, enzymes are immobilized after the synthesis of the support through adsorption, to avoid enzyme denaturation caused by harsh conditions or chemical reagents that are detrimental to the enzyme. A simple immobilization method through physical adsorption under a phosphate buffer has been reported [14,90,91]. Controlling surface charge is important, which is explained by the fact that sodium ions act as counterions for reducing the electrical double layer of the silica surface and thereby decreasing the k-potential via the screening effect. Stronger silica–lipase interaction occurs in the presence of a crosslinker such as glutaraldehyde. It was reported that immobilized Candida Ragusa lipase (ICRL) on fibrous silica nanoparticles KCC-1 maintained above 81% of the initial activity after 28 days, and 80% of the activity was maintained after 8 repeated cycles [92]. Other crosslinkers include sodium dodecyl sulfate (SDS), ethylene diamine tetra acetate (EDTA), and polyethylene glycol (PEG), even though they generally lower the activity.

The catalyst activity of immobilized enzymes should be through the mechanism of interfacial activation, in which the active site of the lipase in aqueous solution is covered by a flexible region of the enzyme molecule, often called the lid. Interaction with a hydrophobic phase can cause opening of the lid to make the active site accessible. The positioning of lipase onto the support is frequently performed through a hydrophobic interaction between the surface and the lid of lipase. For this, covalent bonding of the linkage to increase the surface hydrophobicity is required. Moderate hydrophobicity/hydrophilicity of the support is the best feature providing lipase activity enhancement. However, too hydrophilic (pure silica) or too hydrophobic (butyl-grafted silica) supports are not appropriate for developing high activity for lipases [24]. As an example, the presence of Triton X-100 in a lipase adsorption system had a detrimental effect, reducing the efficiency more than twofold. The competition for surface adsorption sites is probably the major reason for this [90]. Figure 4 describes the expression of lipase immobilization over adsorption, covalent bonding, and cross-linking methods.

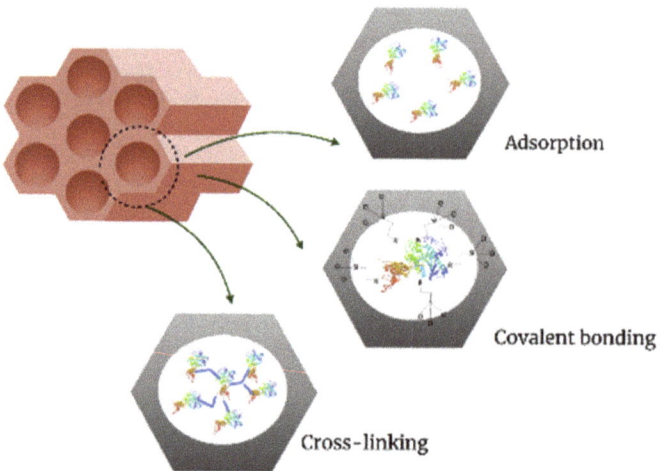

Figure 4. Schematic representation of mesoporous silica lipase immobilized through adsorption, covalent bonding, and cross-linking methods.

Moreover, to provide easy handling and recoverability of the catalytic system, hybrid nanocomposites were created by applying magnetic mesoporous silica materials. The nanocomposites were designed as silica-coated magnetic nanoparticles as a host for lipase. Fe_3O_4 nanoparticles with magnetic properties were the core of the nanocomposite shells to host *Burkholderia* sp. lipase, leading to a biodiesel yield above 90% [93,94]. Greater reusability and thermal stability were achieved through the combination of lipase immobilization onto amine and aldehyde surface-modified mesoporous silica [95]. Superior activity for olive oil transesterification was demonstrated by Rhizopus oryzae lipase (ROL) immobilized onto aminopropyl triethylenesilane (AP) and glutaraldehyde (GA) magnetic mesoporous silica. Silica was coated onto Fe_3O_4 before functionalization followed by ROL immobilization, and the material still showed magnetism (~20 emu/g). Using the Michaelis–Menten kinetics calculation, the composite shows a reduced K_M parameter, suggesting the substrate's affinity to the enzyme [95].

3.4. Mesoporous Silica-Supported Ionic Liquids

Ionic liquids (ILs) are classified as molten salts with several unique properties and characteristics that can be applied in various fields of application, including catalytic processes. One of the main properties of ILs is their ability to act as phase transfer catalysts (PTCs). This unique property facilitates the transfer of reactants or products between two immiscible phases, for example, between liquid and gas or solid and liquid phases. The PTC characteristics of ILs also provide several advantages compared to traditional solvents because ILs have good chemical stability, can dissolve various reactants, are thermally stable, and have low volatility. Several studies have shown that ILs containing pyridinium cations, phosphonium, imidazolium, and ammonium act as PTCs for benzoin condensation, nucleophilic substitution, fluorination, and esterification [96,97]. Santiago et al. (2017) reported ILs for enhancing the epoxidation reaction for the selective synthesis of β-O-glycosides [98]. The results showed that the presence of imidazolium ILs could facilitate the transfer migration of reaction between two different phases, aqueous and organic phases. Moreover, Szepiński et al. (2020) investigated surface-active amino acids ILs as PTCs for several different reactions [99].

Their study found that their effectiveness and efficiency as a PTC could be increased quickly and significantly impact their catalytic characteristics, such as changing the cation structure and extending the structural chain. PTCs in different phase systems, for example, in the liquid/liquid phase, facilitate the migration of anionic reagents between phases so that the process drives the reaction to run faster and improves the percentage of yield product. Although ILs have shown exemplary performance in the PTC process, some limitations of ILs as PTCs, especially on a large scale, still need to be overcome, such as low compatibility, difficult recovery and reuse, low stability, or easy degradation [99]. Therefore, ILs can be heterogeneous with solid materials such as mesoporous materials to overcome these limitations. These supported ILs (SILs) provide new findings and opportunities, especially in several reactions in biodiesel production. In addition, SILs allow for more straightforward and efficient catalyst recovery.

Mesoporous material-supported ILs have been widely studied, especially in the conversion and production of biodiesel. These materials provide many advantages, such as dissolving various organic compounds and good stability at high temperatures and pressures. SBA-15 is a mesoporous silica-based material widely developed as a catalyst because it has uniform hexagonal pores with an average diameter of around 15 nm. Besides being cheap, this material also has a narrow pore size distribution and a variety of chemical functional groups, making it easy to modify to increase efficiency. Yuan et al. (2012) reported the heterogenous catalyst-based basic IL-supported mesoporous SBA-15 for the epoxidation of olefins [100]. The basic ILs, 1-methyl-3-(chloropropyltriethoxysilane) imidazolium chloride (CIL), were attached to the surface of SBA-15 through the surface grafting reaction method. They found that the CIL could be loaded onto mesoporous SBA-15 until 0.39 mmol/g. The CIL-supported SBA-15 showed good catalytic activity for the epoxidation of various olefins compounds, with the highest conversion and epoxide yields until 94% for the cyclooctene compound. This type of catalyst is quite reactive for epoxidation reactions on cyclic olefins and less reactive for open and linear chain olefins. A related study was also reported by Karimi and Vafaeezadeh (2012); they investigated acidic IL-supported SBA-15 for solvent-free esterification [101]. Differences in the structure and functionalization of the IL groups assigned to the mesoporous material SBA-15 have been shown to affect the catalyst's phase and mass transfer properties. Acidic hydrophobic [OMIm] [HSO$_4$] is believed to have efficient mass transfer properties for esterification processes, especially in biodiesel production. Furthermore, this type of IL can also have increased Brønsted acid strength and high reusability because it does not interact with the water produced from the primary reaction process. Figure 5 shows an example of a modified cationic ionic liquid process on mesoporous silica. When cationic ionic liquids are introduced or immobilized onto the surface of mesoporous silica, they can change the properties and functionality of silica materials. The immobilization of cationic ionic liquids onto the surface of silica can increase its catalytic activity, selectivity, stability, and several surface characteristics, as shown in Figure 5. Different types of ILs modified on mesoporous silica can produce various surface areas. Thus, selecting the kind of IL directly affects the characteristics of the resulting mesoporous silica. In addition, the type of modification method used also affects it directly. Various synthesis methods have been studied, but grafting and impregnation are the primary methods for supporting ILs in mesoporous silica materials [102,103]. Both are general methods that are representative of chemical and physical processes. Impregnation is a method based on physical interaction and is more straightforward; namely, it involves the adsorption of ILs on the surface of mesoporous silica. In contrast, the grafting method uses chemical interactions between ILs and the surface of mesoporous materials. As a result, the grafting method shows better stability but has drawbacks such as complicated and complex preparation methods, requiring a long time, and high costs. These differences in synthesis methods directly affect the morphological properties of the catalyst, such as pore size and surface area. In its application as a catalyst in the production of biodiesel through a variety of different reactions, the specific surface area of the IL-supported mesoporous silica determines the percentage of product produced. However, the pore diameter and the reaction time applied during the process also influence the ratio of products [104].

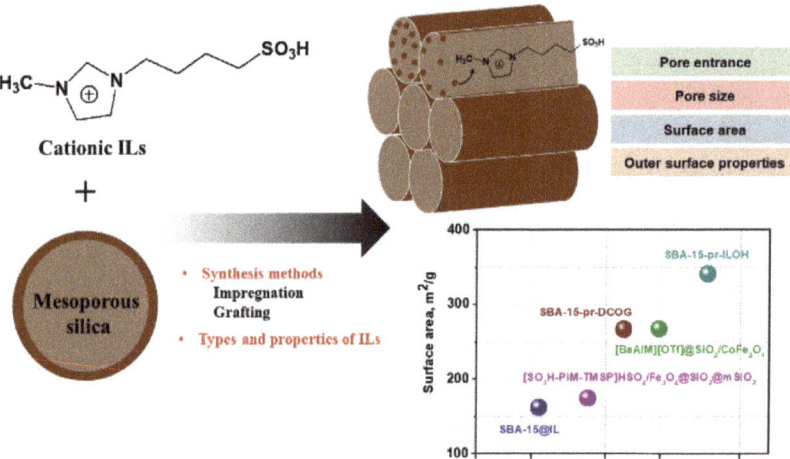

Figure 5. Illustration of heterogenous IL-supported mesoporous silica synthesis and the surface area properties with different types of IL-supported mesoporous silica.

4. Environmental Impact and Circular Economy Analysis

Using mesoporous silica catalysts in biodiesel production can cause several environmental impacts from the catalyst synthesis process and biodiesel production approaches. The production process of silica catalysts generally involves energy-intensive methods for the synthesis, purification, and activation processes. The energy needed in each process contributes directly to the resulting emissions and environmental impact. However, the amount of energy required depends on several factors, especially the synthesis method chosen and used in the catalyst production. Therefore, to minimize excessive energy consumption during the synthesis process of silica mesopore-based catalysts, several strategies can be applied, such as exploring alternative synthesis pathways and optimizing the conditions for the synthesis reaction. Development and synthesis approaches that are more environmentally friendly and sustainable are one of the most efficient steps in helping to reduce the energy requirements for catalyst production. However, it should be noted that the specific energy requirements may vary depending on the desired characteristics of the catalyst and the scale of production.

In addition to the synthesis method, the raw material used is another factor that influences the environmental impact of using mesoporous silica catalysts. These two factors are closely related to the amount of energy required because they are interconnected with the selection of the synthesis and purification methods. Watanabe and co-workers (2021) reported the synthesis of mesoporous silica from geothermal water [105]. They concluded that using geothermal water as a source of silica provides an advantage during the synthesis process because the high water temperature can supply sufficient thermal energy for the catalyst synthesis process. Thus, this process will encourage more efficient or lower energy use. El-Nahas et al. (2020) reported a facile synthesis route for zeolite based on waste raw materials such as disposed silica gel as a source of silica and various types of aluminum waste (cans, aluminum foil, scrap wire cables, bottles, etc.) [106]. Studies have proved that using waste as a raw material in synthesizing catalyst materials could reduce production costs by up to 70% compared to commercial raw materials. In addition, selecting appropriate and economical raw materials can reduce the use of chemicals and multi-step procedures in the synthesis process so that production costs are lower than for commercial raw materials.

Furthermore, if we look at the biodiesel production process, the transesterification process has been widely reported to produce several types of by-products, such as solid residue, methanol, biodiesel washing wastewater, and glycerol [107]. However, among the various types of by-products, glycerol is a by-product with a relatively large quantity produced from the catalytic process. Disposal of this waste must be properly managed to prevent environmental pollution and minimize its impact. The conversion and reuse of biodiesel by-products, especially glycerol, has been studied and continues to be developed. For example, glycerol has been widely used to produce high-value biotechnology products such as biosurfactants, 1,3-propanediol, citric acid, and ethanol. This biodiesel production process can lead to an abundant glycerol yield in the future to reduce the price of glycerol sales. Various techniques have been studied to convert glycerol into biotechnology products or more useful renewable energy sources, such as ammoxidation, esterification, acetylation, pyrolysis, gasification, and steam reforming [108].

Therefore, the application and circular economy approach for the synthesis of mesoporous silica for biodiesel production can be viewed in terms of several essential aspects, namely: (1) selection of waste raw materials that are rich in silica sources such as industrial by-products (fly ash), agricultural residues (dregs sugarcane, rice husk), or household post-consumer waste. (2) Selection of resources such as developing efficient synthesis methods with high purity and low energy use. Optimizing the synthesis process to minimize energy consumption, waste generation, and the use of hazardous chemicals can reduce the resulting environmental impact. (3) Emphasis on recycling and reuse of catalyst materials. Developing catalyst materials that are easy to separate and purify is a challenging area of study for the present and the future. Reusing catalyst materials can reduce waste generation and promote a circular approach. (4) Cultivating collaboration between waste feedstock suppliers, catalyst producers, and biodiesel producers to build closed-loop systems and sustainable production is an essential step in adopting a circular economy approach in the future.

Based on the mapping and bibliometric analysis results, as shown in Figure 6, converting waste biomass into biodiesel and bioethanol is still an exciting topic and supports the renewable energy transition policy. The development of alternative synthesis methods that are more environmentally friendly is still a topic that is continuously being studied to reduce the environmental impact of high energy consumption. Servicing raw materials and their availability is an essential factor in developing a sustainable and abundant system for future biodiesel and bioethanol production. Currently, fast pyrolysis involving enzymatic reactions and microorganisms as the main raw material is considered to have the potential to be developed for the production of bioenergy [109].

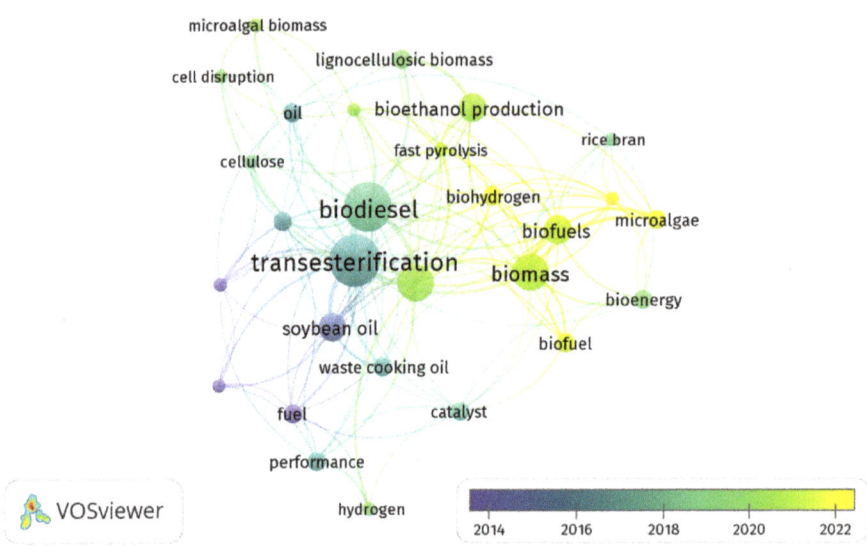

Figure 6. Bibliometric analysis for biodiesel production using mesoporous catalysts (data source: Web of Science, webofscience.com, accessed on 6 June 2023).

5. Conclusions and Outlook

Various catalysts and methods have been proposed for biodiesel production to achieve the optimized conditions and a total low-cost process. Considering the green technology and green chemistry perspective, heterogeneous catalyzed reactions are known as a promising method. Processes with low energy consumption and reusable catalysts are highlighted to fit with the requirements. The present study revealed that the physical and chemical characteristics of the catalyst (such as basicity and acidity) play a pivotal role in biodiesel production. Mesoporous silica materials have been shown to be feasible heterogeneous catalyst supports due to their high specific surface area and tunable properties. Varied surface functionalization methods of mesoporous silica surfaces have been reported with specific features in order to meet the prerequisites of industrial applications. Some issues that can be noted for development include the intensive design of the nanocomposite by applying nanotechnology perspectives, the development of intensified methods, magnetically separable catalysts, ionic liquid modified materials, and the photocatalytic conversion of biodiesel. Immobilized biocatalysts with ultra-selectivity in non-solvent reaction using mesoporous silica are still a promising and relevant scheme of material development. Nevertheless, some nanomaterials from mesoporous silica are still too expensive to be commercialized and to be used for biodiesel production. Silica precursors and other chemicals need to be replaced with feasible materials, and for this issue, some explorations into biogenic silica resources need to be looked at. Sustainable sources of silica such as rice husk, wheat straw, salacca peel waste, and bamboo leaves have been reported to have potential for functionalized silica [110–114]. As the design of catalyst preparation could be inclusively combined with the agricultural industrial cycle, the concept of circular economy to provide biodiesel as a renewable energy resource can be more intensified. It is believed that several newly introduced functionalizations of mesoporous silica along with the intensification of biodiesel production help in producing eco-friendly and economically viable biodiesel. Developing innovative process intensification techniques, such as microreactors, continuous flow systems, or advanced reactor designs that can facilitate increased efficiency and productivity of biodiesel production using mesoporous silica catalysts, still needs further study. In addition, the synthesis of mesoporous silica for biodiesel production at the industrial level is also of interest to consider. Conducting a comprehensive life cycle assessment (LCA) to evaluate the environmental impact of mesoporous silica catalysts on biodiesel production is essential. Overall, future research should aim to advance the understanding of mesoporous silica catalysts in biodiesel production, addressing challenges related to stability, catalyst design, feedstock flexibility, and process optimization, seeking sustainable and efficient development.

Author Contributions: I.F.: Methodology, writing original draft; G.F.: Resources, software; S.S.: validation, review and editing, W.-C.O.: validation, review and editing; K.L.A.: validation, review and editing. All authors have read and agreed to the published version of the manuscript.

Funding: This research was funded by Jurusan Kimia FMIPA Universitas Islam Indonesia.

Data Availability Statement: No new data were created.

Acknowledgments: The authors thank Jurusan Kimia FMIPA Universitas Islam Indonesia for supporting this research.

Conflicts of Interest: The authors declare no conflict of interest.

References

1. Millward-Hopkins, J.; Zwirner, O.; Purnell, P.; Velis, C.A.; Iacovidou, E.; Brown, A. Resource Recovery and Low Carbon Transitions: The Hidden Impacts of Substituting Cement with Imported 'Waste' Materials from Coal and Steel Production. *Glob. Environ. Chang.* **2018**, *53*, 146–156. [CrossRef]
2. Xu, H.; Ou, L.; Li, Y.; Hawkins, T.R.; Wang, M. Life Cycle Greenhouse Gas Emissions of Biodiesel and Renewable Diesel Production in the United States. *Environ. Sci. Technol* **2022**, *56*, 7512–7521. [CrossRef] [PubMed]
3. Chanthon, N.; Ngaosuwan, K.; Kiatkittipong, W.; Wongsawaeng, D.; Appamana, W.; Assabumrungrat, S. A Review of Catalyst and Multifunctional Reactor Development for Sustainable Biodiesel Production. *ScienceAsia* **2021**, *47*, 531–541. [CrossRef]
4. Fazal, M.A.; Haseeb, A.S.M.A.; Masjuki, H.H. Biodiesel Feasibility Study: An Evaluation of Material Compatibility; Performance; Emission and Engine Durability. *Renew. Sustain. Energy Rev.* **2011**, *15*, 1314–1324. [CrossRef]
5. Jothiramalingam, R.; Wang, M.K. Review of Recent Developments in Solid Acid, Base, and Enzyme Catalysts (Heterogeneous) for Biodiesel Production via Transesterification. *Ind. Eng. Chem. Res.* **2009**, *48*, 6162–6172. [CrossRef]
6. Parangi, T.; Mishra, M.K. Solid Acid Catalysts for Biodiesel Production. *Comments Inorg. Chem.* **2020**, *40*, 176–216. [CrossRef]
7. Zhao, X.; Qi, F.; Yuan, C.; Du, W.; Liu, D. Lipase-Catalyzed Process for Biodiesel Production: Enzyme Immobilization, Process Simulation and Optimization. *Renew. Sustain. Energy Rev.* **2015**, *44*, 182–197. [CrossRef]
8. Wang, C.; Hong, H.; Lin, Z.; Yuan, Y.; Liu, C.; Ma, X.; Cao, X. Correction: Tethering Silver Ions on Amino-Functionalized Mesoporous Silica for Enhanced and Sustained Antibacterial Properties. *RSC Adv.* **2016**, *6*, 8329. [CrossRef]
9. Essamlali, Y.; Amadine, O.; Larzek, M.; Len, C.; Zahouily, M. Sodium Modified Hydroxyapatite: Highly Efficient and Stable Solid-Base Catalyst for Biodiesel Production. *Energy Convers. Manag.* **2017**, *149*, 355–367. [CrossRef]
10. Zhang, M.; Jun, S.-H.; Wee, Y.; Kim, H.S.; Hwang, E.T.; Shi, J.; Hwang, S.Y.; Lee, J.; Kim, J. Activation of Crosslinked Lipases in Mesoporous Silica via Lid Opening for Recyclable Biodiesel Production. *Int. J. Biol. Macromol.* **2022**, *222*, 2368–2374. [CrossRef]
11. Al-Jammal, N.; Al-Hamamre, Z.; Alnaief, M. Manufacturing of Zeolite Based Catalyst from Zeolite Tuft for Biodiesel Production from Waste Sunflower Oil. *Renew. Energy* **2016**, *93*, 449–459. [CrossRef]
12. Jamil, F.; Al-Haj, L.; Al-Muhtaseb, A.H.; Al-Hinai, M.A.; Baawain, M.; Rashid, U.; Ahmad, M.N.M. Current Scenario of Catalysts for Biodiesel Production: A Critical Review. *Rev. Chem. Eng.* **2018**, *34*, 267–297. [CrossRef]
13. Li, L.; Cani, D.; Pescarmona, P.P. Metal-Containing TUD-1 Mesoporous Silicates as Versatile Solid Acid Catalysts for the Conversion of Bio-Based Compounds into Valuable Chemicals. *Inorg. Chim. Acta* **2015**, *431*, 289–296. [CrossRef]
14. Sharma, A.; Sharma, T.; Meena, K.R.; Kanwar, S.S. Physical Adsorption of Lipase onto Mesoporous Silica. *Int. J. Curr. Adv. Res.* **2017**, *6*, 3837–3841. [CrossRef]
15. Fu, Z.; Zhang, G.; Tang, Z.; Zhang, H. Preparation and Application of Ordered Mesoporous Metal Oxide Catalytic Materials. *Catal. Surv. Asia* **2020**, *24*, 38–58. [CrossRef]
16. Wang, X.; Zhu, H.; Yang, C.; Lu, J.; Zheng, L.; Liang, H.-P. Mesoporous Carbon Promoting the Efficiency and Stability of Single Atomic Electrocatalysts for Oxygen Reduction Reaction. *Carbon* **2022**, *191*, 393–402. [CrossRef]
17. Xu, W.; Thapa, K.B.; Ju, Q.; Fang, Z.; Huang, W. Heterogeneous Catalysts Based on Mesoporous Metal–Organic Frameworks. *Coord. Chem. Rev.* **2018**, *373*, 199–232. [CrossRef]
18. Gabriel, R.; De Carvalho, S.H.; da Silva Duarte, J.L.; Oliveira, L.M.; Giannakoudakis, D.A.; Triantafyllidis, K.S.; Soletti, J.I.; Meili, L. Mixed Metal Oxides Derived from Layered Double Hydroxide as Catalysts for Biodiesel Production. *Appl. Catal. A Gen.* **2022**, *630*, 118470. [CrossRef]
19. Sha, X.; Dai, Y.; Song, X.; Liu, S.; Zhang, S.; Li, J. The Opportunities and Challenges of Silica Nanomaterial for Atherosclerosis. *Int. J. Nanomed.* **2021**, *16*, 701–714. [CrossRef]
20. Cheah, W.K.; Sim, Y.L.; Yeoh, F.Y. Amine-Functionalized Mesoporous Silica for Urea Adsorption. *Mater. Chem. Phys.* **2016**, *175*, 151–157. [CrossRef]

21. Telalović, S.; Ramanathan, A.; Mul, G.; Hanefeld, U. TUD-1: Synthesis and Application of a Versatile Catalyst, Carrier, Material. *J. Mater. Chem.* **2010**, *20*, 642–658. [CrossRef] [PubMed]
22. Van Der Voort, P.; Mathieu, M.; Mees, F.; Vansant, E.F. Synthesis of High-Quality MCM-48 and MCM-41 by Means of the GEMINI Surfactant Method. *J. Phys. Chem. B* **1998**, *102*, 8847–8851. [CrossRef]
23. Wang, S.; Wu, D.; Sun, Y.; Zhong, B. The Synthesis of MCM-48 with High Yields. *Mater. Res. Bull.* **2001**, *36*, 1717–1720. [CrossRef]
24. Galarneau, A.; Mureseanu, M.; Atger, S.; Renard, G.; Fajula, F. Immobilization of Lipase on Silicas. Relevance of Textural and Interfacial Properties on Activity and Selectivity. *New J. Chem.* **2006**, *30*, 562–571. [CrossRef]
25. Sankar, S.; Sharma, S.K.; Kim, D.Y. Synthesis and Characterization of Mesoporous SiO_2 Nanoparticles Synthesized from Biogenic Rice Husk Ash for Optoelectronic Applications. *Int. J. Eng. Sci.* **2016**, *17*, 353–358.
26. Viscardi, R.; Barbarossa, V.; Maggi, R.; Pancrazzi, F. Effect of Acidic MCM-41 Mesoporous Silica Functionalized with Sulfonic Acid Groups Catalyst in Conversion of Methanol to Dimethyl Ether. *Energy Rep.* **2020**, *6*, 49–55. [CrossRef]
27. Ravi, S.; Roshan, R.; Tharun, J.; Kathalikkattil, A.C.; Park, D.W. Sulfonic Acid Functionalized Mesoporous SBA-15 as Catalyst for Styrene Carbonate Synthesis from CO_2 and Styrene Oxide at Moderate Reaction Conditions. *J. CO2 Util.* **2015**, *10*, 88–94. [CrossRef]
28. Kohns, R.; Meyer, R.; Wenzel, M.; Matysik, J.; Enke, D.; Tallarek, U. In Situ Synthesis and Characterization of Sulfonic Acid Functionalized Hierarchical Silica Monoliths. *J. Solgel Sci. Technol.* **2020**, *96*, 67–82. [CrossRef]
29. Luštická, I.; Vrbková, E.; Vyskočilová, E.; Paterová, I.; Červený, L. Acid Functionalized MCM-41 as a Catalyst for the Synthesis of Benzal-1,1-Diacetate. *React. Kinet. Mech. Catal.* **2013**, *108*, 205–212. [CrossRef]
30. Shah, K.A.; Parikh, J.K.; Maheria, K.C. Biodiesel Synthesis from Acid Oil over Large Poresulfonic Acid-Modified Mesostructured SBA-15: Process Optimization and Reaction Kinetics. *Catal. Today* **2014**, *237*, 29–37. [CrossRef]
31. Sherry, L.; Sullivan, J.A. The Reactivity of Mesoporous Silica Modified with Acidic Sites in the Production of Biodiesel. *Catal. Today* **2011**, *175*, 471–476. [CrossRef]
32. Musterman, M.; Placeholder, P. Study About Naphthoquinone Schiff What Is So Different About Was Ist so Anders Am Neuroenhancement? *Open Chem.* **2018**, *1*, 2–7.
33. Zuo, D.; Lane, J.; Culy, D.; Schultz, M.; Pullar, A.; Waxman, M. Sulfonic Acid Functionalized Mesoporous SBA-15 Catalysts for Biodiesel Production. *Appl Catal B* **2013**, *129*, 342–350. [CrossRef]
34. Melero, J.A.; Bautista, L.F.; Morales, G.; Iglesias, J.; Sánchez-vázquez, R. Biodiesel Production from Crude Palm Oil Using Sulfonic Acid-Modified Mesostructured Catalysts. *Chem. Eng. J.* **2010**, *161*, 323–331. [CrossRef]
35. Mostafa Marzouk, N.; Abo El Naga, A.O.; Younis, S.A.; Shaban, S.A.; El Torgoman, A.M.; El Kady, F.Y. Process Optimization of Biodiesel Production via Esterification of Oleic Acid Using Sulfonated Hierarchical Mesoporous ZSM-5 as an Efficient Heterogeneous Catalyst. *J. Env. Chem. Eng.* **2021**, *9*, 105035. [CrossRef]
36. Alrouh, F.; Karam, A.; Alshaghel, A.; El-Kadri, S. Direct Esterification of Olive-Pomace Oil Using Mesoporous Silica Supported Sulfonic Acids. *Arab. J. Chem.* **2017**, *10*, S281–S286. [CrossRef]
37. Wang, Y.; Wang, D.; Tan, M.; Jiang, B.; Zheng, J.; Tsubaki, N.; Wu, M. Monodispersed Hollow SO3H-Functionalized Carbon/Silica as Efficient Solid Acid Catalyst for Esterification of Oleic Acid. *ACS Appl. Mater. Interfaces* **2015**, *7*, 26767–26775. [CrossRef] [PubMed]
38. Bandyopadhyay, M.; Tsunoji, N.; Bandyopadhyay, R. Comparison of Sulfonic Acid Loaded Mesoporous Silica in Transesterification of Triacetin. *React. Kinet. Mech. Catal.* **2019**, *126*, 167–179. [CrossRef]
39. Uchoa, A.F.; do Valle, C.P.; Moreira, D.R.; Bañobre-López, M.; Gallo, J.; Dias, F.S.; Anderson, M.W.; Ricardo, N.M.P.S. Synthesis of a KIT-6 Mesoporous Sulfonic Acid Catalyst to Produce Biodiesel from Cashew Nut Oil. *Braz. J. Chem. Eng.* **2022**, *39*, 1001–1011. [CrossRef]
40. Jackson, M.A.; Mbaraka, I.K.; Shanks, B.H. Esterification of Oleic Acid in Supercritical Carbon Dioxide Catalyzed by Functionalized Mesoporous Silica and an Immobilized Lipase. *Appl. Catal. A Gen.* **2006**, *310*, 48–53. [CrossRef]
41. Jaroszewska, K.; Nowicki, J.; Nosal-kovalenko, H.; Grzechowiak, J.; Lewandowski, M.; Kaczmarczyk, J. Selected Acid and Basic Functionalized Ordered Mesoporous Materials as Solid Catalysts for Transesterification of Canola Oil: A Comparative Study. *Fuel* **2022**, *325*, 124902. [CrossRef]
42. Ghoreishi, K.B.; Asim, N.; Yarmo, M.A.; Samsudin, M.W. Mesoporous Phosphated and Sulphated Silica as Solid Acid Catalysts for Glycerol Acetylation. *Chem. Pap.* **2014**, *68*, 1194–1204. [CrossRef]
43. Pirez, C.; Caderon, J.; Dacquin, J.; Lee, A.F.; Wilson, K. Tunable KIT-6 Mesoporous Sulfonic Acid Catalysts for Fatty Acid Esteri Fi Cation. *ACS Catal.* **2012**, *2*, 1607–1614. [CrossRef]
44. Brahmkhatri, V.; Patel, A. Biodiesel Production by Esterification of Free Fatty Acids over 12-Tungstophosphoric Acid Anchored to MCM-41. *Ind. Eng. Chem. Res.* **2011**, *50*, 6620–6628. [CrossRef]
45. Sakthivel, A.; Komura, K.; Sugi, Y. MCM-48 Supported Tungstophosphoric Acid: An Efficient Catalyst for the Esterification of Long-Chain Fatty Acids and Alcohols in Supercritical Carbon Dioxide. *Ind. Eng. Chem. Res.* **2008**, *48*, 2538–2544. [CrossRef]
46. Jiménez-Morales, I.; Santamaría-González, J.; Maireles-Torres, P.; Jiménez-López, A. Zirconium Doped MCM-41 Supported WO3 Solid Acid Catalysts for the Esterification of Oleic Acid with Methanol. *Appl. Catal. A Gen.* **2010**, *379*, 61–68. [CrossRef]
47. Esmi, F.; Masoumi, S.; Dalai, A.K. Comparative Catalytic Performance Study of 12-Tungstophosphoric Heteropoly Acid Supported on Mesoporous Supports for Biodiesel Production from Unrefined. *Catalyst* **2022**, *12*, 658. [CrossRef]

48. Singh, S.; Patel, A. 12-Tungstophosphoric Acid Supported on Mesoporous Molecular Material: Synthesis, Characterization and Performance in Biodiesel Production. *J. Clean Prod.* **2014**, *72*, 46–56. [CrossRef]
49. Yu, Z.; Zhang, Y.; Duan, J.; Chen, X.; Piao, M.; Hu, J.; Shi, F. One-Pot Synthesis of Propyl-Sulfonic Phosphotungstic Dual-Acid Functionalized Mesoporous Silica in Non-Hydrochloric Acid Solution: Reusable Catalyst for Efficient Biodiesel Production. *ChemistrySelect* **2020**, *5*, 14666–14669. [CrossRef]
50. Pandurangan, P.S.A. Heteropolyacid ($H_3PW_{12}O_{40}$)—Impregnated Mesoporous KIT-6 Catalyst for Green Synthesis of Bio-Diesel Using Transesterification of Non-Edible Neem Oil. *Mater. Renew. Sustain. Energy* **2019**, *8*, 22. [CrossRef]
51. Meloni, D.; Monaci, R.; Zedde, Z.; Cutrufello, M.G.; Fiorilli, S.; Ferino, I. Transesterification of Soybean Oil on Guanidine Base-Functionalized SBA-15 Catalysts. *Appl. Catal. B* **2011**, *102*, 505–514. [CrossRef]
52. de Lima, A.L.; Mbengue, A.; San, R.A.S.; Ronconi, C.M.; Mota, C.J.A. Synthesis of Amine-Functionalized Mesoporous Silica Basic Catalysts for Biodiesel Production. *Catal. Today* **2014**, *226*, 210–216. [CrossRef]
53. Guerrero, V.V.; Shantz, D.F. Amine-Functionalized Ordered Mesoporous Silica Transesterification Catalysts. *Ind. Eng. Chem.* **2009**, *48*, 10375–10380. [CrossRef]
54. Ngo, H.L.; Zafiropoulos, N.A.; Foglia, T.A.; Samulski, E.T.; Lin, W. Mesoporous Silica-Supported Diarylammonium Catalysts for Esterification of Free Fatty Acids in Greases. *J. Am. Oil Chem. Soc.* **2010**, *87*, 445–452. [CrossRef]
55. Moradi, G.; Mohadesi, M.; Hojabri, Z. Biodiesel Production by CaO/SiO2 Catalyst Synthesized by the Sol–Gel Process. *React. Kinet. Mech. Catal.* **2014**, *113*, 169–186. [CrossRef]
56. Elimbinzi, E.; Nyandoro, S.S.; Mubofu, E.B.; Osatiashtiani, A.; Jinesh, C. Synthesis of Amine Functionalized Mesoporous Silicas Templated by Castor Oil for Transesterification. *MRS Adv.* **2018**, *3*, 2261–2269. [CrossRef]
57. Chermahini, A.N.; Azadi, M.; Tafakori, E.; Teimouri, A.; Sabzalian, M. Amino-functionalized mesoporous silica as solid base catalyst for regioselective aza-Michael reaction of aryl tetrazoles. *J.Porous Mater.* **2016**, *23*, 441–451. [CrossRef]
58. Xie, W.; Hu, L. Mesoporous SBA-15 Silica-Supported Diisopropylguanidine: An Efficient Solid Catalyst for Interesterification of Soybean Oil with Methyl Octanoate or Methyl Decanoate. *J. Oleo Sci.* **2016**, *65*, 803–813. [CrossRef]
59. Xie, W.; Yang, X.; Fan, M. Novel Solid Base Catalyst for Biodiesel Production: Mesoporous SBA-15 Silica Immobilized with 1, 3-Dicyclohexyl-2-Octylguanidine. *Renew. Energy* **2015**, *80*, 230–237. [CrossRef]
60. Mary Anjalin; Kanagathara, N.; Baby Suganthi, A.R. A Brief Review on Aniline and Its Derivatives. *Mater. Today Proc.* **2020**, *33*, 4751–4755. [CrossRef]
61. Santos, E.C.S.; dos Santos, T.C.; Guimarães, R.B.; Ishida, L.; Freitas, R.S.; Ronconi, C.M. Guanidine-Functionalized Fe3O4 Magnetic Nanoparticles as Basic Recyclable Catalysts for Biodiesel Production. *J. Mater. Chem. C* **2015**, *3*, 10715–10722. [CrossRef]
62. Abdullahi, M.; Panneerselvam, P.; Imam, S.S.; Ahmad, L.S. Removal of Free Fatty Acids in Neem Oil Using Diphenylamine Functionalized Magnetic Mesoporous Silica SBA-15 for Biodiesel Production. *J. Pet. Technol. Altern. Fuels* **2016**, *7*, 31–37. [CrossRef]
63. Albuquerque, C.G.; Jime, I.; Me, J.M.; Jime, A.; Maireles-torres, P. CaO Supported on Mesoporous Silicas as Basic Catalysts for Transesterification Reactions. *Appl. Catal. A Gen.* **2008**, *334*, 35–43. [CrossRef]
64. Méndez, J.C.; Arellano, U.; Solis, S.; Asomo, M.; Lara, H.; Padilla, A.J.; Wang, J.A. Synthesis of Hybrid Materials, Immobilization of Lipase in SBA-15 Modified with CaO. *J. Appl. Res. Technol.* **2018**, *16*, 498–510. [CrossRef]
65. Kolo, L.; Firdaus; Taba, P.; Zakir, M.; Soekamto, N.H. Selectivity of the New Catalyst ZnO-MCM-48-CaO in Esterification of Calophyllum Inophyllum Oil. *Automot. Exp.* **2022**, *5*, 217–229. [CrossRef]
66. Li, E.; Rudolph, V. Transesterification of Vegetable Oil to Biodiesel over MgO-Functionalized Mesoporous Catalysts. *Energy Fuels* **2008**, *22*, 145–149. [CrossRef]
67. Qu, S.; Chen, C.; Guo, M.; Lu, J.; Yi, W.; Ding, J.; Miao, Z. Synthesis of MgO/ZSM-5 Catalyst and Optimization of Process Parameters for Clean Production of Biodiesel from Spirulina Platensis. *J. Clean Prod.* **2020**, *276*, 123382. [CrossRef]
68. Sani, Y.M.; Alaba, P.A.; Raji-Yahya, A.O.; Abdul Aziz, A.R.; Daud, W.M.A.W. Facile Synthesis of Sulfated Mesoporous Zr/ZSM-5 with Improved Brønsted Acidity and Superior Activity over SZr/Ag, SZr/Ti, and SZr/W in Transforming UFO into Biodiesel. *J. Taiwan Inst. Chem. Eng.* **2016**, *60*, 247–257. [CrossRef]
69. Peruzzolo, T.M.; Stival, J.F.; Baika, L.M.; Ramos, L.P.; Grassi, M.T.; Rocco, M.L.M.; Nakagaki, S. Efficient Esterification Reaction of Palmitic Acid Catalyzed by WO3-x/Mesoporous Silica. *Biofuels* **2022**, *13*, 383–393. [CrossRef]
70. Vardast, N.; Haghighi, M.; Dehghani, S. Sono-Dispersion of Calcium over Al-MCM-41Used as a Nanocatalyst for Biodiesel Production from Sunflower Oil: Influence of Ultrasound. *Renew. Energy* **2018**, *132*, 979–988. [CrossRef]
71. Sahel, F.; Sebih, F.; Bellahouel, S.; Bengueddach, A.; Hamacha, R. Synthesis and Characterization of Highly Ordered from Bentonite as Efficient Catalysts for the Production. *Res. Chem. Intermed.* **2020**, *46*, 133–148. [CrossRef]
72. Li, Y.; Zhang, W.; Zhang, L.; Yang, Q.; Wei, Z.; Feng, Z.; Li, C. Direct Synthesis of Al—SBA-15 Mesoporous Materials via Hydrolysis-Controlled Approach. *J. Phys. Chem. B* **2004**, *108*, 9739–9744. [CrossRef]
73. Fedyna, M.; Jaroszewska, K.; Leszek, K.; Trawczy, J. Procedure for the Synthesis of AlSBA-15 with High Aluminium Content: Characterization and Catalytic Activity. *Microporous Mesoporous Mater.* **2020**, *292*, 109701. [CrossRef]
74. Kumar, P.; Mal, N.K.; Oumi, Y.; Sano, T.; Yamana, K. Synthesis and Characterization of Al-MCM-48 Type Materials Using Coal Fly Ash. *Stud. Surf. Catal.* **2002**, *142*, 1229–1236.
75. Juwono, H.; Wahyuni, E.T.; Ulfin, I.; Kurniawan, F. Production of Biodiesel from Seed Oil of Nyamplung (Calophyllum Inophyllum) by Al-MCM-41 and Its Performance in Diesel Engine. *Indones. J. Chem.* **2017**, *17*, 316–321. [CrossRef]

76. Dehghani, S.; Haghighi, M. Structural/Texture Evolution of CaO/MCM—41 Nanocatalyst by Doping Various Amounts of Cerium for Active and Stable Catalyst: Biodiesel Production from Waste Vegetable Cooking Oil. *Int. J. Energy Res.* **2019**, *43*, 3779–3793. [CrossRef]
77. Guliani, D.; Sobti, A.; Toor, A.P. Titania Impregnated Mesoporous MCM-48 as a Solid Photo-Catalyst for the Synthesis of Methyl Palmitate: Reaction Mechanism and Kinetics. *Renew. Energy* **2022**, *191*, 405–417. [CrossRef]
78. Dehghani, S.; Haghighi, M. Sono-Sulfated Zirconia Nanocatalyst Supported on MCM-41 for Biodiesel Production from Sunflower Oil: Influence of Ultrasound Irradiation Power on Catalytic Properties and Performance. *Ultrason. Sonochemistry* **2017**, *35*, 142–151. [CrossRef]
79. Mathimani, T.; Rene, E.R.; Sindhu, R.; Al-Ansari, M.M.; Al-Humaid, L.A.; Jhanani, G.K.; Chi, N.T.L.; Shanmuganathan, R. Biodiesel Production and Engine Performance Study Using One-Pot Synthesised ZnO/MCM-41. *Fuel* **2023**, *336*, 126830. [CrossRef]
80. Barros, S.D.T.; Coelho, A.V.; Lachter, E.R.; San, R.A.S.; Dahmouche, K.; Isabel, M.; Souza, A.L.F. Esteri Fi Cation of Lauric Acid with Butanol over Mesoporous Materials. *Renew. Energy* **2013**, *50*, 585–589. [CrossRef]
81. Kazemian, H.; Turowec, B.; Siddiquee, M.N.; Rohani, S. Biodiesel Production Using Cesium Modified Mesoporous Ordered Silica as Heterogeneous Base Catalyst. *Fuel* **2013**, *103*, 719–724. [CrossRef]
82. Corro, G.; Sánchez, N.; Pal, U.; Cebada, S.; Fierro, J.L.G. *Solar-Irradiation Driven Biodiesel Production Using Cr/SiO$_2$ Photocatalyst Exploiting Cooperative Interaction between Cr^{6+} and Cr^{3+} Moieties*; Elsevier: Amsterdam, The Netherlands, 2017; Volume 203, ISBN 2295500729.
83. Carlucci, C.; Degennaro, L.; Luisi, R. Titanium Dioxide as a Catalyst in Biodiesel Production. *Catalysts* **2019**, *9*, 75. [CrossRef]
84. Huang, J.; Jian, Y.; Zhu, P.; Abdelaziz, O.; Li, H. Research Progress on the Photo-Driven Catalytic Production of Biodiesel. *Front. Chem.* **2022**, *10*, 904251. [CrossRef] [PubMed]
85. Welter, R.A.; Santana, H.; de la Torre, L.G.; Robertson, M.; Taranto, O.P.; Oelgemöller, M. Methyl Oleate Synthesis by TiO$_2$ Photocatalytic Esterification of Oleic Acid: Optimisation by Response Surface Quadratic Methodology, Reaction Kinetics and Thermodynamics. *ChemPhotoChem* **2022**, *6*, e202200007. [CrossRef]
86. Bhuyan, M.S.U.S.; Alam, A.H.M.A.; Chu, Y.; Seo, Y.C. Biodiesel Production Potential from Littered Edible Oil Fraction Using Directly Synthesized S-TiO2/MCM-41 Catalyst in Esterification Process via Non-Catalytic Subcritical Hydrolysis. *Energies* **2017**, *10*, 1290. [CrossRef]
87. Trisunaryanti, W.; Larasati, S.; Triyono, T.; Santoso, N.R.; Paramesti, C. Selective Production of Green Hydrocarbons from the Hydrotreatment of Waste Coconut Oil over Ni- And NiMo-Supported on Amine-Functionalized Mesoporous Silica. *Bull. Chem. React. Eng. Catal.* **2020**, *15*, 415–431. [CrossRef]
88. Suryajaya, S.K.; Mulyono, Y.R.; Santoso, S.P.; Yuliana, M.; Kurniawan, A.; Ayucitra, A.; Sun, Y.; Hartono, S.B.; Soetaredjo, F.E.; Ismadji, S. Iron (II) Impregnated Double-Shelled Hollow Mesoporous Silica as Acid-Base Bifunctional Catalyst for the Conversion of Low-Quality Oil to Methyl Esters. *Renew. Energy* **2021**, *169*, 1166–1174. [CrossRef]
89. Bhan, C.; Singh, J. Role of Microbial Lipases in Transesterification Process for Biodiesel Production. *Environ. Sustain.* **2020**, *3*, 257–266. [CrossRef]
90. Nikolić, M.P.; Srdić, V.V.; Antov, M.G. Immobilization of Lipase into Mesoporous Silica Particles by Physical Adsorption. *Biocatal. Biotransform.* **2009**, *27*, 254–262. [CrossRef]
91. Katiyar, M.; Ali, A. Immobilization of Candida Rugosa Lipase on MCM-41 for the Transesterification of Cotton Seed Oil. *J. Oleo Sci.* **2012**, *61*, 469–475. [CrossRef]
92. Ali, Z.; Tian, L.; Zhao, P.; Zhang, B.; Ali, N.; Khan, M.; Zhang, Q. Immobilization of Lipase on Mesoporous Silica Nanoparticles with Hierarchical Fibrous Pore. *J. Mol. Catal. B Enzym.* **2016**, *134*, 129–135. [CrossRef]
93. Thangaraj, B.; Jia, Z.; Dai, L.; Liu, D.; Du, W. Effect of Silica Coating on Fe3O4 Magnetic Nanoparticles for Lipase Immobilization and Their Application for Biodiesel Production. *Arab. J. Chem.* **2019**, *12*, 4694–4706. [CrossRef]
94. Tran, D.T.; Chen, C.L.; Chang, J.S. Immobilization of Burkholderia Sp. Lipase on a Ferric Silica Nanocomposite for Biodiesel Production. *J. Biotechnol.* **2012**, *158*, 112–119. [CrossRef] [PubMed]
95. Esmi, F.; Nematian, T.; Salehi, Z.; Khodadadi, A.A.; Dalai, A.K. Amine and Aldehyde Functionalized Mesoporous Silica on Magnetic Nanoparticles for Enhanced Lipase Immobilization, Biodiesel Production, and Facile Separation. *Fuel* **2021**, *291*, 120126. [CrossRef]
96. Shinde, S.S.; Chi, H.M.; Lee, B.S.; Chi, D.Y. Tert-Alcohol-Functionalized Imidazolium Ionic Liquid: Catalyst for Mild Nucleophilic Substitution Reactions at Room Temperature. *Tetrahedron Lett.* **2009**, *50*, 6654–6657. [CrossRef]
97. Bender, J.; Jepkens, D.; Hüsken, H. Ionic Liquids as Phase-Transfer Catalysts: Etherification Reaction of 1-Octanol with 1-Chlorobutane. *Org. Process Res. Dev.* **2010**, *14*, 716–721. [CrossRef]
98. Santiago, C.C.; Lafuente, L.; Bravo, R.; Díaz, G.; Ponzinibbio, A. Ionic Liquids as Phase Transfer Catalysts: Enhancing the Biphasic Extractive Epoxidation Reaction for the Selective Synthesis of β-O-Glycosides. *Tetrahedron Lett.* **2017**, *58*, 3739–3742. [CrossRef]
99. Szepiński, E.; Smolarek, P.; Milewska, M.J.; Łuczak, J. Application of Surface Active Amino Acid Ionic Liquids as Phase-Transfer Catalyst. *J. Mol. Liq.* **2020**, *303*, 112607. [CrossRef]
100. Yuan, C.; Huang, Z.; Chen, J. Basic Ionic Liquid Supported on Mesoporous SBA-15: An Efficient Heterogeneous Catalyst for Epoxidation of Olefins with H 2O 2 as Oxidant. *Catal. Commun.* **2012**, *24*, 56–60. [CrossRef]
101. Karimi, B.; Vafaeezadeh, M. SBA-15-Functionalized Sulfonic Acid Confined Acidic Ionic Liquid: A Powerful and Water-Tolerant Catalyst for Solvent-Free Esterifications. *Chem. Commun.* **2012**, *48*, 3327–3329. [CrossRef]

102. Yao, J.; Sheng, M.; Bai, S.; Su, H.; Shang, H.; Deng, H.; Sun, J. Ionic Liquids Grafted Mesoporous Silica for Chemical Fixation of CO2 to Cyclic Carbonate: Morphology Effect. *Catal. Lett.* **2022**, *152*, 781–790. [CrossRef]
103. Xu, Q.-Q.; Yin, J.-Z.; Zhou, X.-L.; Yin, G.-Z.; Liu, Y.-F.; Cai, P.; Wang, A.-Q. Impregnation of Ionic Liquids in Mesoporous Silica Using Supercritical Carbon Dioxide and Co-Solvent. *RSC Adv.* **2016**, *6*, 101079–101086. [CrossRef]
104. Gholami, A.; Pourfayaz, F.; Maleki, A. Recent Advances of Biodiesel Production Using Ionic Liquids Supported on Nanoporous Materials as Catalysts: A Review. *Front. Energy Res.* **2020**, *8*, 144. [CrossRef]
105. Watanabe, Y.; Amitani, N.; Yokoyama, T.; Ueda, A.; Kusakabe, M.; Unami, S.; Odashima, Y. Synthesis of Mesoporous Silica from Geothermal Water. *Sci. Rep.* **2021**, *11*, 23811. [CrossRef]
106. El-Nahas, S.; Osman, A.I.; Arafat, A.S.; Al-Muhtaseb, A.H.; Salman, H.M. Facile and Affordable Synthetic Route of Nano Powder Zeolite and Its Application in Fast Softening of Water Hardness. *J. Water Process Eng.* **2020**, *33*, 101104. [CrossRef]
107. Aneu, A.; Pratika, R.A.; Hasanudin; Gea, S.; Wijaya, K.; Oh, W.-C. Silica-Based Catalysts for Biodiesel Production: A Brief Review. *Silicon* **2023**. [CrossRef]
108. Kaur, J.; Sarma, A.K.; Jha, M.K.; Gera, P. Valorisation of Crude Glycerol to Value-Added Products: Perspectives of Process Technology, Economics and Environmental Issues. *Biotechnol. Rep.* **2020**, *27*, e00487. [CrossRef]
109. Osman, A.I.; Qasim, U.; Jamil, F.; Al-Muhtaseb, A.H.; Jrai, A.A.; Al-Riyami, M.; Al-Maawali, S.; Al-Haj, L.; Al-Hinai, A.; Al-Abri, M.; et al. Bioethanol and Biodiesel: Bibliometric Mapping, Policies and Future Needs. *Renew. Sustain. Energy Rev.* **2021**, *152*, 111677. [CrossRef]
110. Ma, Y.; Chen, H.; Shi, Y.; Yuan, S. Low Cost Synthesis of Mesoporous Molecular Sieve MCM-41 from Wheat Straw Ash Using CTAB as Surfactant. *Mater. Res. Bull.* **2016**, *77*, 258–264. [CrossRef]
111. Rangaraj, S.; Venkatachalam, R. A Lucrative Chemical Processing of Bamboo Leaf Biomass to Synthesize Biocompatible Amorphous Silica Nanoparticles of Biomedical Importance. *Appl. Nanosci.* **2017**, *7*, 145–153. [CrossRef]
112. Fatimah, I.; Taushiyah, A.; Najah, F.B.; Azmi, U. ZrO_2/Bamboo Leaves Ash (BLA) Catalyst in Biodiesel Conversion of Rice Bran Oil. In Proceedings of the IOP Conference Series: Materials Science and Engineering, the 12th Joint Conference on Chemistry, Semarang, Indonesia, 19–20 September 2017; Volume 349.
113. Tadjarodi, A.; Haghverdi, M.; Mohammadi, V. Preparation and Characterization of Nano-Porous Silica Aerogel from Rice Husk Ash by Drying at Atmospheric Pressure. *Mater. Res. Bull.* **2012**, *47*, 2584–2589. [CrossRef]
114. Fatimah, I.; Purwiandono, G.; Sahroni, I.; Sagadevan, S.; Chun-Oh, W.; Ghazali, S.A.I.S.M.; Doong, R. Recyclable Catalyst of ZnO/SiO_2 Prepared from Salacca Leaves Ash for Sustainable Biodiesel Conversion. *S. Afr. J. Chem. Eng.* **2022**, *40*, 134–143. [CrossRef]

Disclaimer/Publisher's Note: The statements, opinions and data contained in all publications are solely those of the individual author(s) and contributor(s) and not of MDPI and/or the editor(s). MDPI and/or the editor(s) disclaim responsibility for any injury to people or property resulting from any ideas, methods, instructions or products referred to in the content.

MDPI
St. Alban-Anlage 66
4052 Basel
Switzerland
www.mdpi.com

ChemEngineering Editorial Office
E-mail: chemengineering@mdpi.com
www.mdpi.com/journal/chemengineering

Disclaimer/Publisher's Note: The statements, opinions and data contained in all publications are solely those of the individual author(s) and contributor(s) and not of MDPI and/or the editor(s). MDPI and/or the editor(s) disclaim responsibility for any injury to people or property resulting from any ideas, methods, instructions or products referred to in the content.

www.ingramcontent.com/pod-product-compliance
Lightning Source LLC
LaVergne TN
LVHW070238100526
838202LV00015B/2150